Developments in Hydrobiology 149

Series editor
H. J. Dumont

Assessing the Ecological Integrity of Running Waters

Proceedings of the International Conference, held in Vienna, Austria, 9–11 November 1998

Edited by

M. Jungwirth, S. Muhar & S. Schmutz

Reprinted from Hydrobiologia, volumes 422/423 (2000)

Kluwer Academic Publishers
Dordrecht / Boston / London

Library of Congress Cataloging-in-Publication Data

A C.I.P. Catalogue record for this book is available from the Library of Congress.

Published by Kluwer Academic Publishers,
P.O. Box 17, 3300 AA Dordrecht, The Netherlands

Sold and distributed in North, Central and South America
by Kluwer Academic Publishers,
101 Philip Drive, Norwell, MA 02061, U.S.A.

In all other countries, sold and distributed
by Kluwer Academic Publishers Group,
P.O. Box 322, 3300 AH Dordrecht, The Netherlands

Cover illustration: Natural reference river, Alaska.
Photo: S. Schmutz

Printed on acid-free paper

Contents

Indicators, Assessment Methods, and Applications

Symposium sponsors

Rector of the University of Agricultural Sciences Vienna

Federal Ministry of Environment, Youth and Family Affairs

Federal Ministry of Forestry and Agriculture

Federal Ministry of Science and Transport

Austrian Network for Environmental Research

Austrian Airlines

Austrian Lotteries

Bank Austria

Vienna Tourist Board

VISA Austria

Oracle

Schneiders

Austrian Association of Cities and Towns: Know-How Transfer Center

List of Authors

BARBOUR Michael T., Tetra Tech, Inc., 10045 Red Run Boulevard, Suite 110, Owings Mills, MD 21117, U.S.A.
E-mail: mbarbour@ccpl.carr.lib.md.us

BAUERNFEIND Ernst, Museum of Natural History, Dept. of Zoology, Bird Collection, Burgring 7, A-1014 Vienna, Austria
E-mail: hans-martin.berg@nhm-wien.ac.at

BEISEL Jean-Nicolas, UPRES Ecotoxicité, Biodiversité et Santé Environnementale (E.B.S.E.), Université de Metz, Ile du Saulcy, F-57045 Metz Cedex 01, France
E-mail: beisel@sciences.univ-metz.fr

BIS Barbara, Department of Applied Ecology, Centre of Ecohydrological Studies, University of Lódz, PL-90-237 Lódz, Banacha 12/16, Poland
E-mail: barbis@biol.uni.lodz.pl

BOON Philip J., Scottish Natural Heritage, 2 Anderson Place, Edinburgh EH6 5NP, U.K.
E-mail: pboon@compuserve.com

BOSCAINI Adriano, Museo Tridentino di Scienze Naturali, Via Calepina 14, I-38100 Trento, Italy
E-mail: boscaini@mtsn.tn.it

BUFFAGNI Andrea, Water Research Institute – C.N.R., Via della Mornera 25, I-20047 Brugherio (MI), Italy
E-mail: buffagni@server-mi.irsa.rm.cnr.it

BUNDI Ulrich, EAWAG, Swiss Federal Institute for Environmental Science and Technology, Limnological Research Center, Überlamdstr. 133, CH-8600 Dübendorf, Switzerland
E-mail: ulrich.bundi@eawag.ch

BUNN Stuart E., Centre for Catchment and In-Stream Research, Faculty of Environmental Sciences, Griffith University, Nathan, Qld-4111 Queensland, Australia
E-mail: s.bunn@mailbox.gu.edu.au

CHOVANEC Andreas, Federal Environment Agency, Spittelauer Laende 5, A-1090 Vienna, Austria
E-mail: chovanec@ubavie.gv.at

FISCHER Sabine, Biological Station Lunz, Austrian Academy of Sciences, A-3293 Lunz/See, Austria
E-mail: sabine.fischer@oeaw.ac.at

GAFNY Sarig, Institute for Nature Conservation Research, George S. Wise Faculty of Life Sciences, Tel-Aviv University, Ramat-Aviv 68878, Israel
E-mail: gafny@post.tau.ac.il

GALAT David L., U.S. Geological Survey, Biological Resources Division, Missouri Cooperative Fish and Wildlife Research Unit, 302 Anheuser-Busch Natural Resources Building, University of Missouri, Columbia, MO 65211-7240, U.S.A.
E-mail: galatd@missouri.edu

GAMMON James R., DePauw University, Department of Biological Science, Greencastle, IN 46135, U.S.A.
E-mail: jrgammon@depauw.edu

HABERSACK Helmut M., Institute for Water Management, Hydrology and Hydraulic Engineering, University of Agricultural Sciences Vienna, Muthgasse 18, A-1190 Vienna, Austria
E-mail: haber@edv2.boku.ac.at

HARPER David M., Ecology Unit, Department of Biology, University of Leicester, Leicester LE1 7RH, U.K.
E-mail: dmh@leicester.ac.uk

HUGHES Robert M., Dynamac, 200 SW 35th St. Corvallis, Oreg., U.S.A.
E-mail: hughesb@mail.cor.epa.gov

INNIS Stephanie A., School of Fisheries, Box 357980, University of Washington, Seattle, WA 98195, U.S.A.
E-mail: sinnis@fish.washington.edu

JANSEN Wolfgang, Institute for Zoology, University of Hohenheim, D-70593 Stuttgart, Germany
E-mail: jansen@uni-hohenheim.de

JUNGWIRTH Mathias Dept. of Hydrobiology, Fisheries and Aquaculture, University of Agricultural Sciences, Vienna, Max Emanuel-Str. 17, A-1180 Vienna, Austria
E-mail: mjungwir@edv1.boku.ac.at

KARR James R., University of Washington, Box 357980, Seattle, WA 98195-7980, U.S.A.
E-mail: jrkarr@u.washington.edu

KESMINAS Vytautas, Institute of Ecology, Akademijos 2, 2600 Vilnius, Lithuania
E-mail: kesminas@aiva.lt

MILNER Alexander M., School of Geography, University of Birmingham, Edgbaston, Birmingham B15 2TT, U.K.
E-mail: a.m.milner@bham.ac.uk

MOOG Otto, BOKU-University of Agricultural Sciences, Max Emanuel-Str. 17, A-1180 Vienna, Austria
E-mail: ottomoog@edv1.boku.ac.at

MUHAR Susanne, University of Agricultural Sciences Vienna, Department of Hydrobiology, Fisheries and Aquaculture, Max Emanuel-Str. 17, A-1180 Vienna, Austria
E-mail: smuhar@edv1.boku.ac.at

PETTS Geoffrey E., School of Geography and Environmental Sciences, University of Birmingham, Edgbaston B15 2TT, U.K.
E-mail: pettsge@novell2.bham.ac.uk

RABENI Charles F., Missouri Cooperative Fish and Wildlife Research Unit, Biological Resources Division, U.S. Geological Survey, 302 Natural Resources Building, University of Missouri, Columbia, MO 65211-7240, U.S.A.
E-mail: charles_rabeni@muccmail.missouri.edu

RAVEN Paul J., Environment Agency, Rio House, Waterside Drive, Aztec West, Almondsbury, Bristol BS32 4UD, U.K.
E-mail: paul.raven@environment-agency.gov.uk

RAWER-JOST Claudia, Institute for Zoology, University of Hohenheim, Garbenstraße 30, D-70593 Stuttgart, Germany
E-mail: rawer@uni-hohenheim.de

ROUSSEAU Alain N., INRS-Eau, 2800 rue Einstein C.P. 7500, Sainte-Foy, G1V 4C7 Québec, Canada
E-mail: alain_rousseau@INRS-EAU.uquebec.ca

SANDIN Leonard, Department of Environmental Assessment, Swedish University of Agricultural Sciences, Box 7050, S-75007 Uppsala, Sweden
E-mail: leonard.sandin@ma.slu.se

SCHIEMER Fritz, Dept. of Limnology, Institute of Zoology, University of Vienna, A-1090 Vienna, Althanstrasse 14, Austria
E-mail: schiemer@zoo.univie.ac.at

SCHMUTZ Stefan, Dept. Hydrobiology, Fisheries and Aquaculture, Univ. of Agricultural Sciences Vienna, Max Emanuel-Str. 17, A-1180 Vienna, Austria
E-mail: schmutz@mail.boku.ac.at

SOLIMINI Angelo G., Department of Biology, University of Rome "Tor Vergata", Via della ricerca scientifica, I-00133 Rome, Italy

USSEGLIO-POLATERA Philippe, UPRES, Ecotoxicité, Biodiversité et Santé Environnementale, Université de Metz, Ile du Saulcy, F-57045 Cedex 01, France
E-mail: usseglio@sciences.univ-metz.fr

VERDONSCHOT Piet F.M. Institute for Forestry and Nature Management, P.O. Box 23, NL-6700 AA Wageningen, The Netherlands
E-mail: p.f.m.verdonschot@ibn.dlo.nl

WAGNER Rüdiger, Limnologische Flußstation Schlitz der MPG, P.O. Box 260, D-36105 Schlitz, Germany
E-mail: rwagner@mpil-schlitz.mpg.de

List of Reviewers

BARDONNET Agnes, Institut National de la Recherche Agronomique – INRA, Laboratoire d'Ecologie Aquatique, 65, rue de Saint Brieuc, F-35042 Rennes Cedex, France

BARBOUR Michael, Tetra Tech, Inc., 10045 Red Run Blvd., Suite 110, Owing Mills, MD 21117, U.S.A.
E-mail: mbarbour@ccpl.carr.lib.md.us

BAUERNFEIND Ernst, Museum of Natural History, Dept. of Zoology, Bird Collection, Burgring 7, A-1014 Vienna, Austria
E-mail: hans-martin.berg@nhm-wien.ac.at

BEISEL Jean-Nicolas, Biodiversité et Santé Environnementale (E.B.S.E.), Université de Metz, Ile du Saulcy, F-57040 Metz, France
E-mail: beisel@sciences.univ-metz.fr

BIS Barbara, University of Lodz, Department of Applied Ecology, Banacha 12/16, PL - 90237 Lodz, Poland
E-mail: barbis@biol.uni.lodz.pl

BÖHMER Jürgen, Institute for Zoology, University of Hohenheim, Garbenstraße 30, D-70593 Stuttgart, Germany
E-mail: boehmer@uni-hohenheim.de

BOON Philip, Scottish Natural Heritage, 2 Anderson Place, EH6 5NP Edinburgh, U.K.
E-mail: pboon@compuserve.com

BRAUKMANN Ulrich, University of Kassel, Fachbereich 13, Fachgebiet Gewässerökologie/Gewässerentwicklung, Nordbahnhofstr. 1a, D-37213 Witzenhausen, Germany
E-mail: braukm@wiz.uni-kassel.de

BRETSCHKO Gernot, Institute of Limnology, Biological Station Lunz/See, Am Seehof 4, A-3293 Lunz/See, Austria
E-mail: gernot.bretschko@oeaw.ac.at

BRITTAIN John E., Norwegian Water Resources and Energy Administration, University of Oslo, Sars Gate 1, N-0562 Oslo, Norway
E-mail: j.e.brittain@toyen.uio.no

BUNDI Ulrich, EAWAG/ETH, Swiss Federal Institute for Environmental Science and Technology, Limnological Research Center, Überlandstraße 133, CH-8600 Dübendorf, Switzerland
E-mail: ulrich.bundi@eawag.ch

BUNN Stuart, Griffith University, Faculty of Environmental Sciences, Centre for Catchment and In-Stream Research, Qld-4111 Queensland, Nathan, Australia
E-mail: s.bunn@mailbox.gu.edu.au

CAIRNS John E. Jr., Virginia Polytech Institute and State University, Department of Biology, Blacksburg, VA 24061-0406, U.S.A.

CAPRA Hervé, CEMAGREF, Laboratoire d'Hydroécologie Quantitative, Division Biologie des Ecosystémes Aquatique, 3 bis quai Chauveau, CP220, F-69336 Lyon Cedex 09, France
E-mail: herve.capra@cemagref.fr

DAVIS Wayne S., U.S. Environmental Protection Agency, 401 M. Street, S.W. (2162), Washington, D.C. 20460, U.S.A.
E-mail: davis.wayne@epa.gov

DOLÉDEC Sylvain, Université Lyon I, ESA CNRS, Ecologie des Eaux Douces et des Grands Fleuves, 43 Bvd du 11 Novembre, F-69622 Villeurbanne Cedex, France
E-mail: sylvain@biomserv.univ-lyon1.fr

DUNBAR Michael, Institute of Hydrology, Crowmarsh Gifford, Wallingford, Oxfordshire OX10 8BB, U.K.
E-mail: m.dunbar@va.nwl.ac.uk

EGGER Gregory, Institute of Ecology and Environmental Planning, Fromillerstraße 40/7, A-9020 Klagenfurt, Austria
E-mail: oekup@darth.carinthia.com

FERREIRA Teresa, Instituto Superior de Agronomia, Departamento de Engenharia Florestal, Tapada da Ajuda, 1399 Lisboa Codex, Portugal
E-mail: terferreira@mail.telepac.pt

FRIEDRICH Günther, Landesumweltamt NRW, Wallneyerstraße 6, D-45133 Essen, Germany

FRUTIGER Andreas, EAWAG/ETH, Swiss Federal Institute for Environmental Science and Technology, Hydrology/Limnology, Überlandstraße 133, CH-8600 Dübendorf, Switzerland
E-mail: andreas.frutiger@eawag.ch

FÜREDER Leopold, University of Innsbruck, Institute of Zoology and Limnology, Technikerstraße 25, A-6020 Innsbruck, Austria
E-mail: leopold.fuereder@uibk.ac.at

GAFNY Sarig, Tel Aviv University, George S. Wise Faculty of Life Sciences, Institute for Nature Conservation Research, Ramat Aviv, 69978 Tel Aviv, Israel
E-mail: gafny@post.tau.ac.il

GALAT David, U.S. Geological Survey, Coop. Fish and Wildlife Research Unit, University of Missouri, 302 Anheuser-Busch N.R. Building, Columbia, MO 65211-7240, U.S.A.
E-mail: galatd@missouri.edu

GAMMON James R., De Pauw University, Department of Biology, Greencastle, Indiana IN-46135, U.S.A.
E-mail: jrgammon@depauw.edu

GRAF Wolfram, University of Agricultural Sciences Vienna, Dept. of Hydrobiology, Fisheries and Aquaculture, Max Emanuel-Str. 17, A-1180 Vienna, Austria
E-mail: grafwolf@edv1.boku.ac.at

HABERSACK Helmut, Institute of Water Management, Hydrology and Hydraulic Engineering, University of Agricultural Sciences Vienna, Muthgasse 18, A-1190 Vienna, Austria
E-mail: haber@edv2.boku.ac.at

HARPER David, University of Leicester, Ecology Unit, Department of Biology, Leicester LE1 7RH, U.K.
E-mail: dmh@leicester.ac.uk

HAWKINS Charles, Dept. of Fisheries and Wildlife, Ecology Center, Utah State University, Logan, UT 84322-521, U.S.A.
E-mail: hawkins@cc.usu.edu

HOFFMANN Andreas, University Bielefeld, Verhaltensphysiologie, Morgenbreede 45, D-33615 Bielefeld, Germany
E-mail: hoffmana@igb-berlin.de

HOLMES Nigel T.H., Alconbury Consultants, 57 Ramsey Road, Warboys, Huntington PE17 2RW, U.K.

HUGHES Robert, Dynamac Corporation, Environmental Services, 200 South West 35th Street, Corvallis, Oregon, OR 97333, U.S.A.
E-mail: hughesb@mail.cor.epa.gov

HUMPESCH Uwe, Austrian Academy of Sciences, Institute of Limnology, Gaisberg 116, A-5310-Mondsee, Austria
E-mail: uwe.humpesch@oeaw.ac.at

JANAUER Georg, University of Vienna, Institute of Plant Physiology, Althanstr. 14, A-1090 Vienna, Austria
E-mail: janauer@pflaphy.pph.univie.ac.at

JUNGWIRTH Mathias, University of Agricultural Sciences Vienna, Dept. of Hydrobiology, Fisheries and Aquaculture, Max Emanuel-Str. 17, A-1180 Vienna, Austria
E-mail: mjungwir@edv1.boku.ac.at

KARR James, University of Washington, School of Fisheries, P.O. Box 357980, Seattle, WA 98195-2200, U.S.A.
E-mail: jrkarr@u.washington.edu

KAUFMANN Phil, Oregon State University, Department of Fisheries and Wildlife, Corvallis, WA 97333, U.S.A.
E-mail: kaufmann@mercury.cor.epa.gov

KECKEIS Hubert, University of Vienna, Dept. of Zoology/Limnology, Althanstraße 19, A-1090 Vienna, Austria
E-mail: hubert.keckeis@univie.ac.at

KESTEMONT Patrick, FUNDP-URBO, Faculté Universitaires N.D. de la Paix, 62, rue de Bruxelles, B-5000 Namur, Belgium
E-mail: patrick.kestemont@fundp.ac.be

KNOFLACHER Helmut, Austrian Research Center, Forschungszentrum Seibersdorf, A-2444 Seibersdorf, Austria
E-mail: knoflacher@arcs.ac.at

KUMMER Helmut, University of Agricultural Sciences Vienna, Dept. of Hydrobiology, Fisheries and Aquaculture, Max Emanuel-Str. 17, A-1180 Vienna, Austria
E-mail: helmut@edv1.boku.ac.at

LAMOUROUX Nicolas, ESA CNRS, 5023 Ecologie des Eaux Douces et des Grands Fleuves, Compagnie Nationale du Rhone, Université de Lyon 1, F-69622 Villeurbanne Cedex, France
E-mail: lamourou@biomserv.univ-lyon1.fr

LEK Sovan, Laboratoire de Biologie Quantitative, Université Paul Sabatier, 118 rote de Narbonne, F-31062 Toulouse, France
E-mail: lek@cict.fr

LYONS John, Wisconsin Department of Natural Resources, Bureau of Research, 1350 Femrite Drive, Monona, WI 21228, U.S.A.
E-mail: lyonsj@dnr.state.wi.us

MILNER Alexander, The University of Birmingham, Department of Geography, Edgbaston, Birmingham B15 2TT, U.K.
E-mail: a.m.milner@bham.ac.uk

MOOG Otto, University of Agricultural Sciences Vienna, Dept. of Hydrobiology, Fisheries and Aquaculture, Max Emanuel-Str. 17, A-1180 Vienna, Austria
E-mail: ottomoog@edv1.boku.ac.at

MUHAR Susanne, University of Agricultural Sciences Vienna, Dept. of Hydrobiology, Fisheries and Aquaculture, Max Emanuel-Str. 17, A-1180 Vienna, Austria
E-mail: smuhar@edv1.boku.ac.at

NAIMAN Robert, School of Fisheries, University of Washington, P.O. Box 357980, Seattle, WA 98195, U.S.A.
E-mail: naiman@u.washington.edu

NEWSON Malcolm, University of Newcastle upon Tyne, Department of Geography, Newcastle upon Tyne NE1 7RU, U.K.
E-mail: m.d.newson@ncl.ac.uk

OBERDORFF Thierry, Muséum National d'Historie Naturelle, Laboratoire d'Ichtyologie Générale et Appliquée, 43, rue Cuvier, F - 75005 Paris, France
E-mail: oberdorf@cimrs1.mnhn.fr

PETER Armin, EAWAG/ETH, Swiss Federal Institute for Environmental Science and Technology, Fisheries Sciences, Seestraße 79, CH-6047 Kastanienbaum, Switzerland
E-mail: peter@eawag.ch

PETTS Geoffrey, University of Birmingham, Environmental Research and Management, Edgbaston, Birmingham B15 3TT, U.K.
E-mail: pettsge@novell2.bham.ac.uk

PSENNER Roland, University of Innsbruck, Institute of Zoology and Limnology, Technikerstraße 25, A-6020 Innsbruck, Austria
E-mail: roland.psenner@uibk.ac.at

RABENI Charles F., Missouri Cooperative Fish and Wildlife Research Unit, University of Missouri, 112 Stephens Hall, Columbia, MO 65211, U.S.A.
E-mail: charles_rabeni@muccmail.missouri.edu

RAVEN Paul, The Environment Agency, Rio House, Waterside Drive, Aztec West, Almondsbury, Bristol BS12 4DU, U.K.
E-mail: paul.raven@environment-agency.gov.uk

REUSCH Herbert, Wellendorf 70, D-29562 Suhlendorf, Germany
E-mail: herbert.reusch@t-online.de

ROUSSEAU Alain, Université de Quebec, INRS Eau, 2800 Rue Einstein, Suite 020, C.p: 7500, Ste Foy, Quebec G1V 4C7, Canada
E-mail:
alain_rousseau@INRS-EAU.uQuebec.ca

SCHIEMER Fritz, Institute of Zoology and Limnology, University of Vienna, Althanstrasse 14, A-1090 Vienna, Austria
E-mail: schiemer@zoo.univie.ac.at

SCHILLHAMMER Harald, Museum of Natural History Vienna, Käfersammlung-Beetle Collection, Burgring 7, A-1014 Vienna, Austria
E-mail: heinrich.schoenmann@nhm-wien.ac.at

SCHMUTZ Stefan, University of Agricultural Sciences Vienna, Dept. of Hydrobiology, Fisheries and Aquaculture, Max Emanuel-Str. 17, A-1180 Vienna, Austria
E-mail: schmutz@mail.boku.ac.at

SCOTT Mark C., The University of Georgia, Institute of Ecology, Athens, GA 30602-2202, U.S.A.

SOUCHON Yves, CEMAGREF, Agricultural and Environmental Engineering, Biology of Freshwater Ecosystem Division, Lyon Regional Centre, 3 Bis Quai Chaveau, CP 220, F-69336 Lyon Cedex 09, France
E-mail: yves.souchon@cemagref.fr

STODDARD John, USEPA, 200 SW 35th Street, Corvallis, OR 97333, U.S.A.
E-mail: stoddard@tsunami.cor.epa.gov

THIEL Ralf, Inst. für Hydrobiologie und Fischereiwirtschaft, Olbersweg 24, D-Hamburg 22767, Germany

TOCKNER Klement, EAWAG/ETH Zürich, Swiss Federal Institute for Environmental Science and Technology, Hydrobiology/Limnology, Überlandstraße 133, CH-Dübendorf 8600, Switzerland
E-mail: klement.tockner@eawag.ch

USSEGLIO POLATERA Philippe, Université de Metz, Biodiversité et Santé Environnementale (E.B.S.E.), Ile du Saulcy, F-57045 Metz Cedex 01, France
E-mail: usseglio@sciences.univ-metz.fr

VERDONSCHOT Piet, Inst. Forestry & Nature Research (IBN-DLO), Department of Aquatic Ecology, P.O. Box 23, NL-6700 AA Wageningen, The Netherlands
E-mail: p.f.m.verdonschot@ibn.dlo.nl

VOGEL Birgit, Vienna Technical University, Institute for Water Quality and Waste Management, Karlsplatz 13/226, A-1040 Vienna, Austria
E-mail: vogel@wgnt.iwag.tuwien.ac.at

WAGNER Rüdiger, Limnologische Flußstation Schlitz der MPG, P.O. Box 260, D-36105 Schlitz, Germany
E-mail: rwagner@mpil-schlitz.mpg.de

WAIDBACHER Herwig, University of Agricultural Sciences Vienna, Dept. of Hydrobiology, Fisheries and Aquaculture, Max Emanuel-Str. 17, A-1180 Vienna, Austria
E-mail: waidbach@edv1.boku.ac.at

WARD James, EAWAG/ETH Zürich, Swiss Federal Institute for Environmental Science and Technology, Hydrobiology/Limnology, Überlandstraße 133, CH-Dübendorf 8600, Switzerland
E-mail: james.ward@eawag.ch

WARINGER Johann, University of Vienna, Coordination Ecology, Althanstraße 14, A-1090 Vienna, Austria
E-mail: waringer@pflaphy.pph.univie.ac.at

WATERSTRAAT Arno, Heinrich Mann-Straße 13, D-2080 Neustrelitz, Germany

WEICHSELBAUMER Peter, Gschwendt 1, A-6060 Tulfes, Austria

WOLTER Christian, Institut für Gewässerökologie und Binnenfischerei, Müggelseedamm 310, D-12587 Berlin, Germany
E-mail: wolter@igb-berlin.de

YOUNG William, CSIRO Land and Water, GPO Box 1666, ACT 2601 Canberra, Australia
E-mail: bill.young@cbr.clw.csiro.au

Hydrobiologia **422/423:** xvii–xviii, 2000.
M. Jungwirth, S. Muhar & S. Schmutz (eds), Assessing the Ecological Integrity of Running Waters.

Preamble

The assessment of ecological integrity is necessary in order to shed light on the effects of human alterations. Evaluating degradation processes also provides key information on how to avoid further negative impacts. The success of future conservation, mitigation and restoration activities will rely on sound assessment methodologies and their ecological relevance and applicability. Assessment methodologies are therefore an integral part of sustainable river management. As an issue, 'ecological assessment' is ubiquitous in modern applied research on running waters. Despite an enormous number of individual scientific publications, little attempt has been made to synthesize and discuss the available information and experience. This was the motivation behind making the assessment of ecological integrity the focus of an international scientific meeting on running waters.

In November 1998 the Department of Hydrobiology at the University of Agricultural Sciences Vienna held an International Conference titled 'Assessing the Ecological Integrity of Running Waters'. The 300 participants from 40 countries represented a wide range of fields including basic limnology, hydromorphology, applied ecology, and governmental administration, underlining the inter- and transdisciplinary nature of the issue.

The present volume, as one result of this conference, has taken the next step after the 1995 'Fourth Conference of the Scottish Natural Heritage'. The earlier conference posed the provocative question 'Water Quality – Defining the Indefinable?' to a scientific forum. Numerous contributions dealing with specific river assessment-examples in various countries attempted to answer this question.

Since then, research has made substantial progress. Currently, international research initiatives are focusing on crucial issues such as levels of scale, reference conditions, key parameters, species guild-traits, etc. An attempt has been made to record and evaluate processes, apply new methods of statistical analysis including multimetric and multivariate techniques, introduce neural networks, improve GIS technologies, and develop integrative evaluation schemes.

Concurrently, the European Commission has developed the Water Framework Directive, which is now ready for ratification. The goal of the Directive is to 'prevent further deterioration and protect and enhance the status of aquatic ecosystems'. This is a call to develop and test methods that will ultimately enable scientists to document and evaluate the complex ecological quality of running waters and changes over time using standardized methods. The specific issues treated include classifying rivers according to characteristic features, establishing type-specific reference conditions, defining quality classes, as well as developing and harmonising assessment methods. The appropriate scales must be given due consideration.

The Vienna Conference, which produced nearly 50 oral presentations and 100 posters, was held against this backdrop. These contributions ultimately gave rise to the 37 papers included in this special volume of *Hydrobiologia*.

The organization of the Proceedings largely mirrors that of the congress. The book is divided into three parts. The first part addresses the state-of-the-art knowledge about abiotic and biotic processes and functions of running waters; this provides a fundamental basis for developing adequate assessment methods. The second part focuses on the indicators, assessment methods, and applications. The third part deals with integrated assessment methods.

The success of the conference is built on the dedicated commitment of many colleagues. To all we are grateful. Particular thanks go to Birgit Vogel for organizing the conference and the Proceedings. She was actively supported in this endeavor by Michael Kaufmann and Eva Alvarez. The reviewers contributed significantly to the quality of the book, and we owe them great thanks. We would also like to thank Michael Stachowitsch for his unflagging effort in linguistically improving the manuscripts.

Finally, a special thank you goes to the sponsors of the conference: Austrian Federal Ministry of Environment, Youth and Family Affairs, Austrian Federal Ministry of Agriculture and Forestry, Austrian Federal Ministry of

Science and Transport, Austrian Network for Environmental Research, Austrian Airlines, Austrian Lotteries, Bank Austria, Oracle, Schneiders, Vienna Tourist Board and VISA Austria.

We would like to express our appreciation to all those who came to Vienna and made presentations; all the participants contributed to the pleasant and constructive atmosphere that was palpable throughout the days spent together. We are convinced that this common spirit will stimulate future collaborations and promote further activities in this field.

Finally, we hope that the conference results will contribute to a more effective preservation, restoration and maintenance of the ecological integrity of running water ecosystems.

THE EDITORS

Hydrobiologia **422/423:** xix–xxi, 2000.
M. Jungwirth, S. Muhar & S. Schmutz (eds), Assessing the Ecological Integrity of Running Waters.

Preface

The Austrian approach

The alpine nature of Austria, with annual precipitation rates of 550 to 2500 mm (average 1170 mm), provides not only favourable hydrological conditions, but also a dense river network. With its more than 100 000 km of running waters, Austria has a 'river density' of 1.2 km per km^2. At the same time, this 'wealth in water' entails extreme precipitation events and natural hazards such as avalanches, landslides and floods.

Protection of settlements and economic areas required precautionary measures that interfered with the water regime. On the other hand the combination of flow and altitude provided a great potential for hydroelectric power, which has now been exploited to a considerable extent. Today about 68% of the hydroelectric power potential is used providing two-thirds of Austria's electricity.

The economic development of valleys, flood protection, as well as hydroelectric power generation has heavily modified Austria's running waters. An assessment of Austria's 52 largest rivers (catchment area >500 km^2, about 5000 km total river length) shows that about 80% (3900 km) of the stretches are moderately to heavily impacted by human activity and thus no longer correspond to their original channel form and hydrological dynamics. Only about 6% (291 km) retained a 'high quality' status and about 15% (774 km) can still be classified as 'good quality' (Muhar et al., 1998). Growing environmental awareness as well as increasing knowledge on the interrelationships between water bodies and their environment were the driving forces behind an early change towards developing running waters to a nature-like status, as well as for comprehensively taking into account their ecological integrity.

In the first step in the early 1970s, the authorities responsible for nature protection and fisheries were integrated into river engineering processes. This was followed by attempts to harmonise the required flood protection and the protection of the water body against negative anthropogenic impacts. This was achieved by developing a new river management tool, the 'Gewässerbetreuungskonzept (GBK)'. The success of this new approach is tightly linked to the increasing integration of ecological integrity as a public interest into Austrian water legislation. Water protection is now based on the maintenance of the natural habit of water bodies and their relevant riparian areas. Today, maintaining the ecological integrity is one of the preconditions for issuing new permits under Austrian water law.

Before ecological integrity was defined in detail, the GBK served already as a pragmatic approach for a targeted and feasible river management. In a first step, the GBK focuses on a 'visionary' objective (Leitbild). These ideal conditions of a river are sectorally defined by relevant disciplines such as river engineering and river ecology. In a second step, these objectives are integrated and adjusted in an interdisciplinary approach to what is feasible by subordinating single interests to the benefit of the common goal. In this manner, long-term, river-specific objectives are defined considering individual regional features. Further steps are the development of a set of adequate measures depending on the defined priorities as well as ensuring the necessary legal and financial basis.

Considering the responsibility for such an integrated development of running waters has largely contributed to the implementation of ecological integrity in river engineering measures as well as, at a later stage, in Austrian water legislation. In this context ecological integrity is defined as the capacity to maintain all internal and external processes and attributes interacting with the environment in such a way that the biotic community corresponds to the natural state of the relevant aquatic habitat (Austrian legislation on funding water-related measures, OJ. 1994/516).

Hence, river management has to be oriented at the natural and, at the same time, river-type-specific characteristics of the water body. This new strategy is a consequence of the coherent development of Austrian water management strategies from nature-like river engineering via ecomorphological concepts assessment toward an orientation aimed at specific river types. This approach has always been requested by the scientific community as well as by river authorities; today, it sets a high standard in assessing the state of ecological integrity of surface waters. In the future, the task will be to increase this knowledge in order to guarantee an optimal development of running waters as 'lifelines'.

European developments

The European Union is currently thoroughly restructuring its water policy. Following a broad consultation process involving all interested parties the European Commission presented in 1997 and 1998 its Proposals for a Water Framework Directive[2]. This proposal is currently being negotiated at the European Parliament and the Council of Ministers. This legislation will have the following main objectives:

- expanding the scope of water protection to all waters, surface waters and groundwater;
- achieving 'good status' for all waters by a certain deadline: good status will have ecological, chemical and morphological elements, as well as – for groundwater – sustainable quantity;
- water management based on river basins, involving a 'combined approach' of emission limit values and quality standards; in each particular case, the more stringent approach will apply;
- comprehensive measuring and monitoring system as planning tools as well as for checking progress;
- getting the prices right; prices for services such as drinking water supply and waste treatment should reflect the true costs;
- getting citizens as well as experts more closely involved;
- streamlining legislation: old directives are to be repealed.

Within this framework the quality elements for rivers will be the following:

1. Biological elements
   - Composition and abundance of aquatic flora
   - Composition and abundance of benthic invertebrate fauna
   - Composition, abundance and age structure of fish
2. Hydromorphological elements supporting the biological elements
   - Hydrological regime: quantity and dynamics of water flow connection to groundwater bodies
   - River continuity
   - Morphological conditions: river depth and width variations; structure and substrate of the river bed; structure of the riparian zone
3. Chemical and physiochemical elements supporting the biological elements
   - General: thermal conditions, oxygenation conditions, salinity, acidification status, nutrient conditions
   - Specific pollutants: pollution by all priority substances identified as being discharged into the body of water; pollution by other substances identified as being discharged in significant quantities into the body of water

The Water Framework Directive will provide a sound basis for water protection across the European Union as it approaches the new millennium, aiming for the first time throughout Europe not only at a protection of all waters, but also at ecological and ecomorphological objectives.

The progress made in negotiating this future European water legislation indicates a final adoption in early 2000.

Europe's waters are in need of more protection, in need of increased efforts to get them clean or to keep them clean. After 25 years of European water legislation this is a demand not only by the scientific community and other experts, but to an ever increasing extent by citizens and environmental organizations.

Let us take up the challenge of water protection, particularly with a view to ecological integrity. This is one of the great challenges for the European Union as it approaches the new millennium. Let us seize the initiative generated by the present political process on the Water Framework Directive for the benefit of all Europe's citizens and waters.

WOLFGANG STALZER
Austrian Ministry for Agriculture and Forestry, Director-General for Water Management Issues, A-1010 Vienna, Austria

HELMUT BLÖCH
European Commission, Head of Water Protection, B-1049 Brussels, Belgium

[1] MUHAR, S. et al., Ausweisung flußtypspezifisch erhaltener Fließgewässerabschnitte in Österreich. Wasserwirtschaftskataster, published by Austrian Ministry for Agriculture and Forestry, Vienna (1998)

[2] EUROPEAN COMMISSION: Proposals for a Water Framework Directive of 26.2.1997, COM(97)49, amendments of 26.11.1997, COM(97)614 and of 17.2.1998, COM(98)76, Official Journal C184 of 17.6.1997, C16 of 20.1.1998 and C108 of 7.4.1998, Brussels 1997 and 1998.

Hydrobiologia **422/423:** 1–14, 2000.
M. Jungwirth, S. Muhar & S. Schmutz (eds), Assessing the Ecological Integrity of Running Waters.

Introduction

Sustaining living rivers

James R. Karr[1] & Ellen W. Chu[2]
[1]*University of Washington, Box 355020, Seattle, WA 98195-5020, U.S.A.*
[2]*Northwest Environment Watch, 1402 Third Ave., Suite 1127, Seattle, WA 98101-2118, U.S.A.*

Key words: biological integrity, IBI, rivers, RIVPACS, urban rivers, water cycle

Abstract

Rivers cannot continue to meet society's needs, or the needs of living things, if humans continue to regard river management as a purely political or engineering challenge. The flow of rivers is part of a greater flow, the planet's water cycle, which sustains not only the flow of water but the entire web of life. Ultimately, the condition, or health, of the aquatic biota is the best means of understanding and controlling humans' impact on the Earth's watercourses and on the whole water cycle. Biological monitoring, especially multimetric approaches such as the index of biological integrity, acknowledges the importance of rivers' biotic integrity and offers one of the strongest available tools for diagnosing, minimizing, and preventing river degradation. The broad perspective offered by biological evaluations stands a better chance than narrow chemical criteria or conventional measures of urban development of sustaining living rivers.

One major, overwhelming reason why we are running out of water is that we are killing the water we have.
WILLIAM ASHWORTH, Nor Any Drop to Drink, 1982

Introduction

Humans could not live without rivers. As long as humans have inhabited the Earth, rivers have provided them with food, drink, and sites for settlement. People have long relied on rivers for cleaning as well as for waste removal and decomposition, not to mention commerce, transportation, and recreation. And rivers feed the soul; "without them we should be spiritually as well as materially deprived" (Macdonald, 1999; Pielou, 1998). So as agricultural, industrial, and domestic consumption of water and associated resources grows, concern about river conservation increases. Will the momentum be enough to reverse decades of river degradation?

For most of human history, human interactions with rivers aimed at creating or maintaining water supplies. These challenges were seen as both engineering and political: How can we haul water from stream to hearth? How do we keep water clean enough to drink? How much waste can rivers absorb? How can we protect homes and property from floods? Where can we find more water than flows within our political boundaries?

For thousands of years, but especially in the last hundred, human engineering prowess has been driven to ensure reliable supplies of water, to limit damage from flooding, and to avoid or clean up pollution. Human intents for water were assumed to be attainable by intensifying human control of water and watercourses. But now more than ever, water supplies are insufficient to meet growing demands, and flood control infrastructure cannot control flood damage. Indeed, our efforts to tame water have heightened our vulnerability to long-term fluctuations in climate and flow patterns and even brought about the unraveling of living aquatic systems.

Our sharp focus on political and engineering challenges has led us to ignore this unraveling and the biological challenges the collapse of living riverine systems brings. Many local and regional fisheries are gone; many others are so contaminated with pollutants that they are not safe to eat. Declining fishery or recreational resources are omens of a much wider crisis, but too few people see them as such. Likewise, too few people – including water scientists, engineers, and policymakers it seems – see the greater flow of which rivers are a part: the planet's water cycle.

Don't forget the water cycle

Every drop of water in rivers flowing across the land once fell to Earth from the atmosphere. Given enough time, every drop will return to the atmosphere as water vapor. Schoolchildren learn this lesson early. But when schoolchildren grow up, they forget about the water cycle. They begin to see rivers as 'surface waters' and 'channels' that 'drain' away 'dirt' to the sea. They see wetlands as swamps to be dried to make way for progress, not as sponges that shelter aquatic life and control floods. They forget that groundwater nourishes rivers; they think only of mining it to water cities and farms. And, ignoring the ancient lesson that a deluge is just compensation for human sins, they still build walls to contain the rivers. Yet without the endless cycle of water, human and other life on Earth would simply cease.

Ultimately, water matters most to living things, for without water, the rest of our Earthly habitat could not sustain us. The converse is that life itself, and biology, offers the best means of telling whether our actions are allowing the water cycle to continue unimpaired. The natural transformations and flows of water support all life, including humans, and it is living aquatic systems that provide the goods and services humans need to survive.

River biota: mark of integrity

The biota of every watershed on Earth is the product of millions of years of geological change and biological evolution; the very existence of living organisms represents the integration of conditions around them. Such a highly evolved, complex living web is the mark of integrity against which we can measure degradation. Living things in a water body 'do the integral,' so to speak, of every biogeochemical process that has shaped that water body. Understanding the interactions that keep water bodies and other places alive holds the secret to sustaining water supplies for human enterprise and, indeed, for sustaining all life.

Retaining the biological parts of riverine ecosystems (populations, species, genes), as well as the processes that nourish them (mutation, selection, fish migration, biogeochemical cycles, and the water cycle itself), is crucial to retaining water supplies and all the goods and services associated with water (Karr & Chu, 1999). By using the word *integrity,* the US 1972 Water Pollution Control Act Amendments (now called the Clean Water Act) recognized this importance of whole hydrobiological systems: "The objective of this Act is to restore and maintain the chemical, physical, and biological integrity of the Nation's waters." This perspective has been carried into the Great Lakes Water Quality Agreement between the United States and Canada and into the Water Framework that now guides the European Union.

Still, despite progress in controlling pollution coming from point sources, aquatic ecosystems have continued to decline worldwide. Altered water flows in dammed streams and rivers; pollution from nonpoint sources such as cities, farms, and feedlots; destruction of habitats above and alongside rivers by development or logging; and invasions by alien species have all taken a heavy toll.

For too long, people have regarded water as a simple fluid to be used; water unused was water wasted. Consequently, water resource management is still dominated by legal doctrines from a less crowded world, weak implementation of good laws, and a focus on water chemistry (Karr, 1995). But such narrow views do not guarantee the well-being of aquatic life, the integrity of water and watersheds, or the continuity of the water cycle. Seeing water only as a fluid – even a clean fluid – protects neither water quality nor the continued supply of water itself. Only protecting the integrity of the water cycle can protect both (Falkenmark, 1997).

The most direct and effective measure of a water body's integrity, and of its place in the water cycle, is the status of life in the water. "Ecosystem integrity is primarily a biological concern" (Reynoldson et al., 1995). Living communities reflect watershed conditions better than any chemical or physical measure because they respond to the entire range of biogeochemical factors in the environment. When something alters the landscape of a river's headwaters, life in lowland reaches feels the effects. Actions that protect the biota tell us directly if we are protecting the water cycle. Protecting the biological integrity of water – the full range of diversity and ecological processes reflecting the evolutionary and biogeographical history of a place – will thus protect human uses of that water, whether for drinking, fishing, washing, flushing, shipping, irrigating, generating electricity, or making money in countless ways. When waters no longer support living things, they will no longer support human affairs (Karr & Chu, 1999).

Biological monitoring: vital for the long term

More than a century ago, people recognized that human activities produced pollution harmful to the biota (for reviews see Davis, 1995; Perry & Vanderklein, 1996). They therefore made an effort to track the extent of biological degradation; biological degradation was even considered an indicator for the presence of human activities. So began biological monitoring (Forbes, 1928; Kofoid, 1903, 1908; Kolkwitz & Marrson, 1908).

Rather than staying focused on biology, however, water quality evaluations shifted for much of the twentieth century to the effects of chemical contaminants; rarely were connections between chemical criteria and ambient biotic condition documented. At the same time, water quantity evaluations measured water volume in light of human needs or wants. Rarely, if ever, was the whole flow of the water cycle seen as limiting the quantity of water supplies or as the environment in which water planning should be taking place. Humans pushed ahead with control, not understanding.

During the past two decades, life in the waters has again come to the fore, and biological monitoring – detecting human-caused biotic changes apart from those occurring naturally – is again part of water managers' tool kits (Chovanec et al., 1995; Davis & Simon, 1995; Hellawell, 1986; Karr, 1981, 1991; Karr & Chu, 1999; Norris & Thoms, 1999; Norris et al., 1995; Reynoldson et al., 1995; Rosenberg & Resh, 1993; Simon, 1999; Wright et al., 1989, 1997).

Two major approaches to biological monitoring have emerged in the past 20 years: the river invertebrate prediction and classification system (RIVPACS) and the index of biological integrity (IBI). The underlying principle of both is that the biota is the ultimate integrator of all human actions. The two approaches overlap in many important ways. Both, for example, (1) focus on biological endpoints to define river health; (2) use a concept of reference condition as a benchmark; (3) organize sites into classes with a select set of environmental characteristics; (4) assess change and degradation caused by human effects; (5) require standardized sampling, laboratory, and analytical methods; (6) score sites numerically to reflect site condition; (7) define "bands," or condition classes, representing degrees of degradation; and (8) furnish needed analyses for selecting high-quality areas as acquisition and conservation priorities.

Despite these shared properties, the details of RIVPACS and IBI also differ in important ways (Table 1). IBI was first developed for use with fish (Karr, 1981; Karr et al., 1986) and has subsequently been adapted for use with invertebrates (Karr, 1998; Karr & Chu, 1999). IBI can be used with algae, plankton, and vascular plants in wetlands, streams, and coastal estuaries, and the index is also being adapted for use in terrestrial environments (Bradford et al., 1998; Karr & Chu, 1999).

IBI and RIVPACS both classify stream segments according to important environmental features in order to restrict comparisons to areas with equivalent natural environmental conditions. Both RIVPACS and IBI define a reference standard as a benchmark of a healthy river, although how those standards are defined differ. RIVPACS analyses select sites of "good or fairly good quality" (Armitage et al., 1983) or that are "free of serious pollution" (Wright et al., 1984). The focus on chemical contamination contrasts with the broader reference standards established for IBI, where reference stream segments or watersheds are chosen to have little or no human influence from any source, not just chemical pollution.

Perhaps the most important difference between RIVPACS and IBI is the biological information used to frame the assessment process. The early goal of RIVPACS was to improve site selection for conservation (Wright et al., 1984). As a result, recognizing patterns of species composition was and continues to be the core of RIVPACS analyses. In contrast, IBI was developed to measure river condition, or river health, and included development of measurements for diagnosing causes of degradation. As a result, the biological signals that make up IBI analyses are broader, including taxa richness and composition, trophic or other aspects of ecological organization, presence and relative abundance of tolerant and intolerant taxa, and presence of diseased individuals or individuals with other anomalies (Karr, 1991; Karr & Chu, 1999). A primary goal of IBI is to define the attributes of living systems that change systematically in diverse situations when exposed to the activity of humans. RIVPACS emphasizes species identity and abundances. IBI searches for a broader array of biological signals from species to higher taxa, including sys-

Table 1. Comparative analysis of the conceptual, sampling, and analytical characteristics of multivariate (RIVPACS) and multimetric (B-IBI) approaches to river monitoring and assessment (based on the most common current applications)

	RIVPACS[a] (multivariate)	IBI[b] (multimetric)
Site classification or characterization	Stream size, geography, substrate particle size, water chemistry	Geology, stream size and temperature, altitude
Reference standard	Sites without serious pollution (RP); other human influences not considered. Broader in recent applications (AR)	Sites with little or no human influence
Model foundations	Multivariate associations between environmental variables and species present	Empirically defined measures; dose–response graphs plotting human influence against biological response
Decision criteria	Species presence or absence (observed vs. expected ratios) from probabilistic models; sometimes tolerance measures	Biological attributes such as taxa richness, relative abundance, taxa composition, tolerance and intolerance
Microhabitat sampled	Multiple (e.g., riffle, edge, pool-rock: RP). Riffles (AR)	Riffles
Subsampling	Varies regionally and with application	Full samples counted
Stream applications available	Benthic invertebrates	Benthic invertebrates, fish, algae
Data set required	Extensive, hundreds of sites Regional foundation; some smaller-scale applications being developed	Fifteen to 20 sites representing a broad gradient of human disturbance; can be developed locally or over larger areas
Transferability	Extensive data sets and new species-specific models needed for each region	Consistency in selected measures and biological responses across regions
Treatment of rare species	Often excluded from analyses	Included in analyses
Sampling period	RP: combined 3-season model AR: model for each season	Defined period for sampling
Analytical basis	Species presence–absence	Diverse dimensions of biological and natural history patterns
Human influence	Largely chemical contamination	Full spectrum of human influences
Diagnostic capability	Not explored	Moderately well developed
Communication	Statistical foundations difficult Breadth of signal narrow (O/E ratio); pollution tolerance	Simple dose–response curves similar to toxicology; broad range of biological signal

[a]RIVPACS (RP) developed in England and cloned as AUSRIVAS (AR) in Australia for rivers and streams, being applied in rivers in other regions.
[b]IBI developed in midwestern United States for rivers and streams; now used internationally. Applications for other habitats in use or development.

tematic shifts in the natural history of assemblages of species. Because RIVPACS analyses emphasize probability statements about species presence, rare taxa are excluded from analyses; rare taxa are not excluded in IBI. Finally, RIVPACS depends on large data sets for its multivariate models, whereas smaller data sets still produce an effective IBI.

These and other differences notwithstanding, both RIVPACS and IBI have changed the way many scientists and water managers think. Developing and testing these integrative approaches to water resource assessment have provided a long-needed counterpoint to the history of monitoring chemical water quality and simple water quantity.

What is IBI?

IBI – like the multifaceted indexes measuring economic and human health – integrates multiple biological indicators to measure and communicate biological condition, and thereby the ecological health of a river and its watershed. Much as a physician relies on a battery of medical tests, not just one, to diagnose illness, anyone can use an IBI to diagnose the condition of a water body. Connecting many previously disconnected dimensions of water science and policy, the index has been tested and refined over nearly two decades and now provides the foundation of US federal programs for biological monitoring. An IBI based on fish has been applied on every continent except Antarctica; in developing as well as developed countries; and in basic science, resource management, engineering, public policy, legal, and community volunteer arenas (Hughes & Oberdorff, 1999; Simon, 1999). An IBI based on benthic invertebrates (B-IBI) has been applied in areas from the United States to Japan.

Based on empirically defined 'dose-response' relationships between particular human influences and the biological responses they provoke, IBI is one of the most commonly used – "and arguably the most effective" (Simon, 1999) – biological monitoring approaches. IBI, along with its component metrics:

- tells us whether we are maintaining water bodies, water supply, and flow through the water cycle, along with the vital resources the water cycle supports;
- works in streams, rivers, lakes, wetlands, estuaries, and coastal marine systems;
- integrates elements that conventionally have been fragmented in water policy and decision making, such as water quality and water quantity, surface water and groundwater;
- stands up to scientific and legal scrutiny and is actually improving the legal and regulatory approaches used by governments to protect the public's interest in water resources;
- directly addresses the central call of the United States' Clean Water Act and the European Union's Water Framework to protect the integrity of water;
- is simple to develop and use, demanding no advanced technologies beyond the reach of developing countries with limited financial means or of citizen groups seeking to understand the condition of their local and regional watersheds;
- defines the health of a water resource system and aids in diagnosing and identifying causes of any detected degradation;
- enables us to identify and protect the places most deserving of conservation (defines places where restoration is possible and practical, and guides development activities to prevent or minimize damage to water resources);
- provides results that can be used to assess the effectiveness of particular resource management decisions and to set funding priorities;
- extends the concept of taking multiple measures to assess health – long central in economics and medicine – to environmental assessments;
- integrates precise biological measurements of the condition of waters and their associated resources into numbers and words that are easily understood by diverse audiences;
- allows us to compare the effects of single acts with the cumulative effects of many activities;
- permits comparisons across time and space: of the effects of different human activities through time at the same site or of watershed condition in different geographic regions;
- allows us to measure and compare the relative impacts of different human land uses, including recreation, farming, logging, and urbanization, and to compare such impacts with those affecting water bodies directly, such as pollution, channelization, or dam building;
- can be applied with a broad range of taxa, from algae and vascular plants to invertebrates and fishes.

Human activities degrade water resources by altering one or more of five principal groups of attributes: water quality, habitat structure, flow regime, energy source, and biotic interactions (Figure 1; Karr & Chu,

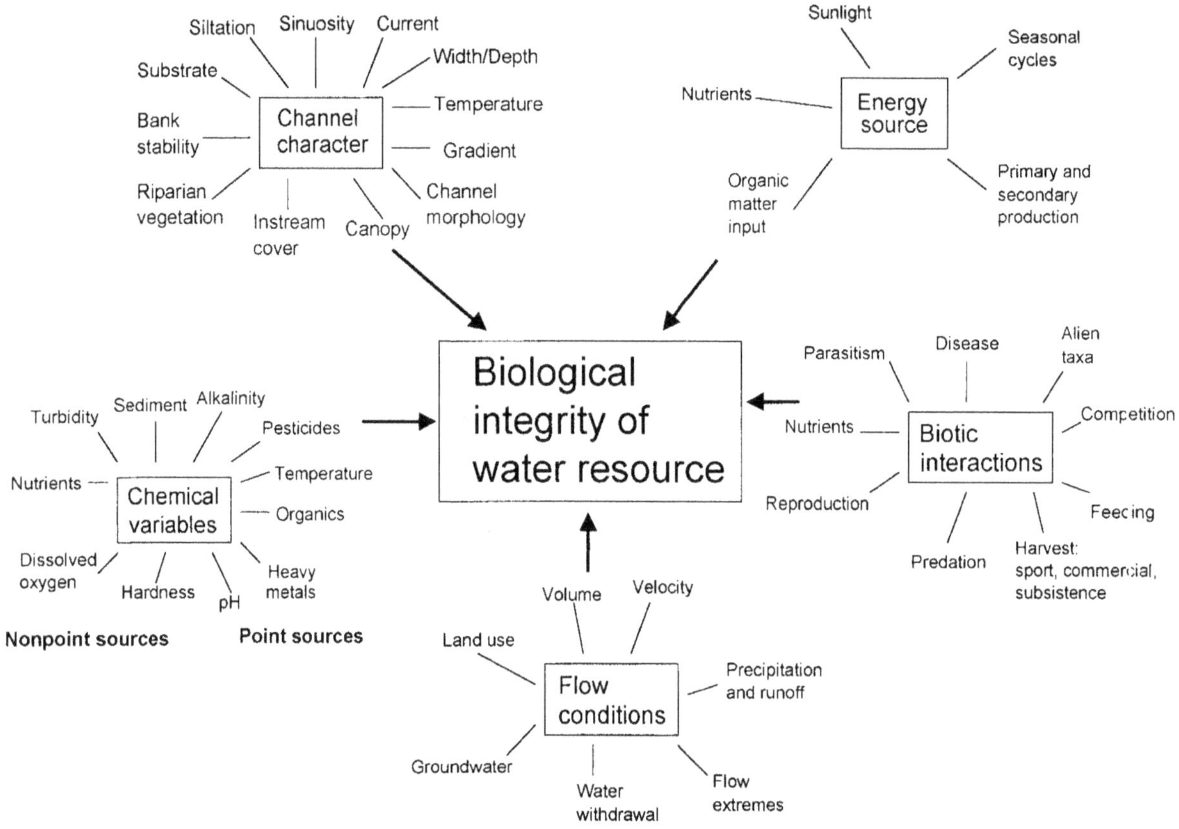

Figure 1. The five principal factors with some of their important chemical, physical, and biological components that are commonly altered by human actions (modified from Karr et al., 1986).

1999). IBI is an effective assessment tool because it incorporates the sensitivity of the biota itself to all five of these factors.

Building a robust and effective IBI requires selecting measurable attributes that provide reliable and relevant signals about the biological effects of human activities. The biological attributes ultimately incorporated into an IBI are called metrics. They are chosen because they reflect specific, predictable responses of organisms to changes in landscape condition. They are sensitive to a range of physical, chemical, and biological factors that alter biological systems, and they are relatively easy to measure and interpret. The index of biological integrity explicitly comprises several attributes of the sampled biota, including taxa richness, indicator groups, health of individual organisms, and ecological processes (Table 2).

The most important criterion for choosing a biological attribute as a metric is whether the attribute responds predictably along a gradient of human influence. Does the attribute vary systematically with varying degrees of human impact? In Japan, the bio-

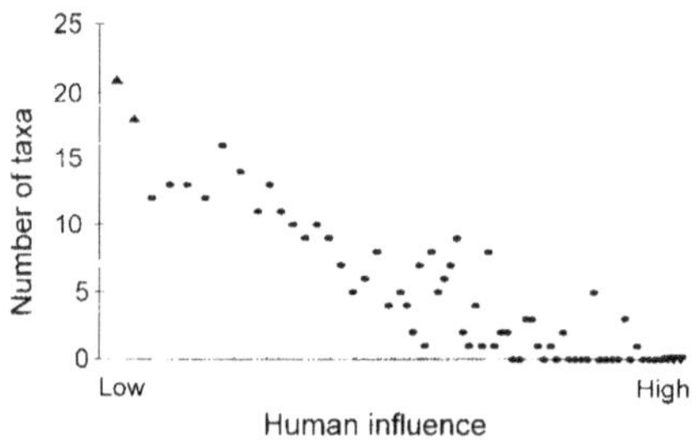

Figure 2. The number of taxa of rock-clinging invertebrates declined with increasing human influence in 65 west-central Japanese rivers (Rossano, 1996) because increasing sediment filled the interstitial spaces in the gravel streambed.

diversity of invertebrates clinging to rocks in a stream goes down as human activities in the landscape increase (Figure 2; Rossano, 1996). Such relationships are best discovered and deciphered with graphs. A multimetric IBI comprising well-chosen metrics integrates information from ecosystem, community, population, and individual levels (Karr & Chu, 1999) and clearly discriminates biological 'signal' – including the effects of human activities – from the 'noise' of

Table 2. Sample biological attributes, in four broad categories, that might have potential as metrics. Actual monitoring protocols have proven some of these attributes effective; other attributes may work but need more testing; still others are difficult to measure or too theoretical. Ideally, an IBI should include metrics in each of these categories, but untested or inadequately tested attributes should not be incorporated into the final index (Modified from Karr & Chu, 1999)

Category	Demonstrated effective	Need more testing	Difficult to measure or too theoretical
Taxa richness	Total taxa richness Richness of major taxa, (e.g., mayflies or sunfish)	Dominance (relative abundance of most-numerous taxa)	Relative abundance distribution, after Preston (1962)
Tolerance, intolerance	Taxa richness of intolerant organisms Relative abundance of green sunfish Relative abundance of tolerant taxa	Number of rare or endangered taxa	Chironomid species (difficult to identify)
Trophic structure	Trophic organization, e.g., relative abundance of predators or omnivores		Productivity
Individual health	Relative abundance of individual fish with deformities, lesions, or tumors Relative abundance of individual chironomids with head-capsule deformities Growth rates by size or age class	Contaminant levels in tissue (biomarkers)	Metabolic rate
Other ecological attributes		Age structure of target species population	

Table 3. Biological monitoring and assessment support the goals of many sections in the US Clean Water Act

Listing of impaired waters (§303d)
Permitting of point-source discharges (§402)
Nonpoint source assessment (§319)
Hazardous waste site assessments (§104e)
Evaluation of habitat modifications (§401/404)
Analysis of total maximum daily loads (TMDLs)
Aquatic life use designations (§305b)
Reporting condition of waters (§305b)
Clean Lakes Program activities (§314)
Thermal effluent and entrainment-impingement (§316a, b)
Wet-weather discharge (stormwater) (§319, §402)
Water quality standards and criteria (§303c)
Comparative risk assessment (multiple sections)
Watershed protection and assessments

natural variation. Moreover, because IBI is founded on empirical data, its use does not require resolution of all the higher-order theoretical debates in contemporary ecology (Miller & Ehnes, in press) or the formal mathematical models of ecological functions.

IBI in action

The strength of IBI is its adaptability – to diverse contexts and geographies and diverse environmental conditions, such as point-source and nonpoint pollution or the effects of urbanization. As the index has been tested worldwide, IBI has proven more sensitive to human impacts than many conventional measures of these impacts. IBI has detected the effects of effluents from a bauxite plant in Guinea (Hugueny et al., 1996) and from salmonid aquaculture on small streams in France (Oberdorff & Porcher, 1994); the effects of channelization and chemical effluents in small Venezuelan rivers (Gutierrez, 1994); metal and organic pollution in central Indian rivers (Ganasan & Hughes, 1998); the cumulative effects of channelization, agricultural runoff, and urbanization in France's Seine River basin; and the impact of diverse land uses on streams in dry west-central Mexico (Lyons et al., 1995). IBIs have been used to reflect the biological effects of chemical contamination (Figure 3) and the impacts of human activities on a watershed or landscape scale (Figure 4). IBI has served as a framework for reinventing biological assessments for water resource investigations (Figure 5), as a foundation for citizen water-monitoring programs (Figure 6), and to

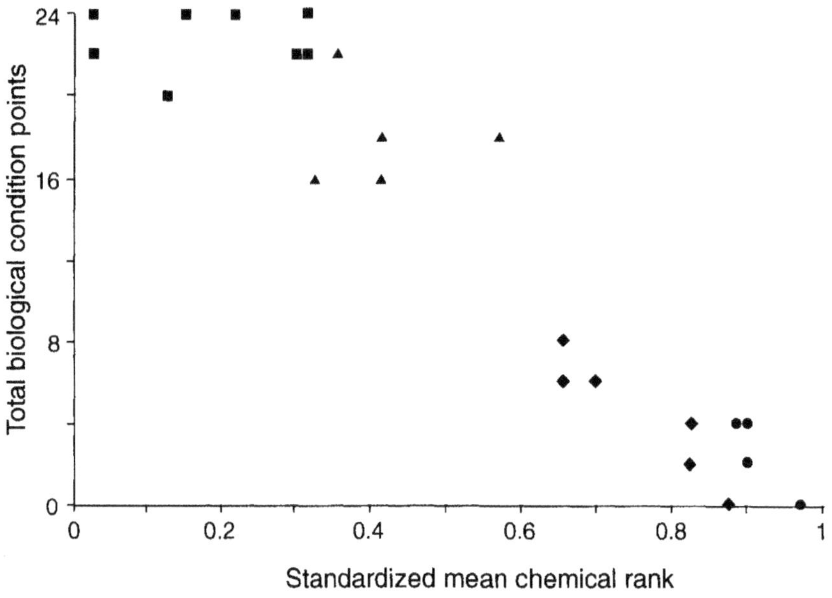

Figure 3. Within stream channels, IBI fully reflects chemical contamination in diverse geographical settings, such as urban point and nonpoint pollution gradients from 'pristine' to 'grossly polluted' in Thailand, Ghana, and Brazil (Thorne & Williams, 1998).

support the goals of the Clean Water Act (Table 3). IBI helped to evaluate watersheds in the entire Sierra Nevada for setting conservation priorities; it was applied in a long-term evaluation of dammed streams and to evaluate a relatively undisturbed stream heavily invaded by alien species (Moyle & Randall, 1998).

Tracking the health of urban rivers

Urbanization, the concentration of more and more humans into densely populated areas, may be the most important trend influencing rivers today. In 1850 only one metropolis boasted a population larger than 1 million. By 1998, 326 cities had populations exceeding 1 million, and 14 megalopolises had populations greater than 10 million (O'Meara, 1999). Globally, urban populations increase by 1 million each week, driven by natural increase, migration from rural areas, and urban sprawl engulfing nearby towns. In the United States, officially designated metropolitan areas have mushroomed from 9% of the nation's land area in 1960 to 19% (Stoel, 1999). Four out of five US citizens live in urban zones, over half in areas with more than a million people. In the twenty-first century, a growing share of people will come to measure environmental quality by conditions in their cities (Stoel, 1999).

Indeed, many US residents believe that urban environmental conditions are deteriorating, and the condition of waterways is high on their list of worries (Karr et al., in press). Rivers in most cities have been straightened, diverted, dammed, contained, and otherwise physically engineered beyond recognition. They have been hydrologically transformed and polluted to the point of catching fire – all this to feed cities' demand for massive concentrations of water, food, energy, and raw materials. Cities also put out huge amounts of organic and synthetic garbage, fully expecting rivers to carry much of it away. As sprawl pushes beyond city centers into neighboring watersheds, it degrades rural watersheds and streams as well. Urban ecological footprints (Wackernagel & Rees, 1996), which extend much farther than urban boundaries, put tremendous pressure on rivers, damaging them more than agriculture, logging, and other land uses. Both within and between midwestern US watersheds, the most degraded streams (as measured by biological condition) are located in urban areas (Karr et al., 1986; see also Karr & Chu, 1999: Figure 6).

In the face of such pressures, looking at the living systems in urban watersheds stands a better chance of protecting long-term human needs and well-being than the narrower management perspectives of the past. Multimetric indexes like IBI can improve both the conservation and restoration of urban rivers even as they guide decision-making about development.

Past urban watershed management concerned itself mostly with stormwater management: managing

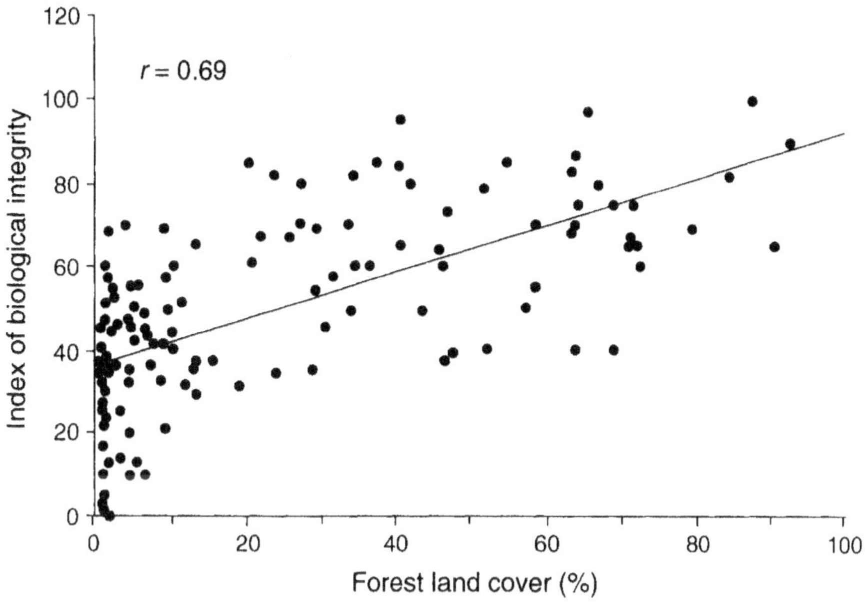

Figure 4. IBIs also reflect impacts of human activities on a watershed scale. For example, an IBI based on fishes declined as forest land cover declined in Wisconsin (Wang et al., 1997).

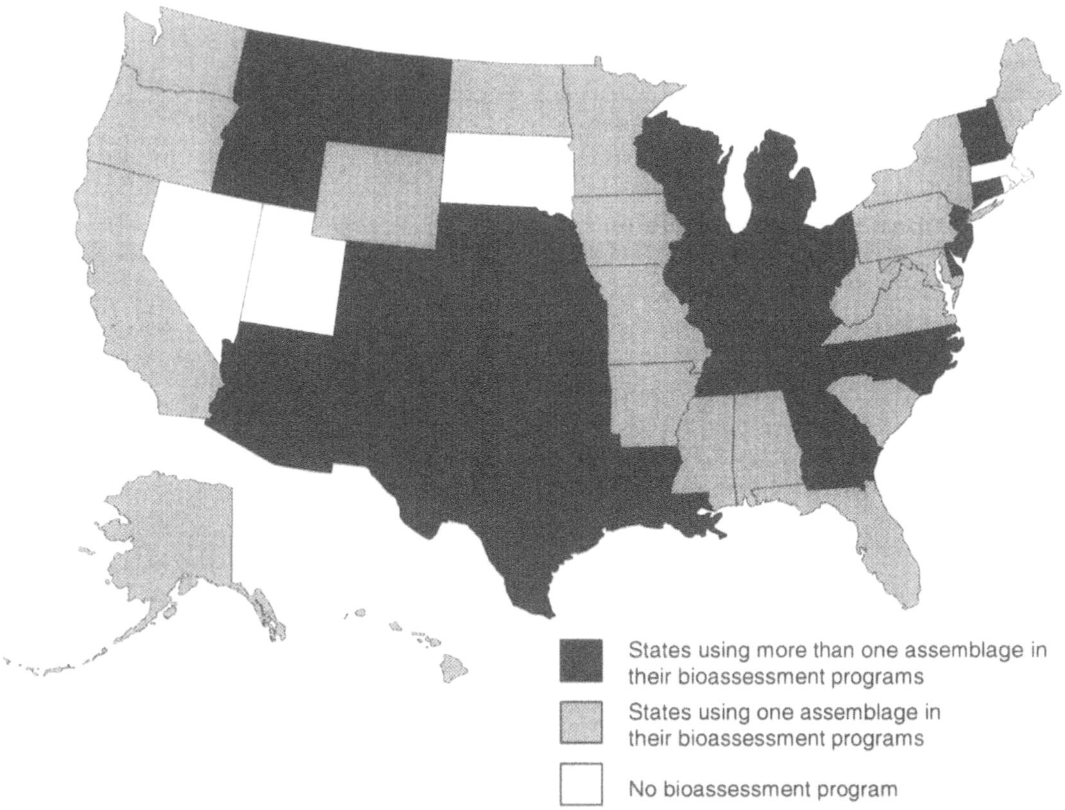

Figure 5. IBI has provided the framework for reinventing biological assessments for water resource investigations, leading many US states to make such assessments central to their water resource programs (from Davis et al., 1996).

increased runoff during storms, a consequence of paving over large segments of a basin. Little attention was directed at measuring or understanding runoff's effects on the biological health of rivers (Roesner, 1997). Fortunately, this perspective is beginning to change, within governments and among professionals and laypersons alike.

One study of Puget Sound lowland streams in some 40 watersheds near Seattle (Washington, USA) sought to identify the linkages between landscape conditions, in-stream environmental factors, and stream biota as well as to develop a set of indexes of stream quality that could help minimize river degradation caused by urban development (May et al., 1997). Among the study's elements was an examination of biological condition (expressed as a benthic IBI) with respect to the percentage of total impervious area (% TIA), a measure that has been widely embraced in regulatory circles because of its reputation as an integrative measure of human activity. The study streams – all similar in size, physiography, and the historical presence of breeding salmon populations – ranged from relatively undeveloped (2–5% TIA) to highly urbanized (60% TIA).

In theory, an integrative measure of urban development should be highly correlated with a stream's biological condition (Figure 7). Sites with little or no development should support a diversity of species and an ecological organization that reflects the history of evolutionary and biogeographic processes for that site and region; such sites have biological integrity. An integrative measure of urbanization's effects on a stream biota should show a drop in biological condition, away from integrity, as urbanization intensifies.

The percentage of total impervious area is indeed correlated with declining stream health, or biological condition as measured by B-IBI (Figure 8a). But the correlation is not tight as it is in Figure 7, suggesting that % TIA alone does not capture all aspects of urbanization (e.g., removal of riparian vegetation) or account for all its biological effects. At low levels of development (low % TIA) some streams remained healthy (B-IBI = 45–50) while others suffered (B-IBI = 15–20); at high development levels, biological condition was uniformly poor.

A line through the plot's top points (Figure 8b) illustrates what Thomson et al. (1986) term a 'factor ceiling distribution,' defining the best biological condition one can expect as a function of % TIA. The scatter of points below the ceiling shows that factors other than % TIA are at work, and that they can have marked effects on the biota even when urbanization is minimal. Yet development permitting is often based on the assumption that the correlation between biological condition and % TIA is direct, tight, and simple. It is not.

Urban development radically alters a watershed's drainage network by changing or destroying many small natural streams and channels and by creating artificial channels across and along roads, in culverts, and through stormwater outfalls. The percentage of impervious area likely takes much of this hydrological transformation into account (producing the 'ceiling' in Figure 8b), but omits other activities associated

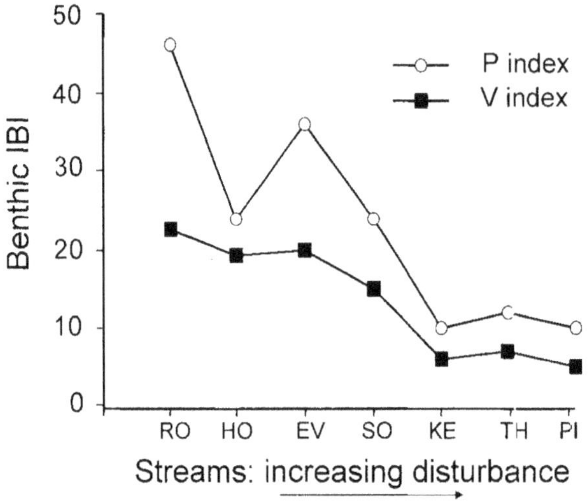

Figure 6. IBI provides a powerful tool that even citizen water-monitoring groups can use. A study in the Puget Sound lowlands (Fore et al., in press) showed volunteer efforts to be in line with professional laboratory studies. Indexes based on volunteer (V) and professional (P) lab analyses were correlated with each other ($P < 0.01$) and with disturbance ($P < 0.01$).

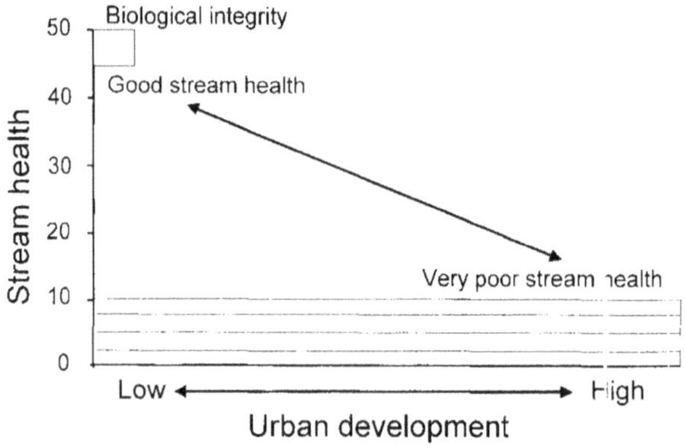

Figure 7. An ideal measurement of urban development (x-axis) that accounts for development's impacts on living aquatic systems should be tightly correlated with actual measures of a stream's biological condition, or health (y-axis). One would expect stream health to fall way from integrity (top left) as development increased.

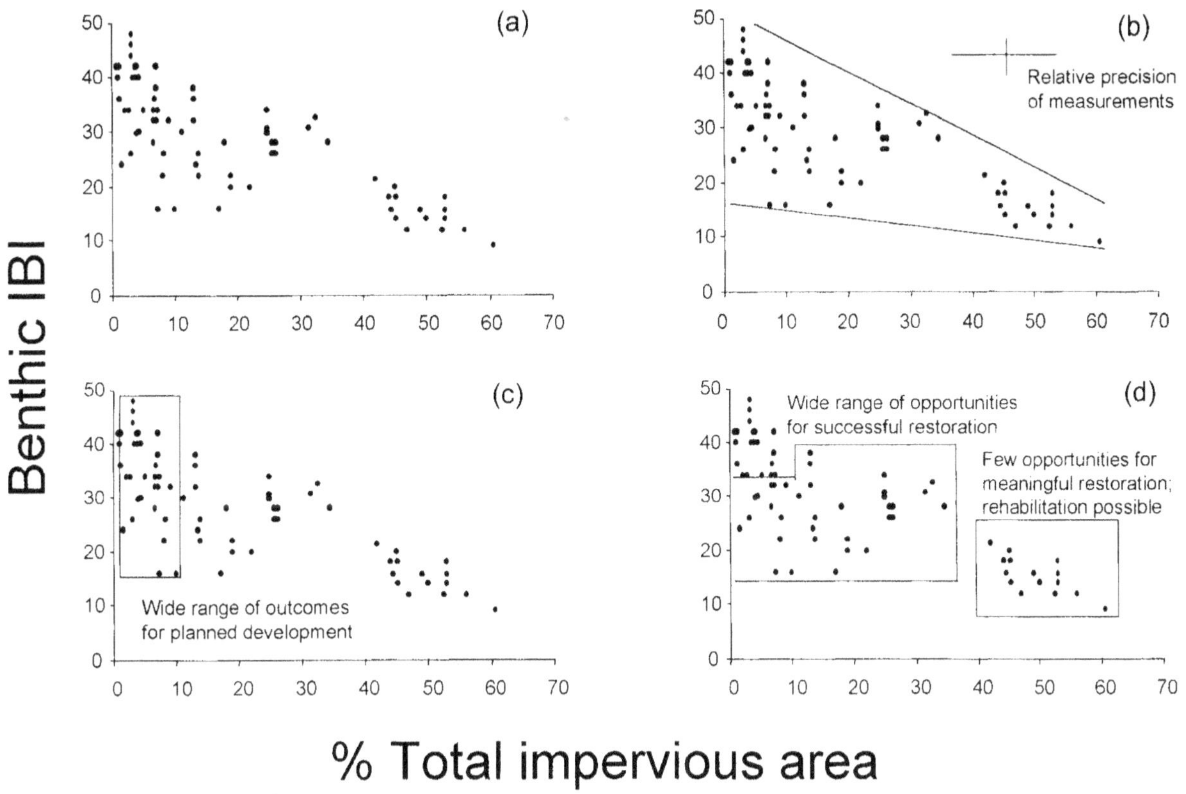

Figure 8. (a) Biological condition, as measured by the benthic index of biological integrity (IBI), declines with increasing urban development expressed as the percentage of total impervious area (% TIA). The correlation is not tight, however, especially at low levels of development. (b) As indicated by the error bars at the top right, B-IBI is a statistically precise measure of biological condition; % TIA, however, does not reflect all of development's effects on a biota, although it does set a 'ceiling' for the biota's response (the distinct upper bound of the plot's wedge shape, termed a factor ceiling distribution). (c) At low levels of imperviousness, other factors not captured in measures such as % TIA, including riparian condition or point sources of pollution, are often responsible for biological decline. (d) At intermediate levels of imperviousness, the index of biological integrity can help managers choose among a variety of potential sites for successful restoration as well as guide managers to the most successful restoration approaches at those sites. At very high levels of imperviousness, IBI can help avoid wasting resources on sites with only limited possibilities for restoration or rehabilitation.

with development, such as point pollution sources or riparian cover. In the Puget Sound lowlands, for example, watersheds with protected riparian corridors and wetlands are in better biological condition, as measured by B-IBI, than watersheds where development is concentrated along streams. A stream with effluent from an old mine seeping into it is in worse condition than a stream with similar levels of development, as measured by impervious area, but no abandoned mine. Similarly, sites on streams adjacent to wastewater treatment plants have lower B-IBIs than streams with the same % TIA but no wastewater plant (J. Adams et al., 1999).

What studies of B-IBI in urban areas has revealed is that an index like IBI better reflects the cumulative influence of human activity on a biota than conventional measures such as % TIA. It also shows that IBI is a better guide than % TIA for planning development. At low levels of development, some activities have relatively little impact on stream biota, while others harm it (Figure 8c). At intermediate and high (Figure 8d) levels of imperviousness, IBI enables managers to pick and choose among a range of opportunities for successful restoration or rehabilitation and to avoid sites where expensive restoration efforts would be wasted. Continuing studies are identifying which development activities do the most damage, ultimately offering planners a guide for development least likely to harm streams and watersheds.

What's next?

Biological monitoring gets at the heart of sustaining water resources because biology – more precisely, the biota – is the best indicator of the health of humans' relationship with water. If we understand this relationship, we can meter our activities to ensure the maintenance of the planet's life-giving water cycle. Without such understanding, we are likely to continue dealing with one problem at a time, out of the larger context. We will pile unintended consequence upon unintended consequence as we have for centuries – and end up robbing the 'watery planet' of its most precious resource.

Inasmuch as the strength of IBI is its adaptability and utility for getting this understanding across to a broad audience, the goal of future IBI work is to adapt it and use it as much as possible. After all, an effective river IBI can be developed within the time frame and budget of a master's thesis. IBI and other biological monitoring tools can:

- heighten awareness that 'it's the biology!' across the broadest possible technical, scientific, and management circles and among citizen groups;
- speed adoption of biological monitoring by growing numbers of local, state, regional, federal, and even international organizations charged with maintaining sustainable supplies of water.

Biological monitoring as exemplified by IBI marks a profound shift in thinking. First, it puts biological attributes, instead of chemical concentrations, at the center of water management. Second, by not emphasizing species – endangered or commodity species, in particular – or population size and ecosystem productivity, it has overturned ideas of which biological attributes are best for biological monitoring and conservation.

So while some scholars continue to argue about the merits of 'health' as an ecological metaphor or how to 'operationalize' the concept of ecological integrity (Schrader-Frechette, 1995), biological monitoring is rapidly being adopted by scientists and groups ranging from local grassroots watershed alliances to local, state, and regional water management agencies and the regulated community. The more we come to use biological monitoring, the more likely we are to keep our rivers and their bounty alive. Our rivers, their biotas, and we who depend on both deserve nothing less.

When it comes to sustaining living rivers and the other life-giving services ecological systems provide, the stakes are too great and the obstacles too large to simply do 'more research.' The results of outstanding research are already piled high on library shelves while human activities unwittingly destabilize our entire planet's water cycle. More research is not enough. We have to put our basic research results – our understanding of *biological* responses – to work in the world's watersheds. Failure to rise to this challenge may spell failure to keep the world's waters alive.

Acknowledgements

This paper was made possible by grants from many agencies and organizations during the past 30 years. Recently, especially important support came from Washington Department of Ecology Centennial Clean Water Fund Grant No. G9400121, Environmental Protection Agency/National Science Foundation Water and Watersheds Program Grant EPA R82-5284-010, and the Consortium for Risk Evaluation with Stakeholder Participation (CRESP) by Department of Energy Cooperative Agreement #DE-FC01-95EW55084.S. Among others, Derek Booth, Stephen Burges, and Sarah Morley contributed significantly to the understanding of urban streams.

References

Adams, J. W., J. R. Karr & T. C. Dewberry, 1999. Evaluating Aquatic Resources in the Clackamas River Basin of Oregon. Report on Contract #CR825041-01, US Environmental Protection Agency, Corvallis, OR, USA.

Armitage, P. D., D. Moss, J. F. Wright & M. T. Furse, 1983. The performance of new biological water quality score system based on macroinvertebrates over a wide range of unpolluted running water sites. Wat. Res. 17: 333–347.

Bradford, D. F., S. E. Franson, A. C. Neale, D. T. Heggem, G. R. Miller, & G. E. Canterbury, 1998. Bird species assemblages as indicators of biological integrity in Great Basin rangeland. Envir. Monit. Assess. 49: 1–22.

Chovanec, A., V. Koller-Kreimel, O. Moog & S. Weiss, 1995. Assessment of the ecological integrity of running waters: the Austrian approach. Proceedings of International Workshop on Assessment and Classification of Rivers, 5–7 November 1995, Düsseldorf.

Davis, W. S., 1995. Biological assessment and criteria: Building on the past. In Davis, W. S. & T. P. Simon (eds), Biological Assessment and Criteria: Tools for Water Resource Planning and Decision Making. Lewis, Boca Raton, FL: 15–29.

Davis, W. S. & T. P. Simon (eds), 1995. Biological Assessment and Criteria: Tools for Water Resource Planning and Decision Making. Lewis, Boca Raton, FL.

Davis, W. S., B. D. Snyder, J. B. Stribling & C. Stoughton, 1996. Summary of State Biological Assessment Programs for Streams and Rivers. EPA 230-R-96-007. Office of Policy, Planning, and Evaluation, US Environmental Protection Agency. Washington, DC.

Falkenmark, M., 1997. Society's interaction with the water cycle: a conceptual framework for a more holistic approach. Hydrol. Sci. J. 42: 451–466.

Forbes, S. A., 1928. The biological survey of a river system - its objects, methods and results. Illinois Nat. Hist. Survey Bull. 17: 277–284.

Fore, L. S., K. Paulsen & K. O'Laughlin, in press. Statistical power of volunteer monitoring protocols. Freshwat. Biol.

Ganasan, V. & R. M. Hughes, 1998. Application of an index of biological integrity (IBI) to fish assemblages of the rivers Khan and Kshipra (Madhya Pradesh), India Freshwat. Biol. 40: 367–383.

Gutierrez, M. A. R., 1994. Utilizacion de la ictiofauna como indicadora de la integridad biotica de los Rios Gouche and Guanare. Estado Portuguesa. Venezuela. M.Sc. thesis, Universidad National Experimental de los Llanos Occidentales 'Ezequiel Zamora,' Guanare, Venezuela.

Hellawell, J. M., 1986. Biological Indicators of Freshwater Pollution and Environmental Management. Elsevier, London.

Hughes, R. M. & T. Oberdorff, 1999. Applications of IBI concepts and metrics to waters outside the United States. In Simon, T. P. (ed.), Assessing the Sustainability and Biological Integrity of Water Resources Using Fish Communities. CRC Press, Boca Raton, FL: 79–93.

Hugueny, B., S. Camara, B. Samoura & M. Magassouba, 1996. Applying an index of biotic integrity based on fish assemblages in a West African river. Hydrobiologia 331: 71–78.

Karr, J. R., 1981. Assessment of biotic integrity using fish communities. Fisheries 6(6): 21–27.

Karr, J. R., 1991. Biological integrity: a long-neglected aspect of water resource management. Ecol. Appl. 1: 66–84.

Karr, J. R., 1995. Clean water is not enough. Illahee 11: 51–59.

Karr, J. R., 1998. Rivers as sentinels: using the biology of rivers to guide landscape management. In Naiman, R. J. & R. E. Bilby (eds), River Ecology and Management: Lessons from the Pacific Coastal Ecoregion. Springer, New York: 502–528.

Karr, J. R. & E. W. Chu, 1999. Restoring Life in Running Waters: Better Biological Monitoring. Island Press, Washington, DC.

Karr, J. R., K. D. Fausch, P. L. Angermeier, P. R. Yant & I. J. Schlosser, 1986. Assessment of biological integrity in running waters: a method and its rationale. Illinois Nat. Hist. Surv. Spec. Publ. 5.

Karr, J. R., J. D. Allan & A. C. Benke, in press. River conservation in the United States and Canada: science, policy, and practice. In Boon, P. J., B. R. Davies & G. E. Petts (eds), River Conservation: Science, Policy, and Practice. John Wiley, Chichester, UK.

Kofoid, C. A., 1903. The plankton of the Illinois River, 1894–1899, with introductory notes upon the hydrography of the Illinois River and its basin. Part I. Quantitative investigations and general results. Bull. Illinois State Lab. Nat. Hist. 8: 1–360.

Kofoid, C. A., 1908. The plankton of the Illinois River, 1894–1899, with introductory notes upon the hydrography of the Illinois River and its basin. Part II. Constituent organisms and seasonal distribution. Bull. Illinois State Lab. Nat. Hist. 8: 1–360.

Kolkwitz, R. & M. Marsson, 1908. Ökologie der pflanzlichen Saprobien. Bericht der Deutschen Botanischen Gesellschaft 26a: 505–519. (Translated 1967). Ecology of plant saprobia. In Kemp, L. E., W. M. Ingram & K. M. Mackenthum (eds), Biology of Water Pollution. Federal Water Pollution Control Administration, Washington, DC: 47–52.

Lyons, J., S. Navarro-Perez, P. A. Cochran, E. Santana C. & M. Guzman-Arroyo, 1995. Index of biotic integrity based on fish assemblages for the conservation of streams and rivers in west-central Mexico. Conserv. Biol. 9: 569–584.

Macdonald, S., 1999. Bishops see God in a river that's restored: letter calls Columbia sacred. Seattle Times, 2 May: B1, B4.

May, C. W., R. R. Horner, J. R. Karr, B. W. Mar & E. B. Welch, 1997. Effects of urbanization on small streams in the Puget Sound lowland ecoregion. Watershed Protect. Tech. 2: 483–494.

Miller, P. & D. Ehnes, in press. Can Canadian approaches to sustainable forest management maintain ecological integrity? In Pimentel, D., L. Westra & R. Noss (eds), Ecological Integrity in the World's Environment and Health. Island Press, Washington, DC.

Moyle, P. & P. J. Randall, 1998. Evaluating the biotic integrity of watersheds in the Sierra Nevada, California. Conserv. Biol. 12: 1318–1326.

Norris, R. H. & M. C. Thoms (eds), 1999. River Health. Freshwat. Biol. 41: 197–479.

Norris, R. H., B. T. Hart, M. Finlayson & K. R. Norris (eds), 1995. Use of biota to assess water quality: an international conference. Aust. J. Ecol. 20(1): 1–227.

Oberdorff, T. & J. P. Porcher, 1994. An index of biotic integrity to assess biological impacts of salmonid farm effluents on receiving waters. Aquaculture 119: 219–235.

O'Meara, M., 1999. Exploring a new vision of cities. In State of the World 1999. Norton, New York: 133–150.

Perry, J. & E. Vanderklein, 1996. Water Quality: Management of a Natural Resource. Blackwell Science, Cambridge, MA.

Pielou, E. C., 1998. Fresh Water. University of Chicago Press, Chicago.

Preston, F. W., 1962. The canonical distribution of commonness and rarity. Ecology 43: 185–215, 410–432.

Reynoldson, T. B., R. C. Bailey, K. E. Day & R. H. Norris, 1995. Biological guidelines for freshwater sediment based on **BE**nthic **A**ssessment of **S**edimen**T** (the BEAST) using a multivariate approach for predicting biological state. Aust. J. Ecol. 20: 198–219.

Roesner, L. A. (ed.), 1997. Effects of Watershed Development and Management on Aquatic Ecosystems. Proceedings of an Engineering Foundation Conference. American Society of Civil Engineers, New York.

Rosenberg, D. M. & V. H. Resh (eds), 1993. Freshwater Biomonitoring and Benthic Macroinvertebrates. Chapman and Hall, NY.

Rossano, E. M., 1996. Diagnosis of Stream Environments with Index of Biological Integrity (in Japanese and English). Museum of Streams and Lakes, Sankaido Publishers, Tokyo.

Shrader-Frechette, K., 1995. Hard ecology, soft ecology, and ecosystem integrity. In Westra, L. & J. Lemons (eds), Perspectives on Ecological Integrity. Kluwer Academic Publishers, Dordrecht, The Netherlands: 125–145.

Simon, T. P. (ed.), 1999. Assessing the Sustainability and Biological Integrity of Water Resources Using Fish Communities. CRC Press, Boca Raton, FL.

Stoel, T. B., Jr., 1999. Reining in urban sprawl. Environment 41: 6–11, 29–33.

Thomson, J. D., G. Weiblen, B. A. Thomson, S. Alfaro & P. Legendre, 1986. Untangling multiple factors in spatial distributions: lilies, gophers, and rocks. Ecology 77: 1698–1715.

Thorne, R. St. J. & W. P. Williams, 1997. The response of benthic invertebrates to pollution in developing countries: a multimetric system of bioassessment. Freshwat. Biol. 37: 671–686.

Wackernagel, M. & W. Rees, 1996. Our Ecological Footprint. New Society Press, Gabriola, BC.

Wang, L., J. Lyons, P. Kanehl & R. Gatti, 1997. Influences of watershed land use on habitat quality and biotic integrity in Wisconsin streams. Fisheries 22(6): 6–12.

Wright, J. F., D. Moss, P. D. Armitage & M. T. Furse, 1984. A preliminary classification of running-water sites in Great Britain based on macro-invertebrate species and the prediction of community type using environmental data. Freshwat. Biol. 14: 221–256.

Wright, J. F., P. D. Armitage & M. T. Furse, 1989. Prediction of invertebrate communities using stream measurements. Regul. Rivers Res. Mgmt 4: 147–155.

Wright, J. F., D. Moss, R. T. Clarke & M. T. Furse, 1997. Biological assessment of river quality using the new version of RIVPACS (RIVPACS III). In Boon, P. J. & D. L. Howell (eds), Freshwater Quality: Defining the Indefinable? Scottish Natural Heritage, Edinburgh: 102–108.

Hydrobiologia **422/423:** 15–27, 2000.
M. Jungwirth, S. Muhar & S. Schmutz (eds), Assessing the Ecological Integrity of Running Waters.

A perspective on the abiotic processes sustaining the ecological integrity of running waters

Geoffrey E. Petts
School of Geography and Environmental Sciences, University of Birmingham, Edgbaston B15 2TT, U.K.

Key words: fluvial hydrosystems, flow regime, channel form, spatial scales

Abstract

Using selected examples of recent research, this paper illustrates the role of abiotic components within running-water ecosystems. The important role of temperature is acknowledged but the paper focuses on another key driver: physical stability, defined in relation to hydrological (frequency, duration and timing of inundation) and substratum parameters (channel dynamics, bedform and sediment size). The importance of this driver is illustrated by reference to four spatial scales. At the scale of the bedform, surface-water and groundwater interactions play an important role not least in driving energy exchanges and determining the temperature dynamics within the ecologically important surface layer of the bed sediments. At the reach scale, bedform development, channel form dynamics, and associated changing hydraulic conditions determine both benthic and riparian community patterns. At the catchment scale, new research has shown that the processes responsible for the formation of islands and divided channels play important roles in the functioning of fluvial hydrosystems. Finally, at the regional scale, the flow regime modified by the geomorphological history of the river over at least the past 16 000 years explains ecological patterns. The integration of hydro-geomorphological knowledge from all four scales of analysis is shown to be fundamental for understanding the ecological characteristics of running waters and for managing ecological integrity.

Introduction

Rivers are three-dimensional systems. Longitudinally, river systems extend from a network of small headwater streams to a single, large lowland river ending in either an estuary or a delta. They are usually considered to be unidirectional, but upstream systems can be influenced by downstream disturbances which disrupt the upstream migration and dispersal of species (Pringle, 1997). Laterally, river systems comprise not only the channel(s) but also the adjacent riparian and floodplain zones which are connected to the channel during high flows (Junk et al., 1989). Vertically, they include the alluvial aquifers with which surface-waters interact (e.g., Stanford & Ward, 1988). The integrity of this three-dimensional *fluvial hydrosystem* (Amoros & Petts, 1993; Petts & Amoros, 1996) depends on the dynamic interactions of hydrological, geomorphological and biological processes over a range of time scales. This paper provides a perspective on the hydrological and geomorphological basis of fluvial hydrosystems and draws conclusions to underpin future interdisciplinary research on the ecological integrity of running waters. First, it provides a brief summary of the development of knowledge on the abiotic factors of running water ecosystems to place contemporary advances in context.

The abiotic basis of river ecosystems

Traditionally, biologists have focused on temperature as the primary abiotic driver of ecosystems. Thus, Rohde (1992) explained latitudinal gradients in biodiversity as a function of temperature-controlled mutation rates and generation times, with higher temperatures accelerating evolution and speciation. In river ecology, Illies (1961), for example, suggested that temperature was the main variable governing changes in faunal structure with altitude and latitude. Other examples are provided by Ward (1985, 1992). More recently, Jacobsen et al. (1997) provided new data to show that the number of aquatic insect orders and families increased with maximum stream temperature and, therefore, decreased with altitude and latitude. This view of temperature as the primary abiotic variable in aquatic ecology probably reflects the early focus on lakes and standing waters (Ryder & Pesendorfer, 1989). Rivers, however, are characterized by strong hydrological and geomorphological gradients which relate to highly dynamic processes of three types: hydrological processes (e.g., flooding, dessication, surface-water and groundwater interactions), geomorphological processes (e.g., channel bed degradation and aggradation, bank erosion, floodplain accretion, channel avulsion), and hydraulic processes (including high shear stresses during spates, and stagnation in deadzones during low flows).

Hynes (1970, 1975) provided the catalyst for recognizing fluvial geomorphology as the basis of freshwater ecology. The roles of disturbance regimes in explaining the composition, structure and function of river ecosystems were proposed by Ward & Stanford (1983). At about the same time, Petts (1984) showed how changing flows and sediment transport processes, and changes of channel morphology, can influence river ecosystems. Building on the seminal work of Schumm & Lichty (1965), geomorphologists have advanced understanding of channel forms and processes over a range of space and time scales focusing on process mechanics, channel dynamics and long-term channel change (e.g., Petts, 1995), and integrating these with studies of hydrological responses to climate change and human impacts. Biologists have provided evidence to illustrate the importance of flow regimes and geomorphological dynamics in sustaining running-water ecosystems (see, e.g., Hildrew 1998; Ward et al., 1999a), and the likely impact upon ecological patterns of both geological history and long-term climate change has been acknowledged (e.g., Jacobsen et al., 1997: Ward, 1992). Furthermore, Milner & Petts (1994) proposed that ecological gradients in glacial rivers can be explained by spatial patterns in temperature and channel stability, and by successional changes (i.e., time since deglaciation) of both biotic communities and abiotic conditions, especially channel form and stability.

The need to consider the river within its drainage basin as the basis for assessing ecological integrity has been firmly estabished. Thus, recovery from disturbance has been shown to be rapid not only where taxa are unaffected by the disturbance because of life-cycle adaptations, such as the terrestrial life stages of aquatic insects or the ocean life stage of many important migratory fishes, but also where there are habitat refugia to allow recolonisation either locally from patches, or catchment-scale from tributaries, unaffected by the disturbance (Hildrew, 1998). Catastrophic events, such as extreme floods, have been shown to benefit the system in the long-term by 'resetting' the system and sustaining a diversity of patches in different stages of succession (see Michener & Haeuber, 1998).

Notwithstanding these advances, full inter-disciplinary collaboration between biologists and geomorphologists has been slow to evolve and the full benefits of such collaboration are yet to be realized. Understanding the ecological integrity of rivers requires a knowledge of the physical processes responsible for sustaining the structural features of fluvial hydrosystems, and their dynamics, over a range of spatial and temporal scales. This paper focuses on (a) flow regime and (b) substratum stability (i.e., susceptibility to erosion or deposition). These are considered at four spatial scales: (i) the regional scale; (ii) a linear sequence of sectors (10^8 to 10^5 m^2) within catchments, (iii) habitat patches (10^4 to 10^2 m^2) within sectors, and (iv) micro-habitats (10 to 10^{-1} m^2) within individual landforms. The strong association between biota and structural habitat features is evident in the 'mesohabitat' (Armitage et al., 1995) and 'functional habitat' (Harper et al., 1995) approaches to river conservation, management and rehabilitation. This paper demonstrates the need to analyse such stuctural features as *dynamic* components of fluvial hydrosystems. It builds from the micro-scale to the region. The paper does not seek to provide a comprehensive treatment but to illustrate the dynamics of the abiotic environment and their importance in assessing the ecological integrity of running waters.

Abiotic components of running water ecosystems

Micro-habitat scale: benthic temperature variations

At the scale of the river bed, river ecologists have often considered their domain of interest to extend down to the light discontinuity: a maximum depth of 4–5 times the mean particle size of the bed sediment. This 'ecologically active' zone has relatively high primary (algae) and secondary (invertebrate) production. However, the influence of the surface stream can extend for a much greater depth and in exceptional cases to several metres (e.g., Stanford & Ward, 1988). Within this hyporheic zone, physico-chemical gradients exist due to the mixing of surface water and groundwater, and to changes caused by biological and geochemical processes, whose strength relates to intra-gravel flow rates (Brunke & Gonser, 1997). Even within sectors, however, differences in bed form (topography, sediment composition and vertical structure) can have marked ecological consequences (Petts et al., 1993).

Of particular importance is the thermal regime at and within the channel bed, as this influences rates of organic matter decomposition, invertebrate life cycles and salmonid egg hatching times. Detailed monitoring of temperatures at two adjacent riffles on the River Blithe, UK (Evans & Petts, 1997) has highlighted the bedform-scale variability of temperatures. The channel is 5 m wide and the riparian zone is rough pasture. Data from 20 miniature data loggers set at the surface and depths of 20 and 40 cm for a 5-day period in July were analysed using Principal Components Analysis and this revealed:

(i) heads of the riffles were dominated by downwelling surface water; they mirror thermal patterns within the surface water but temperatures decreased with depth (1.5°C cooler, on average, at a depth of only 20 cm) and diurnal variations increasingly lagged behind surface variations;

(ii) Tails of the riffles were dominated by upwelling groundwater; on average, temperatures were 4.3°C cooler than surface temperatures and showed no significant diurnal variation;

(iii) Both riffles showed the same general pattern but one riffle had warmer temperatures than the other, reflecting differences in bed topography and sediment composition. One riffle had higher thermal conductivity and diffusivity values, and coarser sediments.

Subsequent research has elucidated the important processes influencing the thermal dynamics of a patch of river bed (Evans et al., 1998). The influence of climatic variables, such as air temperature, solar radiation, cloud cover and wind speed, on both gains and losses of energy at the air–water interface are well understood, but the influence of the water column and the nature of channel bed-water exchanges (Figure 1A) have been largely ignored. Detailed hydrometeorological monitoring in a lowland gravel-bed river in central England showed, as expected, that total energy gains were dominated by net short-wave radiation (97.6%). However, the amount of short-wave radiation attenuated in the water column as measured at the channel bed by a pyranometer varied markedly from that calculated using a standard coefficient (Figure 1B). Recorded values were, on average, 44% lower than calculated values and this was explained by the physics of radiant energy transfer in water.

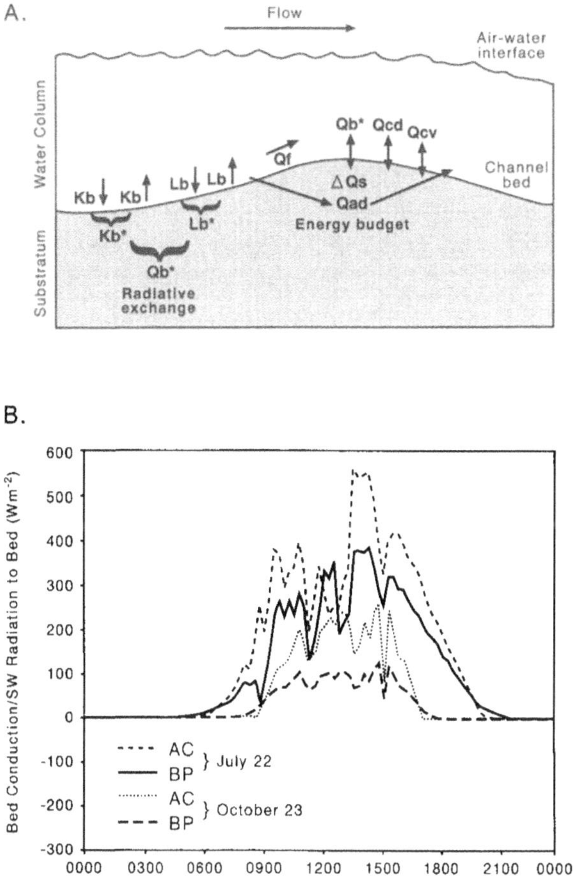

Figure 1. Radiation and energy transfers at the channel bed (after Evans et al., 1998). (A) Conceptual model. radiative exchanges: Kb, short-wave radiation: Lb, long-wave radiation. v, incoming, = reflected/emitted; *, net transfers: Qb*, net bed radiation. Heat fluxes: Qf, due to friction; Qcv, due to convective transfers; Qcd, due to bed conduction: Qb*, due to net bed radiation; Qad, due to advective transfers; •Qs, heat storage within the bed. (B) Differences between short-wave radiation (Kb) measured by a pyranometer (BP) and calculated values using a standard attenuation coefficient (AC) for two days in summer and autumn.

Energy exchanges at the channel bed, resulting from radiative, conductive, convective and advective processes, were shown to account for 15% of the total energy exchanges, on average, with maximum contributions on some days of up to 24% and minima below 1%. The bed acted as a thermal sink in summer with a peak heat loss from the channel water to the substratum of -155 W m^{-2}. By the end of October, however, the bed had become a heat source with energy gains to the channel water of up to 5 W m^{-2}

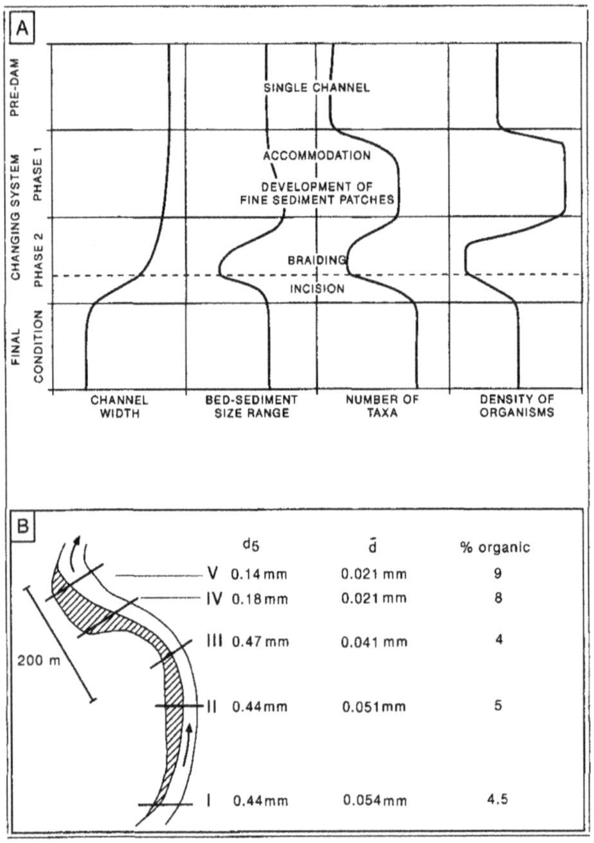

Figure 2. Spatial and temporal patterns of channel form at the reach scale. (A) Changes of a reach on the River Rheidol, Wales, following flow regulation including responses of benthic macroinvertebrates (after Greenwood et al., 1999). (B) Changes in sediments within the riparian zone of a meander on the River Dee, Wales (d5, the 5th percentile particle size; d, mean particle size, and % organic is the percentage organic matter (after Changxing et al. 1999).

and this was found to be strongest in the riffle tail compared with the riffle head areas. Values of bed conduction varied in relation to bed sediment composition and structure, groundwater flux, and surface cover levels of sand, silt, periphyton and macrophytes. Changes to the albedo of the channel bed significantly impacted conductive heat fluxes, indicating that seasonal changes in periphyton cover may exert an important influence on bed conduction rates. Such research not only demonstrates the complex micro-scale structure of the benthic habitat but also establishes the scientific basis for advancing models of this dynamic three-dimensional environment.

Habitat patches within sectors

Channel-bed sediment size and stability are key drivers of lotic ecosystems. This is well illustrated by the succession of changes to flow regulation which are driven by the adjustment of channel form to the imposed flow and sediment transport regimes. Figure 2A illustrates the changes experienced by a reach of a regulated upland stream in Wales over a period of 35 years (Greenwood et al., 1999). Channel adjustment to the imposed conditions involved three stages. First, the regulated flows were accommodated within the existing channel form. The natural channel with a gravel–cobble–boulder bed became stable; mosses developed on boulders and fine organic material accumulated along the channel margins, a process which was accelerated by vegetation encroachment into the channel. Secondly, sediments introduced during major storms in an unregulated tributary basin were deposited over the channel bed and an unstable braided system developed within the former channel. During aggradation, sediment sorting processes reduced the range of bed materials. Boulders and large cobbles were buried and the channel bed became dominated by medium gravels. Thirdly, the main channel incised into the braided channel gravels and the single, narrow, channel cut through to the former bed, exhuming some of the boulders. This new channel has a relatively high flow velocity.

The sequence of channel changes produced a succession of responses within the benthic invertebrate community. In the natural channel the community comprised only 12 taxa with a density of about 400 animals m^{-2}, dominated by chironomids and simuliids. During the accommodation phase, the number of taxa increased to 23 and the density to about 650 animals m^{-2}, including all taxa found in the unregulated stream. The 'new' taxa were mainly associated with slow-flow habitats. The adjusted channel had a mixed community of 27 taxa with an average density of 425 animals m^{-2}. Chironomidae again dominated the community. New species included caddis flies and other taxa found in the lower river and a number of predatory taxa. The data show the sensitivity of the benthic community to changes in channel bed conditions: bedmaterial heterogeneity (and heterogeneity of hydraulic conditions) and bed stability.

Over time, channel sectors establish a quasi-equilibrium between form, i.e., habitat patch structure, and process, with predictable patterns of the physical variables associated with geomorphological units determining the spatial pattern of biota. Such patterns are well illustrated by the ecology of the river margin. Detailed studies of the River Dee, UK, have elucidated the physical and ecological patterns at a scale of half a meander wavelength (Changxing et al., 1999;

Table 1. Comparison of the flora and carabid fauna of two different patch types on the River Dee, north Wales (see also Figure 2B)

	Drier, sandier patches, with lower organic matter content	Damp, silty patches with higher organic matter content
Duration of flooding	10–15%	15–25%
Sediment size (d5)	>0.4 mm	<0.2 mm
Characteristics plants	*Alliaria petiolata*	*Phalaris arundinacia*
	Galium aparine	*Plantago major*
		Myosotis scorpioides
		Rorippa sylvestris
Typical carabids	*Calathus fuscipes*	*Trechus discus*
	Amara montivaga	*Bembidium biguttatum*
	Bembidion lampros	*Bembidion lunulatum*
	Agonum dorsale	

Gurnell et al., 1998). Flow regulation in the headwaters of the Dee has led to channel adjustment, creating a channel of reduced dimensions within the existing channel form by bench development. At the reach-scale the bench is characterised by a continuum from convex- to concave-bank environments characterised by deposits of decreasing grain-size and increasing organic matter content (Figure 2B). Spatial patterns of animals and plants within this berm reflect both changes of substrate and of the frequency/duration of inundation (Table 1). Such studies provide the basis for models that relate spatial patterns of plant and animal communities to predictable patterns of inundation and sediment composition within geomorphological units which are themselves related to channel style and regional setting.

Catchment-scale patterns

Much attention has been directed to catchment scale patterns and processes (e.g. Amoros & Petts, 1993; Giller et al., 1994; Hildrew 1996; Vannote et al., 1980). One widely adopted classification involves the aggregation of sequences of reach types into three zones: headwater streams (source-area catchments), middle-order rivers (the 'transfer zone') and large, lowland floodplain rivers (the depositional zone). This simple division has a sound basis in process geomorphology (Schumm, 1977) and explains major changes in fluvial ecosystems at the basin scale (Figure 3A). However, this fundamental concept is based largely on the geomorphological structure of temperate rivers, which have a long history of manipulation and management for navigation, power, land-drainage, and flood control. Recent research on a geomorphologically intact river in north-east Italy has allowed the development of this model (Ward et al., 1999a,b).

The Fiume Tagliamento flows for 172 km from the limestone Alps (highest peak 2781 m a.s.l) to the Adriatic. This seventh-order river has a drainage area of 2580 km^2 and is characterised by a flashy pluvio-nival regime with a maximum flood of some 4000 m^3 s^{-1}. Using data extracted from maps, the downstream pattern of channel forms has been objectively classified into six styles (Figure 3B). Single-thread reaches (Class 1 and 2) dominate the headwaters and local variations relate to the degree of channel confinement, Class 2 wandering channel reaches developing where the valley floor widens. Thus, the headwater channels show repeated sequences of styles (confined-wandering-braided) with basins separated by confined sections, rock gorges or sections confined by major alluvial fans. The lower river is a single-thread meandering river which in its lowest section has been fixed by flood embankments (Class 0). These classes are common to most channel classification systems. The major difference is within the braided river section where four classes have been defined, forming a clear downstream sequence: Class 3 bar-braided, Class 4 bar-braided with occasional islands (island area/gravel area < 0.25), Class 5 island braided (island area/gravel area 0.25 < 0.5), and Class 6 extensively island braided (island area/gravel area >0.5).

The study highlights the importance of islands and divided channels in downstream patterns within fluvial systems; they increase the habitat heterogeneity, the influence of riparian zones, and the diversity of suc-

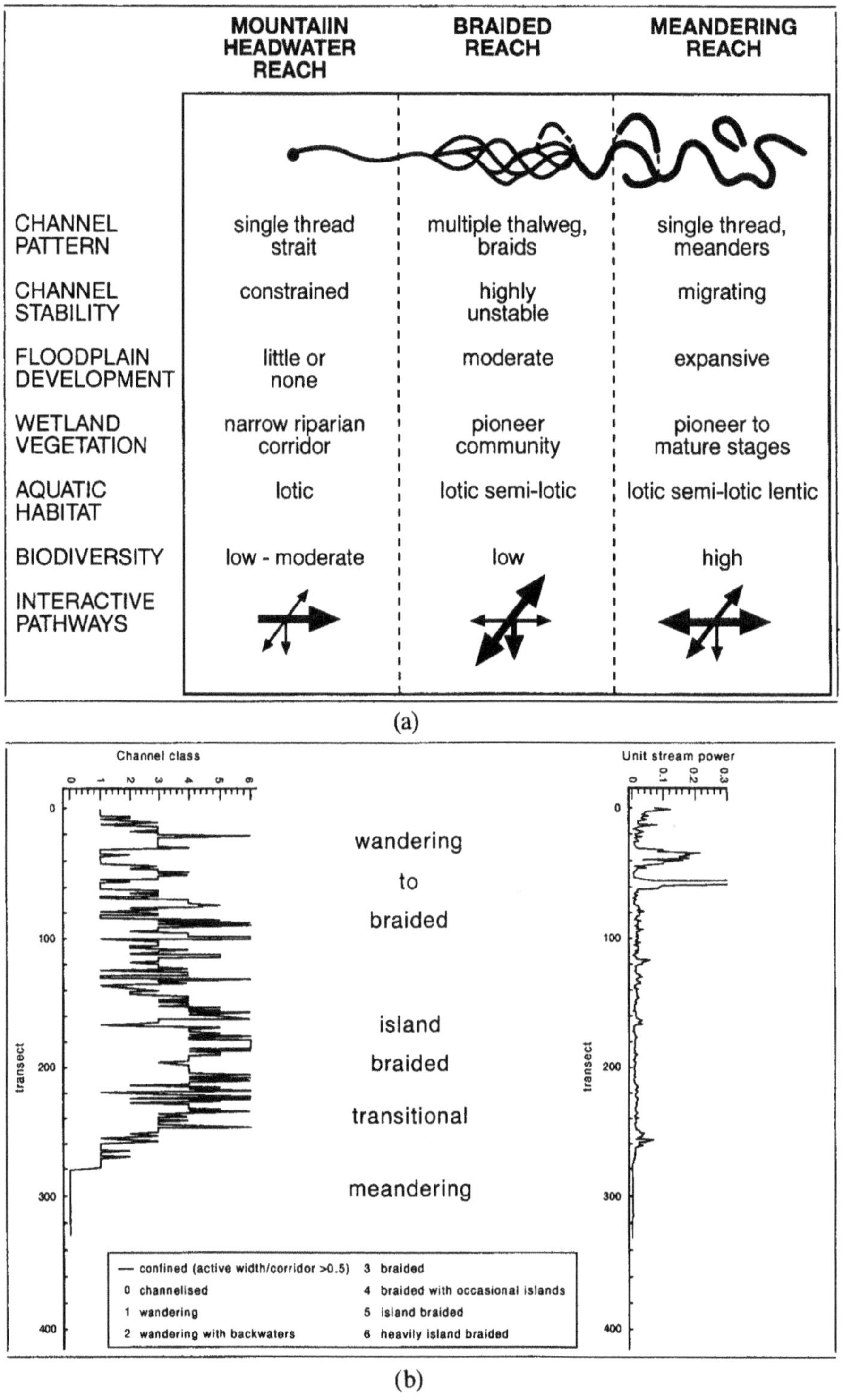

Figure 3. Catchment-scale patterns within fluvial systems. (A) The three-sector model (after Ward & Stanford, 1995). (B) Patterns of channe style and stream power along the River Tagliamento, north-east Italy (after Gurnell et al., 1999).

cessional stages. It is proposed (Ward et al., 1999a,b) that islands increase the retention of nutrients and organic matter, provide 'hot spots' of biodiversity, and contribute to increasing the resistance and resilience of the river to both natural and anthropogenic disturbances. Furthermore, the transition from braided to meandering sectors is associated with a peak of stream power (Figure 3B) which Statzner & Higler (1986) hypothesised as a location of high biodiversity.

The traditional model underestimates the importance of (a) divided channels and (b) islands in large rivers of the temperate zone. The former existence of these features has been widely reported not only from Alpine rivers but also from lower relief environments (e.g., Large & Petts, 1996). The importance of these features has been overlooked in rehabilitation programmes and the re-creation of secondary channels may provide a practicable solution for river management, with benefits for conservation as well as for sustaining the flood control and land drainage functions of rivers.

Regional relationships

At the regional scale, it has become popular to relate ecological patterns to simple gradients, such as a latitudinal temperature gradient, modified by altitude and oceanic influences. However, flow pattern, especially the timing and predictability of the flood season, and geomorphological setting are also important in determining aquatic and riparian processes and biotic communities.

Across Europe, large rivers are dominated by two contrasting, climate-controlled flow regimes (Figure 4A) whose distribution reflects a strong oceanic influence. These two regimes are:

(i) The pluvial oceanic regime characterised by maximum discharge during the period of combined maximum precipitation and minimum evapotranspiration, normally January to March, with minimum flow in July to September;

(ii) The continental regime characterised by the dominance of snowmelt at the beginning of the warm season generating high water in April to June, with low flow in late winter. In northern rivers high spring flows combine with ice-jams during break-up to determine flood levels.

The geomorphological setting reflects the Quaternary and Holocene histories of landscape change. At the maximum extent of the last glacial (ca. 18 000 BP), western Europe was divided into three broadly east-west zones (Figure 5): glaciated to the north, a tundra zone extending to southern France, around the Alpine foothills and approximately along the course of the Danube (Bell & Walker, 1992), and a southern Steppe zone. After the glaciation all rivers experienced a complete, and relatively sudden, transformation: first of flow regime (16 000 to 10 000 BP) and secondly of baselevel (sea level) rise (16 000 to 6000 BP) at rates of more than 1 cm year^{-1} averaged over 1000 years. However, the hydrological transformation was most dramatic in western areas where ice- and snow-melt-dominated periglacial regimes were transformed into low-energy, maritime damped regimes. In basins with high relief, the change in flow regime and reduction in sediment sources associated with the establishment of the forest cover led to a change from braided (bed-load dominated) to meandering (suspended-load dominated), with a reduction in channel width and an increase in sinuosity (Starkel, 1991). Lowland basins showed less dramatic changes. Rivers with continental flow regimes today appear to have experienced a more variable evolution during the Holocene, with dramatic shifts from degrading to aggrading conditions, often reflecting tectonic controls (see Starkel et al., 1991). The 'young' rivers of the glaciated zone, especially in northern latitudes (above 60°N), today transport relatively low sediment loads, and even large channels are often bedrock-lined. In part, this relates to glacio-isostacy.

Differential rates of uplift associated with isostatic recovery following the last glaciation are continuing to affect rivers of north west Europe. Rivers draining to the Gulf of Bosnia in the northern Baltic continue to experience baselevel rise of 0.8–0.9 cm year^{-1} (Figure 5). Thus, the profile of the Vindelälven is tilting because the rate of rise at the tidal limit is about double that in the headwaters. In contrast, the baselevel of the Helge in southern Sweden is more or less static. Farther south, the North Sea has been a focus of crustal downwarping. Subsidence over the past 6500 years at the mouth of the River Ouse in eastern England has averaged 0.9 cm year^{-1} in addition to sea-level rise at a rate of 0.1 cm year^{-1} since 3000 years BP. Thus, the lower reaches of European rivers have different forms: in the north, bed-rock channels have narrow riparian zones dominated by sands whereas mid-latitude and southern floodplain rivers, are strongly connected to major alluvial aquifers.

The results of the EU-funded European River MArgin Systems (ERMAS) programme (Table 2)

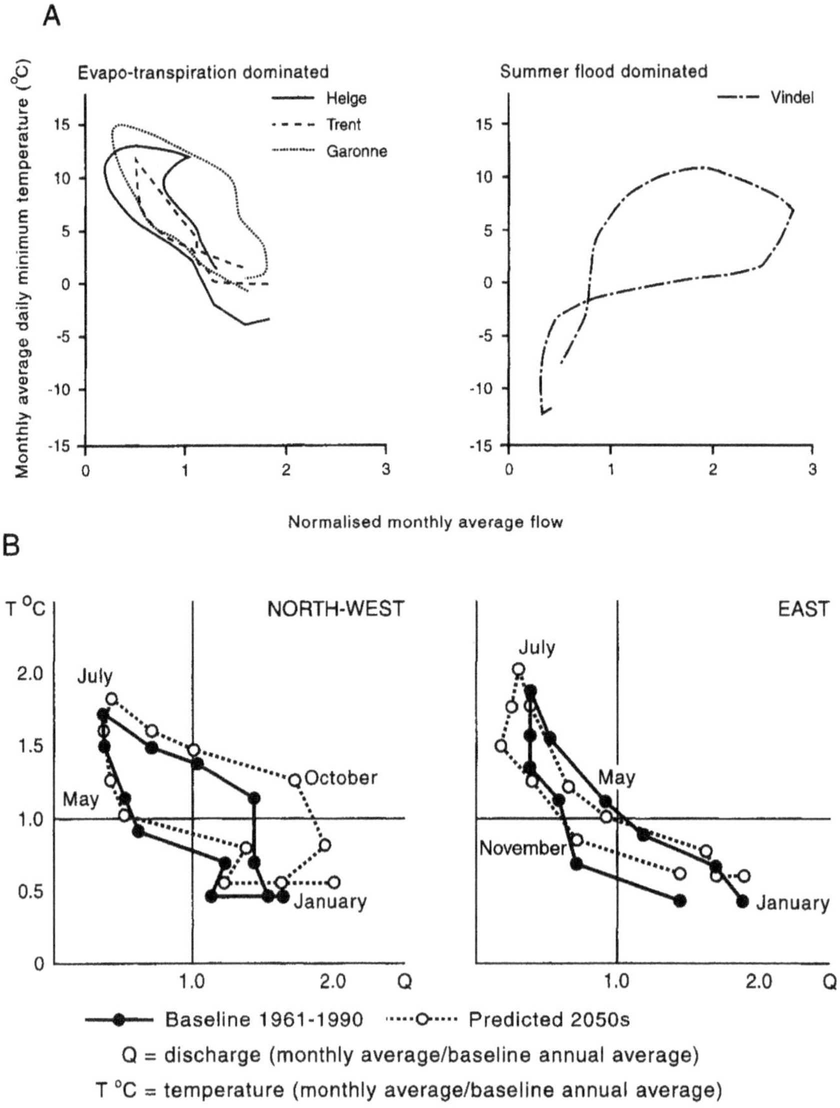

Figure 4. Regional and temporal variations in river flows. (A) The two primary flow regime types in western Europe (rivers are located on Figure 5). (B) Impacts of climate change on flow (Q) and temperature (T) regimes across England.

highlight the role of temperature and river flow regimes, and geomorphological setting as reflected by soil/sediment characteristics, in the functioning of river margin ecosystems. Simple climatic gradients are disrupted by geomorphological discontinuities reflecting the history of the region. An analysis of the soil parameters clearly distinguished the alluvial soils of the Trent and Garonne from the non-fluvial, valley bottom soils of the Helge and Vindel, a separation that reflects the glacial history of the northern and southern areas. The sandy soils along the Helge and Vindel, with high C/N ratio and accumulated organic matter, contrast with the low C/N, silt-clay soils of the other rivers.

The north-south divide is also manifest within data on rates of leaf litter decomposition for alder *Alnus glutinosa* Gaernt (Table 2). The gradient of decomposition rates in river margin sites was significantly related to the hydro-climatic setting, with lowest values in northern sites and highest rates at the warm temperate sites ($P<0.01$). Along the northern rivers (Vindel and Helge) flooding slightly increased the breakdown rate. The average percentages of leaf AFDM lost after 200 days in the terrestrial and frequently flooded sites on the Vindel were 32 and 50%,

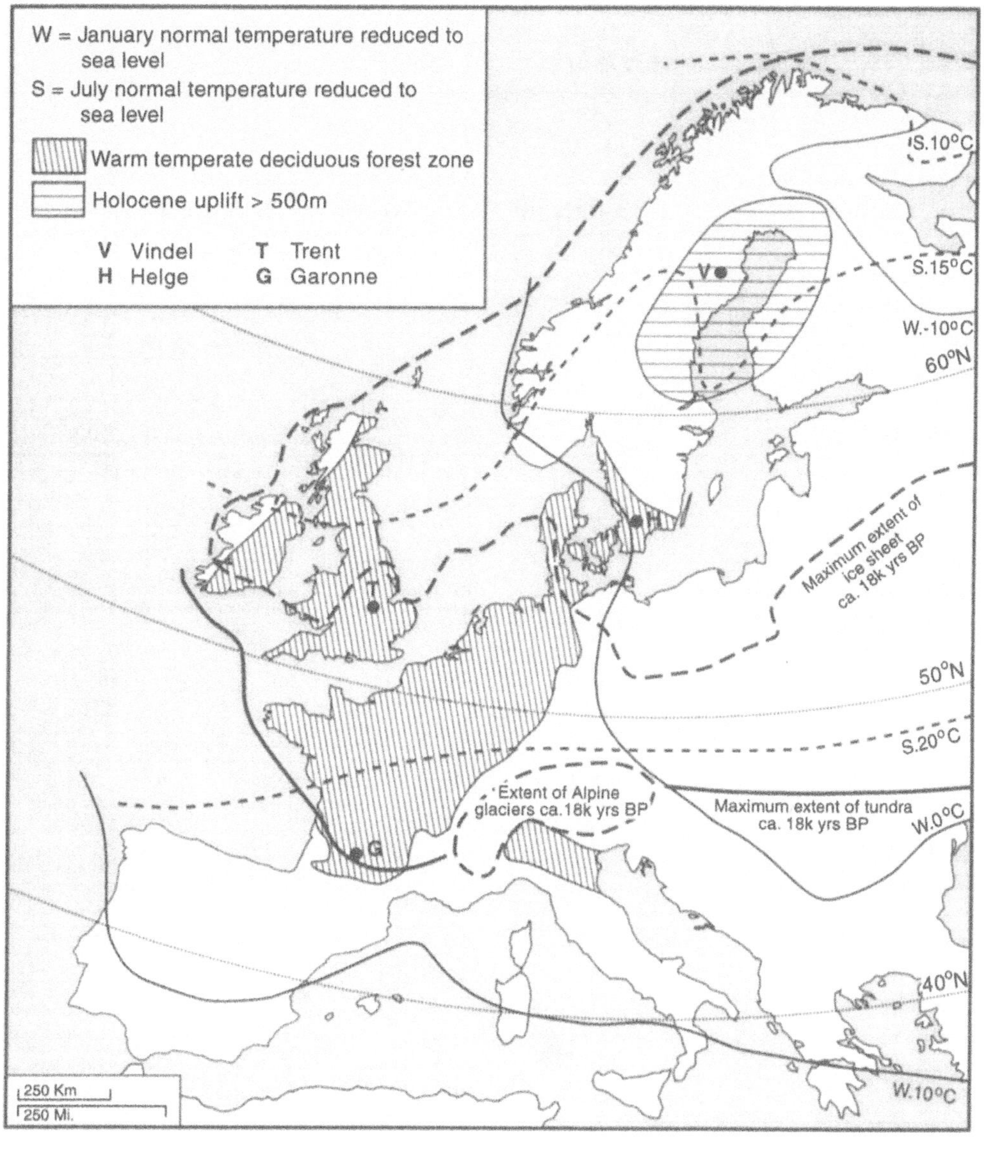

Figure 5. Regional patterns within western Europe. Simple latitudinal gradients are shown to be complicated by the oceanic influence and the legacy of the last Glacial Period.

respectively; values for the Helge were 67 and 80%. In contrast, along the southern rivers flooding markedly reduced the rate of decomposition; on the Trent for example, the percentage leaf AFDM remaining after 200 days increased from 2 to 30% from the terrestrial to frequently flooded, riparian patches.

Flooding plays an important role in soil processes within the riparian zone. Highest rates of *in situ* denitrification on the Helge and Trent occurred in patches that are typically flooded for more than 3 weeks each year. On the Garonne and Vindel, however, the terrestrial sites consistently showed the highest rates. On the Garonne, the duration of flooding of the river margin sites was relatively short, and this may explain the failure to differentiate between patches and to define a peak of denitrification in the frequently flooded sites. On the Vindel, denitrification rates are clearly depressed in relation to terrestrial rates by inundation, suggesting the importance of flood timing for the denitrification process.

ERMAS has also shown that vegetation zonation and macroinvertebrate community patterns relate to

Table 2. Comparison of the ecological characteristics of river margins on a latitudinal gradient across western Europe. Sites are part of the European Commission funded ERMAS network (ENV4-CT95-0061). Field data collected over 2 years (1993–94) at multiple sites on each river. See also Figure 4A and 5

	Vindel	Helge	Trent	Garonne
River characteristics				
Latitude (° N)	64	56	52	44
Landscape 10 500 BP[1]	Ice	Tundra	Tundra	Tundra
Landscape 6000 BP[1]	Birch conifer forest	Mixed deciduous forest	Mixed deciduous forest	Mixed deciduous forest
Mean annual temperature (°C, post 1950)	1	7	10	13
Continentality[2]	27	17	13	16
Month of maximum discharge	June	April	February	April
Growing season (months)	4	5	7	9
Predicted litter production (g m^{-2} $year^{-1}$)	250	400	450	550
Riparian zone				
Soil				
% sand	56	67	40	29
Denitrification (ng N g dry soil h^{-1})	2.7	6.7	36.7	7.5
Time flooded (%) in 1993–4				
Terrestrial sites (norm=0%)	0	10	0	0
Riparian sites (norm= 5<15%)	8	44	14	19
Litter decomposition rates[3]				
Terrestrial sites	0.003	0.005	0.018	0.018
Riparian sites	0.004	0.008	0.011	0.014
Average available resource[4] (g m^{-2})				
Terrestrial sites	42	303	261	802
Riparian sites	242	309	289	1039
Average macro-invertebrate abundance (m^{-2})				
Terrestrial sites	2257	3316	2056	4408
Riparian sites	921	2016	3456	1600

[1]Based upon pollen data in Huntley & Birks (1983).
[2]Continentality = average degree difference between warmest month and coldest month (post 1950).
[3]The breakdown coefficient (k) where $m_t = m_0 \cdot e^{-kt}$ and, m_t is the leaf mass remaining at time t; m_0 is the initial leaf mass.
[4]Average available resource = tree leaves+herbs and forbs at time of maxium productivity.

both intra-regional variations (catchment history) and local factors (hydrological regime) (Bell et al., 1999). Here, two quantitative indicators of the ecological characteristics of the different patches have been included (Table 2). The difference between rivers in average total available plant biomass in terrestrial patches reflects the variation of temperature. The influence of inundation is to markedly increase productivity along the Vindel, but the results for the other rivers are less clear, with a 25% increase in biomass from the terrestrial to most frequently inundated patches on the Trent and Garonne. The invertebrate data show a contrasting picture: inundation reduces invertebrate abundance – markedly so on the Vindel and Helge. The relatively high numbers in the marginal environment of the Trent probably reflect the reduced fauna at the terrestrial sites due to past and present management practices.

Impacts of climate change

Global warming has been one of the major scientific issues of the past two decades, driven by concerns about the 'greenhouse effect'. Recent changes must be set in the longer-term context, with the warmest period since the end of the last glacial period being between 5000 and 6000 years BP when temperatures of the mid latitudes were 1–2°C warmer than at present. In contrast, the Little Ice Age of Europe, which peaked between 1650 and 1750, witnessed temperatures about 0.6–0.8°C lower than present. However, the hydrological consequences of past and future changes of climate also reflect catchment characteristics.

By 2050, global warming is estimated to increase average temperatures by 1.63°C against the 1961–1990 baseline (CCIRG, 1996), but the changes of climate and river flows will have marked seasonal and regional variations. In snowmelt-dominated catchments, the rise in temperatures will cause a change from snowfall to rainfall, leading to reduced flows in spring and increased flows in winter. In boreal landscapes, climate warming with constant precipitation will lead to an extended ice-free season, increased evapotranspiration and an increased threat of forest fires; groundwater levels, stream flows and lake levels will all decline (Schindler, 1997).

Arnell (1996) reports predicted changes in climate created for the 1996 Climate Change Impacts Review Group for 2050. Winter temperatures rise by 2°C in the south-east and by around 0.8°C in the north. Summer temperatures rise by between 1.8 and 1.2°C. The regional pattern results from the slow response of the North Atlantic to increasing temperatures. Rainfall changes show an even greater regional variation with an increase of less than 5% in the south rising to more than 15% in the north. Winter rainfalls are predicted to increase throughout Britain but to be greatest (<10%) in the south. Summer rainfalls will increase by up to 10% in the north but will decrease by 10% in the south-east. The regional effects of these changes are to increase the differences in the physical drivers of stream ecosystems (Figure 4B). In the north and west, change will be manifest by a marked increase in flows in autumn and winter, with warmer temperatures thoughout the year. In the east, the dominant effects of climate change will be reduced summer flows, which may fall by more than 50%, and increased temperatures during late autumn/ early winter. Thus, climate change is likely to exaggerate regional differences.

Perspectives for future research

Assessing and managing the ecological integrity of running waters are complex problems requiring an understanding of the dynamics of abiotic processes over a range of spatial scales. At the *regional scale*, latitudinal, altitudinal and oceanic-continental gradients are modified by the geomorphological legacy of climate change, and by variations of flow and temperature regimes. Within *catchments*, ecological patterns are related to upstream–downstream gradients interrupted by transition zones associated with tributary confluences, knick-points in the long profile, valley constrictions, changes of channel pattern and large-scale, groundwater–surface-water interactions. Gradients also appear at the scale of the geomorphological unit (e.g., meander wavelength) and through lateral (aquatic to terrestrial) and vertical (surface to groundwater) ecotones, giving a characteristic structure to rivers at the *sector* scale. However, these gradients can be disturbed over short timescales, such as by changes in bedform structure and sediment type, reflecting the real-time sequence of erosional and depositional processes. These processes also influence process dynamics at the *patch* and *micro-habitat* scales, for example by influencing surface water and groundwater interactions.

The scientific search for explanations to empirical observations of biological patterns within fluvial hydrosystems has been influenced by two constraints. First, there remains a lack of *truly* inter-disciplinary research involving hydrologists, geomorphologists and biologists. Experience with such programmes is proving enormously beneficial but there is still much to do to establish focused, integrated research unhindered by the prejudices and perceptions of the different specialisms. Establishing unifying objectives that cross disciplines is the key, but this often requires each specialist to stand back from his or her own research frontier in order to put in place effective interfaces. Such projects can be seen as technically unsophisticated and they have been viewed within disciplines as having less prestige than single-discipline, reductionist research. Nevertheless, such programmes (e.g., ERMAS) are making fundamental advances in developing a comprehensive understanding of the dynamics of fluvial hydrosystems. They also provide effective communication channels for the transfer of advances within disciplines to cognate areas.

Secondly, there is a need for longer-term studies addressing the *dynamic interactions* of the physical, chemical and biological processes. Functional approaches have focused upon the sector scale, usually involving studies of representative reaches within which average or steady state conditions have been defined. Whilst empirical approaches, if poorly conceived and weakly designed, can produce generalisations which are "the product of the sampling and experimental design rather than intrinsic to nature" (Lane & Richards, 1997), rigorously designed empirical studies have generated valuable insights into the behaviour of fluvial hydrosystems (Petts, 1995). They have also established the basis for detailed mechanistic and experimental research to elucidate and model the mechanisms of channel, habitat and micro-habitat response in time and space to temporal variations in the governing variables (temperature, discharge, water quality, sediment load, and the supplies of particulate organic matter and large woody debris). The advance of this approach towards a 'full explanation' (Lane & Richards, 1997), however, requires a complementary approach to that used by many scientists who, for practical reasons, have collapsed the time- and space-scales of enquiry to consider variations in specific components of the system over short periods of time. Approaches need to be advanced to integrate such short-term studies with those of species lifecyles, and again with those of system recovery to disturbance and successional change. Biologists must work within interdisciplinary teams including hydrologists, geomorphologists and environmental chemists in order to advance new models integrating process–structure relationships over the range of scales. Only then will we be able to assess the ecological integrity of running water ecosystems with confidence. A key challenge is for the different scientific discipines to improve communication between themselves.

Acknowledgements

This synthesis paper reflects on-going research in collaboration with many colleagues across Europe. I am pleased to acknowledge their contribution to the development of the research; any errors in interpretion or use of the data are mine! Thanks especially to Christer Nilsson, Gilles Pinay, Eric Chauvet, Lena Vought, James Ward, Peter Edwards, Angela Gurnell, Mel Bickerton, Malcolm Greenwood and Henri Decamps. Thanks also to Stan Gregory whose honest advice is always appreciated!

References

Amoros, C. & G. E. Petts, 1993. Hydrosystemes Fluviaux. Masson, Paris.

Armitage, P. D., 1995. Faunal community change in response to flow manipulation. In: Harper, D. M. & A. J. D. Ferguson (eds), The Ecological Basis for River Management. Wiley, Chichester, UK: 59–78.

Arnell, N., 1996. Global Warming, River Flows and Water Resources. Wiley, Chichester, UK.

Bell, M. & M. J. C. Walker, 1992. Late Quaternary Environmental Change. Longman, Harlow, UK.

Bell, D., G. E. Petts & J. P. Sadler, 1999. The distribution of spiders in the wooded riparian zone of three rivers in western Europe Regulated Rivers 15: 141–158.

Brunke, M. & T. Gonser, 1997. The ecological significance of exchange processes between rivers and groundwater. Freshwat. Biol. 37: 1–33.

CCIRG, 1996. The potential effects of climate change in the United Kingdom. Climatic Change Impacts Review Group. HMSO. London.

Changxing Shi, G. E. Petts & A. M. Gurnell, 1999. Berm development along the regulated, lower Dee, UK. Earth Surf. Processes Landf. 24: 135–149.

Evans, E. C. & G. E. Petts, 1997. Hyporheic temperature patterns in riffles Hydrol. Sci. J. 2: 199–213.

Evans, E. C., G. McGregor & G. E. Petts, 1998. River energy budgets with special reference to bed processes. Hydrol. Processes 12: 575–596.

Giller, P. S., A. G. Hildrew & D. G. Raffaelli, 1994. Aquatic Ecology: Scale, Pattern and Process. Blackwell Scientific, Oxford. UK.

Greenwood, M. T., M. A. Bickerton, A. M. Gurnell & G. E. Petts, 1999. Channel changes and inverterate faunas below Nant-y-Moch dam, River Rheidol, Wales, UK: 35 years on. Regulated Rivers, in press.

Gurnell, A. M., M. Bickerton, P. Angold, D. Bell, I. Morrisey, G. E. Petts & J. Sadler, 1998. Morphological and ecological change on a meander bend: the role of hydrological processes and the application of GIS. Hydrol. Processes 12: 981–993.

Gurnell, A. M., G. E. Petts, N. Harris, J. V. Ward, K. Tockner, P. J. Edwards & J. Kollman, 1999. Large wood retention in river channels: the case of the Fiume Tagliamento, Italy. Earth Surf. Processes Landf. 24: 1–21.

Harper, D. M., C. Smith, P. Barham & R. Howell, 1995. The ecological basis for the management of the natural river environment. In Harper, D. M. & A. J. D. Ferguson (eds), The Ecological Basis for River Management. Wiley, Chichester, UK: 219–238.

Hildrew, A. G., 1996. Whole river ecology: spatial scale and heterogeneity in the ecology of running waters. Arch. für Hydrobiol. Suppl. 113, 10: 25–43.

Hildrew, A. G., 1998. Physical habitat and the benthic ecology of streams and rivers. In: Bretschko, G. & J. Helesic (eds), Advances in River Bottom Ecology. Backhuys, Leiden: 3–22.

Huntley, B. & H. J. B. Birks, 1983. An Atlas of Past and Present Pollen Maps of Europe 0–13,000 years ago. Cambridge University Press, London, UK.

Hynes, H. B. N., 1970. The Ecology of Running Waters. University of Toronto Press, Toronto.

Hynes, H. B. N., 1975. The stream and its valley. Verhandlungen der Internationalen Vereinigung für theoretische und angewandte Limnologie 19: 1–15.

Illies, J. 1961. Versuch einer allgemein biozönotischen Gliederung der Fliessgewässer. Int. Rev. ges. Hydrobiol. 46: 205–213.

Jacobsen, D., R. Schultz & A. Encalada, 1997. Structure and diversity of stream invertebrate assemblages: the influence of temperature with altitude and latitude. Freshwat. Biol. 38: 247–261.

Junk, W. J., P. B. Bayley & R. E. Sparks, 1989. The flood pulse concept in river floodplain systems. Can. Spec. Publ. Fish. aquat. Sci. 106: 110–127.

Lane, S. N. & K. S. Richards, 1997. Linking river channel form and process: time, space and causality revisited. Earth Surf. Processes Landf. 22: 249–260.

Large, A. R. L. & G. E. Petts, 1996. Historical channel-floodplain dynamics along the River Trent: implications for river rehabilitation. Applied Geography 16: 191–209.

Michener, W. K. & R. A. Haeuer, 1998. Flooding: natural and managed disturbances. BioScience 48, 9: 677–680.

Milner, A. M. & G. E. Petts, 1994. Glacial rivers: physical habitat and ecology. Freshwat. Biol. 32: 295–307.

Petts, G. E., 1984. Impounded Rivers. Wiley, Chichester.

Petts, G. E., 1995. Changing river channels: the geographical tradition. In: Gurnell, A. M. & G. E. Petts (eds), Changing River Channels. Wiley, Chichester, UK.

Petts, G. E. & C. Amoros (eds), 1996. Fluvial Hydrosystems. Chapman and Hall, London, UK.

Petts, G. E., H. Moller & A. L. Roux (eds), 1989. Historical Change of Large Alluvial Rivers: Western Europe. Wiley, Chichester, UK.

Petts, G. E., P. D. Armitage & E. Castella, 1993. Physical habitat change and macroinvertebrate response to river regulation: the River Rede, UK. Regulated Rivers 8: 167–178.

Pringle, C. M., 1997. Exploring how disturbance is transmitted upstream: going against the flow. J. n. am. Benthol. Soc. 16: 425–438.

Rohde, K., 1992. Latitudinal gradients in species diversity. Oikos 65: 514–527.

Ryder, R. A. & J. Pesendorfer, 1989. Large rivers are more than flowing lakes: a comparative review. In Dodge, D. P. (ed.), Proc. International Large River Symposium. Can. Special Publ. Fisheries aquat. Sci. 106: 65–85.

Schindler, D. W., 1997. Widespread effects of climatic warming on freshwater ecosystems in North America. Hydrol. Processes 11: 1043–1067.

Schumm, S. A., 1977. The Fluvial System. Wiley, New York.

Schumm, S. A. & R. W. Lichty, 1965. Time, space and causality in geomorphology. Am. J. Sci. 263: 110–119.

Stanford, J. A. & J. W. Ward, 1988. The hyporheic habitat of river ecosystems. Nature 335: 64–66.

Starkel, L., 1991. Characteristics of the temperate zone and fluvial palaeohydrology. In: Starkel, L., K. J. Gregory & J. B. Thornes (eds), Temperate Palaeohydrology. Wiley, Chichester, UK.

Starkel, L., K. J. Gregory & J. B. Thornes (eds), 1991. Temperate Palaeohydrology. Wiley, Chichester, UK.

Statzner, B. & B. Higler, 1986. Stream hydraulics as a major determinant of benthic invertebrate zonation patterns. Freshwat. Biol. 16: 127–139.

Vannote, R. L., G. W. Minshall, K. W. Cummins, J. R. Sedell & C. E. Cushing, 1980. The river continuum concept. Can. J. Fish. aquat. Sci. 37: 130–7.

Ward, J. V., 1985. Thermal characteristics of running waters. Hydrobiologia 125: 31–46.

Ward, J. V., 1992. A mountain river. In Calow, P. & G. E. Petts (eds), The Rivers Handbook. Blackwell Scientific, Oxford, UK: 493–510.

Ward, J. V. & J. A. Stanford, 1983. The intermediate disturbance hypothesis: an explanation of biotic diversity patterns in lotic ecosystems. In: Fontaine, T. D. & S. M. Bartell (eds), Dynamics of Lotic Ecosystems. Ann Arbor Science, Ann Arbor, USA: 347–356.

Ward, J. V. & J. A. Stanford, 1995. The serial discontinuity concept: extending the model to floodplain rivers. Regulated Rivers 10: 159–168.

Ward, J. V., K. Tockner, P. J. Edwards, J. Kollmann, G. Bretschko, A. M. Gurnell, G. E. Petts & B. Rossaro, 1999a. A reference system for the Alps: The Fiume Tagliamento. Regulated Rivers 15: 125–140.

Ward, J. V., K. Tockner, P. J. Edwards, J. Kollmann, G. Bretschko, A. M. Gurnell, G. E. Petts & B. Rossaro, 1999b. Potential role of island dynamics in river ecosystems. Verh. int. Verein. Limnol., in press.

Hydrobiologia 422/423: 29–48, 2000.
M. Jungwirth, S. Muhar & S. Schmutz (eds), Assessing the Ecological Integrity of Running Waters.

Restoring ecological integrity of great rivers: historical hydrographs aid in defining reference conditions for the Missouri River

David L. Galat[1] & Robin Lipkin[2]
[1] *U.S. Geological Survey, Biological Resources Division, Missouri Cooperative Fish and Wildlife Research Unit, 302 Anheuser-Busch Natural Resources Building, University of MO, Columbia, MO 65211-7240, U.S.A.*
[2] *The School of Natural Resources, 302 Anheuser-Busch Natural Resources Building, University of Missouri, Columbia, MO 65211-7240, U.S.A.*

Key words: ecological reference, flood pulse, flow regulation, hydrology, Range of Variation Approach

Abstract

Restoring the ecological integrity of regulated large rivers necessitates characterizing the natural flow regime. We applied 'Indicators of Hydrologic Alteration' to assess the natural range of variation of the Missouri River's flow regime at 11 locations before (1929–1948) and after (1967–1996) mainstem impoundment. The 3768 km long Missouri River was divided into three sections: upper basin least-altered from flow regulation, including the lower Yellowstone River; middle basin inter-reservoir, and lower basin channelized. Flow regulation was associated with a reduction in magnitude and duration of the annual flood pulse, an increase in magnitude and duration of annual discharge minima, a reduction in frequency of annual low-flow pulses, earlier timing of March–October low-flow pulses, and a general increase in frequency of flow reversals with a reduction in the rate of change in river flows. Hydrologic alterations were smallest at two least-altered upper-basin sites and most frequent and severe in inter-reservoir and upper-channelized river sections. The influence of reservoir operations on depressing the annual flood pulse was partially offset by tributary inflow in the lower 600 km of river. Reservoir operations could be modified to more closely approximate the 1929–1948 flow regime to establish a simulated natural riverine ecosystem. For inter-reservoir and upper channelized-river sections, we recommend periodic controlled flooding through managed reservoir releases during June and July; increased magnitude, frequency and duration of annual high-flow pulses; and increased annual rates of hydrograph rises and falls. All of the regulated Missouri River would benefit from reduced reservoir discharges during August–February, modified timing of reservoir releases and a reduced number of annual hydrograph reversals. Assessment of ecological responses to a reregulation of Missouri River flows that more closely approximates the natural flow regime should then be used in an adaptive fashion to further adjust reservoir operations.

Introduction

So paramount is a river's flow regime as an underpinning to ecological integrity that its protection or restoration has been accorded 'the natural flow paradigm' (Poff et al., 1997; Richter et al., 1997). Alteration of the natural flow regime through impoundment and flow regulation has severely compromised the ecological health (*sensu* Karr & Chu, 1999) of most of the world's rivers. This is pervasive for large and great rivers in developed countries because of their long association with human activities (Davies & Walker, 1986; Benke, 1990; Bravard & Petts, 1996). Rehabilitation of the hydrological and ecological integrity of large rivers is therefore a major thrust of contemporary fluvial ecology (Gore, 1985; Boon et al., 1992; National Research Council, 1992; Calow & Petts, 1992).

Assessment is a fundamental aspect of characterizing, conserving or recovering the ecological integrity of fluvial hydrosystems and benchmarks or reference conditions are necessary to quantify what constitutes 'healthy' or 'integrity.' Hughes (1995) gave six approaches for determining reference conditions. These include: regional reference sites, historical data, paleoecological data, experimental laboratory data, quantitative models and best professional judgement. Assessing and restoring ecological integrity of large rivers precludes use of regional reference systems as there is only one Danube or Missouri River. Additionally, biomonitoring, experimental studies and quantitative models of large rivers are limited relative to wadeable streams due to their size and sampling difficulties (Johnson et al., 1995; Reash, 1998). However, long-term hydrological records are often available for large rivers (Sparks, 1992).

An initial consideration to realize fluvial restoration is to reregulate the natural flow regime (National Research Council, 1992; Gore & Shields, 1995; Stanford et al., 1996; Poff et al., 1997), since attempts to reestablish a river's biological integrity are doomed without recreating the underlying physical template. Consequently, 'managed flooding' is now recommended to naturalize river flows and their associated ecological processes in some large rivers (Michener & Haeuber, 1998).

Our objective was to assess ecologically relevant components of the Missouri River's flow regime and their longitudinal variability before and after mainstem regulation. We use this information to provide initial guidelines for restoring a more natural hydrograph to enhance the ecological integrity of the river's imperiled biota.

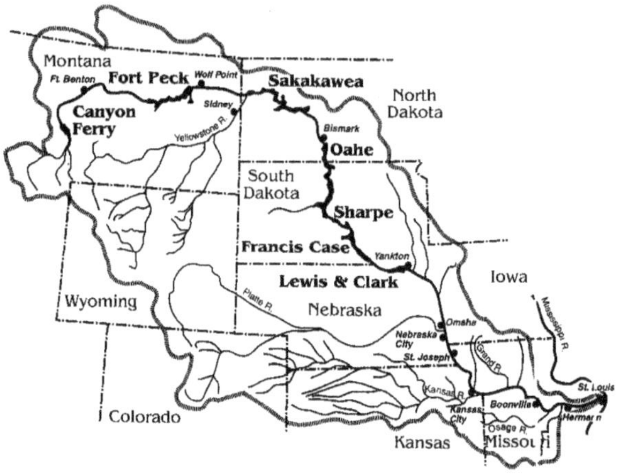

Figure 1. Map of the Missouri River basin (thick broken line), the Missouri River and its major tributaries, mainstem reservoirs (names in **bold**), and locations of gauging stations where historical flow records were analyzed.

Missouri River hydrosystem

The Missouri River is the longest river in the conterminous United States. It extends 3768 km from southwest Montana to the Mississippi River, near St. Louis, Missouri (Figure 1). Its drainage basin encompasses about one-sixth of the conterminous United States (1 371 000 km^2) and is largely semi-arid, resulting in a low discharge relative to basin area. Mean annual Missouri River discharge per unit drainage area from 1951 to 1980 was about four times less (0.0016 m^3 s^{-1} km^{-2}) than for the adjacent upper Mississippi River (0.0065 m^3 s^{-1} km^{-2}; data from Hedman & Jorgensen, 1990).

Public demands to improve navigation, irrigate the arid Great Plains, control devastating floods and generate electricity began in earnest on the Missouri River in the early 1900s. Today, this highly regulated river can be divided into three approximately equal length sections. The upper 1241 km represents a 'least-altered' section relative to the remaining river. Although four mainstem dams and reservoirs are present – Helena, Hauser, Holter and Canyon Ferry – their usable capacity (2.7 km^3) is only 3% of the downriver mainstem reservoirs. The 1233 km long middle or 'inter-reservoir' section was impounded between 1937 and 1963 by a cascade of six large mainstem reservoirs (total gross volume: 90.7 km^2; total average annual discharge: 100.5 km^3 yr^{-1}). Flows in the 1212 km long lower section are regulated by upstream reservoirs. Channel-floodplain morphology in the lower section from Sioux City, Iowa (1178 km), to the mouth was also altered by channelization, bank stabilization and levee construction and will be referred to as the 'channelized' section.

System impoundment, flow regulation, channelization, levee construction, and other basin developments and their impacts on the system's ecology have been considerable and are chronicled elsewhere (see Hesse et al., 1989; Schmulbach et al., 1992; Galat & Frazier, 1996 and references therein). Collectively, Missouri River basin development has contributed to listing as endangered, threatened, or rare by state or federal agencies seven species of plants, six insects, two mussels, 16 fishes, four reptiles, 14 birds and three mammals (Whitmore & Keenlyne, 1990). The conservation organization, American Rivers, listed the Missouri River as North America's most endangered river in 1997 (American Rivers, 1997). Re-establishing a semblance of the natural flow regime is an essential element to restore the Missouri River's ecological integrity; characterizing pre-regulation hydrologic variability is an initial step.

Methods

The Range of Variation Approach, or RVA, was developed by The Nature Conservancy (Richter et al., 1997) to assess and define river ecosystem management targets based on a statistical characterization of ecologically relevant hydrologic parameters (Richter et al., 1996). The hydrologic parameters used in the RVA comprise the 'Indicators of Hydrologic Alteration,' or IHA, method (Richter et al., 1996). These parameters reflect five fundamental attributes of river flow that have profound ecological significance: magnitude, timing, frequency, duration and rate of change of discharge (Richter et al., 1996, 1998; Poff et al., 1997; Scott et al., 1997). The IHA method calculates 32 hydrologic parameters for each year of flow record (Table 1). Measures of central tendency (mean, median) and dispersion (range, standard deviation, percentiles, coefficient of variation (CV) and coefficient of dispersion (CD)) are used to characterize inter-annual variation before (reference period) *versus* after the system has been altered by human activities (Richter et al., 1996). A fundamental concept of the RVA is that post-regulation river flows should be managed to fall within the targeted range of IHA values at the same frequency as the pre-regulation flows (Richter et al., 1997).

Richter et al. (1997) recommended six steps for setting, implementing and refining flow management targets for a specific river or reaches within a river. We applied the first two steps to evaluate contemporary Missouri River hydrology relative to historical conditions. First, we characterized the natural range of flow variation (Step 1) for the Missouri River using IHA before and after mainstem flow regulation, referred to hereafter as pre- and post-regulation, respectively. Second, initial flow management targets were identified for these hydrologic parameters using the RVA approach.

River-flow data were analyzed from gauging stations above (least-altered), between (inter-reservoir) and below (channelized) the large mainstem reservoirs. Comparing hydrologic parameters before and after flow regulation at stations above major reservoirs provides an estimate of natural temporal variability between the two time periods. It also enabled us to evaluate the applicability of relatively least-altered sites as spatial references for distant locations below large mainstem reservoirs.

The IHA software uses daily flow data input by water year (October 1–September 30). The first year of continuous daily flow records for most mainstem Missouri River stations was water year 1929 (October 1928–September 1929). We searched the U.S. Geological Survey's (USGS) stream-flow records to locate mainstem Missouri River stations meeting three criteria: 1. a minimum of 13 years of continuous flow records both before and after reservoir construction; 2. location of stations within each of the least-altered, inter-reservoir and channelized-river sections; 3. concentration of stations in river sections where input from major tributaries was highest. Ten of the 24 mainstem gauging stations met these criteria.

Gauging stations are identified by their location (km) upstream from the mouth of the Missouri River followed by a two-letter abbreviation of the station name. Least flow-altered sites are identified in **bold** typeface. Inter-reservoir stations and station 1297YT, just below the most downstream reservoir (Lewis and Clark), are identified by normal type face and stations in the channelized section are distinguished by *italics*. We also included one tributary, least-altered station in our analysis: Sidney, Montana (**2592SN**), 47 km upstream from the Yellowstone River's confluence with the Missouri. The Yellowstone is the longest free-flowing, high-quality, large river in the conterminous U.S. (Benke, 1990; White & Bramblett, 1993), and has a greater discharge than the mainstem Missouri at their confluence (Table 2). Stations **3336FB** and **2592SN** are affected by human activities (e.g. small impoundments and water diversions), but their flow regimes are little altered relative to other mainstem Missouri River sites (White & Bramblett, 1993; Scott et al., 1997; Shields et al., 1997).

Construction of Ft. Peck (2850 km), the first dam on the mainstem Missouri, began in 1937. Hesse & Mestl (1993) reported that its early operation did not affect the hydrograph at Bismarck, North Dakota (2115BM), until 1948 and they defined the pre-regulation period for the entire river as 1929–1948. We adopted this interval, but recognize that filling and operation of Ft. Peck Dam may have influenced river flow at the Wolf Point station (2738WP) between Ft. Peck Dam and 2115BM. The remaining mainstem reservoirs were constructed between 1953 and 1963 and the six-dam system commenced operation in 1967 (Ferrell, 1993); we defined the post-regulation period as 1967–1996. Flow data were incomplete over the 50 year record for three of the 11 stations. One year of pre-regulation flows was absent from *905NC* (1929) and two years each from **2592SN** (1932–1933) and 1297YT (1929–1930).

Table 1. Summary of the 32 hydrologic parameters used in the Indicators of Hydrologic Alteration (IHA) method for the Missouri and lower Yellowstone rivers. Sources: Richter et al. (1996, 1998)

IHA statistics group		Hydrologic parameters (number)
Group 1:	Magnitude of monthly discharge conditions	Median discharge for each calendar month (12)
Group 2:	Magnitude and duration of annual extreme discharge conditions	Annual median discharge maxima and minima averaged over 1-day, 3–day, 7-day, 30-day and 90-day intervals (10)
Group 3:	Timing of annual extreme discharge conditions	Median Julian date of each annual (Jan–Dec) 1-day maximum and minimum discharge and median Julian date of each seasonal (Mar–Oct) 1-day minimum discharge (3)
Group 4:	Frequency and duration of high- and low-flow pulses	Median number and median duration of high and low discharge pulses each year (4)
Group 5:	Rate and frequency of hydrograph changes	Mean of all positive and negative differences between consecutive daily discharge values and mean number of flow reversals (3)

Indicators of hydrologic alteration calculations

We added an additional variable to the Group 3 statistics (Table 1): Julian date (JD) of the vegetation 'growing season' 1-day minimum flow where the growing season was defined as March 1–October 31 (JD 122–305). Minimum flows during this period are relevant to reproductive success of the federally endangered least tern (*Sterna antillarum*) and threatened piping plover (*Charadrius melodus*), which nest on exposed sand islands along the Missouri and its major tributaries (Smith, 1996; Bacon & Rotella, 1998). Timing of sand island exposure is also important to nesting success of softshell turtles (*Apalone* spp., pers. comm., R. Bodie, Dept. Biol. Sci., University of Missouri), as is exposure of mud flats to germination of annual moist-soil plants (Galat et al., 1998). Additionally, most Missouri River fishes reproduce during the March–October interval (Galat et al., 1998) and shallow-water habitats are important nursery areas for many large-river fishes (Copp, 1991; Poizat & Pont, 1996). It was not necessary to add a new variable to reflect the date of the growing season 1-day discharge maximum, because the maximum for the whole year at all stations always occurred between March and October.

Specific high and low discharge thresholds are user-defined in the IHA to compute the number and duration of high and low pulses relative to these flows. Richter et al. (1997) suggest a default definition of high pulses as >75th percentile (%ile) of all pre-dam flows and low pulses as <25th%ile of all pre-dam flows. We adopted this approach, but applied a more conservative criterion, defining the annual high-flow threshold at each station as the 75th%tile daily discharge for the month with the highest pre-regulation monthly median discharge. Conversely, the low-discharge threshold was set as the 25th%ile daily discharge for the month with the lowest pre-regulation monthly median discharge.

Range of variability calculations

Once IHA parameter values are calculated (Step 1),

Table 2. Location (kilometers upstream from Missouri River mouth), drainage area, pre- (Oct 1929–Sep 1948) and post-flow (Oct 1967–Sep 1996) regulation mean annual discharge, and percent change in mean annual discharge from pre- to post-flow regulation at two stations on the Missouri and lower Yellowstone (**2592SN**) rivers upstream from large mainstem reservoirs (**bold**), three stations immediately below dams or between reservoirs, and six stations downstream from reservoirs in the channelized river (*italics*). Pre-flow regulation period is missing 1932–33 at km **2592SN**, 1929–1930 at km 1297YT, and 1929 at km *905NC*

Name (USGS station number)	Station location (km) and ID	Mean annual discharge ($m^3 s^{-1}$)			
		Drainage area (km^2)	1929–1948	1967–1996	Percent change
Fort Benton, MT (06090800)	**3336FB**	64 100	175.1	228.4	30.4
Wolf Point, MT (06177000)	2738WP	213 131	212.0	300.1	41.6
Sidney, MT YSR (06329500)	**2592SN**	178 977	332.6	358.4	7.8
Bismarck, ND (06342500)	2115BM	482 776	583.9	698.5	19.6
Yankton, SD (06467500)	1297YT	723 905	692.8	812.8	17.3
Omaha, NE (06610000)	*991OM*	846 306	771.5	1012.2	31.2
Nebraska City, NE (06807000)	*905NC*	1 072 154	913.7	1193.8	30.6
St. Joseph, MO (06818000)	*721SJ*	1 088 577	1007.4	1386.4	37.6
Kansas City, MO (06893000)	*589KC*	1 256 668	1240.9	1673.0	34.8
Boonville, MO (06909000)	*317BV*	1 299 403	1487.0	2020.4	35.9
Hermann, MO (06934500)	*158HM*	1 357 678	1955.6	2629.6	34.5

the RVA recommends that flow management 'targets' for each hydrologic parameter be based on a river management team's selected ranges of natural variation for that parameter (Step 2). In the absence of specific ecological information to identify these targets, Richter et al. (1997) recommend using ±1 standard deviation (SD) of pre-development hydrologic parameters as initial targets; we follow this general approach. However, hydrologic data are often skewed so that ±1 SD falls outside the range of observed values. This occurred for various parameters within Groups 1–4 of the IHA statistics for the Missouri River data set. Consequently, we report median (50th%ile) values rather than means for parameters within Groups 1–4 and use the 25th and 75th%iles as our initial flow management targets. The mean and ±1 SD were used for Group 5 parameters (rise rate, fall rate and number of flow reversals). We report each IHA parameter as the percent change in medians or means and CDs or CVs from the pre-regulation to post-regulation period:

$$\%CHG = ((\text{post-regulation value}) - (\text{pre-regulation value}))/(\text{pre-regulation value}) * 100.$$

See Galat & Lipkin (1999) for numerical values of each variable.

Using measures of natural dispersion based on pre-regulation data as flow 'targets' evaluates if post-regulation hydrologic conditions occur at the same frequency as before regulation (Richter et al., 1998). Pre-regulation annual values for hydrologic parameters fall within the 25–75th %ile values 50% of the time and within ±1 SD about 67% of the time. Thus, for post-regulation IHA observations to meet target criteria, only one-half to two-thirds of annual values are expected to fall within the pre-development flow regime. The degree to which the selected pre-regulation RVA measure of dispersion is not attained is an estimate of 'hydrologic alteration' (Richter et al., 1998). We follow Richter et al.'s (1998) example and report the percent of hydrologic alteration as:

$$\%HA = ((\text{Observed} - \text{Expected})/\text{Expected}) * 100.$$

'Observed' is the count of post-regulation years the hydrologic parameter was observed within the pre-regulation 25–75th %iles or ±1 SD of the pre-regulation mean. 'Expected' is the count of post-regulation years the hydrologic parameter is expected within the pre-regulation 25–75th %iles or ±1 SD of the mean, which by definition, is either 50 or 67%, respectively. Thus, %HA = 0, when the observed frequency of post-regulation years falling within the pre-regulation target range is the same as expected during the pre-regulation period. When %HA >0, post-regulation annual parameter values fell within the RVA target window *more* often than expected, while %HA <0 for a post-regulation parameter indicates annual values fell within the RVA target window *less* often than expected. We further abstract hydrologic alteration among stations by dividing %HA values (absolute) into four classes of equal range: 0 = 0–25%, low flow alteration; 1= 26–50%, moderate flow alteration; 2 = 51–75%, high-flow alteration; and 3 = 76–100%, extreme flow alteration. These ranks were first averaged over the hydrologic indicators within each of the five IHA groups and then the group means averaged to yield an index of overall hydrologic alteration for each station. Caution is advised not to overinterpret this summary index, as combining offsetting variables might yield similar overall %HAs among stations which exhibit widely different causes of impairment. Management actions should rely on consideration of each of the 32 individual indicators.

Results

Mean annual discharge over the study period (1929–1996) increased gradually from **3336FB** to *991OM*, as only a few major tributaries contribute flow in this section (e.g. Yellowstone River, Table 2). Downriver from *991OM*, flow increased more steeply with the input of several tributaries (Platte, 957 km; Kansas, 591 km; Grand, 402 km; Chariton, 366 km; Osage, 209 km; and Gasconade, 168 km). Although the lower 991 km of river drains only about 38% of the total Missouri River catchment, it contributed 61% of the 50 year mean annual discharge at *158HM*.

Mean annual discharge for the 30 year post-regulation period was higher at all stations than for the 20 years before mainstem dams operated as a complex (Table 2). This increase in discharge was <10% at **2592SN** (Yellowstone River), but about 30% higher at **3336FB**. The discharge increase at the six channelized stations on the lower Missouri River ranged from about 30 to 38%. The greatest increase in post-regulation discharge was observed at 2738WP, below Ft. Peck Dam. This appeared to be an artifact of including reservoir filling in the 1929–1948 pre-regulation interval.

Magnitude of monthly discharge

The general pattern of mean monthly discharge at all stations before mainstem impoundments were operational was an extended period of low flow from August through February (Figure 2). Mean discharge increased beginning in March at most stations, showed a small peak in April between 2115BM and *589KC*, and was highest during June throughout the river continuum. The annual flood pulse was unimodal at

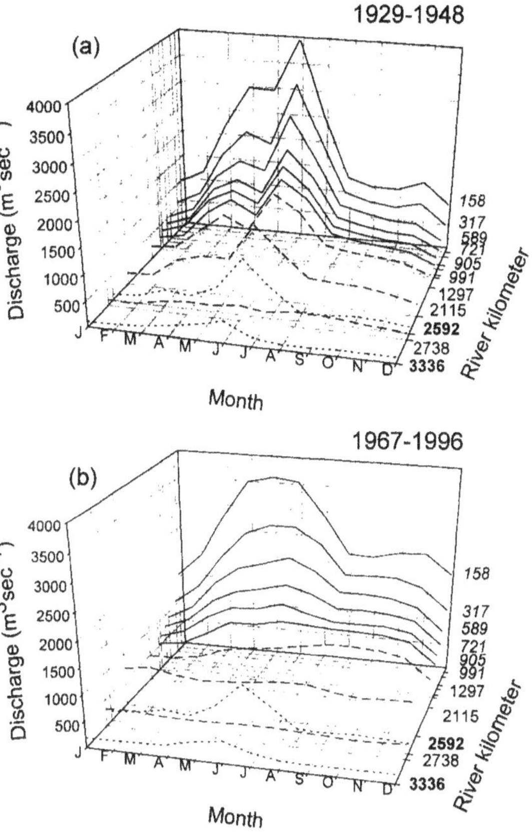

Figure 2. Mean monthly discharge before (a, Oct 1929–Sep 1948) and after (b, Oct 1967–Sep 1996) flow regulation along the mainstem Missouri and lower Yellowstone rivers. Stations above large mainstem reservoirs are indicated by dotted lines, those between and immediately below reservoirs by dashed lines, and those below reservoirs in the channelized Missouri River are identified by solid lines. River kilometer is distance above the Missouri River mouth. Station 2592 (Sidney, MT) is on the undammed lower Yellowstone River.

least-altered site **3336FB** (June peak), weakly bimodal at 2738WP and 2115BM (April, June peaks), strongly bimodal at **2592SN** (March, June peaks) and also at 1297YT, *991OM* and *905NC*, but with April and June peaks. The bimodal flood-pulse pattern weakened downriver from *991OM*, gradually becoming nearly unimodal again at the lowermost station (*158HM*). A small November flow pulse was also observed before flow regulation at the two lowermost stations (*317BV*, *158HM*). The absence of a distinct flood pulse at 2738WP relative to **3336FB** and **2592SN** (Figure 2) again appeared to be a result of filling Ft. Peck reservoir in the 'pre-impoundment' period.

The general pattern of post-regulation flows was a stabilization of mean monthly discharge by an increase in August through February low flows and a reduction in June and July high flows (Figure 2, Table 3). This change was absent at least-regulated stations (**3336FB**, **2592SN**), most pronounced at inter-reservoir and upper-channelized-river sites, and less extreme at lower-channelized river sites (*589KC*–*158HM*). Once the mainstem dams became operational, the naturally bimodal flood-pulse became unimodal at all sites below reservoirs and the small fall pulse at sites *317BV* and 158HM disappeared because of constantly high summer-autumn reservoir water releases.

August through February median monthly discharges at most stations were above the pre-regulation 75th%ile flow interval more often than expected for post-regulation years (Table 3). April and July %HA was often positive following flow regulation, indicating that more years were within the pre-regulation 25–75th%iles than expected for many stations, but there was no consistent trend in %HA for March, May or June. Flow variability (% change in CD) was lower following impoundment for most months and gauging stations.

Median flows for the peak discharge month of June decreased 16% at least-altered station **2592SN** between the two time intervals, the CD increased by 50%, and 27% fewer years were within the 1967–1996 target window for June discharge (Figure 3). In contrast, median June discharge increased 37% at **3336FB**, the CD was less, and 20% fewer years than expected were within the 1929–1948 25–75th%iles. Decreases in median June discharge attributed to flow regulation were highest at inter-reservoir sites 2115BM and 1297YT and – moving downriver through the channelized reach – became progressively smaller (Figure 3). Median June flows decreased <10% following impoundment at channelized-river stations *721SJ*, *589KC*, *317BV* and *158HM*, while variability in June discharge decreased by ≥45% (Figure 3, Table 3). Fewer post-regulation years than expected were within the pre-regulation 25–75th%iles for June discharge at most stations above *905NC* (except 2738WP). The number of years where June discharge was outside the pre-regulation target window generally decreased downriver until there was no difference between the two time intervals at *905NC* (Figure 3). Below *905NC*, a higher number of post-regulation years were within the pre-regulation June flow 25–75th%iles than expected.

Table 3. Percent change in median, coefficient of dispersion, and hydrologic alteration of monthly discharge between pre- (Oct 1929–Sep 1948) and post-flow regulation (Oct 1967–Sep 1996) periods along the Missouri and lower Yellowstone (**2592SN**) rivers. Station locations are kilometers above Missouri River mouth. Station numbers in **bold** type are the least flow-altered, those in regular type are inter-reservoir or immediately below reservoirs, and those in *italics* are below reservoirs in the channelized river. See text for how hydrologic alteration was calculated

Station	Jan	Feb	Mar	Apr	May	Jun	Jul	Aug	Sep	Oct	Nov	Dec
Median												
3336FB	77	72	45	21	21	37	105	85	51	36	34	48
2738WP	285	366	90	−12	30	−6	64	47	100	77	92	230
2592SN	49	68	−12	24	11	−16	−7	17	42	35	26	70
2115BM	314	272	18	9	−6	−54	−21	43	43	56	118	201
1297YT	128	97	−15	5	19	−43	−18	62	88	157	143	126
991OM	106	108	13	14	31	−27	−7	65	83	120	128	165
905NC	63	52	14	23	32	−19	−15	55	59	67	97	128
721SJ	71	61	22	32	47	−7	2	63	90	75	103	123
589KC	55	50	10	22	44	−3	−4	68	88	98	95	129
317BV	70	63	0	27	50	−4	−6	64	79	80	107	117
158HM	47	81	19	37	48	0	5	55	67	76	101	93
Coefficient of dispersion (CD)												
3336FB	−61	−29	18	−54	4	−17	10	63	15	46	2	−50
2738WP	−84	−81	−55	28	−56	−48	−52	−68	−75	−50	−27	−81
2592SN	−25	17	−6	−28	−4	50	−21	−19	−39	−26	25	−44
2115BM	−67	−56	−52	−60	−36	9	−49	−47	−49	−12	−8	−48
1297YT	−25	−51	−69	−79	−63	−55	−72	−17	−48	−65	−37	46
991OM	−8	−35	−38	−67	−47	−43	−45	−23	−52	−53	−12	54
905NC	17	−17	73	−44	−32	−28	−27	−35	−41	−36	−9	45
721SJ	40	−8	27	−46	−19	−46	−46	−23	−48	−53	−14	−7
589KC	25	32	44	−9	−20	−60	−29	−27	−10	−44	−39	−15
317BV	−15	45	25	2	−42	−63	−15	−27	20	−24	−37	−17
158HM	−6	8	93	−1	−25	−45	−10	−34	18	−19	0	−20
Hydrologic alteration (HA)												
3336FB	−87	−93	−47	67	−53	−20	−27	−80	−67	−33	−40	−73
2738WP	−100	−100	−53	33	53	60	−7	93	87	27	−60	−100
2592SN	−93	−87	40	27	−7	−27	20	0	7	−13	−47	−53
2115BM	−100	−100	27	33	13	−60	60	−47	0	−13	−47	−100
1297YT	−100	−80	60	67	47	−80	87	−87	−87	−87	−87	−100
991OM	−100	−87	60	87	0	−13	80	−100	−100	−100	−73	−100
905NC	−93	−47	−40	33	7	0	73	−87	−100	−100	−73	−93
721SJ	−93	−40	−27	47	−13	27	67	−80	−87	−100	−80	−87
589KC	−80	−47	−20	20	−7	40	60	−60	−87	−100	−73	−80
317BV	−67	−53	−13	33	13	40	40	−67	−87	−73	−67	−73
158HM	−13	−33	−47	7	7	47	40	−67	−67	−67	−53	−53

Magnitude and duration of discharge extremes

Patterns of annual discharge maxima and minima were similar among the 1-, 3-, 7-, 30- and 90-day averaging windows so we report only the 1-, 7- and 30-day results (Table 4). Post-regulation median discharge maxima for 1-, 7-, and 30-day intervals were between 12 and 23% lower than before regulation at **2592SN** on the Yellowstone River, but about 30% higher at

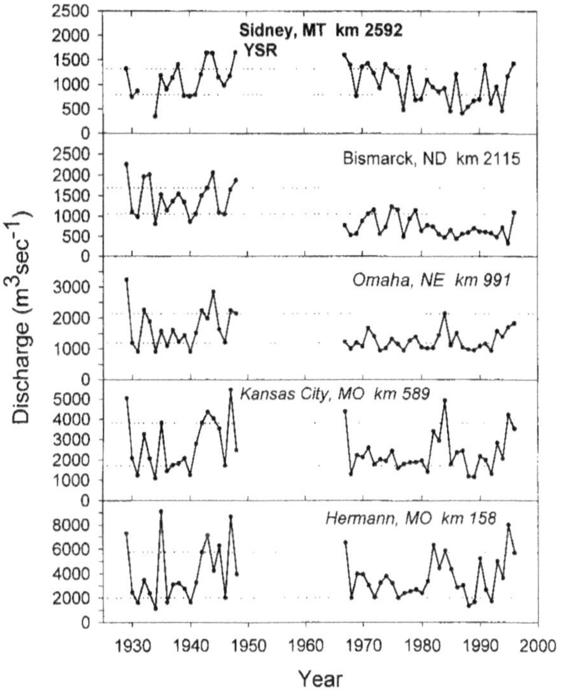

Figure 3. Median June discharge at five stations along the Missouri and lower Yellowstone rivers (YSR) before (1929–1948) and after (1967–1996) flow regulation. No flow records were available for Sidney, MT, during 1932–33. The horizontal dotted lines are the 25th (lower) and 75th (upper) percentiles for the pre-regulation interval and define a target range of acceptable hydrologic variability for post-regulation years. Numbers within panels identify station locations in kilometers above Missouri River mouth. Least-altered station names are in **bold** typeface, inter-reservoir stations are in normal typeface, and channelized river stations are in *italics*.

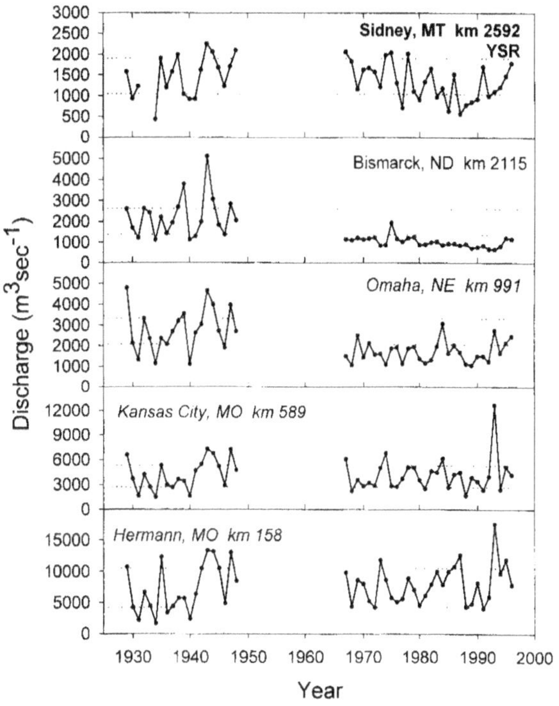

Figure 4. Annual 7-day median discharge at five stations along the Missouri and lower Yellowstone rivers (YSR) before (1929–1948) and after (1967–1996) flow regulation. The horizontal dotted lines identify a target range of variability for the post-regulation period. Refer to Figure 3 for further explanation.

the least-regulated Missouri River station **3336FB** (Table 4). Median annual maximum flows for the three averaging durations were nearly all less after regulation for inter-reservoir and channelized-river stations 2738WP to *905NC*; station *721SJ* was transitional, and maximum flows increased from *589KC* downstream to the Missouri's terminus. Post-impoundment variability of annual maximum flows decreased at **3336FB**, all inter-reservoir stations and the uppermost channelized-river site (*991OM*), increased at stations *905NC* and *721SJ* in the channelized section, but then decreased again at *589KC* and downriver. Hydrologic alteration in annual maximum flows at least-altered stations ranged from 27% more years within the pre-regulation 25–75th%ile than expected to 20% less years than expected. Most regulated river stations exhibited a greater alteration in medians of annual maximum flows than least-altered sites. Fewer years than expected were within the pre-regulation 25–75th%ile target range for 1- and 7-day (Figure 4) averaging periods of median annual maxima for inter-reservoir stations. The river from Kansas City (*589KC*) to the mouth showed a different trend with a higher number of post-regulation years falling within the target 25–75th%iles of the 1–30-day averaging windows of maximum discharge than before flow regulation.

Post-regulation medians of annual minimum flows for 1-, 7- and 30-day averaging intervals at all study gauges were higher than pre-dam medians, over 100% at many stations (Table 4, Figure 5). Locations with the smallest increases in annual minimum flows (although still ranging between 41 and 136% higher) were the two least-altered sites and lowermost stations *317BV* and *158HM*. Over 50% of post-dam years at all stations had fewer 1-, 7- and 30-day annual minimum flows within the pre-dam 25–75th%iles than expected; all 30 post-dam years were above the target window for one or more of the annual minimum-flow durations at six of the inter-reservoir and channelized-river stations (Table 4, Figure 5).

Table 4. Percent change in nonparametric (median, coefficient of dispersion) and parametric (mean, coefficient of variation) statistics and hydrologic alteration for five groups of hydrologic variables between pre- (Oct 1929–Sep 1948) and post-flow regulation (Oct 1967–Sep 1996) periods along the Missouri and lower Yellowstone (**2592SN**) rivers. All hydrologic parameters are summarized by nonparametric statistics except rise rate, fall rate and number of reversals, which are summarized by parametric statistics. Refer to Table 2 for station locations in relation to reservoirs. See Table 1 for definitions of hydrologic terms and text for how hydrologic alteration was calculated. JD = Julian date

Station	Group 2 – Magnitude & duration of annual discharge extremes						Group 3 – Timing of annual discharge extremes			Group 4 – Frequency & duration of high & low pulses				Group 5 – Rate & frequency of hydrograph changes		
	Max			Min			JD annual		JD Mar–Oct	High pulse		Low pulse				
	1d	7d	30d	1d	7d	30d	Max	Min	Min	Count	Dur.	Count	Dur.	Rise rate	Fall rate	No. flow reversals
	Median/mean															
3336FB	30	27	33	69	41	51	3	34	14	100	−21	−11	−77	−15	−17	−6
2738WP	−27	−28	−21	370	346	176	−31	−38	22	50	−48	−40	−72	−39	−27	51
2592SN	−12	−23	−12	126	102	51	4	8	−3	0	24	−17	−45	−33	−34	−4
2115BM	−61	−52	−41	256	260	177	−38	−55	5	25	−41	0	−74	−62	−48	145
1297YT	−59	−52	−34	144	148	149	95	48	−48	−50	−3	−75	76	−76	−62	56
991OM	−36	−40	−17	207	204	157	12	13	−51	−33	−42	−75	156	−53	−44	40
905NC	−21	−29	−12	174	186	124	1	12	−29	−17	−31	−78	101	−42	−40	30
721SJ	6	−8	−1	212	250	140	3	9	−28	40	5	−63	46	−13	−17	24
589KC	15	1	5	163	179	126	−1	7	−30	0	17	−63	16	−6	−11	19
317BV	49	30	25	117	136	97	4	5	−18	50	19	−56	−34	13	11	16
158HM	21	38	53	91	94	100	−1	7	−21	60	20	−33	−34	14	13	14
	Coefficient of dispersion/coefficient of variation (CD/CV)															
3336FB	−20	−28	−11	−48	−8	−5	94	−27	48	25	68	−39	152	37	35	−21
2738WP	−5	−27	−40	−50	−53	−39	−10	406	49	50	168	483	38	−23	0	−22
2592SN	−25	5	29	48	18	−2	−15	159	25	33	2	100	0	−35	−39	−26
2115BM	−51	−47	−45	−38	−27	8	71	144	212	260	37	100	−31	−9	14	−30
1297YT	−16	−22	−14	28	−12	−46	−12	187	70	167	125	33	97	18	58	75
991OM	−1	−13	−15	30	2	7	7	60	84	250	315	60	314	−4	−11	−28
905NC	39	52	19	15	−4	10	14	−14	106	140	292	200	133	16	18	−7
721SJ	25	50	5	−28	−45	−3	20	17	197	25	42	33	107	34	7	−8
589KC	−36	−16	−8	16	−18	−15	149	−14	252	25	−58	100	81	15	9	10
317BV	−67	−50	−32	17	11	76	53	0	140	17	−49	80	43	−10	−3	−4
158HM	−46	−47	−51	34	46	1	168	125	115	−22	−45	−200	60	−20	−20	−6
	Hydrologic alteration (HA)															
3336FB	−7	7	−20	−53	−73	−87	−53	−40	−60	−60	−33	47	−40	−35	−35	15
2738WP	−40	−40	−33	−100	−100	−100	13	80	−67	−27	−47	−80	−53	−30	−15	−70
2592SN	27	13	−7	−87	−67	−73	−20	13	−13	13	27	−13	−33	15	5	20
2115BM	−93	−93	−40	−100	−100	−100	−40	80	−87	−73	7	−27	−67	−85	−80	−100
1297YT	−100	−100	−47	−67	−80	−87	−93	−73	−93	−33	−53	−67	−73	−100	−95	−95
991OM	−73	−60	7	−87	−100	−100	−40	−60	−47	−40	−73	−60	−73	−80	−75	−100
905NC	−47	−40	0	−87	−100	−100	−27	−67	−67	−60	−73	−87	−60	−50	−55	−90
721SJ	−33	−27	0	−73	−87	−100	−20	−13	−67	−20	−40	−53	−53	−20	0	−80
589KC	47	33	20	−73	−80	−87	−47	47	−80	−20	47	−67	−47	0	5	−75
317BV	67	53	20	−93	−100	−100	−7	67	−73	−47	27	−33	−7	10	10	−35
158HM	60	53	47	−73	−73	−73	−20	73	−40	−27	53	−53	−33	15	20	−25

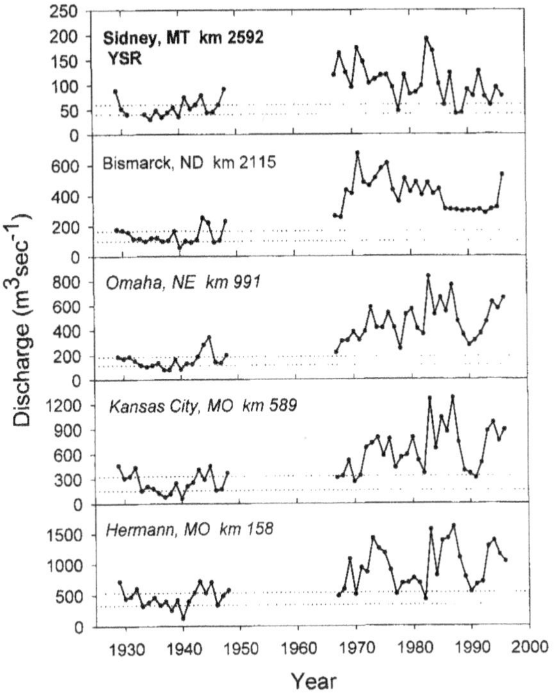

Figure 5. Annual minimum 7-day median discharge at five stations along the Missouri and lower Yellowstone rivers (YSR) before (1929–1948) and after (1967–1996) flow regulation. The horizontal dotted lines identify a target range of variability for the post-regulation period. Refer to Figure 3 for further explanation.

Timing of annual discharge extremes

Median date of the annual maximum daily discharge before flow regulation occurred within the same three weeks (21 May – 14 June, JD 142–166) among all stations except 1297YT (10 April). There were only small differences (<4%) in timing of the median Julian date of annual daily flow maxima following regulation at above-reservoir stations and also in the channelized river from *905NC* downriver (Table 4). However, variability in the date of annual maximum daily flow among years was generally higher in the channelized section following river impoundment (Table 4); fewer years than expected were within the pre-regulation 25–75th%iles. Annual peak daily discharges occurred between 56 and 70 days earlier at inter-reservoir stations 2738WP and 2115BM, but 173 days later at station 1297YT below Gavins Point Dam, an important nesting area for federally threatened least terns (Smith, 1996).

Prior to flow regulation, annual daily discharge minima occurred between mid-December and early January (JD 345–4) at all stations except site **3336FB**, where the median date of annual minimum daily discharge was 12 August. Median Julian date of annual flow minima occurred much earlier following dam operation at inter-reservoir stations 2738WP and 2115BM, and 80% more post-regulation years than expected were within the pre-regulation 25–75th%iles, even though variability in the timing of annual daily flow minima was much higher (Table 4). The gauge below Gavins Point Dam (1297YT) was again different from other inter-reservoir sites, as the timing of annual daily flow minima was delayed after flow regulation from Julian day 350 (15 December) to JD 71 (11 March) and 73% fewer years than expected fell within the pre-regulation 25–75th%iles. Variability in CD at channelized-river sites ranged from +60 to −14%, except at *158HM* where it increased to 125%. Also, the percentage of post-regulation years within the pre-regulation 25–75th%iles was higher than expected at the three lowermost channelized stations (Table 4).

Pre-dam median date of lowest daily flow between May and October occurred from JD 223 to 283 (10 August–9 October) at all stations, was most common in September (7 of 11 stations), and generally occurred later in the season farther downriver (October in the three lowermost stations). Timing of the 1967–1996 May–October daily flow minima was generally later than the 1929–1948 25–75th%iles for stations **3336FB** and 2738WP, was erratic at inter-reservoir station 2115BM, and occurred much earlier in the growing season at station 1297YT and all sites downriver (Table 4).

Frequency and duration of annual high- and low-flow pulses

Discharges selected as the minimum threshold to define annual high pulses occurred in June at all Missouri River stations (Galat & Lipkin, 1999). Least-altered and inter-reservoir stations exhibited fewer (3–4 yr^{-1}) high-flow pulses before flow regulation than channelized-river stations (5–6 yr^{-1}). The median number of high-flow pulses per year doubled at the Missouri River least-altered station **3336FB** (from 3 to 6 yr^{-1}) between the two time intervals, but remained constant at the Yellowstone River least-altered station (**2592SN**). Frequency of high-flow pulses increased following flow regulation at inter-reservoir stations 2738WP and 2115BM; decreased at inter-reservoir and channelized stations 1297YT, *991OM* and *905NC*; and increased at three of the four lower-basin channelized stations (*721SJ*, *317BV*, *158HM*).

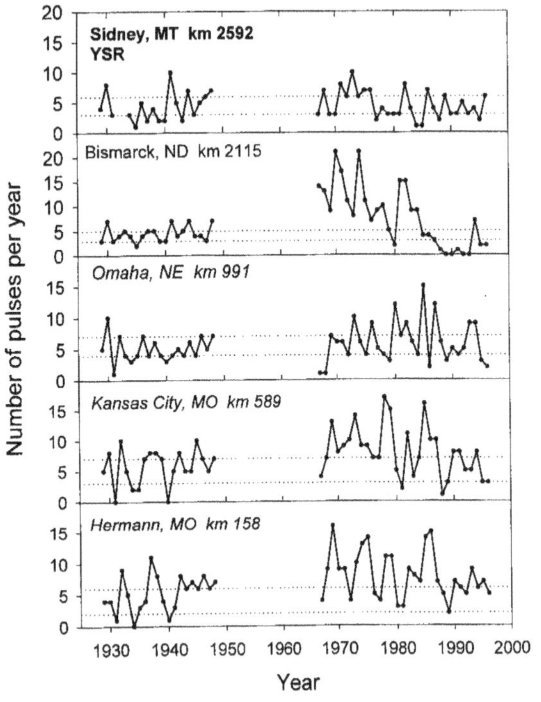

Figure 6. Number of high-flow pulses per year at five stations along the Missouri and lower Yellowstone rivers (YSR) before (1929–1948) and after (1967–1996) flow regulation. The horizontal dotted lines identify a target range of variability for the post-regulation period. Refer to Figure 3 for further explanation.

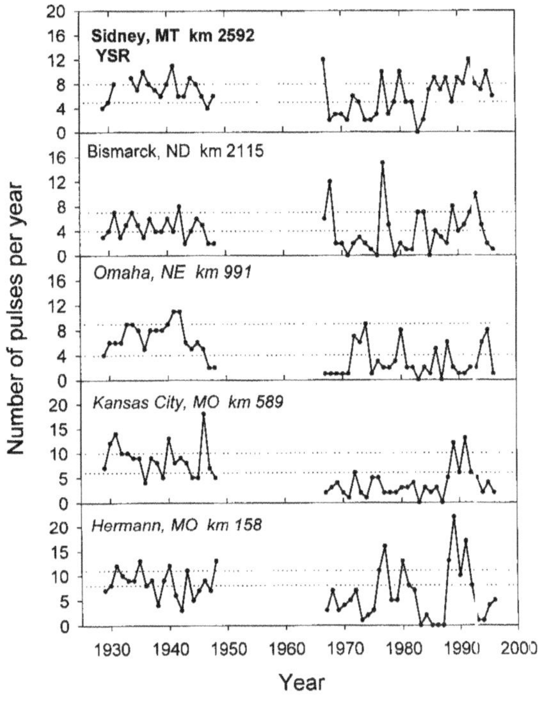

Figure 7. Number of low-flow pulses per year at five stations along the Missouri and lower Yellowstone rivers (YSR) before (1929–1948) and after (1967–1996) flow regulation. The horizontal dotted lines identify a target range of variability for the post-regulation period. Refer to Figure 3 for further explanation.

Variability in the number of high-flow pulses per year increased by over 100% following flow regulation at inter-reservoir and channelized-river stations 2115BM to *905NC*, compared to less than 35% at least-altered upper basin sites (Figure 6).

Median duration of high-flow pulses before impoundment ranged from 12 to 17 days yr^{-1} at stations **3336FB** to 1297YT and decreased downriver to 7–9 days yr^{-1} at stations *991OM–158HM*. After impoundment, the median duration of high-flow pulses decreased (−3 to −48%) at 6 of the 7 stations between **3336FB** and *905NC*, except station **2592SN** where it increased by 24%. In contrast, the length of high-flow pulses increased from 5 to 20% at the four lowermost channelized-river stations (*721SJ–158HM*) during 1967–1996. The number of days per year of high-flow pulses at the Yellowstone River site (**2592SN**) and the three lowermost channelized-river stations (*589KC*, *317BV* and *158HM*), were within the pre-regulation 25–75th%iles for over 60% of post-regulation years. In summary, the number of high-flow pulses per year generally increased between the two time periods, but their length was shorter at the two least-regulated stations and the two upper river, inter-reservoir sites. Both the number and duration per year of high-flow pulses were reduced below Gavins Point Dam at station 1297YT, while their values generally increased between 1929–1948 and 1967–1996 at the four farthest downriver gauges (*721SJ–158HM*).

The number of low-flow pulses per year decreased at 10 of the 11 stations following flow regulation (Table 4). This reduction was smallest at the two above-reservoir stations and at the lowermost river gauge (*158HM*). The post-dam decrease in number of low-flow pulses per year was highest (−40 to −78%) at inter-reservoir and channelized-river stations, with the exception of 2115BM (Figure 7). Variability in the number of low-flow pulses per year generally increased during the reservoir operation period except at *158HM* where the CD decreased. Fewer post-regulation years than expected for most stations were within the 25–75th%ile target range for the number of low-flow pulses, and the decrease in%HA was lowest at least-altered site **2592SN**.

Changes in the duration of annual low-flow pulses were recorded between pre- and post-regulation intervals, but were variable along the river continuum (Table 4). Duration following regulation was 45–77%

shorter at the two above-reservoir stations and the two uppermost inter-reservoir gauges. Below Gavins Point Dam the pattern was reversed, duration increasing by over 75% at stations 1297YT, *991OM* and *905NC*. This increase was dampened downriver until it again decreased at the two lowermost stations (*317BV*, *158HM*). So, while there was a general basin-wide reduction in the number of low-flow pulses between the pre- and post-dam intervals, their duration showed a complex longitudinal pattern: decreasing, increasing and then decreasing again.

Rate and frequency of change in discharge

Rates of change in river flow were the only hydrologic parameters analyzed using parametric statistics and the pattern of results were similar for both discharge rises and falls (Table 4). Mean rates of discharge rises and falls were between 15 and 34% lower and the mean number of flow reversals per year decreased slightly (4–6% less) during 1967–1996 at the two least-altered stations. There was a more pronounced post-regulation reduction in rise and fall rates at inter-reservoir and upper channelized-river sites than at least-altered stations, and between 50 and 100% fewer years were observed within the pre-regulation %HA target window at stations 2115BM to *905NC* (Table 4). The post-regulation reduction became progressively less proceeding downriver from 1297YT to 721SJ–589KC, while below 589KC (*317BV* and *158HM*) the mean rate of discharge change increased (Table 4). Additionally, 10–20% of post-regulation years were within the pre-regulation target windows at the two lowermost sites, while between 50 and 100% fewer years than expected were within the target window at most regulated river stations upstream from *721SJ*.

More flow reversals per year were observed following regulation at stations downstream from dams, but this increase gradually diminished to only 14% more pulses per year after regulation at station *158HM*. Similarly, the reversal rate per year between the two time periods was highest at inter-reservoir stations and decreased linearly downriver to *158HM*.

Basin-wide summary

Major differences in hydrologic indicators between pre- and post-regulation for the 11 stations are highlighted by reporting %HA for hydrologic indicators ranked high or extreme (Table 5).

These include variables and stations where >50% more post-regulation years than expected (>22 of the 30 post-regulation years for Groups 1–4) were within or outside of the RVA target window (25–75th%ile for Groups 1–4, ±1 SD for Group 5).

Mean annual discharge for all stations along the Missouri River was higher from 1967 to 1996 than between 1929 and 1948, but noticeably less so at **2592SN**, the only station with no upriver impoundments. Flow regulation of the mainstem Missouri River was associated with significant alterations in many of the 32 hydrologic indicators (Table 5). Most notably, these were: 1. a reduction in the magnitude (i.e. lower high flows) and duration of the annual flood pulse; 2. an increase in the magnitude (i.e. higher low flows) and duration of annual discharge minima; 3. a reduction in the frequency of annual daily low-flow pulses and earlier timing of growing season daily low-flow pulses; and 4. a general increase in the frequency of discharge reversals per year coupled with a reduced rate of change in river flows. Collectively, hydrologic alterations were lowest at least-altered station **2592SN**, most severe at inter-reservoir and upper channelized-river sections, and intermediate at the lowermost channelized-river stations (Figure 8). Least-altered station **3336FB**, below Canyon Ferry Reservoir, showed a composite hydrologic alteration much higher than least-altered site **2592SN** and similar to that of the four lowermost channelized-river stations. Higher post-regulation discharge throughout most of the Missouri River catchment, but to a much lesser extent in the Yellowstone River basin, appeared to interact with reservoir operations to influence measures of hydrologic alteration between the two time intervals.

Discussion

Alterations in seasonal flow patterns and ecological effects

Our characterization of the Missouri River's flow regime between 1929 and 1948 shows the system to be more complex than previously reported (Hesse et al., 1989; Hesse & Mestl, 1993). The protracted pre-regulation annual flood pulse in the lower Missouri River was a temporally cumulative result of complex precipitation and climatic patterns throughout the catchment: spring ice-out in the upper and middle basins, spring snowmelt in the middle basin, runoff from spring rains in the lower basin, and runoff from June snowmelt in the upper basin. We suspect the

Table 5. Summary of flow alterations between pre- (Oct 1929–Sep 1948) and post- (Oct 1967–Sep 1996) regulation intervals for five groups of hydrologic variables at 11 stations along the Missouri and lower Yellowstone (**2592SN**) rivers. Hydrologic variables included are where >50% of post-regulation years than expected were within or outside of the RVA target window (25–75th percentile for Groups 1–4, ±1 SD for Group 5). The direction of change for the post-regulation median or mean relative to pre-regulation conditions is shown as + or –. An "o" indicates that the post-impoundment change was ≤±20% of the pre-impoundment median or mean. JD = Julian date

Station	Group 1 – Magnitude of monthly discharge extremes[1]		Group 2 – Magnitude & duration of annual discharge extremes		Group 3 – Timing of annual discharge extremes			Group 4 – Frequency & duration of high & low pulses				Group 5 – Rate & frequency of hydrograph changes		
					JD annual		JD Mar–Oct	High pulse		Low pulse		Rise	Fall	No. flow
			Max (d)[2]	Min (d)[2]	Max	Min	Min	Count	Dur.	Count	Dur.	rate	rate	reversals
Least-altered														
3336FB	1,2,4,5,8,9,12	+		1,7,30 +	o		o	+						
2592SN	1,2,12	+		1,7,30 +										
Inter-reservoir														
2738WP	1,2,3,5,8,9,11,12	+		1,7,30 +		–	+			–	–			+
	6	o												
2115BM	1,2,12	+	1,7 –	1,7,30 +		–	o	+			–	–	–	+
	6,7	–												
1297YT	1,2,4,8,9,10,11,12	+	1,7 –	1,7,30 +	+	+	–		o	–	+	–	–	+
	3,7	o												
	6	–												
Below reservoir, channelized														
991OM	1,2,8,9,10,11,12	+	1,7 –	1,7,30 +		o	–		–	–	+	–	–	+
	3,4	o												
905NC	1,8,9,10,11,12	+		1,7,30 +		o	–	–	–	–	+		–	+
	7	o												
721SJ	1,8,9,10,11,12	+		1,7,30 +			–			–	+			+
	7	o												
589KC	1,8,9,10,11,12	+		1,7,30 +			–			–				+
	7	o												
317BV	1,2,8,9,10,11,12	+	1,7 +	1,7,30 +		o	o							
158HM	8,9,10,11,12	+	1,7 +	1,7,30 +		o			–	–				

[1] Numbers indicate months: 1=Jan, 2=Feb, etc.; [2] d=day (1-day, 7-day or 30 day averaging interval).

pre-regulation annual flow regime in the middle-river section was more bimodal than elsewhere because a spring rainy season is absent in the Great Plains Province.

The importance of a predictable annual flood pulse to reproduction of fishes in large floodplain rivers is well documented (Junk et al., 1989; Ward, 1989; Welcomme et al., 1989; Bayley & Li, 1992) and is the basis of the 'flood-pulse advantage' observed in the high production of fishes in large river-floodplain systems (Bayley, 1991). If tenets of the flood-pulse concept are applicable to the Missouri River hydrosystem, we hypothesize that modifications in the magnitude, frequency, timing and duration of the annual flood pulse have reduced recruitment and production of native river-floodplain fishes in the inter-reservoir and channelized-river sections. Galat et al. (1998) showed that inundation for $\geq$10 days of the Missouri River floodplain in central Missouri occurred during the optimal spawning temperature ranges of native fishes for only 4 of the 15 post-regulation years between 1983 and 1996.

There was historically a small fall increase in discharge along the lowermost Missouri River, similar to that reported by Sparks et al. (1990) and Sparks (1995) for the nearby upper Mississippi River. This small pulse is regional and due to the onset of autumn rains in the well-watered Central Lowlands and Interior Highlands physiographic provinces (Galat et al., 1998). A fall flow pulse is important to provide fishes access to wintering habitats on the floodplain and in backwaters before cold-water temperatures reduce their swimming ability (Bodensteiner & Lewis, 1992). Furthermore, it inundated annual moist-soil vegetation in floodplain wetlands, providing forage to fall migrating waterfowl. This historically small autumn flow pulse inundated floodplain wetlands to a shallower depth than the subsequent large June pulse, with the result that forage remained available for waterfowl on their return migration the following spring (Sparks, 1992; Galat et al., 1998). Flow releases for navigation and levees that disconnect the lower Missouri River from its floodplain wetlands have largely eliminated this fall river rise and the benefits to fish and wildlife it once provided.

Assessment of intra-annual flow patterns also identified alteration of several ecologically important summer low-flow variables and showed that changes are widespread over the regulated Missouri River. Sustained reservoir water releases during the naturally low-water season cause protracted flooding of about two-thirds of the Missouri River and may be as pervasive and damaging a disturbance as reduction of the annual June flood pulse. Circumstantial evidence for this hypothesis comes from the native fish community that reproduces in or along the Missouri and lower Yellowstone rivers. Based on adult habitat use and reproductive requirements described in Pflieger (1971, 1997), we identified as 'fluvial specialists' (∗) seven of the nine fishes federally listed as candidate, threatened, or endangered (U.S. Fish & Wildlife Service, 1994, 1995): *Acipenser fluvescens*, ∗*Scaphirhynchus albus*, *Polyodon spathula*, ∗*Hybognathus argyritus*, ∗*H. placitus*, ∗*Macrhybopsis gleida*, ∗*M. meeki*, ∗*Platygobio gracilis* and ∗*Cycleptus elongatus*. Fluvial specialists use flowing-water habitats throughout life (Bain, 1992; Kingsolving & Bain, 1993). The remaining two species, *A. fluvescens* and *P. spathula*, are considered 'fluvial dependent,' requiring flowing water at some point in their life cycle (Bain, 1992; Kingsolving & Bain, 1993). Both fishes reproduce in flowing waters and may migrate into tributary streams to spawn. All nine fishes are capable of completing their entire life-cycle within the channel complex and are included in Pflieger's (1971) 'Big River Faunal Group.' Loss of a river-floodplain connection due to a reduced annual flood pulse should have less direct affect on spawning success of these fishes than on floodplain-dependent species like centrarchids. Therefore, the disproportionate decline in Missouri River's native fluvial fishes appears most associated with protracted summer-fall high flows and in-channel habitat loss.

Loss of the lower Missouri River's braided channel geometry through channelization has eliminated most sand island and shallow in-channel habitats used by riverine fishes for spawning and nursery (Funk & Robinson, 1974; Latka et al., 1993). The few low-elevation sand islands and associated shoals that remain are now flooded or their surface area reduced during much (July–September) of the reproductive season for many riverine fishes, as well as for birds and turtles (Galat et al., 1998) that make similar use of these critical habitats. Additionally, protracted summer-fall high flows prevent germination of early-successional tree species (Johnson, 1992) and moist-soil annual vegetation in suitable habitats that remain along the lower Missouri River.

This hypothesis regarding the ecological effects of altered summer-fall low flows is not intended to discount the importance of restoring the annual flood pulse, but to emphasize the need for the full range of seasonal flow variability to re-establish a naturalized flow regime. The annual flood pulse is a critical cue to initiate spawning migrations for many fluvial-specialist and fluvial-dependent fishes, it is essential to transport floodplain-derived organic matter and nutrients to the main channel, and to maintain channel-floodplain geomorphic diversity (Welcomme, 1985; Junk et al., 1989).

Natural factors contribute to spatio-temporal variability in river flows

Station **2592SN** on the Yellowstone River was the only site evaluated where no upstream mainstem impoundments were present and it showed the lowest degree of hydrologic alteration between the two time intervals (Figure 8). Proceeding downriver from the last mainstem dam, Gavins Point (km 1305), overall flow alteration between 1967 and 1996 declined from that observed at inter-reservoir sites. Hydrologic variability was less from *721SJ* to the mouth than between the mainstem reservoirs and was similar to, or lower than, the upstream least-altered Missouri River site at **3336FB**. However, sources of hydrologic alteration differed somewhat among the channelized-river stations and between them and **3336FB** (Table 5). Notably, post-regulation magnitude and duration of annual maximum flows were higher at *317BV* and *158HM* than at other channelized sites because *317BV* and *158HM* were more affected by climatically driven flooding in 1993, 1995 and 1996 (Parrett et al., 1993; Galat et al., 1998).

Flow differences were moderate to extreme comparing the 1929–1948 interval to 1967–1996 at all sites examined, including least-altered site **2592SN** (Table 5). Consequently, caution must be used in attributing hydrologic alteration among sites and between the two time intervals solely to flow regulation. While mainstem impoundment, flow regulation, and changes in land use greatly influenced hydrologic variability of the Missouri River, at least two natural factors also contributed to these spatio-temporal differences. Basin-wide precipitation and runoff differed between the two time intervals, and hydrology is inherently variable along the Missouri River continuum.

The 'dust bowl' droughts of the 1930s occurred during the pre-regulation period and recurrent flooding in the 1990s was prevalent during the post-regulation interval. Qi Hu et al. (1998) analyzed interdecadal variations in precipitation from the 1890s through the 1990s in the central U.S., including the lower Missouri River basin states of Kansas, Nebraska, Iowa and Missouri. They concluded that there has been a trend in the region's annual mean precipitation and that it has changed from a decrease before the mid-1960s to an increase thereafter. Thus, the increase in low-flow discharge between 1967 and 1996, and most pronounced at inter-reservoir and channelized-river sites, appears to be an additive effect of mainstem flow regulation operating within the context of a catchment-wide increase in precipitation. This conclusion illustrates the importance of including least-altered sites in analyses of historical flow variability to normalize or filter out temporal climatic variation, even if such sites are not entirely anthropogenically unaltered.

A paradigm of the flood-pulse concept is that hydrologic buffering of a large catchment area results in smooth and predictable flooding and that effects of seasonal climatic changes are observed downstream only after several weeks or months in unaltered large-floodplain rivers in temperate and tropical regions (Junk et al., 1989). This has been shown graphically for the pre-regulation upper Mississippi (Sparks, 1992; Sparks et al., 1998) and Illinois rivers (Sparks, 1995). In contrast, timing of the pre-regulation median annual discharge maxima at 10 of 11 locations over the entire Missouri and lower Yellowstone rivers all occurred within 24 days and there was no longitudinal time lag. Additionally, IHA variables along the Missouri River continuum did not show a longitudinal decrease in predictability before flow regulation. Indices of dispersion (CD and CV, Tables 3 and 4) and the relative ranges of the pre-regulation 25–75th%iles among stations (Figures 3–7) either showed no longitudinal pattern or upper – and lowermost – stations exhibited the greatest flow variability for most hydrologic indicators.

The Missouri River does not fit neatly into the flood-pulse concept because it arises in the well-watered Rocky Mountains and then flows over 1000 km through the semi-arid Great Plains. Consequently, its middle section is largely allogenic, analogous to a dryland river as described by Walker et al. (1995). Tributary influx to the Missouri River is greatest in the lowermost section, and variability in frequency and duration of high-flow pulses and discharge is also high here. Our assessment of 50 years of Missouri River hydrology illustrates that the influence of reservoir operations on the annual flood pulse was partially offset by tributary influx downriver from Kansas City (*589KC*) during the wet period of 1967–1996. Therefore, the flood pulse retains a more

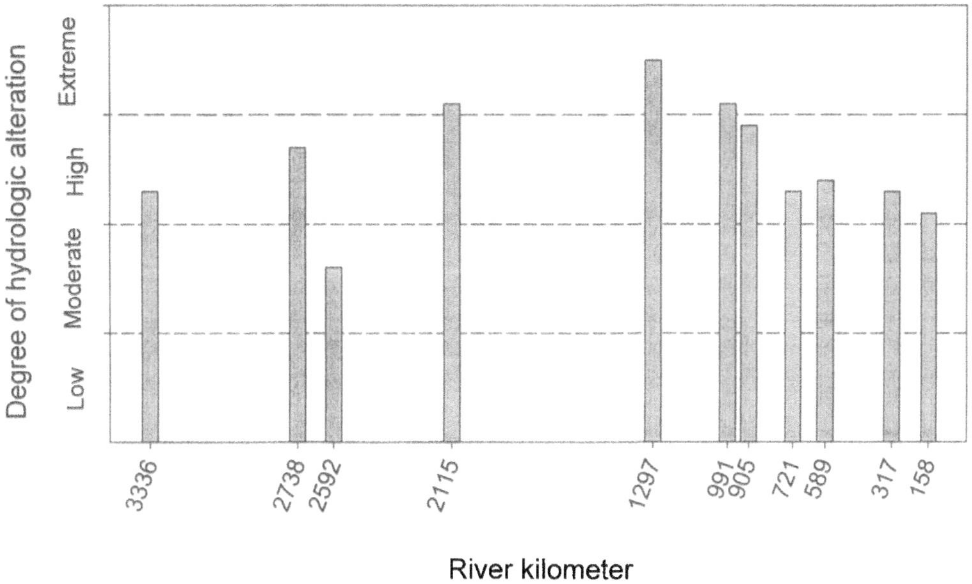

Figure 8. Ranked percentages of hydrologic alteration for the 1967–1996 post-regulation period averaged over five groups of hydrologic indicators for 11 stations along the Missouri and lower Yellowstone rivers. Hydrologic alteration occurs when the post-regulation median or mean of a hydrologic variable falls within the observed pre-regulation 25th to 75th percentiles or ± 1SD, respectively, more or less often than expected. Low alteration indicates that annual values for the average of all groups of post-regulation hydrologic indicators fell within the observed pre-regulation range of hydrologic alteration 0–25% more or less often than expected, moderate: 26–50%, high: 51–75%, and extreme: 76–100%.

natural character here than at upriver channelized and inter-reservoir locations.

Hughes (1995) recommended regional or ecoregional references to develop biological criteria for rivers. Our results support use of 'regional' rather than basinwide references, since distant locations were shown to be inappropriate spatial references for large rivers where natural longitudinal and geographic variability are great. Additionally, spatio-temporal differences in precipitation can confound the applicability of historical flow data for establishing baseline conditions. Prescribing initial flow targets from least-altered locations and historical discharge data at the landscape scale for large rivers maximizes use of available information. However, these sources are not without their shortcomings and, therefore, should be applied with caution, and not be the sole criteria for designing flow guidelines.

Preliminary flow recommendations within an adaptive management framework

The IHA and RVA methods are comprehensive techniques to assess the hydrological and, by inference, ecological integrity of running waters. A particular value of these methods is that they identify both increases and decreases in hydrologic variability and give high – and low – flow alterations equal weight. The idea of a human-induced disturbance as an alteration from the natural range of flow variation is bi-directional. River regulation that produces high flows during the historical low-flow season can have ecological consequences as harmful as an imposed reduction in the annual flood pulse. Reduced seasonal discharge variability through impoundment and flow regulation is evident today along much of the Missouri River (Figure 2). Additionally, channelization, bank stabilization, levee construction and changes in land use and land cover greatly influence the discharge-stage relationship, river-floodplain hydrology, and habitat quality and availability. These factors also need to be considered when designing, implementing and refining flow-management guidelines. Restoration of a river's ecological integrity often requires rehabilitating and managing both flow and habitat.

Operational flexibility within the existing reservoir complex could be used to more closely approximate the 1929–1948 flow regime of the Missouri River in order to establish a 'simulated natural ecosystem' (*sensu* Schmidt et al., 1998). For example, target guidelines for Groups 1–4 variables are that 50% of years should fall between the pre-regulation 25th and 75th%ile values (Figures 3–7), and for Group 5 variables, 67% of the years should fall between the pre-regulation −1SD and +1SD values. Preliminary flow guidelines based on this approach are detailed in Galat & Lipkin (1999) for each of the 32 hydrologic variables and for the nine flow-altered stations. Our results indicate that overall ecological structure and function of the inter-reservoir and upper channelized-river sections would benefit by: 1. controlled flooding through managed reservoir releases during June and July of some years; 2. increasing the magnitude, frequency, and duration of annual high-flow pulses; and 3. increasing the annual rate of hydrograph rises and falls. All of the regulated Missouri River would benefit from: 1. reducing reservoir discharges from August through February, 2. modifying the timing of reservoir releases, and 3. reducing the number of annual hydrograph reversals. These actions would increase the annual flood pulse, increase frequency and reduce monthly and annual low flows, delay timing of the growing season daily discharge minima and reduce the frequency of flow reversals per year. Assessment of ecological responses to a reregulation of Missouri River flows that more closely approximates the natural flow regime should then be used in an adaptive fashion to further adjust reservoir operations. Results summarized here and detailed in Galat & Lipkin (1999) provide a first approximation of flow recommendations that approach the 1929–1948 range of variability of mainstem river flows throughout the Missouri River catchment.

Aspects of these ecologically based flow-management guidelines conflict with contemporary Missouri River reservoir management objectives of maximizing mid-summer power production in the inter-reservoir river and providing summer-autumn flow releases for navigation in the channelized river. Agency and public support for a more naturalized Missouri River hydrograph is increasing as the Missouri River Natural Resources Committee (Risland, 1999) is recommending a trial 'spawning rise' below Ft. Peak Reservoir. Flow releases from Lewis and Clark Reservoir to enhance the spring flood pulse are also under review by the U.S. Army Corps of Engineers (1998). Impoundment and flow regulation have provided large economic benefits by reducing annual flow extremes and increasing predictability of intermediate flows. In contrast, ecological systems benefit most from the full natural range of seasonal flows and associated uncertainty. We believe that the total economic value of the Missouri River will be higher when the traditional products of agriculture, electric-power generation and transportation are integrated with the socio-ecological benefits of a naturalized flow regime. We hope that consideration of the range of flow variability approach presented here and elsewhere (Richter et al., 1998) will stimulate discussion among the various beneficiaries within and outside the Missouri River basin to reconcile differences.

Analyses of hydrologic variability are needed on other river systems to better define the geographic diversity of natural and altered flow regimes. Such assessments will assist development of integrated and adaptive hydro-ecological models to predict a range of structural and functional responses of river-floodplain biota to various flow management scenarios within a framework of broader policy issues and societal values.

Acknowledgements

Chuck Smythe and Brian Richter assisted in application of the IHA and RVA methods. Thai Ngu contributed to the Figures, Pam Haverland assisted with data preparation, and Sandy Clark edited the manuscript. Lee Bergsted, Doug Dieterman, Robb Jacobson, Tom Parks, Mark Pegg, Brian Richter, Mike Ruggles and Robert White provided information and reviewed the manuscript. Funding was provided by the Missouri Department of Conservation to the University of Missouri. This article is a contribution from the Missouri Cooperative Fish and Wildlife Research Unit (U.S. Geological Survey-Biological Resources Division, Missouri Department of Conservation, University of Missouri and Wildlife Management Institute cooperating). We dedicate this paper to the memory of Tom Parks; we've lost a great friend to the Missouri.

References

American Rivers, 1997. North America's most endangered and threatened rivers of 1997. American Rivers, Washington, D.C.

Bacon, L. M. & J. J. Rotella, 1998. Breeding ecology of interior least terns on the unregulated Yellowstone River, Montana. J. field Ornithol. 69: 391–401.

Bain, M. L., 1992. Study designs and sampling techniques for community-level assessment of large rivers. In Cuffney, T. F. & M. E. Gurtz (eds), Biological Assessments in Large Rivers. 5th Annual Technical Information Workshop, N. am. Bentho. Soc., Louisville, KY: 63–74.

Bayley, P. B., 1991. The flood pulse advantage and the restoration of river-floodplain systems. Reg. Riv.: Res. Manage. 6: 75–86.

Bayley, P. B. & H. W. Li, 1992. Riverine fishes. In Calow, P. & G. E. Petts (eds), The Rivers Handbook. Vol. 1. Blackwell Scientific Publications, London, England: 251–281.

Benke, A. C., 1990. A perspective on America's vanishing streams. J. n. am. Benthol. Soc. 9: 77–88.

Bodensteiner, L. R. & W. M. Lewis, 1992. Role of temperature, dissolved oxygen and backwaters in the winter survival of freshwater drum (*Aplodinotus grunniens*) in the Mississippi River. Can. J. Fish. aquat. Sci. 49: 173–184.

Boon, P. J., P. Calow & G. E. Petts, 1992. River conservation and management. Wiley & Sons, Chichester.

Bravard, J.-P. & G. E. Petts, 1996. Human impacts on fluvial hydrosystems. In Petts, G. E. & C. Amoros (eds), Fluvial Hydrosystems. Chapman & Hall, London, England: 242–262.

Calow, P. & G. E. Petts, 1992. The rivers handbook. Blackwell Scientific Publications, London, England.

Copp, G. H., 1991. Typology of aquatic habitats in the Great Ouse, a small regulated lowland river. Reg. Riv.: Res. Manage. 6: 125–134.

Davies, B. R. & K. F. Walker (eds), 1986. The Ecology of River Systems. Dr W. Junk Publishers, Dordrecht.

Ferrell, J. R., 1993. Big dam era. Missouri River Division, U.S. Army Corps of Engineers, Omaha, NE.

Funk, J. L. & J. W. Robinson, 1974. Changes in the channel of the lower Missouri River and effects on fish and wildlife. Aquatic Series 11, Missouri Department of Conservation, Jefferson City, MO.

Galat, D. L. & A. G. Frazier (eds), 1996. Overview of River-Floodplain Ecology in the Upper Mississippi River Basin. Vol. 3 of J. A. Kelmelis (ed.), Science for Floodplain Management into the 21st Century. U.S. Government Printing Office, Washington, D.C.

Galat, D. L. & R. Lipkin, 1999. Characterizing the natural flow regime of the Missouri River using historical variability in hydrology. Missouri Cooperative Fish & Wildlife Research Unit, Univ. of Missouri, Columbia.

Galat, D. L. & 16 coauthors, 1998. Flooding to restore connectivity of regulated, large river wetlands. BioScience 48: 721–733.

Gore, J. A. (ed.), 1985. The Restoration of Rivers and Streams. Butterworth Publishers, Boston, MA.

Gore, J. A. & F. D. Shields, 1995. Can large rivers be restored? BioScience 45: 142–152.

Hedman, E. R. & D. G. Jorgensen, 1990. Surface and ground-water interactions and hydrologic budget of the Missouri River valley aquifer between Yankton, South Dakota, and St. Louis, Missouri. Atlas HA-721, U.S. Geological Survey, Denver, CO.

Hesse, L. W. & G. E. Mestl, 1993. An alternative hydrological cycle for the Missouri River based on the pre-control condition. N. am. J. fish. Manage. 13: 360–366.

Hesse, L. W., J. C. Schmulbach, J. M. Carr, K. D. Keenlyne, D. G. Unkenholz, J. W. Robinson & G. E. Mestl, 1989. Missouri River fishery resources in relation to past, present and future stresses. Can. Spec. Publ. Fish. aquat. Sci. 106: 352–371.

Hughes, R. M., 1995. Defining acceptable biological status by comparing with reference conditions. In Davis, W. S. & T. P. Simon (eds), Biological Assessment and Criteria: Tools for Water Resource Planning. Lewis Publ., Boca Raton, FL: 31–47.

Johnson, B. L., W. B. Richardson & T. J. Naimo, 1995. Past present and future concepts in large river ecology. BioScience 45: 134–141.

Johnson, W. C., 1992. Dams and riparian forests: case study from the upper Missouri River. Rivers 3: 229–242.

Junk, W. J., P. B. Bayley & R. E. Sparks, 1989. The flood pulse concept in river-floodplain systems. Can. Spec. Publ. Fish. aquat. Sci. 106: 110–127.

Karr, J. R. & E. W. Chu, 1999. Restoring life in running waters: better biological monitoring. Island Press, Washington, D.C.

Kingsolving, A. D. & M. B. Bain, 1993. Fish assemblage recovery along a riverine disturbance gradient. Ecol. Appl. 3: 531–544.

Latka, D. C., J. Nestler & L. W. Hesse, 1993. Restoring physical habitat in the Missouri River: a historical perspective. In Hesse, L. W., C. B. Stalnaker, N. G. Benson & J. R. Zuboy (eds), Restoration Planning for the Rivers of the Mississippi River Ecosystem. Biol. Rep. 19, National Biological Survey, Washington, D.C.: 350–359.

Michener, W. K. & R. A. Haeuber, 1998. Flooding: natural and managed disturbances. BioScience 48: 677–680.

National Research Council, 1992. Restoration of Aquatic Ecosystems. National Academy Press, Washington, D.C.

Parrett, C., N. B. Melcher & R. W. James Jr., 1993. Flood discharges in the upper Mississippi River basin, 1993. U.S. Geological Survey Circular 1120-A, Denver, CO.

Pflieger, W. L., 1971. A distributional study of Missouri fishes. Univ. Kansas Pubs., Mus. Nat. Hist. 20: 225–570.

Pflieger, W. L., 1997. The fishes of Missouri. Missouri Dept. Conservation, Jefferson City. Poff, N. L. & 7 coauthors, 1997. The natural flow regime. BioScience 47: 769–784.

Poizat, G. & D. Pont, 1996. Multi-scale approach to species habitat relationships: juvenile fish in a large river section. Freshwat. Biol. 36: 611–622.

Qi, Hu, C. M. Woodruff & S. E. Mudrick, 1998. Interdecadal variations of annual precipitation in the central United States. Bull. am. Meteorol. Soc. 79: 221–229.

Reash, R. J., 1998. Considerations for characterizing midwestern large-river habitats. In Simon, T. P. (ed.), Assessing the Sustainability and Biological Integrity of Water Resources Using Fish Communities. CRC Press, Boca Raton, FL: 463–473.

Richter, B. D., J. V. Baumgartner, D. P. Braun & J. Powell, 1998. Assessing hydrological connectivity within river ecosystems. Reg. Riv.: Res. Manage. 14: 329–340.

Richter, B. D., J. V. Baumgartner, J. Powell & D. P. Braun, 1996. A method for assessing hydrological alteration within ecosystems. Conserv. Biol. 10: 1163–1174.

Richter, B. D., J. V. Baumgartner, R. Wigington & D. P. Braun, 1997. How much water does a river need? Freshwat. Biol. 37: 231–249.

Risland, S., April, 1999. Summary of the Kansas City MRBA Conference. In The Missouri River Report. Missouri River Basin Association, Lewistown, MT: 2–8.

Schmidt, J. C., R. H. Webb, R. A. Valdez, G. R. Marzolf & L. E. Stevens, 1998. Science and values in river restoration in the Grand Canyon. BioScience 48: 735–747.

Schmulbach, J. C., L. W. Hesse & J. E. Bush, 1992. The Missouri River – Great Plains thread of life. In Becker, C. D. & D. A. Nietzel (eds), Water Quality in North American River Systems. Batelle Press, Columbus, OH: 137–158.

Scott, M. L., G. T. Auble & J. Friedman, 1997. Flood dependency of cottonwood establishment along the Missouri River, Montana, U.S.A. Ecol. Appl. 7: 677–690.

Shields, R. R., M. K. White, P. N. Ladd, C. L. Chambers & K. A. Dodge, 1997. Water resources data, Montana, Water Year 1997, Water-data Report MT-97-1. U.S. Geological Survey, Helena, MT.

Smith, J. W., 1996. Wildlife use of the Missouri and Mississippi River basins – an ecological review. In Galat, D. L. & A. G. Frazier (eds), Overview of River-Floodplain Ecology in the Upper Mississippi River Basin. Vol. 3 of J. A. Kelmelis (ed.), Science for Floodplain Management into the 21st Century. U.S. Government Printing Office, Washington, D.C.: 91–112.

Sparks, R. E., 1992. Risks of altering the hydrologic regime of large rivers. In Cairns Jr., J. B. R. Niederlehner & D. R. Orvos (eds),

Predicting Ecosystem Risk. Vol. XX: Advances in Modern Environmental Toxicology. Princeton Scientific Publ. Co., Princeton: 119–152.

Sparks, R. E., 1995. Need for ecosystem management of large rivers and their floodplains. BioScience 45: 168–182.

Sparks R. E., J. C. Nelson & Y. Yin, 1998. Naturalization of the flood regime in regulated rivers. BioScience 48: 706–720.

Sparks, R. E., P. B. Bayley, S. L. Kohler & L. L. Osborne, 1990. Disturbance and recovery of large floodplain rivers. Envir. Manage. 14: 699–709.

Stanford, J. A., J. V. Ward, W. J. Liss, C. A. Frissell, R. N. Williams, J. A. Lichatowich & C. C. Coutant, 1996. A general protocol for restoration of regulated rivers. Reg. Riv.: Res. Manage. 12: 391–413.

U.S. Army Corps of Engineers, 1998. Missouri River Master Water Control Manual. Review and Update, Preliminary Revised Draft Environmental Impact Statement. U.S. Army Corps of Engineers, Northwestern Division, Missouri River Region, Omaha, NE.

U.S. Fish & Wildlife Service, 1994. Endangered and threatened wildlife and plants; animal candidate review for listing as endangered or threatened species; Proposed rule. Dept. Interior, Fish & Wildlife Service, Federal Register, 50 CFR Part 17, 50(219): 58982-59028.

U.S. Fish & Wildlife Service, 1995. Endangered and threatened wildlife and plants. Dept. Interior, Fish & Wildlife Service, Federal Register, 50 CFR Parts 17.11 & 17.12: 1–44.

Walker, K. F., F. Sheldon & J. T. Pukridge, 1995. A perspective on dryland river ecosystems. Reg. Riv.: Res. Manage. 11 85–104

Ward, J. V., 1989. Riverine-wetland interactions. In Sharitz. R. & J. Gibbons (eds), Freshwater Wetlands and Wildlife. CONF-8603101, DOE Symposium Series No. 61, U.S. Dept. of Energy, Office of Scientific and Technical Information, Oak Ridge, TN: 385–399.

Welcomme, R. L., 1985. River fisheries. FAO Fisheries Technical Paper 262. Food & Agr. Organization of the United Nations, Rome.

Welcomme, R. L., R. A. Ryder & J. A. Sedell, 1989. Dynamics of fish assemblages in river systems – a synthesis. Can. Spec. Publ. Fish. aquat. Sci. 106: 569–577.

White, R. G. & R. G. Bramblett, 1993. The Yellowstone River: its fish and fisheries. In Hesse, L. W., C. B. Stalnaker, N. G. Benson & J. R. Zuboy (eds), Restoration Planning for the Rivers of the Mississippi River Ecosystem. Biol. Rep. 19, National Biological Survey, Washington, D.C.: 396–414.

Whitmore, S. B. & K. D. Keenlyne, 1990. Rare, threatened and endangered endemic species of the Missouri River floodplain Report MRC-90-1, U.S. Fish & Wildlife Service, Missouri River Coordinator's Office, Pierre, SD.

Hydrobiologia **422/423:** 49–60, 2000.
M. Jungwirth, S. Muhar & S. Schmutz (eds), *Assessing the Ecological Integrity of Running Waters.*

The river-scaling concept (RSC): a basis for ecological assessments

Helmut M. Habersack
Institute for Water Management, Hydrology and Hydraulic Engineering, University of Agricultural Sciences, Muthgasse 18, A-1190 Vienna, Austria
Fax: [+43] 1 36006-5549; E-mail: haber@edv2.boku.ac.at

Key words: scaling, morphodynamics, river morphology, ecology, ecological integrity

Abstract

The aim of this paper is to discuss the river-scaling concept (RSC) as a basis for ecological assessments. Since river morphology is a result of two major boundary conditions – transport of water and sediments – the size of project areas and the analysis procedure were found to be critical. Restricting the assessment of abiotic and biotic river components and its variability to a certain scale neglects the fact that ecological integrity depends on the process scale of boundary conditions. A newly developed two-step procedure for assessing the ecological integrity at various temporal and spatial scales is presented. During the so-called downscaling phase, abiotic and biotic components are analysed at the regional–continental (only for special questions), catchment-wide, sectional, local, and point scales. Catchment-wide analyses are based on digital elevation models (DEMs) and geographic information systems (GISs). Mass balance analysis, the application of fractals, and self-similarity studies for channel developments gain increasing importance. The application of linear and non-linear theories allows analytical tools to be derived to describe abiotic components like morphodynamics at sectional and local scales. Point scale investigations are based on deterministic models, where input data can be measured directly in nature. As the scale varies dramatically, the results obtained from various analysis tools are significantly different but interdependent. Biotic analysis are also performed at the same scales, so that the interrelations between morphodynamics and habitat quality can be derived at the end of this first phase. In a second phase an upscaling integrates and aggregates the results from the first step in order to yield overall conclusions on ecological integrity.

Introduction

Assessing ecological integrity requires performing abiotic as well as biotic investigations. The issue of scales has recently gained more scientific importance (Bult et al., 1998; Frisell et al., 1986; Giller et al., 1994; Habersack, 1998; Holling, 1992; Horne & Schneider; 1995; Levin, 1992; Menge & Olsen, 1990; Naiman et al., 1992; Rossi et al., 1992; Wiens, 1989). Physical processes have emerged as architects of heterogeneity in all aquatic systems (Raffaelli et al., 1994). These physical influences operate in a hierachy across a complete range of spatio-temporal scales, largely encompassing water movement (hydraulics) or resulting from forces generating such movement. The description of pattern is the description of variation, and the quantification of variation requires determining scales. Patterns and processes may be different depending on the resolution with which they are observed (Levin, 1992).

Studies designed to understand the ecological integrity of certain rivers attempt to understand and evaluate the processes affecting ecosystems. Process is a more dynamic term and refers to a series of actions or operations leading to a particular river configuration and distribution of organisms.

The integrity of the processes can be understood by analysing the heterogeneity or patchiness of patterns in relation to environmental features and habitats. In the past many of the involved processes were discussed as if they were dimensionless or of unspecified dimension, although many of the processes and derived patterns are relevant to specific ranges of space and time (Fisher, 1994). As the scale of the investigations varies dramatically, the results obtained from various analysis tools are very different. Several attempts to incorporate scales in river classifications have been made in the past (Frisell et al., 1986, Naiman et al., 1992) and are taken into consideration herein. This paper aims to:

- define scales distinguished in this paper,
- analyse abiotic and biotic processes and patterns at various scales,
- present the river scaling concept (RSC) as a basis for assessing ecological integrity,
- describe existing models in scale-variant parts of catchments, allocate links between different scales (down-, upscaling),
- suggest future research demands.

Definition of scales and scaling

The term scale refers to a characteristic time or length, and the process of scaling means to change the scale in space or time. Upscaling means increasing the scale and downscaling decreasing the scale. The definitions of different scales are given in Table 1.

Table 1. Sedimentological, morphological scales (after Habersack, 1997)

Scale	In space (typical lengths, km)	In time (morphological changes)
Regional–continental	>1000	Geological time units
Catchment-wide	100–10000	Geological time units
Sectional	1–100	10–1000 years
Local	0.01–1	0.1–10 years
Point	0.0001–0.01	<1 year

Abiotic, morphodynamic scaling: processes and patterns

Morphodynamic processes occur over a wide range of scales, from continental/regional, long-term changes of the Earth's crust to the initiation of motion of individual particles (Figure 1a). From the morphodynamic perspective switching between different scales requires describing the major processes and related patterns on each individual scale:

In general the boundary conditions for a certain landscape or river pattern are given by the large scale, regional–continental or even worldwide acting processes. Plate tectonics lead to the rise of mountain ranges, and precipitation in combination with climatic variations is responsible for the onset of denudation processes. These processes determine channel network patterns, splitting the land surface into concave and convex parts.

To a certain extent the patterns derived from catchment scale processes give the boundary condition for processes at the sectional scale, e.g., meander migration, river bed aggradation or degradation. Consequently a special channel type may develop within a shorter time period than the larger scale processes act and form a certain catchment pattern. Local scale processes like channel braiding or meander cutoffs or gravel bar development cause typical bedforms, bars, meander loops or pool-riffle-pool sequences.

At the point scale, shear stress fluctuations, the existence or non-existence of sweeps/bursts or grain sorting processes yield characteristic point scale patterns (e.g., patchy structure of the river bed surface).

Biotic scaling

Different biological processes inhabit different spatio-temporal scales, depending on the organism's generation time and life-time space. The set of processes will be nested within each life-time range. Combining the life-time range and the acting radius, various biota more or less reflect specific scales (Figure 2). Whereas the vegetation communities typically reflect larger scales, individuals reflect the local scale. Birds and mammals reflect more the regional/continental, catchment-wide scale than the point scale. For example otters live along several-kilometer-long river sections, and are reported to wander between different catchments and use aquatic as well as terrestrial habitats (Michor et al., 1993). Fish populations predominantly reflect the sectional scale, although, of course, both larger and smaller scales are relevant. Benthic invertebrates show the shortest life span and therefore reflect the smallest spatial scale, with dominance at the point scale.

Organisms respond to environmental change at a range of scales. The time scale of response by a population depends on whether the response is behavioral, physiological, or genetic. The time scale of response by communities ranges from hours (daily variation in primary production) to millions of years (faunal changes due to vicariant effects of plate tectonics on the biogeography of species). Therefore the levels of biological organization and related ranges of scales, including the individual, the population, the community and the ecosystem, must be examined for

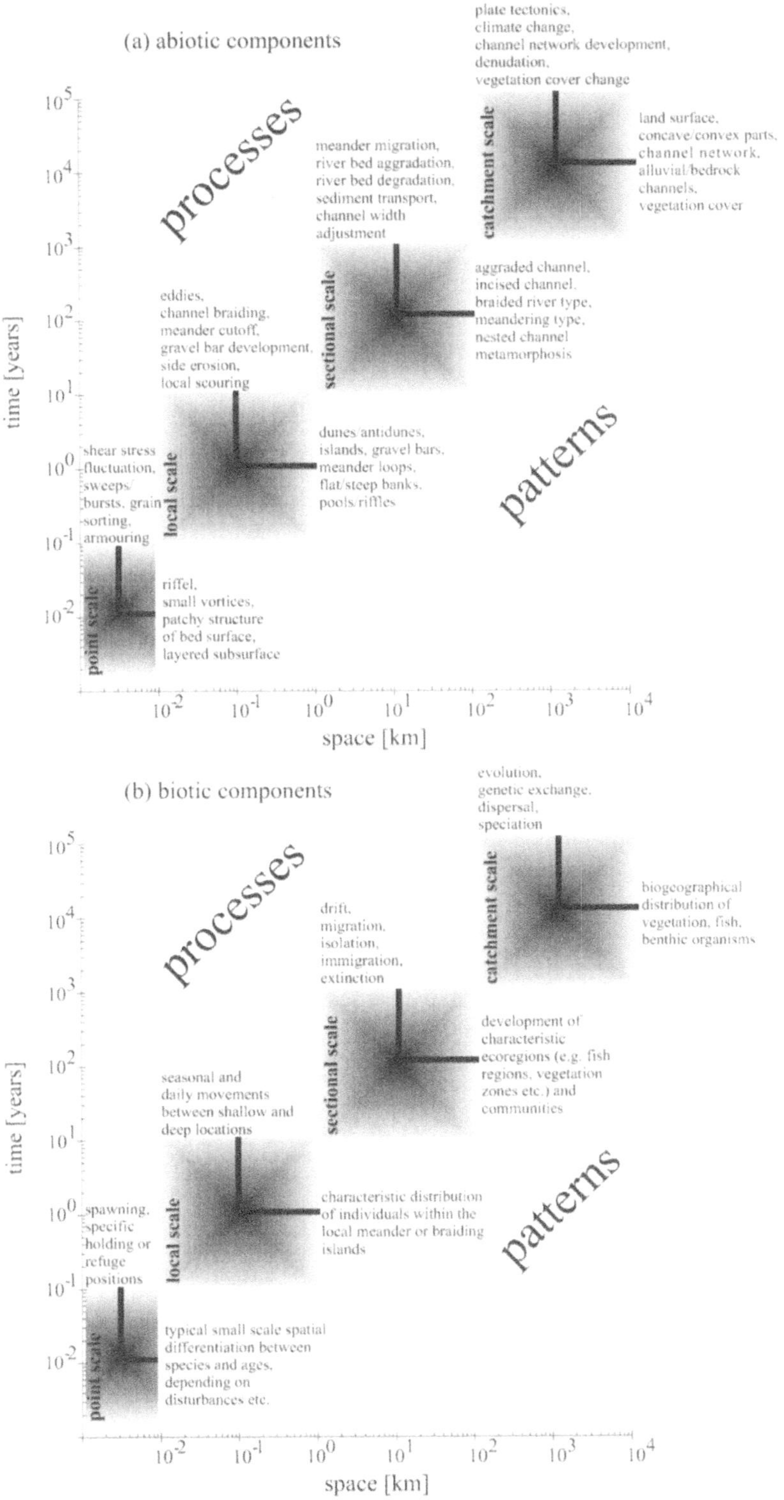

Figure 1. Morphodynamic (a) and biotic (b) processes and patterns at various scales.

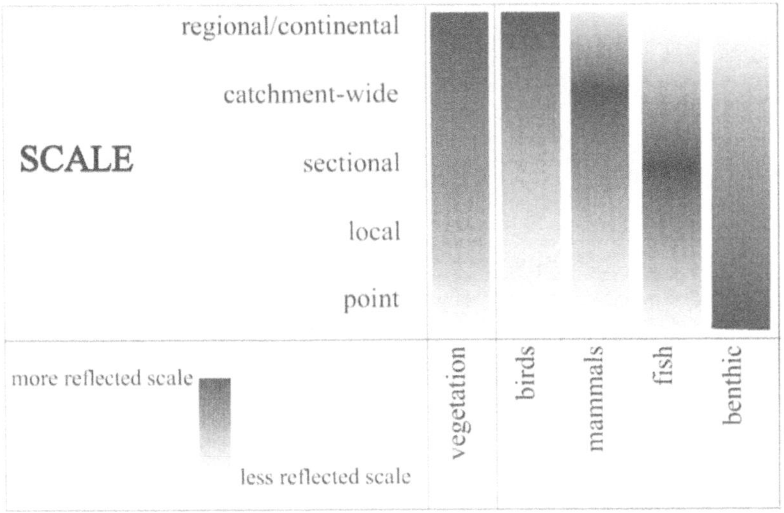

Figure 2. Scales reflected by various biota.

each group (Beeby, 1993). The biotic processes and resulting patterns in relation to scales are shown in Figure 1b.

Evolution and genetic exchange or non-exchange lead to characteristic biogeographic distributions of vegetation, mammals, birds, fish and benthic organisms over thousands of kilometers and years. This situation establishes boundary conditions for processes at the sectional scale, like drift and migration, followed by a development of characteristic ecoregions. At the local scale, seasonal and daily movements result in a typical pattern of individuals within local meanders or braiding islands. Point scale processes of spawning or the specific selection of holding or refuge positions yield characteristic small-scale spatial differentiation between species and age-classes, depending on disturbances, etc.

Interdependences between abiotic, morphodynamic and biotic scaling

According to the morphodynamics the adjustments of river channels generally have a direct influence on ecological integrity. With few exceptions, large areas have more species than small areas (Huston, 1994). The species/area curve, which shows the relation between increase of diversity with increasing sample size, can provide insight into the total number of species in a region. In order to describe these relations, equations of the following type were defined (Lomolino, 1989):

$$S = kA^z, \tag{1}$$

with S as number of species, k as constant, A as area of the sample, z as an exponential parameter. One potential explanation for the general increase of diversity with area is that it is a result of environmental heterogeneity (Huston, 1994).

Concerning interdependences between morphodynamics and biota, Bult et al. (1998) demonstrate the necessity of spatial scaling for the investigation of habitat selection by fish in riverine habitats. Fish act over a wide range of spatial and temporal scales (Figure 2), in which interdependences between the abiotic and biotic processes occur (Jungwirth, 1998). Often an overlap or coexistence of various scales can be observed, where, e.g., salmonids, select specific holding positions (small-scale) with relatively low snout velocities in areas of high current conditions (larger scale), where drift is concentrated (Bult et al., 1998).

Based on the strong interdependences and the above-discussed scale-dependent processes and patterns, it can be concluded that one important step in assessing ecological integrity is scaling. The following section describes a newly developed conceptual model and procedure based on scaling.

River scaling concept (RSC)

The aim of the river scaling concept (RSC) is to be a basis for the assessment of ecological integrity. As far as morphodynamics are concerned integrity is achieved when a system functions in a process-oriented, self-organised manner. If this is the case

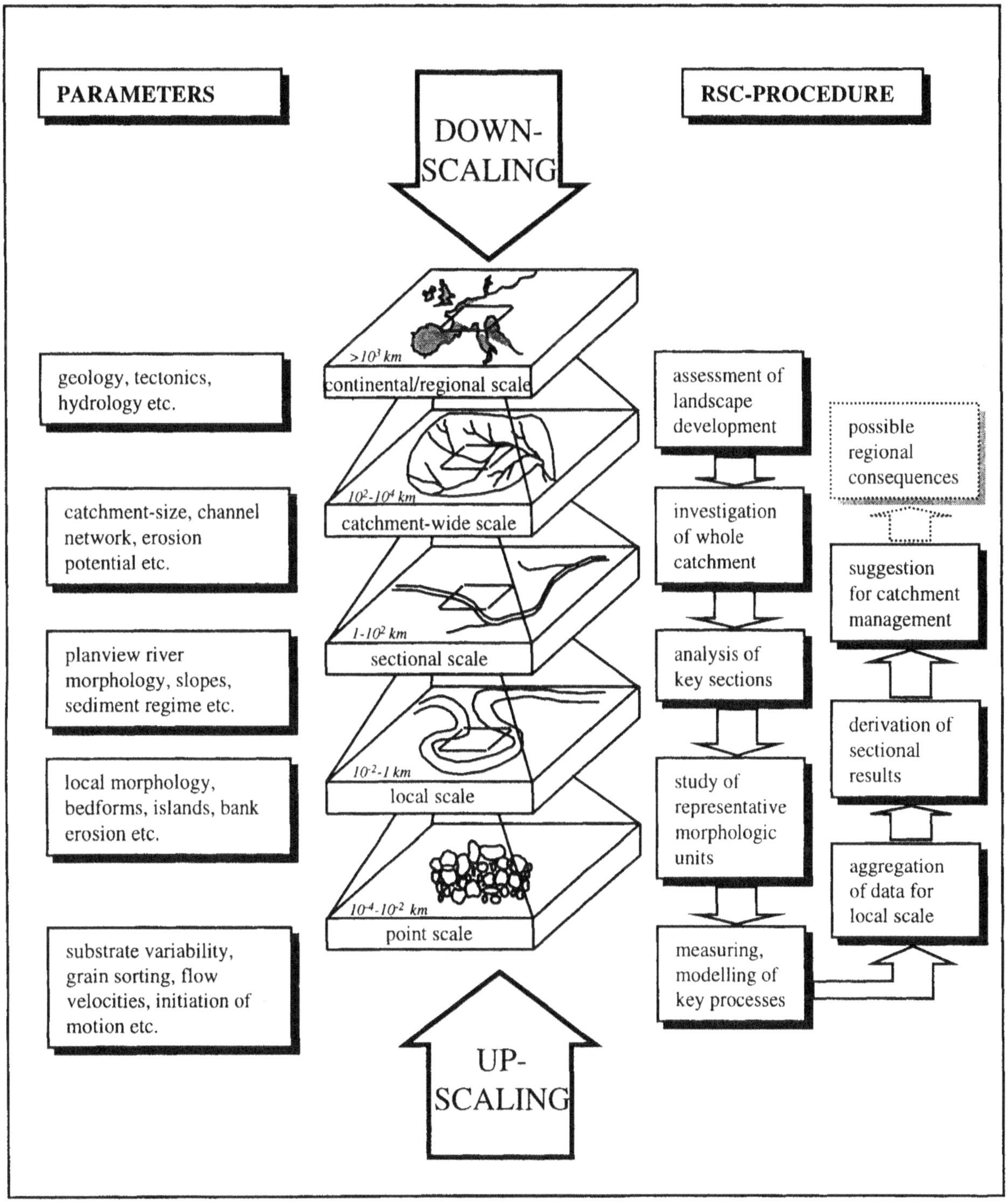

Figure 3. Abiotic parameters and procedure of the RSC.

a positive biological response can be expected and ecological integrity is consequently positive.

The RSC proposes a two-phase (down- and up-scaling) procedure (Figure 3). As already discussed, morphodynamics, which includes hydrology, hydraulics, etc., can be seen as the boundary condition for the ecosystem. Consequently the assessment procedure should start with the analysis of abiotic conditions. During the downscaling phase biotic investigations have to be performed subsequently or parallel to the abiotic analysis at each scale and interdependences of abiotic and biotic results have to be studied. Furthermore, scaling relations between subsequent scales are established using techniques described below. Conclusions about the ecological integrity of a given area with the aim of evaluating or suggesting management measures can only be derived after an upscaling of point and local scale information obtained from the previous downscaling process. Basically the RSC provides a hierachical framework for the systematic investigation of processes and patterns of running waters.

As a central idea the RSC suggests performing the following two-step procedure (Figure 3):

Step 1: downscaling phase

At each of the following scales the process-oriented (without anthropogenic influences) and existing situation have to be analysed and compared in order to evaluate ecological integrity.

Catchment-wide scale

The downscaling phase starts at the catchment-wide (continental–regional) scale. At this scale the possible development of the morphodynamics in the covered area is determined. Tectonics, geological processes, uplift, denudation rates, climate change, vegetation cover and relief energy are the essential processes (Figure 4).

The analysis of the network structure using digital elevation models (DEMs) and fractal techniques allows us to differentiate the whole catchments into subcatchments and to determine key locations as well as key variables. For example, major sources of sediments and the potential amount of mobilized materials determine key locations which separate key sections.

At this level the biogeographic situation determines the potential species richness and has to be analysed for the given catchment. The role of abiotic parameters for the special biotic conditions is examined and deficits of the existing versus the potential situation are defined.

Sectional scale

Based on key variables and locations, river reaches of several kilometers are investigated at the sectional scale (Habersack & Nachtnebel, 1997). At this level, mainly the balance of sediments (sediment input–output comparison) – leading to stable, aggrading or degrading river beds – the statistical evaluation of gravel bars, and the influence of bends and knick-points in the longitudinal profile are studied (Figure 5). Numerical models, linear and non-linear theories are analytical tools for the description and simulation of morphodynamics.

Within the section these data again allow key variables to be derived for the local scale. At the same time key locations for the investigation at the local scale are defined. At the sectional scale, biological studies include studies of fish migration, large-scale fish investigations, vegetation surveys on the community level, mammals, etc.

The comparison with biogeographic data reveals differences. Abiotic data may help to explain the differences, leading to habitat suitabilities in comparison with suitability curves. The assessment of ecological integrity requires additionally analysing the detailed situation at the local scale.

Local scale

From the morphodynamic point of view, riffle–pool–riffle sequences, step–pool–step systems, bend–straight–bend situations or sequences of alternate bars are analysed. The investigation of the possible variability (Figure 6) gives an indication of the potential ecological situation. Subsequent measurements of cross-sections at short intervals and over subsequent floods provide a good insight into the morphological dynamics of a local river site (Habersack & Nachtnebel, 1995). Point scale habitats vary dramatically within the different locations, depending on flow depth, velocity and shear stress fluctuations.

Parallel biotic investigations lead to a direct comparison and relation between habitats and biotic populations. The characteristic distribution of individuals within the local meander or braiding channel is studied, taking into consideration seasonal and daily movements between shallow and deep locations, etc.

Here, the spatial scaling and habitat selection processes in riverine habitats have to be determined (Bult

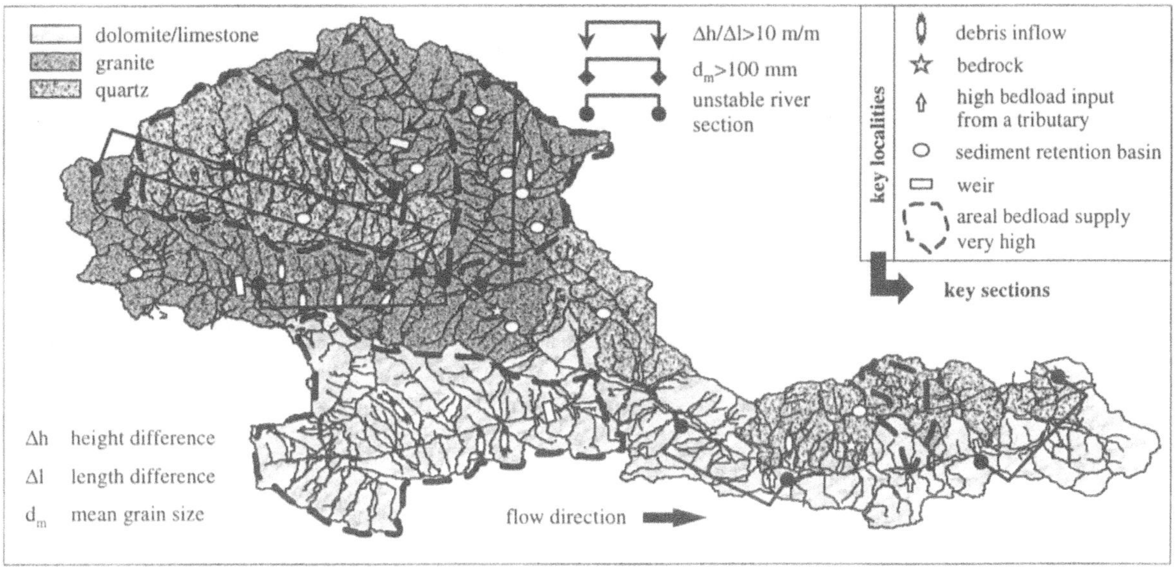

Figure 4. Illustration of catchment-wide morphodynamic investigations.

et al., 1998) and characteristic locations for necessary point scale measurements defined.

Point scale

This scale is restricted to very small areas (several m^2), yielding a high short-term variability of morphodynamics. The morphodynamics can be observed by small-scale monitoring of patches at the river bed, down to the initiation of motion of individual grains. Due to the small spatial and short temporal variations, deterministic models, calibrated by direct measurements, dominate.

Associated with the point scale morphodynamics are point measurements of the related biotic communities, like fish and benthic organisms. Emphasis is based on investigating the quality of spawning places and specific holding or refuge positions. During the downscaling a general switch from stochastic to more deterministic models is assumed because, compared to the catchment-wide scale, at the point scale all relevant calibration parameters can be measured directly (Habersack, 1997). Relevant scaling models are described in the next section.

Step 2: upscaling phase

The downscaling phase, leading to the hierachical assessment of ecological integrity, is followed by an aggregation (upscaling) of the information gathered at the small scale, based on a multiscaling approach (Figure 3). The detailed deterministic knowledge at the point scale, derived at key localities and measuring points, is interpreted in relation to the meaning for the larger scale.

This transformation is possible via the morphodynamic relationships between different scales. One way is to analyse different habitat types and then to extrapolate the results to the local, sectional and even catchment-wide scale. Especially the planning and realisation of measures for improving ecological integrity have to be expanded from the small to the large scale. Beside the analysis of the situation within a defined scale it is necessary to analyse the relation between the different scales, for which the following methods are suggested.

Scaling methods in the RSC

In the RSC a variety of scaling methods may be used. The selection of appropriate models depends on the individual running water system, the data availability and the special boundary conditions.

Abiotic, morphodynamic investigations In order to investigate at a certain scale or to switch to another scale a variety of methods is available:

- deterministic models based on differential equations,
- stochastic models based on distribution functions, covariance functions and moments,
- dimensional analysis,

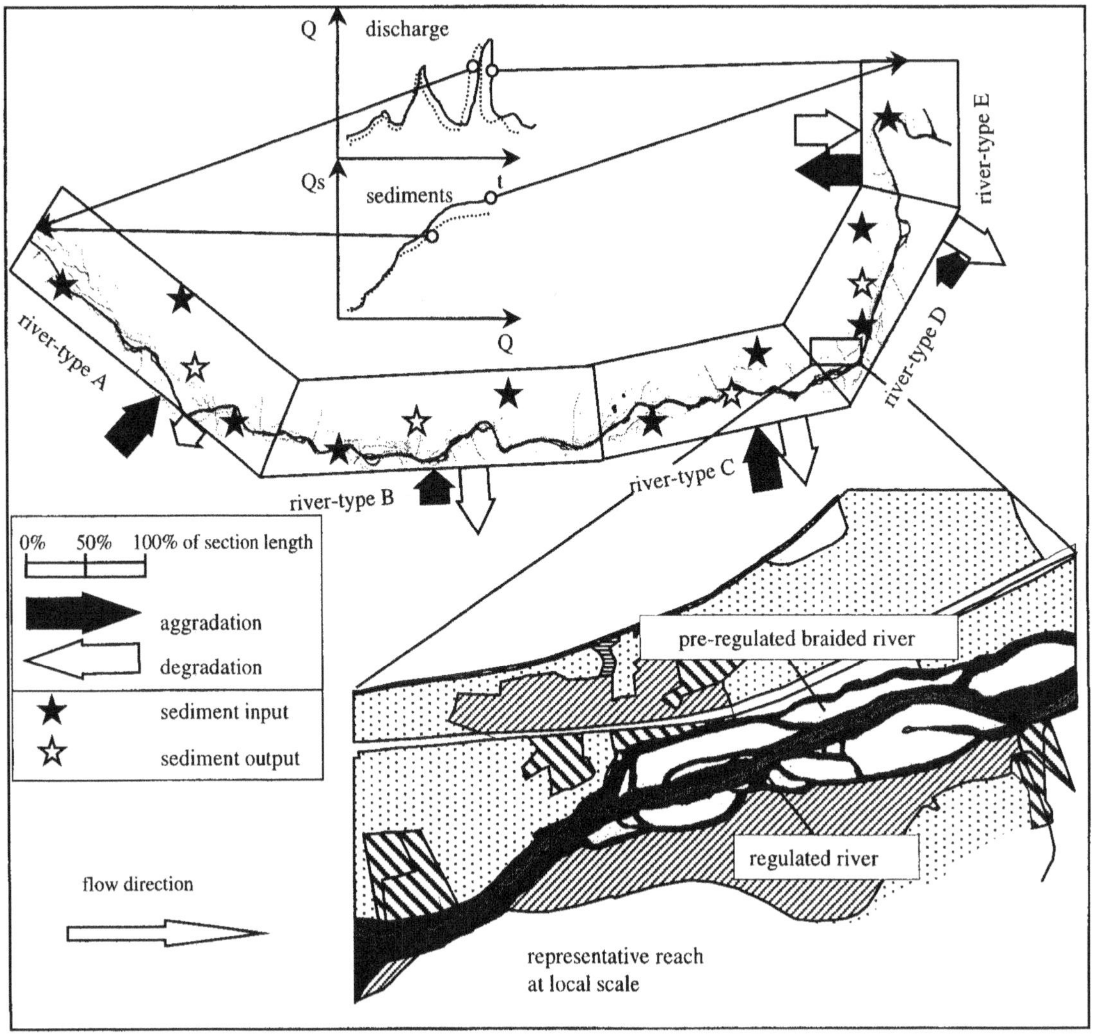

Figure 5. Morphodynamics at the sectional scale.

- self similarity analysis, simple fractals and multifractals,
- wavelets.

Linkage between scales In deterministic models the linkage between scales is reached by distributing small-scale modeled values, resulting in a spatial or temporal pattern. Stochastic models use distribution and covariance functions, often characterised by their moments, for distributing results (Blöschl, 1996). Both deterministic and stochastic models aggregate the results in order to derive avarage values for larger time and space scales.

Concerning the dimensional analysis of catchment form and dynamics, Strahler (1964) showed that a significant geometric similarity exists in catchments with common underlying geology, climate and lithology. Horton's empirical laws of network morphology (Horton, 1945; Strahler, 1957) are relevant here. Fractals are also used to analyse the structure of stream networks. This technique allows the discrimination of hillslope and channel processes and the definition of a critical scale at which the situation deviates from the fractal relationship (Montgomery & Foufoula-Georgiou, 1993; Tarboton et al., 1989). Both deterministic (geometric) fractals and random (statistical) fractals exist, which generally are sets with certain properties independent of scale (Cutler, 1993; Voss, 1985). Very often, power law equations like

$$v \propto Q^{m}; d \propto Q^{f}; w \propto Q^{b}; Q^{b}; S \propto Q^{z}; n \propto Q^{y} \quad (2)$$

with v as flow velocity, d as mean flow depth, w as surface flow width, S as energy slope, n as the Manning

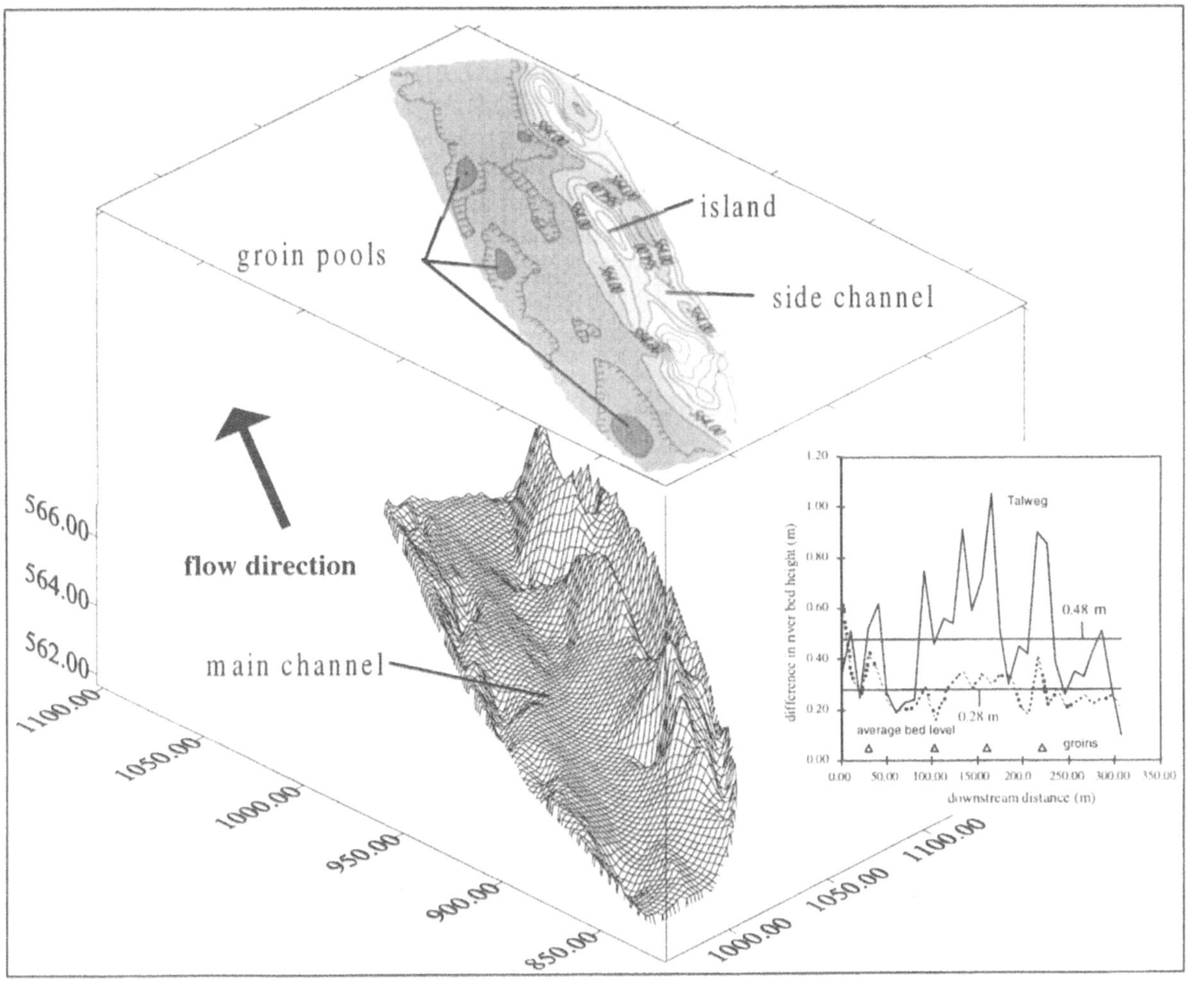

Figure 6. Morphodynamics at the local scale: variation of a river bed over a period of several years.

friction coefficient and the exponents m, f, z and y as constants, can be viewed as scaling relationships (Hey, 1988; Knighton, 1984; Leopold & Maddock, 1953; Richards, 1982). Furthermore, the issue of energy expenditure plays a central role in deriving key structural characteristics of stream networks, like downstream hydraulic geometry variaton, variation of stream slope as a function of catchment area and Horton order ratios (Habersack, 1997; Rodriguez-Iturbe et al., 1992).

Biotic and combined investigations The following tools and models are available:

fixed scale: (a) Morisita's index, Lloyd's index of mean crowding, exponent of Taylor's power law (Lloyd, 1967; Morisita, 1959; Taylor, 1961); (b) correlation, regression, frequency analyses

range of scales: (a) pattern analysis (Greig-Smith, 1983); (b) correlograms and variograms (Rossi et al., 1992; Sokal & Oden, 1978); (c) variance analysis on hierachical models (Downes et al., 1993); (d) second-order neighbourhood analysis (Getis & Franklin, 1987; Muotka & Penttinen, 1994); (e) spectral analysis (Platt & Denman, 1975); (f) coherence analysis (Chatfield, 1980); (g) habitat models (Allan et al., 1997; Lewis et al., 1996) (h) multiscale technique based on frequency analysis and randomisation (Bult et al., 1998; Poizat & Pont, 1996; Richards et al., 1996; Syms, 1995).

Scale-dependent processes such as dispersal or energy transformation can be visualized in fractal dimensions rather than in just two or three Euclidean dimensions (Schneider, 1994). Fractal geometry (Mandelbrot, 1977) is a promising method for quantifying scale-dependent patterns in the shape and form of organisms, the geography of populations, and the physical structure of habitats (Sugihara & May, 1990; Williamson & Lawton, 1991). Other examples of fractal objects include root and branch systems of trees, habitat boundaries, and the nesting of eddies within eddies of the fluid environments.

In the RSC the selection of adequate models is essential for the quality of the assessment of ecological integrity and is determined by available data, size of the study area and type of site-specific problems.

The importance of the RSC as a multiscale analysis concept Multiscale analysis is analysis with respect to multiples of a unit of measurement (Schneider, 1994). If the interaction of biological with physical processes is the norm, then a shifting rather than characteristic scale can be expected (Levin, 1992). Shifting scales of biological interaction with the environment suggests a permanent role for multiscale analysis in ecology. Multiscale analysis (zooming) and sequential analysis (panning) can be applied to space and time. The importance of the RSC as a multiscale analysis concept lies in the defined two-step procedure and sequential or parallel abiotic and biotic scaling approach.

Conclusions and future research demands

Large-scale events often propagate to smaller scales within a fluid environment (Stommel, 1963; Mackas et al., 1985), with important consequences for the organisms inhabiting the fluids (Dayton & Tegner, 1984). Processes and patterns are characteristic for defined scales and are very often scale dependent. Nevertheless, it is necessary to use multiscaling approaches, fractals or other modeling concepts in order to analyse the existing abiotic and biotic situation. This fact yields the river scaling concept (RSC) involving a two-step down- and upscaling suggested in this paper. The downscaling process starts at the large scales like the catchment-wide scale and ultimately leads to the point scale, where especially deterministic concepts can be used to interpret the observed processes and patterns. During every step, the analysis of morphodynamics allows the derivation of basic abiotic processes, key localities and key sections. Biotic investigations must also be performed at various scales so that – knowing the response function – ecological integrity can be assessed. Based on the necessity of scaling in morphodynamics and biological responses, a variety of techniques for the scaling process is described. Ecological integrity is assessed by comparing the existing situation with a potential self-organizing and self-regulating, morphologically dynamic system without anthropogenic influences.

Additionally, multiscaling techniques are used to analyse the relation between different scales. Furthermore, the aggregation of information during the second step of upscaling may lead to overall conclusions on the ecological situation in a larger spatial and temporal dimension. This upscaling is crucial with regard to management consequences because the ecological situation can often only be improved by measures being realized in the catchment (boundary conditions for smaller scales).

Future research must be directed at improving scaling methods in general. The development of detailed down- and upscaling procedures is necessary. Through the investigation of the interdependences between abiotic and biotic scaling, objective criteria for defining ecological integrity have to be found. A variety of examples in Austria have already demonstrated the practible applicability of parts of the river scaling concept (Habersack & Nachtnebel, 1995). The systematic use of the RSC is currently under way. An application of the RSC for different climatic, geological and biogeographical conditions will lead to a fine-tuning of the concept.

Acknowledgements

Thanks to the Austrian Ministry of Agriculture and Forestry for funding several projects through which the author received the opportunity to experience morphodynamics in nature. The author is also thankful to the editors and reviewers for commenting on this paper.

References

Allan, J.D., D.L. Erickson & J. Fay, 1997. The influence of catchment land use on stream integrity across multiple spatial scales Freshwat. Biol. 37: 149–161.

Beeby, A. (ed.), 1993. Applying Ecology. Chapman & Hal , London, 441 pp.

Blöschl, G., 1996. Scale and scaling in hydrology. Wiener Mitteilungen 132.

Bult, T.P., R.L. Haedrich & D.C. Schneider, 1998. New technique describing spacial scaling and habitat selection in riverine habitats. Regul. Rivers: Res. Mgmt. 14: 107–118.

Chatfield, C., 1980. The Analysis of Time Series: an Introduction, 4th edn. Chapman and Hall, London, 241 pp.
Cutler, C.D., 1993. A review of theory and estimation of fractal dimension. Tech. Rep. Ser. STAT-93-06. Dept. of Stat. and Act. Science, Univ. of Waterloo, Ontario, 107 pp.
Dayton, P.D. & M.J. Tegner, 1984. The importance of scale in community ecology: a kelp forest example with terrestrical analogs. In Price, P.W., C.M. Slobodchikoff & W.S. Gaud (eds), A New Ecology: Novel Approaches to Interactive Systems. John Wiley & Sons, New York: 457–481.
Downes, B.J., P.S. Lake & E.S.G. Schreiber, 1993. Spatial variation in the distribution of stream invertebrates: implications of patchiness for models of community organization. Freshwat. Biol. 30: 119–132.
Fisher, S.G., 1994. Pattern, process and scale in freshwater systems: some unifying thoughts. In Giller, P.S., A.G. Hildrew & D.G. Raffaelli (eds), Aquatic Ecology: Scale, Pattern and Process. Blackwell, Oxford: 575–591.
Frisell, C.A., W.J. Liss, C.E. Warren & M.D. Hurley, 1986. A hierachical framework for stream classification: viewing streams in a watershed context. Envir. Mgmt. 10: 199–214.
Getis, A. & J. Franklin, 1987. Second-order neighborhood analysis of mapped point patterns. Ecology 68: 473–477.
Giller P.S., A.G. Hildrew & D.G. Raffaelli (eds), 1994. Aquatic Ecology: Scale, Pattern and Process. Blackwell, Oxford, 649 pp.
Greig-Smith, P., 1983. Quantitative Plant Ecology. Blackwell Scientific Publications, Oxford, 359 pp.
Habersack, H., 1998. Numerical sediment transport models-theoretical and practical aspects. IAHS Publ. 249: 299–308.
Habersack, H., 1997. Catchment-wide, sectional and local aspects in sediment transport modelling and monitoring. J. Sediment Res. 12: 120–130.
Habersack, H. & H.P. Nachtnebel, 1995. Short term effects of local river restoration on morphology, flow field, substrate and biota. Regul. Riv. Res. Mgmt. 10: 291–301.
Habersack, H. & H.P. Nachtnebel, 1997. Changes in sediment transport and river engineering concepts, case study of the river Drau in Austria, UNESCO – IHP-V/Tech. Doc. Hydrol. 10: 277–286.
Hey, R.D., 1988. Mathematical models of channel morphology. In Anderson, M.G. (ed.), Modelling Geomorphological Systems. Wiley, Chichester: 99–125.
Holling, C.S., 1992. Cross-scale morphology, geometry and dynamics of ecosystems. Ecol. Monogr. 62: 447–502.
Horne, J.K. & D.C. Schneider, 1995. Spatial variance in ecology. Oikos 74: 18–26.
Horton, R.E., 1945. Erosional development of streams and their drainage basin: hydrophysical approach to quantitative morphology. Geol. Soc. am. Bull. 38: 275–370.
Huston, M.A., 1994. Biological Diversity, the Coexistence of Species on Changing Landscapes. Cambridge University Press, Cambridge, 681 pp.
Jungwirth, M., 1998. River continuum and fish migration. In Jungwirth, M., S. Schmutz & S. Weiss (ed.), Fish Migration and Bypass Channels. Fishing News Books, Blackwell Science, Oxford: 19–32.
Knighton, D., 1984. Fluvial Forms and Processes. Edward Arnold, London, 218 pp.
Leopold, L.B. & T. Maddock, 1953. The hydraulic geometry of stream channels and some physiographic implications. US Geol. Surv. Prof. Pap. 252: 57 pp.
Levin, S.A., 1992. The problem of pattern and scale in ecology. Ecology 73: 1943–1967.
Lewis, C.A., N.P. Lester, A.D. Bradshaw, J.E. Fitzgibbon, K. Fuller, L. Hakanson & C. Richards, 1996. Consideration of scale in habitat conservation and restoration. Can. J. Fish. aquat. Sci. 53 (suppl. 1): 440–445.
Lloyd, M., 1967. Mean crowding. J. anim. Ecol. 36: 1–30.
Lomolino, M.V., 1989. Interpretations and comparisons of constants in the species–area relationship: an additional caution. Am. Nat. 133: 277–280.
Mackas, D.L., H.L. Denman & M.R. Abbott, 1985. Plankton patchiness: biology in the physical vernacular. Bull. mar. Sci. 37: 652–674.
Mandelbrot, B.B., 1977. Fractals: Form, Chance and Dimension. Freeman, San Francisco, 365 pp.
Menge, B.A. & A.M. Olson, 1990. Role of scale and environment factors in regulation of community structure. Trends Ecol. Evol. 5: 52–57.
Michor, K., A. Zedrosser & M. Unterlercher, 1993. Mammal investigations at the river Drau, special report of the interdisciplinary project Gewässerbetreuungskonzept Obere Drau. Lienz, Austria.
Montgomery, D.R. & E. Foufoula-Georgiou, 1993. Channel network source representation using digital elevation models. Wat. Resour. Res. 29: 3925–3934.
Morisita, M., 1959. Measuring of the dispersion of individuals and analysis of the distributional patterns. Mem. Fac. Sci., Kyushu Univ. 2: 215–235.
Muotka, T. & A. Penttinen, 1994. Detecting small-scale spatial patterns in lotic predator-prey relationship: statistical methods and a case study. Can. J. Fish. aquat. Sci. 51: 2210–2218.
Naiman, A.J., D.G. Lonzarich, T.J. Beechie & S.C. Ralph, 1992. General principles of classification and the assessment of conservation potential in rivers. In Boon, Calow, Petts, G.E. (eds), River Conservation and Management. Wiley & Sons, Chichester, Sussex: 93–123.
Platt, T. & K.L. Denman, 1975. Spectral analysis in ecology. Ann. Rev. Ecol. Syst. 6: 189–210.
Poizat, G. & D. Pont, 1996. Multi-scale approach to species–habitat relationship – juvenile fish in a large river section. Freshwat. Biol. 36: 611–622.
Raffaelli, D.G., A.G. Hildrew & P.S. Giller, 1994. Scale, pattern and process in aquatic systems: concluding remarks. In Giller, P.S., A.G. Hildrew & D.G. Raffaelli (eds), Aquatic Ecology: Scale, Pattern and Process. Blackwell, Oxford: 601–606.
Richards, C., L.B. Johnson & G.E. Host, 1996. Landscape-scale influences on stream habitats and biota. Can. J. Fish. aquat. Sci. 53 (suppl. 1): 295–311.
Richards, K., 1982. Rivers – Form and Process in Alluvial Channels. Methuen, London, 361 pp.
Rodríguez-Iturbe, I., A. Rinaldo, R. Rigon, R.L. Bras, Marani & E. Ijjász-Vásquez, 1992. Energy dissipation, runoff production and the 3-dimensional structure of river basins. Wat. Resour. Res. 28: 1095–1103.
Rossi, R.E., D.J. Mulla, A.G. Journel & E.H. Franz, 1992. Geostatistical tools for modeling and interpreting ecological spatial dependence. Ecol. Monogr. 62: 277–314.
Schneider, D.C., 1994. Quantitative Ecology: Spacial and Temporal Scaling. Academic Press, CA, 395 pp.
Sokal, R.R. & N.L. Oden, 1978. Spatial autocorrelation in biology 1. Methodology. Biol. J. linn. Soc. 10: 199–228.
Stommel, H., 1963. The varieties of oceanographic experience. Science 139: 572–576.
Strahler, A.N., 1957. Quantitative analysis of watershed geomorphology. Trans. am. Geophys. Union 38: 913–920.
Sugihara, G. & R.M. May, 1990. Application of fractals in ecology. Trends Res. Ecol. Evol. 5: 79–87.
Syms, C., 1995. Multi-scale analysis of habitat association in a guild of blennioid fishes. Mar. Ecol. Prog. Ser. 125: 31–43.

Tarboton, D.G., R.L. Bras & I. Rodríguez-Iturbe, 1989. Scaling and elevation in river networks. Wat. Resour. Res. 25: 2037–2051.
Taylor, L.R., 1961. Aggregation, variance and the mean. Nature 189: 732–735.
Voss, R.F., 1985. Random fractals: characterisation and measurement. In Pynn, R. & A. Skjeltorp (eds), Scaling Phenomena in Disordered Systems. Plenum Press, New York: 1–11.
Wiens, J.A., 1989. The Ecology of Bird Communities. Cambridge University Press, Cambridge, 276 pp.
Williamson, M.H. & J.H. Lawton, 1991. Fractal geometry of habitats. In Bell, S.S., E.D. McCoy & H.R. Mushinsky (eds), Habitat Structure. Chapman and Hall, London: 69–86.

Hydrobiologia **422/423:** 61–70, 2000.
M. Jungwirth, S. Muhar & S. Schmutz (eds), Assessing the Ecological Integrity of Running Waters.

Biological processes in running waters and their implications for the assessment of ecological integrity

Stuart E. Bunn[1] & Peter M. Davies[2]
[1]*Centre for Catchment and In-Stream Research, Faculty of Environmental Sciences, Griffith University, Nathan, Queensland, 4111, Australia*
[2]*Department of Zoology, The University of Western Australia, Nedlands, Western Australia, 6907, Australia*

Key words: biological monitoring, streams, benthic communities, ecosystem processes, recruitment

Abstract

Although biomonitoring approaches are being increasingly used in the measurement of stream and river health, critical assumptions about the nature of biological populations and communities that underpin them are often ignored. Many approaches based on pattern detection in plant and animal communities assume high temporal persistence in the absence of anthropogenic disturbances. However, this has been rarely tested with long-term data sets and there is evidence that this assumption is not true in some river systems. Biological processes, such as predation and recruitment, can account for considerable spatial and temporal variation in the structure of some stream communities. These processes may prevent the development of robust predictive models or indices based on pattern detection. Measurements of population or community attributes also are often used to infer ecosystem processes, yet the link between pattern and process has rarely been demonstrated. Many goals of river management relate to the maintenance of natural ecological processes and ecosystem function; direct measurement of these processes is, however, often neglected in assessment programs. Such measures are often sensitive to causal factors that are known to affect river health and it is possible to develop simple but powerful predictive models. Perhaps more importantly, should an impact to be detected, strategies for remediation are more obvious as the causal processes are generally better known. The ultimate success of biomonitoring approaches depends on how well we understand the biophysical processes that influence the structure and dynamics of stream and river systems, and the way they function.

Introduction

Assessing the ecological 'health' of rivers and streams is a fundamental and increasingly important water management issue worldwide and, as a consequence, a vast array of biological monitoring approaches has been developed. These range from measurements of subtle changes at the sub-organism level to those directly measuring changes at the population, community and ecosystem level (see Norris & Norris, 1995 and references therein). Although selection of appropriate assessment methods is often the subject of considerable debate (Bunn, 1995; Wallace et al., 1996; Wicklum & Davies, 1995), there is no doubt that such biological monitoring approaches are essential to achieving the goals of ecological sustainability, which espouse protection of biodiversity and the maintenance of essential ecological processes and life support systems (Council of Australian Governments, 1992; Karr, 1991).

Biomonitoring approaches, especially those based on pattern detection, make several critical assumptions about the nature of biological populations and communities. In particular, they assume that, in the absence of anthropogenic disturbance, spatial and seasonal patterns in stream animal (and plant) communities will be consistent from year to year. This assumption of high temporal persistence (sensu Grossman et al., 1982) has rarely been demonstrated with long-term data sets. Furthermore, they assume that the structure of stream communities is primarily determined by the nature of the physical habitat (Huryn & Wallace, 1987; Pringle et al., 1988; Schofield & Davies, 1996; Townsend, 1989) and often ignore the possibility that marked variation in natural community patterns can occur as a direct result of biological processes (e.g., competition, recruitment, predation). The ultimate success of biomonitoring approaches depends on how well we understand the biophysical processes that influence the structure and dynamics of stream and river communities, and the way in which they function.

The aim of this paper is to explore some of the assumptions that underpin predictive models that are widely used to assess the ecological integrity of streams and rivers. Rather than attempt to provide a broad overview, we wish to draw on recent experiences in developing strategies for monitoring river health in Australia. We begin by looking at whether high temporal persistence is a feature of stream macroinvertebrate communities and, if not, what are the implications for the development of predictive biomonitoring models based on pattern detection? We then consider some of the biological processes that may be responsible for marked changes in the dynamics of stream populations and ultimately patterns in community structure, both in space and in time. We briefly examine whether measures of patterns (at the community or population level) can tell us much about essential ecosystem processes (an important component of river health), and conclude with an example of how direct measurements of these processes can be used to indicate river health.

Definitions of terms such as ecological 'health' and 'integrity' are often the subject of much contention (e.g., Karr, 1991; Schaeffer et al., 1988; Suter, 1993; Wallace et al., 1996). Ecosystem health has been recently defined in terms of system organisation, vigour and resilience, as well as the absence of ecosystem stress (Rapport et al., 1998). The latter emphasises essential ecosystem functions and life support systems, which are key elements of ecological sustainability (Council of Australian Governments, 1992). Throughout this paper, we use the term 'integrity' to refer to minimal deviation from a natural reference condition, and 'healthy' as shorthand for good condition as per Karr (1991) and Karr & Chu (2000). In agreement with Rapport et al. (1998), we also consider that ecosystem processes are key indicators of river health and integrity (Bunn et al., 1999).

Materials and methods

Much of the original data presented in this paper is derived from ongoing comparative studies in three different forested biomes within Australia: the Mary River catchment (sub-tropical rainforest) in south-east Queensland (152° 40′ E, 26° 45′ S), the Johnstone River catchment (tropical rainforest) in far north Queensland (145° 39′ E, 17° 30′ S), and from the northern jarrah forest (Mediterranean, dry sclerophyll) in southwestern Western Australia (116° 05′ E, 32° 50′ S). A full description of the study streams can be found in Bunn et al. (1986), Davies (1994), Hughes et al. (1996), Hancock & Bunn (1997), Mosisch & Bunn (1997) and Bunn et al. (1999).

Persistence in benthic community structure

Temporal persistence of stream benthic fauna was determined from a long-term monitoring programme of macroinvertebrates collected in southwestern Australia. Quantitative samples were collected with a modified Surber sampler (250-μm mesh, area of base 0.0625 m^2) (Surber, 1970) from two sites in the North Dandalup catchment (ND1: Foster Brook and ND3: North Dandalup River) and two in the Canning catchment (CD2: Death Adder Creek and CD3: Poison Gully). Sites were situated in first and second order upland streams in relatively undisturbed forest and were sampled seasonally from 1984 to 1988 (inclusive). During this time, sites in the North Dandalup catchment were permanently flowing and sampled four times a year and the Canning catchment sites were seasonally intermittent and sampled twice a year (winter and autumn). For details on collection and laboratory processing of samples see Storey et al. (1990).

At each site during each time, six random Surber samples were collected from riffles. This number of samples was sufficient to collect all but the rarest taxa (i.e., those comprising less than 0.5% of the total abundance of fauna) (Storey et al., 1990). Macroinvertebrates were identified to the lowest taxon possible (in most cases genus or species) and community data were ordinated by detrended correspondence analysis (DECORANA: Hill, 1979) using the mean abundance of the total fauna collected from each site during each year. Data were transformed using $\ln(x+1)$ and an option to downweight rare species in proportion to their abundance was utilised (i.e., Hill, 1979).

Community patterns versus ecosystem processes

Community metabolism is a fundamental ecological process describing the movement of carbon (e.g., Bott et al., 1985; Bunn et al., 1999). At seven sites in southwestern Australia, benthic metabolism was measured by monitoring dissolved oxygen using dataloggers within Perspex chambers over 24 h (see Bunn et al., 1999). The amount of carbon produced (benthic gross primary production; GPP) and consumed (respiration; R_{24}) at the stream reach scale was estimated by patch-weighting the habitat-level metabolism values (e.g., Naiman, 1983).

At the same sites and during the same seasons, the 'health' of these streams was determined by the AUstralian RIVer Assessment Scheme (AusRivAS; Schofield & Davies, 1996), which is a national protocol based on the RIVPACS approach used in the UK (Wright, 1995; Wright et al., 1984). AusRivAS uses family-level identifications of benthic macroinvertebrates and associated measurements of environmental variables to derive predictive models of river health for different biomes. The model then uses the observed (O) families of taxa in the collected sample to determine if the faunal composition is what would be expected (E) from an undisturbed site in the same region. The ecological condition of a site is tested by calculating the O/E ratio number of families. The resulting O/E values are banded to provide a classification of the health of a study site (for further details of methodology, see Kay et al., 1999; Smith et al., 1999).

Relationships between benthic community patterns (O/E scores) and ecosystem processes (GPP, R_{24}) were then examined using simple correlations. The relationships between these biological variables and water quality parameters, including turbidity and nutrients, were also examined.

Spatial and temporal patterns in community structure

Many strategies for assessing ecological integrity of streams and rivers are based on direct measures of populations and communities (Bunn, 1995). Examples are the National Monitoring River Health Initiative in Australia (AusRivAS, Schofield & Davies, 1996) and the RIVPACS approach in the UK (Wright, 1995). The general approach is:

- selection of control or reference sites which are deemed to be largely free from human disturbance, taking care to distinguish between broad ecoregions (e.g., climate, vegetation, biogeographical differences);
- development of predictive models that describe spatial patterns and relate these to abiotic variables (e.g., classification and ordination techniques that underpin AusRivAS and RIVPACS); and
- use of the models to predict patterns at other sites. Deviations from expected reference condition are assumed to reflect changes in ecological integrity and provide an assessment of river 'health'.

Biogeographical considerations

Recognizing the existence of natural geographic variation in species distribution and abundance is an important consideration in the development of biomonitoring programmes, particularly at large spatial scales (Karr, 1991; Resh et al., 1995). At regional or continental scales, biogeographical differences are likely to give rise to inherent differences in biota despite similar local attributes (such as climate and geology). Developing a national framework to biomonitor streams and rivers in a country the size of Australia is clearly no small undertaking (Schofield & Davies, 1996). There are considerable variations in climate and hydrology (Keast, 1981), ranging from cool temperate systems with winter rain (i.e., the southern regions) to arid regions of highly episodic and unpredictable rainfall events (i.e., much of northern and central Australia).

Comparison of streams in southwestern and southeastern Australia provides a striking example of the difficulties in developing national guidelines. Both regions share a similar climate and common Gondwanan history but there are major differences in the composition and biodiversity of the fauna (Bunn & Davies, 1990). Some groups (e.g., Plecoptera and many families of algal-grazing insects) are naturally absent or poorly represented in the southwest compared with similar streams in southeastern Australia. This has been attributed, in part, to the unique geological history of southwestern Australia (Bunn & Davies, 1990). As a consequence, indices based on species richness of particular groups would give a poor ranking of southwestern Australian streams if based on seemingly similar reference sites (based on climate, vegetation, physical habitat) from the southeast. For example, mean EPT taxa (Ephemeroptera, Plecoptera, Trichoptera; see Resh et al., 1995) for forest streams in southwestern Australia was only 33 $\pm$ 1 compared with 122 $\pm$ 8 in southeastern Australia (mean $\pm$ 1S.E.; see Bunn & Davies, 1990, for details). Using the SIGNAL rapid assessment procedure proposed by Chessman (1995), which incorporates pollution grades for each family, the mean score (out of a maximum of 10) for benthic invertebrate families in southwestern Australia forest streams (6.5 ± 0.1) was significantly lower than that observed in similar streams in the southeast (7.4 ± 0.03). Perhaps more disturbing is that the SIGNAL value for streams in southwestern Australia is close to the 'doubtful water quality' status (SIGNAL = 5–6; see Chessman, 1995), even though the sites are in undisturbed forest. These biogeographical distinctions emphasise the need for careful selection of reference sites and the futility of developing nationwide indices of stream health (see Resh et al., 1995).

Temporal patterns and community persistence

Biological monitoring studies typically emphasise the importance of spatial differences and tend to neglect temporal scales. However, an underlying assumption is that the communities monitored show high persistence. High persistence indicates a community is deterministic or regulated and therefore has an underlying structure (Grossman et al., 1982; Zaret, 1982; however, see Ebeling et al., 1990), and low persistence describes communities that are more likely produced by stochastic processes (Moyle & Vondracek, 1985; Sousa, 1979).

High persistence indicates that the community structure changes little between years. If we return to the same sites, year after year, we would expect the biota to be essentially the same unless some disturbance (either natural or anthropogenic) has occurred. For example, if the benthic fauna of two reference streams is sampled each summer and winter over several years, we can represent the spatial and temporal variation in community structure in ordination space (Figure 1). If the community showed high temporal persistence with either little seasonality (Figure 1a) or a strong seasonal pattern (Figure 1b), we can easily develop a robust predictive model for biomonitoring. However, the spatial pattern could also change from year to year due to low persistence. Both sites could show a similar temporal trend in community structure, perhaps in response to inter-annual variation in environmental conditions (Figure 1c), or vary completely independently of each other with no obvious causal explanation (Figure 1d). In the former case (1c), observed in some streams in northern Australia (Faith et al., 1995), dissimilarity between the two sites would be constant over time because the communities were changing in parallel rather than not changing at all. We would need to continually sample reference sites to account for natural temporal shifts in assemblage structure. In the latter example (1d), we would have little confidence in the predictive capacity of our model based on reference sites. Unfortunately, there are few published long-term data sets to provide convincing evidence that stream benthic community patterns are persistent over time.

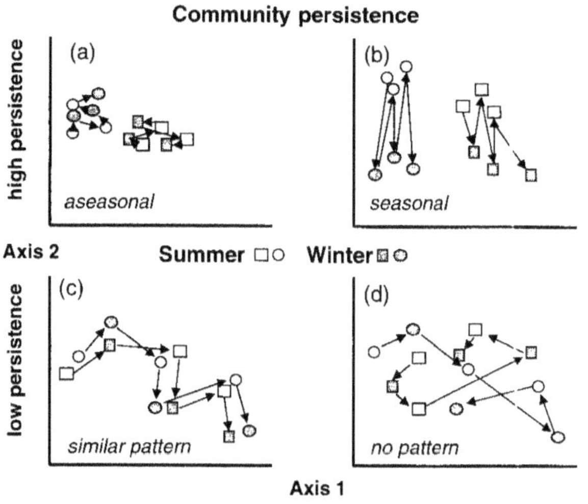

Figure 1. Temporal variation in benthic community structure from two hypothetical streams (plotted in ordination space), where sites show: (a) high temporal persistence and no seasonal pattern; (b) high temporal persistence and strong seasonal pattern; (c) low temporal persistence with similar trends among sites; and (d) low temporal persistence and no consistent trends among sites (from Bunn, 1995).

Ordinations of benthic community data, collected over a 5-year period from perennial and intermittently flowing streams in southwestern Australia, provide a good example of the range of temporal patterns we might find (Figure 2). These reference streams flow through relatively undisturbed sclerophyll forest in one of the most predictable climatic regions of the country (Bunn et al., 1986). The two perennial streams in the North Dandalup catchment (ND1 and ND3) show little inter-annual variation in benthic community structure over the 5-year period (Figure 2). In contrast, the two intermittently flowing streams in the adjacent Canning River catchment (CD2 and CD3) show marked variation over time with no obvious pattern, similar to that suggested in Figure 1d.

As longer-term data sets are gathered, it is becoming increasingly apparent that not all assemblages show high persistence. In some instances, this may be a function of inter-annual variation in environmental conditions (e.g., stream benthic communities in the wet-dry tropics of Australia, Faith et al., 1995).

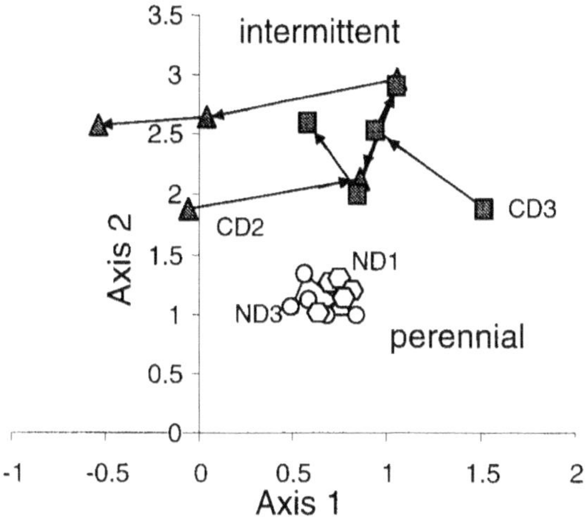

Figure 2. Ordination of annual benthic macroinvertebrate samples from two seasonally intermittent (CD2 and CD3) and two perennial (ND1 and ND3) forest streams in southwestern Australia, collected over five years.

Temporal changes in community composition, however, may be more stochastic and seemingly unrelated to inter-annual variation in environmental parameters. For example, species turnover between years in some arctic and subarctic streams (Miller & Stout, 1989) and cold desert spring-streams (Cushing & Gaines, 1989) is very high and the process of species extinctions and immigrations is thought to be largely stochastic. Similar stochastic patterns in space and time have been observed in subtropical streams in southeast Queensland (Bunn & Hughes, 1997) and are also expected throughout drought-prone regions of warm-temperate eastern Australia.

High temporal variability in community structure has the potential to limit the sensitivity of biomonitoring approaches such as AusRivAS and RIVPACS. Low persistence in benthic community structure makes it extremely difficult to construct robust predictive models and will increase the likelihood of failure of the models (i.e., falsely conclude that observed changes in patterns are a consequence of changes in ecological health).

Influence of biological processes on community structure

There are many possible explanations for high levels of spatial and temporal variation in benthic community structure in some streams. One obvious conclusion is that such variation simply reflects a direct response of the fauna to changes in abiotic conditions (e.g., associated with floods and droughts), and therefore it may be possible to factor this into predictive models. The seemingly overriding importance of abiotic processes and habitat has been discussed by Petts (this volume). However, we must also acknowledge that marked spatial and temporal variation can also be the result of several important biological processes, including competition, predation and recruitment. Of these, we will focus attention on the latter two. There is little evidence to suggest that interspecific competition can result in major changes in the structure of stream benthic assemblages (Hildrew & Townsend, 1987). At most, competition may result in shifts in abundance of interacting pairs of species (e.g. Allan, 1983), but cascading effects are unlikely.

Predation

Predators are common in streams, although their role in structuring aquatic communities is the subject of considerable debate (Cooper et al., 1990; Hildrew, 1992). Spectacular evidence of strong predation effects comes from work on Californian streams (Power, 1990; Power et al., 1985). These studies have demonstrated that predatory fish can have a 'cascading' effect on the stream biota, through direct effects on small fish and predatory invertebrates, to primary consumers and ultimately both the composition and biomass of benthic algae. In the presence of predatory fish, smaller predators were reduced, tube-weaving chironomid larvae proliferated, and the benthic substrate was reduced to a midge-infested residue (Power, 1990). Predation effects of this kind clearly can be a major cause of spatial and temporal variation in stream community structure. Biomonitoring models based entirely on abiotic variables would be unable to predict such marked changes in the nature of the stream. Predator-mediated shifts in benthic community structure would be classified as a change in ecological integrity and mistakenly attributed to some form of anthropogenic disturbance.

As an aside, the notion that 'all species are not equal' is an important consideration in selecting taxa for bioassay studies or as indicators (see Bunn, 1995). Selection of species on the basis of their high functional importance (i.e., those that provide important links in food webs or are important in organizing the community – see Hurlbert, 1997) should be a primary criterion.

Recruitment and dispersal

The important role of population-level processes such as dispersal and recruitment is not new to ecology. The 'supply' of recruits is a well-known process in marine systems and can be of over-riding importance in determining community structure and dynamics (Sale, 1990; Underwood & Fairweather, 1989). However, the influence of recruitment and dispersal on the structure and dynamics of stream communities has been downplayed in the past (Bunn & Hughes, 1997). Stream and river animals are generally considered to have high dispersal capabilities and, given the extensive geographic distributions of some species, it is apparent that viable mechanisms for dispersal either do exist or at least must have existed in the past (e.g., Mackay, 1992; Wallace, 1990).

Recent work on the genetic structure of stream and river invertebrates, however, suggests that this is not always the case (Bunn & Hughes, 1997). For fully aquatic animals, such as atyid shrimps, there appeared to be limited movement even on a small spatial scale (e.g., streams within the same subcatchment) (see Hughes et al., 1995, 1996). Although widespread adult dispersal of stream insects was apparent, movement of larvae between reaches was also limited. The realisation that larval insects sampled at the reach scale were the offspring of only a few matings implies that populations can be accounted for by the chance oviposition of only a few females (Bunn & Hughes, 1997). Such genetic studies provide strong evidence that dispersal capability and the stochastic effects of recruitment can easily explain the observed marked spatial and temporal variability in community structure.

The role of biological processes

In light of the above, underlying biological processes cannot be ignored when developing predictive multivariate models or even simple metrics for biomonitoring that are based on measures of populations or communities. Selecting reference sites for predictive models requires considerable care to ensure there are no inherent biogeographical constraints to the distribution of biota. Although stream benthic organisms generally respond in a predictable fashion to variations in the physical habitat, this may be over-ridden by marked variation resulting from biological processes such as predation and recruitment. Observed deviations from reference condition could result from stochastic processes such as larval recruitment and be misinterpreted as changes in ecological integrity.

Pattern and process

These limitations aside, what else can we infer about the health of rivers from measurements of patterns alone (i.e., populations, communities)? Many goals relating to river management and protection refer to ecosystem-level processes, with the notion that streams and rivers should function in the same way as they do in undisturbed catchments. This is an important element of the principles of ecological sustainable development (ESD), which consider not only the protection of biodiversity, but also the maintenance of essential ecological processes and life support systems (Commonwealth of Australia, 1996; Council of Australian Governments, 1992).

Often, we use patterns of species richness and composition as surrogate measures of these fundamental processes. However, pattern and process are not necessarily linked (see Bunn, 1995): changes in patterns can occur without any detectable change in processes (e.g., functional redundancy). Even if pattern and process are linked, there may be a lag before a modification in an ecological process is manifest as a change in pattern. Alternatively, changes in processes may occur without any detectable change in population or community attributes. There is little doubt that simple structural indices of stream health (e.g., EPTs) can be used to track ecosystem processes (e.g., secondary production, leaf processing rates) in the case of extreme disturbance (e.g., Wallace et al., 1996). However, this may not be the case at lower levels of disturbance.

Comparisons of measures of benthic metabolism (GPP, R_{24}) and AusRivAS scores from small forest streams in southwestern Australia provide an example of the latter (Figure 3). AusRivAS scores (O/E) reflect the deviation from reference condition and it is worth noting that all of these sites score in the 'undisturbed' category. Although there was a marked (10-fold) spatial variation in GPP and R_{24}, this was clearly not correlated with O/E scores (i.e., pattern and process were not related – Figure 3). The observed variation in GPP and R_{24}, however, was clearly related to spatial differences in water quality (Figure 4a). Specifically, these measures of benthic metabolism were correlated with nitrogen concentration and turbidity – two variables likely to affect river health. Importantly, there was no correlation between any of the water quality parameters tested and O/E scores based on benthic macroinvertebrate communities (Figure 4b). In this case, predictive models based on patterns were

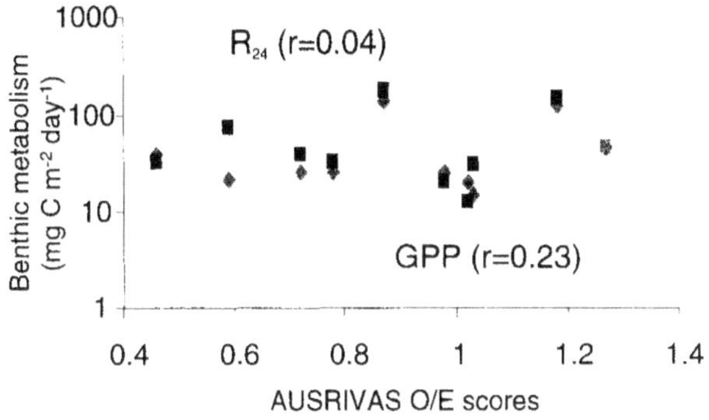

Figure 3. Relationships between benthic community metabolism (GPP = diamond symbols; R_{24} = square symbols) and AusRivAS scores (O/E) for small forest streams in southwestern Australia.

not sensitive enough to detect changes in ecological integrity.

Ecosystem processes and river health

Because of the above issues (lack of persistence and unpredictability of fauna, and the lack of obvious links between patterns and process), we have argued that direct measures of ecosystem attributes should form an integral component of river health assessment in Australia (Bunn, 1995; Bunn et al., 1999). In our recent work, we have used replicated and patch-weighted measures of benthic metabolism, and stable isotope analysis to track the sources and fate of organic carbon and nitrogen. These are direct measures of 'vigor' and 'organization', considered to be key indicators of ecosystem health (sensu Rapport et al., 1998).

A particular focus of our work has been to determine whether such indicators are influenced more by changes in riparian condition than by land-use activities elsewhere in the catchment. Using data from 20 streams in the Mary River catchment in southeast Queensland, it has been possible to develop simple but powerful predictive models to describe the response of these indicators to changes in both aspects (Bunn et al., 1999). We have shown that riparian canopy cover is by far the best predictor of benthic metabolism. Values of GPP and R_{24} from undisturbed forest catchments provided reference values to assess changes in ecological integrity. Stimulation of GPP was shown to occur when the canopy cover was reduced to below approximately 75% (= average cover for forest streams); however, major changes in health occurred when the canopy was lower than 40–50% and GPP exceeded R_{24}. At this point, much of the aquatic plant production shifted from palatable microalgae to filamentous algae and macrophytes, which do not appear to enter the food web (Bunn et al., 1999). The proliferation of these aquatic plants is also associated with a major decline in water quality (Bunn et al., 1998).

Conclusions

If we are to rely solely on measures of ecological patterns to assess the ecological integrity of streams and rivers, then we must use caution developing and applying predictive models and simple biotic indices. Changes in patterns (abundance, richness, species composition) do not always equate to changes in ecological integrity. Marked changes can and do occur as a result of natural biological processes and may falsely lead to conclusions that impacts have occurred when they have not (i.e., Type I statistical error). Conversely, patterns may not change even when important ecosystem processes do. Biomonitoring approaches reliant solely on pattern detection may be unable to detect changes in ecological integrity (i.e., Type II statistical error). Measurement of patterns alone often tells us very little about the essential life support systems that are an integral component of ESD. Direct measurements of ecosystem processes are often neglected in river health assessment programs (Bunn, 1995). However, they are sensitive to factors that are known to directly influence river health and it is possible to develop simple but powerful predictive models. Importantly, should changes in ecological integrity be detected, it is possible to recommend strategies for remediation because the causal processes are generally well known.

Acknowledgements

We are extremely grateful to the Department of Hydrobiology, Fisheries and Aquaculture, University of Agricultural Sciences (BOKU), Vienna, for the opportunity to attend the International Conference on *Assessing the Ecological Integrity of Running Waters*, and for their generous hospitality. Much of the original data presented in this paper has been put together with the assistance of staff from the Queensland Department of Natural Resources and the Water and Rivers Commission of Western Australia. In particular, we thank John Amprimo, Heather Hunter, Damon Telfer,

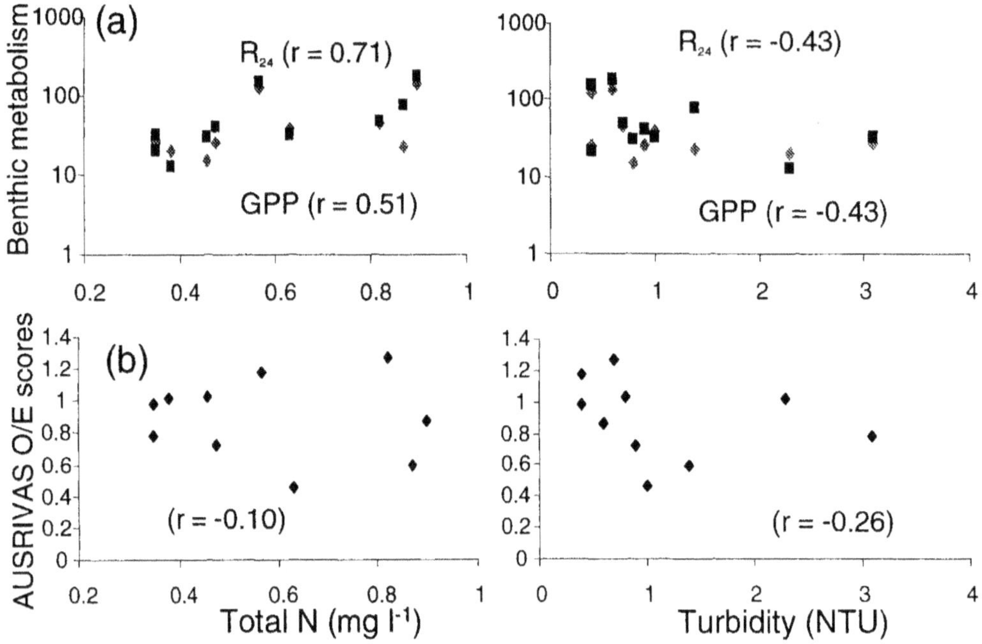

Figure 4. Relationships between water quality (total nitrogen concentration and turbidity) in small forest streams in southwestern Australia and (a) benthic community metabolism (GPP = diamond symbols; R_{24} = square symbols) and (b) AusRivAS scores.

Steve Kelly and Jeff Kite. Permission to collect in State Forests and National Parks was given by the Department of Conservation and Land Management (Western Australia), and the Queensland Departments of Natural Resources and Environment. Chris Marshall, Sue Creagh, Thorsten Mosisch and Michelle Jenkins assisted with the collection of field data and laboratory analysis of samples. This work has been largely funded by the Australian Research Council, and the Land and Water Resources Research and Development Corporation under a National Riparian Lands Program grant to SB (GRU17), and a National River Health Program grant to PD (UWA14).

References

Allan, J. D., 1983. Predator-prey relationships in streams. In: Barnes, J. R. & G. W. Minshall (eds), Stream Ecology: Application and Testing of General Ecological Theory Plenum Press, New York: 191–229.

Bott, T. L., J. J. Brock, C. S. Dunn, R. J. Naiman, R. W. Ovink & R. C. Peterson, 1985. Benthic community metabolism in four temperate stream systems: an inter-biome comparison and evaluation of the river continuum concept. Hydrobiologia 123: 3–45.

Bunn, S. E., 1995. Biological monitoring of water quality in Australian: workshop summary and future directions. Aust. J. Ecol. 20: 220–227.

Bunn, S. E. & P. M. Davies, 1990. Why is the stream fauna of southwestern Australia so impoverished? Hydrobiologia 194: 169–176.

Bunn, S. E. & J. M. Hughes, 1997. Dispersal and recruitment in streams: Evidence from genetic studies. J. N. am. Benthol. Soc. 16: 338–346.

Bunn, S. E., D. H. Edward & N. R. Loneragan, 1986. Spatial and temporal variation in the macroinvertebrate fauna of streams of the northern jarrah forest, Western Australia: community structure. Freshwat. Biol. 16: 67–92.

Bunn, S. E., P. M. Davies, D. M. Kellaway & I. Prosser, 1998. Influence of invasive macrophytes on channel morphology and hydrology in an open tropical lowland stream, and potential control by riparian shading. Freshwat. Biol. 39: 171–178.

Bunn, S. E., P. M. Davies & T. D Mosisch, 1999. Ecosystem measures of river health and their response to riparian and catchment degradation. Freshwat. Biol. 41: 333–345.

Chessman, B. C., 1995. Rapid assessment of rivers using macroinvertebrates: a procedure based on habitat-specific sampling, family level identification and a biotic index. Aust. J. Ecol. 20: 122–129.

Commonwealth of Australia, 1996. National Principles for the Provision of Water for Ecosystems. Occasional paper SWR No 3, Agriculture and Resource Management Council of Australia and New Zealand and Australian and New Zealand Environment and Conservation Council. ISBN 0 7310 2382 X.

Cooper, S. D., S. J. Walde & B. L. Peckarsky, 1990. Prey exchange rates and the impact of predators on prey populations in streams. Ecology 71: 1503–1514.

Council of Australian Governments, 1992. National Strategy for Ecologically Sustainable Development. Australian Government Printing Service, Canberra.

Cushing, C. E. & W. L. Gaines, 1989. Thoughts on recolonization of endorheic cold desert spring-streams. J. N. am. Benthol. Soc. 8: 277–287.

Davies, P. M., 1994. Ecosystem ecology of upland streams of the northern jarrah forest, Western Australia. Unpublished Ph.D. Thesis, Department of Zoology, The University of Western Australia, 236 pp.

Ebeling, A. W., S. J. Holbrook & R. J. Schmitt, 1990. Temporally concordant structure of a fish assemblage: bound or determined? Am. Nat. 135: 63–73.

Faith D. P., P. L. Dostine & C. L. Humphrey, 1995. Detection of mining impacts on aquatic macroinvertebrate communities: Results of a disturbance experiment and the design of a multivariate BACIP monitoring program at Coronation Hill, Northern Territory. Aust. J. Ecol. 20: 167–180.

Grossman, G. D., P. B. Moyle & J. O. Whitaker Jr., 1982. Stochasticity in structural and functional characteristics of an Indiana stream fish assemblage: a test of community theory. Am. Nat. 120: 423–454.

Hancock, M. A. & S. E. Bunn, 1997. Population dynamics and life history of *Paratya australiensis* Kemp, 1917 (Decapoda: Atyidae) in upland rainforest streams, southeast Queensland. Mar. Freshwat. Res. 48: 361–369.

Hildrew, A. G., 1992. Food webs and species interactions. In: Petts, G. & P. Calow (eds), The Rivers Handbook, Volume 1. Blackwell Science, London: 309–330.

Hildrew, A. G. & C. R. Townsend, 1987. Organization in freshwater benthic communities. In: Gee, J. H. & P. S.Giller (eds), Organization of Communities: Past and Present, Symposia of the British Ecological Society. Blackwell Scientific, Oxford: 347–371.

Hill, M. O., 1979. DECORANA — A FORTRAN Program for Detrended Correspondence Analysis and Reciprocal Averaging. Ecology and Systematics, Cornell University, Ithaca, New York.

Hughes, J. M., S. E. Bunn, D. Hurwood & M. Kingston, 1995. Genetic differentiation of populations of *Paratya australiensis* (Decapoda: Atyidae) in rainforest streams in south-east Queensland. J. N. am. Benthol. Soc. 14: 158–173.

Hughes, J. M., S. E. Bunn, D. A. Hurwood, S. Choy & R. Pearson, 1996. Genetic differentiation among populations of *Caridina zebra* (Decapoda: Atyidae) in tropical rainforest streams, northern Australia. Freshwat. Biol. 36: 289–296.

Hurlbert, S. H., 1997. Functional importance *vs* keystoneness: reformulating some questions in theoretical biocenology. Aust. J. Ecol. 22: 369–382.

Huryn, A. D. & J. B. Wallace, 1987. Local geomorphology as a determinant of macrofaunal production in a mountain stream. Ecology 68: 1932–1942.

Karr, J. R., 1991. Biological integrity: a long-neglected aspect of water resource management. Ecol. Appl. 1: 66–84.

Karr, J. R. & E. W. Chu, 2000. Sustaining living rivers. Hydrobiologia 422/423: 1–13.

Kay, W. R., M. J. Smith, A. M. Pinder, J. M. McRae, J. A. Davis & S. A. Halse, 1999. Patterns of distribution of macroinvertebrate families in rivers of north-western Australia. Freshwat. Biol. 41: 299–316.

Keast, A., 1981. Ecological Biogeography of Australia. Dr W. Junk Publishers, The Hague.

Mackay, R. J., 1992. Colonization by lotic macroinvertebrates: a review of processes and patterns. Can. J. Fish. aquat. Sci. 49: 617–628.

Miller, M. C. & J. R. Stout, 1989. Variability of macroinvertebrate community composition in an arctic and subarctic stream. Hydrobiologia 172: 111–127.

Mosisch, T. D. & S. E. Bunn, 1997. Temporal patterns in stream epilithic algae in response to discharge regime. Aquat. Bot. 58: 181–193.

Moyle, P. B. & B. Vondracek, 1985. Persistence and structure of the fish assemblage in a small California stream. Ecology 66: 1–13.

Naiman, R. J., 1983. The annual pattern and spatial distribution of aquatic oxygen metabolism in boreal forest watersheds. Ecol. Monogr. 53: 73–94.

Norris, R. H. & K. R. Norris, 1995. The need for biological assessment of water quality: Australian perspective. Aust. J. Ecol. 20: 1–6.

Petts, G. E., 2000. A perspective on the abiotic processes sustaining the ecological integrity of running waters. Hydrobiologia 422/423: 15–27.

Power, M. E., 1990. Effects of fish in river food webs. Science 250: 811–814.

Power, M. E., W. J. Matthews & A. J. Stewart, 1985. Grazing minnows, piscivorous bass, and stream algae: dynamics of a strong interaction. Ecology 66: 1448–1456.

Pringle, C. M., R. J. Naiman, G. Bretschko, J. R. Karr, M. W. Oswood, J. R. Webster, R. L. Welcomme & M. J. Winterbourn, 1988. Patch dynamics in lotic systems: the stream as a mosaic. J. N. am. Benthol. Soc. 7: 503–524.

Rapport, D. J., R. Costanza & A. J. McMichael, 1998. Assessing ecosystem health. Trends Ecol. Evol. 13: 397–402.

Resh, V. H., R. H. Norris & M. T. Barbour, 1995. Design and implementation of rapid assessment approaches for water resource monitoring using benthic macroinvertebrates. Aust. J. Ecol. 20: 108–121.

Sale, P. F., 1990. Recruitment of marine species: is the bandwagon rolling in the right direction? Trends Ecol. Evol. 5: 25–27.

Schaeffer, D. J., E. E. Henricks & H. W. Kerster, 1988. Ecosystem health: 1. Measuring ecosystem health. Envir. Manage. 12: 445–455.

Schofield, N. J. & P. E. Davies, 1996. Measuring the health of our rivers. Water (May/June): 39–43.

Smith, M. J., W. R. Kay, D. H. D. Edward, P. J. Papas, K. S. J. Richardson, J. C. Simpson, A. M. Pinder, D. J. Cale, P. H. J. Horwitz, J. A. Davis, F. H. Yung, R. H. Norris & S. A. Halse, 1999. AusRivAS: using macroinvertebrates to assess ecological condition of rivers in Western Australia. Freshwat. Biol. 41: 269–282.

Sousa, W. P., 1979. Disturbance in marine intertidal boulder fields; the non equilibrium maintenance of species diversity. Ecology 60: 1225–1239.

Storey, A. W., S. E. Bunn, P. M. Davies & D. H. Edward, 1990. Classification of the macroinvertebrate fauna of two river systems in south-western Australia in relation to physical and chemical parameters. Regul. Riv. 5: 217–232.

Surber, E. W., 1970. Procedure in taking stream bottom samples with the stream square foot bottom sampler. Proceedings of the 23rd Annual Conference of South East Game and Fisheries Commission 23: 587–591.

Suter, G. W., 1993. A critique of ecosystem health concepts and indexes. Envir. Toxicol. Chem. 12: 1533–1539.

Townsend, C. R., 1989. The patch dynamics concept of stream community ecology. J. N. am. Benthol. Soc. 8: 36–50.

Underwood, A. J. & P. G. Fairweather, 1989. Supply-side ecology and benthic marine assemblages. Trends Ecol. Evol. 4: 16–20.

Wallace, J. B., 1990. Recovery of lotic macroinvertebrate communities from disturbance. Envir. Manage. 15: 605–620.

Wallace, J. B., J. W. Grubaugh & M. R. Whiles, 1996. Biotic indices and stream ecosystem processes: results from an experimental study. Ecol. Appl. 6: 140–151.

Wicklum, D. & R.W. Davies, 1995. Ecosystem health and integrity. Can. J. Bot. 73: 997–1000.

Wright, J. F., 1995. Development and use of a system for predicting the macroinvertebrate fauna in flowing waters. Aust. J. Ecol. 20: 181–197.

Wright, J. F., D. Moss, P. D. Armitage & M. T. Furse, 1984. A preliminary classification of running water sites in Great Britain based on macroinvertebrate species and the prediction of community type using environmental data. Freshwat. Biol. 14: 221–256.

Zaret, T. M., 1982. The stability/diversity controversy: a test of hypotheses. Ecology 63: 721–731.

Hydrobiologia **422/423:** 71–83, 2000.
M. Jungwirth, S. Muhar & S. Schmutz (eds), Assessing the Ecological Integrity of Running Waters.

Mayflies (Insecta: Ephemeroptera) and the assessment of ecological integrity : a methodological approach

Ernst Bauernfeind[1] & Otto Moog[2]
[1]*Naturhistorisches Museum Wien, I.Zool.Abteilung, Burgring 7, A-1014 Vienna, Austria*
[2]*Department of Hydrobiology, Fisheries and Aquaculture, University of Agricultural Sciences Vienna, Max Emanuel-Straße 17, A-1180 Vienna, Austria*

Key words: Ephemeroptera, mayflies, ecological integrity, natural habitats, sampling strategies, typical assemblages, identification

Abstract

Mayflies (Ephemeroptera) play an important role in almost all undisturbed freshwater communities and their larvae frequently form a considerable part of the material sampled during biomonitoring procedures. Mayfly taxa are widely accepted as bioindicators for water quality and form an integral part of standardized systems for the saprobic evaluation of running waters. Nevertheless, successful interpretation of data depends heavily on sampling strategies and the technical concepts adopted therein. The following paper discusses current problems affecting the correct assessment of mayfly community structure and proposes basic requirements to improve the meaningfulness of limnological routine sampling with regard to Ephemeroptera. The topics covered include: mayfly diversity in different types of freshwater communities in Europe; natural habitats and sampling sites / sampling strategies; life cycles and minimum sampling frequency; level of identification and interpretation of data.

Introduction

The assessment of ecological integrity provides information for environmental managers and decision-makers to take accurate and justifiable actions as well as to evaluate the effectiveness of legislative regulations already in force. While the political mandate to strive for ecological integrity has already been formally confirmed in legislation, the monitoring and assessment of the environmental status of surface water-resources has sometimes been hampered by a lack of conceptual clarity. For the benefit of both human society and nature itself, an effort must be made to ensure that national environmental policies are based upon the most comprehensive scientific information available.

Focussing on the insect order Ephemeroptera, this paper raises technical issues based on the present understanding of mayfly ecology, proposing an appropriate methodology for the establishment of standard operation procedures.

Technical questions addressed in this study include: (1) mayfly assemblages and diversity in different types of Central European freshwater communities; (2) natural mayfly habitats and determination of habitats to be sampled (spatial sampling regime); (3) sampling equipment and sampling techniques; (4) life cycles and temporal sampling regime regarding minimum sampling frequency; (5) level of identification and identification problems in Europe, including a short review of the taxonomic situation in Ephemeroptera world-wide.

It is the main goal of this paper to walk the line between the needs of science and the financial and technical means of national monitoring projects. Applying the experience gained by taxonomic and faunistic research on Ephemeroptera will no doubt improve the results of limnological routine sampling and thereby contribute considerably towards a better understanding of the ecological integrity of aquatic systems.

Numerous studies demonstrate that mayfly community structure effectively reflects the environmental situation of watercourses. Since Kolkwitz & Marsson (1902) initially proposed the use of organismic indicators (including Ephemeroptera) to assess water pollution, their concept has proved highly successful, with mayflies forming an integral part of the taxonomic groups currently considered to be especially valuable for biomonitoring (Moog, 1993; Moog et al., 1997). When Illies & Botosaneanu established their river zonation concept in 1963, two species of Ephemeroptera (*Heptagenia sulphurea, Potamanthus luteus*) had been used to differentiate between the potamon and rhithron, together with Trichoptera (Lepneva, 1949), Plecoptera, and Coleoptera. Subsequent research led to a broader understanding of mayfly biocoenoses, revealing significant changes in diversity and community structure in connection with ecological changes along watercourses (e.g., Landa & Soldán, 1991; Krno, 1987, 1990; Sowa, 1975a). The application of modern theoretical concepts including trophic levels and functional feeding guilds (Cummins & Klug, 1979; Townsend, 1996; Vannote et al., 1980) is also supported by findings based on the distributional differentiation among Ephemeroptera.

Curiously, the significance of mayflies in indicating the ecological integrity of running waters has been widely neglected in the past (Karr, 1991), especially when compared with traditionally popular environmental indicators such as fish and bird communities (Errington, 1945; Kendeigh, 1934; Utschick, 1976). During the last decade, however, the so-called EPT-philosophy has successfully emphasized the outstanding importance of Ephemeroptera, Plecoptera, and Trichoptera in describing environmental conditions (Lenat & Barbour, 1994; Plafkin et al., 1989; Resh & Jackson, 1993). Besides water temperature, oxygenation and current, the availability of suitable mesohabitat structures (stratotopes, choriotopes) is clearly one of the most important factors influencing the occurrence and distribution of Ephemeroptera larvae (Cogerino et al., 1995; Krno, 1990, 1991). Ecological degradation of running waters caused by human impact (channelization, irrigation, impoundments) generally leads to deficiencies in or complete loss of characteristic natural structures, creating environmental uniformity and thereby significantly affecting mayfly diversity and abundance. Furthermore, interrelationships and possible cumulative effects between pollution and structural deterioration for mayflies are only poorly understood at present. The much higher drop in mayfly diversity in the potamon region is probably not due only to the (assumed) greater pollution in the lower courses of rivers or to technical problems in collecting. The availability of different mesohabitats in comparatively still natural (though moderately polluted) rivers in eastern Europe versus in the heavily impaired watercourses in industrial Central Europe strongly suggests that channelization and impoundments are even more important than pollution. Under pristine conditions river dynamics create a mosaic of different, closely adjoining habitats. Changing currents and instable river banks allow for ramification, erosion, and gradual sedimentation. This in turn provides a rich variety of mesohabitats suitable for both rheophilic (riffle-dwelling) as well as stillwater forms inhabiting pools and stagnant backwaters.

Although the amount and frequency of monitoring studies have increased considerably within the last 20 years, great gaps still exist even in our basic knowledge. Autecological and faunistic studies of Ephemeroptera are rather limited at present and further co-operation between taxonomists and limnologists should prove especially fruitful. The urgent need for intensified research in the near future necessitates improving the meaningfulness of limnological routine sampling data. Although some of the problems dealt with in the present paper may appear rather trivial and have in fact already been discussed by Macan as early as 1957(a), experience shows that Ephemeroptera are still underrepresented in many monitoring programs. Understanding and considering group-specific requirements should perceptibly extend the basis for future data analysis.

Mayfly diversity in Central Europe

Under natural conditions mayfly diversity is directly correlated with habitat variety and therefore a modified 'Species–Deficit–Concept' ('Artenfehlbetrag'; Kothé, 1962) should effectively reflect the state of overall ecological integrity for a defined sampling area. The greatest impediment for practical application is the supposedly insufficient knowledge of the species inventory (as well as of the target list) for different sampling stations due to unknown distribution models and, subsequently, data incomparability. The outstanding importance of reliable faunistic baseline data and their value for future conservation projects is amply documented from long-term studies (Landa et al., 1997; Sartori & Landolt, 1998). In Europe at least, these difficulties may be overcome with regard to Ephemeroptera, providing that sampling standards are well defined and comparisons are restricted to the same river zone within the zoogeographical units defined by Illies (1978). The data given in Bauernfeind et al. (1995) are a first step in that direction: they summarize records from Austrian watercourses in connection with limnosaprobity and zonation for all Ephemeroptera species. Table 1 shows as an example the zonational distribution of *Rhithrogena* species in ecologically intact streams and rivers in Austria. The average number of families and species in ecologically pristine to moderately disturbed conditions shows an increase of ephemeropteran taxa from the springs to the potamal sections (Figure 1).

The spring and springbrook zone (krenon)

No strictly steno–krenobiont taxa exist in Central Europe (Zollhöfer, 1997), but mayfly larvae may be numerous and at least some species are found in most types of springs. The occurrence of Ephemeroptera is apparently restricted to the meta- and hypocrenon, where species composition is similar to that in the upper courses of rivers (epirhithral), including euryzonal species as well. *Rhithrogena fonticola*, known so far only from the type locality, has been found in a small spring in southwestern France and may represent a krenophilic taxon. The suspected significance of *Acentrella sinaica** for the hypokrenal subzone (Zollhöfer, 1997) remains to be confirmed. Haybach & Fischer (1997) suggest that, along the northwestern border of its distribution area, *Ecdyonurus subalpinus* is restricted to springs. In Italy, *Electrogena gridellii* has also been found under similar conditions (Moog, 1995, unpubl.). These observations fit well into Kühnelt's (1943) concept of 'Regionale Stenözie', explaining the preference for different habitats under changing external conditions throughout the geographical distribution area of a species.

With regard to Ephemeroptera, diversity in the spring brook zone is comparatively low, seldom exceeding about five to seven species (one to three families) in Austrian karst springs (Weigand, 1998; Weigand & Tockner, 1996). In Switzerland a total of 19 mayfly species (five families) has been found in various types of springs, most of them rhithrobiontic or euryzonal (Zollhöfer, 1997). Whether all of the taxa recorded from the metakrenon finish their developmental cycle there, or whether they undergo a drift phase before emerging, is not known so far due to the lack of emergence trap data.

Species composition is mostly influenced by substratum structure (especially by the occurrence of leaf litter and mosses), with minor human impact having little effect. More radical habitat changes like tapping of a spring or piping usually destroy the community altogether.

Streams and rivers (rhithron)

The mayfly community structure of Central European mountain torrents differs significantly above and below approximately 400 m NN (coinciding roughly with the lower border of the montane vegetation zone), presumably because it is dependent on characteristics of the surrounding terrestrial ecosystem as well as on current velocity/gradient. The autecological basis for this phenomenon is not well understood, but food resources may be a partial explanation. At present this interesting vertical faunistic cline cannot be explained convincingly with historical distribution patterns or the temperature regiment alone (Moog & Wimmer, 1994), nor does it correspond exactly with the conventional understanding of the meta– and hyporhithron.

In Central Europe the upper zone is strictly characterized by *Baetis alpinus*, *Epeorus alpicola*, *Rhithrogena loyolaea*, *Rhithrogena alpestris*, and *Ecdyonurus helveticus*, the latter usually accompanied by at least one of the very closely allied taxa *E. austriacus*, *E. zelleri*, and *E. picteti*. *Electrogena lateralis* is also a typical member of the community throughout the Alps, whereas western populations occur at lower elevation. Species richness is highest in the genus *Rhithrogena*, of which five to eight species may be present at a single sampling station; unfortunately, species–specific ecological amplitude and distributional patterns remain poorly understood. Breitenmoser-Würsten & Sartori (1995) found a negative correlation between the number of species and the slope (and most probably altitude as well). Usually a total of about 15–20 mayfly species (five to six families) may be expected. Structural deficiencies are not always clearly reflected by reduced mayfly diversity, perhaps due to compensational immigration of azonal species, although the rhithron environment may be threatened by fragmentation in the future (Zwick, 1992).

Below 400 m NN in Central Europe, *Electrogena lateralis* is replaced by *E. ujhelyii*; *Ecdyonurus venosus* usually follows *E. helveticus* and its allies, sometimes forming a very short range of overlap. *Rhithrogena picteti* and *Ecdyonurus starmachi* clearly prefer streams, whereas *Rh. semicolorata* and *E. venosus* are usually found in rivers. Species diversity is

* Presumably based on a misidentification. Material received by Graf (pers. comm.) was identified as *Baetis alpinus*.

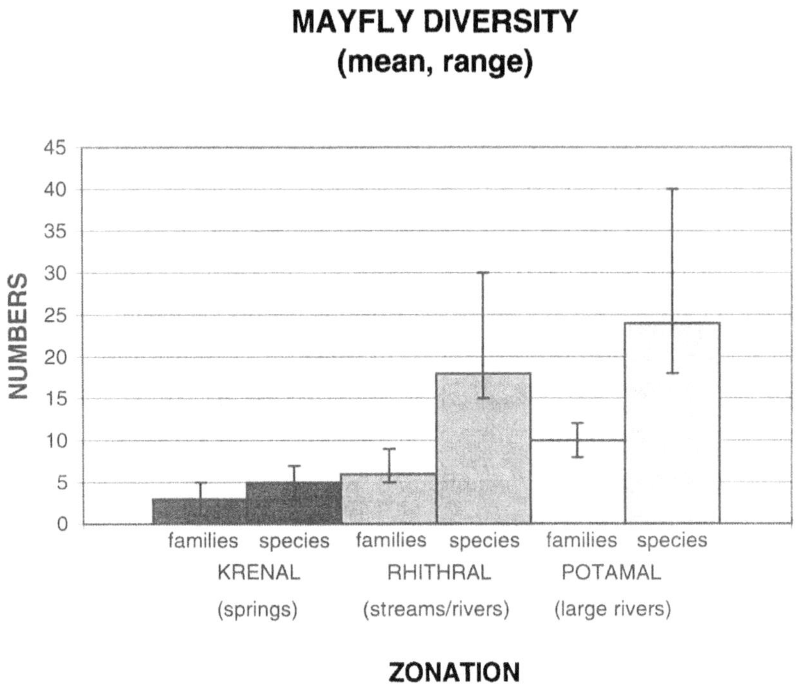

Figure 1. Average number of families and species in Austria. Ecologically pristine to moderately disturbed sampling stations

highest in the genera *Baetis* and *Ecdyonurus*. The occurrence of up to 30 species (seven to nine families) of mayflies reflects pristine environmental conditions or ecological integrity. Eutrophication, increasing water temperature due to extensive clearings of bank vegetation, the loss of riparian structures (roots, dead wood, sandy/silty back-currents, leaf litter) as well as a smaller variety of current/substratum types significantly reduce mayfly diversity, especially in the genera *Ecdyonurus, Rhithrogena,* and *Leptophlebiidae*. The phenomenon is also well reflected in a quantitative index (ratio Baetidae:Heptageniidae), which decreases progressively with decreasing altitude and/or increasing disturbance (Devan & Mucina, 1986).

The lower courses of rivers (potamon)

In ecologically intact potamalic rivers the extremely high diversity in Ephemeroptera is based on the variety of different habitat types, which provides a wide amplitude of temperature, current, oxygenation, and substratum conditions. More than 30 species (10–12 families) may be found in an undisturbed river section including connected backwaters and ramifications. Diversity is especially high on the genus and family level, the region being clearly characterized by the simultaneous occurrence of three or more species of *Heptagenia*. The replacement of *Ecdyonurus venosus* by *E. dispar* (*E. aurantiacus* in southern and southeastern Central Europe) is also significant. Structural losses are common, considerably affecting community structure and allowing for a quick and sound assessment even on a less precise taxonomic level. A sample containing members of *Rhithrogena* (*Rh. beskidensis, Rh. semicolorator*) and *Ecdyonurus* (*E. dispar, E. aurantiacus, E. ruffii*) together with at least two species of *Heptagenia* (or of the typical potamobiont families Isonychiidae, Oligoneuriidae, Potamanthidae, Polymitarcyidae, Prosopistomatidae, Palingeniidae) evidently indicates an ecologically intact environment. Loss of river dynamics disturbs riffle/pool sequences and is immediately reflected by the combined absence of *Rhithrogena* (riffle structures), *Ecdyonurus* (shingle banks) and Leptophlebiidae (pools/roots/leaf barriers). *Choroterpes picteti* is usually dependent on deep pools with macrophytic vegetation; its occurrence in backwaters may signalize the influx of cool groundwater, whereas *Paraleptophlebia werneri* (and *Siphlonurus aestivalis*) prefer leaf barriers in riverine flood-ponds or dense vegetation in more stagnant waters. *Electrogena fascioculata* and *E. affinis* are mostly found on dead wood or submerged roots, preferably along the edge of currents bordering still water. Steep clay banks were once colonized by the burrowing larvae of *Palingenia longicauda* (Sartori et al., 1995) and *P.*

Table 1. Species-specific distribution of the Austrian species of *Rhithrogena* according to biocoenotic region (Rhithral-Potamal-Concept), (10-point system; Moog, 1995)

	Spring	Springbrook	Epirhithral	Metarhithral	Hyporhithral	Epipotamal
R. taurisca	1	6	3	–	–	–
R. loyolaea	1	5	4	-	-	-
R. nivata	1	3	6	–	–	–
R. alpestris	–	3	7	+	–	–
R. degrangei	–	2	5	3	–	–
R. rolandi	–	1	7	2	–	–
R. hybrida	–	1	6	3	–	–
R. endenensis	–	1	5	4	+	–
R. picteti	–	+	7	3	–	–
R. puytoraci	–	+	5	5	–	–
R. austriaca	–	–	5	5	–	–
R. puthzi	–	–	5	5	–	–
R. carpatoalpina	–	+	4	5	1	–
R. allobrogica	–	–	4	5	1	–
R. circumtatrica	–	–	2	5	3	–
R. podhalensis	–	–	2	4	4	–
R. zelinkai	–	–	4	4	2	–
R. beskidensis	–	+	+	2	5	3
R. landai	–	–	3	4	3	+
R. savoiensis	–	–	3	4	3	+
R. vaillanti	–	–	1	5	4	–
R. gratianopolitana	–	–	1	3	5	1
R. iridina	–	–	1	4	4	1
R. semicolorata	–	+	1	3	4	2
R. hercynia	–	–	+	4	4	2
R. germanica	–	–	–	2	5	3

fuliginosa, both species now threatened in Europe and restricted to a limited area in the east.

Special types of streams and rivers

In some cases low mayfly diversity is the result of extreme ecological conditions in the natural environment.

The alpine glacial brooks and streams

Such streams are typically inhabited by the highly specialized *Rhithrogena nivata*, almost invariably accompanied by *Baetis alpinus* and *Ecdyonurus picteti*. The joint occurrence of *Rh. loyolaea, Rh. alpestris, Rh. degrangei,* and *Epeorus alpicola* clearly indicates increasing pluvial influences. Most heptageniid species are absent in North European glacial streams. Human impacts such as minor eutrophication, artificial bank stabilization, or even channelization are not reflected by immediate and obvious changes in the mayfly community, as long as key environment features (discharge regime, temperature, and bed-load situation) remain unimpaired.

Acidic streams and highland brooks

The high vulnerability of mayflies to acidification (compared with Trichoptera and some genera of Plecoptera) is well documented and commonly used for bioindication (Braukmann, 1992; Engblom & Lingdell, 1987; Fjellheim & Raddum, 1990; Guerold et al., 1991, 1993). The naturally low pH-value of brooks draining *Sphagnum* bogs and fens is reflected by an extremely species-poor Ephemeroptera community consisting mainly of *Leptophlebia vespertina* (in slow current only) and potentially *Ameletus inopinatus* in higher altitudes. *Leptophlebia marginata, Baetis niger,* and *B. buceratus* may also be found when conditions are favourable, but their ecological amplitude remains poorly understood. The same applies to *Heptagenia fuscogrisea*, which is common in some limestone lakes and slow-flowing brooks in Ireland as well as in some slightly acidic brooks in the Bohemian Forest and northern Europe.

Rivulets and ditches
The inconspicuous and very small watercourses draining wetlands and traditionally farmed meadows have rarely been included in biomonitoring programs. Dramatic changes in agricultural land use threaten these often neglected habitats and their characteristic mayfly assemblage, consisting typically of *Habrophlebia fusca* and sometimes *Paraleptophlebia cincta* (known among fly fishermen under the characterizing name 'Ditch Fly'). *Baetis niger* (replaced by *B. muticus* in more alkaline water) is also abundant, accompanied by more ubiquitous taxa (*B. rhodani, B. vernus, Centroptilum luteolum, Cloeon simile, Caenis luctuosa* etc.). Eutrophication by agricultural sewage and channelization may significantly reduce diversity in favour of the tolerant ubiquitous taxa, whereas piping destroys the community altogether.

Intermittent streams
Lowland brooks and streams that dry up during summer are sometimes inhabited by mayfly species that are only rarely discovered elsewhere but are abundant where they do occur, e.g., *Metreletus balcanicus* and *Siphlonurus armatus*. Suitable habitats are extremely scarce in Central Europe and should be of concern for nature conservation. Under similar conditions, species of *Caenis*, mostly *C. robusta* and *C. luctuosa*, may also be found along with the ubiquitous *Cloeon dipterum*. All of them are very tolerant to eutrophication and relatively high water temperature and may quickly colonize slow-flowing or stagnant shallows that fall dry in autumn.

Stretches within karst streams that regularly fall dry during summer are recolonized by downstream drift and presumably from within the substrate; their vernal aspect in species usually does not differ significantly from their upper reaches, although only little systematic research has been conducted.

Large lowland rivers with sandy bottom
The river bed of large lowland rivers, consisting mainly of shifting sands, is extremely thinly inhabited by highly specialized mayfly species which are only rarely encountered by limnologists. Typical examples for eastern and southeastern Europe are *Behningia ulmeri, Baetopus wartensis, Brachycercus harrisellus, B. minutus, Oligoneuriella pallida*, and *O. borysthenica. Ametropus fragilis* and *Electrogena affinis* are found almost exclusively among dead wood or rotting logs partly embedded in sand.

Different habitats and sampling sites

Due to high substrate specificity the mayfly diversity at a given sampling station is easily underestimated if certain (meso)habitats are unintentionally neglected; seasonal changes in habitat preference further complicate matters. In most *Ecdyonurus* species, nymphs accumulate in the mesolithal along shallows (Macan, 1957b), where they find suitable conditions for emergence in water less than 5 cm deep. Similar movements occur in most *Leptophlebiidae* (independent of water depth), whereas in *Oligoneuriella rhenana* emergence usually takes place in midstream (Reisinger, 1995). Nymphs of many species of *Baetis* and *Rhithrogena* inhabit the riffle section of streams and rivers, but may also emerge further downstream in relatively strong currents. Habitats with the same substrate composition but different water velocity or water depth often harbour different species or different developmental stages. Incomplete sampling will therefore seriously influence the results, not only with regard to the number of species found, but also by affecting the potential level of identification and by increasing the amount of time necessary for identification. Careful selection of sampling sites (including all visually different (meso)habitat types) is therefore crucial for correctly assessing ecological integrity by means of Ephemeroptera diversity. The comparison between different stations within the same river reach may be misleading or impossible if a different set of habitats has been sampled or if the samples have not been taken at approximately the same time of year. For longitudinal studies, sampling for all stations should be completed within 1 week.

The interpretation of limnological routine sampling data is often hampered by the incomplete record of mayfly taxa due to inadequate sampling strategies, frequently combined with poor taxonomic standards or insufficient level of determination. When Danecker (1992) analysed a 16-year water quality survey based on 1017 samples from more than 50 sampling stations in southeastern Austria (Burgenland), single samples on average contained 10–20% of the total species number found, but some samples covered as little as 0.6% of the species later found at the same station. Clearly, the basic requirements for successfully monitoring mayfly diversity include a spatial and temporal sampling regime as well as suitable sampling equipment and techniques. Sampling strategies must be carefully adapted to the individual watercourses, as described for instance in the Austrian Standard ÖNORM M 6232 (1997). The final establishment of sampling sites and sampling locations requires at least one preliminary site survey, preferably during low water discharge in autumn.

Sampling equipment and sampling techniques

Although numerous sampling techniques and devices have been described and successfully used in the field (see review in Elliott & Tullett, 1978, 1983), only relatively few have proved to effectively collect representative mayfly larvae samples. The most versatile piece of equipment is a simple triangular handnet (ISO 7828) with a side length of approximately 20 cm that may be attached to a long handle for use in deeper water. A relatively short, shallow net bag (not drawn out acutely) about 20 cm long and made of stiff plastic or wire allows for quick and easy sorting out of live specimens, which is often preferable to fixing of the total net contents. The advantages of sorting live samples in the field include better direct control of collecting success, better preservation, less damaged specimens and a considerable saving of time. For monitoring purposes a mesh size between 0.5 and 1.0 mm is most suitable – young larvae usually cannot be identified with any reliability, and debris held back by a finer mesh unnecessarily complicates sorting. The same net should be used for kick-sampling (all substrates), sweeping through roots and submerged plants, scraping the moss-covered surface of boulders, and sieving rotten leafs. This method yields comparable semi-quantitative results when applied in a standardized manner, e.g., 10 'sweeps', 10 'kicks' (disturbing approx. 400 cm^2 of substrate) each for the different mesohabitats. Sorting must be done after every single kicking or sweeping action and therefore a fixed number of actions is preferable to given time standards for collecting. Especially in Ephemeroptera the equipment and techniques described above have many advantages over (supposedly) more quantitative sampling techniques (Pescador et al., 1997): the small net is fit for use even under restricted circumstances like very low water, along the broken edge of shingle banks, among submerged roots, or in close conditions and patchily distributed small habitats. The timesaving qualities allow a greater number of samples to be taken per unit time, increase efficiency and comparability, and lower costs. Sampling large rivers efficiently is always difficult (Petto et al., 1991) and no special recommendations can be given for Ephemeroptera. Standard methods are described in Clifford et al. (1989), Humpesch & Elliott (1990) and Ofenböck (1998). Under extreme conditions, acceptable results can be obtained by scuba-diving. Even in the most frequently investigated rivers, some substrates/habitats are far more difficult to sample than others. Leptophlebiidae, Siphlonuridae, Ephemeridae, and some Caenidae are often underrepresented or even missing in routine samples because they prefer habitats that are neglected or insufficiently/ineffectively treated. Big boulders as well as barriers of leaf litter, accumulated dead leaves, sand and mud are major obstacles for the collector, and a far greater number of samples must be taken to allow for these difficulties. When sampling among leaves, sand and mud, only very small quantities of substrate should be taken into the net, otherwise cryptic species are easily overlooked. Larvae of the potamobiont genus *Baetopus* inhabit the thin layer of silt or F–POM on boulders in the riprap section of channelized rivers, whereas *Ephemerella mesoleuca* may be found hidden in small crevices of the same habitat. Under more natural conditions both have been found among submerged vegetation and gravel in relatively swift current. Nymphs of *Oligoneuriella* and *Epeorus* may cling persistently to the rough surface of big boulders in deeper water and strong current, where they are very rarely obtained by mere 'kicking'. The same applies to the small larvae of the nearly extinct potamobiont *Prosopistoma foliaceum*. Shifting sands are extremely thinly inhabited. They require taking a great number of substrate samples over a longer period as well as the use of artificial substrates, drift nets, and light traps. Uniform stretches of sandy bottom in smaller streams may be effectively sampled using a shovel sampler (Jazdzewska, 1997).

The aim of ultimately obtaining a complete species inventory for a given locality is greatly enhanced if light traps are used parallel to the more conventional bottom sampling techniques. This applies especially to large rivers or localities otherwise difficult to sample. The additional use of light traps is highly recommended for biomonitoring projects and is considered to be state-of-the-art for Trichoptera, Ephemeroptera and Plecoptera (Reusch, 1995; Usseglio-Polatera, 1997).

This technique considerably reduces costs by reducing sampling frequency, yielding an increased range of taxa, and simplifying determinations.

Life cycles and minimum sampling frequency

Life history parameters are part of the framework underlying 'ecological plasticity' in mayflies (Brittain, 1991) and basically limit the ability to withstand both natural and man-made ecological disturbances. The different life cycles of Ephemeroptera (Landa, 1968; Sowa, 1975b) clearly influence both the completeness and comparability of samples. Furthermore, among the Heptageniidae (which comprise about 45% of the mayfly fauna of Central Europe) only fully grown nymphs can be reliably identified on the species level. Investigations that continue to neglect these considerations impair ecological and faunistic interpretations (Bauernfeind, 1998). The expense of field work makes the issue of minimum sampling frequency crucial for any biomonitoring program. As Reusch (1985, 1995) and others have pointed out, monthly samples over at least 1 year are the only means to obtain objective data on insect community structure in streams and rivers. For economical reasons, restrictions towards two to four samples a year, depending on altitude or insect groups, have been proposed (Finck, 1998; Finck et al., 1992; Holm, 1989; Peissner, 1992). In the case of Ephemeroptera the minimum sampling frequency for compiling an accurate species inventory at a given locality should be determined individually by considering the range of theoretically expected taxa based on altitude, river zonation, and discharge dynamics. A rough guideline for choosing sampling dates is provided in Bauernfeind (1994, 1995) and Marrer (1992) for most Central European species and information on zonal distribution is listed in Moog (1995). A longitudinal survey covering more than one river zone requires a higher sampling frequency than a more restricted local investigation. A minimum sampling scheme suitable for many rivers in Central Europe should at least consider the following sampling dates:

Krenon
planar–submontane (0–500 m): April and July
montane–alpine (500–2000 m): May and August

Rhithron
planar–submontane: March, May, July
montane–alpine: May and August

Potamon
March, May, July, October

Nevertheless, a higher sampling frequency is desirable whenever possible. In the planning phase, monitoring projects should consult local faunistic and taxonomic papers for pertinent information on life cycles. For qualitative aspects a 3-year study period and monthly samples are almost indispensable.

Level of identification

From the scientific point of view indicators must be accurately identified to species (De Pauw & Vanhooren, 1983; Furse et al.,1984; Hilsenhoff, 1987; Marten & Reusch, 1992; Moog et al., 1997; Resh & Unzicker, 1975; Rosenberg et al., 1986). Most biological monitoring techniques are dependent on correct identifications. Better taxonomy produces more accurate results and clearly increases the precision of site classifications, improving the ability to detect subtle changes in environmental quality (Lenat & Barbour, 1994). Although higher taxonomic units have occasionally been proposed either for rapid assessment of water quality (Alba-Tercedor & Sanchez-Ortega, 1988; Hilsenhoff, 1988) or to overcome difficulties of identification ('morpho-taxonomic groups'; Buffagni, 1997), the loss of biological information is a serious drawback, even if the method itself is adequate for the purpose stated. The ready availability of published data on ecological and faunistic research is an additional argument for identification on the species level.

Species identification in Ephemeroptera does present certain difficulties, but this is balanced by the relatively small number of taxa and comparatively good knowledge of both larval and imaginal stages.

Requirements for species level identification

Modern regional keys and revisions are available for most of Europe (see detailed review in Studemann et al., 1992; Bauernfeind, 1994; Engblom, 1996), whereas the taxonomic basis for identifying Ephemeroptera from the Balkans and south-central to southern Europe remains unsatisfactory. For Russia and adjacent countries, the identification of most species was recently summarized by Kluge (1997). Keys and descriptions published before 1980 contain valuable information for the specialist but should be used with caution by the less experienced: the evaluation of diagnostic characters has changed considerably and many species have since been newly described.

For North and Central America, regional keys on the species level were discussed by Edmunds et al., (1976) and the environmental requirements and pollution tolerance of Ephemeroptera have been dealt with in detail (Hubbard & Peters, 1978a). A more recent checklist for North America is now available (McCafferty, 1997), along with an online database described by McCafferty (1996). For Central and South America the taxonomic basis is still rather incomplete and keys exist only for small groups or geographically limited areas. Checklists including nominal species have already been prepared by Hubbard (1982) and McCafferty & Lugo-Ortiz (1996). A recent generic level key that also includes some species is now available for Argentina (Dominguez et al., 1994).

With the exception of Japan (Gose, 1979–1980) no modern keys on the species level are available for any major area in Asia, although checklists have been provided for India (Hubbard & Peters, 1978b), China (Gui Hong, 1985) and Korea (Bae et al., 1994).

Comprehensive information on mayfly species in Australia was fully summarized by Campbell (1988), and recently a key to genera for mayfly nymphs was provided by Dean & Suter (1997). Identification of Ephemeroptera remains rather difficult in Africa; since the annotated list by Demoulin (1970), no comprehensive or regional keys for more general use have been published. Revisional work, however, is progressing rapidly and a provisional checklist of South African Ephemeroptera has recently been published (McCafferty & de Moor, 1995).

High species diversity in some genera and close morphological similarities between certain species still impede the more general use of Ephemeroptera as indicators for ecological integrity, even in Central Europe. Nevertheless, experience shows that many identification problems actually arise from inadequate sampling strategies, inefficient collecting, or poor preservation. All of these can be easily avoided.

Larval identifications require much care because most key characters only apply to fully grown mature nymphs. No reliable keys are available at present for early instars, and misidentifications even at the genus or family level are possible.

Basic technical requirements for preservation and preparation are discussed in Edmunds et al. (1976) and Bauernfeind (1994). For documentation and future research the material collected should be preserved permanently to allow for comparison and re-evaluation. A sufficient number of undamaged voucher specimens should be housed in an institution where curatorial care is guaranteed.

Selected groups and identification pitfalls (Central Europe)

Heptageniidae probably represent the most difficult family and instruction by trained taxonomists is recommended. Especially in *Rhithrogena*, problems are numerous and benthos samples should be accompanied by light trap catches. The use of reference collections is often necessary and keys should be used with discretion. Egg chorionic structures provide valuable additional information for identifying mature female nymphs and most of the discriminating characters are already visible at a magnification of about 400×, provided a suitable mounting medium is used. In *Ecdyonurus*, mature nymphs and male imagines can usually be correctly identified (Bauernfeind, 1997; Hefti et al., 1989), but some training is essential. To acquire the necessary experience the use of reference specimens is advisable. *Electrogena* comprises only a few species in Central Europe, most of them easily recognisable in the nymphal and imaginal stage; the separation of *E. fascioculata* and *E. affinis* may prove more difficult (Malzacher, 1996*; Sowa, 1984). The genus *Heptagenia* should provide no difficulties and even half-grown larvae are readily identifiable with the aid of current keys.

Oligoneuriidae are represented by the single species *O. rhenana* in Central Europe, but the geographic range of closely allied taxa from eastern and southeastern Europe may show some overlap. For dubious specimens, consult Mol (1984). Discriminating characters are difficult to interpret and comparison with reference material is advisable.

In Siphlonuridae, only fully grown mature nymphs and imagines of both sexes are identifiable with certainty and slide preparation may be necessary. Distinct genital characters are discernible in male nymphs.

Most Baetidae can be identified in the larval stage, whereas male imagines require more experience. Although very time consuming, careful slide preparation is a prerequisite to become sufficiently acquainted with variations in the characteristic pronotum pattern in *Baetis*, which usually allows for quick separation of most of the material. Larvae of *Centroptilum* s.l.

* *Electrogena rivuscellana* Sartori & Landolt is currently considered a junior synonym of *E. ujhelyii* Sowa. See Belfiore & Desio (1995).

(usually included in *Procloeon* Bengtsson or *Pseudocentroptilum* Bogoescu, see McCafferty & Waltz, 1990) are poorly characterized morphologically and only incompletely dealt with in most keys; the descriptions in Keffermüller & Sowa (1984) should be consulted. Egg chorionic structures are usually distinctive. Fully grown nymphs and imagines of Leptophlebiidae are readily identifiable to species, but the interpretation of gills and abdominal colour pattern requires some experience and well-preserved, undamaged specimens. The shape of spines on the surface and border of femora provide valuable diagnostic characters, but correct interpretation is dependent on adequate slide preparation and reference material is desirable for the less experienced.

Identification of most Ephemerellidae species poses no difficulties; the gill arrangement and the shape of gill II are characteristic features. Note, however, that the well-known ventral abdominal markings in *E. notata* are not exclusively distinctive and variation in colour patterns is generally extremely high in this family.

Ephemeridae include only five species in Central Europe, but differentiating morphological characters are poorly developed; also, abdominal markings in nymphs are less distinctive and more blurred in nymphs than in most imagines. Larvae of the rare potamobiont *Ephemera lineata*, for example, may easily be confused with the common rhithrobiont *E. danica*.

Although most Caenidae can be readily identified in the larval, subimaginal, and imaginal stage, some experience and microscopical preparations are necessary to avoid misidentifications. Morphological characters for separating nymphs of the more rhithrobiont *Caenis beskidensis* from the potamobiont *C. pseudorivulorum* are sometimes misleading and identification of single specimens may remain doubtful.

The remaining families Isonychiidae, Potamanthidae, Polymitarcyidae, Palingeniidae and Prosopistomatidae are each represented only by a single species in Central Europe and identification of nymphs as well as of half–grown larvae should cause no special problems with current keys. Their status, however, is precarious in Europe, and any records would be of especial interest. Forthcoming specimens should at any rate be permanently preserved.

Summary

Successful assessment of ecological integrity by means of Ephemeroptera diversity data, obtained by biomonitoring or more specialized studies, depends on a careful evaluation of the basic criteria influencing the spatial and temporal distribution as well as the abundance of mayfly larvae.

A thorough study design should include the following aspects:

(1) Evaluation of the potential species inventory of a given river section with regard to zoogeographic distribution, longitudinal zonation, and typological aspects which affect the regional patterns of species assemblages. Major river types and their characteristic mayfly communities are discussed for Central Europe, concentrating on longitudinal and altitudinal typologies.

(2) The distribution and abundance of mayfly communities is strongly dependent on habitat composition and mesohabitat structure. Compiling a nearly complete species list therefore requires sampling all habitat types available.

(3) Seasonal variations are reflected not only in different abundance and distribution but may also influence identification considerably. Problems may arise especially when comparing samples with a theoretically derived target list or with samples taken at different seasons or stations. Guidelines for a suitable temporal sampling scheme are discussed.

(4) Successful sampling depends on suitable equipment and sampling techniques, adapted to allow for the patchy distribution of mayfly nymphs as well as for the different structure in the various habitats. A simple handnet is cost effective and can be used with maximum versatility in medium-sized streams and rivers.

(5) The correct use of ephemeropteran assemblages to assess ecological integrity must be based on identification on the species level. The taxonomic situation world-wide and possible identification pitfalls in Central Europe are discussed in brief.

Acknowledgements

We want to express our sincere gratitude to W. Reisinger (Weißenberg) and F. Stich (Ferlach) for providing material and photographs and Dr Michael

Stachowitsch for linguistic help. K. Foxall (Torquay), Mag. Dr M. D. Larsen (Baden), Dr H. Reusch and Dr. P. Weichselbaumer gave valuable comments on the draft.

References

Alba-Tercedor, J. & A. Sanchez-Ortega, 1988. Un método rápido y simple para evaluar la calidad biológica de las aguas corrientes basado en el de HELLAWELL, 1978. Limnetica 4: 51–56.

Bae, Y. J., I. B. Yoon & D. J. Chun, 1994. A Catalogue of the Ephemeroptera of Korea. Entomol. res. Bull. 20: 31–50.

Bauernfeind, E., 1994. Bestimmungsschlüssel für die österreichischen Eintagsfliegen (Insecta: Ephemeroptera), 1. Teil. Wasser Abwasser Suppl. 4: 1–92.

Bauernfeind, E., 1995. Bestimmungsschlüssel für die österreichischen Eintagsfliegen (Insecta: Ephemeroptera), 2. Teil. Wasser Abwasser Suppl. 4: 1–96.

Bauernfeind, E., 1997. Discriminating characters in Central European species of *Ecdyonurus* EATON. In Landolt P. & M. Sartori (eds), Ephemeroptera & Plecoptera: Biology–Ecology–Systematics (= Proc. 8th Int. Conf. Ephemeroptera, Lausanne, 1995), Fribourg: 418–426.

Bauernfeind, E., 1998. Ergänzungen und Berichtigungen zu den Eintagsfliegen des Lavanttales (Insecta: Ephemeroptera). Linzer Biol. Beitr. 30 (1): 293–298.

Bauernfeind, E., P. Weichselbaumer & O. Moog, 1995. Ephemeroptera (Eintagsfliegen). In Moog O. (Ed.), Fauna Aquatica Austriaca, Lief. Mai 1995. Wasserwirtschaftskataster, Bundesministerium f. Forstwirtsch., Wien: 17 pp.

Belfiore, C. & F. Desio, 1995. Taxonomy and distribution of *Electrogena ujhelyii* (Sowa, 1991) (Insecta: Ephemeroptera: Heptageniidae). Ann. Naturhistor. Mus. Wien 97 [B]: 151–154.

Braukmann, U., 1992. Biological indication of stream acidity in Baden-Württemberg. In Böhmer, J. & H. Rahmann (eds), Bioindikationsverfahren zur Gewässerversauerung, Literaturstudie zur Erarbeitung von Bioindikationsverfahren zur Gewässerversauerung. Landesanstalt für Umweltschutz Baden-Württemberg, Karlsruhe: 58–71.

Breitenmoser-Würsten, Ch. & M. Sartori, 1995. Distribution, diversity, life cycle and growth of a mayfly community in a prealpine stream (Insecta, Ephemeroptera). Hydrobiologia 308: 85–101.

Brittain, J., 1991. Life history characteristics as a determinant of the response of mayflies and stoneflies to man-made environmental disturbance (Ephemeroptera and Plecoptera). Proc. VIth Int. Conf. Ephemeroptera & Plecoptera, Granada: 539–545.

Buffagni, A., 1997. Mayfly community composition and the biological quality of streams. In Landolt P. & M. Sartori (eds), Ephemeroptera & Plecoptera: Biology–Ecology–Systematics (= Proc. 8th Int. Conf. Ephemeroptera Lausanne, 1995). Fribourg: 235–246.

Campbell, I., 1988. Ephemeroptera. In Walton, D. W. & W. W. K. Houston (eds), Zoological Catalogue of Australia 6. Bureau Flora Fauna, Victoria: 1–22.

Clifford, H. F., V. Gotceitas & R. J. Casey, 1989. Roughness and color of artificial substratum particles as possible factors in colonization of stream invertebrates. Hydrobiologia 175: 89–95.

Cogerino, L., B. Cellot & M. Bournaud, 1995. Microhabitat diversity and associated macroinvertebrates in aquatic banks of a larger European river. Hydrobiologia 304: 103–115.

Cummins, K. W. & M. J. Klug, 1979. Feeding ecology of stream invertebrates. Ann. rev. ecol. Syst. 10: 147–172.

Danecker, E., 1992. Makrozoobenthos und Gewässergüte burgenländischer Flüsse 1974–1990. Wiss. Arb. Burgenland 91, Burgenl. Landesmus. Eisenstadt, 1–88.

Dean, J. C. & P. J. Suter, 1997. Mayfly Nymphs of Australia: A Guide to Genera. Coop. Res. Centre Freshwat. Ecol., Albury, Victoria.

Demoulin, G., 1970. Ephemeroptera des fauna Éthiopienne et Malgache. South African Anim. Life 14: 24–170.

De Pauw, N. & G. Vanhooren, 1983. Method for biological quality assessment of water courses in Belgium. Hydrobiologia 100: 153–168.

Devan, P. & L. Mucina, 1986. Structure, zonation and species diversity of the mayfly communities of the Bela River, Slovakia. Hydrobiology 135: 155–165.

Dominguez, E., M. D. Hubbard & M. L. Pescador, 1994. Los Ephemeroptera en Argentina. Fauna Agua Dulce República Argentina 33 (1): 1–120.

Edmunds, G. F. Jr., S. L. Jensen & L. Berner, 1976. The Mayflies of North and Central America. Univ. Minnesota Press, Minneapolis: 1–330.

Elliott, J. M. & P. A. Tullett, 1978. A Bibliography of Samplers for Benthic Invertebrates. Freshwat. Biol. Ass., Occas. Publ. 4: 1–61.

Elliott, J.M. & P.A. Tullett, 1983. A Supplement to a Bibliography of Samplers for Benthic Invertebrates. Freshwat. Biol. Ass., Occas. Publ. 20: 1–26.

Engblom, E., 1996. Ephemeroptera, Mayflies. In Nilsson, A.N. (ed.), Aquatic Insects of North Europe – A Taxonomic Handbook, Vol. 1. Apollo Books, Stenstrup: 13–53.

Engblom, E. & P.-E. Lingdell, 1987. Vilked skydd har de vattenlevande smadjuren i landets naturskyddomraden? En studie av förurnings- och föroreningsförhallanden. Naturvardsverket Rapport N 3349, Solna: 1–274.

Errington, P. L., 1945. Some contributions of a fifteen-year local study of the northern Bobwhite to a knowledge of population phenomena. Ecol. Monogr. 15: 1–34.

Finck, P., 1998. Der Einfluß von Probenahmenzeitpunkt und -häufigkeit auf die Erfassung der Makroinvertebraten in Mittelgebirgsbächen. Lauterbornia 34: 245–254.

Finck, P., D. Hammer, M. Klein, A. Kohl, U. Riecken, E. Schröder, A. Ssymank & W. Völkl, 1992. Empfehlungen für faunistisch-ökologische Datenerhebungen und ihre naturschutzfachliche Auswertung im Rahmen von Pflege- und Entwicklungsplänen für Naturschutzgroßprojekte des Bundes. Natur Landschaft 67: 329–340.

Fjellheim, A. & G. G. Raddum, 1990. Acid precipitation: biological monitoring of streams and lakes. Sci. Tot. Envir. 96: 57–66.

Furse, M. T., D. Moss, J. F. Wright & P. D. Armitage, 1984. The influence of seasonal and taxonomic facts on the ordination and classification of running water-sites in Great Britain and on the prediction of their macro-invertebrate communities. Freshwat. Biol. 14: 257–280.

Gose, K., 1979–1980. The Mayflies of Japan. Aquabiol. 1–11.

Guerold, N., V. Vein & G. Jaquemin, 1991. Les peuplements d'éphémeroptères de plécoptères et de trichoptères des ruisseaux acides et non acides du massiv vosgien: premièr approche. Rev. Sci. L'eau 4: 299–314.

Guerold, N., Vein V., Jaquemin G. & J.-C. Moreteau, 1993. Impact de l'acidification des russieux vosgiens sur la biodiversité de la macrofaune benthique. C.R. Acad. Sci. Paris, Sci. Vie 316: 1388–1392.

Gui Hong, 1985. A Catalogue of the Ephemeroptera of China (in Chin.). J. Nanjing Teacher's Univ. 4: 79–97.

Haybach, A. & T. Fischer, 1997. Ein Beitrag zur Kenntnis der Heptageniidae-Fauna im nördlichen Hessen mit einem Nachweis von *Rhithrogena germanica* Eaton (Ephemeroptera: Heptageniidae). Lauterbornia 31: 41–48.

Hefti, D., I. Tomka & A. Zurwerra, 1989. Revision of morphological and biochemical characters of the European species of the *Ecdyonurus helveticus*-group (Ephemeroptera, Heptageniidae). Bull. Soc. Entomol. Suisse 62: 329–344.

Hilsenhoff, W. L., 1987. An improved biotic index of organic stream pollution. Great Lakes Entomol. 20 (1): 31–39.

Hilsenhoff, W. L., 1988. Rapid field assessment of organic pollution with a family-level biotic index. J. N. Am. Benthol. Soc. 7: 65–68.

Holm, A., 1989. Ökologischer Bewertungsrahmen Fließgewässer (Bäche). Landesamt Natursch. Landschaftspflege Schleswig-Holstein, Kiel, 1–46 & Anhang.

Hubbard, M. D., 1982. Catálogo abbreviado de Ephemeroptera da América do Sul. Papéis Avulsos Zool., S. Paulo 34 (14): 257–28F2.

Hubbard, M. D. & W. L. Peters, 1978a. Environmental Requirements and Pollution Tolerance of Ephemeroptera. USEPA Report EPA–600/4–78–061, Cincinnati, VI+461 pp.

Hubbard, M. D. & W. L. Peters, 1978b. A catalogue of the Ephemeroptera of the Indian subregion. Oriental Insects (Suppl.) 9: 1–43.

Humpesch U. H. & J. M. Elliott (eds), 1990. Methods of biological sampling in a large deep river – the Danube in Austria. Was. Abwass. Suppl. 2: 1–83.

Illies, J. (ed.), 1978. Limnofauna Europaea. 2nd edn. Fischer, Stuttgart: 1–532.

Illies, J. & L. Botosaneanu, 1963. Problèmes et méthodes de la classification et de la zonation écologique des eaux courantes, considerées surtout du point de vue faunistique. Mitt. int. Verein. Theor. Angew. Limnol. 12: 1–57.

ISO 7828, 1985. Water quality – Methods of biological sampling – Guidance on handnet sampling of aquatic benthic macroinvertebrates.

Jazdzewska, T., 1997. Mayflies (Ephemeroptera) of the sandy bottom of the River Grabia (Central Poland). In Landolt, P. & M. Sartori (eds), Ephemeroptera & Plecoptera. Biology–Ecology–Systematics (= Proc. 8th Int. Conf. Ephemeroptera, Lausanne, 1995). Fribourg: 157–166.

Karr, J. R., 1991. Biological integrity: a long-neglected aspect of water resource management. Ecol. Appl. 1: 66–84.

Keffermüller, M. & R. Sowa, 1984. Survey of Central European species of the genera *Centroptilum* Eaton and *Pseudocentroptilum* Bogoescu (Ephemeroptera, Baetidae). Polsk. Pis. Entomol. 54: 309–340.

Kendeigh, S. C., 1934. The role of environment in the life of birds. Ecol. Monogr. 4: 299–417.

Kluge, N. Y., 1997. Ephemeroptera. In Key to Freshwater Invertebrates of Russia and Adjacent Lands (in Russ.) 3. Zool. Inst. Acad. Sci., St. Petersburg: 176–220.

Kolkwitz, R. & M. Marsson, 1902. Grundsätzliches für die biologische Beurteilung des Wassers nach seiner Flora und Fauna. Mitt. K. Prüfanst. Wasserversorg. Abwasserbes. Berlin-Dahlem 1: 33–72.

Kothé, P., 1962. Der 'Artenfehlbetrag' ein einfaches Gütekriterium und seine Anwendung bei biologischen Vorfluteruntersuchungen. Dt. Gewässerkundl. Mitt. 6: 60–65.

Krno, I., 1987. Classification of streams of the Upper Váh River Basin (West Carpathians). Acta Fac. Rer. Nat. Univ. Comenianae Zool. 29: 33–51.

Krno, I., 1990. Longitudinal Changes in the Structure of Macrozoobenthos and its Microdistribution in Natural and Moderately Eutrophicated Waters of the River Rajcianka (Strázovské vrchy). Acta Fac. Rer. Nat. Univ. Comen. Zoologia 33: 31–48.

Krno, I., 1991. Podenky (Ephemeroptera) a posvatky (Plecoptera) dunaja (Palkovicovo-Chlaba) a prilahlych vodnych biotopov. Biologia (Bratislava) 46 (2): 139–145.

Kühnelt, W., 1943. Die Leitformenmethode in der Ökologie der Landtiere. Biol. Gen. 17: 106–146.

Landa, V., 1968. Developmental cycles of Central European Ephemeroptera and their interrelations. Acta Entomol. Bohemoslov. 65: 276–284.

Landa, V. & T. Soldán, 1991. The possibility of mayfly faunistics to indicate environmental changes of large areas. In: Alba-Tercedor, J. & A. Sanchez-Ortega (eds), Overview and Strategies of Ephemeroptera and Plecoptera (=Proc. VIth Int. Conf. Ephemeroptera, Granada, 1989). Sandhill Crane Press, Gainesville, Fl: 559–565.

Landa, V., S. Zahrádková, T. Soldán & J. Helesic, 1997. The Morava and Elbe river basins, Czech Republik: A comparison of long term changes in mayfly (Ephemeroptera) biodversity. In Landolt, P. & M. Sartori (eds), Ephemeroptera & Plecoptera: Biology–Ecology–Systematics (= Proc. 8th Int. Conf. Ephemeroptera, Lausanne, 1995). Fribourg: 219–226.

Lepneva, S. G., 1949. K voprosu ob ekologitcheskoi klass fikatzij ruceinikov tekutchih vod. Entomol. Obozr. 30: 253–265.

Lenat, D. R. & M. T. Barbour, 1994. Using benthic macroinvertebrate community structure for rapid, cost-effective, water quality monitoring: rapid bioassessment. In Loeb, S. L. & A. Spacie (eds), Biological Monitoring of Aquatic Systems. Lewis Publishers, Boca Raton FL: 187–215.

Macan, T. T., 1957a. The Ephemeroptera of a stony stream. J. anim Ecol. 26: 317–342.

Macan, T. T., 1957b. The life-histories and migrations of the Ephemeroptera of a stony stream. Trans. Soc. Br. Entomol. 12: 129–156.

Malzacher, P., 1996. Genitalmorphologische Merkmale zur Unterscheidung der in Baden-Württemberg vorkommenden Electrogena-Arten (Heptageniidae, Ephemeroptera). Lauterbornia 25: 81–93.

Marrer, H., 1992. Gewässerschutz im Einzugsgebiet des Vierwaldstättersees. Die Invertebratenfauna der Fließgewässer. EAWAG, Solothurn, No. 4752, 1–370 (+ appendices).

Marten, M. & H. Reusch, 1992. Anmerkungen zum DIN 'Saprobienindex' (38 410 Teil 2) und Forderung alternativer Verfahren. Natur Landschaft 67 (11): 544–547.

McCafferty, W. P., 1996. The Mayflies of North America online. Entomol. News 107: 61–63.

McCafferty, W. P., 1997. Ephemeroptera. In Poole, R. W. & P. Gentili (eds), Nomina Insecta Nearctica: A Checklist of Insects of North America 4. Entomol. Inf. Serv., Rockville, MD: 89–117.

McCafferty, W. P. & F. C. de Moor, 1995. South African Ephemeroptera: Problems and priorities. In Corkum, L. D. & J. H. Ciborowski (eds), Current Directions in Research on Ephemeroptera (= Proc. 7th Int. Conf. Ephemeroptera, Orono, 1992). Canadian Scholar's Press, Toronto: 463–476.

McCafferty, W. P. & C. R. Lugo-Ortiz, 1996. Los Efemerópteros (Ephemeroptera) de América Central. Rev. Nicaragua Entomol. 35: 19–28.

McCafferty, W. P. & R. D. Waltz, 1990. Revisionary synopsis of the Baetidae (Ephemeroptera) of North and Middle America. Trans. am. entomol. Soc. 116 (4): 769–799.

Mol, A. W. M., 1984. *Oligoneuriella polonica* n.sp., and a note on *Oligoneuriella pallida* (Hagen, 1833) (Ephemeroptera: Oligoneuriidae). Entomol. Ber. Amsterdam 44 (8): 126–128.

Moog, O., 1993. Makrozoobenthos als Indikator bei ökologischen Fragestellungen. In: 14 Seminar Landschaftswasserbau an der Technischen Universität Wien: 102–139.

Moog, O. (ed.), 1995. Fauna Aquatica Austriaca. Katalog zur autökologischen Einstufung aquatischer Organismen Österreichs. Lieferung Mai 1995, Wasserwirtschaftskataster, Bundesministerium f. Land- und Forstwirtschaft. Wien.

Moog, O. & R. Wimmer, 1994. Comments to the water temperature based assessment of biocoenotic regions according to Illies and Botosaneanu. Verh. int. Verein. Limnol. 25: 1667–1673.

Moog, O., E. Bauernfeind & P. Weichselbaumer, 1997. The use of Ephemeroptera as saprobic indicators in Austria. In: Landolt, P. & M. Sartori (eds), Ephemeroptera & Plecoptera: Biology–Ecology–Systematics (= Proc. 8th Int. Conf. Ephemeroptera, Lausanne, 1995). Fribourg: 254–260.

Önorm, M., 1997. Guidelines for the ecological study and assessment of rivers. Austrian Standards M 6232: 1–84.

Ofenböck, G., 1998. Entwicklung einer praxisorientierten Methode zur (semi)quantitativen Benthosentnahme an großen Flüssen. Diploma Thesis, BOKU, Vienna: 1–169.

Peissner, T., 1992. Erfassung und Eignung des Makrozoobenthos für die Gütebestimmung und Beurteilung von Gewässern. In: Trautner, J. (ed.), Arten- und Biotopschutz in der Planung – Methodische Standards von Tierartengruppen. BDLV-Tagung Bad Wurznach 1991 (= Ökologie in Forschung und Anwendung, Bd. 5). J. Margraf Verl., Weikersheim: 75–96.

Pescador, M. L., M. D. Hubbard, A.K. Rasmussen, J. Jones & S. Leong, 1997. Influence of sampling methods on the determination of Mayfly (Ephemeroptera) species richness in a coastal plain stream. In: Landolt, P. & M. Sartori (eds), Ephemeroptera & Plecoptera: Biology–Ecology–Systematics (= Proc. 8th Int. Conf. Ephemeroptera, Lausanne, 1995). Fribourg: 395–401.

Petto, H., U. H. Humpesch & P. H. Anderwald, 1991. Güte des Wassers der Donau im Bereich der Staustufe Altenwörth (Stromkilometer 1980–2007). 1.Teil: Ist-Zustand im Stauwurzelbereich in den Jahren 1986 und 1997. Österr. Wasserwirtsch. 43: 17–23.

Plafkin, J. L., M. T. Barbour, K. D. Porter, S. K. Gross & R. M. Hughes, 1989. Rapid bioassessment protocols for use in streams and rivers. Benthic macroinvertebrates and fish. EPA/444/4–89/001. Office of Water Regulation and Standards: US EPA, Washington, DC.

Reisinger, W., 1995. Die Rheinmücke. Der Fliegenfischer. Verl. J. Schück, Nürnberg: 118: 32–36, 119: 34–38.

Resh, V. H. & J. K. Jackson, 1993. Rapid assessment approaches to biomonitoring using benthic macroinvertebrates. In: Rosenberg, D. M. & V. H. Resh (eds), Freshwater Biomonitoring and Benthic Macroinvertebrates. Chapman and Hall, New York: 159–194.

Resh, V. H. & J. D. Unzicker, 1975. Water quality monitoring and aquatic organisms: the importance of species identification. J. Wat. Pollut. Cont. Fed. 47 (1): 9–19.

Reusch, H., 1985. Limnofaunistische Untersuchungen über die Eintags-, Stein- und Köcherfliegen (Ephemeroptera, Plecoptera, Trichoptera) der Örtze (Niedersächsisches Tiefland). Jb. Naturw. Verein Fstm. Lüneburg 37: 117–139.

Reusch, H., 1995. Planungsrelevante Aufbereitung und Bewertung faunistisch-ökologischer Daten vom Makrozoobenthon in Fließgewässern. Sch.-R. Landschaftspfl. Natursch. 43: 31–42.

Rosenberg, D. M., H. V. Danks & D. M. Lehmkuhl, 1986. Importance of insects in environmental impact assessment. Envir. Manage. 10: 773–783.

Sartori, M. & P. Landolt, 1998. SEG & CSCF (ed.), Atlas de distribution des Ephemeres de Suisse (Insecta: Ephemeroptera). Fauna Helvetica 2: Neuchâtel, 1–220.

Sartori, M., P. Landolt, V. Lubini & L. Ruffieux, 1995. Biological studies of *Palingenia longicauda* (Olivier) (Ephemeroptera: Palingeniidae) in one of its last European refuges – Abiotic characteristics and description of the habitat. In Corkum, L. D. & J. H. Ciborowski (eds), Current Directions in Research on Ephemeroptera (= Proc. 7th Int. Conf. Ephemeroptera, Orono, 1992). Canadian Scholar's Press, Toronto: 263–272.

Sowa, R., 1975a. Ecology and biogeography of mayflies (Ephemeroptera) of running waters in the polish part of the carpathians. 1. Distribution and quantitative analysis. Acta hydrobiol. 17: 223–297.

Sowa, R., 1975b. Ecology and biogeography of mayflies (Ephemeroptera) of running waters in the polish part of the carpathians. 2. Life cycles. Acta hydrobiol. 17: 319–359.

Sowa, R., 1984. *Ecdyonurus fascioculatus* sp.n., espèce voisine d'*E. affinis* Eaton du midi de la Pologne (Ephemeroptera, Heptageniidae). Bull. Acad. Pol. Sci., Sér. Sci. Biol., II 22: 315–323.

Studemann, D., P. Landolt, M. Sartori, D. Hefti & I. Tomka, 1992. Soc. Entomol Suisse (ed.), Ephemeroptera. Insecta helvetica, Fauna 9. Fribourg: 1–174.

Townsend, C. R., 1996. Concepts in river ecology: pattern and process in the catchment hierarchy. Arch. hydrobiol. Suppl. 113, Large Rivers 10: 1–4, Stuttgart: 3–21.

Usseglio-Polatera, P., 1997. Long-term changes in the Ephemeroptera of the River Rhone at Lyon, France, assessed using a fuzzy coding aproach. In: Landolt, P. & M. Sartori (eds), Ephemeroptera & Plecoptera: Biology–Ecology–Systematics (= Proc. 8th Int. Conf. Ephemeroptera, Lausanne, 1995). Fribourg: 227–234.

Utschick, H., 1976. Die Wasservögel als Indikatoren für den ökologischen Zustand von Seen. Verh. Orn. Ges. Bayern 22: 395–438.

Vannote, R. V., G. W. Minshall, K. W. Cummins, J. R. Sedell & C.E. Cushing, 1980. The River Continuum Concept. Can. J. Fish. aquat. Sci. 37: 130–137.

Weigand, E., 1998. Limnologisch-faunistische Charakterisierung von Karstquellen, Quellbächen und unterirdischen Gewässern nach Choriotopen und biozönotischen Gewässerregionen (Nationalpark o.ö. Kalkalpen, Österreich). Nationalpark o.ö. Kalkalpen, unpubl. Report AZ: 1603–7.6/1996–1997. Obergrünburg, Wien: 1–173.

Weigand, E. & K. Tockner, 1996. Limnologische Charakterisierung ausgewählter Karstquellen im Nationalparkgebiet o.ö. Kalkalpen, unpubl. Report 1603–7.6/95. Nationalpark Kalkalpen, Wien: 1–105.

Zollhöfer, J., 1997. Quellen, die unbekannten Biotope im Schweizer Jura und Mittelland. Bristol-Schriftenreihe 6, Zürich: 1–153.

Zwick, P., 1992. Stream habitat fragmentation – a threat to biodiversity. Biodiversity Conservation 1: 80–97.

Hydrobiologia 422/423: 85–97, 2000.
M. Jungwirth, S. Muhar & S. Schmutz (eds), Assessing the Ecological Integrity of Running Waters.

Fundamentals of fish ecological integrity and their relation to the extended serial discontinuity concept

M. Jungwirth, S. Muhar & S. Schmutz
Dept. of Hydrobiology, Fisheries and Aquaculture, University of Agricultural Sciences, Vienna, Max-Emanuel Str. 17, A-1180 Vienna, Austria

Key words: ecological integrity, running waters, fish ecology, extended serial discontinuity concept, connectivity

Abstract

Current ecological theories and concepts describe running waters as four-dimensional systems, their longitudinal, lateral and vertical linkages, interactions and exchange processes varying over time and over different scales. According to Ward & Stanford's (1983) extended serial discontinuity concept (ESDC), along a three-reach river, model the relative strength of the longitudinal pathways is highest in the constrained headwaters, vertical interactions reach their maximum in the braided middle course and lateral connectivity plays the major role in alluvial floodplain rivers. The present paper examines the general tenets of the ESDC from a fundamental fish ecology perspective. Specifically, it focuses on the degree to which the spatial/temporal connectivity requirements of fish communities or key fish species along a schematic longitudinal river course are compatible with the underlying principles of the ESDC and whether these requirements provide basic criteria for assessing the ecological integrity of running waters. From the fish ecological perspective, the examples provided here demonstrate a principle agreement with the concept of the four-dimensional nature of running waters, whose key functional and structural elements are spatial/temporal fluvial dynamics, disturbances, connectivity, succession and ecotones. Fish are particularly useful indicators of the temporally variable connectivities on the full range of scales at the three spatial dimensions. As far as the habitat requirements of typical species/developmental stages are concerned, two major modifications of the ESDC are needed. In constrained headwaters, in addition to longitudinal connectivity, the crucial importance of the vertical pathway river/bed sediments for reproduction must be emphasized. Also, in addition to lateral connectivity, longitudinal connectivity remains a vital basis for potamal fish communities in alluvial braided and meandering zones.

Introduction

The highly complex functional and structural elements of running waters are largely based on the system-inherent, dynamic genesis and development of those systems. Because of the very intricate interrelations with the surrounding environments, running waters are among the most distinctive landscape elements. Especially in alluvial river floodplain systems, the high spatio-temporal heterogeneity and, therefore, the great diversity of complex habitats and ecotones in successional stages, represent key features. These diverse environments support species-rich fish communities that contribute to the overall high biodiversity of river ecosystems (Schiemer & Waidbacher, 1992, 1994; Arthington & Welcomme, 1995; Ward & Stanford, 1995a, b; Tockner et al., in press; Ward et al., in press).

Theoretical constructs, such as the river continuum concept (Vannote et al., 1980) and the serial discontinuity concept (Ward & Stanford, 1983), focused much too strongly on longitudinal functions and processes. In fact, a number of well-known concepts of running water ecology demonstrate that rivers and streams are highly complex, four-dimensional systems whose dynamics during flood events involve intensive exchange processes with their surroundings that are vital for the wide range of communities found along their course (Jungwirth, 1998). Seminal contributions include the flood pulse concept (Junk et al., 1989), the ecotone concept (Naiman & Decamps, 1990) and the concept of ecological connectivity (Amoros &

Roux, 1988). According to the conceptual framework of these authors and especially to the extended serial discontinuity concept (ESDC, Ward & Stanford, 1995a), river systems are interactive along three spatial dimensions: the longitudinal (river/river or river/tributaries), vertical (river bed/aquifer) and lateral (river bed/floodplains) dimensions. The fourth dimension – the temporal scale – is also critical; this encompasses the whole range of temporal variability from the short-term processes related to annual hydrological cycles to the long-term development of river systems.

River systems are heavily degraded by various human uses and impacts on a global level (Benke, 1990; Dynesius & Nilsson, 1994; Arthington & Welcomme, 1995; Raven et al., 1998; Muhar et al., 2000). It is, therefore, highly relevant for both practical river conservation and river management in general to determine which basic processes, functions and structures make up the ecological integrity of running waters (Karr & Dudley, 1981; Boon, 2000). In this context, fish assume a major indicator function. Based on their high vagility, on their complex species- and stage-specific movement patterns, as well as on their distinct habitat requirements, fish prove to be sensitive indicators especially for assessing the highly variable connectivity conditions of running waters over space and time (Jungwirth, 1998; Northcote, 1998). The protection and conservation, as well as restoration and management of intact running water environments, require exact knowledge and consideration of the autecological demands of the most important species. A precondition for successful management is that the different connectivity requirements of the various ecological guilds, species and developmental stages from headwaters to the sea are recognized and factored into measures taken.

The ESDC approach to connectivity and interactive pathways

Today, the highly complex four-dimensional nature of running water systems is a generally recognized, fundamental principle (Ward, 1989). Spatial/temporal fluvial dynamics and disturbances are considered to be the key prerequisites for the development of type-specific sets of functional and structural elements, which are largely defined by connectivity, succession and ecotones (Stanford & Ward, 1992; Ward & Stanford, 1995a,b).

The strength of the interactive pathways – *sensu* complex linkages, interactions and exchange processes – in the three spatial dimensions represented in the ESDC point to characteristic changes over the longitudinal course (Figure 1): 1. In constrained headwaters, longitudinal processes clearly dominate. Vertical exchange processes between river and river bed sediments are of limited extent. The lateral pathway is comparatively insignificant. 2. In the braided middle reach, vertical exchange processes as well as lateral connectivity are stronger, whereas the strength of interactions along the longitudinal pathway is weaker than in the other reaches. 3. In the meandering river course of alluvial river/floodplain systems, the ESDC specifies that the vertical pathway is weaker than longitudinal and especially lateral connectivity, which becomes the central ecological element, both functionally and structurally: here, a multitude of dynamic interactive pathways link the river channel and the diverse aquatic and riparian communities of the floodplain.

Against this backdrop, we examine the fundamental aspects, demands and connectivity requirements of healthy fish communities in temperate running waters of Western Europe, from forested headwaters of the trout region over the braided middle reaches of the grayling and barbel region down to the alluvial floodplain systems inhabited by potamal species. This is followed by a synoptic discussion of the degree to which the three spatial dimensions over time – as outlined in the ESDC – are valid from a fish ecological point of view and can serve as a general model for assessing ecological integrity.

Traits and requirements of fish communities along temperate rivers

Constrained headwaters

Alpine headwaters are summer-cold and frequently (due to groundwater sources) winter-warm lotic environments. Usually they are characterized by steep gradients, single thread constrained channels with coarse bed material, high structural diversity, diverse patterns of currents and substrates, accumulations of woody debris and narrow riparian corridors. Erosive processes predominate and the steep banks and adjoining terrestrial areas leave little or no room for floodplains. This, combined with often strong shading, means that – based on the river continuum concept –

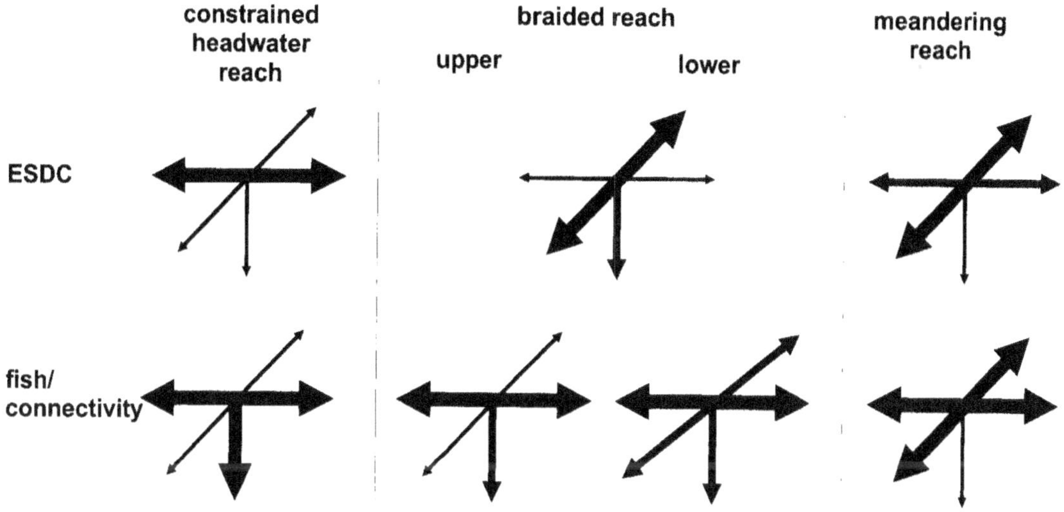

Figure 1. Interactive pathways along a three-reach model river system. Arrows indicate interactions along longitudinal (horizontal arrows), lateral (oblique arrows) and vertical (vertical arrows) dimensions. Thickness of the arrows indicates the relative strength of interactions according to the ESDC (top) and their importance from the fish ecological point of view (bottom).

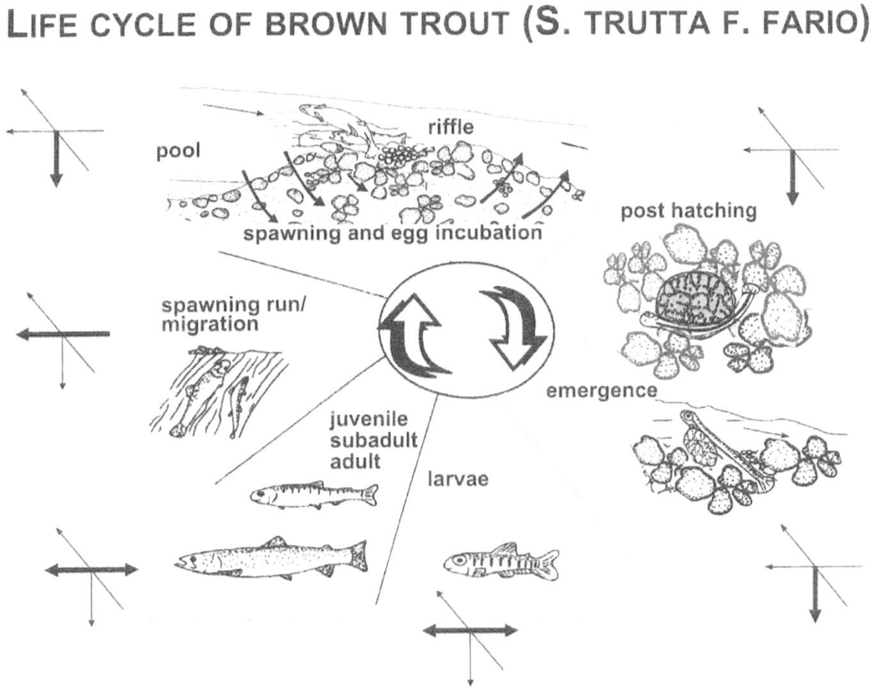

Figure 2. Brown trout life cycle: the importance of longitudinal connectivity and vertical interactions (arrow thickness indicates the importance of the interactions along longitudinal [horizontal arrows], vertical [vertical arrows] and lateral [oblique arrows] dimensions from the fish ecological point of view; (compare Figure 1 and text).

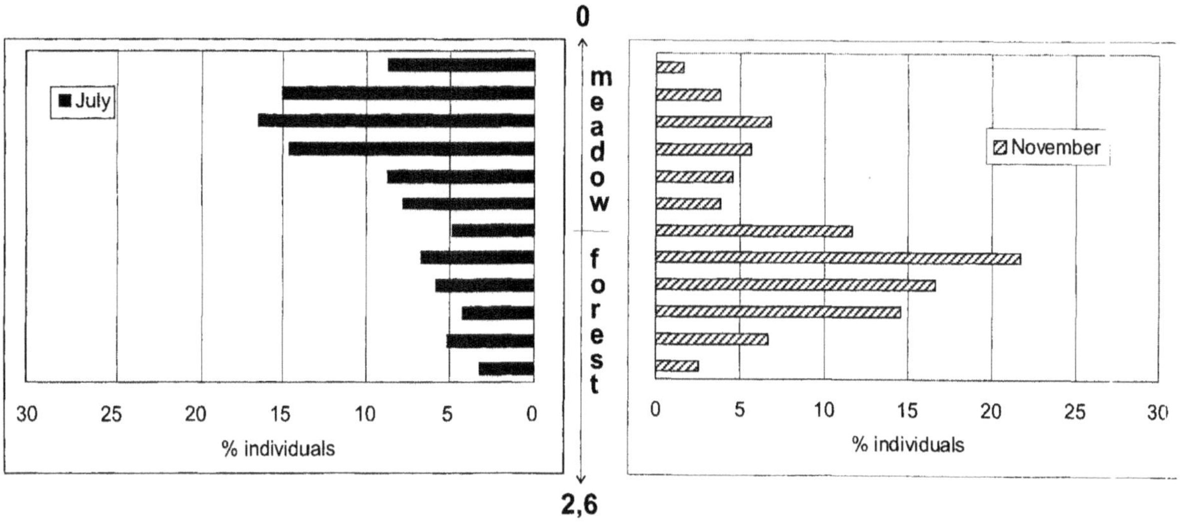

Figure 3. Different summer/winter-distributions of adult brown trout in six test stretches of a flat, meandering meadow stream with fine sediment and six adjoining, steeper forest stream sections with coarse substrate suitable for spawning (after Bohl & Lehmann, 1989).

Table 1. Different age class structure of the brown trout population in a naturally structured and a straightened section of the Lower Lunzer Seebach: due to the monotony of the straightened section, the relative proportion of juveniles dropped by half and the age class of large adults disappeared entirely (from Jungwirth, 1984b)

	Structured	Monotonous
age class 1 (1 – 3+ juneniles)	39,4	19,6
age class 2 (4 – 7+ adults)	52,3	80,4
age class 3 (>7+ adults)	8,3	0

allochthonous energy input is higher than autochthonous production (Vannote et al., 1980).

The brown trout (*Salmo trutta forma fario* L.) is the dominating fish species. In many cases it is the sole representative, but it also often co-occurs with the bullhead (*Cottus gobio* L.) and various small-sized fish species. The life cycle of the brown trout (Figure 2) clearly demonstrates that the longitudinal dimension is crucial both for spawning migrations of adult individuals, as well as for seasonal and/or stage-specific shifts. Migrations between various habitats (Figure 3) and stage-specific niche occupancy in relation to structural features or the supply/abundance of instream structures (Table 1) are obligatory elements over the course of the life cycle. Depending on the winter temperature conditions, it takes 1.5–6 months from the moment of spawning and egg incubation, over the subsequent vertical dispersal of the yolk-sac fry into the substrate, up to the emergence of the larvae from the gravel bed. The dependence of various developmental stages on the gravel habitat (Figure 2) along with the fundamental role that the hyporheic zone plays as a habitat for the benthic food resource of brown trout underlines the eminent significance of vertical connectivity.

Recent studies on the bullhead demonstrate that highly complex shifts, above all at the micro- and mesohabitat level, are also characteristic for this species. Stage-specific structure and substrate preferences induce longitudinal migrations on the micro- und mesoscale, which may be interrupted by bed falls of only 10 cm (Bless, 1990; Peter, 1998); more importantly, vertical distribution patterns within the stream bed related to developmental stage and substrate quality characterize this species (Fischer & Kummer, 2000).

The complex life cycle requirements of the brown trout and bullhead, as well as of other fish and lamprey species (e.g. Waterstraat, 1989), show that factors beyond longitudinal connectivity (along with the necessary complement of instream and shore structures) are essential in constrained headwaters of temperate running waters. As far as reproduction, early life stages and benthic food resources are concerned, the vertical pathway and exchange processes at the interface between river and bed sediments are crucial.

Braided middle course

Braided reaches with less steep gradients and moderate floodplain development represent highly unstable lotic to semi-lotic alluvial channels. The dominating coarse material of bed and banks is transported and deposited by fluvial action. Diverse current and substrate patterns dominate the aquatic environment. A characteristic feature is the very long shoreline due to numerous channels. The range of different morphological types within braided systems is very broad. In the following, only two main types are distinguished: the upper reaches, in which re-deposition of bed load takes place several times a year or even more frequently, are characterized by a particularly varied network of arms with gravel banks that are largely free of plants and/or bear only pioneer vegetation. In Austria's Alpine rivers, these braided reaches frequently correspond to the grayling zone (Muhar et al., 2000). In the subsequent downriver braided sections with first transitions to river meanders, instability is often somewhat reduced. Branching into few yet larger arms is characterized by a trend toward shorter shore lengths; more importantly, it gives rise to island systems that are stable for up to several years and colonized by woody vegetation. Other features include more extensive floodplains and an increasing proportion of oxbows and other lenitic elements. In Austria, these sections of the braided zone often correspond to the barbel region, in which indifferent and stagnophilous species appear alongside the still dominating rheophilous guild.

The life cycle (Figure 4) of the grayling (*Thymallus thymallus* L.) – the dominant fish species of the upper braided zone – clearly demonstrates that longitudinal connectivity once again plays a key role. Both the 5–15 km long spawning migrations, which go beyond the reach scale, as well as the adult migrations between summer and winter habitats and distinctive shifts of larvae and juveniles at the micro- and mesohabitat scale are the norm here. This is particularly true of the early life stages directly after larval emergence from the substrate; after a brief passive drift, they congregate at distinct microhabitats with very shallow depth, low flow velocity and sandy substrate. Two to three weeks later the juveniles already settle in distinctly deeper habitats with more rapid flow velocities and gravelly substrate; from here, they continue to switch back to the early fry habitats at night (Sempeski & Gaudin, 1995a,b,c; Guthruf, 1996). The extensive shoreline length along the upper braiding zone thus proves to be a key prerequisite for early larvae and juveniles.

Beyond longitudinal migrations and shifts, vertical connectivity is also important. Above all, the grayling depends on the superficial layer of the gravel beds for egg incubation and larval development. Intact gravel beds are equally crucial for the rheophilous representatives of those cyprinids that dominate in the barbel region. As demonstrated in Figures 5 and 6, spawning of the nase (*Chondrostoma nasus* L.) takes place in early spring (March/April), largely in riffle zones with high flow velocities in the range of 100 cm/s and shallow depths of 30 cm (Schmutz & Melcher, in prep.). In addition, the grayling and many rheophilous cyprinids also depend on the gravel layer of the river bed as a habitat for the benthic invertebrates on which they feed.

At the same time, very detailed investigations on the nase serve to underline the importance of the longitudinal river corridor. Tagging experiments have shown that the adults of this species, which was once very abundant in those river sections of the Danube catchment that correspond to the barbel region, undergo annual migrations on a scale of up to several hundred kilometers (Steinmann et al., 1937). On the other hand, Figures 5 and 6 also show distinct seasonal habitat preferences of different juvenile age classes. Specific seasonal shifts between various instream and shore habitats and niche occupancy prove to be obligatory phenomena in the life cycle.

Especially in the lower braided stretches, lateral connectivity gains importance alongside longitudinal migrations and shifts. Side arms are only one essential element for juvenile fish and various small-sized species (Figure 7). Oxbows in the increasingly extensive floodplains are an additional important element of the overall system (Muhar, 1996). The species involved here are basically the same indifferent and stagnophilous forms that contribute to broadening the species spectrum toward potamal communities in the extensive alluvial flood plain systems of meandering rivers (compare Figure 8).

Alluvial floodplain rivers

According to the ESDC, meandering zones within alluvial river/floodplain systems show very high lateral interactions with the surrounding landscape. Meandering reaches commonly have sinuous single thread channels whose relative slow lateral development into the alluvial floodplain erodes a concave slope along

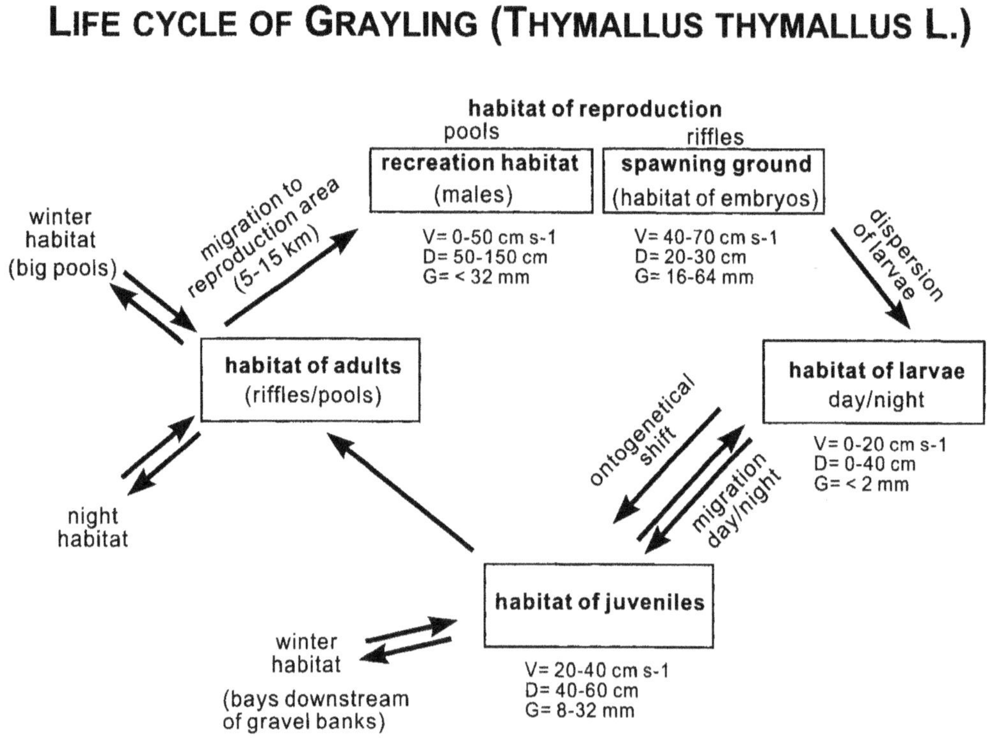

Figure 4. Grayling life cycle: For the importance of longitudinal connectivity and bed sediments for reproduction, and of micro- to mesoscale habitats for early life stages, compare text (V = flow velocity; D = depth; g = grain size; after Sempeski & Gaudin, 1995a, b, c).

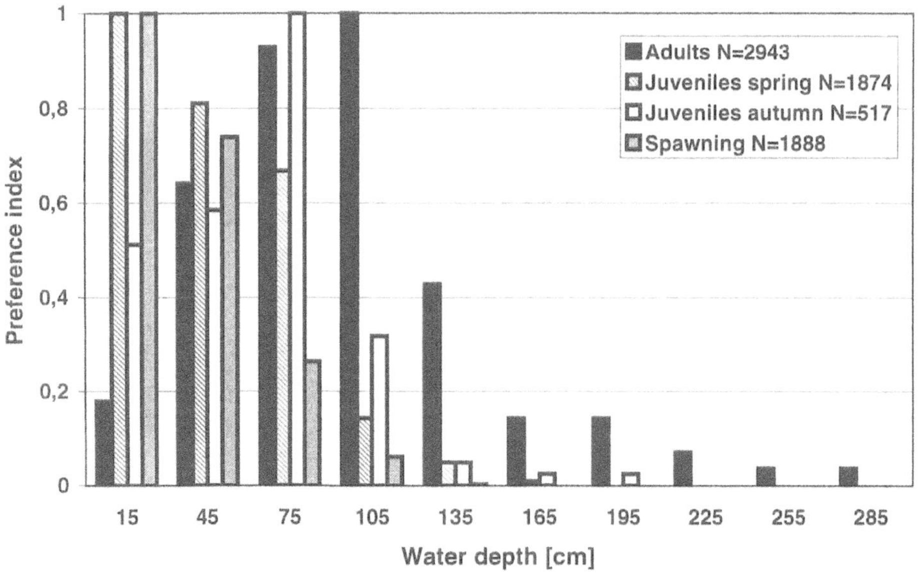

Figure 5. Habitat preference of nase: Depth preference indices of 0+ juveniles in spring and autumn as well as of adults in late autumn and at spawning (after Schmutz & Melcher, in prep.; compare text and Figure 6).

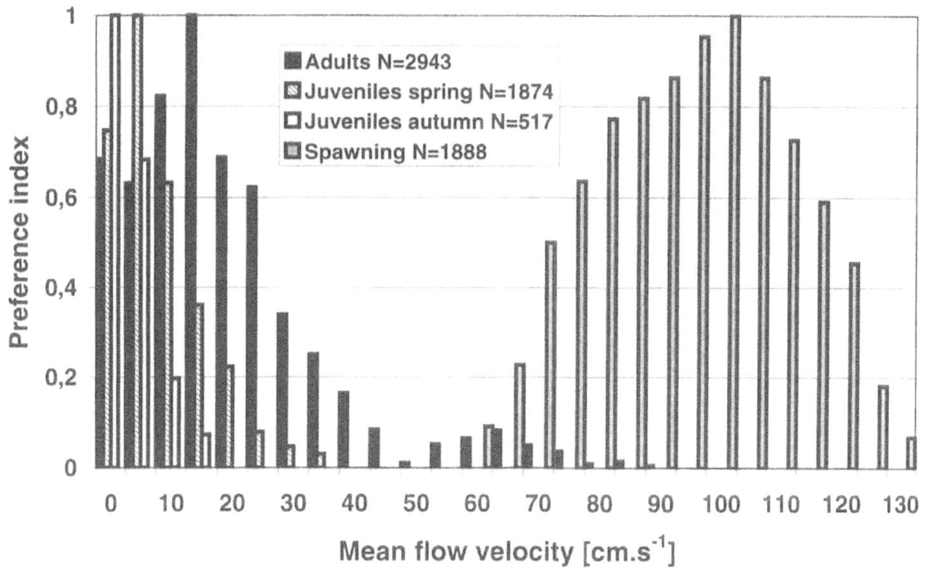

Figure 6. Habitat preference of nase: Flow velocity preference indices of 0+ juveniles in spring and in autumn as well as of adults in late autumn and at spawning (after Schmutz & Melcher, in prep.; compare text and Figure 5).

the outer bank while depositing sediment along the point bar. As part of the fluvial dynamic, this meander development leads to a broad range of backwaters and oxbow habitats in various successional stages spanning from 'eupotamon' to 'paleopotamon' types (Amoros et al., 1982). Together with the rich mosaic of pioneer vegetation and riverine forest types, they provide the basis for extremely species-rich aquatic, semiaquatic and terrestrial communities, which are further promoted by the high spatial and temporal variability of the temperature regime.

High instability and regular disturbances are further distinguishing features that result in a wealth of transitional and border zones involving many communities (Junk et al., 1989; Naiman & Decamps, 1990) and a multitude of interactions and exchange processes. The highest intensity of lateral connectivity according to the concept of ecological connectivity (Amoros & Roux, 1988) is reached in riverine backwaters of meandering systems.

The flood pulse concept (Junk et al., 1989) underlines the enormous importance of the hydrology of such systems with respect to the adaptation of the inhabiting communities. In this context, floodplains are termed 'aquatic/terrestrial transition zones' (ATTZ); they are flooded on a regular, annual basis and are, therefore, responsible for the very high biodiversity and productivity of such river/floodplain systems (see also Ward & Stanford, 1995b). Specific fish assemblages and, according to movement patterns, ecological guilds, can be clearly documented for the different types of water bodies in floodplains (Welcomme, 1985; Ward & Stanford, 1995b).

In the following, we describe only a few selected results of specific investigations over the last few years along the Austrian Danube east of Vienna that relate to fundamental aspects of fish ecology in temperate alluvial river/floodplain systems. Details on the assessment and evaluation of the fish ecological integrity of such systems can be found in the contribution of Schiemer (2000).

The numerous investigations of the ichthyofauna of the Danube east of Vienna (Schiemer & Waidbacher, 1992, 1994) show that of the 62 species currently present (six of these being exotic species), 27 are rheophilous, 25 indifferent and only 10 stagnophilous (Figure 8). In former times, the group of rheophilous species was enlarged by four anadromous acipenserids to yield a total of 31 species. Among these four anadromous acipenserids, the great sturgeon (*Huso huso* L.) for example undertook two separate, seasonal upstream runs from the Black Sea into the Danube. Important fishing and probably also spawning sites were located in former braided sections of the Danube's upper course (e.g. Tullnerfeld upstream of Vienna). Sporadic catches were even documented from the Alpine Danube tributaries (e.g. river Salzach). The construction of the Iron Gate power

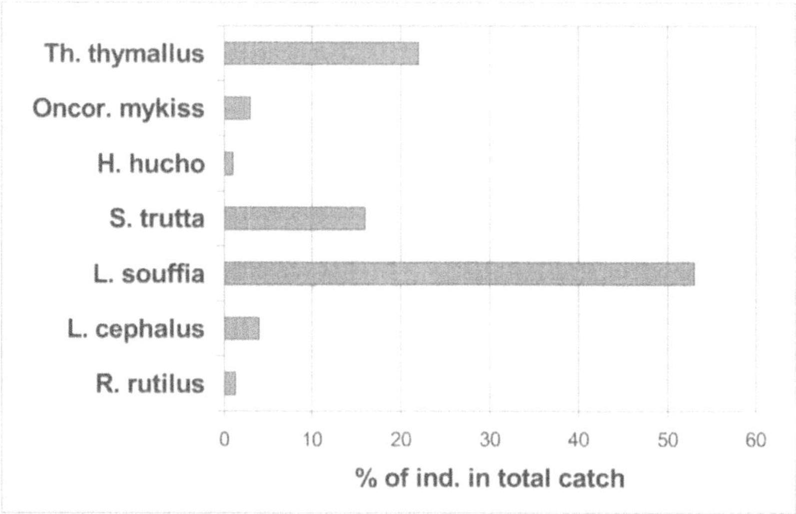

Figure 7. Newly created side arm of Drau River at Kleblach-Lind, Carinthia, in the framework of a large-scale river restoration project. Only a few months after completion, numerous juveniles of the typical former fish community, including highly endangered species such as Danube salmon (Hucho hucho) and souffie (Leuciscus souffia), inhabit the newly created environment (after Muhar, 1996).

plant in the 1960s created an unsurmountable barrier for the anadromous species (Waidbacher & Haidvogl, 1998).

Along with the originally widely distributed anadromous acipenserids, numerous rheophilous species as adults migrated over distances averaging several hundred kilometers annually (i.e. nase, barbel) or at least underwent regular and significant seasonal migrations between the Danube and its tributaries or between different habitats within the Danube itself; this demonstrates that the longitudinal river corridor is still crucial in the potamal region of temperate rivers. This is accompanied by significant migrations or ontogenetic shifts with regard to microhabitat requirements (in relation to flow velocity, substrate type and food) in the early fry or juvenile stages of many rheophilous species (Schiemer et al., 1991). Furthermore, different stages of many of the 25 fish species that are classified as being indifferent show active and/or passive migrations and shifts along the longitudinal corridor (Mader et al., 1998; Schmutz & Melcher, in prep.).

Without a doubt, lateral connectivity also plays a key role in potamal fish communities. At rising water levels in spring, various rheophilous species

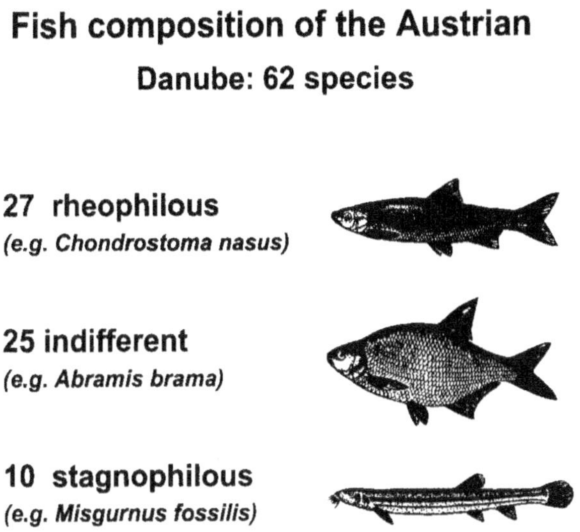

Figure 8. The present fish diversity of the Austrian Danube within the three main ecological guilds.

preferentially reproduce at lotic sites along the laterally displaced shoreline of the main channel, or in sidearms and reunited backwaters (Schiemer et al., 1991; Schiemer & Waidbacher, 1992, 1994). With highly elevated water levels in late spring and summer, however, many indifferent and most of the stagnophilous species migrate from the main river and from backwaters and sidearms into shallowly inundated floodplain areas in order to spawn and feed (Figure 9). In winter, when the water level drops, they retreat and concentrate in the deepest sites of the river or its backwaters.

Therefore, especially for the stagnophilous community inhabiting the meandering zone of temperate lowland rivers, the interplay between the lateral dimension and hydrology is crucial. On the one hand, suitable spawning and breeding sites, as well as juvenile and adult habitats, require ongoing natural reconstitution and restructuring of backwaters and other floodplain elements in the framework of long-term hydrological processes. On the other hand, the annual inundation and resulting regular alternation between connection and isolation of habitat components on various spatial scales remains a prerequisite for the functioning of such river/floodplain systems.

Conclusion

Current theories and concepts basically view running waters as complex four-dimensional systems (*sensu* Ward, 1989). Especially in the lower braided and in the meandering zones of alluvial river/floodplain systems, the relevant system-inherent features and processes include high instability and dynamic development. These phenomena, triggered and driven by disturbances and/or cyclical hydrological processes, lead to a wealth of structural and functional habitats and ecotones. The temporally and spatially variable linkages and connectivities of habitat subsystems are viewed as a vital basis for a wide range of exchange processes, migration opportunities, connection, isolation and recombination of communities.

As introduced above, the ESDC describes the relative strength of interactions along the three spatial dimensions/pathways over time for a temperate three-reach scheme river based on well-known papers on running water ecology. Summarizing that concept, the longitudinal pathway attains its greatest relative strength in constrained headwaters and the vertical pathway in the middle course. Lateral connectivity is clearly strongest in braided and meandering zones of alluvial river/floodplain systems.

The following picture emerges if one examines the three schematic river sections in the ESDC with respect to selected, fundamental fish ecological criteria (compare Figure 1): In constrained headwaters of Europe, brown trout as the key fish species is heavily dependent on the longitudinal (river/river and river/tributaries) migration possibilities and shifts between different reaches, mesohabitats and instream or riparian structures. The same holds true for resident representatives of the rhithral in North America (e.g. rainbow trout, *Oncorhychus mykiss* WALBAUM; brook trout, *Salvelinus fontinalis* MITCHILL). In anadromous species, the demands on the longitudinal river corridor expand to include the whole catchment scale, since their habitats comprise the entire reach between the headwaters and the sea (Northcote, 1998).

From a fish ecology perspective, the ESDC does not place sufficient emphasis on the vertical pathway. Although vertical exchange is minor compared to the longitudinal movement of water, the small scale interactions between the surface water and the sediment is crucial. Practically all salmonids, but also bullhead, strongly depend on both the vertical and longitudinal dimension. This is because the hyporheic zone serves as the vital niche for incubated eggs and early fry stages for several months of every year and because benthic invertebrates are the most important food items.

From the fish ecological standpoint, evaluating the relative importance, rather than the relative strength

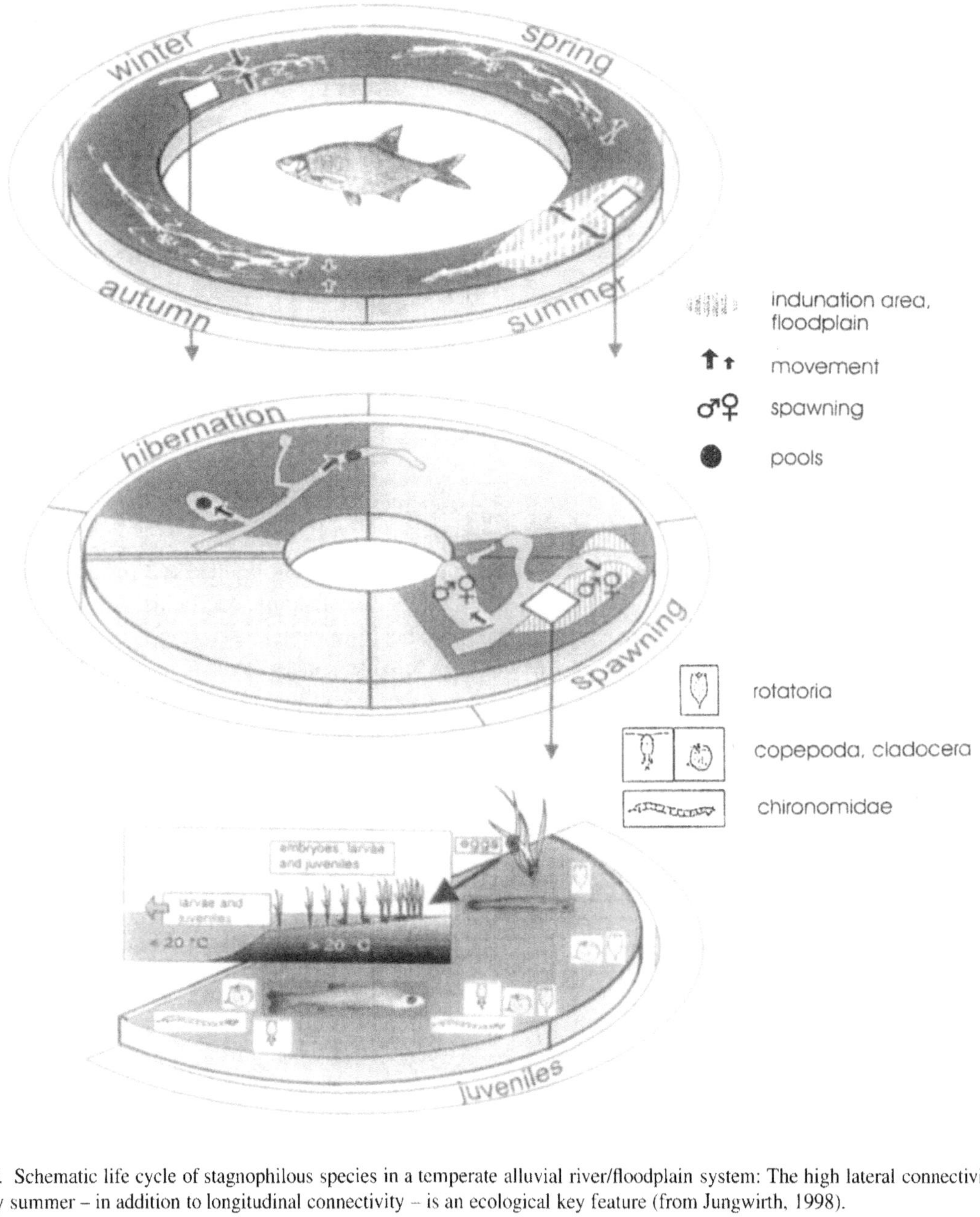

Figure 9. Schematic life cycle of stagnophilous species in a temperate alluvial river/floodplain system: The high lateral connectivity in spring and early summer – in addition to longitudinal connectivity – is an ecological key feature (from Jungwirth, 1998).

of interactions, is crucial. This is a different approach, which takes the individual situation of the aquatic habitat and the occurring species into consideration. River sections whose longitudinal continuum is naturally or anthropogenically fragmented but which are richly structured may support viable brown trout or bullhead populations if the hyporheic zone of the bed sediments enables sufficient reproduction and provides enough food. Populations of anadromous species, however, are excluded from such stretches. On the other hand, river sections with an open continuum often harbor only small fish populations when adequate structures and bed sediments are missing.

In the braided middle course, our model (Figure 1) distinguishes between the upper zone frequently dominated by grayling and the downstream (barbel) zone which, due to increasingly complex systems of sidearms, oxbows etc., shows a broader species spectrum including certain stagnophilous representatives. In agreement with the ESDC, the vertical pathway is essential for reproduction and feeding (invertebrates), especially in the upstream braided zone. Nevertheless, the assertion that the relative strength of the vertical pathway reaches its maximum in the braiding middle course is not necessarily relevant from the fish ecological perspective. During the reproductive phase, long use of the hyporheic zone, combined with comparatively deep penetration depths of larvae, are more typical of fall and winter spawners (brown trout, brook trout, various salmon species, etc.) in constrained headwaters. On the other hand, many rheophilous representatives of the frequently braided grayling and barbel region incubate their eggs in relatively superficial layers or, as spring spawners, for a comparatively short period. The early larval stages of these species are restricted to the uppermost layer or surface of the bed sediments.

From the fish ecological perspective, the relative strength of interactions described by the ESDC for the middle course do not indicate the importance of such pathways. The above examples of the grayling and nase are representative for many other species as well: seasonal and stage-specific longitudinal migrations and shifts are obligatory phenomena in the life cycle of virtually all rheophilous species, but also of many indifferent species. Therefore, longitudinal connectivity over different scales is a crucial part of ecological integrity here as well.

As far as lateral connectivity in the braided middle course is concerned, agreement with the ESDC exists insofar as its relative importance increases strongly compared with constrained headwaters. This is particularly true for the lower zone corresponding to the barbel region, where the alluvial floodplain often becomes extensive. The broadening spectrum of side arms and, above all, oxbows, which increasingly represent lotic/lenitic or lenitic habitats functionally and structurally, also tends to enlarge the fish species spectrum in the direction of potamal communities.

The significance of lateral linkages in meandering reaches of alluvial river/floodplain systems was mentioned quite early for large European rivers (Antipa, 1910), but was subsequently largely unappreciated until more recent work on temperate rivers in the northern hemisphere (Petts, 1989; Copp, 1989) confirmed the complex interrelationships and the importance of flooding cycles (Arthington & Welcomme, 1995). In the meandering zone, the ESDC quite clearly stresses lateral connectivity. The fish ecological perspective also attaches great importance to lateral connectivity (compare Schiemer et al., 1991; Schiemer & Waidbacher, 1992, 1994; Jungwirth, 1998) in that, for example, the productivity of temperate potamal systems is dependent on annual flooding and the associated exchange processes, migrations and reproductive conditions. If these functions become disturbed or if local exchange processes become disrupted by river engineering measures, fish stocks and catches strongly decline, as was the case in the Danube in the 20th century (Busnita, 1961; Jungwirth 1984a, 1998; Bacalbasa-Drobovici, 1998). Nonetheless, other interactive pathways are also important from a fish ecological perspective. A case in point is the fact that, beyond the four species of anadromous acipenserids formerly inhabiting the Upper Danube, an additional 27 rheophilous species with in part considerable adult migration distances along the main channel systems, but also with obligatory seasonal and stage-specific longitudinal short migrations or shifts, make up nearly half of the entire species spectrum. Once again, the longitudinal river corridor proves to play a central role for fish communities in the meandering zone of alluvial river/floodplain systems.

Overall, we can confirm that the main points outlined in the ESDC – the highly complex and four-dimensional nature of running waters with spatial/temporal dynamics, disturbances, connectivity, succession and ecotones as essential functional and structural elements – correspond with the fundamental aspects and demands of healthy fish communities. Nevertheless, along a temperate three-reach model river the relative strength of the single interactive pathways described in the ESDC do not always correspond to their importance for the connectivity requirements of the fish communities. Without a doubt, the greatest discrepancy between interaction strength and importance lies in the fact that, as far as reproduction in fish communities of constrained headwaters is concerned, the small-scale vertical exchange processes between river and bed sediments play a key role. Moreover, as Figure 1 shows, the longitudinal river corridor also represents a crucial fish ecological element for migrations and shifts at different scales over the entire river course, including all zones of alluvial river/floodplain systems.

As far as the assessment of the ecological integrity of running waters is concerned, the principle of connectivity clearly does play a central role in fish communities. The latest studies on a wide range of fish species indicate that stage-specific movements, drift, ontogenetic shifts and/or migrations over various scales are obligatory phenomena in the life cycles of almost every species (Schiemer et al., 1991; Mader et al., 1998; Schmutz & Melcher, in prep.). Once again, the exact determination of individual species and sound knowledge of their autecological requirements with regard to the various spatial dimensions etc. proves to be a key prerequisite for selecting and applying adequate methods and criteria to evaluate the ecological integrity and, therefore, ultimately to protect, sustainably manage, and restore running waters (Sedell et al., 1994; Muhar 1996; Tockner & Schiemer, 1997; Jungwirth, 1998; Tockner et al., 1998; Muhar & Schwarz, in prep.; Tockner et al., in press; Ward et al., in press).

References

Amoros, C. & A. L. Roux, 1988. Interaction between water bodies within the floodplains of large rivers: function and development of connectivity. In Schreiber, K. F. (ed.), Connectivity in Landscape Ecology. Münstersche Geographische Arbeiten 29: 125–130.

Antipa, G. P., 1910. Regiunea inudabila a Dungrii. Starea ei actuala si mijloacele de a o pue in valoare. Bucharest Cited in Welcomme 1992.

Arthington, A. H. & R. L. Welcomme, 1995. The Condition of Large River Systems of the World. In Armantrout, N. B. & R. J. Wolotira Jr. (eds), Condition of the World's Aquatic Habitats. Proceedings of the World Fisheries Congress, Theme 1. 44–75.

Bacalbasa-Drobovici, N., 1982. Anthropogene Einwirkungen auf Fischbestände. Schweiz Z. Hydrobiol. 44 (2): 243–51.

Benke, A. C., 1990. A perspective on America's vanishing streams. J. north am. Benthol. Soc. 9: 77–88.

Bless, R., 1990. Die Bedeutung von gewässerbaulichen Hindernissen im Raum-Zeit-System der Groppe (Cottus gobio, L.). Natur und Landschaft 65: 120.

Bohl, E. & R. Lehmann, 1989. Zur Bedeutung der Struktur von Fließgewässern für das Fischleben. Arbeiten des Deutschen Fischereiverbandes 46: 27–41.

Boon, 2000. The development of integrated methods for assessing river conservation value. Hydrobiologia 422/423: 413–428.

Busnita, T., 1961. Die Wandlung der Fischfauna der unteren Donau während der letzten 100 Jahre. Verh. int. Ver. Limnol. 14: 381–385.

Copp, G. H., 1989. The habitat diversity and fish reproductive function of floodplain ecosystems. Envir. Biol. Fishes 26: 1–26.

Dynesius, M., & C. Nilsson, 1994. Fragmentation and flow regulation of the river systems in the northern third of the world. Science 266: 753–762.

Fischer, S. & H. Kummer, 2000. Effects of residual flow and habitat fragmentation on distribution and movement of bullhead (*Cottus gobio* L.) in an alpine stream. Hydrobiologia 422/423: 305–317.

Guthruf, J., 1996. Populationsdynamik und Habitatwahl der Äsche (*Thymallus thymallus* L.) in drei verschiedenen Gewässern des schweizerischen Mittellandes. Diss. ETH Zürich: 175 pp.

Jungwirth, M., 1984a. Die fischereilichen Verhältnisse in Laufstauen alpiner Flüsse, aufgezeigt am Beispiel der österreichischen Donau. Oesterreichische Wasserwirtschaft 36 (5/6): 103–111.

Jungwirth, M., 1984b. Auswirkungen von Fließgewässerregulierungen auf Fischbestände, Teil II. BMLuF. Wasserwirtschaftsamt – Wasservorsorge: 188 pp.

Jungwirth, M., 1998. River continuum and fish migration – going beyond the longitudinal river corridor in understanding ecological integrity. In Jungwirth, M., S. Schmutz & S. Weiss (eds), Fish Migration and Fish Bypasses. Fishing News Books, Blackwell Science, Oxford: 19–32.

Junk, W. J., P. B. Bayley & R. E. Sparks, 1989. The Flood Pulse Concept in River-Floodplain Systems. In Dodge, D. P. (ed.), Proceedings of the Int. River Symposium, Can. Spec. Publ. Fish. aquat. Sci. 106: 110–127.

Karr, J. R. & D. R. Dudley, 1981. Ecological perspective on water quality goals. Envir. Manage. 5: 55–68.

Mader, H., G. Unfer & S. Schmutz, 1998. The effectiveness of nature-like bypass channels in a lowland river, the Marchfeldkanal. In Jungwirth, M., S. Schmutz & S. Weiss (eds), Fish Migration and Fish Bypasses. Fishing News Books. Oxford: 384–402.

Muhar, S., 1996. Habitat Improvement of Austrian Rivers with regard to different scales. Regul. Riv. Res. Manage. 12: 471–482.

Muhar, S. & M. Schwarz, (in prep.). Assessing rivers with high and good habitat integrity in the Alpine region as the foundation for conservation programmes: applications in Austria. SIL Proceedings.

Muhar, S., M. Schwarz, S. Schmutz & M. Jungwirth, 2000. Identification of rivers with high and good habitat quality: methodological approach and applications in Austria. Hydrobiologia 422/423: 343–358.

Naiman, R. J. & H. Décamps, 1990. The Ecology and Management of Aquatic Ecotones. Man and the Biosphere Series 4: 316 pp.

Northcote, T. G., 1998. Migratory behaviour of fish and its significance to movement through to riverine fish passage facilities. In Jungwirth, M., S. Schmutz & S. Weiss (eds), Fish Migration and Fish Bypasses. Fishing News Books, Blackwell Science. Oxford: 3–18.

Peter, A., 1998. Interruption of the River Continuum by Barriers and the Consequences for Migratory Fish. In Jungwirth, M., S. Schmutz & S.Weiss (eds), Fish Migration and Fish Bypasses. Fishing News Books, Blackwell Science, Oxford: 99–1 2.

Petts, G. A., 1989. Perspectives for ecological management of regulated rivers. In Gore, J. A. & G. E. Petts (eds), Alternatives in Regulated River Management. CRC Press Boca Raton, Florida: 3–24 pp.

Raven, J., N. T. H. Holmes, F. H. Dawson, & M. Everard, 1998. Quality assessment using river habitat survey data. Aquat. Cons. 8 (4): 477–500.

Schiemer, F., 2000. Fish as indicators for the assessment of the ecological integrity of large rivers. Hydrobiologia 422/423: 271–278.

Schiemer, F. & H. Waidbacher, 1992. Strategies of conservation of a Danubian fish fauna. In Boon, P. J., P. Calow & G. E. Petts (eds), River Conservation and Management. John Wiley & Sons: 363–382.

Schiemer, F. & H. Waidbacher, 1994. Naturschutzerfordernisse zur Erhaltung einer typischen Donau-Fischfauna. In Kinzelbach (ed.), Biologie der Donau. Gustav Fischer Verlag, Stuttgart. Limnologie aktuell 2: 247–265.

Schiemer, F., T. Spindler, H. Wintersberger, A. Schneider & A. Chovanec, 1991. Fish fry associations: important indicators for the ecological status of large rivers. Verh. int. Ver. Limnol. 24: 2497–2500.

Schmutz, S. & A. Melcher, (in prep.). Habitat preference of juvenile and adult Nase (Chondrostoma nasus, L.) in the Pielach river.

Sedell, J. R., G. H. Revees & K. M. Burnett, 1994. Development and evaluation of aquatic conservation strategies. J. Forestry 92 (4): 28–31.

Sempeski, P. & P. Gaudin, 1995a. Habitat selection by grayling. I. Spawning habitats. J. Fish Biol. 47: 256–265.

Sempeski, P. & P. Gaudin, 1995b. Habitat selection by grayling. II. Preliminary results on larval and juvenile daytime habitats. J. Fish Biol. 47: 345–349.

Sempeski, P. & P. Gaudin, 1995c. Size-related cheanges in diel distribution of young *grayling (Thymallus thymallus)*. Can. J. Fish. aquat. Sci. 52: 1842–1848.

Stanford, J. A. & J. V. Ward, 1992. Management of aquatic resources in large catchments: recognizing interactions between ecosystem connectivity and environmental disturbance. In Naiman, R. J. (ed.), Watershed Management. Springer-Verlag, New York: 91–124.

Steinmann, P., W. Koch & L. Scheuring, 1937. Die Wanderung unserer Süßwasserfische. Dargestellt aufgrund von Markierungsversuchen. Zeitschrift f. Fischerei u. d. Hilfswissenschaften 35: 369–467.

Tockner, K. & F. Schiemer, 1997. Ecological aspects of the restoration strategy for a river-floodplain system on the Danube River in Austria. Global Ecology and Biogeography Letters 6: 321–329.

Tockner, K., F. Schiemer & J. V. Ward, 1998. Conservation by restoration: the management concept for a river – floodplain system on the Danube River in Austria. Aquat. Conserv: Mar. Freshw. Ecosyst. 8: 71–86.

Tockner, K., F. Schiemer, C. Baumgartner, G. Kum, E. Weigand, I. Zweimüller & J. V. Ward, in press. The Danube Restoration Project: Species Diversity Patterns across Connectivity Gradients in the Floodplain System. Regul. Riv.

Vannote, R. L., G. W. Minshall, K. W. Cummins, J. R. Sedell & C. E. Cushing, 1980. The River Continuum Concept. Can. J. Fish. aquat. Sci. 37: 130–137.

Waidbacher, H. & G. Haidvogl, 1998. Fish migration and fish passage facilities in the Danube: past and present. In Jungwirth, M., S. Schmutz & S. Weiss (eds), Fish Migration and Fish Bypasses. Fishing News Books, Blackwell Science, Oxford: 85–98.

Ward, J. V., 1989. The four-dimensional nature of lotic ecosystems. J. n. am. Benthol. Soc. 8: 2–8.

Ward, J. V. & J. A. Stanford, 1983. The serial discontinuity concept of lotic ecosystems. In Fontaine T. D. & S. M. Bartell (eds), Dynamics of Lotic Ecosystems. Ann. Arbor. Science: 29–42.

Ward, J. V. & J. A. Stanford, 1995a. The serial discontinuity concept: Extending the model to floodplain rivers. Regul. Riv. Res. Manage. 10: 159–168.

Ward, J. V. & J. A. Stanford, 1995b. Ecological connectivity in alluvial river ecosystems and its disruption by flow regulation. Regul. Riv. Res. Manage. 11: 105–119.

Ward, J. V., K. Tockner & F. Schiemer, in press. Biodiversity of floodplain river Ecosystems: Ecotones and Connectivity. Regul. Riv.

Waterstraat, A., 1989. Einfluß eines Gewässerausbaus auf eine Population des Bachneunauges *Lampetra planeri* (Bloch, 1784) in einem Flachlandbach im Norden der DDR. Fischökologie 1 (2): 29–44.

Welcomme, R. L., 1985. River Fisheries. FAO Fisheries Technical Paper 262.

Hydrobiologia **422/423:** 99–109, 2000.
M. Jungwirth, S. Muhar & S. Schmutz (eds), Assessing the Ecological Integrity of Running Waters.

Assessing the ecological integrity of rivers: walking the line among ecological, political and administrative interests*

Otto Moog & Andreas Chovanec
Otto Moog, BOKU-University of Agricultural Sciences, Max-Emanuel-Str. 17, A-1180 Vienna, Austria
Andreas Chovanec, Federal Environment Agency, Spittelauer Lände 5, A-1090 Vienna, Austria

Key words: aquatic systems, assessment, benthic invertebrates, ecological integrity, functional feeding guilds, longitudinal zonation, running waters, taxonomic precision

Abstract

This article considers the interactions among science, politics, public administration and the general public within the framework of the assessment of biological condition as it pertains to ecological integrity. Strong public concern for environmental protection and health helps to shape policy. The legislative mandate of "maintaining and restoring the ecological integrity of waters" is both a statement of political will and a commitment. The public administration bears the responsibility for efficient water management under the auspices of maintaining a high quality of environmental health. The increasing complexity of perspectives on environmental topics increases the input of ecological sciences to the administrative process. Ecologists play an essential role in advancing scientific knowledge, developing methods to implement legislative mandates and educating the public (and public officials) about natural processes and interactions, as well as the ecological consequences of human activities. A variety of questions posed at the administration level concerning both pollution and physical impairments can be answered by biological assessment. With emphasis on benthic macroinvertebrates and levels of taxonomic resolution, examples from Austria are provided to show the efforts necessary to maintain scientific accuracy whilst maintaining practicality. Some examples of species-centred and community-based surrogate parameters used in Austria to assess the ecological integrity of rivers are presented (e.g. saprobiological aspects, functional feeding groups, longitudinal zonation analysis).

Introduction

Many countries now have the political will to manage their natural resources in a sustainable way (e.g. UN, 1992; Austrian Federal Government, 1995); this has been summarized by Karr (1996), who stated that "*there is a strong public concern for environmental protection throughout the world*". This fact is reflected by numerous strong legislative regulations which explicitly point out aspects of ecological integrity, including the U.S. Clean Water Act in 1972, the Austrian Water Act in 1990 and the 1998 draft version of the EU Water Framework Directive.

A gap between the political will and reality, however, remains. In the aquatic environment, for example, impaired rivers dominate the landscape on a worldwide scale (Dynesius & Nilsson, 1994). With respect to Austria, these impairments can be quantified to a certain degree. The study of Muhar et al. (1998) – including 56 rivers with a catchment area larger than 500 km^2 – comes to the conclusion that only 6% of the river stretches assessed have maintained their more or less natural character.

In addition to economic pressures, which have obviously led to this gap, another potential reason for the relatively weak enforcement is the poor input of ecological knowledge into water management decisions. Most of the **practical measures on aquatic environments** are undertaken **without asking ecologists**. It will be the duty of ecologists to argue for a better

* This article is dedicated to all who are engaged in ensuring the survival of our remaining natural aquatic resources and the successful restoration of those riverine landscapes impaired by human uses.

understanding of ecological integrity. To strengthen this aspect in the frame of decision-making processes, reflections on the interactions among the main societal interests within this field of applied ecology are essential: science, politics, administration and the public. The struggle for ecologically healthy systems must walk the line among them.

To activate this process of incorporating ecological aspects, ecologists have to interact with politicians, the public and particularly with administrators. Including the public in this scheme underscores the fact that public awareness helps to shape policy. Availability of information, education, democratic freedom and an improved transformation of bureaucratic tradition has enhanced public input to policy development. The 'political greening' (Steedman, 1994) is a result of a strong public concern for environmental protection and health.

Ecologists can play a critical role in this debate by educating the public about the ecological consequences of human activities. They can and must act as opinion leaders for politicians. In Austria, for example, a handful of limnologists were the driving force behind the incorporation of ecological integrity into legislation. Politicians are responsible for the legal framework guiding the environmental policy. The legislative mandate of "maintaining and restoring the ecological integrity of waters" is a political will – a promise.

The public administration in Austria is charged with ensuring efficient water management under the auspices of environmental health. Changing perspectives and the increasing complexity of environmental topics have boosted the input of ecological sciences to the administrative process. Ecologists are called upon to provide answers to various questions put by the administration: The most important questions concern:

1. the fundamentals of environmental decision-making;
2. the creation and enforcement of laws;
3. fundamentals for outcome control;
4. scientific answers to various topics (e.g. sustainability, ecological integrity);
5. the development of environmental indicators.

Supported by international programmes and collaborations, environmental indicators have become key elements of environmental policies. Indicators are used to streamline information on the environment and to enhance related communication, thus providing the basis for environmental decision-taking. Furthermore, indicators help formulate environmental goals and to verify successful implementation and compliance. And, finally, indicators serve to gauge the success of remedial measures. In this context, the search for indicators of sustainable forms of development plays an important role.

The sustainable use of water comprises two objectives: to conserve water in its manifested forms as a resource and to conserve water in its function as a habitat. This implies that type-specific biotic and environmental characteristics of waters shall be maintained at their highest condition or subjected to only minor alterations.

A prerequisite for this is an integrated assessment of waters that allows as many potential impacts as possible to be evaluated in an interrelated manner. Modern applied limnological research, therefore, focuses on ecosystem-related, integrated assessment procedures, with biological indicators playing a central role. The importance of an integrated approach in limnology and practical water management is reflected in descriptive terms such as biological integrity, ecosystem health, environmental health and ecological integrity.

The Austrian approach to assessing ecological integrity

The ecological integrity has been established in the Austrian Water Laws since 1985, reflecting the new philosophy of water management of Austrian rivers. A theoretical definition of ecological integrity into the Austrian Standards M 6232 (1997) can be expressed as – *the maintenance of all internal and external community processes and attributes, interacting with their environment in such a way that the biotic community corresponds to the natural state of the relevant aquatic habitat*. The ecological integrity of an aquatic system, though a holistic concept, is fundamentally based on the ability of all naturally occurring aquatic species to maintain healthy and reproducing populations. An acceptable and sustainable use of an aquatic environment would not overburden its capacity to maintain itself and, hence, its ecological integrity would remain intact.

The widely accepted definition of biological integrity given by Karr & Dudley (1981) "*ability of an ecosystem to support and maintain a balanced, integrated, adaptive community of organisms having a species composition, diversity, and functional organisation comparable to that of a natural habitat*

Table 1. Categories of disturbance, or disruptions in the ecological integrity of the benthic invertebrate communities (four main classes and three transition classes)

Level 1: **Pristine** – The species richness, dominance structure, abundances, longitudinal zonation and composition of functional feeding guilds all conform to the site-specific reference state. The saprobity index is below 1.25.

Level 1–2: **Slightly disturbed** – The species richness conforms to the site-specific reference state but the dominance structure can deviate slightly. The abundances of individual species can likewise deviate either up or down from the expected conditions. The longitudinal and functional feeding group distributions for the most part meet the ideal expectations. The saprobity index is between 1.26–1.75.

Level 2: **Moderately disturbed** – The species composition of the gregarious stock corresponds to the site-specific reference state but the occurrence of additional species is possible. The dominance structure is altered but, in form, resembles the reference conditions. The abundances of individual species can deviate considerably from the reference state. With some exceptions, seasonal deviations in the longitudinal zonation and functional feeding guild distributions fundamentally resemble the reference state. A moderate flattening of the longitudinal zonation curve or a shift in the regions – not more than one region – can also be observed. The saprobity index is between 1.76–2.25.

Level 2–3: **Significantly disturbed** – The species richness is altered but generally comparable to that found in the site-specific reference state. The most sensitive species will not be found and the occurrence of unexpected species is possible. The abundances of individual species as well as the dominance structure of the community compared with the reference state is clearly altered. The longitudinal zonation curve does not conform to the reference state and a shift in regions – not more than two – is witnessed. The composition of functional feeding groups is altered, although their natural relation is still maintained. The saprobity index is between 2.26–2.75.

Level 3: **Heavily disturbed** – The species richness is altered from that found in the site-specific reference state. Site-specific species disappear and unexpected species adapted to the altered conditions occur. The abundances of individual species do not conform to the reference conditions. Major alterations in the dominance structure are seen and the longitudinal zonation and the distribution of feeding guilds deviate considerably from expectations. The distributional curve is flattened considerably or its focal point is shifted up to two regions from the site-specific reference state. The saprobity index is between 2.76–3.25.

Level 3–4: **Very heavily disturbed** – The species richness is noticeably altered from that found in the site-specific reference state. Site-specific species disappear and species atypical to this site occur. The dominance structure is significantly changed and bears little resemblance to the reference state. Abundances of most species do not conform to the reference state. The longitudinal zonation and the composition of functional feeding groups clearly deviate from the reference condition. Remaining assemblages may include animals and plants adapted to specific conditions promoted in unnatural habitats such as hygropetric animals on the walls of outflow channels and thermally tolerant fauna near heating or cooling outflows. The distribution curve of longitudinal zonation is extremely flat and scarcely reveals a known community. The saprobity index is between 3.26–3.75.

Level 4: **Completely disrupted** – Few if any species are found that would conform to the reference state. Completely new assemblages, monocultures, and species especially adapted to extreme saprobic conditions may be all that is found. Accordingly, the dominance structure and abundances of individual species in no way correspond to the reference state. The distribution curve depicting longitudinal zonation relates to no known natural community. The saprobity index is above 3.76.

of a region" closely parallels the Austrian meaning. As Karr (1996) pointed out, the sum of physical, chemical and biological integrity is 'the ecological integrity'; the meaning behind 'integrity' and 'Funktionsfähigkeit' is therefore quite similar. This semantic compatibility is also confirmed by the fact that 'integrity' implies an unimpaired condition of the quality or state in correspondence with certain original conditions (Frey, 1977; Karr & Dudley, 1981; Karr et al., 1986; Angermeier & Karr, 1994), which is equivalent to the "natural state of the relevant aquatic habitat" as formulated in the Austrian definition.

A first step toward blending science and application was put forth by Chovanec et al. (1994), who expressed this deviation in a seven-category system of levels of disturbance (four main classes plus three transition classes) (Table 1). Any classification of ecological integrity is a tool or a vehicle to introduce the concept of viewing rivers on an ecosystem level to administrative authorities. Rivers *are* functional entities, and, for proper management or preservation, they require decision-makers to perceive them as integrated systems. The seven levels of disturbance or disruption (Table 1) of the ecological integrity of an aquatic community provide a comprehensive rating scheme that can be used as an administrative tool to assist both legal and logistic attempts to progressively manage activities influencing the systems.

Thus far, the Austrian method of monitoring ecological integrity focuses on a set of environmental criteria relating to hydrologic conditions, stream morphology and structure, water chemistry, indication of pollution (saprobity), toxicity and biotic indicators like periphyton, protozoans, benthic invertebrates, fish and other vertebrates (Jungwirth, 1986; Austrian Standards M 6232, 1997; Berger et al., 1997; Rott et al., 1997; Moog & Chovanec, 1998). Because of their high indicative value, benthic invertebrates are the most widely used indicators for water quality assessment (see Metcalfe, 1989; Moog, 1991, 1995; Danecker, 1992; Rosenberg & Resh, 1993).

Based on the Austrian Standards M 6232 (1997), the assessment of the benthic invertebrate fauna includes the analysis of the species richness of a site, abundance, dominance, functional feeding guilds and longitudinal zonation patterns. The additional use of other metrics such as diversity indices and EPT-ratios are recommended by the new federal "Guidelines for the saprobiological investigation of running waters", although the threshold values remain to be defined.

With respect to the biological aspect, a proposal is set forth to define the variations in characteristics that would allow one to classify a reach of river within the above-suggested classes of ecological disturbance (Table 1). The criteria used include the species inventory and the dominance structure – including at least relative abundance – and three community-level response characteristics, namely the composition of functional-feeding guilds, shifts in the longitudinal zonation of biocoenotic communities and the saprobity index.

The Austrian approach is in line with the European Union philosophy in which the monitoring results and classification of ecological status are to be reported in five levels (Table 2).

Table 2. Classification of ecological status of surface waters (EU Water Framework Directive, Draft 6/98)

Ecological status classification	Colour code
High	Blue
Good	Green
Moderate	Yellow
Poor	Orange
Bad	Red

The EU aspect of the 'ecological state' of the Water Framework Directive

For surface water categories, the ecological status classification for a waterbody shall be represented by the lower of the values for the biological and physico-chemical monitoring results for the relevant quality elements classified in accordance with the first column of Table 2. Member States shall provide a map for each River Basin District illustrating the classification of the ecological status for each waterbody, colour-coded in accordance with the second column of Table 2 to reflect the ecological status classification of the waterbody.

According to the Water Framework Directive, the assessment of the 'ecological status' of rivers is fundamentally based on biological indicators. The seven levels of ecological disturbance in Austria comprise a multitude of factors, all of which can significantly influence the ecological integrity of an aquatic system. They also necessarily include a mixture of causes, proximate physical effects and community-level disturbances. Additional criteria or indicator groups such as macrophytes, dragonflies, amphibians, birds, etc. may be used depending on specific conditions or questions (e.g. Chovanec, in print; Kohler & Janauer, 1996).

Based on an evaluation of the above-listed components (either by a classification process or by verbal description and discussion), the disturbance of the ecological integrity is assessed and classified within the seven-level system (Table 1). The components disturbed most may have the strongest influence on the final assessment: component-specific ranks or levels are not simply averaged to obtain an integrated rank. The final assessment is based on a comparison between the current conditions and a natural reference situation. The development of a reference state is based on the condition that existed prior to significant human disturbances.

Taxonomic resolution

From the scientific point of view, there is no doubt that the accurate identification of the indicators to species is necessary (Resh & Unzicker, 1975; Hilsenhoff, 1982; De Pauw & Vanhooren, 1983; Furse et al., 1984; Rosenberg et al., 1986; Hilsenhoff, 1987). Most biological monitoring techniques are dependent on the correct identification of the organisms collected. Better taxonomy can be expected to produce more accurate results and clearly increases the precision of site classifications, with a better ability to detect subtle changes in environmental quality (Lenat & Barbour, 1994).

The principal underlying scientific assumption in using indicator organisms for water quality assessment is that the presence of the indicator species is a reflexion of its environment (Cairns & Pratt, 1993; Johnson

Table 3. Species-specific distribution of saprobic valences and saprobic indices of *Hydropsyche* species

Species	x	o	β	α	p	SI
Hydropsyche tenuis	4	6				0.6
H. dinarica	1	7	2			1.1
H. fulvipes	1	7	2			1.1
H. instabilis	1	4	5			1.4
H. saxonica	1	4	3	2		1.6
H. siltalai		2	6	2		2.0
H. incognita		2	5	3		2.1
H. pellucidula		2	5	3		2.1
Cheumatopsyche lepida		1	6	3		2.2
Hydropsyche angustipennis		1	5	4		2.3
H. bulgaromanorum			8	2		2.2
H. bulbifera			6	4		2.4
H. contubernalis			2	8		2.8
H. modesta			2	8		2.8

Legend: x: xenosaprobic, o: oligosaprobic, β: β-mesosaprobic, α: α-mesosaprobic, p: polysaprobic, SI: Saprobic index.

et al., 1995). An indicator species has particular requirements with regard to a known set of variables, and the tolerances of species may vary widely even within genera (Resh & Unzicker, 1975; Cranston, 1990). Accepting this, the use of biological indicators on a generic or higher taxonomic level is clearly combined with a loss of sensitivity.

Saprobic analysis

In deriving biotic indices, consideration must be given to variation in sensitivity or tolerance values between genera or species in the same family (Hilsenhoff, 1987; Lenat, 1993). The saprobiologic analysis of nearly 1600 aquatic species in Austria indicates that nearly none of the different species within a genus was ranked with the same saprobic index (Moog, 1995). The principal loss of information by analysing taxonomic units higher than species is a given fact. Stubauer & Moog (in press) demonstrated that the use of family-based biotic indices and scores can result in systematic misinterpretations of environmental disturbances, thereby rendering these assessment methods all but meaningless for water management purposes. A case in point is the Austrian genera of the Trichopteran family Hydropsychidae (data from Graf et al., 1995) (Table 3). Analogous examples of other systematic groups are presented in Moog (in press) and Bauernfeind & Moog (2000).

Concentrating on a species-level taxonomy – and focusing on 'walking the line' – the Austrian Ministry for Agriculture and Forestry, in co-operation with the University of Agricultural Sciences in Vienna (BOKU), along with a staff of ecological or taxonomic experts, provides facilities for practitioners to improve their taxonomic and ecological knowledge. In a three-year cycle, taxonomic courses are offered in which, within about 25 days, the participants can upgrade their practical knowledge.

Functional feeding guilds

Among those concepts that allow ecological questions at the functional level to be addressed, the "functional feeding group – philosophy" (Cummins, 1973; Cummins & Klug, 1979) is among the most widely applied concepts for lotic ecosystems. The analysis of functional feeding guilds (FFG) seems to provide sufficient information on a taxonomic level higher than species. We can only partly agree with this commonly held opinion and refer to Merritt & Cummins (1984), who state that although functional feeding group assignements should be based on morphology and behaviour, different levels of taxonomic identification are required to classify aquatic invertebrates to functional groups. Whereas the ordinal level may be adequate to describe functional trophic relations of the Odonata, even the generic level may be insufficient in some others groups. From a pragmatic point of view – if potential errors are acceptable – the family and genus level may sufficiently describe the functional feeding habits of a taxonomic group. More detailed analyses require using the species level. Even those taxonomic groups whose members appear to belong to comparable trophic guilds need more attention. A few examples demonstrate this point.

Trichoptera

Mackay & Wiggins (1979) presented an analysis of trichopteran functional feeding guilds that indicates a reasonable functional consistency at the generic level. These findings are confirmed for the Austrian fauna by Graf et al. (1995). Nonetheless, special attention should be given to certain genera. In applied studies, the FFG-categorisation of the world-wide distributed genus *Rhyacophila* commonly regards the members as predaceous. This is true for most, but not all, of the species: *R. laevis* for instance feeds exclusively by shredding leaves and CPOM. The fact that this spring- and spring-brook-dwelling rhyacophilid sometimes significantly contributes to the dominance structure of the lotic invertebrate community could easily distort the interpretation of trophic interactions. At the level of applied lotic ecological work, however, the specimens of this species can easily be identified by the distinct lateral protuberances on their abdominal segments (Waringer & Graf, 1997).

Likewise, *Drusus discolor*, a caddis-fly species which contributes significantly to the faunal composition of alpine mountain streams, does not support an FFG assignment based on a generic level. Whereas most *Drusus*-species feed by grazing, shredding or detritivory, the members of the *discolor*-group show a quite different behaviour. They are predaceous and filter-feeding; special head-bristles enable these larvae to filter the water and also to catch drifting insects. Again, even practitioners performing rapid field assessments can easily identify these species, which are characterised by the typical catching area that somewhat resembles a punk haircut.

Simuliidae

The larvae of black flies are generally classified as passive filter-feeders; their impressive filtering combs are classic demonstration objects in laboratory courses. Some reputable categorisation-lists indicate that certain genera (e.g. *Twinnia*) lack these apparati and feed on algae (e.g. Merritt & Cummins, 1984; Moog & Car, 1995). The fact that the first larval instars of the genus *Prosimulium* also feed by grazing is neglected in many studies. Because members of this genus sometimes dominate lotic communities of mountain streams, misinterpretations of the autotrophically induced energy pathways can occur. Again, from the cost/benefit side of biomonitoring, the members of this very common genus in Austrian mountain brooks (Car & Moog, 1993) can easily be identified in the field: with the exception of *Simulium monticola*, the *Prosimulium* specimens can be distinguished from other black fly species by their very dark head capsules.

Chironomidae

Orendt (1993) was able to show that detritivorous feeding is obligatory in younger instars among the mostly carnivorous subfamily Tanypodinae. A generic classification is therefore possible (Janecek et al., 1995). Within the other members of the family Chironomidae, even the generic level is often insufficient to describe the functional feeding habits, due to the variability of feeding types of different species of a genus. A species-based classification of FFG of most of the chironomid fauna in Austria is provided by the authors of the "Fauna Aquatica Austriaca".

Table 4. Longitudinal stream zonation assemblages

Zone	Abbreviation	Region
Eucrenal	EUC	Mountain spring
Hypocrenal	HYC	Mountain stream
Epirhithral	ER	Upper-Trout Region
Metarhithral	MR	Lower-Trout Region
Hyporhithral	HR	Grayling Region
Epipotamal	EP	Barbel Region
Metapotamal	MP	Bream Region, connected backwaters
Hypopotamal	HP	Brackish-Water Region
Littoral	LIT	Lake shorelines, lentic sites, ponds, stagnant backwaters etc.
Profundal	PRO	Lake bottom

The contribution of the longitudinal stream zonation concept to the assessment of ecological integrity

The analysis of stream zonation patterns is based on the fact that typical stream communities change along the longitudinal course of a river section when trends in physiographic and physico-chemical conditions change. This spatial succession is driven by both physiographic and biological inputs along the longitudinal gradient. Spatial succession has been recognized for almost 140 years, and in Europe it has led to a subdivision of rivers into typical fish regions. Between 1952 and 1961, Illies expanded upon this notion and developed a concept of biocoenotic regions – the Rhithron-Potamon-concept – regarding fish, invertebrates and environmental criteria (Illies & Botosaneanu, 1963). To cover the entire stream system, Moog (1994) included the spring and spring-brook region as well as the littoral and profundal areas into this system (Table 4).

Numerical ranking of stream zonation patterns

To numerically describe the stream zonation patterns, a 10-point system has been used. The allotment method follows Zelinka & Marvan (1961) and scores individual species according to their occurrence within the stream zones described in Table 4. The sum of

Table 5. Ranking of longitudinal zonation patterns of *Hydropsyche* species. For legend consult Table 4

Species	EUC	HYC	ER	MR	HR	EP	MP	HP	LIT	PRO
Hydropsyche fulvipes	1	7	2							
H. saxonica		2	6	2						
H. tenuis			8	2						
H. dinarica			6	4						
H. instabilis			4	4	2					
H. siltalai			1	5	4					
H. pellucidula			1	2	4	3				
H. incognita				5	4	1				
H. silvenii				4	6					
H. bulbifera				1	4	5				
H. contubernalis				1	4	4	1			
H. bulgaromanorum				1	3	3	3			
H. guttata					5	4	1			
Cheumatopsyche lepida					3	6	1			

valency values for each species across the longitudinal zonation spectrum (10 regions, Table 4) is set to 10. A species which exclusively lives in one zone of a river would have 10 valency points in this zone. A species occurring throughout the longitudinal profile (i.e. from the headwaters to the potamal zones) would have one valency point in each of these regions. The valency values of individual species for stream zonation patterns and for functional-feeding guilds were not calculated quantitatively. A team of some 30 taxonomic specialists, as well as other field-practitioners, drafted a list of Austrian species and their valencies based upon the integration of available descriptive information, educated assumptions and professional judgement. These draft catalogs were reviewed by various institutions in Europe. Criticisms and comments were incorporated before this now standardized list was agreed upon. All contributors to this effort fully recognize that this descriptive and empirical valence system is a compromise between the standards of basic and applied science; nonetheless, these authors jointly acknowledge that this information is currently the best available. As a valuable help for both scientists and practitioners, the Austrian Ministry for Agriculture and Forestry, in co-operation with the University of Agricultural Sciences in Vienna (BOKU), included these categorization lists in the "Fauna Aquatica Austriaca". The first issue (May 1995) comprises nearly 2000 benthic invertebrates. The following example demonstrates the assignments of longitudinal distribution valency numbers to species of the mayfly genus *Hydropsyche* (Table 5). Each species of the family Hydropsychidae shows typical longitudinal zonation patterns ranging from spring – and headwater – dwelling species (*H. fulvipes* and *H. saxonica*) to the inhabitants of big streams and slowly flowing lowland rivers (*H. bulgaromanorum*).

Calculation measures

The calculation of indices for longitudinal zonation (biocoenotic regions) includes the formula for determining a weighted mean; the operation process follows the well-known 'Zelinka-Marvan'-method modified by Moog (1992). Again, to fullfill the concept of 'walking the line' between science and cost/benefit-aspects, a computer software application (Ecoprof) that considerably eases the analysis process was developed. The result is a modern, user-friendly, flexible and scientifically sound tool for environmental quality assessment and monitoring (Moog et al., 1997).

By comparing a reference stretch with the actual state of a disturbed system, certain predictable patterns or shifts in the longitudinal zonation become visible. As the following examples will demonstrate, pulse releases, channel straightening and the subsequent influence that these measures have on the sediment balance, organic inputs and residency time often lead to the 'rhithralization' of a community. On the other hand, retaining organic matter, impounding, heating and bed-load suspension all lead to the 'potamalization' of the faunal structure. Thus the phenomena described in the River Continuum Concept, the Serial

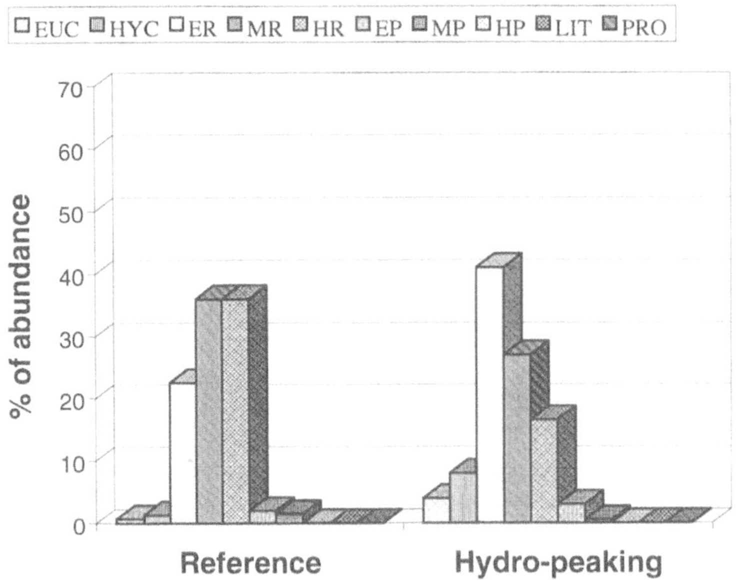

Figure 1. Longitudinal zonation patterns of benthic macroinvertebrates in a 'natural' (reference) and a disturbed river stretch (Salzach River; 28 Dec. 1990; 68 resp. 47 taxa; Surber Sampler, 6 replicates each; percentage of individual abundances). For legend consult Table 4.

Discontinuity Concept, the Mosaic Dynamic Aspects and others can be more clearly demonstrated in a qualitative, inexpensive yet sensitive manner. Two human measures that alter the community into more rhithralic or more potamalic assemblages are discussed below.

The effects of hydro-peaking on longitudinal zonation

The changes in a benthic invertebrate community below a power station operated with hydro-peaking is exemplified in the tail waters of the power plant Urreiting in the Salzach River (Salzburg). The Salzach is a 7th-order sub-montane river, partially glacier-fed, with a snow-fed discharge regime flowing through a mountain range of mixed limestone and crystalline composition. The term hydro-peaking refers to a special type of reservoir-operated hydroelectric power generation. Water is stored in base load run-off river plants or retired in storage basins to generate electricity at times of peak demand. The water released leads to diurnally variable water pulses in the river section below the power station. The artificial discharge variations affect the riverine biota immensely.

Besides an 85% reduction of the benthic invertebrate biomass in the most affected section, the benthic assemblage below the power plant has clearly shifted to one dominated by an epirhithralic fauna; for comparison, a reference stretch in this reach contains faunal characteristics of the meta- and hyporhithral zones (Figure 1).

The effects of river channel straightening on longitudinal zonation

The Mattig at Uttendorf is a 4th-order, lowland river with a pluvio-nivalic discharge regime flowing through the basin landscape of the Flachgau district of Salzburg. The reference stretch, meandering through a natural floodplain area depicted below, is clearly dominated by a hyporhithralic fauna (although moderate channel straightening many decades ago has already degraded this stretch).

In a 'disturbed' site (straightened river channel) this community composition is further shifted from hyporhithralic to one dominated by epi- to metarhithralic elements, clearly indicating an artificial 'rejuvenation' of this river (Figure 2).

Conclusions

Growing concerns about environmental quality have given ecological science a prominent role within water management. Rivers were formerly regarded as mere water transport channels, but are now viewed as ecological systems. An integrative view of ecological 'health' is the goal of the current science-based

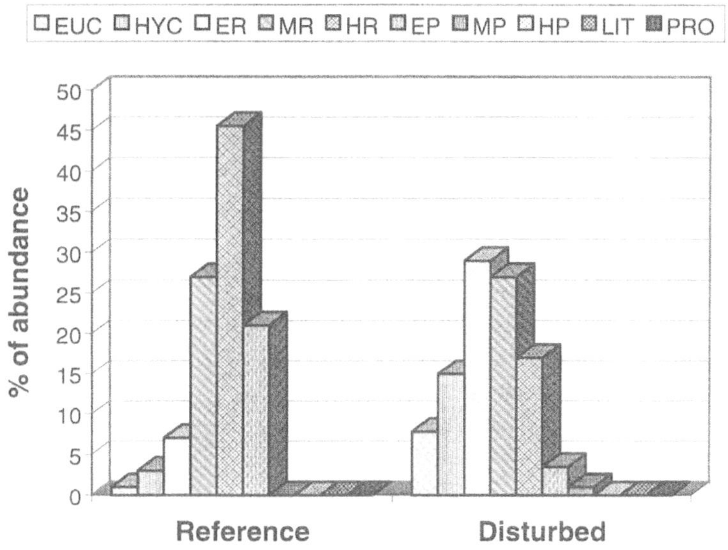

Figure 2. Longitudinal zonation patterns of benthic macroinvertebrates following in a near-natural (reference) and a channelized section. (Mattig River, 27th Aug., 1989; 107 resp. 92 taxa; Surber Sampler, 5 replicates each; percentage of individual abundances).

management of aquatic resources. Biological indicators play an increasingly crucial role in monitoring the ecological status (Davis & Simon, 1995). In Europe, the assessment criteria laid down in the draft of the Water Framework Directive (WFD) primarily embrace bio-indicative approaches with physico-chemical and hydro-morphological aspects assuming a supporting function.

Although there is a consensus that a nation's environmental policy must be based upon the best science, the design of (biological) monitoring programmes has to factor in the cost-benefit aspect. From a scientific point of view, however, the accurate determination of indicators on the species level is essential for the sound evaluation of environmental conditions, e.g. 'ecological integrity'. We argue that there is a strong need to bridge the gap between the principles of strong science and the demands of administration for an efficient and cost-effective assessment procedure.

The principle of 'walking the line' among these different interests is founded on a close co-operation among scientists, administrators and practitioners. In Austria, this collaboration comprises the following aspects of a sound framework of quality assurance to guarantee comparable results and international consistency of methods:

1. The political will to manage the natural resources in a sustainable way is explicitly described in a National Environmental Plan (Austrian Federal Government, 1995).
2. The protection of aquatic systems is based on a strong legal foundation (Austrian Water Act, 1990).
3. Technical support and scientific methods have been developed for applied uses
   - with respect to taxonomic resolution
     - providing applicable determination keys
     - offering taxonomic courses
     - compiling check lists of the species inventory (Fauna Aquatica Austriaca, FAA)
   - with respect to standardised sampling and analysis procedures (FAA)
     - describing the numerical transformation of ecological information (FAA)
     - providing a computer software application (Ecoprof)
   - with respect to sampling design
     - establishing a countrywide national monitoring network
     - establishing a provincial monitoring network
     - developing type-specific reference assemblages
4. In compliance with the requirements of the EU Water Framework Directive assessment approaches, a current study has established an Austrian river classification and typology which serves

as a basis for generating type-specific benthic invertebrate reference communities.

Acknowledgements

The authors wish to express their gratitude to Astrid Schmidt-Kloiber, Ilse Stubauer and Thomas Ofenböck for graphic assistance, Michael Stachowitsch and Steve Weiss for linguistic help and Michael Barbour for his valuable comments on the manuscript. We also acknowledge the Ministry for Agriculture and Forestry and the Federal Environment Agency in Vienna for initiating and funding many of the activities mentioned in the article.

References

Angermeier, P. L. & J. R. Karr, 1994. Relationships between woody debris and fish habitat in a small warmwater stream. Trans. am. Fish. Soc. 113 (6): 716–726.

Austrian Federal Government, 1995. National Environmental Plan. Ministry for Environmental Affairs, Vienna: 324 pp.

Austrian Standards M 6232, 1997. Önorm M 6232: Guidelines for the ecological study and assessment of rivers, Vienna: 1–84.

Bauernfeind, E. & O. Moog, 2000. Mayflies (Insecta: Ephemeroptera) and the assessment of ecological integrity – a methodological approach. Hydrobiologia 422/423: 71–83.

Berger, H., W. Foissner & F. Kohmann, 1997. Bestimmung und Ökologie der Mikrosaprobien nach DIN 38410. G. Fischer, Stuttgart: 1–291.

Cairns, J. Jr. & J. R. Pratt, 1993. A history of biological monitoring using benthic macroinvertebrates. In Rosenberg D. M. & V. H. Resh (eds), Freshwater Biomonitoring and Benthic Invertebrates. Chapman & Hall, New York: 10–27.

Car, M. & O. Moog, 1993. Höhen- und längenzonale Verteilung der Simuliidenfauna (Diptera) Österreichs und Liechtensteins. In Timm, T. & W. Rühm (eds), Beiträge zur Taxonomie, Faunistik und Ökologie der Kriebelmücken in Mitteleuropa. Essen: Westarp-Wiss. ISBN 3-89432-0788. Essener Ökologische Schriften, Bd. 2: 63–79.

Chovanec, A., in press. Dragonflies (Insecta: Odonata) as indicators of the ecological integrity of aquatic systems – a new assessment approach. Verh. Int. Verein. Limnol. 27.

Chovanec A., H. Heger, V. Koller-Kreimel, O. Moog, T. Spindler & H. Waidbacher, 1994. Anforderungen an die Erhebung und Beurteilung der ökologischen Funktionsfähigkeit von Fließgewässern – eine Diskussionsgrundlage. ÖWAW 11/12: 257–264.

Cranston, P. S., 1990. Biomonitoring and invertebrate taxonomy. Envir. Monit. Assess. 14: 265–273.

Cummins, K. W., 1973. Trophic relations of aquatic insects. Ann. Rev. Ent. 18: 183–206.

Cummins, K. W. & M. J. Klug, 1979. Feeding ecology of stream invertebrates. Ann. Rev. Ecol. Syst. 10: 147–172.

Danecker, E., 1992. Phototrophe und Heterotrophe Organismen als Indikatoren der Verunreinigung von Fliessgewässern. Wiener Mitteilungen 105: 26 pp.

Davis, W. S. & T. P. Simon, 1995. Biological assessment and criteria: tools for water resource planning and decision making. CRC Press, Boca Raton, FL: 1–415.

De Pauw, N. & G. Vanhooren, 1983. Method for biological quality assessment of water courses in Belgium. Hydrobiologia 100: 153–168.

Dynesius, M & C. Nilsson, 1994. Fragmentation and flow regulation of river systems in the northern third of the world. Science 266: 753–762.

Frey, D. G., 1977. Biological integrity of waters – an historical approach. In Ballentine, R. K. & L. J. Guarraia (eds), The Integrity of Water. Proceedings of a symposium. US Environmental Protection Agency. Washington, D.C.: 127–140.

Friedrich, G., 1992. Ökologische Bewertung von Fließgewässern – eine unlösbare Aufgabe? In Friedrich, G. & J. Lacombe (eds), Ökologische Bewertung von Fließgewässern. Limnologie aktuell 3, G. Fischer Vlg., Stuttgart-New York: 1–7.

Furse, M. T., D. Moss, J. F. Wright & P. D. Armitage, 1984. The influence of seasonal and taxonomic facts on the ordination and classification of running water-sites in Great Britain and on the prediction of their macro-invertebrate communities. Freshwat. Biol. 14: 257–280.

Graf, W., U. Grasser & J. Waringer, 1995. Trichoptera. In Moog, O. (ed.), Fauna Aquatica Austriaca – A Comprehensive Species Inventory of Austrian Aquatic Organisms with Ecological Notes. Federal Ministry for Agriculture and Forestry. Wasserwirtschaftskataster Vienna: loose-leaf binder.

Hilsenhoff, W. L., 1982. Using a biotic index to evaluate water quality in streams. Technical Bulletin No. 132, Wisconsin Dept. of Natural Resources: 1–22.

Hilsenhoff, W. L., 1987. An improved biotic index of organic stream pollution. The Great Lakes Entomologist 20,1: 31–39.

Illies, J. & L. Botosaneanu, 1963. Problèmes et methodes de la classification et de la zonation ecologique des eaux courantes, considerées surtout du point de vue faunistique. Int. Verein. theor. und angew. Limnologie 12: 1–57.

Janecek, B., C. Orendt & O. Moog 1995. Tanypodinae. In Moog, O. (ed.), Fauna Aquatica Austriaca – A Comprehensive Species Inventory of Austrian Aquatic Organisms with Ecological Notes. Federal Ministry for Agriculture and Forestry, Wasserwirtschaftskataster Vienna: loose-leaf binder.

Johnson, B. L., W. B. Richardson & T. J. Naimo, 1995. Past, present, and future Concepts in Large River Ecology. BioScience 45(3): 134–141.

Jungwirth, M., 1986. Flußbau und Fischerei. Quantitative Untersuchungen über die Auswirkungen unterschiedlicher Flußregulierungen auf Fischbestände. Wr. Mitt. Bd. 70, Institut für Wasserwirtschaft, Universität für Bodenkultur, Wien.

Karr, J. R., 1996. Ecological integrity and ecological health are not the same. In Schulze, P. (ed.), Engineering Within Ecological Constraints. National Academy of Science, Washington D.C.

Karr, J. R. & I. J. Dudley, 1981. Ecological perspective on water quality goals. Envir. Manage. 5: 55–68.

Karr, J. R., K. D. Fausch, P. D. Angermeier, P. R. Yant & I. J. Schlosser, 1986. Assessing biological integrity in running waters: a method and its rationale. Illinois Natural History Survey Special Publication 5, Champaign, Illinois.

Kohler, A. & G. A. Janauer, 1996. Zur Methodik der Untersuchungen von aquatischen Makrophyten in Fließgewässern. In Steinberg, C., H. Bernhardt & H. Klapper (eds), Handbuch Angewandte Limnologie. Grundlagen - Gewässerbelastung - Restaurierung - Aquatische Ökotoxikologie - Bewertung - Gewässerschutz. ecomed, Landsberg am Lech.

Lenat, D. R., 1993. A biotic index for the southeastern United States: derivation and list of tolerance value, with criteria for assigning water-quality ratings. J. n. am. Benthol. Soc. 12 (3): 279–290.

Lenat, D. R. & M. T. Barbour, 1994. Using Benthic Macroinvertebrate Community Structure for Rapid, Cost-effective, Water Quality Monitoring: Rapid Bioassessment. In Loeb, S. L. & A. Spacie (eds), Biological Monitoring of Aquatic Systems, Lewis Publishers: 187–215.

Mackay, R. J. & G. B. Wiggins, 1979. Ecological diversity in Trichoptera. Ann. Rev. Ent. 24: 185–208.

Mauch, E., 1990. Ein Verfahren zur gesamtökologischen Bewertung der Gewässer. Wasser + Boden 11/1190): 763–767.

Merritt, R.W. & K. W. Cummins (eds), 1984. An Introduction to the Aquatic Insects of North America, 2nd edn. Kendall/Hunt Publishing Company.

Metcalfe, J. L., 1989. Biological water quality assessment of running waters based on macroinvertebrate communities: history and present status in Europe. Envir. Pollut. 60: 101–139.

Moog, O., 1991. Biologische Parameter zum Bewerten der Gewässergüte von Fließgewässern. Wien. Landschaftswasserbau 11: 235–266.

Moog, O., 1992. Das Konzept der biozönotischen Regionen - ein Hilfsmittel zur Charakteristik anthropogener Einflüsse auf benthische Fließgewässerzönosen. Erw. Zusammenfassung der DGL Jahrestagung 1992, Bd. II: 622–626.

Moog, O., 1994. Ökologische Funktionsfähigkeit des aquatischen Lebensraumes. Wiener Mitt. 120: 15–59.

Moog, O. (ed.), 1995. Fauna Aquatica Austriaca – A Comprehensive Species Inventory of Austrian Aquatic Organisms with Ecological Notes. Federal Ministry for Agriculture and Forestry, Wasserwirtschaftskataster Vienna: loose-leaf binder.

Moog, O., in press. Biological River Water Quality Assessment in Austria. Proceedings of the "Internat. Workshop on Assessment and Classification of Rivers", Düsseldorf 5–7.11.1996.

Moog, O. & A. Chovanec, 1998. Die ökologische Funktionsfähigkeit – ein Ansatz der integrierten Gewässerbewertung in Österreich. Münchener Beiträge zur Abwasser-, Fischerei- und Flußbiologie 51, Oldenburg Verlag München, Wien: 57–118.

Moog, O. & M. Car, 1995. Simuliidae. In Moog, O. (ed.), Fauna Aquatica Austriaca – A Comprehensive Species Inventory of Austrian Aquatic Organisms with Ecological Notes. Federal Ministry for Agriculture and Forestry, Wasserwirtschaftskataster Vienna: loose-leaf binder.

Moog, O., A. Schmidt-Kloiber & R. Vogel, 1997. Ecoprof – Entwicklung und Erprobung eines Auswertungsmodelles zur Beurteilung der ökologischen Funktionsfähigkeit auf Basis benthischer Fließgewässerzönosen. Report BMLF, Vienna. The computer software Ecoprof is available at the Ministry of Agriculture and Forestry, Sektion IV, A-1030 Vienna, Marxerg. 2, Austria.

Muhar, S., M. Kainz, M. Kaufmann & M. Schwarz, 1998. Erhebung und Bilanzierung flußtypspezifisch erhaltener Fließgewässerabschnitte in Österreich. ÖWAW 50: 119–127.

Orendt, C., 1993. Vergleichende Untersuchungen zur Ökologie litoraler, benthischer Chironomidae und anderer Diptera (Ceratopogonidae, Chaoboridae) in Seen des nördlichen Alpenvorlandes, mit einem Beitrag zur Eignung und Verwendung von Chironomiden als Trophie-Indikatoren. Thesis Univ. München: 1–274, 18 tables.

Resh, V. H. & J. D. Unzicker, 1975. Water quality monitoring and aquatic organisms: the importance of species identification. J. Wat. Pollut. Cont. Fed. 47, 1.

Rosenberg, D. M. & V. H. Resh, 1993. Freshwater biomonitoring and benthic invertebrates. Chapman & Hall, New York: 1–488.

Rosenberg, D. M., H. V. Danks & D. M. Lehmkuhl, 1986. Importance of insects in environmental impact assessment. Envir. Manage. 10: 773–783.

Rott, E., K. Hofmann, P. Pall, P. Pfister & E. Pipp, 1997. Indikatorlisten für Aufwuchsalgen – Teil 1: Saprobielle Indikation. Bundesministerium für Land- und Forstwirtschaft, Wien: 1–73.

Steedman, R. J., 1994. Ecosystem health as a management goal. J. n. am. Benthol. Soc. 13(4): 605–610.

Stubauer, I. & O. Moog, in press. Taxonomic sufficiency versus need of information – comments based on the Austrian experiences in biological water quality monitoring. Verh. int. Verein. Limnol 27.

UN (United Nations), 1992. Report of the Conference on Environment and Development 'The Earth Summit'. Rio de Janeiro, June 3–14, 1992, New York.

Waringer, J. & W. Graf, 1997. Atlas der österreichischen Köcherfliegenlarven. Facultas-Universitätsverlag, Vienna: 1–286.

Zelinka, M. & P. Marvan, 1961. Zur Präzisierung der biologischen Klassifikation der Reinheit fließender Gewässer. Arch. Hydrobiol. 57, 3: 389–407.

Hydrobiologia 422/423: 111–131, 2000.
M. Jungwirth, S. Muhar & S. Schmutz (eds), Assessing the Ecological Integrity of Running Waters.

Indicators and assessment methods for measuring the ecological integrity of semi-aquatic terrestrial environments

Stephanie A. Innis[1], Robert J. Naiman[1] & Scott R. Elliott[2]
[1]*School of Fisheries, Box 357980, University of Washington, Seattle, WA 98195, U.S.A.*
Tel.: [+1] 206 616-9421; E-mail: sinnis@fish.washington.edu
[2]*College of Forest Resources, Box 352100, University of Washington, Seattle, WA 98195, U.S.A.*

Key words: riparian, wetland, assessment, integrity, review

Abstract

Our objective is to summarize scientific knowledge concerning assessment methods addressing ecological integrity in wetlands and riparian zones, with an emphasis on riparian areas. This article examines which indicators (abiotic parameters, species, faunistic and floristic communities and functional assemblages) are used, how they are applied (single or integrative indicator), and which assessment algorithms and models have been successful to date. Overall, our review shows that despite the relatively recent emergence of riparian ecology, riparian assessments are better developed than the wetland functional assessments currently employed. In general, it is recommended that useful methods be updated and cross-calibrated, that new rapid assessment methods provide reasonable levels of accuracy for a variety of users in a variety of situations, that assessment be developed for specific applications (with identified users), that uncertainty be explicitly acknowledged, that the policy implications of specific assessments methods be openly discussed, and that methods be formally tested for accuracy, cost and practicality. In addition, we offer a revised protocol for the effective and rapid assessment of functional integrity in riparian environments associated with freshwater ecosystems. This protocol encourages the use of terrestrialization, canopy development, biodiversity, microclimate and seston as integrative indicators of integrity.

Introduction

Riparian forests and wetlands associated with rivers are among the most biologically rich systems on earth. Though they constitute only a fraction of the total land area, they exhibit greater biotic diversity and biomass per unit area than the remainder of the land base and perform well-known critical functions for maintaining the environmental vitality of many aquatic systems (Naiman & Décamps, 1997). Swift (1984) estimates that before extensive human settlement in the United States, riparian ecosystems may have once occupied between 30 and 40 million hectares (about 5% of the total land mass). At present, riparian forests occupy between 10 and 14 million hectares, a loss of 66%. In addition, wetland losses have exceeded 50% in the United States (NRC, 1995), while Europe's wetlands have decreased by more than 66% since the beginning of the 20th century (Commission of the European Community, 1995).

This has occurred while remaining semi-aquatic habitat suffers from a variety of human impacts: pollution from point and non-point sources, exotic species devastating native populations and erosion from cultivation and grazing in combination with altered hydrological regimes causing substantial changes to habitat (Naiman et al., 1995). For example, Wilcove et al. (1998) report that in the USA habitat destruction by water development (including dams and agricultural diversion) threatens 91% of endangered fish and 99% of endangered mussels and more than 50% of endangered invertebrate, amphibian and crustacean species, posing a substantial threat to biodiversity.

These changes are crucial issues for resource managers. A key component of any management issue, whether for habitat preservation or restoration, is the assessment of current ecological integrity. A broad array of assessments are currently employed by various agencies and institutions. But how appropriate or successful are these methods? How compatible? Kusler

& Niering (1998) address whether wetland assessment methods are being used effectively by resource managers. Based on a symposium, workshops and extensive interviews with wetland professionals, they report three sobering conclusions: (1) There is overall dissatisfaction with existing wetland assessment approaches and little use of formal, rapid assessment techniques. (2) Qualitative assessments based on a 'little common sense' provide a more satisfactory assessment than the application of detailed models. (3) Regulators and managers repeatedly use the terms 'unrealistic' and 'out of touch' when discussing assessment methods. Even though there has been no similar evaluation for riparian assessment methods, one must assume that the problems of implementation are similar.

In this article, we examine current riparian and wetland assessment methods. We summarize which indicators are used, how they are applied and provide a critical evaluation of the relative advantages and disadvantages of each method. We then discuss the fundamental components of successful assessments and offer a novel, generalized strategy to assess riparian functional integrity.

Approach

We conducted literature searches of the Aquatic Sciences and Fisheries Abstracts, BIOSIS, GeoRef, Current Contents and GeoBase databases using the following keywords singly and in combinations: riparian, wetland, river, semi-aquatic, ecosystem, ecological, integrity, assessment, evaluation, and health. Reference lists from papers selected for review supplied additional relevant references. This search produced hundreds of potential documents. Therefore, we imposed criteria to maintain objectivity in the selection of papers for review:

1. Document selection was limited to classifications and assessments that focused specifically on riparian or wetland systems. Later, we included some methods with wider scope (e.g., river habitats, in-stream conditions and catchment condition) if riparian assessment was a significant component of the design.
2. Methods and descriptions of indicators were included if they reflected riparian and wetland condition, regardless of their success.
3. In addition, documents were excluded in a few extenuating circumstances. For example, in cases where there was no citation of either the scientific or the technical literature, the document was excluded from our analysis.

The approaches used in individual studies were then summarized (Tables 1, 2, and 3) and assessed for strengths, limitations, and validation. These critiques were primarily based on how well each method treated the fundamental components of ecological assessment of integrity, which are addressed in a later section of this article.

Current state of knowledge

Semi-aquatic ecosystems

Semi-aquatic ecosystems encompass both wetlands and riparian zones. Both are fundamentally characterized by an intimate association with the hydrologic setting and exist as ecotones between terrestrial and aquatic systems. In addition, both wetlands and riparian zones are areas of rich biodiversity and generally comprise a minor proportion of the landscape. However, the inherent differences between riparian and wetland ecosystems are also significant; they differ substantially in spatial context, disturbance regimes, the nature of the hydrologic setting and ecological organization. For example, terrestrial and aquatic aspects are intimately associated in wetlands, while riparian zones are more terrestrial and generally do not include the in-stream environment. In addition, riparian ecosystems exist as highly connected, linear ecosystems in space whereas freshwater wetlands exist as lightly connected, more circular or oblong entities in space. Natural riparian ecosystems are dynamic and are often shaped by high-energy, system-resetting disturbance regimes; in contrast, freshwater wetlands are normally less dynamic in space and time and natural disturbances are generally less catastrophic in nature. Therefore, riparian processes may be generally limited by disturbance while wetlands may be limited by competition.

Despite these fundamental ecological differences, riparian zones historically were considered as one class of wetland so that early classifications and assessments of wetlands and riparian zones overlapped considerably (e.g. Cowardin et al., 1979). The emergence of riparian ecology as a focus area independent of wetland ecology has led to increasingly separate treatment of the two. Many current publications, however, persist in treating these systems as commensurate entities. Ecological patterns are often best described by lumping information from similar ecosystems. Riparian and wetland ecosystems are strongly related by their semi-aquatic nature; in our opinion, despite this relationship, the inherent differences call for their distinction in ecological terms.

Application of ecological information

What are the fundamental components of an assessment and how are these used to formulate a decision about ecological integrity? Clearly, specific assessments involve unique conditions, objectives, parameters, reference conditions, synthesis, and interpretation. The literature has treated these components to varying degrees and has generated a great deal of ecological information which has taken many forms such as inventory, classification, indicator, and assessment. These are distinct in their nature and therefore the application(s) for which they are appropriate. In general, inventories generate comprehensive lists of attributes, classifications generate groups based on co-occurring attributes, indicators are attributes that respond in a known way to disturbance, and assessments or evaluations compare attributes or indicators in relation to criteria specified for a certain class (see Figure 1).

Inventory

In an ecological context, an inventory is a list of observable or measurable physical, chemical and biological features. The dynamic nature of ecosystems is characterized by spatial and temporal variability of attributes; therefore, the results of an inventory apply only to a narrow window of space and time, in essence providing a detailed snapshot. As these snapshots provide essential ecological information in the development of an assessment, some inventory procedures have been standardized.

One the most widely-used references in riparian inventories, classifications and assessments is Platts et al. (1987), which was developed to evaluate riparian habitats by the United States Department of Agriculture's (USDA) Forest Service. This publication provides a well-organized, comprehensive list of potential parameters (many tested for accuracy and precision), includes detailed instructions in the proper measurement of these parameters and discusses statistical considerations in sampling methods. The manual, however, does not address selection and integration of parameters for building an inventory, but has been used in many subsequent publications that do (Table 1).

The United States Department of the Interior (USDI) Bureau of Land Management (BLM) has generated two widely used riparian inventory manuals. Myers (1989) presents a two-stage approach that consists of a broad-scale, extensive inventory that can be followed up by a site-specific, intensive inventory. Leonard (1992) builds on Myers' approach and previous BLM manuals to generate a multi-stage, interdisciplinary approach for ecological site inventories (Table 1). Both methods make recommendations toward the determination of ecological status (comparing the present vegetative community with the potential natural community, PNC). However, this is distinct from addressing ecological integrity, a subject which will be developed in a later section of this paper.

A recent inventory by Briggs et al. (1997) demonstrates how a well-developed inventory generates a detailed snapshot of a system with careful consideration of the next steps in the ecological evaluation process. This inventory employs well-established methods to gather quantitative information and to choose attributes with some discrimination so that a relatively limited number of ecological attributes are measured. Although developed specifically for a study area in Arizona, this inventory provides a solid framework for the development of other ecological inventories.

Classification

Classification provides a means of appropriate application of the ecological 'snapshots' generated by an inventory. This is accomplished by grouping areas based on important common attributes. The attributes used and the categories created depend on the objective of the classification; the same area could be classified differently using different sets of attributes (Kondolf, 1995). This is significant because wetlands are highly complex ecosystems and are distinct not only from the surrounding landscape but are usually distinct from one another. Additionally, riparian systems exhibit changing characteristics along the length of a river (Naiman & Décamps, 1997). This natural variation makes comparison within and between wetland and riparian systems problematic unless a clear classification is provided.

Indicators

Indicators represent the next highest level toward ecological assessment (Figure 1). Indicators and multi-

Table 1. Summary of major riparian inventories

Method	Inventory items	Approach	Strengths	Limitations	Testing/validation	Selected references
Methods for evaluating riparian habitats. Platts et al. (1987)	Hydrology, Includes detailed geomorphology streambanks, vegetation, water column, soils, benthic invertebrates.	• Comprehensive; includes information for quantitative measurement of riparian vegetation, statistical analyses, and a community classification linking vegetation to soils and environmental properties.	• Does not prescribe the use statistics and experimental design considerations. • Provides detailed instructions for techniques used.	Many variables were tested, of reference conditions. • No discussion on synthesis of data/information obtained.	Briggs et al., 1997 and their precision and accuracy were reported.	Myers, 1989
Inventory and monitoring of riparian areas. Myers (1989)	Hydrology, geomorphology, vegetation, soils, benthic invertebrates.	US Bureau of Land Management method. Two-stage approach: extensive inventory provides broad classification and identifies sites for detailed intensive inventory. Some parameters quantitative, some qualitative.	• Clear tabular presentation of parameters, reference(s) and relative merits of each. • Distinguishes between ecological status and management conditions.	• Two-stage approach uses consistent methods applied to either level of details.	None specified.	Briggs et al., 1997 Leonard et al., 1992 USDI, 1993
Procedures for ecological site inventory: riparian-wetland sites. Leonard et al. (1992)	Landscape, climate, soil, wildlife, vegetation.	US Bureau of Land Management ESI adapted for riparian wetlands. Presents a comprehensive list of attributes to fully describe an ecosystem using categories listed to the left.	• Designed with future application for classification and assessment in mind.	• To minimize labor, often relies on use of pre-existing data that may be of uncertain quality.	None specified; all are well-established descriptors of river and riparian systems.	All USDI, BLM Riparian Area Management Procedures e.g. USDI, 1993 Cagney, 1993
Inventory of riparian habitat. Briggs et al. (1997)	Hydrology: discharge, sediment movement, groundwater levels. Vegetation: diversity, density, stem diameter, percent cover.	Initial inventory of riparian habitat to create a baseline for comparison with future changes. Includes quantitative inventory as well as a detailed understanding of landscape features and history.	• Parameters allow for integration of physical and biotic information to examine cause and effect. • Could easily be adapted for a robust ecological assessment.	• Case study; only examines a single reach in catchment. • Uses site-specific historical information rather than rregional reference conditions.	None specified. Uses well-established methods for measurements.	None located

Table 2. Summary of classifications for semi-aquatic environments

Method	Class basis	Approach	Strengths	Limitations	Testing/validation	Selected references
Classification of wetlands and deepwater habitats of the United States. Cowardin et al. (1979)	System: water source Subsystem: water regime. Class: broad substr Subclass: specific substrate/vegetation.	Qualitative. Hierarchically groups wetland habitat. Main classification units include systems (5), subsystems (10), classes (54). Further description possible using subclass, dominance and modifier categories.	• High degree of objectivity responsible for observed precision.	Classification units do not reflect ecological processes.	Most extensively used and peer-reviewed method of its kind. Many other classifications based on this approach.	Basis for all US Fish and Wildlife National Wetlands Inventory (NWI) Information. Adamus et al., 1991 Adamus, 1992 Leibowitz et al., 1991
Hydrogeomorphic classification for wetlands (HGM). Brinson (1993)	Characteristics geomorphic setting, water source and transport, hydrodynamics Indicators: water characteristics, color, pH, nutrients, soils.	Functional-based classification with emphasis on hydrology and geomorpology. Qualitative and quantitative evidence describes each class. Develops functional description for each class (physical and biotic associations).	• Basis on physical forces enhances global application. • Distinguishes between wetland functions and values. • Prescribes use of long-term reference sites.	Does not provide a means to synthesize gathered information.	Widely accepted, but very little testing and validation. Ainslie (1994) reports positive results from a field test using creeks.	Wilson & Mitsch, 1996 Ainslie, 1994 Adamus, 1992
A hierarchical framework for stream classification: viewing streams in a watershed context. Frissel et al. (1986)	Capacity time, boundaries, substrate, water depth and velocity, overhead cover.	Streams classified hierarchically using several spatiotemporal scales, from watershed to microhabitat and from tectonic to seasonal variation.	• Considers spatial and temporal components. • Broad range of applications.	None.	Widely used especially by agencies. Design very flexible, so likely to work anywhere.	
River habitat survey (RHS). Raven et al. (1998)	Physical attributes: solid geology, altitude, slope, and size (flow category).	Rivers classified based on habitat quality. Focuses on physical characteristics of the stream and adjacent areas (banks and floodplain), and uses data from reference sites to generate a predictive model associating physical structure to habitat quality.	• Provides a standardized method for UK classifications and future habitat assessments. • Demonstrates excellent use of reference sites.	Use of physical structure as a surrogate measure of biodiversity may not be optimal.	Presently being widely tested in the U.K.	Boon et al., 1997 Brevin et al., 1998 Naura & Robinson, 1998 Wilkinson et al., 1998
Riparian community type classification of eastern Idaho–western Wyoming. Youngblood et al. (1985)	Vegetation.	Ecological sites classified by vegetation community type. Includes key, successional, and management information. Associated physical features also listed for each community type.	• Establishes a highly organized systematic approach for vegetative classification.	Regional specificity limits widespread application.	This method is the prototype for many subsequent riparian classifications.	Hansen et al., 1995 Padgett et al., 1989 Cagney, 1993

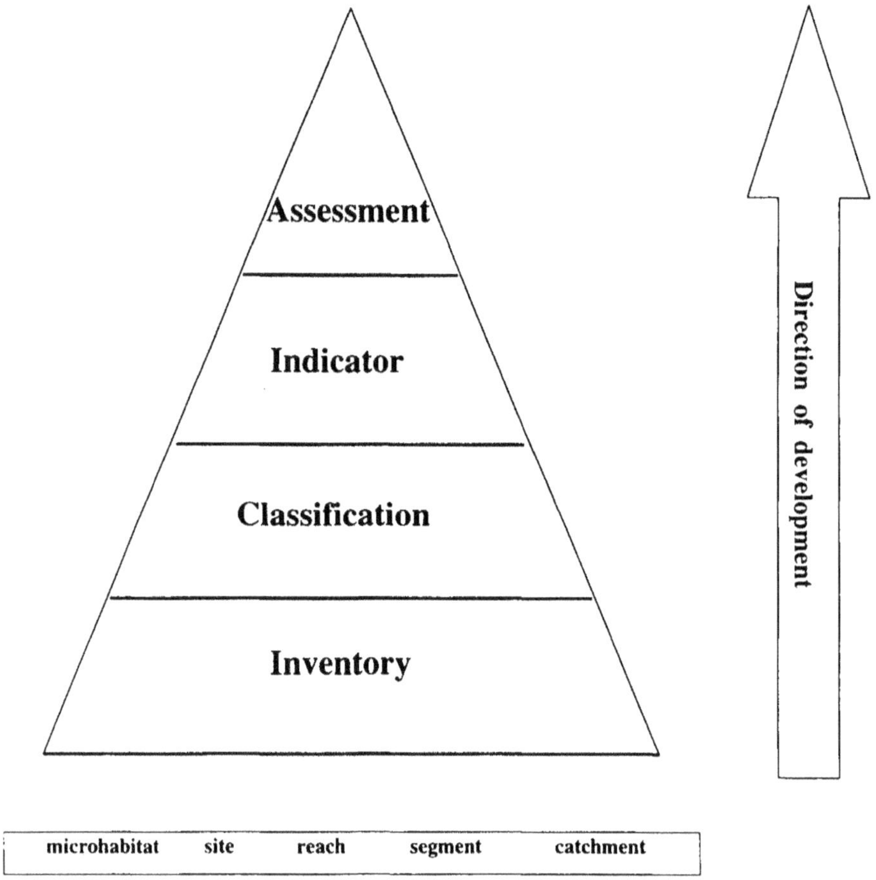

Figure 1. Hierarchical relationship between various forms of ecological information. Each level provides necessary information for the development of the next higher level. Assessment represents the best format for information to determine ecological information.

metric indices are highly integrative, allowing one to address broad-scale questions using a few carefully selected parameters (Adamus & Brandt, 1990; Croonquist & Brooks, 1991; Chovanec & Raab, 1997). However, the use of indicators requires a comprehensive understanding of the structural components of an ecosystem and the interactions, response to various stresses and spatial and temporal variability of these components. Rigorous testing and validation are necessary to establish such understanding and this requires time and resources to develop (Adamus & Brandt, 1990; Leibowitz et al., 1991).

The selection of indicators in formulating multimetric methods is critical. Strategic selections which cover appropriate spatial and temporal ranges and which reflect multiple environmental factors (physical, chemical and biotic considerations) provide robust results. Figure 2 illustrates the scale addressed by some widely used indicators and demonstrates how one can cover a wide range of spatial and temporal settings by using overlapping indicators. The use of carefully selected indicators maximizes the amount and quality of information about the ecological integrity of a system while minimizing the time and expense involved. Current indicators of riparian and wetland integrity are often vertebrate and invertebrate communities, although other biotic and physical indicators are also employed (Table 4).

Assessment

The assessment represents the highest level of ecological information for making decisions concerning the management of semi-aquatic environments (Figure 1). An assessment is an integrated statement regarding the current state and the factors that contribute to that state. It follows then that assessing ecological integrity requires consideration of physical, chemical and biological factors, an understanding of how these factors exist and interact in space and time and why they exist as they do in their present state. Much of this information is obtained through ecological inventories and

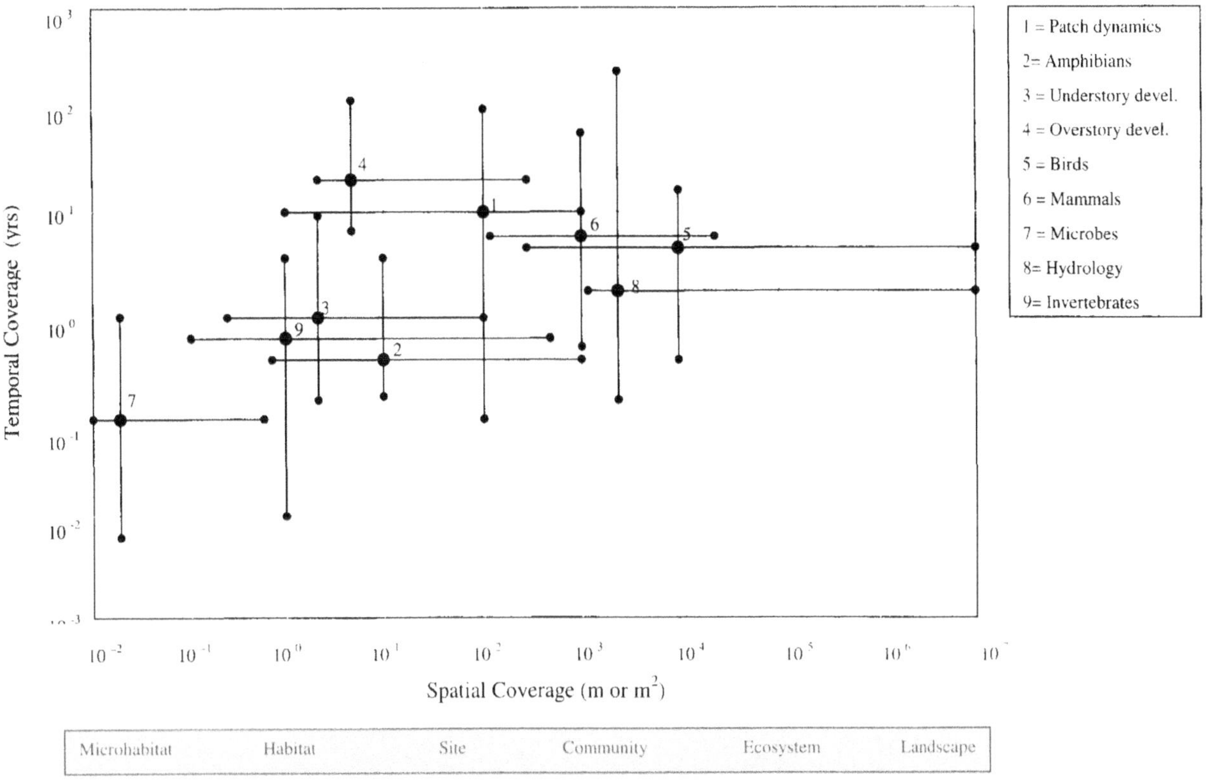

Figure 2. Spatial and temporal coverage of commonly used indicators in semi-aquatic ecosystems.

classifications; however, until inventories and classifications are analyzed and conclusions are drawn, they are simply tabular lists.

Assessing the assessments

Wetlands

Wetland assessment is a topic of current interest and the various approaches available have been extensively reviewed (e.g. Gebhardt et al., 1990; Adamus et al., 1992; Larson & Mazzarese, 1994; NRC, 1995). Other recent reviews have focused on the shortcomings of wetland assessments (Maltby et al., 1994; Kusler & Niering, 1998; Karr, 1999). Kusler & Niering (1998) report that over 40 rapid bioassessment methods were developed within the last decade, but found that most were not being used. Table 3 summarizes the principal riparian and wetland assessments currently available. Wetland assessments include the Wetland Evaluation Technique (WET: Adamus et al., 1991), the hydrogeomorphic classification (HGM: Brinson, 1993), and a wetland IBI (Index of Biotic Integrity) which is currently being developed and is the focus of considerable attention in the United States.

WET is the first technique to integrate broad ecological considerations such as landscape condition and hydrology in addition to vegetation and fauna. It synthesizes over 80 qualitative indicators into categories (high, medium, low) reflecting the probability that a given function or value is present. Each indicator is presented along with a discussion of its measurement, ranking, rationale, confidence, directness and importance. The method is usually applied consistently to all wetlands, which makes it broadly applicable, but unweighted consideration of all functions may cloud the importance of some functions that are especially important to certain types of wetlands. Nonetheless, the comprehensive list of indicators and their rationale enhance the adaptability of the method for users who might compromise a bit of consistency to gain more site-specific information. Another limitation of WET is the lack of distinction between ecological functions performed by the wetland and the values (i.e. those functions of value to human society). This is one of the major points of departure that led to the development of the HGM system.

Table 3. Summary of wetland and riparian assessments

Method	Parameters	Approach	Strengths	Limitations	Testing/validation	Selected references
			WETLANDS			
Wetland Evaluation Technique (WET). Adamus et al. (1991)	Indicators: climate, landscape descriptors, hydrology, soils, vegetation, geomorphology, chemical environment, invertebrates.	Functional assessment. Uses yes/no responses regarding ~ 80 qualitative indicators to categorize a group of 11 integrative functions and values as High, Medium, or Low.	• Simple approach. Integrates landscape conditions into evaluation. • Criteria are sensitive to different wetland classes.	• Allows focus to be limited to only those functions of societal value. • Not aimed at assessing ecological integrity.	Widely used by US government agencies. Extensively tested by over 100 users and Vol. II incorporated feedback.	Adamus, 1992
Hydrogeomorphic (HGM) Assessment. Brinson (1996)	Hydrologic and structural variables used in model. Specific choices dependent on function of interest.	Functional assessment. Models functions specific to an HGM subclass of wetland to determine if functioning differently from reference wetland.	• Builds on HGM classification, which has been applied to many wetlands.	• Does not address ecological integrity. • Lack of quantification.	Widely used, but not yet validated.	Wilson & Mitsch, 1996 Ainslie, 1994 USEPA, 1998
Wetlands Index of Biological Integrity (WIBI). USEPA (1998)		Wetlands IBI still in development. IBI uses metrics which demonstrate an empirical, predictable response to human impacts to generate an overall score.	• Utilizes quantifiable metrics for statistical robustness. • Compares conditions to regional reference wetlands. • Focus is biological integrity.	• Not yet fully developed for wetland application. • Requires initial investment to develop and test metrics for an area.	Based on Karr & Dudley's IBI (1981) which has been widely used and validated in many aquatic settings.	Karr, 1999
			RIPARIAN			
Integrated Riparian Evaluation Guide: Intermountain Region. USDA (1992)	Aquatic habitat, soils, hydrology, geomorphology, vegetation, terrestrial habitat, riparian resource rating.	Semi-quantitative, interdisciplinary approach to stratify, classify, and evaluate riparian areas. Uses three levels of evaluation: office procedures (level I), extensive field work (II), and site-specific intensive field work (III).	• Recognizes importance of watershed and landscape types, characteristics. • Advises integration and synthesis of data into a written interpretation.	• Presentation is complicated. • Approach allows for inconsistency.	Uses several validated methods.	None located

Continued on p. 119

Table 3. Continued

Method	Parameters	Approach	Strengths	Limitations	Testing/validation	Selected references
Riparian Evaluation and Site Assessment (RESA). Fry et al. (1994)	RE-flow regime SA- vegetation, channel morphology, erosion, wildlife, local land use, groundwater recharge, surface water quality, recreation, upland conditions.	Semi-quantitative. Based on SCS land evaluation and site assessment (LESA). Adapted to rank riparian areas based on functions, values and benefits they provide and to prescribe buffer widths for individual sites.	• Weighted scoring stresses most important features. • Approach is rapid and index generates an overall score.	• Does not prescribe use of reference conditions. • Focus on maintaining valuable functions rather than ecological integrity. • Subjective scoring may lead to inconsistency.	None indicated. However, based on LESA, which is widely used for rangelands in the USA.	None located
Assessment of Habitat of Wildlife Communities. Schroeder & Allen (1992)	Percent vegetative cover type compared to 1956, channel and floodplain complexity, percent reach subject to human disturbance.	Generates a quantitative model whose output is a Habitat Suitability index (HSI) for riparian vertebrates. Output predicts proportion 1956 wildlife species richness. Vegetation weighted over other attributes in the model.	• Provides temporal and spatial flexibility. • Concisely addresses three components of habitat integrity: hydrology, morphology, and vegetation.	• Criteria site-specific, so wide applicability uncertain. • Assumes 1956 vegetative communities undisturbed. • Does not look at historical state of other attributes (channel and foodplain complexity and human disturbance).	None indicated. Need to validate predictions of the index with actual species richness data.	None located
An assessment of riparian environmental quality by using butterflies and disturbance susceptibility scores. Nelson & Anderson (1994)	DSS ranked using: Species mobility, larval host-plant form and specificity, and riparian dependency.	Developed an index of butterfly riparian quality (BRQ) for vegetative habitat condition. Index based on species richness and disturbance susceptibility scores for individual taxa (DSS based on semi-quantitative scoring of species attributes related to riparian quality).	• Field component quick and inexpensive. • Relatively few, easily identifiable species.	• Not as effective in regions with less distinction between riparian and surrounding landscape • Only one pair of sites were compared (disturbed and undisturbed).	None indicated: the index score, however, reflected intuitive observable conditions.	None located

Continued on p. 120

Table 3. Continued

Method	Parameters	Approach	Strengths	Limitations	Testing/validation	Selected references
Avian and mammalian guilds as indicators of cumulative impacts in riparian-wetland areas. Croonquist & Brooks (1991)	Guilds based and scored for wetland dependency, trophic level, species status, habitat specificity, and seasonality (birds only).	Develops avian and mammalian response guilds based on life history and status for all Pennsylvania mammals and birds. Correlates responses to land use patterns at landscape level and investigates hierarchical patterns at the catchment level.	• Addressed landscape changes. Wide spatial coverage (catchment as well as riparian segments)	• Does not address alteration leading to the whole impact response. • Assessment endpoint is restoration for specific wildlife needs rather than ecological integrity.	Statistically supports intuitive expectations of observed changes.	None located
Dragonflies (Insecta. Odonata) and the ecological status of newly created wetlands. Chovanec & Raab (1997)	Dragonflies. Also discussed, much more generally, other common indicator species: frogs, fish, birds.	Assessed created wetlands using single indicator group. Sampled two created wetlands on the river Danube: one along a manmade island and another bordering a new channel.	• Developed using frequent sampling over several years. (Could yield a fine degree of temporal resolution.)	• Instead of collecting data at a reference site, used published data. • Clearly artificial structures could appear to have high integrity if they provide habitat.	Extensive knowledge of dragonfly life history requirements validates rating of habitat quality, but not ecosystem integrity.	None located
A qualitative procedure for the assessment of the habitat integrity. Kleynhans (1996)	Water abstraction, flow modification, channel modification, water wuality, inundation, indigenous vegetation removal, exotic vegetation encroachment, bank erosion.	Assesses relative significance of factors degrading river habitat quality. Scored instream and riparian separately then combined for overall evaluation. Used aerial survey and ground level observations.	• Provides a wide, general survey of modifications that affect instream habitats. • Allows separate or combined evaluation of instream and riparian information.	• Use of largely remotely sensed data and descriptive classes limits sensitivity. • Many riparian attributes scored were based on function with respect to instream environment rather than stand-alone riparian integrity.	Intentionally excluded an undisturbed tributary, rather than scoring it to test the method. Integrity class rating did not correlate well to observed ratio of intolerant to tolerant fish species.	None located

Continued on p. 121

Table 3. Continued

Method	Parameters	Approach	Strengths	Limitations	Testing/validation	Selected references
The RCE: a Riparian, Channel, and Environmental inventory for small streams in the agricultural landscape. Petersen (1992)	Upland use, riparian zone width and completeness, vegetation, retention devices, channel structure and sediments, bank undercut.	Describes habitat quality of stream sites for comparison with other sites in same region. Rapid, semi-quantitative scoring of 16 weighted metrics covering spatial scales from landscape to macrobenthos. Examines both instream and adjacent riparian zone, with emphasis on riparian metrics.	• Output easily understandable to non-scientists (color-coded maps reflect condition and management recommendations). • Quite rapid.	• Somewhat subjective. • High degree of correlation between many metrics. • Does not recommend the use of refence conditions.	Developed for lowland agricultural setting, but authors report that tested well in alpine streams and cold mountain interbasin desert. However, Stauffer and Goldstein (1997) report poor results in another agricultural setting.	None located
SERCON: System for evaluating rivers for conservation. Boon et al. (1997)	Scored: physical diversity, naturalness, representativeness, rarity, species richness, special features, human impacts.	Hierarchical approach to assess conservation status of U.K. rivers. Uses site, evaluated catchment sections (ECS), or subcatchment levels of spatial scale. Integrates riparian characteristics into scoring of ecological attributes of river as a whole.	• Results presented in a form easily understood by non-scientists. • Easy interpretation at different spatial scales. • Standardizes evaluations across U.K.	• Does not recommend the use of reference conditions. • Criteria chosen for comprehensives rather than careful integration. • Riparian-related attributes would poorly represent the riparian integrity if stood alone.	Recently developed, but has recently been applied in the U.K. widely. Still no definitive information regarding overall success.	Wilkinson et al., 1998 Boon et al., 1998

HGM is a classification scheme for wetlands based on hydrologic and geomorphic settings. HGM has also been increasingly applied as an assessment approach for wetlands. The classification is applied to assessment by associating ecological properties with the wetlands classes generated by the classification module. Therefore, the hydrogeomorphic class can be associated with certain characteristic functions and the presence, absence, or performance of functions can then be evaluated for functional assessment. Karr (1999) discusses some limitations to the application of HGM as an assessment method. Briefly, these include a lack of quantification and an overall score, insufficient attention to naturalness or ecological integrity, undemonstrated validation and the inability to address all functions of importance based on limited knowledge of fundamental wetland ecology. Despite these concerns, the US Army Corps of Engineers indicated an intention to implement HGM in 80% of Section 404 permits in the United States by 1998 (Federal Register, 1996).

Since its first publication, the index of biological integrity, or IBI (Karr & Dudley, 1981), has gained widespread use for the biological assessment of instream conditions in the United States. Presently, a Wetlands IBI is being developed in response to some of the criticisms of other available wetland assessments. The IBI is a multi-metric index that employs biological metrics (such as the number of sensitive taxa present, or relative proportion of functional groups) to formulate an overall score for a site in relation to reference conditions for that class of wetland. The choice of biological parameters represents an effort to use a small number of integrative indicators, as living organisms respond to their physical, chemical and biotic environments.

Riparian zones

Several methods have been developed for assessing riparian integrity, while many in-stream assessments include a strong riparian component. Table 3 reflects the following discussion of the more commonly employed riparian assessments, summarizing each method and suggesting the strengths and weaknesses of each.

The USDA Forest Service developed the Integrated Riparian Evaluation Guide which presents a three-level procedure including analytical methods, extensive field methods, and intensive site characterization (USDA, 1992). An inventory of dozens of physical (e.g. hydrologic, geomorphic), chemical (e.g. soil attributes) and biological parameters sets the stage for various classifications: stream and valley types, soils and vegetative communities. The actual ecological evaluation portion of the method is limited to a riparian rating of instream fish habitat, comparison to the vegetative potential natural community (PNC) and a descriptive discussion of management implications for an area.

The Riparian Evaluation and Site Assessment (RESA) method (Fry et al., 1994) was modeled after the US Soil Conservation Service land evaluation and site assessment (LESA) (SCS, 1987), which was developed to determine appropriate land use in agricultural areas. Similarly, this riparian counterpart ranks sites for appropriate management activities and prescribes riparian buffer widths. The RESA approach utilizes a semi-quantitative index that incorporates both physical and biological information and integrates upland conditions with site characteristics. Although the method neither prescribes the use of reference conditions nor tests the rating system using a reference site, a relatively undisturbed site that was evaluated did receive a high score. RESA is not explicitly designed to assess ecological integrity (the index is quite general), but the results do go beyond the simple prescription of buffer widths and describe the ecosystem rather well.

Schroeder & Allen (1992) present managers with a community-level habitat suitability index (HSI) for riparian wildlife. Their approach relies on remotely sensed aerial information and assesses three criteria determined to influence species richness of riparian wildlife: percent vegetative cover type compared to 1956 values, channel and floodplain complexity and percent of stream reach subject to human disturbance. The attributes are scored using a detailed descriptive index and the scores are incorporated in a model whose output is the HSI. The HSI then reflects the proportion of wildlife that could potentially exist in the study area in relation to the 1956 wildlife community. Such an output may be useful for managers; however, there is no indication of testing or validation that the score reflects actual conditions, or if the community in 1956 could actually be considered a reference (undisturbed) state.

Nelson & Anderson (1994) developed a novel assessment method using butterflies as indicators of riparian condition in the southwestern United States. They developed an index of butterfly riparian quality (BRQ) based on species richness and disturbance susceptibility scores for individual taxa (DSS; Croonquist & Brooks, 1991). DSS are generated for each species using carefully selected attributes that reflect changes in the entire riparian ecosystem. These attributes include species mobility, larval host-plant form and specificity and riparian dependency. The authors stress the use of reference sites when applying the BRQ since the method generates relative scores. The success of this method lies in the fact the indicator is restricted in its habitat due to the unique nature of the riparian zone in the arid landscape.

Kleynhans (1996) presents a descriptive, qualitative description of habitat integrity for the Luvuvhu River (South Africa) in order to assess the relative significance of various factors responsible for habitat degradation. Habitat integrity is assessed separately for the riparian zone and instream habitat, although riparian attributes assessed are generally those which influence the instream habitat, while the state of the riparian habitat itself is not explicitly addressed. The approach generates a weighted score for a given reach based on the degree of impact or modification. The integrity class rating did not always correlate well with observed indicators such as intolerant species, but such discrepancies may be attributed to the limited sensitivity of indices based on qualitative description. Still, this approach represents a solid first step toward the development of a more data-intensive assessment that addresses ecological integrity.

The Riparian, Channel and Environmental inventory (RCE: Petersen, 1992) was developed to assess physical and biological condition of small streams in a lowland, agricultural setting. This multi-metric index generates a score for a stream by integrating 16 characteristics of the stream channel and the adjacent riparian zone. Although the index places strong emphasis on physical attributes of the riparian zone, these attributes do not reflect riparian integrity so much as they reflect riparian functions important to instream integrity. This stands to reason, as the RCE was developed to describe instream conditions, which it seems to accomplish fairly adequately. The authors report that the method tested well in a variety of stream systems, although Stauffer & Goldstein (1997) report poorer results.

The System for Evaluating Rivers for Conservation (SERCON; Boon et al., 1997) represents another method for river evaluation that integrates riparian attributes into an overall river evaluation. SERCON is a weighted index which scores various attributes within the 'conservation criteria' of physical diversity, naturalness, representativeness, rarity, species richness and several special features. These criteria represent ecological integrity quite well and tightly integrate the various physical, chemical and biological attributes of the river system as a whole. The list of attributes assessed for each criteria is comprehensive, requiring a knowledgeable administrator and substantial time investment, but yields considerable information and resolution. While this approach does not employ reference conditions to test the index each time, other testing did assess high quality rivers and report positive results (Boon et al., 1998). SERCON is expected to gain in popularity in the United Kingdom, as it is closely associated with the widely used River Habitat Survey (RHS: Raven et al., 1997).

Attributes of successful approaches

Based on the previous review of individual methods, it became obvious that there was a specific need to articulate the fundamental components of a successful ecological assessment. The following discussion outlines some of the attributes that were absent or insufficiently developed in many methods.

Choice of endpoint

Ecological integrity is the most appropriate measure of ecosystem performance, rather than functional success or ecological status. Ecological integrity is defined as the capacity to support and maintain a balanced, integrated and adaptive biological system having the full range of elements and processes expected in a region's natural habitat (Karr & Dudley, 1981).

Of the wetland assessments, only the Wetland Index of Biotic Integrity (WIBI: currently under development) addresses ecological integrity; both WET and HGM are functional assessments. Only five of nine riparian assessments reviewed focused on ecological integrity as an endpoint. Further, riparian zones were also subject to a sort of functional assessment in some of the river assessments examined. For example, although the RCE (Petersen, 1992) presents an acceptable river assessment, only riparian functions that are of value to the instream environment (bank stability, sediment retention, shading) are considered, while little attention is given to riparian ecosystem integrity itself.

While functional success and ecological integrity may not seem mutually exclusive, their distinction can be illustrated with an example. Consider a swimming pool; this structure retains water and provides habitat for aquatic plants and invertebrates so could, therefore, be considered a pond or wetland from a functional standpoint. Would this system represent ecological integrity? Such questions not only clarify the distinction between functionality and ecological integrity, but also emphasize the value of reference conditions and naturalness.

Determining successional stage of a vegetative community compared to the potential natural community assesses 'ecological status' at a site (Myers, 1989; Leonard et al., 1992; USDA, 1992; USDI, 1993). Due to the dynamic nature of rivers, various successional stages will be observed along their length and over time (Naiman & Décamps, 1997). Disturbance plays an important ecological role and is important in maintaining ecological integrity. Therefore, early ecological status alone does not indicate an unhealthy system. In fact, a highly regulated river that exists for several years in the absence of disturbance may be characterized by late successional vegetation along much of its length. Common riparian colonizers and early seral species may decline, decreasing riparian species diversity and subsequently ecological integrity.

Reference conditions

A successful assessment considers the natural potential of an ecosystem, described by regional reference sites or conditions. Reference sites illustrate the spatial and temporal dynamics of natural ecosystems and are needed for comparisons between and within wetland and riparian systems. Most are based on 'natural' areas or, if unavailable, historical information may be used to make a 'best guess' about the ecological status of an undisturbed site within the appropriate climatic, geologic and landscape contexts. Also, a range of reference conditions from highly disturbed to pristine may be used for comparing and testing indicators, as well as training field crews (Brinson & Rhinehardt, 1996). While reference conditions are prescribed in two of three wetland assessments (HGM and WIBI), this is true for only two of the nine riparian assessments. Using undisturbed reference sites tests the performance of an assessment each time it is employed to ensure that it is appropriate for that region or system.

Scale

Appropriate spatial and temporal scales will vary with the question being asked, but should be clearly addressed. Functional and structural parameters should be carefully selected to provide information that is suited to the scale(s) addressed (Conquest & Ralph, 1998). In addition, interpretation of results should take into account natural spatiotemporal variation. Of all the methods reviewed, temporal considerations of seasonal or long-term dynamics were prescribed only by the Wetland Evaluation Technique (WET: Adamus et al., 1991). We assume that the resolution of the other methods reviewed would probably be on the scale of a season to several years. Fortunately, most methods can be adapted and applied over the long term for monitoring, so they have no upper limit.

On the other hand, the spatial scales covered by most techniques are broad, clustering around 10^3m and ranging over two to three orders of magnitude (Figure 3). Solid lines in Figure 3 represent the scale directly addressed by each assessment, while dashed lines extending beyond the solid lines indicate that the information gathered can be extrapolated to cover other spatial scales. To illustrate, the RESA method (Fry et al., 1994) used transects at several sites throughout a catchment. This yields detailed information about each site and an overall picture of the catchment, so points were placed at the transect and catchment scale where information is readily interpretable. In addition, dashed lines extend both below this scale to what we determined was the smallest scale of resolution possible for the information they gathered and above to the regional scale.

Unfortunately, many of the methods that have a wide geographical application are not well designed in their treatment of scale. Most methods have seemingly limitless catchment and regional applicability (Figure 3). However, in order to gather ecologically meaningful information over multiple spatial scales, a hierarchical approach is optimal. Frissell et al. (1986) provide an excellent approach to describe and evaluate stream habitat based on spatial hierarchies. This concept is perhaps easier to apply to river systems, which are naturally hierarchically ordered, although Mitsch (1992) does apply a hierarchical framework to wetlands.

Integration

An essential element for articulating the state of an ecological system is the careful integration of information into a statement of overall integrity. High-quality data are of little use if not interpreted in an ecologically relevant way. Integration aids in determining un-

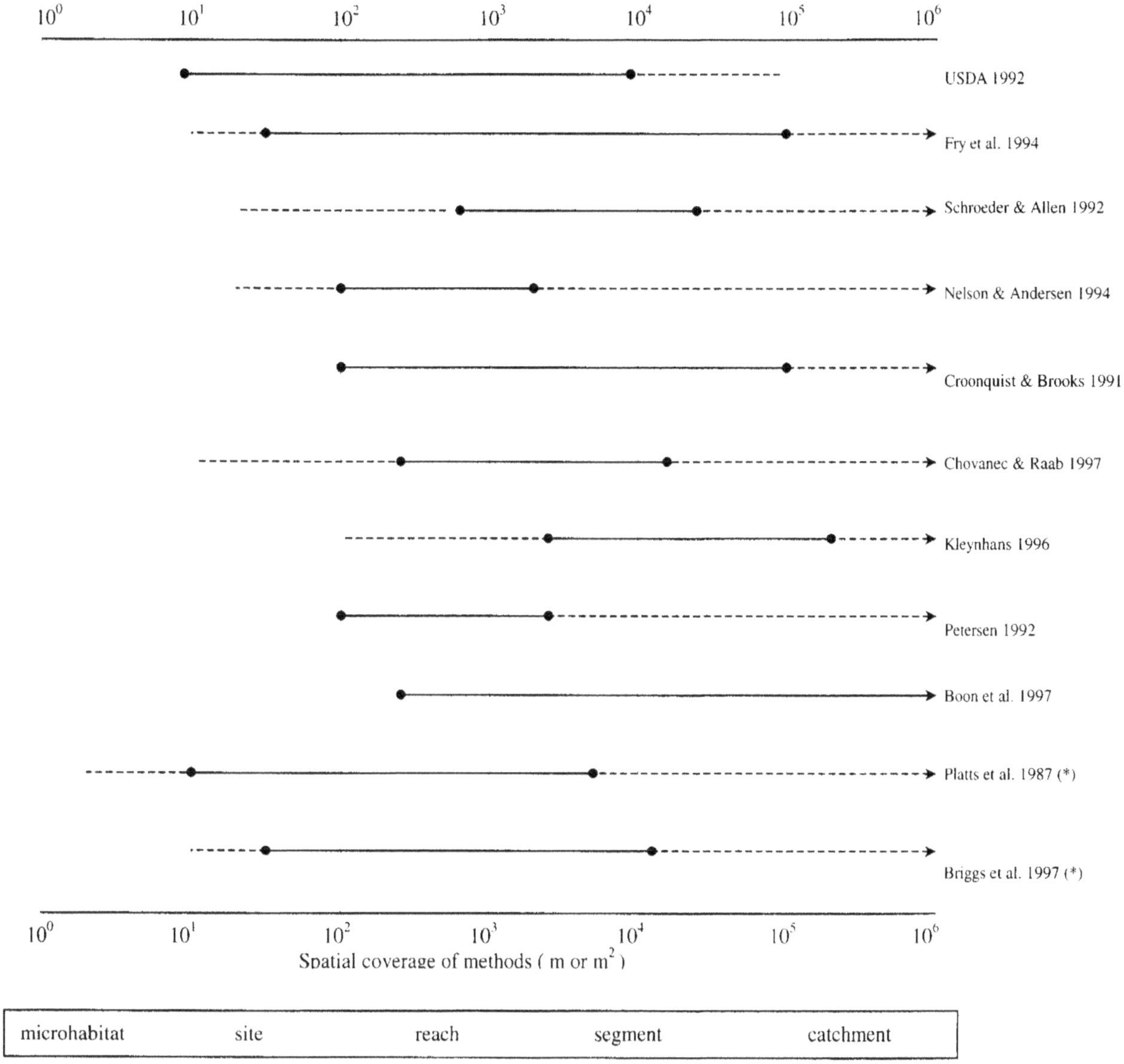

Figure 3. Spatial coverage of major assessment methods for semi-aquatic environments, including two inventory methods (*) with clearly defined spatial extent. Solid lines between points represent spatial scale each method directly addresses. Dashed lines represent scale at which the information obtained can be inferred. Arrows imply that a method could potentially be applied to an unlimited geographical range.

derlying causes for departures from desired conditions and for taking appropriate management actions.

Creating an integrated approach begins at the method design stage so that parameters and scale are chosen based on a clear understanding of their relationship to each other and to the questions being addressed (Conquest & Ralph, 1998). Indicators are useful integrators, and living organisms 'whose very existence represents the integration of conditions around them,' are the optimal indicator of ecosystem integrity (Karr, 1999). The issue of integration is well treated for riparian assessments, with seven of nine methods utilizing indices: these not only require ecological information to be integrated, but generate an overall score useful for management purposes. In addition, all three wetland assessments reviewed used an index in some capacity.

Assessing streamside integrity: a new strategy for riparian zones

Assessing the degree to which a riparian corridor has maintained or re-established its ecological integrity

Table 4. Indicators used in riparian and wetland ecological assessment

Indicator	Advantages	Disadvantages	Selected references
Invertebrates	• often highly specialized and therefore sensitive • short generation time allows for rapid response easily identified • life histories and ecology often well understood	• low perceived societal value • communities may be foraged selectively by fish/birds	Karr, 1991 Adamus & Brandt, 1990 Dragonflies: Chovanec & Raab, 1997 Butterflies: Nelson & Andersen, 1994 Hoverflies: Murphy et al., 1994
Birds and mammals	• high societal value • ecology and life history well worked-out • good landscape integrator present in all riparian and wetland classes	• mobility and seasonality create spatial spatial and temporal variability • difficult to assess causes of impacts for migratory or wide-ranging species	Cronquist & Brooks, 1991 Adamus & Brandt, 1990
Amphibians	• few species, many easily identifiable • especially sensitive to hydrologic and chemical alteration • most highly dependent on riparian or wetland areas for at least some portion of life history	• great inherent variation in dispersal • seasonal variation in abundance	Adamus & Brandt, 1990
Vegetation	• life history requirements and successional pathways well-known for most species • large amount of existing data on many riparian and wetland vegetative communities • sessile; cannot avoid negative impacts	• long recovery times for some species following disturbance • successional stage of vegetative community naturally variable; in dynamic systems, endpoints are not always realistic or optimal	Wetland plants: Murphy et al., 1994 Native aquatic plants: Medina, 1996 Adamus & Brandt, 1990
Hydrologic patterns	• long-term data often available from guaging stations • hydrologic alteration often an underlying cause for observed changes in biotic indicators	• does not provide information about biotic response to alterations; may not be sensitive to acute impacts from short-term events	Richter et al., 1996 Nestler & Long, 1997 Poff et al., 1997

is a formidable task considering the dynamic nature of these boundary ecosystems. Riparian zones are shaped by hydrologic and geomorphic forces (Malanson, 1994; Naiman & Décamps, 1997; Naiman et al., 1998). Few other ecological systems are so completely defined by change and disruption. Recognizing a healthy ecological template against a background of spatially and temporally diverse hydrologic, geomorphic and faunistic processes as well as upland erosional forces requires a unique suite of sampling parameters that effectively reveal the structural and functional integrity of the system in question. While

Table 5. Limitations and recommendations for future riparian and wetland assessments*

Limitation	Recommendation
• Scientific information may be improved	• Better attention to fundamental components (see this paper) • Recognize the 'principle of place' (e.g. Dale et al., 1999)
• Recognize gaps and limits in scientific knowledge, including uncertainty	• Formally field test techniques and indicators
• Deliver too little information for time and cost involved	• Develop rapid techniques • Test techniques for cost and practicality
• Not accurate enough for broad application	• Develop hierarchical techniques applicable at multiple levels of detail and expertise
• Inadequate consensus building for multi-agency and multi-cultural settings	• Openly discuss policy aspects of specific assessment methods
• Too many methods	• Retire out-of-date methods • Strengthen and cross calibrate useful ones
• Inadequate training and support for users	• Consider a wide variety of users and situations • Provide regular training opportunities by developers

*Adapted from Karr (1999) and Kusler & Neiring (1998).

each of these shaping forces can be directly measured, effective assessments should measure and predict integrated biological responses to a biophysical template. The parameters we propose as indicators reflect this viewpoint:

Microclimate

Given their position between terrestrial and aquatic systems, riparian zones typically maintain unique microclimates and microclimatic gradients, so that conditions within the riparian zone reflect conditions in adjacent systems (Brosofske et al., 1997; Chen et al., 1999). These seemingly subtle gradients have much stronger interactions with biological communities than previously thought. Anthropogenic disturbances in or around the riparian corridor typically impact microclimatic gradients; indeed, riparian disturbance has been shown to alter pre-existing levels of relative humidity, air temperature, soil temperature and solar radiation (Brosofske et al., 1997).

Patch Heterogeneity

Most natural riparian forests typically support a mosaic of patches, within which species with different life history strategies or at different successional stages may become successful. Spatial patterns of tree species, height and diameter reflect the maturity of a site following disturbance. Typical riparian species which are adapted to withstand periodic inundation (e.g. willow) will be established in varying sizes close to the channel, while large old-growth members will be situated on protected microsites. Patch heterogeneity largely reflects the integrity of the discharge regime but may also integrate upland impacts such as logging and altered landform dynamics.

Biodiversity

In addition to structural diversity within the riparian corridor (patch heterogeneity), species diversity is an important index of riparian integrity. The biological community (e.g. herbaceous or woody plants, small mammals, amphibians, soil invertebrates, birds) supported within a patch in the landscape is an integrated response to the biophysical template. In riparian and wetland systems, flood frequency, productivity and spatial heterogeneity have all been observed in positive correlation with plant species richness (Pollock, 1998; Pollock et al., 1998). Although this topic requires further investigation prior to identifying species diversity as an interpretable recovery parameter, other assessment methods (e.g. IBI) have successfully established the species richness of selected taxa as a useful metric of biological integrity.

Terrestrialization

The increasing relative abundance of terrestrial species in the riparian corridor (terrestrialization) may signal the degree to which a riparian forest has been impacted by water regulation. Hydrologically regulated streams typically do not experience the highly variable discharge fluctuations that characterize most natural systems. In addition, hydrologic regulation may reduce seasonal or annual flows. In the absence of such physical and biological constraints, species that are typically excluded from growth along the stream by physical factors may become successfully established. For example, most coniferous species are ill-equipped to survive within frequently flooded areas. Less variable discharges in regulated systems may allow for such terrestrial species to encroach within the riparian corridor while hindering the recruitment and growth of typical riparian species.

Seston

The nature and quantity of suspended particulate matter (seston) in streams is largely a function of various physical and biological parameters of the stream channel and the surrounding riparian forest. Ecological factors such as woody debris, canopy cover and discharge regime all play significant roles in the dynamics of seston. Seston analysis shows promise as a method for assessing the condition of the riparian habitat. At baseflow, for instance, seston chlorophyll concentrations and C:N ratios are higher and lower, respectively, in more impacted systems. Hysteresis curves which plot the relationship between seston and discharge during storm events may also be markedly different across a range of conditions and may be valuable in assessing ecological integrity. We suggest the above parameters for developing a more robust assessment scheme based on biophysical characteristics and processes associated with riparian zones. All parameters relate back to the fundamental notion that riparian corridors are comprised of a mosaic of patches and microsites, created by a blend of abiotic and biotic processes (Pringle et al., 1988; Malanson, 1994). In addition, the proposed parameters overlap in the biophysical forces they integrate (e.g. both terrestrialization and patch heterogeneity largely reflect the integrity of the flow regime) as well as in their spatial and temporal resolution.

Concluding remarks and recommendations

Currently employed wetland and riparian assessments are generally based on good science and provide useful information at some levels; however, the approaches are not being used effectively. We feel that careful treatment of endpoint, reference conditions, scale and integration will improve scientific rigor. Kusler & Niering's (1998) finding that most wetland assessment methods are not even being employed should indicate that good and useful science is not enough. Karr (1999) and Kusler & Niering (1998) have identified some of the key issues limiting the utility of wetland assessments which also apply to current riparian assessments (Table 5).

We suggest that there are immediate actions to be taken in order to make wetland and riparian assessment methods more available to practicing professionals (Table 5). This should begin with the retirement of out-of-date methods while strengthening and cross-calibrating useful ones. Aquatic scientists and managers need to collaborate on continued development of useful methods. The 'principle of place' (Dale et al., 2000) is an especially important scientific consideration in that every habitat has a relationship with the surrounding landscape and interpretation of site-specific data needs to be conducted within the larger spatial and temporal perspective. Of course, any technique needs to be formally tested for accuracy, cost and practicality before widespread implementation.

Truly rapid assessment methods that provide a reasonable level of accuracy for everyday use by a variety of users and in a variety of situations are needed to cope with the thousands of assessments required annually to meet legal obligations. Although many of the methods in use today are scientifically rigorous, they are not effective in stemming environmental degradation because of the time and expertise required for evaluations. At the same time, it is important to recognize that there are a wide variety of specific applications and users and individual assessment methods may need to be crafted for specific situations.

The diversity of assessments employed as well as the variation that characterizes semi-aquatic ecosystems may result in disparate information. In addition, there will always be gaps and limits in scientific knowledge and this uncertainty should be clearly stated. Finally, every method will have inherent policy implications which should be openly discussed between scientists, managers, decision-makers and the public. The development and implementation of each of these recommendations is beyond the scope of this review; indeed, each one presents a challenge to future scientists and managers. It is our hope that a better understanding of the current state of assessments will set the stage for this next critical step.

Acknowledgements

We thank James R. Karr and Robert C. Wissmar for discussions and comments that substantially improved the manuscript. Financial support for this review was provided by the Andrew W. Mellon Foundation and the Pacific Northwest Research Station of the U.S. Forest Service.

References

Adamus, P. R., 1992. Wetlands data sources and collection methods. In Statewide wetland strategies: a guide to protecting and managing the resource. World Wildlife Fund, Island Press, Washington, D.C.: 171–224.

Adamus, P. R. & K. Brandt, 1990. Impacts on quality of inland wetlands of the United States: a survey of indicators, techniques and applications of community-level biomonitoring data. U.S. Environmental Protection Agency, Environmental Research Laboratory, Corvalis, OR, U.S.A.: 406 pp.

Adamus, P. R., L. T. Stockwell, E. J. Clairain Jr., M. E. Morrow, L. P. Rozas & R. D. Smith, 1991. Wetland Evaluation Technique (WET) Vol. I: Literature Review and Evaluation Rationale. WRP-DE-2. U.S. Army Corps of Engineers Waterways Experiment Station, Vicksburg, MS. U.S.A.: 299 pp.

Ainslie, W. B., 1994. Rapid wetland functional assessment: its role and utility in the regulatory arena. Wat. Air Soil Pollut. 77: 433–444.

Boon, P. J., J. Wilkenson & J. Martin, 1998. The application of SERCON (system for evaluating rivers for conservation) to a selection of rivers in Britain. Aquat. Cons. 8: 597–616.

Boon, P. J., N. T. H. Holmes, P. S. Maitland, T. A. Rowell & J. Davies, 1997. A system for evaluating rivers for conservation (SERCON): development, structure and function. In Boon, P. J. & D. L. Howell (eds), Freshwater Quality: Defining the Indefinable. Her Majesty's Stationary Office, Edinburgh: 299–326.

Brevin, P. A., S. T. Buckton & S. J. Omerod, 1998. River habitat surveys and biodiversity in acid-sensitive rivers. Aquat. Cons. 8: 501–514.

Briggs, M. K., M. K. Schmid & W. L. Halvorson, 1997. Monitoring riparian ecosystems: an inventory of riparian habitat along Rincon Creek near Tucson, AZ. Technical Report No. 58. USGS, Tucson, AZ: 78 pp.

Brinson, M. M., 1993. A hydrogeomorphic classification for wetlands. Wetlands Research Program Technical Report WRP-DE-4. Final Report. U.S. Army Corps of Engineers Waterways Experiment Station, Vicksburg, MS: 101 pp.

Brinson, M. M., 1996. Assessing wetland functions using HGM. National Wetlands Newsletter 18: 10–16.

Brinson, M. M. & R. Rhinehardt, 1996. The role of reference wetlands in functional assessment and mitigation. Ecol. Applic. 6: 69–76.

Brosofske, K. D., J. Chen, R. J. Naiman & J. F. Franklin, 1997. Harvesting effects on microclimatic gradients from small streams to uplands in western Washington. Ecol. Applic. 7: 1188–1200.

Cagney, J., 1993. Riparian area management: Greenline riparian-wetland monitoring. Bureau of Land Management /SC/ST-93/001+1737, Denver, CO: 45 pp.

Chen, J., S. Saunders, T. Crow, R. J. Naiman, K. Brosofske, G. Mroz, B. Brookshire, & J. F. Franklin, 1999. Microclimate in forest ecosystem and landscape ecology. BioScience 49: 288–297.

Chovanec, A. & R. Raab, 1997. Dragonflies (Insecta, Odonta) and the ecological status of newly created wetlands – examples for long term bioindication programmes. Limnologica 27: 381–392.

Commission of the European Communities, 1995. Wise use and conservation of wetlands. Communication from the Commission to the Council and the European Parliament COM (95) 189 final, 29.05.95, Brussels: 54 pp.

Conquest, L. L. & S. C. Ralph, 1998. Statistical Design and Analysis Considerations for Monitoring and Assessment. In Naiman, R. J. & R. E. Bilby (eds), River Ecology and Management: Lessons from the Pacific Coastal Ecoregion. Springer-Verlag, New York: 455–475.

Cowardin, L. M., V. Carter, F. C. Golet & E. T. LaRoe, 1979. Classification of wetlands and deepwater habitats in the United States. FWS/OBS-79/31, US Fish and Wildlife Service, Washington, D.C: 103 pp.

Croonquist, M. J. & R. P. Brooks, 1991. Use of avian and mammalian guilds as indicators of cumulative impact in riparian-wetland areas. Envir. Man. 15: 701–714.

Dale, V. H., S. Brown, R. Haeuber, N. T. Hobbs, N. Huntley, R. J. Naiman, W. E. Riebsame, M. G. Turner & T. Valone, 2000. Ecological principles and guidelines for managing the use of land. Ecol. Applic.

Federal Register, 1996. National action plan to develop the hydrogeomorphic approach for assessing wetland functions. Federal Register 61: 42593–42603.

Frissell, C. A., W. J. Liss, C. E. Warren & M. D. Hurley, 1986. A hierarchical framework for stream habitat classification: viewing streams in a watershed context. Envir. Man. 10: 199–214.

Fry, J., F. R. Steiner & D. M. Green, 1994. Riparian evaluation and site assessment in Arizona. Landscape & urban Planning 28: 179–199.

Gebhardt, K., S. Leonard, G. Staidl & D. Prichard, 1990. Riparian area management: riparian and wetland classifiction review. US Department of the Interior, Bureau of Land Management, TR-1737-5, Denver, CO: 56 pp.

Hansen, P. L., R. D. Pfister, K. Boggs, B. J. Cook, J. Joy & D. K. Hinkley, 1995. Classification and management of Montana's riparian and wetland sites. Montana Forest and Conservation Experiment Station, School of Forestry, University of Montana. Missoula, MT: 646 pp.

Karr, J. R., 1991. Biological integrity: A long neglected aspect of water resource management. Ecol. Applic. 1: 66–84.

Karr, J. R., 1999. Seeking suitable endpoints: biological monitoring for wetland assessment. Report to U.S. Environmental Protection Agency, Region 10, Seattle: 55 pp.

Karr, J. R. & D. R. Dudley, 1981. Ecological perspectives on water quality goals. Envir. Man. 5: 55–68.

Kleynhans, C. J., 1996. A qualitative procedure for the assessment of the habitat integrity of the Luvuvuhu River (Limpopo System, South Africa). J. Aquat. Ecosystem Health 5: 41–54.

Kondolf, G. M., 1995. Geomorphological stream channel classifiction in aquatic habitat restoration: uses and limitations. Aquat. Cons.: mar. & fresh. Ecosystems 5: 127–141.

Kusler, J. & W. Niering, 1998. Wetland assessment: Have we lost our way? National Wetlands Newsletter 20: 9–20.

Larson, J. S. & D. B. Mazzarese, 1994. Rapid assessment of wetlands: history and application to management. In Mitsch, W. J. (ed.), Global Wetlands: Old World and New. Elsevier, Amsterdam: 625–636.

Leibowitz, N. C., L. Squires & J. P. Baker, 1991. Research plan for monitoring wetland ecosystems. EPA/600/3-91/010. U.S. Environmental Protection Agency, Environmental Research Laboratory. Corvalis, OR: 168 pp.

Leonard, S., G. Staidl, J. Fogg, K. Gebhardt, W. Hagenbuck & D. Pritchard, 1992. Riparian area management: procedures for ecological site inventory-with special reference to riparian-wetland sites. US Department of the Interior, Bureau of Land Management TR-1737-7. Denver: 135 pp.

Malanson, G. P., 1994. Riparian Landscapes. Cambridge University Press, Cambridge: 296 pp.

Maltby, E., D. V. Hogan, C. P. Immirzi, J. H. Tellam & M. J. Van Der Peijl, 1994. Building a new approach to the investigation and assessment of wetland ecosystem functioning. In Mitsch, W. J. (ed.), Global Wetlands: Old World and New. Elsevier, Amsterdam: 637–658.

Medina, A. L., 1996. Native aquatic plants and ecological condition of southwestern wetlands and riparian areas. In Desired Future Conditions for Southwestern Riparian Ecosystems: Bringing Interests and Concerns Together. U.S. Department of Agriculture, Forest Service. RM-GTR-272. Albuquerque, NM: 359 pp.

Myers, L. H., 1989. Riparian area management: Inventory and monitoring riparian areas. US Department of the Interior, Bureau of Land Management TR-1737-7. Denver, CO: 79 pp.

Mitsch, W. J., 1992. Combining ecosystem and landscape approaches to Great Lakes wetlands. J. Great Lakes Res. 18: 552–570.

Mosley, M. P., 1987. The classification and characterization of rivers. In Richards, K. (ed.), River Channels, Environment and Process. Basil Blackwell, Oxford: 295–320.

Murphy, K. J., E. Castella, B. Clément, J. M. Hills, P. Obrdlik, I. D. Pulford, E. Schneider & M. C. D. Speight, 1994. Biotic indicators of riverine wetland ecosystem functioning. In Mitsch, W. J. (ed.), Global Wetlands: Old World and New. Elsevier, Amsterdam: 659–682.

Naiman, R. J. & H. Décamps, 1997. The ecology of interfaces – riparian zones. Ann. Rev. Ecol. Syst. 28: 621–658.

Naiman, R. J., D. G. Lonzarich, T. J. Beechie & S. C. Ralph, 1992. General principles of classification and the assessment of conservation potential in rivers. In Boon, P. J., P. Calow & G. E. Petts (eds), River Conservation and Management. John Wiley & Sons, New York: 92–123.

Naiman, R. J., J. J. Magnuson, D. A. McKnight & J. A. Stanford, 1995. The Freshwater Imperative: A Research Agenda. Island Press, Washington, D.C: 165 pp.

Naiman, R. J., K. L. Fetherston, S. J. McKay & J. Chen, 1998. Riparian Forests. In Naiman, R. J. & R. E. Bilby (eds), River Ecology and Management: Lessons from the Pacific Coastal Ecoregion. Springer-Verlag, New York: 289–323.

National Research Council, 1995. Wetlands: Characteristics and Boundaries, 10. National Academy Press, Washington, D.C.: 307 pp.

Naura, M. & M. Robinson, 1998. Principles of using river habitat survey to predict the distribution of aquatic species: an example applied to the native white-clawed crayfish *Austropotamobius Pallipes*. Aquat. Cons. 8: 515–527.

Nelson, S. M. & D. C. Andersen, 1994. An assessment of riparian environmental quality by using butterflies and disturbance susceptibility scores. The Southwestern Naturalist 39: 137–142.

Nestler, J. M. & K. S. Long, 1997. Development of hydrological indices to aid cumulative impact analysis of riverine wetlands. Regul. Rivers: Research & Management 13: 317–334.

Padgett, W. G., A. P. Youngblood & A. H. Winward, 1989. Riparian community type classification of Utah and southeastern Idaho. R4-Ecol-89-01. U.S. Department of Agriculture, Forest Service. Intermountain Region. Ogden, UT: 191 pp.

Petersen, R. C. Jr., 1992. The RCE: a riparian, channel, and environmental inventory for small streams in the agricultural landscape. Freshwat. Biol. 27: 295–306.

Platts, W. S., C. Armour, G. D. Booth, M. Bryant, J. L. Bufford, P. Culpin, S. Jensen, G. W. Lienkaemper, G. W. Minshall, S. B. Monsen, R. L. Nelson, J. R. Sedell & J. S. Tuhy, 1987. Methods for evaluating riparian habitats with applications to management. U.S. Department of Agriculture, Forest Service. General Technical Report INT-221. Ogden, UT: 177 pp.

Poff, N. L. & 7 co-authors, 1997. The natural flow regime. BioScience 47: 769–784.

Pollock, M. M., 1998. Biodiversity. In Naiman, R. J. & R. E. Bilby (eds), River Ecology and Management: Lessons from the Pacific Coastal Ecoregion. Springer-Verlag, New York: 430–454.

Pollock, M. M., R. J. Naiman & T. A. Hanley, 1998. Plant species richness in riparian wetlands – a test of biodiversity theory. Ecology 79: 94–105.

Pringle, C. M., R. J. Naiman, G. Bretschko, J. R. Karr, M. W. Oswood, J. R. Webster, R. L. Welcomme & M. J. Winterbourn, 1988. Patch dynamics in lotic systems: The stream as a mosaic. J. N. am. Benthol. Soc. 7: 503–524.

Raven, P. J., P. Fox, M. Everard, N. T. H. Holmes & F. H. Dawson, 1997. River habitat survey: a new system for classifying rivers according to their habitat quality. In Boon, P. J. & D. L. Howell (eds), Freshwater Quality: Defining the Indefinable. Her Majesty's Stationary Office, Edinburgh, U.K.: 215–234.

Richter, B. D., J. V. Baumgartner, J. Powell & D. P. Braun 1996. A method for assessing hydrologic alteration within ecosystems Cons. Biol. 10: 1163–1174.

Rosgen, D. L., 1994. A classification of natural rivers. Catena 22: 169–199.

Schroeder, R. L. & A. W. Allen, 1992. Assessment of habitat of wildlife communities on the Snake River, Jackson, Wyoming. U.S. Department of the Interior, Fish and Wildlife Service, Resource Publication 190. Washington, D.C.: 21 pp.

Soil Conversation Service, U.S. Dept. of Agriculture, 1987. National agricultural land evaluation and site assessment handbook 310:1. Washington, D.C.

Stauffer, J. C. & R. M. Goldstein, 1997. Comparison of three qualitative habitat indices and their applicability to prairie streams. N. Am. J. Fish. Manage. 17: 348–361.

Swift, B. L., 1984. Status of riparian ecosystems in the United States. Wat. Res. Bull. 20: 223–227.

U.S. Department of Agriculture, 1992. Integrated riparian evaluation guide: intermountain region. Forest Service. Ogden, UT.

U.S. Department of the Interior, 1993. Riparian area management: Process for assessing proper functioning condition. TR-1737-3. Bureau of Land Management. Denver, CO: 51 pp.

U.S. Environmental Protection Agency, 1998. Wetland Bioassessment Fact Sheets. EPA 843-F-98-001. Office of Wetlands, Oceans, and Watersheds, Office of Water, USEPA, Washington, D.C.

Wilcove, D. S., D. Rothstein, J. Dubrow, A. Phillips & E. Losos, 1998. Quantifying threats to imperiled species in the United States. BioScience 48: 607–615.

Wilkinson, J., J. Martin, P. J. Boon & N. T. H. Holmes, 1998. Convergence of field survey protocols for SERCON (system for evaluating rivers for conservation) and RHS (river habitat survey). Aquat. Cons. 8: 579–596.

Wilson, R. F. & W. J. Mitsch, 1996. Functional assessment of five wetlands constructed to mitigate wetland loss in Ohio, U.S.A. Wetlands 16: 436–451.

Youngblood, A. P., W. G. Padgett & A. H. Winward, 1985. Riparian community type classification of eastern Idaho – western Wyoming. R4-Ecol-85-01. U.S. Department of Agriculture, Forest Service, Intermountain Region. Ogden, UT: 78 pp.

Hydrobiologia **422/423**: 133–142, 2000.
M. Jungwirth, S. Muhar & S. Schmutz (eds), Assessing the Ecological Integrity of Running Waters.

Towards the assessment of 'ecological integrity' in running waters of the United Kingdom

David M. Harper[1], Joanna L. Kemp[1], Birgit Vogel'[2] & Malcolm D. Newson[3]
[1]*Department of Biology, University of Leicester, Leicester LE1 7RH, England, U.K.*
[2]*Department of Hydrobiology, Fisheries and Aquaculture, University of Agricultural Sciences, Vienna, Austria*
[3]*Department of Geography, University of Newcastle, Newcastle, England, U.K.*

Key words: ecological quality, ecological integrity, BMWP, macroinvertebrates, river conservation, functional habitats, flow biotopes

Abstract

The development of methods to assess ecological integrity of running waters requires integrating physical and chemical parameters, as well as their effects upon biological structure, diversity and processes. River management in the U.K. has seen the development of a series of quality assessment methodologies with little integration between them. Methods which assess habitat structure in a biologically meaningful way offer the best approach to integrating measures of the effects of physical and chemical processes upon the river environment. Simultaneous measurement of surface flow types ('flow biotopes') and habitat structure ('functional habitats') showed clear links. These links could form the first stage in the development of a rapid, cost-effective measure of ecological integrity in rivers of the U.K.

Introduction

In many countries of the European Union, there has been rapid development of practical methods for evaluating ecological integrity in preparation for the implementation of the EU Water Framework Directive (Bloch, 1999). Outside Europe, the concept of 'biological integrity' has been a cornerstone of legislation in the U.S.A. for over 20 years (Karr & Chu, 2000). In the U.K., however, there has been considerable discussion but little practical development of methods for assessing 'integrity' as opposed to 'quality' in running waters. Why is this?

The probable answer lies in the strong, but parallel growth of an array of methods for assessing quality from several different perspectives. A simple classification of the approach developed in the U.K. over the last decade is provided in Figure 1.

Assessment of water quality

The best known is water quality assessment directed at the impact of organic pollution, known as the BMWP method (from Biological Monitoring Working Party); it uses hand-net sampling and family-level taxonomy of macroinvertebrates to generate a site's total and average score (ASPT: Average Score per Taxon) (Furze et al., 1981). This was introduced in the early 1980s to supersede the Chandler Biotic Score and before that the Trent Biotic Index, both of which (though successful) were seen as too biased towards upland river communities (Chapman, 1992). A more recent development of this methodology is the use of the predictive power of a model, RIVPACS, which identifies deviation from expected score and invertebrate assemblage, based upon an extensive dataset of invertebrate and morphological/hydrological data from clean rivers throughout Britain (Wright et al., 1994, 1997, 1998).

Additional water quality assessment is provided by aquatic macrophytes, whose distribution has been extensively studied over the past 25 years. These now provide a ten-class typology of British rivers and a developing methodology for quality assessment, known as 'mean trophic ranking' (MTR); it is indicative more of nutrient contamination than of organic pollution

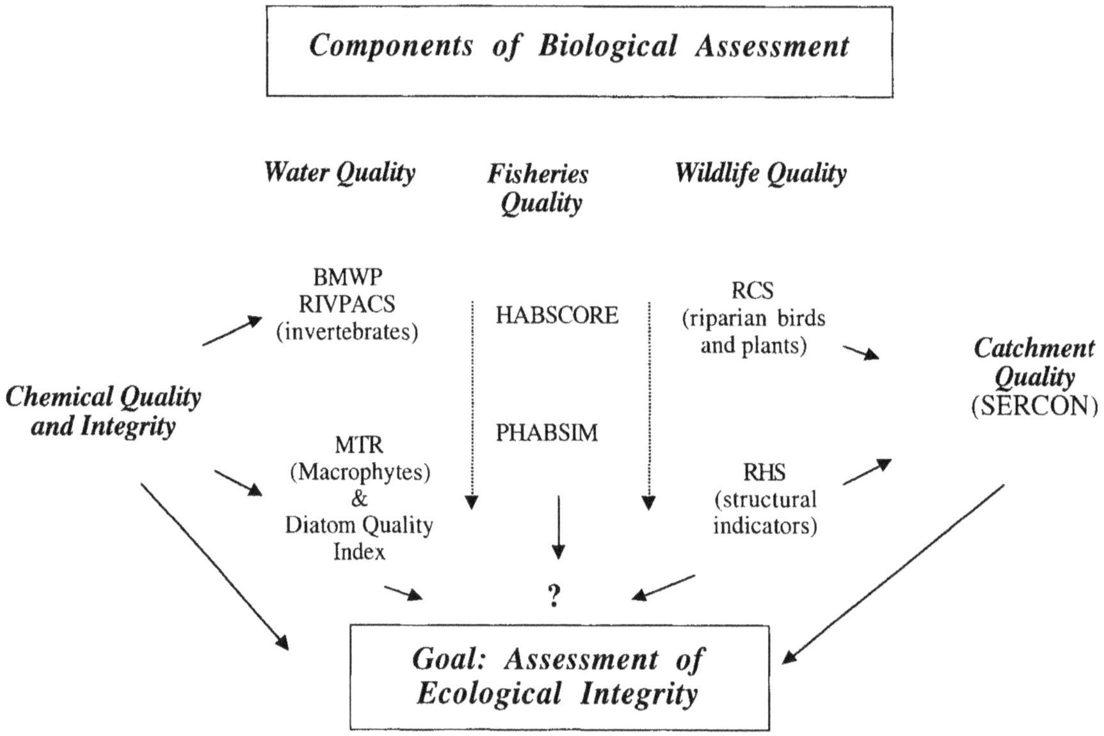

Figure 1. A schematic diagram of the components of biological quality assessment of rivers developed in the U.K. over the past 20 years. Three, more or less independent disciplines, have developed their own systems for measuring 'quality' as defined by the discipline (vertical downwards), with chemical influence strongest from left to right. Although several of these systems may also be used for wider ecological assessment (e.g. PHABSIM, HABSCORE) the goal of 'ecological integrity' requires an integrated approach which has not yet been achieved.

(Holmes et al., 1998). Finer detail is provided by the use of diatoms, whose index parallels the plant MTR. This index can also be used (through the proportions of particular species) to calculate a probable influence of organic pollution at a site whose flora demonstrates nutrient contamination. An example of its use would be downstream of treated sewage effluent where the impact of organic pollution gives way to nutrient pollution as 'self-purification' proceeds (Kelly & Whitton, 1998).

The development of such indices has relied upon the analyses of large datasets using multivariate methods, such as TWINSPAN (Hill, 1979) in the case of the RIVPACS method. Multivariate analyses have also been used in their own right, particularly methods which allow simultaneous analyses of environmental data and biological (species/abundance/sites) data. Probably most widely used has been CANOCO and its derivatives (Ter Braak, 1988). Both methods have been used to detect community changes arising from water quality deterioration (Gower et al., 1995) or physical habitat deterioration.

Fish have been used as water quality indicators in the U.K. in a rough four-class system (e.g. salmonids; coarse fish dominated by species such as the chub *Leuciscus leuciscus*; coarse fish dominated by the roach *Rutilus rutilus*; no fish), probably because there are so few species in U.K. rivers compared with continental rivers. Moreover, the greatest damage to fish populations in Britain may well have been caused by engineering (for agriculture, drainage and flood control), altering discharge regime and/or channel structure. Fisheries quality is predominantly assessed by a habitat-assessment system known as HABSCORE, reflecting the fact that physical, rather than chemical, damage is the main threat to quality (Milner et al., 1998). Over the past five years the U.K. has also tested the US-developed model, PHABSIM, for predicting the habitat requirements of individual species, with mainly fish but also some invertebrates as the target species (Milner et al., loc cit.; Johnson et al., 1994).

The methods for assessing aspects of biological water quality have been paralleled by the development of methods for assessing aspects of ecosystem stress. These have almost all focussed upon stress posed by chemical contaminants, and have usually involved single species. Notable developments in this field of ecotoxicology (Calow, 1989) have been:

(a) 'scope for growth': a physiological measure of the amount of energy an organism has left for growth after it has dealt with the stress of detoxifying or excreting contaminants (Maltby et al., 1990), and
(b) 'biomarkers': the (mainly) biochemical measurement of stress, such as induction of protein or enzyme synthesis (McCarthy & Shugart, 1990).

The most recent developments are those measuring response of populations (of single species) to stress. This is a logical extension of the 'scope for growth' concept, examining how non-lethal stress has affected the pattern of population natalities, mortalities and growth. Such studies are more common in the marine environment than the freshwater one, although planktonic organisms such as *Daphnia* are well studied (Jones et al., 1991).

Assessment of conservation value

Conservation of the river environment has, until recently in England and Wales, been a separate responsibility within the statutory agency (the National Rivers Authority from 1989 and the Environment Agency since 1995), as well as partially coming under the government agency responsible for all wildlife conservation issues (originally the Nature Conservancy Council for the U.K. but now separate agencies for England, Wales and Scotland). This has resulted in the development of conservation quality indices independently of those for fish or invertebrates. The Nature Conservancy Council has relied primarily on the classification of river plant communities (Holmes et al., 1998), whilst both NCC and the NRA/EA also used River Corridor Surveys, map-based descriptions of river stretches (500 m) recording above-water structure and mainly terrestrial plants and birds (Nature Conservancy Council, 1985; National Rivers Authority, 1992). In the 1990s, the lack of a consistent, fully integrated, computer-based evaluation system became acute and the NRA/EA developed the River Habitat Survey. This has recorded physical structural features of U.K. rivers on an annually increasing number of sites since 1994, maintaining a computerised database which provides national presentation and analysis of data (Fox et al., 1998; Raven et al., 2000). The conservation agencies, led by Scottish Natural Heritage, have now developed a broader-based catchment-scale assessment, SERCON, which incorporates all available information on biotic and structural components (Boon et al., 1998).

The U.K. thus has developed a range of assessment methods (Raven et al., 1998), each serving individual quality functions but none, other than SERCON, at present integrating the structure and values of a river (despite the fact that macroinvertebrates, for example, have the potential to indicate a wider range of environmental conditions than just water quality). Moreover, none of them address the processes of the river.

Links between the scales of river assessment

Only the most recent methods – RHS and SERCON – consider the hierarchical scales of river systems. However, recognising scales within a recording system is one thing (Raven et al., 2000): integrating the scales so that recording at one scale provides valuable information about another is entirely different and has not yet been achieved. We argue that this should be the goal in measuring 'ecological integrity' (see other papers in this volume).

The hydrological and geomorphological processes, which chiefly operate at catchment scale and downwards, are shown in schematic form in Figure 2; linked to the biological scales, which operate largely at the microhabitat scale and upwards. The rectangles represent the important scales, their contents the important processes. Each rectangle contains elements which are measured by existing assessment systems – from river discharge at catchment scale to species richness at microhabitat scale. The circles between the rectangles represent the possibilities for linking the scales in a more holistic system for assessing ecological integrity.

The left-hand circle, 'evaluation of surface flow biotopes', represents a field assessment system for the bankside measurement of surface flow character in a stream and river. This system provides a rapid yet accurate way of measuring the link between a river's hydrological regime and its physical habitat pattern (Padmore, 1997; Newson et al., 1998; Padmore et al., 1999).

The middle circle represents a field assessment system for measuring habitat structure of river channels. Sixteen 'functional habitats' (usually not all are found in one surveyed river stretch) have been shown to accommodate the river's overall macro-invertebrate biodiversity (Harper et al., 1992; Kemp et al., 1999). Many other studies demonstrate that the habitat fea-

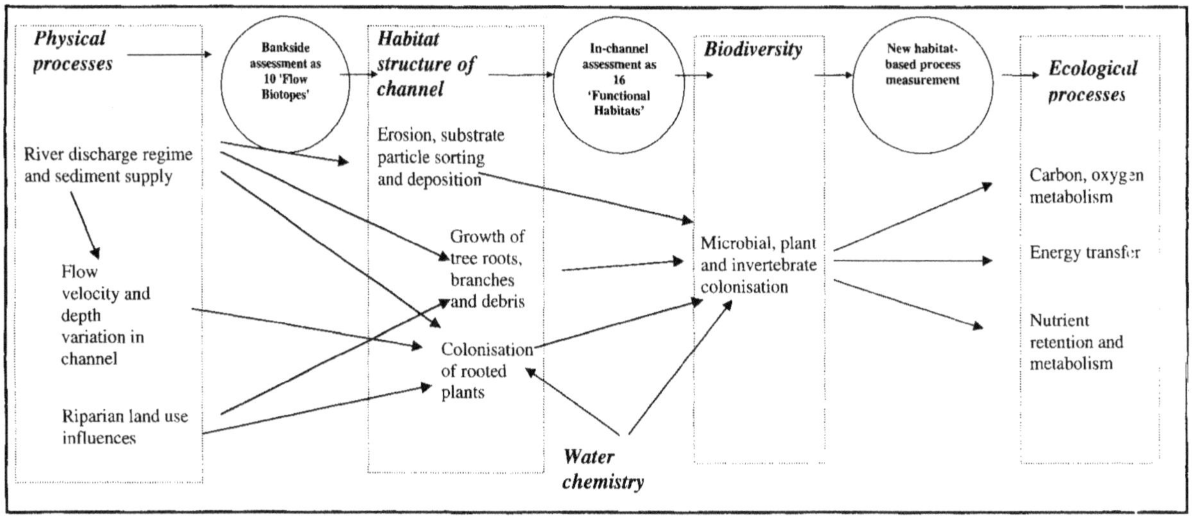

Figure 2. A schematic diagram of the hierarchy of scales and processes important in river ecology (rectangular boxes, scale decreasing from left to right), and the proposed development of methods (circles) used to integrate the assessment of these processes and thus assess 'ecological integrity'.

tures represented by functional habitats are critical for invertebrate communities (Brooker, 1982; Bournaud & Cogerino, 1986; Barmuta, 1989; Armitage et al., 1995; Pardo & Armitage, 1997).

The right-hand circle represents a potential system for assessing ecological processes, which are dependent upon the physical and chemical environment but driven by the biological community. Such processes are normally considered part of the functioning of the ecosystem, and important ones are the cycling of carbon, oxygen, nitrogen and phosphorus together with the transfer of energy (Edwards, 1995). Whilst many studies have measured these processes in rivers, relatively few have done so in order to understand the scale links – the extent to which physical habitat affects chemical processes. Elosegui et al. (1995) showed the importance of cobble-type habitats for processing carbon in organically polluted streams. Several recent studies have addressed the role of macrophytes and in-channel habitats in altering sediment and nutrient retention (e.g. Aub-Robinson et al., 1995; Sand-Jensen & Mebus, 1996; Hill, 1997; Sand-Jensen, 1998). Crosa & Buffagni (1996) demonstrated a clear difference in secondary productivity between functional habitats and proposed that this measure of energy processing could be a useful tool for comparing quality of Italian streams.

Habitats, and habitat structure of river channels, already play important roles in the existing U.K. monitoring and assessment framework. Water quality, fisheries quality and wildlife quality assessments are all partially or completely habitat-based:

1. The BMWP index for water quality assessment, initially derived from single-habitat sampling ('stony riffles'), is now calculated from field collections made from all habitats in proportion to their abundance; one index, the 'Lincoln Index' takes 'habitat-rich' and 'habitat-poor' sites into account in its derivation.
2. The Habscore assessment of fisheries is a means of evaluating the habitat features cricial to successful fish growth and recruitment in a reach.
3. Both River Conservation Surveys in the 1970s and 1980s and River Habitat Surveys of the 1990s, identify habitat features of the river and highlight rare or unusual ones.

Elsewhere in this volume are papers from Australia (Bunn & Davies, 2000), Austria (Chovanec et al., 2000) and the U.S. (Karr & Chu, 2000) which integrate measures of physical and biological processes. It is clearly necessary for the U.K. to move towards better integration of its existing assessment methods to be able to accurately monitor ecological integrity. The hypothesis advanced here is that it is possible to do this practically at habitat scale, and that this provides an effective way of integrating both 'up-scale' and 'down-scale'. This hypothesis is tested by measuring the strengths of links between the circles shown in Figure 2.

Linking approaches between flow biotopes and functional habitats

The meso-scale is the essential fulcrum for moving information to both larger and smaller scales on the hierarchy (Kershner & Snyder, 1992; Pardo & Armitage, 1997). Recognising this, 'flow biotopes' and 'functional habitats' were independently developed by geomorphologists and ecologists, respectively, (see above) but although each system is incorporated within the protocol of the River Habitat Survey, there was initially no linkage developed between them. Newson et al. (1998) and Newson & Newson (in press) stressed the need for such a linkage and some has been shown indirectly through relationships with hydraulic factors such as Froude Number (Kemp et al., in press). Preliminary analysis of the 1995 dataset for lowland rivers from the RHS showed that there was a relationship between functional habitats and their equivalent flow biotopes (Newson et al., loc. cit.). However, field work which recorded the two simultaneously had not been conducted.

Methods

Four rivers – two lowland and two upland – were chosen for a comparative study. The methods for recording 'functional habitats' (Kemp et al., 1999) and 'flow biotopes' (Padmore, 1997) were applied simultaneously at 1 m intervals on a series of transects. The transects were 5 m apart (10 m apart in the Wharfe), for at least 70 m stretches of each river. The list of flow biotopes and functional habitats used is shown in Table 1. All measurements were carried out during low flow conditions in late summer.

The association between 'functional habitats' and 'flow biotopes' was calculated as the number of times the habitat was observed in a particular biotope, divided by the total number of times that 'flow biotope' was recorded at the site. In this way, bias from more common biotopes being recorded more frequently was avoided. The output was then standardised, to give a value of 1 to the strongest association for each 'functional habitat' type.

Table 1. The categories of functional habitats and flow biotopes used, together with their abbreviations

Functional habitat name
Boulder (BR)
Cobble (CO)
Gravel (G)
Sand (SA)
Silt (SI)
Tree roots (R)
Trailing vegetation (TV)
Marginal plants (MP)
Emergent macrophytes (ME)
Submerged, fine-leaved macrophytes (MSF)
Submerged, broad-leaved macrophytes (MSB)
Moss (M)
Macroalgae (MA)
Woody debris (WD)
Leafy debris (LD)
Flow Biotope name
No flow (NF)
No perceptible flow (NP)
Smooth Boundary Turbulent flow (SM)
Rippled flow (RP)
Unbroken Standing Wave UW)
Broken Standing Wave (BW)
Chute (CH)
Free Fall (FF)

Results

A total of 864 data points were recorded from the four rivers. The distribution of biotopes is shown in Figure 3.

Individual rivers showed differences in habitat frequency, following an expected pattern (Figures 4 and 5). The lowland rivers are plant habitat-dominated. The Welland – a eutrophic, low-gradient river on clay geology, predominantly unshaded – is dominated by marginal and emergent plants, with macro-algae in slow flows and by macro-algae with fine-leaved plants in faster flow, giving way to gravel and cobble habitats in the fastest flow. The Gwash – a more shaded, groundwater and reservoir-supported, low-gradient river on limestone – is dominated more by trees (roots and trailing vegetation) and marginal plants in slower flows, and submerged broad-leaved macrophytes and moss in the faster flow types. The upland rivers il-

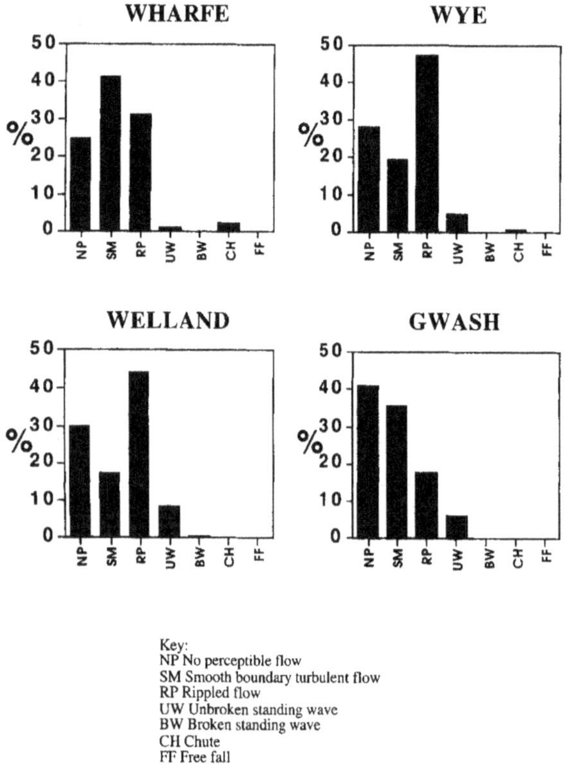

Figure 3. The percentage of different 'flow biotopes' in the four English study rivers.

lustrate both differences and similarities. The Wharfe – a high-gradient upland moorland, granite stream – contains few plant habitats except moss and is dominated by cobbles. In contrast the Wye – an upland limestone, groundwater-supported stream – is richer in plant habitats; submerged fine-leaved macrophytes and moss associated with cobbles are characterisitc of the faster flow types.

The 13 most common functional habitats recorded display these patterns more clearly when expressed as an association with flow type. Silt, tree roots, trailing vegetation, marginal plants and emergent vegetation all showed strong association with the slowest flow types. Coarser substrate habitats were, in order of increasing size, associated with increasingly energetic flow types (except for 'rock' habitat). Mosses were associated with unbroken standing waves, fine-leaved submerged macrophytes with rippled flow and broad-leaved submerged macrophytes with smooth boundary turbulent flow (Figure 6). These results are consistent with previous work which linked functional habitats, depth and velocity (Kemp et al., 1999) and Froude number (Kemp et al., in press).

Discussion

The link between species and physical variables has been demonstrated by many authors (e.g. Statzner & Higler, 1986; Statzner et al., 1988; Barmuta, 1989, 1990; Wetmore et al., 1990). Few however, have found any easily-measurable variable clearly related to biodiversity, yet useful in assessment, management or restoration. Perhaps the best example is Jungwirth et al. (1995), who quantified the bed variability of 15 Austrian epipotamal streams by calculating the variance of maximum depths. They found this to be significantly related to the richness and diversity of fish species. Physical restoration work, to increase the spatial heterogeneity of the channel, was carried out on one, low-scoring reach. Fish species richness and diversity increased accordingly (within a 3-year period). A greater diversity and abundance of invertebrates were also found (Jungwirth et al., 1993).

Here we have shown that an easily-measured physical variable – surface flow type – can be related to an easily-measured channel structural variable – functional habitat type/frequency – which has itself been shown to be related to biodiversity in the UK. The relationship between them indicates their interdependence – flow types under low flow conditions are influenced not just by discharge but also by depth and substrate type; factors themselves controlled by high channel-forming discharges (Newson & Newson, in press). Submerged and emergent plants alter both flow characteristics and sediment distribution as they grow and senesce. Since they operate at the same scale, we may tentatively put forward the assessment of flow biotope diversity as a rapid, cost-effective method of estimating potential biodiversity.

Each technique – flow biotopes and functional habitats – has already independently proved valuable in assessing and remediating physical damage (Newson et al., 1998; Raven et al., 1998; Harper et al., 1998). There is good reason to believe that both techniques are applicable elsewhere in Europe. Studies of the functional habitats of two Italian rivers, which were of very different character to the English river sites, support this belief. In the river Ticino, a large, meandering piedmont/lowland river, five functional habitats were identified. These matched individual English functional habitats (although were designated differently). 'River margins with macrophytes' matched with 'submerged, broad-leaved macrophytes'; 'margins without macrophytes' with 'silt' and 'sand'; 'backwater' with 'leaf litter' habitat; 'run-riffle' with 'cobbles' and

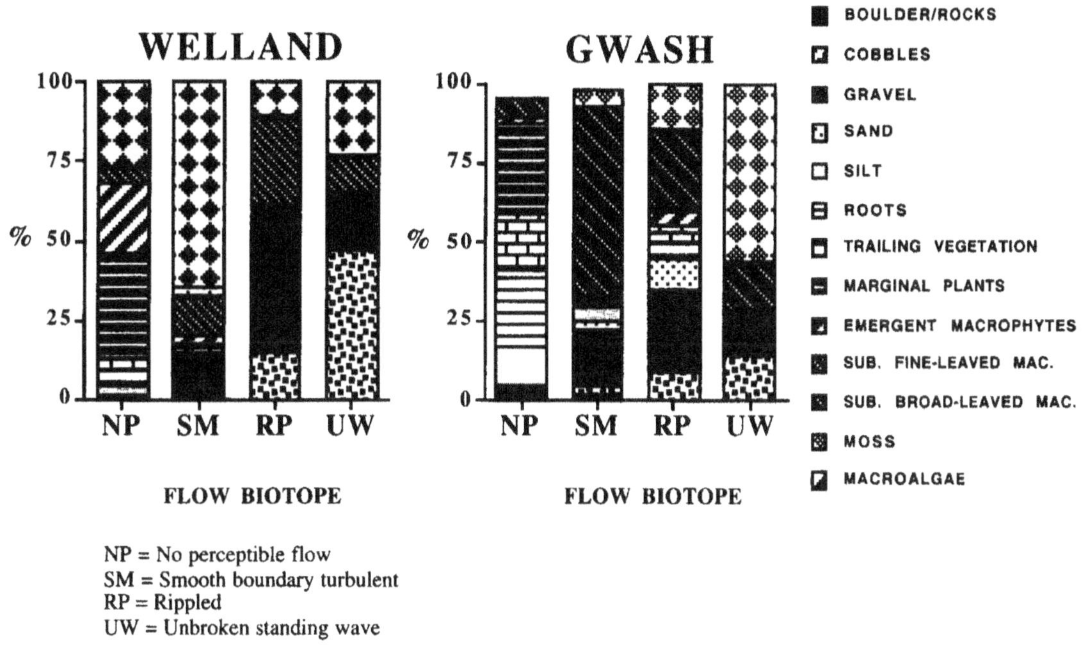

Figure 4. The proportion of different 'functional habitats' found in the different 'flow biotopes' of the two lowland rivers.

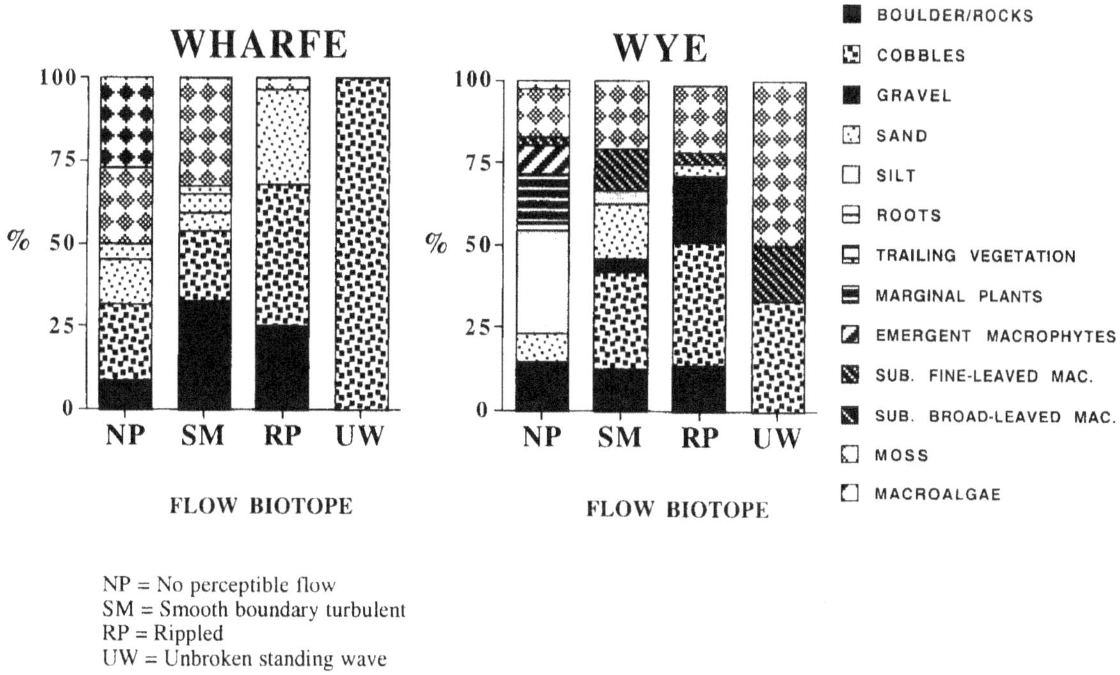

Figure 5. The proportion of different 'functional habitats' found in the different 'flow biotopes' of the two upland rivers.

'macrophytes in flow' with 'submerged, fine-leaved macrophytes' (Buffagni et al., 2000). In a typical North Italian mountain stream, the Pioverna, four functional habitats were found. These were 'riffle', 'pool', 'transition' and 'bedrock' habitats (Crosa & Buffagni, 1996; Buffagni & Comin, 2000). They were identifiable by their flow type and their substrate type, which reflects the greater relative impact of flow on habitat structure in an upland stream. Nevertheless, the Pioverna study demonstrates that the functional habitat concept may be transferred across European streams.

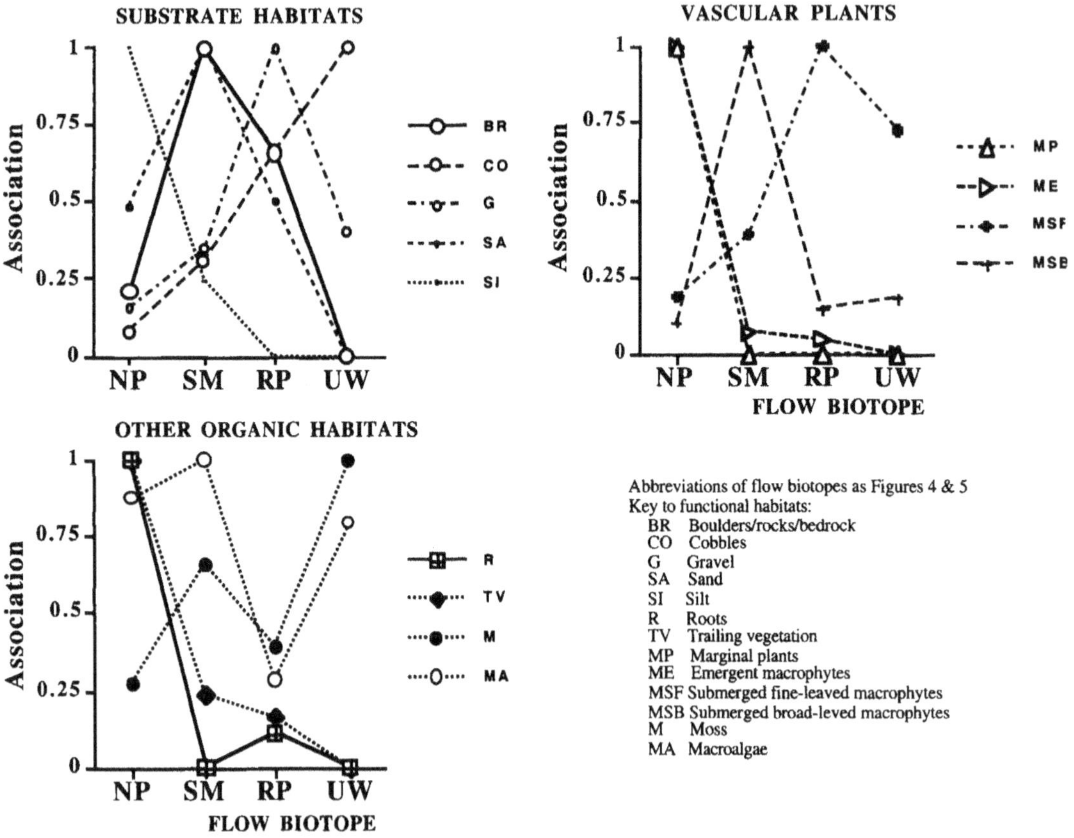

Figure 6. The association between 'functional habitats' and 'flow biotopes' in the overall dataset.

The differences in habitat distribution between the four different U.K. rivers suggest that further work in analysing distribution patterns could be fruitful in identifying damage from chemical quality changes. Plant functional habitats are groups of species responding in different ways to their chemical and physical environment. For example the nutrient status of a river affects the proportion of macroalgae (Figure 3), which increases usually at the expense of other plant habitats but also at the expense of coarse particle habitats as these become covered in algal growth.

An effective method for the assessment of ecological integrity must integrate physical and chemical measures of integrity, and must include ecological processes as well as static parameters such as chemical concentrations. Most continental European approaches to ecological integrity are trying to incorporate the evaluation of ecological processes rather than merely ecological structure, moving forward from a habitat-based structural classification (Muhar et al., 1993, 1996). This paper demonstrates that the U.K., starting from a strong base of measures of water quality coupled with measures of physical habitat quality, is now beginning to develop rapid yet accurate methods for linking the measurement of physical processes with biologically meaningful measures of habitat structure. Refinement of these methods is underway. The challenge for the future will be to find efficient yet cost-effective measures of ecological processes at the meso-habitat scale. This will provide a comprehensive and accurate tool to help assess 'ecological integrity' (see Figure 2).

References

Armitage, P. D., I. Pardo & A. Brown, 1995. Temporal constancy of faunal assemblages in mesohabitats – application to management. Archiv für Hydrobiologie 133: 367–387.

Aub-Robinson, C., O. Falkum, C. D. Hansen & B. Kronvang, 1995. Retention of particulate nutrients in macrophytes and on the floodplain. In Kronvang B. & L. M. Svendsen (eds), Erosion and Delivery, Transport and Fate of Sediments and Sediment-associated Nutrients in Watersheds. N.E.R.I., Silkeborg, Denmark: 130–134.

Barmuta, L. A., 1989. Habitat patchiness and macrobenthic community structure in an upland stream in temperate Victoria Australia. Freshwat. Biol. 21: 223–236.

Barmuta, L. A., 1990. Interaction between the effects of substratum, velocity and location on stream benthos – an experiment. Aust. J. Mar. Freshwat. Res. 41: 557–573.

Bloch, H., 1999. European water policy facing the new millenium: The EU Water Framework Directive. In Assessing the Ecological Integrity of Running Waters. Vienna, 9–11 November 1998, Austria.

Bournaud, M. & L. Cogerino, 1986. Les microhabitats aquatiques des rives d' un grand cours d' eau: approche faunistique. Annales de Limnologie 22: 285–294.

Brooker, M. P., 1982. Conservation of wildlife in river corridors, Part I. Methods of survey and classification, Part II. Brecon: Welsh Water Authority.

Buffagni, A. & E. R. Comin, 2000. Secondary production of benthic communities at the habitat scale as a tool to assess ecological integrity in mountain streams. Hydrobiologia 422/423: 183–195.

Buffagni, A., G. A. Crosa, D. M. Harper & J. L. Kemp, 2000. A test of the functional habitats concept in a large, unpolluted Italian river (River Ticino, Northern Italy). Paper presented at the Vienna Conference 1998.

Bunn, S. E. & P. M. Davies, 2000. Biological processes in running waters and their implications for the assessment of ecological integrity. Hydrobiologia 422/423: 61–70.

Calow, P., 1989. Physiological ecotoxicology: theory, practice and application. In Lokke, H., H. Tyle & F. Bro-Rasmussen (eds), Proceedings of the First European Conference on Ecotoxicology. Polyteknisk, Forlag, Odense, Denmark: 23–35.

Chovanec, A., P. Jager, M. Jungwirth, V. Koller-Kreimel, O. Moog, S. Muhar & S. Schmutz, 2000. The Austrian way of assessing the ecological integrity of running waters: a contribution to the EU Water Framework Directive. Hydrobiologia 422/423: 445–452.

Chapman, D., 1992. Water Quality Assessments. UNESCO/WHO/ UNEP & Chapman & Hall, London.

Crosa, G. A. & A. Buffagni, 1996. L'habitat idraulico quale elemento per la gestione degli ambienti fluviali. Atti della Societe Italiana di Ecologia: 17: 581–583.

Edwards, R. W., 1995. The ecological basis for the management of water quality. In Harper, D. M. & A. J. D. Ferguson (eds), The Ecological Basis for the River Management. John Wiley & Sons, Chichester: 135–146.

Elosegui, A., X. Arana, A. Basaguren & J. Pozo, 1995. Self-purification processes along a medium-sized stream. Envir. Manage. 19: 931–939.

Fox, P. J. A., M. Naura & P. Scarlett, 1998. An account of the derivation and testing of a standard field method: River Habitat Survey. Aquati. Cons.: Mar. Freshwat. Ecosystems 8: 455–476.

Furze, M. T., J. F. Wright, P. D. Armitage & D. Moss, 1981. An appraisal of pond-net samples for biological monitoring of lotic macroinvertebrates. Wat. Res. 15: 679–689.

Gower, A. M., G. Myers, M. Kent & M. E. Foulkes, 1995. The use of macroinvertebrate assemblages in the assessment of metal-contaminated streams. In Harper, D. M. & A. J. D. Ferguson (eds), The Ecological Basis for the River Management. John Wiley & Sons, Chichester: 181–191.

Harper, D. M., C. D. Smith & P. Barham, 1992. Habitats as building blocks for river conservation. In Boon, P. J., P. Calow & G. Petts (eds), River Conservation and Management. John Wiley & Sons, Chichester: 311–319.

Harper, D. M., C. Smith, J. L. Kemp & G. A. Crosa, 1998. The use of functional habitats in the conservation, management and rehabilitation of rivers. In Bretschko G. & J. Helesic (eds), Advances in River Bottom Ecology. Backhuys Publishers, Leiden: 315–326.

Hill, A. R., 1997. The potential role of in-stream and hyporheic environments as buffer zones. In Haycock, N. E., T. P. Burt, K. W. T. Goulding & G. Pinay (eds), Buffer Zones: their Processes and Potential in Water Protection. Quest Environmental, Harpenden, U.K.: 115–127.

Hill, M. O., 1979. TWINSPAN – A Fortran programme for arranging multivariate data in an ordered two-way table by classification of the individuals and attributes. Cornell University, Ithaca, New York.

Holmes, N. T., P. J. Boon & T. A Rowell, 1998. A revised classification system for British plants based on their aquatic plant communities. Aquat. Cons.: Mar. and Freshwat. Eco. 8: 555–578.

Jones, M., C. Folt & S. Guarda, 1991. Characterising individual, population and community effects of sublethal aquatic toxicants: an experimental case study using *Daphnia*. Freshwat. Biol. 26, 35–41.

Johnson, I. W., C. R. N. Elliot & A. Gustard, 1994. Data requirements and collection procedures for application of the In-stream Flow Incremental Methodology in the U.K. Institute of Hydrology, Wallingford, U.K.

Jungwirth, M., O. Moog & S. Muhar, 1993. Effects of river bed restructuring on fish and benthos of a 5th-order stream, Melk, Austria. Reg. Riv.: Res. Man. 8: 195–204.

Jungwirth, M., S. Muhar & S. Schmutz, 1995. The effects of re-created instream and ecotone structures on the fish fauna of an epipotamal river. Hydrobiologia 303: 195–206.

Karr, J. R. & E. W. Chu, 2000. Introduction: Sustaining living rivers. Hydrobiologia 422/423: 1–14.

Kelly, M. G. & B. A. Whitton, 1998. Biological monitoring of eutrophication in rivers. Hydrobiologia 384: 55–67.

Kemp, J. L., D. M. Harper & G. A. Crosa, 1999. Use of 'functional habitats' to link ecology with morphology and hydrology in river rehabilitation. Aquat. Cons.: Mar. and Freshwat. Eco. 9: 159–178.

Kemp, J. L., D. M. Harper & G. Crosa, in press. The habitat scale ecohydraulics of rivers. Ecological Engineering.

Kershner, J. L. & W. M. Snyder, 1992. Importance of a habitat-level classification systsem to design instream flow studies. In Boon, P. J., P. Calow & G. E. Petts (eds), River Conservation and Management. John Wiley & Sons, Chichester, U.K.: 179–192.

McCarthy, J. F. & L. R. Shugart, 1990. Biomarkers of Environmental Contamination. Lewis Publishers, Baton Rouge, Lousiana.

Maltby, L., C. Naylor & P. Calow, 1990. Field deployment of a 'scope for growth' assay involving Gammarus pulex, a freshwater benthic invertebrate. Ecotoxicol. Envir. Safety 19: 292–300.

Milner, N. J., R. J. Wyatt & K. Broad, 1998. HABSCORE – Applications and future development of related habitat models. Aquat. Cons.: Mar. Freshwat. Eco. 8: 633–644.

Muhar, S., A. Muhar, G. Imhof, M. Jungwirth, S. Schmutz, R. Wimmer, H. Wiesbauer, B. Hozang & P. Tschemernig, 1993. Ausweisung naturnaher Fließgewässerabschnitte in Österreich – Vorstudie. Universität für Bodenkultur, Abteilung für Hydrobiologie, Fischereiwirtschaft und Aquakultur und Institut für Landschaftsgestaltung. Blaue Reihe des Bundesministeriums für Umwelt, Jugend und Familie (BMLF), Bd 1, Wien, 175 pp.

Muhar, S., M. Kainz, M. Kaufmann & M. Schwarz, 1996. Ausweisung flußtypspezifisch erhaltener Fließgewässerabschnitte in Österreich, Bundesministerium für Umwelt, Jugend und Familie (BMLF), Bd 1, Wien, 167 pp.

Nature Conservancy Council, 1985. Surveys of Wildlife in River Corridors. NCC, Peterborough, U.K.

National Rivers Authority, 1992. *River Corridor Surveys*. Conservation Technical Handbook 1, N.R.A., Bristol, U.K.

Newson, M. D., D. M. Harper, C. L. Padmore, J. L. Kemp & B. Vogel, 1998. A cost-effective approach for linking habitats, flow types and species requirements. Aquat. Cons.: Mar. Freshwat. Eco. 8: 431–446.

Newson, M. D. & C. L. Newson, in press. Geomorphology, ecology and river channel habitat: mesoscale approaches to basin-scale challenges. Progress in Physical Geography.

Padmore, C. L., 1997. Biotopes and their hydraulics: a methods for defining the physical component of freshwater quality. In Boon, P. & D. L. Howell (eds), Freshwater Quality, Defining the Indefinable. The Stationery Office, Edinburgh: 251–257.

Padmore, C. L., M. D. Newson & M. E. Charlton, 1999. Instream habitat in gravel-bed rivers: identification and characterisation of biotopes. In Klingeman, P. C., R. L. Beschta, P. D. Komar & J. B. Bradley (eds), Gravel-bed rivers in the Environment. Water Resource Publications, Englewood, Colorado, U.S.A.: 345–364.

Pardo, I. & P. D. Armitage, 1997. Species assemblages as descriptors of mesohabitats. Hydrobiologia 344: 111–128.

Raven, P. J., P. J. Boon, F. H. Dawson & A. J. D. Ferguson, 1998. Towards and integrated approach to classifying and evaluating rivers in the U.K. Aquat. Cons.: Mar. and Freshwat. Eco. 8: 383–393.

Raven, P. J., N. T. H. Holmes, M. Naura & F. H. Dawson, 2000. Using river habitat survey for environmental assessment and catchment planning in the U.K. Hydrobiologia 422/423: 359–367.

Statzner, B. & B. Higler, 1986. Stream hydraulics as a major determinant of benthic invertebrate zonation patterns. Freshwat. Biol. 16: 127–139.

Statzner, B., J. A. Gore & V. H. Resh, 1988. Hydraulic stream ecology – observed patterns and potential applications. J. N. Am. Benthol. Soc. 7: 307–360.

Ter Braak, C. J. F., 1988. CANOCO – A Fortran Programme for Canonical Community Ordination by (Partial) (Detrended) (Canonical) Correspondence Analysis. TNO, Wageningen, The Netherlands.

Wetmore, S. H., R. J. Mackay & R. W. Newbury, 1990. Characterization of the hydraulic habitat of *Brachycentrus occidentalis*, a filter-feeding caddisfly. J. N. Am. Benthol. Soc. 9: 157–169.

Wright, J. F., M. T. Furze & P. D. Armitage, 1994. Use of macroinvertebrate communities to detect environmental stress in running waters. In Sutclifee, D. W. (ed.), Water Quality and Stress Indicators in Marine and Freshwater Ecosystems: Linking Levels of Organisation (Individuals, Populations and Communities). Freshwater Biological Association, Ambleside: 15–34.

Wright, J. F., D. Moss, R. T. Clarke & M. T. Furze, 1997. Biological assessment of river quality using the new version of RIVPACS (RIVPACS III). In Boon P. J. & D. L. Howell (eds), Freshwater Quality: Defining the Indefinable? The Stationery Office, Edinburgh: 102–108.

Wright, J. F., M. T. Furse & D. Moss, 1998. River classification using invertebrates: RIVPACS applications. Aquat. Cons.: Mar. Freshwat. Eco. 8: 617–631.

Hydrobiologia **422/423:** 143–152, 2000.
M. Jungwirth, S. Muhar & S. Schmutz (eds), Assessing the Ecological Integrity of Running Waters.

The influence of environmental variables on the abundance of aquatic insects: a comparison of ordination and artificial neural networks

R. Wagner[1], T. Dapper[2] & H.-H. Schmidt[1]
[1]*Limnologische Flußstation Schlitz der MPG, P.O. Box 260, D-36105 Schlitz, Germany*
[2]*Forschungsgruppe Neuronale Netzwerke der Universität GH Kassel, Heinrich-Plett-Straße 40, D-34132 Kassel, Germany*

Key words: artificial neural networks, ordination, aquatic insect emergence, prediction, discharge pattern

Abstract

Two methods to predict the abundance of the mayflies *Baetis rhodani* and *Baetis vernus* (Insecta, Ephemeroptera) in the Breitenbach (Central Germany), based on a long-term data set of species and environmental variables were compared. Statistic methods and canonical correspondence analysis (CCA) attributed abundance of emerged insects to a specific discharge pattern during their larval development. However, prediction (specimens per year) is limited to magnitudes of thousands of specimens (which is outside 25% of the mean). The application of artificial neural networks (ANN) with various methods of variable pre-selection increased the precision of the prediction. Although more than one appropriate pre-processing method or artificial neural networks was found, R^2 for the best abundance prediction was 0.62 for *B. rhodani* and 0.71 for *B. vernus*.

Introduction

Biodiversity, species richness, density or biomass of populations are results of a multitude of environmental variables. The dependence of a species or a community on its habitat is a crucial hypothesis in ecology. Thus, the forecast of abundance or biomass of species or populations, or even the community structure based on habitat characteristics is an interesting task in basic and applied ecology (Baran et al., 1996; Chon, 1996; Chon et al., 1996, Cisneros Mata et al., 1996; Whitehead et al., 1997). Such studies are of high interest in particular for managers of fish and wildlife, and for engineers dealing with stream and river channels (Giske et al., 1998; Guegan et al., 1998; Lek & Baran, 1997; Lek et al., 1996a; Mastorillo et al., 1997a; Mastorillo et al., 1997b).

During the past decades human-caused effects have altered the interactions between running waters and their environment, with implications on major portions of the community. A proper evaluation of human impact on ecosystems, however, depends on the availability of data from undisturbed reference areas. Intact systems need not only to be characterised in terms of average species abundance or the mean of selected environmental factors, but also in natural variability and its effects on populations and communities. This requires the availability of long-term data sets to generate models of the magnitude of the dependence of populations or communities on environmental variables. The results of such studies can generate features to characterise the ecological integrity of ecosystems. Regression and correlation models have repeatedly been used to explain patterns of ecosystem attributes and they have provided useful insights on environmental control of ecosystems, but their predictive power is low (Paruelo & Tomasel, 1997; Ter Braak & Verdonschot, 1995; Walley & Fontama, 1998).

Since 1969, the Limnologische Flußstation Schlitz (Germany) has collected data on emerging insects at the Breitenbach and on environmental variables that are believed to influence species abundance. Using traditional statistical methods and canonical correspondence analysis (CCA) (Ter Braak, 1988, 1990), the abundance of adults of individual species was attributed mainly to discharge patterns during larval development (Wagner & Schmidt, in press); this indicates that if the discharge pattern is known, at least the magnitude of a species' abundance (emerging adults) is predictable. Due to the necessity to recognise patterns and not just single events, artificial neuronal networks (ANN) are believed to be an alternative method to model species abundance (Colasanti, 1991; Lek et al., 1996b). The applicability of ANNs to this set of problems is tested below.

Study site

The Breitenbach is a small perennial stream in Central Germany, (50° 40′ N, 9° 45′ E) in an area of Bunter Sandstone. The stream flows mainly through meadows and has a drainage of approximately 9 km^2. The main spring in the middle course has an elevation of 310 m a.s.l., the stream flows into the Fulda River at 220 m a.s.l., and stream length is 2 km. Monitoring of insect emergence, water temperature and discharge was carried out at the 'classical site', 660 m below the spring (Illies, 1971), precipitation was measured at about 2 km distance from the drainage area, at an official site.

Material and methods

Species and environmental variables

Aquatic insects have been collected in emergence traps since 1969 (Illies, 1971). Collecting methods have changed: until 1986, insects were collected daily with a vacuum device, but in 1987 the traps were altered to collect specimens automatically. The mayflies *Baetis vernus* (Curtis, 1834) and*Baetis rhodani* (Pictet, 1843) are among the most numerous aquatic insects in the Breitenbach. Larvae of both species are grazers on stones and avoid sandy substratum (Wagner, 1989). *B. rhodani* is typically bivoltine, but *B. vernus* is univoltine with an egg diapause until winter (Bohle, 1969; Clifford, 1982; Schmidt, 1984). Mortality during the terrestrial phase of the life cycle was estimated to be 90–99% (Wernecke & Zwick, 1992). We compared the accuracy for predicting the abundance (A) of adult *B. vernus* and *B. rhodani* in the Breitenbach at one monitoring site with two methods described below. Water temperature (T) and discharge (D) were measured at the Breitenbach, and precipitation (P) was measured at a station close to the catchment. Monthly maxima of water temperature and discharge, and the monthly amount of precipitation, were used as predictors in the models.

Statistical methods

Correlation, regression (SPSS, 1997), and ordination (Ter Braak, 1988, 1990) are traditional methods to relate environmental variables with species abundance. They calculate abundance difference and provide methods to test significance. However, they involve data reduction to a smaller number of vectors or site points, with a loss of information.

As an alternative, artificial neural networks (ANN) use all available data of precipitation (P), discharge (D), water temperature (T), and abundance (A) of the 12 preceding months to predict species abundance in the target month (12 in Table 1).

ANNs consist of interconnected layers of simple processing elements called neurons. In feed forward networks — the most common type of ANNs — an input vector provided at the first layer (the input layer) is propagated step by step through all layers, resulting in an output vector at the last layer (output layer). If the neurons use a non-linear function to map their input (a weighted sum of all outputs from neurons in the previous layer) the network as a whole represents a non-linear function. The more neurons an ANN has, the more capable and complex the networks and its function becomes.

One of the most useful features of ANNs is the ability to learn relationships from examples, i.e., to adapt their weighted connections so that the network represents a model fitting the training data (usually a set of pairs of input vectors and corresponding target outputs). The rules for the adaptation process are put together in the learning algorithm. Given representative examples during training, ANNs also have the ability to generalise, i.e., provide sensible outputs for new (untrained) input vectors. The available data are split into a set of input/target pairs; these are used for the training and as a test set solely to check the generalisation performance of the trained net. This involves applying the input vectors and comparing the network's output with the target output. From that, an overall error value E is calculated by

$$E = \frac{1}{2\mathrm{PAT}} \sum_{p=0}^{\mathrm{PAT}-1} \sum_{i=0}^{\mathrm{OUT}-1} (\mathrm{tar}_{pi} - \mathrm{out}_{pi})^2,$$

where PAT is the number of patterns, OUT the number of output neurons, out_{pi} the value of the i-th output neuron for the p-th pattern, and tar_{pi} the target value of the i-th output neuron for the p-th pattern. For the experiments, we also calculated the determina-

Table 1. Structure of the 52-dimensional variable vectors. Abundance 12 is the target vector to be predicted by the remaining variable vectors

Variable	Index (13 subsequent months)												
Abundance	0	1	2	3	4	5	6	7	8	9	10	11	12
Discharge	0	1	2	3	4	5	6	7	8	9	10	11	12
Precipitation	0	1	2	3	4	5	6	7	8	9	10	11	12
Water temperature	0	1	2	3	4	5	6	7	8	9	10	11	12

Table 2. Number (rank on the *x*-axis in Figures 2, 4, 5, 6, 7) and date (year/month) of the test data set (7008 = 1970 August)

Date	No.	Date	No.	Date	No.	Date	No.	Date	No.
7008	1	7503	12	7910	23	8405	34	8812	45
7101	2	7508	13	8003	24	8410	35	8905	46
7106	3	7601	14	8008	25	8503	36	8910	47
7111	4	7606	15	8101	26	8508	37	9003	48
7204	5	7611	16	8106	27	8601	38	9008	49
7209	6	7704	17	8111	28	8606	39	9101	50
7302	7	7709	18	8204	29	8611	40	9106	51
7307	8	7802	19	8209	30	8704	41	9111	52
7312	9	7807	20	8302	31	8709	42	9204	53
7405	10	7812	21	8307	32	8802	43	9209	54
7410	11	7905	22	8312	33	8807	44		

tion coefficient R^2 for all output/target pairs in the test set. We used senso networks (Dapper, 1997), a simple but effective variation of feed-forward networks that implement the well-known back-propagation learning algorithm (Rumelhart et al., 1986). Senso nets contain an additional layer of weights, each corresponding to exactly one input neuron. These weights are included in the adaptation process (training) and provide a direct way to measure the relevance of the input variables in respect to the output variables. In addition, senso nets use some improvements of the standard back-propagation algorithm. We also carried out experiments with another kind of ANN called feature map (Kohonen, 1982).

Every model calculated is termed an experiment. Original data were scaled sigmoidally, exponentially, or logarithmically (minimum 0, maximum 1) to optimise the accuracy of the models. Furthermore, modelling with the entire database was compared with methods of a preceding reduction of vector dimensions by correlation, regression or neural sensitivity analysis (to reduce computing time). Eighty percent of the data sets were used in the training session; the remaining data were used as test data. A key for the year and months of the test data set is provided in Table 2.

Results

Ordination

The results of ordination (CCA) indicated a strong dependence of species abundance on the discharge pattern (Wagner & Schmidt, in press). Abundance (specimens per 5 m^{-2} $year^{-1}$) between patterns was significantly different (Figure 1). However, predictions can be made only in magnitudes of thousands of specimens per year.

Artificial neural networks (ANN)

Modelling the abundance of B. rhodani *(experiment 1)*

The precision of the model was relatively high (R^2=0.56) with the original data (April 1969 to December 1992). Almost all months with any abundance ($n > 0$) were predicted correctly. However,

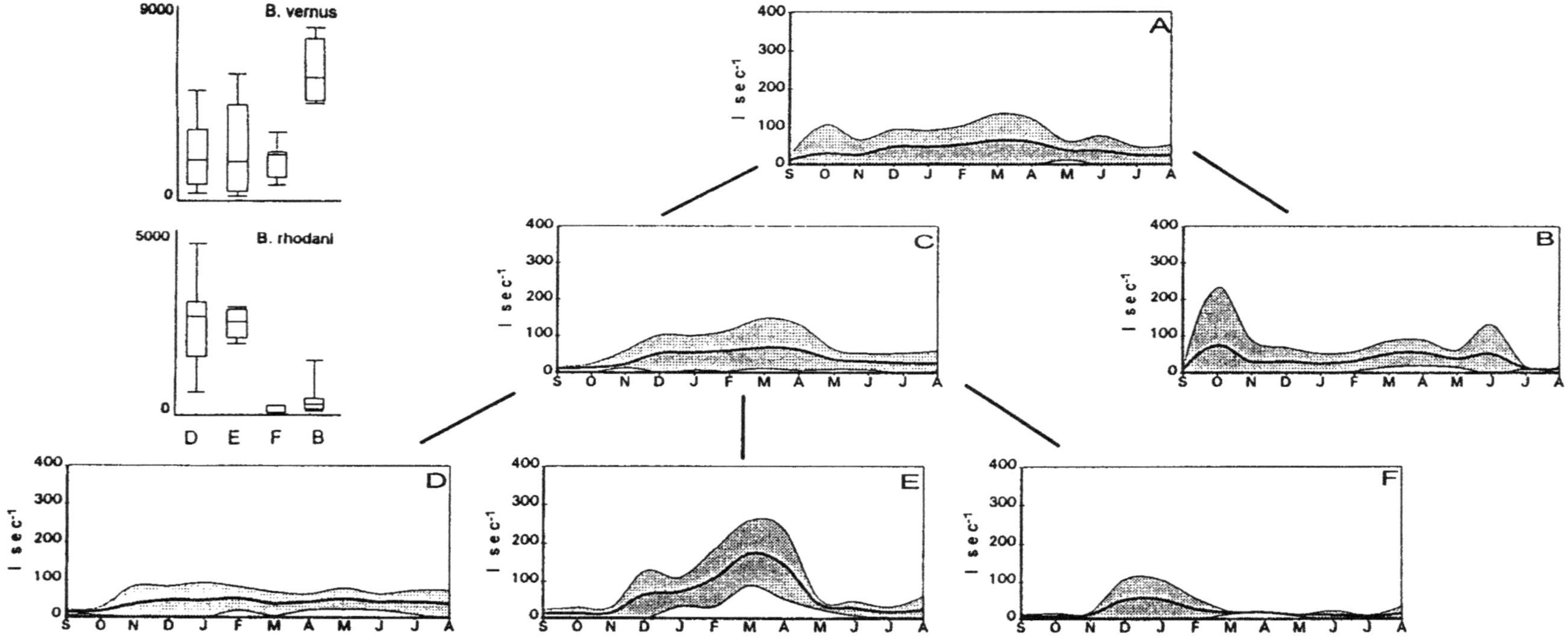

Figure 1. Discharge patterns of the Breitenbach discriminated with traditional statistical methods: (A) 25-year mean of monthly maximum discharge; (B) years with 'non-seasonal' events; (C) the seasonal pattern; (D) permanent, 'good' discharge; (E) winter and spring floods; (F) long-term low flow [mean of within group monthly maximum (line) ±1SD (stippled)]. The abundance of *B. vernus* and *B. rhodani* at patterns D, E, F and B, and significant differences between patterns ($P < 0.05$, Wilcoxon) on left top.

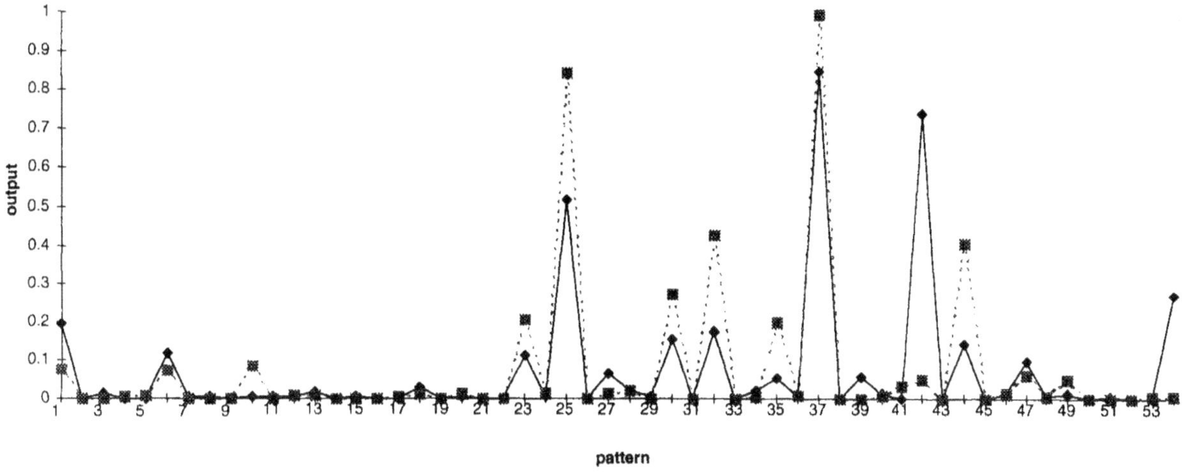

Figure 2. Abundance prediction of *B. rhodani*. Model: all input variables, best generalisation (lowest error): 51–10–1–senso net, R^2=0.56. (Model, full line, square; observed data, dotted line, quadrate).

Table 3. Overview of the different ANN models to predict the abundance of *B. rhodani*. Determination coefficient, minimum and mean generalisation errors of the models

Senso nets	Minimum	R^2	Mean
All variables	0.0082	0.56	0.0156
Correlation	0.0074	0.62	0.0108
Regression	0.0111	0.40	0.0151
Sensitivity analysis	0.0085	0.55	0.0139

Table 4. Overview of ANN models of *B. rhodani* abundance based on discharge measures. Determination coefficient, minimum, mean and maximum generalisation errors of the models

Method	Minimum	R^2	Mean	Maximum
Max	0.01600	0.11	0.02323	0.03453
Min	0.01853	0.18	0.02171	0.02993
MitMax	0.01423	0.37	0.01487	0.01556
Diff	0.01627	0.28	0.02305	0.03095
Quo	0.02166	0.01	0.02190	0.02256
Minus	0.02117	0.03	0.02649	0.03696
Div	0.02221	0.03	0.02850	0.03955

the magnitude of abundance differed between prediction and actual data (Figure 2). Pre-selection of five variables ($abundance_0$, $abundance_1$, $abundance_{11}$, $temperature_0$, $temperature_5$, compare Table 1) with correlation analysis increased the accuracy of the model to R^2=0.62 (Figures 3 and 4). Cross correlation indicated almost no influence of precipitation on *B. rhodani* abundance, but some influence of temperature and discharge. The abundance of the grandparent generation, low discharge and high temperature 12 months before, and high discharge with low temperature 6–8 months before emergence provide sufficient conditions for the success of the population. Pre-selection by regression analysis found other relevant variables ($abundance_1$, $abundance_{10}$, $abundance_{11}$, $discharge_6$, $precipitation_9$), yet decreased the accuracy of the model to R^2=0.40. A sensitivity analysis selected the variables $abundance_0$, $abundance_{11}$, $discharge_9$, $precipitation_9$, $precipitation_{10}$ and had an accuracy of R^2=0.55 (Figure 5). An overview of these experiments indicates that the best model was computed with a pre-selection of the 'best five' variables by correlation, and the next best models included all variables or a pre-selection by sensitivity analysis. A pre-selection by regression resulted in a low accuracy of the model (Table 3).

Altering the discharge data (experiment 2)

We also tested the prediction quality of the *B. rhodani* model using various measures of discharge only: maximum monthly discharge (Max), minimum monthly discharge (Min), the long-term mean pattern of monthly discharge (MitMax), the deviation of actual from long-term mean monthly discharge (Diff = Max minus MitMax), the quotient (Quo = Max by MitMax), difference of Max and Min (= Minus), and Div (= Max by Min). The highest R^2 of 0.37 was

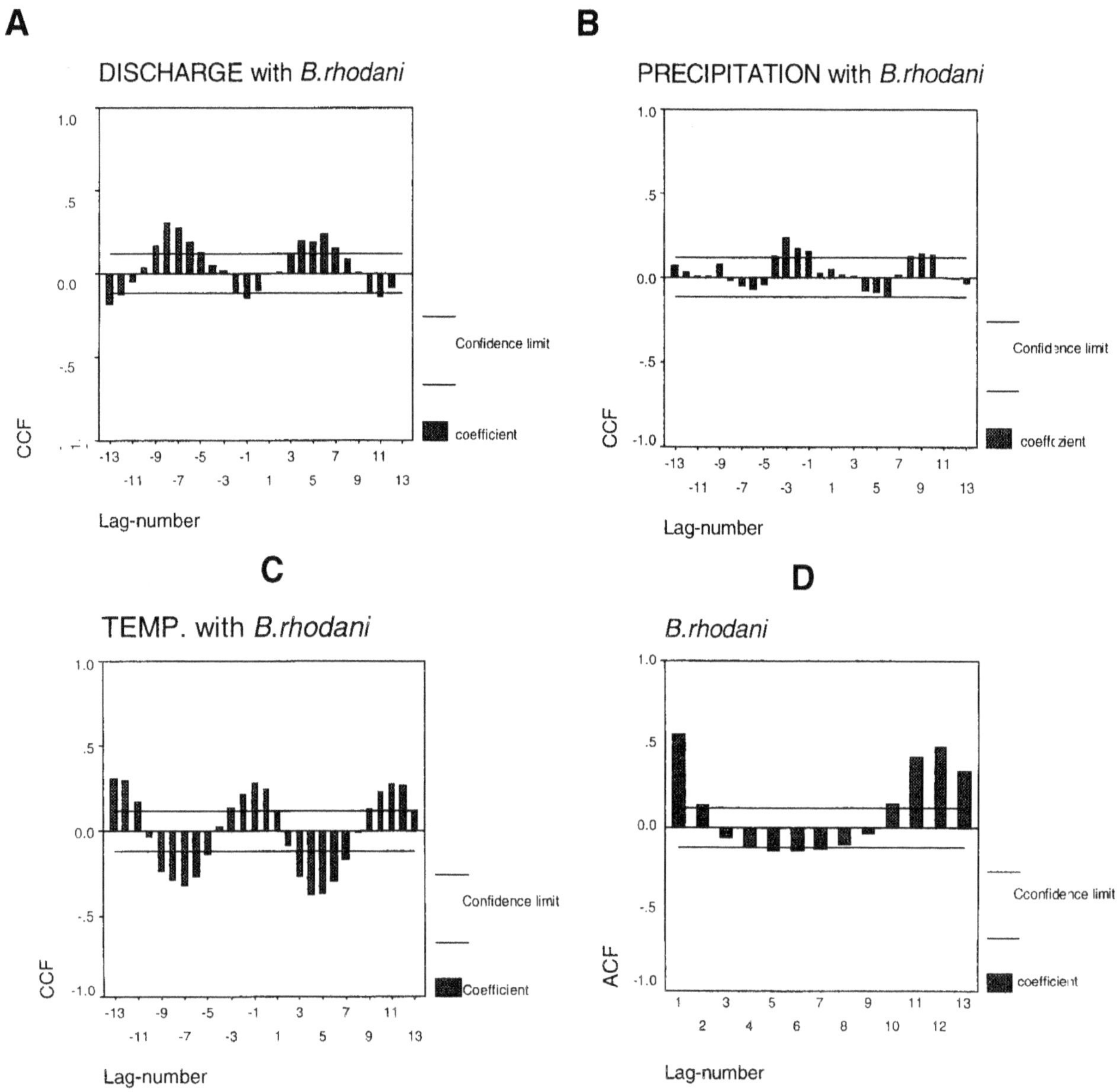

Figure 3. Auto- (ACF) and cross-correlation (CCF) of *B. rhodani* and environmental variables.

found with the long-term mean monthly discharge pattern, but the determination coefficient declined with discharge (Table 4).

Actual discharge or long-term mean discharge (experiment 3)

In a further experiment the actual discharge data (Max) of experiment 1 were replaced by the long-term mean discharge (MitMax of experiment 2) in combination with *P*, *T* and *A* to predict the abundance of *B. rhodani*. The determination coefficient increased to 0.63 for the model with all variables, but decreased to 0.57 for the correlation model.

Altering the scaling options (experiment 4)

The scaling options of the original data may contribute to variation of a model's precision. Thus, in experiment 4 the data were scaled logarithmically, exponentially, or sigmoidally. Variable pre-selection by correlation showed the best generalisation, although, after re-scaling the determination coefficient of the models decreased to 0.40–0.42 (Table 5).

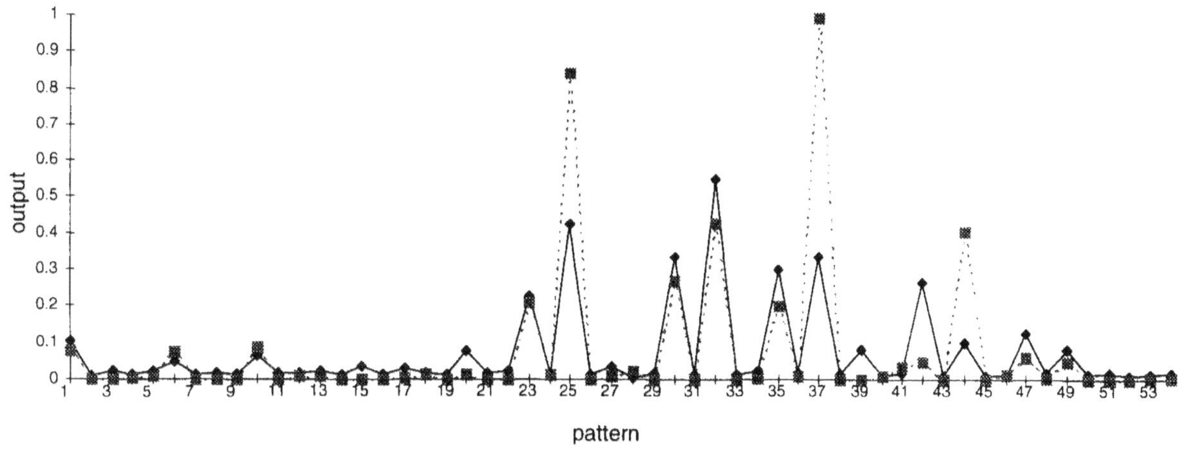

Figure 4. Abundance prediction of *B. rhodani*. Model: pre-selection by correlation, best generalisation (lowest error): 5–3–1–senso net, R^2=0.56. (Model, full line, square; observed data, dotted line, quadrate).

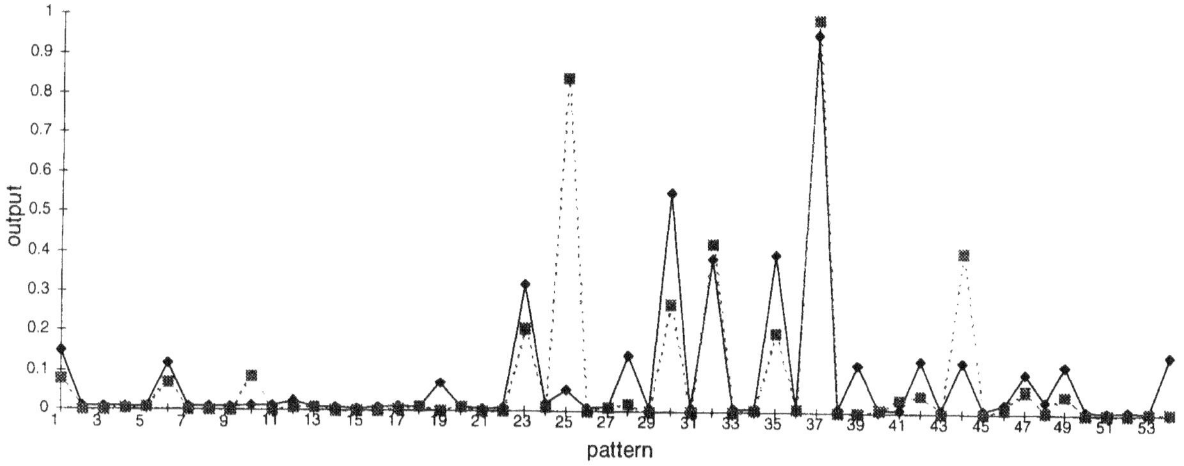

Figure 5. Abundance prediction of *B. rhodani*. Model: pre-selection by sensitivity analysis, best generalisation (lowest error) 5–3–1–senso net: R^2=0.55. (Model, full line, square; observed data, dotted line, quadrate).

Table 5. Overview of the ANN models with logarithmically scaled data. Minimum and mean generalisation errors of the models, and determination coefficient of the scaled and re-scaled data

Senso nets	Minimum	R^2	Mean	R^2 (re-scaled)
All variables	0.015	0.63	0.0164	0.40
Correlation	0.013	0.68	0.0147	0.42
Regression	0.016	0.59	0.0173	0.42

Table 6. Overview of ANN models of *B. vernus* abundance based on various pre-selection methods. Determination coefficient, minimum and mean generalisation errors of the models

Senso nets	Minimum	R^2	Mean
All variables	0.0057	0.54	0.0110
Correlation	0.0060	0.45	0.0087
Regression	0.0054	0.63	0.0077
Sensitivity analysis	0.0036	0.71	0.0064

Deviation of the actual data from the long-term mean (experiment 5)

We also tested the influence of the differences of the actual abundance and discharge values from the long-term mean. The model was not significant (R^2=0.01) with actual discharge and abundance, but with actual discharge and abundance of the parent generation the R^2 increased to 0.44.

Modelling the abundance of B. vernus *(experiment 6)*

The same models used for *B. rhodani* were applied to predict the abundance of *B. vernus*. A

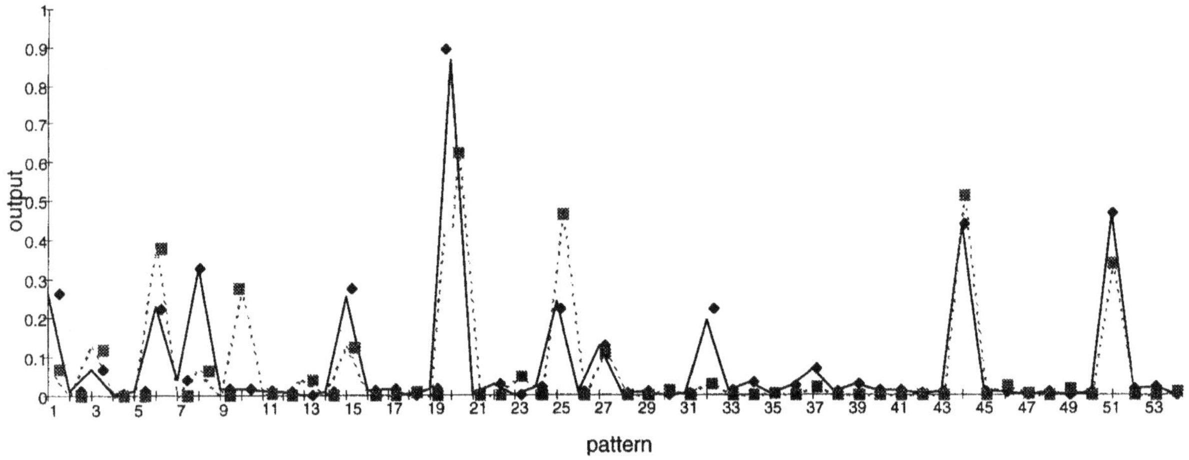

Figure 6. Abundance prediction of *B. vernus*. Model: pre-selection by sensitivity analysis, best generalisation (lowest error) 5–3–1–senso net: R^2=0.71. (Model, full line, square; observed data, dotted line, quadrate).

neural sensitivity analysis that selected the variables abundance$_0$, abundance$_{10}$, abundance$_{11}$, discharge$_0$, and precipitation$_0$ provided the best solution. The best model (5–3–1–senso net) had an accuracy of R^2=0.71 (Figure 6). The determination coefficient of other methods was lower, and lowest with pre-selection by correlation. An overview of the coefficient of other models for *B. vernus* is provided in Table 6.

Using only the months of larval development (experiment 7)

It was interesting to test whether the precision of the model increased if only months relevant for larval development (i.e., excluding the period of egg diapause until January) were used with the *B. vernus* data. We used previous abundance, discharge, precipitation and temperature data from six consecutive months to forecast the abundance in the last month by a 23–5–1 network. The resulting coefficient of this model was lower (R^2=0.60) compared with the regression model or the sensitivity analysis, but higher compared with a model that used all variables for an entire year.

Discussion and conclusion

The abundance differences of *B. rhodani* and *B. vernus* among years were related to discharge patterns using CCA and traditional statistics. However, only two classes with high and two classes with low abundance were discriminated for *B. rhodani*, whereas three classes of low and one class of high abundance were discriminated for *B. vernus*. Further discrimination was senseless because the 50% boxes of mean abundance per pattern widely overlapped (Figure). In combination with species traits, it can be deduced that *B. rhodani* has high success at permanently high discharge and at a seasonal discharge pattern (D, E), but low success at permanent low flow or a non-seasonal discharge pattern (F, B). *B. vernus*, on the other hand, reaches highest abundance values only at the non-seasonal pattern B. This is an example of discharge-mediated competitive exclusion because all probable competitors reach only low abundance levels as an effect of high discharge, but the *B. vernus* population remains more or less unaffected because of diapausing eggs.

ANN models are based on monthly data of species and environmental variables. The amount of data is larger in the original data set and computing is time consuming. The dimension reduction techniques provided improved generalisation performance of the ANNs in many, but not all cases. The selection of proper pre-processing methods is important for the success of the neural modelling (Dapper, 1997).

No single pre-processing method to model the abundance of both *Baetis* species was found. Pre-selection by correlation was optimal in *B. rhodani* but was the least sensitive method in *B. vernus*. *B. vernus* was best modelled by sensitivity pre-selection. Predictors for the best model of both species were abundance of the grandparent or parent generation and temperatures during the emergence and oviposition period of the grandparents and parents. Thus, abundance predictions can be made quite precisely even several months before emergence.

In both species, the manipulation of some or all data by different scaling options decreased the precision of the ANN models (comparison of R^2 values is useful only after re-scaling). A similar phenomenon was observed in CCA analyses (Wagner & Schmidt, in press). Several combinations of long-term and actual data also did not improve the model. Only one small but negligible improvement was observed in the *B. rhodani* model, if the actual discharge was replaced by the long-term mean, but the previously best (correlation) model then decreased in precision. Ecologically, a probable link of the species' life cycle with repetitive seasonal patterns of the environmental variables exists and probably it explains the high determination coefficients of the ANN models.

Based also on the results of more traditional statistical methods, increased experience with ANNs will help optimise the models to predict the abundance of aquatic insects. These methods can be further developed as a tool for better understanding the interrelations of environmental variables impacting individual species and the entire community of streams. The forecast of environmental effects on habitats, species, populations and communities is crucial in basic as well as in applied science.

Acknowledgements

This study was supported by a grant of the Deutsche Forschungsgemeinschaft 'Modellierung ökosystemarer Zusammenhänge in Fließgewässern mit Neuronalen Netzen' (BO 1012). We thank the numerous people who have contributed for many years in the collection of the data. In recent years E. Döring, G. Stüber, and B. Landvogt-Piesche contributed most to these data sets. They, like many assistants and students, were involved in this study by collecting and determining emerged insects, or by recording environmental variables. T. Horvath provided linguistic help. We sincerely acknowledge the comments of two anonymous reviewers on the manuscript.

References

Baran, P., S. Lek, M. Delacoste & A. Belaud, 1996. Stochastic models that predict trout population density or biomass on a mesohabitat scale. Hydrobiologia 337: 1–9.

Bohle, H. W., 1969. Untersuchungen über die Embryonalentwicklung und die embryonale Diapause bei *Baetis vernus* Curt. und *Baetis rhodani* (Pict.). Zool. Jb. Anat. Ontog. Tiere 86: 493–575.

Chon, T. S., 1996. A self-organizing mapping of community patterns for benthic macroinvertebrates in streams in Korea through artificial neural networks. 1996 Annual Combined Meeting of the Ecological Society of America on Ecologists/Biologists as Problem Solvers, Providence, Rhode Island, USA, August 10–14, 1996. Bull. Ecol. Soc. Am. 77 (3 Suppl. Part 2): 81.

Chon, T. S., Y. S. Park, K. H. Moon & E. Y. Cha, 1996. Patternizing communities by using an artificial neural network. Ecol. Model. 90: 69–78.

Cisneros Mata, M. A., T. Brey, A. Jarre Teichmann, W. Garcia Franco & G. Montemayor Lopez, 1996. Artificial neural networks to forecast biomass of Pacific sardine and its environment. Ci. Mar. 22: 427–442.

Clifford, H. C., 1982. Life cycles of mayflies, with special reference to voltinism. Quest. Ent. 18: 15–90.

Colasanti, R. L., 1991. Discussions of the possible use of neural network algorithms in ecological modelling. Binary Comput. Microbiol. 3: 13–15.

Dapper, T., 1997. Dimensionsreduzierende Vorverarbeitungen für Neuronale Netze mit Anwendungen in der Gewässerökologie. Dissertation im Fachbereich Mathematik/Informatik der Universität Gesamthochschule Kassel, 243 pp.

Giske, J., G. Huse & O. Fiksen, 1998. Modelling spatial dynamics of fish. Rev. Fish Biol. Fisher. 8: 57–91.

Guegan, J. F., S. Lek & T. Oberdorff, 1998. Energy availability and habitat heterogeneity predict global riverine fish diversity. Nature 391: 382–384.

Illies, J., 1971. Emergenz 1969 im Breitenbach. Arch. Hydrobiol. 69: 14–59.

Kohonen, T., 1982. Self-organized formation of topologically correct feature maps. Biol. Cyber. 43: 59–69.

Lek, S. & P. Baran, 1997. Estimations of trout density – a neural networks approach. Nonlinear Anal. Theory Methods Appl. 30: 4985–4990.

Lek, S., A. Belaud, P. Baran, I. Dimopoulos & M. Delacoste, 1996a. Role of some environmental variables in trout abundance models using neural networks. Aquat. Liv. Res. 9: 23–29.

Lek, S., M. Delacoste, P. Baran, I. Dimopoulos, J. Lauga & S. Aulagnier, 1996b. Application of neural networks to modelling nonlinear relationships in ecology. Ecol. Model. 90: 39–52.

Mastrorillo, S., S. Lek & F. Dauba, 1997a. Predicting the abundance of minnow *Phoxinus phoxinus* (Cyprinidae) in the river Ariège (France) using artificial neural networks. Aquat. Liv. Res. 10: 169–176.

Mastrorillo, S., S. Lek, F. Dauba & A. Belaud, 1997b. The use of artificial neural networks to predict the presence of small-bodied fish in a river. Freshwat. Biol. 38: 237–246.

Paruelo, J. M. & F. Tomasel, 1997. Prediction of functional characteristics of ecosystems – a comparison of artificial neural networks and regression models. Ecol. Model. 98: 173–186.

Rumelhart, D. E., G. E. Hinton & R. J. Williams, 1986. Learning internal representations by error propagation. In: Rumelhart, D.E. & J.L. McClelland (eds), Parallel Distributed Processing, Vol. I. MIT Press: Cambridge, MA.

Schmidt, H.-H., 1984. Einfluß der Temperatur auf die Entwicklung von *Baetis vernus* Curtis. Arch. Hydrobiol. (Suppl.) 69: 364–410.

SPSS, 1997. SPSS Professional Statistics™ 7.5.1. Chicago, SPSS Inc., 1997.

Ter Braak, C. J. F., 1988. CANOCO – a Fortran program for canonical community ordination by [partial] [detrended] [canonical] correspondence analysis, principal component analysis and redundancy analysis (version 2.1). Agricultural Mathematics Group, Wageningen, The Netherlands, 95 pp.

Ter Braak, C. J. F., 1990. Update Notes, CANOCO version 3.10. Agricultural Mathematics Group, Wageningen, The Netherlands, 35 pp.

Ter Braak, C. J. F. & P. F. M. Verdonschot, 1995. Canonical correspodence analysis and related multivariate methods in aquatic ecology. Aquat. Sci. 57: 255–289.

Wagner, R., 1989. The influence of artificial stream bottom siltation on Ephemeroptera in emergence traps. Arch. Hydrobiol. 115: 71–80.

Wagner, R. & H.-H. Schmidt, in press. Relationship of aquatic insects emergence (Ephemeroptera, Plecoptera, Trichoptera) to environmental variables, a 25 years analysis of the Breitenbach. Freshwat. Biol.

Walley, W. J. & V. N. Fontama, 1998. Neural network predictors of average score per taxon and number of families at unpolluted river sites in Great Britain. Wat. Res. 32: 613–622.

Wernecke, U. & P. Zwick, 1992. Mortality of the terrestrial adult and aquatic nymphal life stages of *Baetis vernus* and *Baetis rhodani* in the Breitenbach (Insecta. Ephemeroptera). Freshwat. Biol. 28: 249–255.

Whitehead, P. G., A. Howard & C. Arulmani, 1997. Modelling algal growth and transport in rivers. A comparison of time series analysis, dynamic mass balance and neural network techniques. Hydrobiologia 349: 39–46.

Hydrobiologia **422/423:** 153–162, 2000.
M. Jungwirth, S. Muhar & S. Schmutz (eds), Assessing the Ecological Integrity of Running Waters.

Biomonitoring through biological traits of benthic macroinvertebrates: how to use species trait databases?

Philippe Usseglio-Polatera[1,2], Michel Bournaud[1], Philippe Richoux[1] & Henri Tachet[1]
[1]*ESA CNRS 5023, Ecologie des Eaux Douces et des Grands Fleuves, Université Lyon I, 43 Boulevard du 11 Novembre, F-69622 Villeurbanne Cedex, France*
[2]*UPRES, Ecotoxicité, Biodiversité et Santé Environnementale, Université de Metz, Ile du Saulcy, F-57045 Cedex 01, France*

Key words: species traits, macroinvertebrates, community structure, indicators, bioassessment

Abstract

The aim of this paper was to investigate the potential use of biological and ecological traits of macroinvertebrates as indicator systems of quality conditions in freshwater ecosystems. To provide a framework for the trait analysis, a data base was developed; it stored biological information about 472 benthic macroinvertebrate taxa. Twenty-two variables describing biological and ecological traits were resolved into a number of simple attributes. A fuzzy coding procedure was used to describe the link between taxa and variables. Three strategies, using simple descriptive statistics or multivariate ordination techniques, were examined in their abilities to demonstrate some of the characteristics of the environment and to bring out clear information on the rate of changes in habitats. We used the Loire River, France, as a specific example. Like faunal analyses, trait analyses showed that a major change occurred in the community in the upper reaches of the river, but transition zones that were not detected with faunal data alone, were discriminated. They corresponded to differences in combinations of biological and ecological traits of communities and were linked to tributary influences (e.g. the Allier–Loire confluence) or anthropogenic disturbances: dams (Grangent, Villerest) or urban zones (Le Puy, Tours). We concluded that the species trait approach had the potential to evaluate the actual state of ecosystems, to discriminate among different types of human impact, and to develop monitoring tools considering the functional diversity of communities.

Introduction

The increasing need for methods to evaluate the actual state of ecosystems and to monitor their rate of change has promoted exploration of how communities of organisms may be used as indicator systems of anthropogenic environmental changes or as the basis for criteria to identify conservation priorities or restoration tools (Cairns & Pratt, 1993; Cao et al., 1996, 1997; DeShon, 1995; Palmer et al., 1996; Pardo & Armitage, 1997; Thorne & Williams, 1997; Verdonschot et al., 1992). Due to their diversity of forms and habits, benthic macroinvertebrates – which potentially respond in many ways to environmental stresses – are good candidates to bioassess aquatic ecosystem quality (Rosenberg & Resh, 1993).

The biological assessment of river quality often used indices based on the abundance and/or richness of various taxonomic groups (Andersen et al., 1984; Barbour et al., 1996; Extence & Ferguson, 1989; Kerans & Karr, 1994; Lenat, 1993; Slàdecek, 1965, 1973; Verneaux et al., 1982; Wallace et al., 1996; Woodiwiss, 1964; and others). However, the generalization of these indices to every type of running water failed because of the natural fluctuations of these groups according to the ecoregional conditions. In addition, the use of such single value indices as a summary of the environmental quality of rivers reduces the available ecological information and may introduce bias into bioassessment (Charvet et al., 1998). Consequently, new synthetic expressions preserving more of the ecological information provided by macroinvertebrates are required.

Spatial and temporal changes in habitats provide a shifting mosaic of abiotic and biotic conditions that play a major role in organizing stream communities. Abiotic factors, in particular those related to disturbance (Poff, 1992; Resh et al., 1988; White & Pickett, 1985) and/or habitat heterogeneity that buffers against disturbances (Minshall & Robinson, 1998; Quinn & Hickey, 1990; Scarsbrook & Townsend, 1993; Sedell et al., 1990), clearly determine the composition of invertebrate communities. If certain species are excluded by their interactions with predators or competitors, what remains is likely to be a subset of those with traits that match the physical environment (Townsend & Hildrew, 1994). Therefore, certain combinations of general biological traits and life history tactics (especially reproductive, physiological and behavioural adaptations for both individual and species survival) are assumed to be selected depending on the environmental characteristics of habitats. Consequently, to explain the local distribution of aquatic organisms, we must study the link between traits of organisms and environmental variability.

Recent studies have proposed various qualitative trends for specific species traits over gradients of streamflow (Poff & Ward, 1989) and habitat heterogeneity, predictability and/or productivity (Minshall, 1988; Minshall & Robinson, 1998; Poff & Ward, 1990; Townsend, 1989; Townsend & Hildrew, 1994; Williams & Feltmate, 1992). Emphasis has turned to individual species and the predominance of biological or ecological traits in relation to differences in environmental characteristics among lotic habitats (Growns & Davis, 1994; Richards et al., 1997; Scarsbrook & Townsend, 1993; Townsend & Hildrew, 1994). Trends of species traits in these habitat templets were usually apparent and often statistically significant (Resh et al., 1994; Scarsbrook & Townsend, 1993; Spence, 1989; Statzner et al., 1997; Townsend et al., 1997; Usseglio-Polatera, 1994; Williams & Feltmate, 1992).

Therefore, if population dynamics can be assumed to be governed essentially by autecological processes, the dynamics of communities may be predicted from the knowledge of the traits of the faunal assemblage. In a complementary way, the knowledge of the traits of a faunal assemblage can demonstrate some of the characteristics of the environment and brings out clear information both on the actual state of freshwater ecosystems and on the rate of spatial or temporal changes in their habitats.

Based on such an approach, the objective of this study was to examine (1) how to synthesize a very large and scattered biological information readily accessible only to specialists into easily interpreted sets of attributes, and (2) how to practically use this information in the course of the evaluation or monitoring processes.

Materials and methods

To provide a framework for the trait analysis of benthic macroinvertebrates, we developed a database storing autecological information on about 472 benthic macroinvertebrate taxa that were known to occupy French freshwater systems. Patterns of traits were derived from a close examination of information gathered from a very large and scattered published expert knowledge and diverse literature sources, ranging from general handbooks to specialized papers. Because of the lack of specific information about many traits for a large set of species (data do not exist or are not published), it was difficult to adopt the species level. As a result, the faunal list included all the genera of benthic macroinvertebrate groups (except for Diptera and Oligochaeta, for which different systematic levels were used: genus but also subfamily or family).

Traits of taxa

We defined eleven biological traits (Table 1) and eleven ecological traits (Table 2) which could be generally assigned across the 472 taxa with the least amount of ambiguity. To structure biological information, we documented features concerning the life cycle ('aquatic stages', 'life cycle duration', 'potential number of reproductive cycles per year', 'maximal size'), the resistance or resilience potential of organisms beyond that of life history traits ('dissemination', 'locomotion and substrate relation', 'resistance stages'), general physiological traits ('respiration') and behavioural aspects of reproduction or nutrition of taxa ('reproduction', 'food', 'feeding habits').

For ecological information, we focused on variables that reflected different spatial characteristics of the habitat used by the taxa (biogeographical zonation (Illies, 1978), longitudinal, altitudinal and transversal distributions). We also coded the microhabitat preferences of taxa for bottom substrate, current velocity, temperature, pH and salinity. In addition, we documented, for each taxon, the 'trophic level' of freshwater it lives in and its value of 'saprobity' (Slàdecek, 1973).

Table 1. Biological traits and modalities (= categories) of aquatic macroinvertebrates defined in this study. Numbers were used as labels in figures

No.	Traits	No.	Modalities
1	Maximal size	1	≤ 0.25 cm
		2	> 0.25–0.5 cm
		3	> 0.5–1 cm
		4	> 1–2 cm
		5	> 2–4 cm
		6	> 4–8 cm
		7	> 8 cm
2	Life cycle duration	1	≤ 1 year
		2	> 1 year
3	Potential number of reproductive cycles per year	1	< 1
		2	1
		3	> 1
4	Aquatic stages	1	egg
		2	larva
		3	pupa
		4	adult
5	Reproduction	1	ovoviviparity
		2	isolated eggs, free
		3	isolated eggs, cemented
		4	clutches, cemented or fixed
		5	clutches, free
		6	clutches, in vegetation
		7	clutches, terrestrial
		8	asexual reproduction
6	Dissemination	1	aquatic passive
		2	aquatic active
		3	aerial passive
		4	aerial active
7	Resistance form	1	eggs, statoblasts
		2	cocoons
		3	cells against desiccation
		4	diapause or dormancy
		5	none
8	Respiration	1	tegument
		2	gill
		3	plastron
		4	spiracle (aerial)
		5	hydrostatic vesicle (aerial)
9	Locomotion and substrate relation	1	flier
		2	surface swimmer
		3	full water swimmer
		4	crawler
		5	burrower (epibenthic)
		6	interstitial (endobenthic)
		7	temporarily attached
		8	permanently attached

Table 1. Continued

No.	Traits	No.	Modalities
10	Food	1	fine sediment + microorganisms
		2	detritus < 1mm
		3	plant detritus > 1mm
		4	living microphytes
		5	living macrophytes
		6	dead animal > 1mm
		7	living microinvertebrates
		8	living macroinvertebrates
		9	vertebrates
11	Feeding habits	1	absorber
		2	deposit feeder
		3	shredder
		4	scraper
		5	filter feeder
		6	piercer (plants or animals)
		7	predator (carver/engulfer/swallower)
		8	parasite

Coding of information

We resolved variables describing biological and ecological traits into a number of modalities (= categories). We used a fuzzy coding procedure (Chevenet et al., 1994; Usseglio-Polatera, 1991) to describe the link between a taxon and each of the variables, providing information both on the amplitude of the taxon preference or tolerance and on the intensity of the link for each modality. We assigned a score to each taxon, describing its affinity to each modality of each trait from '0' indicating 'no affinity' to '5' indicating 'high affinity'. Thereby, the relevant features of a taxon were described by the relative distribution of the information among the modalities of each trait. This procedure was a pertinent and simple way to 'translate' the available knowledge about traits of organisms into numerical values. It also avoided the obligate assignment of a taxon to a single category, which can lead to inaccurate characterization of biological or ecological taxa profiles. In the case of missing information for a taxon and a variable, we coded '0' for all modalities. This ensured that in multivariate analyses this 'not documented' taxon was treated with the average profile of all other taxa for the corresponding variable, i.e., its discriminative weight for this particular variable was zero (Chevenet et al., 1994).

As a result, we obtained an array of 472 taxa by traits with 22 variables and 118 modalities. For brevity, this fuzzy coded basis is not included in this paper.

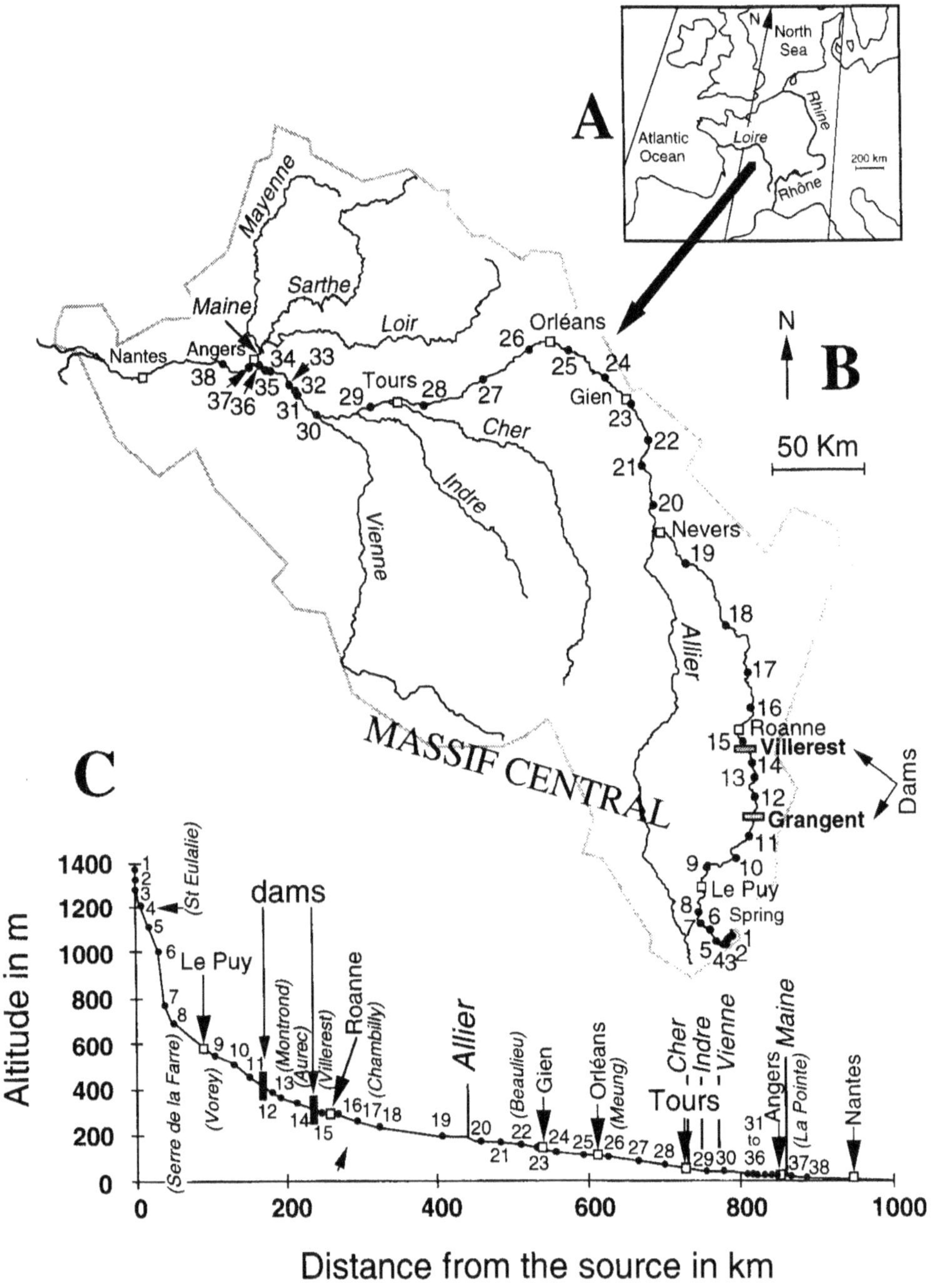

Figure 1. (A) The Loire River in Western Europe, (B) Map of the Loire River Basin, (C) Longitudinal profile of the Loire River. The location of the main tributaries (names in italics), the main urban areas (open squares), the two dams (full rectangles) and the 38 macroinvertebrate sampling sites (full circles with numbers) are indicated. Full labels of sampling sites (names in italics and in brackets) are indicated only if sites were explicitly cited in text.

Table 2. Ecological traits and modalities (= categories) of aquatic macroinvertebrates defined in this study. Numbers were used as labels in figures

No.	Traits	No.	Modalities
1	Transversal distribution	1	river channel
		2	banks, connected side-arms
		3	ponds, pools, disconnected side-arms
		4	marshes, peat bogs
		5	temporary waters
		6	lakes
		7	groundwaters
2	Longitudinal distribution[1]	1	crenon
		2	epirhithron
		3	metarhithron
		4	hyporhithron
		5	epipotamon
		6	metapotamon
		7	estuary
		8	outside river system
3	Altitude	1	lowlands (< 1000 m)
		2	piedmont level (1000–2000 m)
		3	alpine level (> 2000 m)
4	Biogeographic regions[2]	1	2:Pyrenees
		2	4:Alps
		3	8:Vosges, Jura, Massif Central
		4	13a:lowlands (Oceanic)
		5	13b:lowlands (Mediterranean)
5	Substrate (preferendum)	1	flags/boulders/cobbles/pebbles
		2	gravel
		3	sand
		4	silt
		5	macrophytes
		6	microphytes
		7	twigs/roots
		8	organic detritus/litter
		9	mud
6	Current velocity (preferendum)	1	null
		2	slow (< 25 cm s^{-1})
		3	moderate (25–50 cm s^{-1})
		4	fast (> 50 cm s^{-1})
7	Trophic level (preferendum)	1	oligotrophic
		2	mesotrophic
		3	eutrophic
8	Salinity (preferendum)	1	freshwater
		2	brackish water
9	Temperature (preferendum)	1	cold (< 15°C)
		2	warm (> 15°C)
		3	eurythermic
10	Saprobity	1	xenosaprobic
		2	oligosaprobic
		3	β-mesosaprobic
		4	α-mesosaprobic
		5	polysaprobic
11	pH (preferendum)	1	≤ 4
		2	> 4–4.5
		3	> 4.5–5
		4	> 5–5.5
		5	> 5.5–6
		6	> 6

[1]Illies & Botosaneanu (1963).
[2]Illies (1978).

However, it will be available in the major revision (in press) of Tachet et al. (1991).

Illustration of some applications of databases of biological traits

To illustrate some practical applications of such biological information, we selected faunal data from the studies of Guinand et al. (1997) and Ivol et al. (1997), who sampled benthic invertebrate communities in 38 lotic sites (Figure 1) distributed over the entire length of the main channel of the Loire River from the spring to the last reach before the estuary influence (about 800 km). The Loire River is about 1012 km long, 'moderately' dammed (Figure 1), with a mean annual discharge at the mouth of about 860 m^3 s^{-1}. Four insect orders were examined: Plecoptera, Ephemeroptera, Trichoptera and Coleoptera.

Results

We proposed three successive strategies, using simple descriptive statistics or multivariate ordination techniques, to illustrate how biological and ecological traits could be used to investigate community structure and understand mechanisms influencing species distributions.

First strategy

Because individual responses of organisms to physical and chemical gradients shape biological and ecological characteristics of populations, each trait is itself a potential candidate to provide valuable information on the functional diversity of communities. As a result, a first approach can consist in the simple calculation of the sum of affinity scores for all the taxa of an assemblage and for each modality of a trait. Each sum of scores (within a trait) may be expressed as a relative frequency per trait. For each trait, relative frequency distributions (i.e., ecological or biological profiles) may be compared among sites.

As an example, Figure 2 illustrates results obtained for five sampling sites of the Loire River and for five traits. At Saint-Eulalie (site 4), a station situated in the upper reach of the river (8 km from the source) without anthropogenic impact, the faunal assemblage was mainly composed of oligosaprobic, rheophilous and monovoltine taxa which were more representative of rhithronic and oligotrophic environments. Comparing the profiles of stations highlighted an increasing

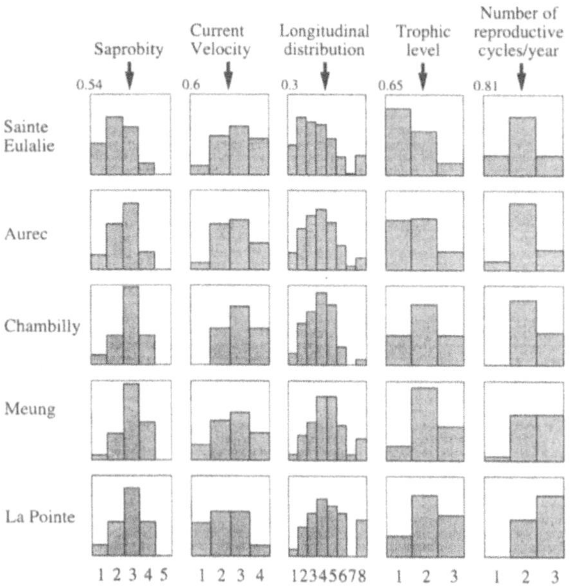

Figure 2. Mean profiles of macroinvertebrate assemblages for five sampling sites of the Loire River and for five traits. Each profile corresponded to the relative frequency distribution of the sums of affinity scores of taxa for modalities (numbers) within a trait. See Tables 1 and 2 for the full label of modalities.

proportion of polyvoltine, β- to α-mesosaprobic organisms along the longitudinal gradient. These taxa displayed more affinities for lentic, mesotrophic or eutrophic environments.

Second strategy

It is also possible, for a particular trait, to compare a typology observed on a set of sampled or tested communities to a reference typology taking into account affinities of the full potential fauna of a given area (e.g., taxa already observed on a set of reference sites of this area). Different scales (e.g., ecoregion, biogeographic zone, several adjacent biogeographic zones) and systematic levels (e.g., species, genus or family) may be envisioned.

To analyse relationships within each set of fuzzy coded trait modalities, we used fuzzy correspondence analysis (Chevenet et al., 1994), which can be regarded as a correspondence analysis of a taxon by modality array where the affinities of the taxa for the modalities are indicated by frequency distributions (e.g., for a given taxon and a trait with five modalities, as 10%/20%/0%/50%/20% for an original code of 1/2/0/5/2). Therefore, the reference typology, for a given trait, was illustrated by the two first factorial axes highlighted from the analysis of the affinity scores for the taxa of the reference fauna. Each taxon and each trait modality exhibited a particular location on this reference plane (F1–F2). Every sample was located on this plane at the weighted average of its taxa locations. Comparing sample locations on this plane potentially provides valuable information about community patterns.

Figure 3A gives the reference typology for the variable 'trophic level' according to the affinity scores of the 472 invertebrate taxa of the reference list. Figure 3B gives the observed typology for 38 sampling sites of the Loire River. Sites were ordinated close to their longitudinal distribution (Figure 3B). Communities demonstrated an increasing affinity for freshwater, exhibiting an elevated trophic level along an upstream–downstream gradient. Some sites, however, as for example 'Montrond' (site 13), despite their geographical location (212 km from the source), corresponded to a macroinvertebrate community with a high affinity for eutrophic environments.

Third strategy

A third way consists in directly matching the faunal structure of a set of samples with the corresponding species trait organization of their communities. All the traits are simultaneously examined. The major objective is to optimally determine the common structure between the two sets of data and then to interpret faunal differences among samples in terms of differences in combinations of biological and ecological traits for communities (Charvet et al., 1998; Usseglio-Polatera, 1997).

In preparation for the simultaneous ordination of the two data sets, both faunal and biological tables were independently and respectively processed by correspondence analysis (CA) and fuzzy correspondence analysis (FCA). Then co-inertia analysis (Dolédec & Chessel, 1994) was performed to assess the relationship between traits and population dynamics.

This approach enabled to observe longitudinal changes in communities for the 38 sampling sites of the Loire River, in terms of both faunal structure and combinations of traits (Figures 4 & 5). The first eight sampling sites were clearly separated from the other sites by their higher F1 scores. Like faunal analyses (Guinand et al., 1996; Ivol et al., 1997), the trait analysis showed that a major change occurred in the community organization in the upper reaches of the river between the Serre de la Fare (site 8) and Vorey (site 9) sampling sites. This first transition zone also

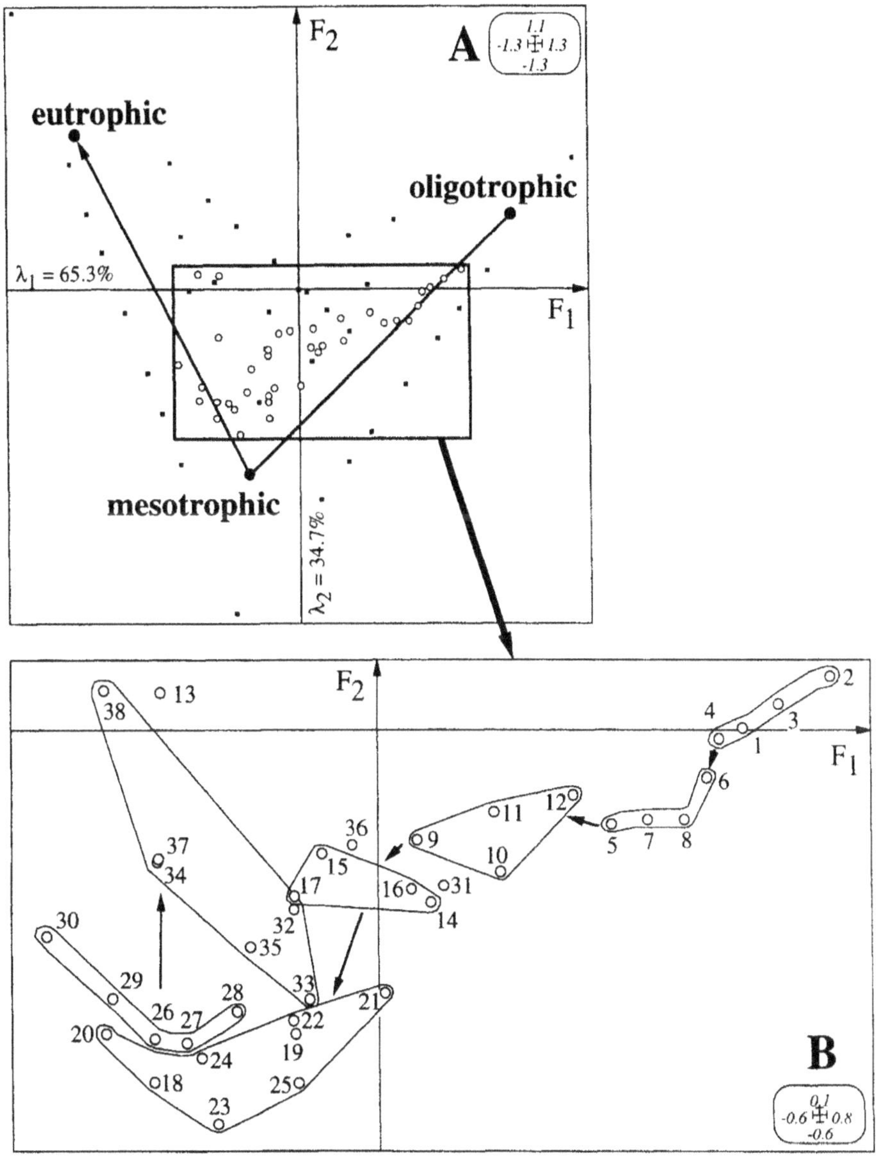

Figure 3. (A) Ordination of modalities (full circles) of the trait 'trophic level' by the fuzzy correspondence analysis of the 472 taxa by three-modality array. The F1–F2 factorial plane was considered as the reference plane. Small squares represented the taxa. 38 sampling sites of the Loire River (open circles) were positioned on this plane at the weighted average of their taxa. (B) For readability, the selected area was enlarged. Sampling sites were indicated by circles with numbers used as labels (see Figure 1 for geographic location of sites).

coincided with the main change of slope, at about 50 km from the source, downstream from site 8 (Figure 1). Two groups of sites (group I for samples from Roanne to Tours and group II for samples from Tours to the last sampling site) had similar negative F1 scores but were separated mainly by their F2 scores. This transition zone hightlighted the impact of the city of Tours and three tributaries draining both urban and agricultural areas (Cher, Indre, Vienne) on the stream quality. Sites 9–11 were also negative on the F1 axis, but closer to the origin than the previous groups. The location of sites 12–15 highly fluctuated on the F1–F2 plane, illustrating the disturbing impact of dams on the assemblage structure. This effect was not so apparent with only faunal data (Guinand et al., 1996; Ivol et al., 1997).

The F1 scores for modalities clearly demonstrated that sites were arranged according to the habitat utilization of their faunal communities. The distribution of modalities for most of the traits corresponded to a gradient from (1) long-lived, xeno- to oligosaprobic taxa that were scrapers and shredders, preferring cold

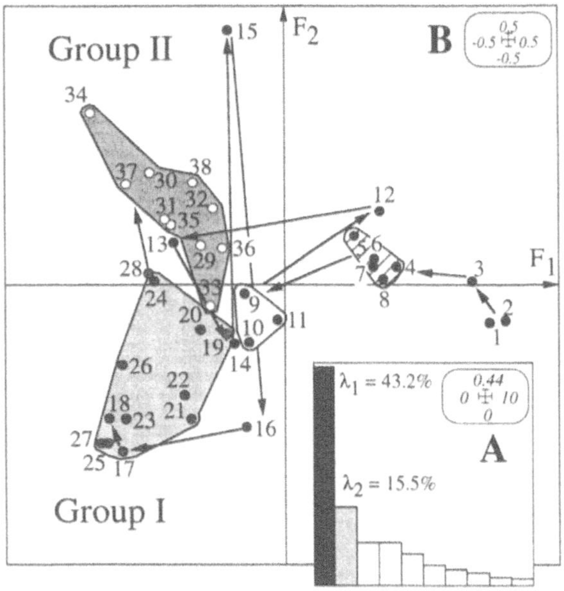

Figure 4. Co-structure between traits and spatial distribution of taxa in the Loire River by co-inertia analysis. (A) Histogram of eigenvalues. The percentage of the total variability explained by each axis is indicated. (B) Ordination of the 38 sampling sites (circles plus their labels) on the F1–F2 factorial plane of the co-inertia analysis.

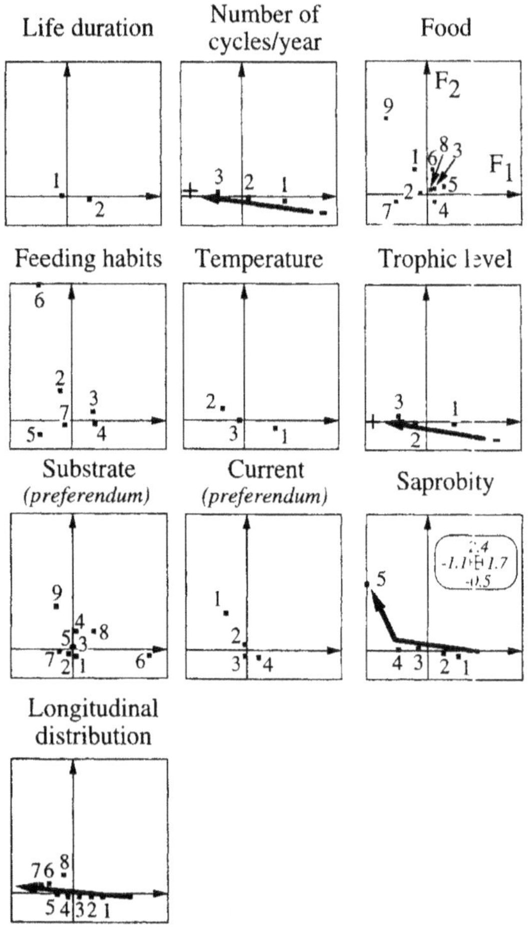

Figure 5. Co-structure between traits and spatial distribution of taxa by co-inertia analysis, in the Loire River. Ordination of the modalities (full squares plus their labels) of 10 traits on the F1–F2 factorial plane of the co-inertia analysis. See Tables 1 and 2 for the full label of traits and modalities.

waters and high current velocities, and living on coarse mineral substrates often covered with epilithic microphytes in the main channel of small oligotrophic rivers, towards (2) short-lived, eurythermic taxa that were filterers or deposit feeders, exhibiting higher affinities for moderately lotic or lentic habitats, with fine mineral sediment, in mesotrophic/eutrophic streams of higher orders (Figure 5).

Differences between groups I and II mainly corresponded to differences among species for food or feeding habits, substrate preferendum, saprobity and tolerance for high temperature. The faunal communities of the lower section had more affinities for muddy substrates and still waters in eutrophic habitats. The corresponding taxa were more often deposit-feeders or piercers (*Hydroptila*), feeding on microorganisms, fine organic matter or filamentous algae.

Discussion

Within the last decade, an increasing effort has been devoted to design a more effective use of macroinvertebrates as monitoring and assessment tools for management of water resources. The most effective use of such tools occurs when there is a clear understanding of the mechanisms which lead to the presence or absence of species in the environment. Biological and ecological traits have the potential to give a pertinent image of the mechanisms structuring stream communities, although trade-offs among traits can sometimes make data analysis and interpretation difficult (Statzner et al., 1997).

To provide a framework for the trait analysis, we developed a database storing autecological information about 472 taxa, mainly genera, that were known to occupy French freshwater ecosystems. Even if some authors have focused on the importance of species identification for water quality monitoring (Resh & Unzicker, 1975), others have shown that higher taxonomic levels were suitable for an ecological study (Bournaud et al., 1996), especially for studying the functional structure of communities (Dolédec et al., 1998). Biological and ecological traits were examined in their abilities to delineate taxa. This data base is usable in France, but also with no or minor adjustments in most of the Western Europe (e.g., England, Belgium, Luxemburg, Netherlands, Western Germany, Northern Spain and Northern Italy).

Each trait has itself the potential to provide valuable information on the structure and the functioning of communities. Furthermore, many modalities of the ecological and the biological traits are well distributed (but with different scores) among taxa and are little or not constrained by biogeographic distributions and thus can be applied across ecoregions.

In this paper, we proposed three strategies that should be useful (1) in evaluating the actual state of ecosystems, (2) in discriminating among different types of human impact, and (3) in developing monitoring tools considering the functional diversity of communities. They allowed changes along a large river to be detected and interpreted more efficiently than with only faunal (and environmental) data.

However, this approach is highly dependent on the availability of autecological information. Despite a thorough examination of the literature, the lack of data forced us to adopt the genus level and to omit potentially important descriptive variables such as, for example, fecundity, dissemination potential (in terms of travelled distances) or P/B ratios. This situation underlines the importance to continue studies on fundamental life history traits and the ecological requirements of species in order to improve our knowledge on such biological and ecological information.

Acknowledgements

We are especially grateful to Jeanne-Marie Ivol for providing some faunal data on the Loire River.

References

Andersen, M.M., F.F. Riget & H. Sparholt, 1984. A modification of the Trent Index for use in Denmark. Wat. Res. 18: 141–151.

Barbour, M.T., J. Gerritsen, G.E. Griffith, R. Frydenborg, E. McCarron, J.S. White & M.L. Bastian, 1996. A framework for biological criteria for Florida streams using benthic macroinvertebrates. J. N. am. Benthol. Soc. 15: 185–211.

Bournaud, M., B. Cellot, P. Richoux & A. Berrahou, 1996. Macroinvertebrate community structure and environmental characteristics along a large river: congruity of patterns for identification to species or family. J. N. am. Benthol. Soc. 15: 232–253.

Cairns, J. & J.R. Pratt, 1993. A history of biological monitoring using benthic macroinvertebrates. In Rosenberg, D.M. & V.H. Resh (eds), Freshwater Biomonitoring and Benthic Macroinvertebrates. Chapmann and Hall, New York: 10–27

Cao, Y., A.W. Bark & W.P. Williams, 1996. Measuring the responses of macroinvertebrate communities to water pollution: a comparison of multivariate approaches, biotic and diversity indices. Hydrobiologia 341: 1–19.

Cao, Y., A.W. Bark & W.P. Williams, 1997. Analysing benthic macroinvertebrate community changes along a pollution gradient: a framework for the development of biotic indices. Wat. Res. 31: 884–892.

Charvet, S., A. Kosmala & B. Statzner, 1998. Biomonitoring through biological traits of benthic macroinvertebrates: perspectives for a general tool in stream management. Arch. Hydrobiol. 142: 415–432.

Chevenet, F., S. Dolédec & D. Chessel, 1994. A fuzzy coding approach for the analysis of long-term ecological data. Freshwat. Biol. 31: 295–309.

DeShon, J.E., 1995. Development and application of the Invertebrate Community Index (ICI). In Davis, W.S. & T.P. Simon (eds), Biological Assessment and Criteria: Tools for Water Resource Planning and Decision Making. Lewis Publishers, Boca Raton, FL: 217–243.

Dolédec, S. & D. Chessel, 1994. Co-inertia analysis: an alternative method for studying species–environment relationships. Freshwat. Biol. 31: 277–293.

Dolédec, S., B. Statzner & V. Frainay, 1998. Accurate description of functional community structure: identifying stream invertebrates to species-level? Bull. N. am. Benthol. Soc. 15: 154–155.

Extence, C.A. & J.D. Ferguson, 1989. Aquatic invertebrate surveys as a water quality management tool in the anglian water region. Regul. Riv. 4: 139–146.

Growns, I.O. & J.A. Davis, 1994. Longitudinal changes in near-bed flows and macroinvertebrate communities in a Western Australian stream. J. N. am. Benthol. Soc. 13: 413–438.

Guinand, B., J.M. Ivol & H. Tachet, 1996. Longitudinal distribution of Trichoptera in the Loire River (France): simple ordination methods and community structure. Hydrobiologia 317: 231–245.

Illies, J. (ed.), 1978. Limnofauna Europaea, 2nd edn. Fischer, Stuttgart, 532 pp.

Illies, J. & L. Botosaneanu, 1963. Problèmes et méthodes de la classification et de la zonation écologique des eaux courantes considérées surtout du point de vue faunistique. Mitt. Intern. Verein. Limnol. 12: 1–57.

Ivol, J.M., B. Guinand, P. Richoux & H. Tachet, 1997. Longitudinal changes in Trichoptera and Coleoptera assemblages and environmental conditions in the Loire River (France). Arch. Hydrobiol. 138: 525–557.

Kerans, B.L. & J.R. Karr, 1994. A benthic index of biotic integrity (B-IBI) for rivers of the Tennessee Valley. Ecol. Appl. 4: 768–785.

Lenat, D.R., 1993. A biotic index for the southeastern United States: derivation and list of tolerance values, with criteria for assigning water-quality ratings. J. N. am. Benthol. Soc. 12: 279–290.

Minshall, G.W., 1988. Stream ecosystem theory: a global perspective. J. N. am. Benthol. Soc. 7: 263–288.

Minshall, G.W. & C.T. Robinson, 1998. Macroinvertebrate community structure in relation to measures of lotic habitat heterogeneity. Arch. Hydrobiol. 141: 129–151.

Palmer, C.G., B. Maart, A.R. Palmer & J.H. O'Keeffe, 1996. An assessment of macroinvertebrate functional feeding groups as water quality indicators in the Buffalo River, Eastern Cape Province, South Africa. Hydrobiologia 318: 153–164.

Pardo, I. & P.D. Armitage, 1997. Species assemblages as descriptors of mesohabitats. Hydrobiologia 344: 111–128.

Poff, N.L., 1992. Why disturbances can be predictable: a perspective on the definition of disturbance in streams. J. N. am. Benthol. Soc. 11: 86–92.

Poff, N.L. & J.V. Ward, 1989. Implications of streamflow variability and predictability for lotic community structure: a regional analysis of streamflow patterns. Can. J. Fish. aquat. Sci. 46: 1805–1818.

Poff, N.L. & J.V. Ward, 1990. Physical habitat template of lotic systems: recovery in the context of historical pattern of spatiotemporal heterogeneity. Envir. Manage. 14: 629–645.

Quinn, J.M. & C.W. Hickey, 1990. Characterisation and classification of benthic invertebrate communities in 88 New Zealand rivers in relation to environmental factors. N.Z. J. Mar. Freshwat. Res. 24: 387–409.

Resh, V.H. & J.D. Unzicker, 1975. Water quality monitoring and aquatic organisms: the importance of species identification. J. Wat. Pollut. Cont. Fed. 47: 9–19.

Resh, V.H., A.V. Brown, A.P. Covich, M.E. Gurtz, H.W. Li, G.W. Minshall, S.R. Reice, A.L. Sheldon, J.B. Wallace & R. Wissmar, 1988. The role of disturbance in stream ecology. J. N. am. Benthol. Soc. 7: 433–455.

Resh, V.H., A.G. Hildrew, B. Statzner & C.R. Townsend, 1994. Theoretical habitat templets, species traits, and species richness: a synthesis of long-term research on the Upper Rhône River in the context of concurrently developed ecological theory. Freshwat. Biol. 31: 539–554.

Richards, C., R.J. Haro, L.B. Johnson & G.E. Host, 1997. Catchment and reach-scale properties as indicators of macroinvertebrate species traits. Freshwat. Biol. 37: 219–230.

Rosenberg, D.M. & V.H. Resh, 1993. Introduction to freshwater biomonitoring and benthic macroinvertebrates. In Rosenberg, D.M. & V.H. Resh (eds), Freshwater Biomonitoring and Benthic Macroinvertebrates. Chapmann & Hall, New York: 1–9.

Scarsbrook, M.R. & C.R. Townsend, 1993. Stream community structure in relation to spatial and temporal variation: a habitat templet study of two contrasting New Zealand streams. Freshwat. Biol. 29: 395–410.

Sedell, J.R., G.H. Reeves, F.R. Hauer, J.A. Stanford & C.P. Hawkins, 1990. Role of refugia in recovery from disturbances: modern fragmented and disconnected river systems. Envir. Manage. 14: 711–724.

Slàdecek, V., 1965. The future of the saprobity system. Hydrobiologia 25: 518–537.

Slàdecek, V., 1973. System of water quality from the biological point of view. Arch. Hydrobiol. 7: 1–218.

Spence, J.R., 1989. The habitat templet and life history strategies of pond skaters (Heteroptera: Gerridae): reproductive potential, phenology, and wing dimorphism. Can. J. Zool. 67: 2432–2447.

Statzner, B., K. Hoppenhaus, M.F. Arens & P. Richoux, 1997. Reproductive traits, habitat use and templet theory: a synthesis of world-wide data on aquatic insects. Freshwat. Biol. 38: 109–135.

Tachet, H., M. Bournaud & P. Richoux, 1991. Introduction à l'étude des macroinvertébrés des eaux douces (systématique élémentaire et aperçu écologique). Université Lyon I & Association Française de Limnologie (eds), CRDP, Lyon, 155 pp.

Thorne, R.S.J. & W.P. Williams, 1997. The response of benthic macroinvertebrates to pollution in developing countries: a multimetric system of bioassessment. Freshwat. Biol. 37: 671–686.

Townsend, C.R., 1989. The patch dynamics concept of stream community ecology. J. N. am. Benthol. Soc. 8: 36–50.

Townsend, C.R. & A.G. Hildrew, 1994. Species traits in relation to a habitat templet for river systems. Freshwat. Biol. 31: 265–275.

Townsend, C.R., S. Dolédec & M.R. Scarsbrook, 1997. Species traits in relation to temporal and spatial heterogeneity in streams: a test of habitat templet theory. Freshwat. Biol. 37: 367–387.

Usseglio-Polatera, P., 1991. Représentation graphique synthétique de la signification écologique d'un peuplement. Application aux macroinvertébrés du Rhône à Lyon. Bull. Ecol. 22: 195–202.

Usseglio-Polatera, P., 1994. Theoretical habitat templets, species traits, and species richness: aquatic insects in the Upper Rhône River and its floodplain. Freshwat. Biol. 31: 417–437.

Usseglio-Polatera, P., 1997. Caractéristiques biologiques, stratégies écologiques et évolution environnementale d'un hydrosystème: l'exemple des Trichoptères du Rhône à Lyon (France). Geobios 21: 151–158.

Verdonschot, P.F.M., L.W.G. Higler, W.F. Van der Hoek & J.G.M. Cuppen, 1992. A list of macroinvertebrates in Dutch water types: a first step towards an ecological classification of surface waters based on key factors. Hydrobiol. Bull. 25: 241–259.

Verneaux, J., P. Galmiche, F. Janier & A. Monnot, 1982. Une méthode pratique d'évaluation de la qualité des eaux courantes. Un indice biologique de qualité générale. Annls Sci. Univ Franche-Comté, Besançon 4: 11–21.

Wallace, J.B., J.W. Grubaugh & M.R. Whiles, 1996. Biotic indices and stream ecosystem processes: results from an experimental study. Ecol. Appl. 6: 140–151.

White, P.S. & S.T.A. Pickett, 1985. Natural disturbance and patch dynamics: an introduction. In Pickett, S.T.A. & P.S. White (eds), The Ecology of Natural Disturbance and Patch Dynamics. Academic Press, Orlando, FL: 3–13.

Williams, D.D. & B.W. Feltmate, 1992. Aquatic Insects. CAB International, Wallingford, 358 pp.

Woodiwiss, L., 1964. The biological system of stream classification used by the Trent River Board. Chem. Ind. (Lond.) 14: 443–447.

Hydrobiologia **422/423:** 163–171, 2000.
M. Jungwirth, S. Muhar & S. Schmutz (eds), Assessing the Ecological Integrity of Running Waters.

The spatial heterogeneity of a river bottom: a key factor determining macroinvertebrate communities

Jean-Nicolas Beisel[1,*], Philippe Usseglio-Polatera[1,2] & Jean-Claude Moreteau[1]
[1]*UPRES Ecotoxicité, Biodiversité et Santé Environnementale (EBSE), Université de Metz, Ile du Saulcy, F-57045 Metz Cedex 01, France*
[2]*UPRESA CNRS 5023 'Ecologie des Eaux Douces et des Grands Fleuves', Université Lyon I, F-69622 Villeurbanne Cedex, France*

Key words: mesohabitats, macroinvertebrate communities, spatial heterogeneity, running waters

Abstract

We examined the relationships between mesohabitat heterogeneity and benthic macroinvertebrate assemblages in a fourth-order stream (Mortagne, North-Eastern France). The heterogeneity of the mesohabitat mosaic was described around each sampling point for eight different surfaces from 0.5 to 4 m radius. Environmental descriptors integrated both the composition and configuration heterogeneity of the substrate mosaic. Faunal data were analysed in terms of biocenotic indices and taxonomic composition. The maximum number of significant Spearman correlations was obtained for a mosaic with a diameter of 2 m around the sampling point and significant correlations were both the highest and the most numerous for this surface. Relationships were established between community structure and substrate heterogeneity. The faunal richness was higher in a heterogeneous environment composed of numerous substrates, an elevated patchiness and with high perimeters. Such a mosaic potentially offers a great number of niches for invertebrates. A reduced distance between two types of substrate favours exchange of species. At the opposite, a very homogeneous mosaic offers a low variety of niches and shelters fewer taxa. Furthermore, in a homogeneous environment we observed that one or two particular taxa dominated the community, probably because competition with taxa coming from neighbouring patches was reduced. The community composition mainly depended on the characteristics of the mesohabitat sampled. Our results showed that the neighbouring environment around such mesohabitats also had a significant influence. Further research must be conducted to specify the influence of the mosaic heterogeneity on biological and ecological traits of invertebrates.

Introduction

Macroinvertebrate communities are widely used to assess the biological integrity of running waters and represent a scale compatible with the definition of criteria for conservation, fisheries management and channel engineering objectives (Harper & Everard, in press). To elaborate or to improve bioindicator systems of environmental changes, considerable research has been devoted to the question of whether and to what extent environmental characteristics influence macroinvertebrate communities at a variety of spatial and temporal scales.

* Author to whom correspondence should be sent.

Recent developments from a micro- or mesohabitat approach (Armitage & Cannan, 1998; Wood, 1998) allowed the biologically important features of a river channel to be identified (Harper & Everard, in press). Mesoscale is potentially perceived by invertebrates; this is determinant for autecological studies. The development of suitable criteria for invertebrates within mesohabitats could potentially help in riverine management operations (Harper et al., 1992; Wood, 1998). As underlined by Armitage & Cannan (1998), there is still a need for small-scale basic studies which examine the distribution of benthos in running waters because such studies are the building blocks of knowledge for wider assessments (Harper & Everard, in press).

Mesohabitat studies have focused on environmental conditions that vary over a distance of a few centimeters or meters, such as substrate composition (Minshall & Minshall, 1977; Reice, 1980), hydraulic conditions (Chutter, 1969; Statzner et al., 1988), or food availability (Dobson & Hildrew, 1992; Drake, 1984). These parameters are well known as determinants for macroinvertebrate assemblages of mesohabitats (Beisel et al., 1998b), mesohabitat being practically defined as a combination of a substrate type and a range of current speed.

Other environmental features such as mesohabitat mosaic heterogeneity are potentially important to explain or predict patterns of invertebrate distribution. The spatial heterogeneity of the mesohabitat mosaic includes for example the number of substrates, their patchiness, their shape and/or their arrangement. However, studies on this topic are scarce.

Preliminary analyses allowed us to demonstrate significant relationships between the spatial heterogeneity of mesohabitat and faunal community structure (Beisel et al., 1998a). This paper analyses a set of faunal data sampled in a restricted area (73 invertebrate assemblages of mesohabitat sampled in the rhithron area of a fourth-order stream) where temporal heterogeneity was the same from a site to another. The lack of spatial heterogeneity within a stream, at least with respect to certain habitat factors, may be compensated for by higher levels of temporal heterogeneity (Minshall & Robinson, 1998). Our main objective was to investigate, at the mesohabitat scale, how macroinvertebrate community assemblages were affected by the local heterogeneity of the substrate mosaic around a sampling point. This implied identifying which ecological processes are key factors.

Material and methods

Study areas

Macroinvertebrate community structure was studied at a mesohabitat scale in a fourth order stream, the Mortagne River, located in the north-eastern part of France (Figure 1). Field work was conducted in four localities in June and September 1993. Sampling sites were selected (1) for easy accessibility, (2) to permit sampling of all the representative habitats of each study site, (3) to have no or little sustained anthropogenic land use, and (4) to exhibit high microscale heterogeneity.

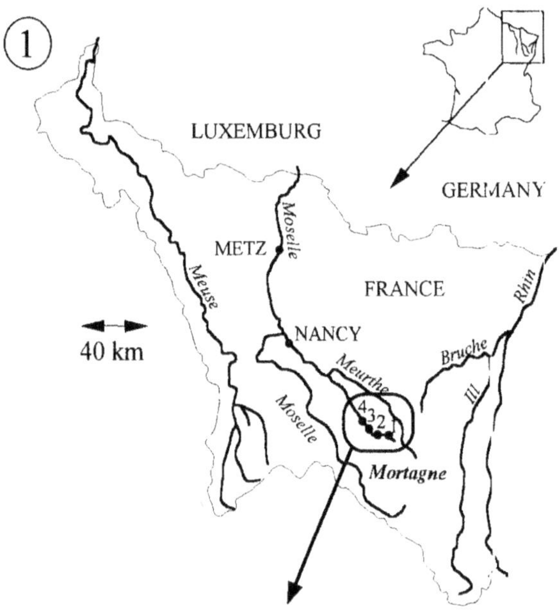

2

Mortagne River	
Mean slope of study sites (‰)	4 to 7
Mean width (m)	4 to 6.5
Dominant substrate	Cobbles
Altitude (m)	315 to 405
Mean depth (cm)	20 to 40
Mean current speed ($cm.s^{-1}$)	35 to 40
pH	6.5 to 7.7
Conductivity at 20°C ($\mu S.cm^{-1}$)	48 to 93

Figure 1. Location of the four stations (black circles) on the Mortagne River (1), and ranges for physical, mesological and chemical variables recorded during field work (2).

The river at the study site was poorly mineralized with a neutral to slightly acid pH (Figure 1). Each site was approximately 30 m in length and comprised a minimum of one pool–riffle sequence. The bottom was composed of a number of organic and mineral mesohabitats. Further details can be found in Beisel (1996).

Sampling design

For each sampling date, the substrate mosaic was mapped with an accuracy of 0.10–0.25 m (Figure 2). Substrate type was selected among 39 modalities (Beisel, 1996). These maps enabled us (1) to identify a maximum number of different mesohabitat types before sampling and (2) to quantify the spatial heterogeneity of the substrate mosaic. One sample was

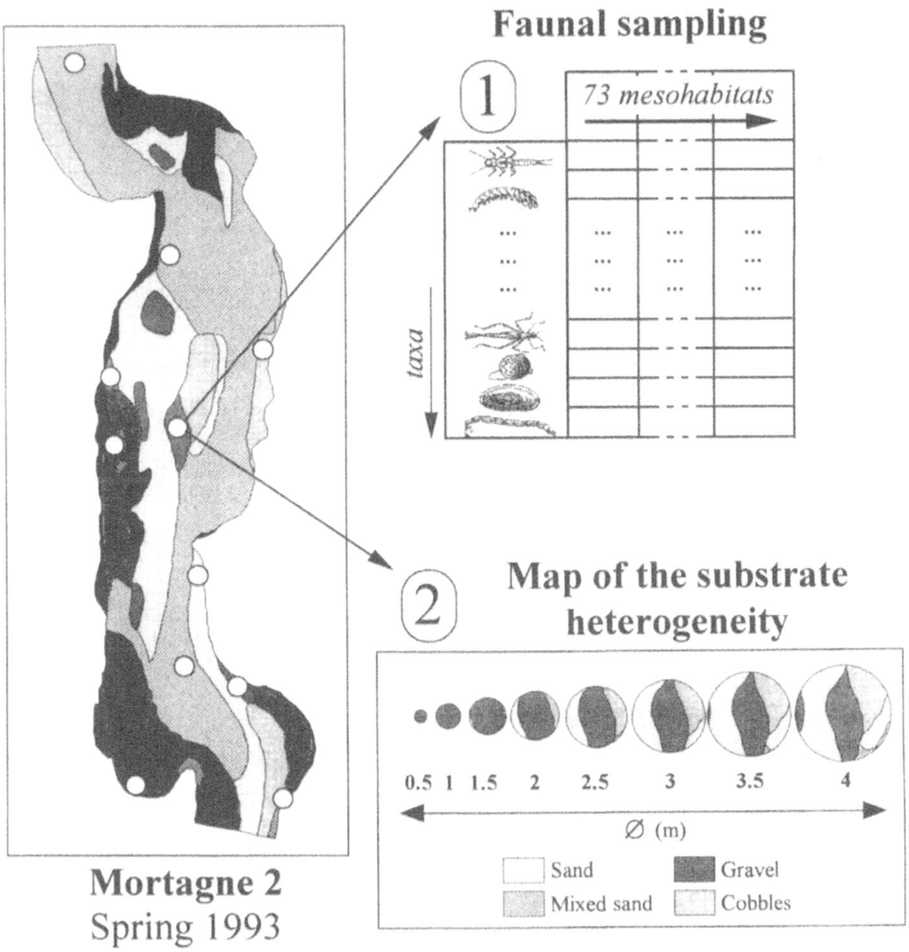

Figure 2. An example of a station map drawn in the field. All the different mesohabitats (substrata type × current velocity) in a station were sampled and yielded a macroinvertebrate list (1). Then, the map allowed us to quantify substrate heterogeneity for eight round surfaces within a radius of 0.5 to 4 m around each sampling point (2).

performed according to mesohabitat type and, depending on station and/or season, eight to 12 different mesohabitat types were sampled per study site. A total of 73 faunal samples and their environments were analysed.

Quantitative macroinvertebrate samples were collected using a 0.05-m^2 Surber sampler (200 × 250 mm frame, mesh size 500 μm). The substrate within the sampler was removed to a depth of about 6 cm and thoroughly brushed and washed to remove all organisms. All samples were fixed in 4% formaldehyde and identified to family level with the exception of Oligochaeta and Hydracarina, which were recorded as such.

Description of the mosaic heterogeneity

Maps drawn in the field provided information on the distribution and proportion of substrates, allowing us to quantify the spatial heterogeneity of the mosaic around a sampling point (Figure 2). In order to test the influence of the area considered we studied eight round surfaces with a radius of 0.5, 1, 1.5, 2, 2.5, 3, 3.5, and 4 m around each sampling point.

To quantify bottom heterogeneity, we used the operational definition proposed by Li & Reynolds (1994, 1995) for categorical maps. Spatial heterogeneity was defined as the complexity in both composition and configuration of elements. Composition implied the number of element types present and the proportion of each type of elements. Configuration mainly referred to spatial arrangement of elements, to element shape, and to contrasts between neighbouring elements.

In this study, elements taken into account for calculations were both the different substrate types and the different patches (a substrate type can be divided into several independent surfaces called patches), which composed the substrate mosaic.

A panel of 14 indices allowed us to measure spatial heterogeneity of the substrate mosaic:

- richness, Hurlbert evenness, Shannon diversity and Simpson dominance calculated on element (i.e. substrates or patches) inventories (number and/or proportions of elements per area);
- surface of the most important element (in m^2 or in%);
- surface of the sampled patch (in m^2 or in%);
- mean of the substrate/perimeter ratio calculated for each inventory of patches and for each inventory of patches excluding macrophytes.

A methodological study showed that biocenotic indices calculated on patch or substrate inventories provided meaningful information on macroinvertebrates and bottom heterogeneity relationships (Beisel et al., 1998a).

Data analysis

Macroinvertebrate community structure was summarized with six biocenotic indices (density for 0.05 m^2, richness, Hurlbert evenness, Shannon–Weaver diversity and dominance-Simpson measure and the proportion of the most dominant taxon) calculated on invertebrate lists.

Nonparametric correlation coefficients (Spearman R, which is only sensitive to the ordinal arrangement of values) enabled us to test the relationship between mosaic heterogeneity descriptors and community structure indices. When a single patch composed the mosaic, evenness of patch surfaces was not considered and correlation was calculated without these points.

A within-station factorial correspondence analysis (CA) was realized on a 48 taxa by a 73-sample array. The identification level of taxa depended on their frequencies: some groups were amalgamated to reduce the number of taxa with a low frequency in the samples.

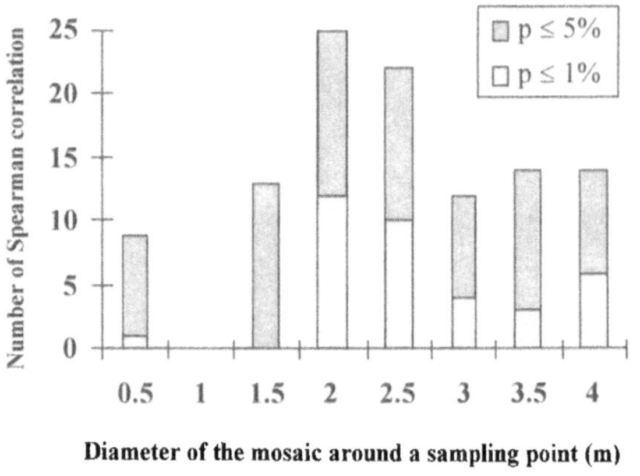

Figure 3. Number of significant Spearman correlations calculated between six macroinvertebrate structure indices and 14 spatial heterogeneity descriptors for each diameter of the mosaic around a sampling point (n=73). Two levels of significance were examined.

Results

Importance of the surface area considered around a sampling point

Preliminary analysis was undertaken to assess which bottom surface diameter influenced the macroinvertebrate assemblage of a mesohabitat. We compared the number of significant Spearman correlations (for $P \leq 5\%$ and $P \leq 1\%$) between environmental heterogeneity (14 measures) and invertebrate assemblage structure (six descriptors) for the eight surfaces considered around each sampling point (Figure 2). The number of significant correlations was maximal for a diameter of 2 m or, to a lesser extent, 2.5 m around a sampling point (Figure 3). Depending on the radius of the surface considered, the mean and the maximum values of correlations were different. Higher values were obtained for mosaics with a diameter of 2 m around a sampling point (Figure 4). Consequently, the results obtained for mosaics of this size were more significant and were detailed below.

Relationships between community structure and mosaic heterogeneity

A detailed analysis allowed us to describe the influence of mosaic heterogeneity on mesohabitat invertebrate assemblages. Twenty-five significant correlations suggested that the community structure was influenced by the mosaic heterogeneity around the respective mesohabitat. None of the environmental descriptors were correlated with the density or evenness of faunal assemblages.

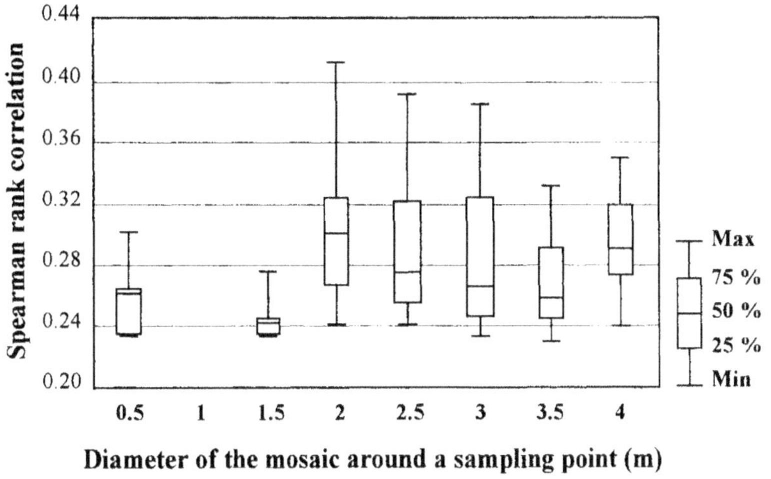

Figure 4. Distribution of the significant Spearman correlations for each diameter of the bottom surface considered.

Mosaic heterogeneity	Faunal richness	Faunal diversity	Faunal dominance
Surface/Perimeter ratio *(All substrates except macrophytes)*	**0.261** *0.026*	**0.318** *0.006*	**-0.311** *0.007*
Surface/Perimeter ratio (*All substrates*)	**0.319** *0.005*	**0.341** *0.003*	**-0.313** *0.006*
Number of patches	**0.341** *0.003*	**0.411** *0.000*	**-0.366** *0.001*
Number of substrates	**0.300** *0.009*	**0.290** *0.013*	**-0.246** *0.035*

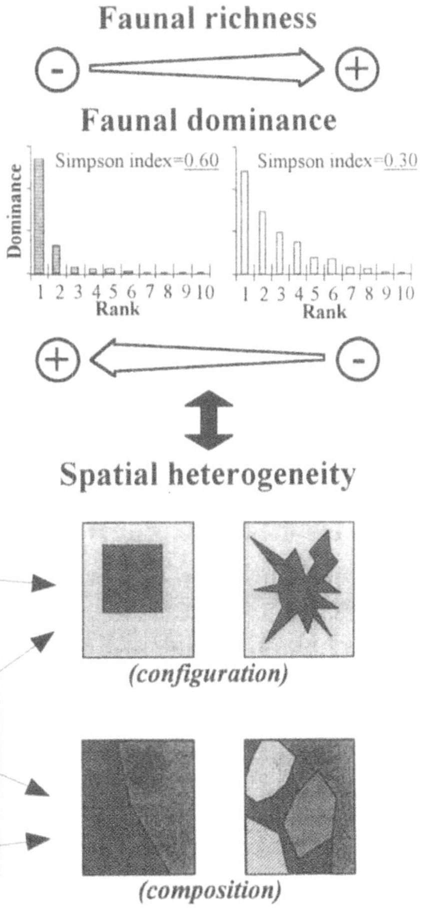

Figure 5. Spearman correlation between community structure indices and mosaic heterogeneity around a sampling point, significant at $P \leq 1\%$, and illustration of the ecological phenomena observed.

We chose to detail correlations significant at $P \leq 1\%$.

Taxonomic richness was positively correlated with the number of patches, the number of substrates and the mean ratio surface/perimeter when considering all substrates (Figure 5).

Numerous significant correlations were found with faunal dominance measures (Simpson index and the relative abundance of the most numerous taxon) that quantify the distribution of individuals among taxa. However, no correlations were obtained with evenness. Dominance seemed to decrease with the mean ratio surface/perimeter (with or without macrophytes), the relative surface of the sampled patch, and the number of patches around a sampling point (Figure 5).

Diversity integrates both the richness and the distribution of individuals among taxa. Correlations with richness and, to a lesser extent, with dominance explained the significant results observed between Shannon index and environmental descriptors. The highest correlation was obtained between patch richness and faunal diversity (r=0.411).

Relationships between macroinvertebrate community composition and mosaic heterogeneity

The first factorial plane of the within-station CA represented 27.45% of the total inertia of the faunal array (the 'station' effect which was eliminated by the within-CA represented 11.24% of the total inertia).

F1 seemed to ordinate taxa (Figure 6) depending on sampling point features (Figure 7). A representation of the gravity centre of the current velocity, resolved into four modalities, showed a clear gradient along the F1 axis. Samples corresponding to mesohabitats with a low current velocity were located at the negative side of the F1 axis. They corresponded to fine mobile substrates. Mesohabitat samples corresponding to fast current velocity occurred at the positive side of this axis. The F2 axis isolated substrates such as mud, silt, gravel and cobble from substrates with a higher organic content or with a coarser granulometry.

Mosaic heterogeneity descriptors like the substrate number, the mean ratio surface/perimeter (all substrates taken into account) or the dominance in patch surface were strongly related to the F2 axis. A clear gradient of mosaic heterogeneity appeared from the negative (homogeneous mosaic) to the positive side (heterogeneous mosaic) of the F2 axis, demonstrating that some substrates (roots, bryophytes, rock, flagstone) more frequently had a heterogeneous environment than others (mud, sand, cobble). Ordination of taxa on the first factorial plane may be interpreted according to these combinations of habitat characteristics (Figure 6). Invertebrates having an inherent need for current or well adapted to shear stress were plotted on the positive side of the F1 axis (*Hydropsyche*, Simuliidae or Heptageniidae), together with taxa requiring sheltered microhabitats in a fast-flowing environment (*Eriocera* or *Atherix*). Taxa adapted to low current velocity and uncohesive substrates were plotted on the negative side of the F1 axis (Tabanidae or *Ephemera danica*). The ordination was more complex on the F2 axis, where preferendi for habitat were mixed with influences of the environmental heterogeneity.

Discussion

Our approach highlighted clear relationships between the macroinvertebrate assemblages and the bottom heterogeneity of their environment at a mesohabitat scale. Even if community assemblages were mainly determined by the specific abiotic characteristics of the sampling habitat, the heterogeneity of the neighbouring mosaic could directly or indirectly have a significant influence.

The number of taxa appeared to be higher when the mosaic was more heterogeneous in terms of patchiness, patch shape or substrate number. Two factors explain this result. Firstly, in a heterogeneous environment, habitats (at the macroinvertebrate scale) were more varied and a higher number of taxa could potentially find ecological niches. Secondly, reduced distances between neighbouring substrates in a complex mosaic favoured the colonisation of a given mesohabitat community with taxa from neighbouring mesohabitats by different mechanisms of dispersal over short distances. A radius of one meter around a sampling point was potentially sufficient to evaluate the complexity of a mosaic and then to evaluate the potential intensity of faunal exchange among mesohabitats. The faunal richness was higher at sites exhibiting such a heterogeneous mosaic. Experience with the colonization of clean substrates has demonstrated the importance of some movement of invertebrates from neighbouring zones (Doeg et al., 1989; Oertli, 1992). Different studies have suggested that individual mobility of invertebrates is not simply a random process driven by disturbance, but is rather an aspect of species' exploitation of their environment and may be important in niche partitioning (Hildrew, 1998). At a

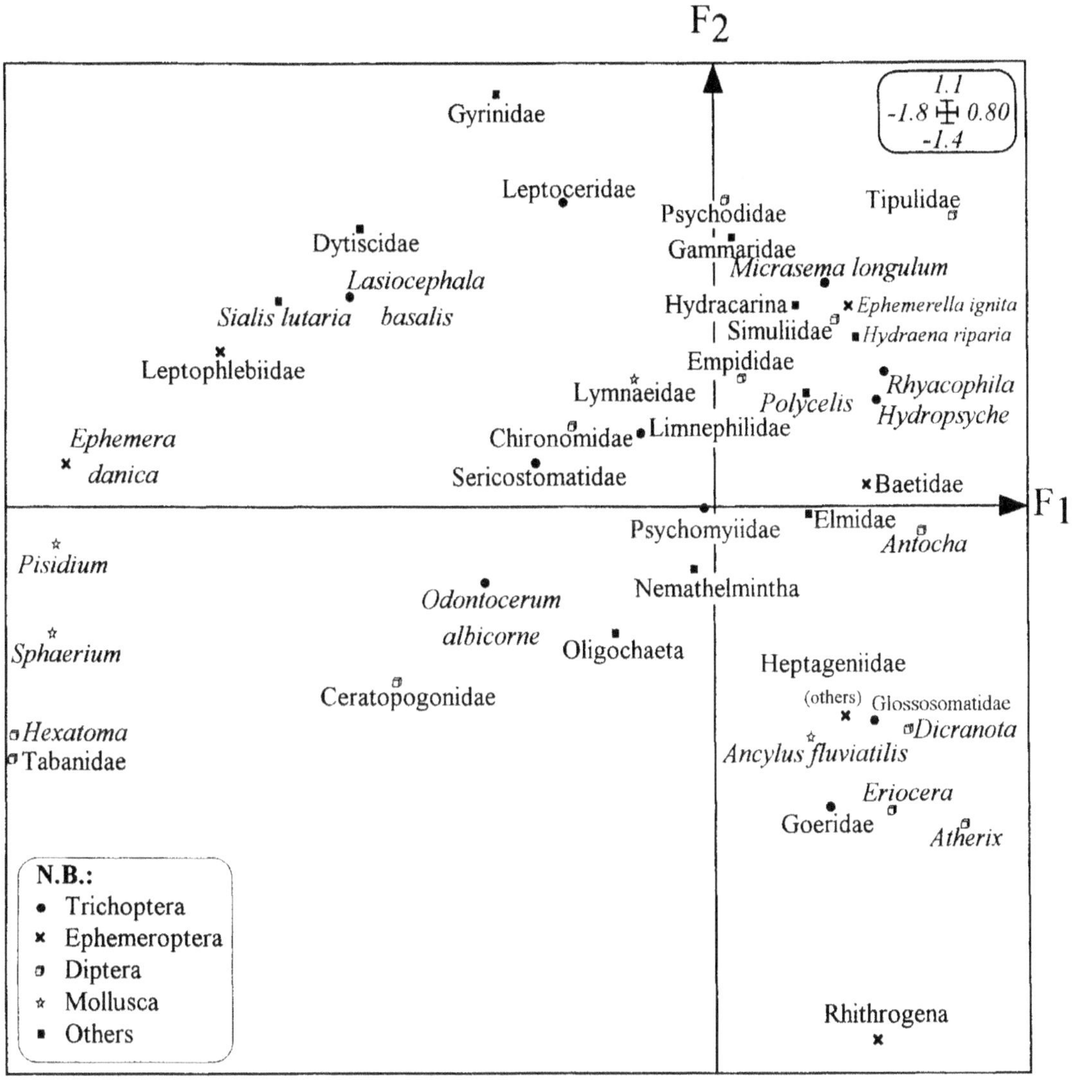

Figure 6. Results of the within-station correspondence analysis of the (48 taxa × 73 samples) array. Ordination of taxa on the first factorial plane.

higher scale, streams having more physically heterogeneous environments should have a greater diversity of species than streams with less heterogeneous habitats (Minshall, 1988; Harper et al., 1992; Minshall & Robinson, 1998).

In this study, the dominance of the most abundant taxa apparently depended on mosaic complexity. In a homogeneous environment a particular type of microhabitat is much better represented and can shelter a high number of individuals of a specific set of species. If exchanges with neighbouring habitats are scarce, then competition with other species (not well adapted) is potentially reduced. Aggregation of very adapted species is favoured in a very homogeneous environment.

Specific faunal communities depend on the features of their mesohabitats (Armitage et al., 1995; Beisel, 1996); these relationships can be used to detect changes associated with modified flow regimes (Armitage & Pardo, 1995). The spatial heterogeneity of the environment around the mesohabitat has a significant effect on the composition of faunal assemblages. Further research is needed to specify how the heterogeneity of the neighbouring environment affects the biological and ecological traits of invertebrates.

In conclusion, faunal diversity was higher in heterogeneous mosaics, even if the complexity of a mosaic and the type of substrate were not always independent (roots alone cannot be found in a round area of 2 m^2). Consequently, faunal diversity was higher both

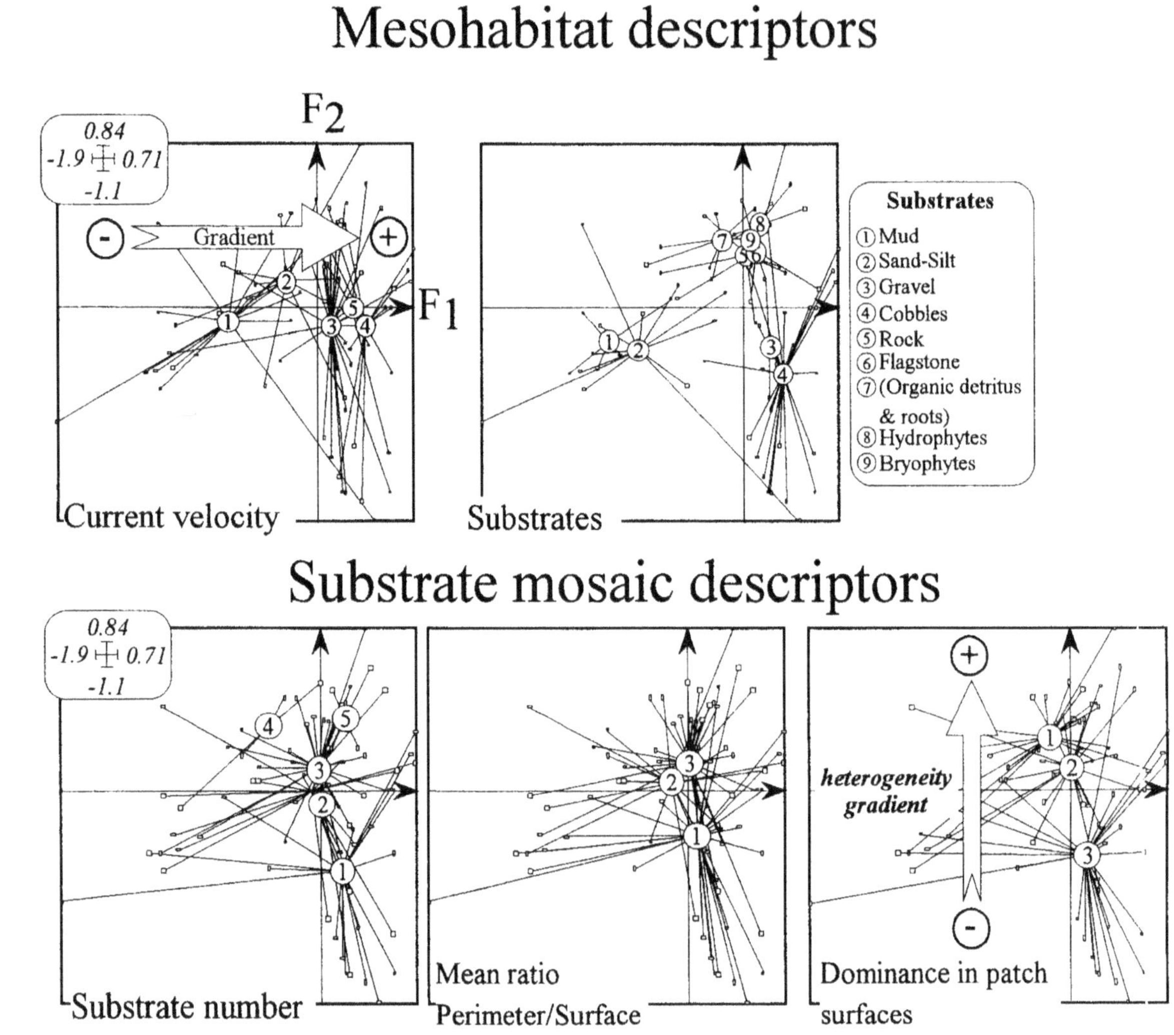

Figure 7. Distribution of modalities (the number in the circle) of two mesohabitat descriptors and three mosaic heterogeneity descriptors on the F1 × F2 factorial plane of the within-station CA. Each modality is positioned at the weighted average of samples representing this modality. Modalities of the four other variables corresponded to a gradient. Modality one was at the lowest side of this gradient.

in complex mesohabitats (Beisel et al., 1998b) and heterogeneous mosaics. These results highlighted that protecting the environmental quality of running waters should primarily focus on the study and the maintenance of habitat diversity (Armitage et al., 1995; Harper et al., 1992).

References

Armitage, P. D. & C. E. Cannan, 1998. Nested multi-scale surveys in lotic systems – tools for management. In: Bretschko, G. & J. Helesic (eds), Advances in River Bottom Ecology. Backhuys Publishers, Leiden, The Netherlands: 293–314.

Armitage, P. D. & I. Pardo, 1995. Impact assessment of regulation at the reach level using macroinvertebrate information from mesohabitats. Reg. Riv. 10: 147–158.

Armitage, P. D., I. Pardo & A. Brown, 1995. Temporal constancy of faunal assemblages in 'mesohabitats' – application to management? Arch. Hydrobiol. 133: 367–383.

Beisel, J. N., 1996. Microrépartition des invertébrés benthiques en eau courante: caractéristiques des microhabitats et organisation de leurs peuplements. Ph.D. Thesis, Univ. Metz: 259 pp.

Beisel, J.-N., P. Usseglio-Polatera, S. Thomas, S. & J. C. Moreteau, 1998a. A method to describe substrate heterogeneity at a microhabitat scale. First results on relationships with the macroinvertebrate community structure. In: Bretschko, G. & J. Helesic (eds), Advances in River Bottom Ecology. Backhuys Publishers, Leiden, The Netherlands: 39–46.

Beisel, J.-N., P. Usseglio-Polatera, S. Thomas & J. C. Moreteau, 1998b. Stream community structure in relation to spatial variation: the influence of mesohabitat characteristics. Hydrobiologia 389: 73–88.

Chutter, F. M., 1969. The distribution of some stream invertebrates in relation to current speed. Int. Rev. ges Hydrobiol. 54: 413–422.

Dobson, M. & A. G. Hildrew, 1992. A test of resource limitation among shredding detritivores in low order streams in southern England. J. anim. Ecol. 61: 69–77. Doeg, T. J., R. Marchant, M. Douglas & P. S. Lake, 1989. Experimental colonization of sand, gravel and stones by macroinvertebrates in the Acheron River, southeastern Australia. Freshwat. Biol. 22: 57–64.

Drake, J. A., 1984. Species aggregation: the influence of detritus in a benthic invertebrate community. Hydrobiologia 112: 109–115.

Harper, D. M. & M. Everard, in press. Why should the habitat-level approach underpin holistic river survey and management? Aquatic Conservation.

Harper, D. M., C. D. Smith & P. J. Barham, 1992. Habitats as the building blocks for river conservation assessment. In: Boon, P. J., P. Calow & G. E. Petts (eds), River Conservation and Management. John Wiley & Sons Ltd, Chichester: 311–319.

Hildrew, A. G., 1998. Physical habitat and the benthic ecology of streams and rivers. In: Bretschko, G. & J. Helesic (eds), Advances in River Bottom Ecology. Backhuys Publishers, Leiden, The Netherlands: 13–22.

Li, H. & J. F. Reynolds, 1994. A simulation experiment to quantify spatial heterogeneity in categorical maps. Ecology 75: 2446–2455.

Li, H. & J. F. Reynolds, 1995. On definition and quantification of heterogeneity. Oikos 73: 280–284.

Minshall, G. W., 1988. Stream ecosystem theory: a global perspective. J. n. am. Benthol. Soc. 7: 263–288.

Minshall, G. W. & J. N. Minshall, 1977. Microdistribution of benthic invertebrates in a rocky mountain (U.S.A.) stream. Hydrobiologia 55: 231–249.

Minshall, G. W. & C. T. Robinson, 1998. Macroinvertebrate community structure in relation to measures of lotic habitat heterogeneity. Arch. Hydrobiol. 141: 129–151.

Oertli, B., 1992. L'influence de trois substrats (*Typha*, *Chara*, feuilles mortes) d'un étang forestier sur la densité, la biomasse et la production des macroinvertébrés aquatiques. Dissertation No. 2557. Université de Genève, Genève, 283 pp.

Reice, S. R., 1980. The role of substratum in benthic macroinvertebrate microdistribution and litter decomposition in a woodland stream. Ecology 61: 580–590.

Statzner, B., J. A. Gore & V. H. Resh, 1988. Hydraulic stream ecology: observed patterns and potential applications. J. n. am. Benthol. Soc. 7: 307–360.

Wood, P., 1998. Reach-scale mesohabitat variations in a small chalk stream under low flow conditions. In: Bretschko, G. & J. Helesic (eds), Advances in River Bottom Ecology. Backhuys Publishers, Leiden, The Netherlands: 31–38.

Hydrobiologia **422/423:** 173–181, 2000.
M. Jungwirth, S. Muhar & S. Schmutz (eds), Assessing the Ecological Integrity of Running Waters.

River ecotones: carabid beetles as a tool for quality assessment

A. Boscaini, A. Franceschini & B. Maiolini*
Museo Tridentino di Scienze Naturali, Via Calepina 14, 38100 Trento, Italy
E-mail: maiolini@mtsn.itc.it

Key words: carabidae, river ecotones, quality assessment, Europe, biodiversity

Abstract

Carabids are frequently used in terrestrial ecology on account of their well-known autecology, wide distribution, their role as predators generally restricted to a given area, their sensitivity to environmental changes and also due to the availability of good identification keys. The EC project ERMAS started in 1993 and involves six European countries; its aim is to study the structure and functioning of riverine ecotones in different environmental conditions. Among the tasks of the project was the evaluation of biodiversity using terrestrial invertebrates. To achieve this a considerable sampling effort was undertaken by all teams. Pitfall traps were chosen as the sampling strategy because they permit greater comparability as opposed to when different operators work in different environments. On each river system we considered five series of transects from the river margin to the edge of the floodplain. All samples were sent to the Museum of Natural Sciences of Trento for identification and data processing. In all, more than 400 000 invertebrates were identified and the results were computed to form a common database. Data on the carabid community are used here for biodiversity trends, classification analysis and similarity indices.

Introduction

The structure and functioning of river ecotones across a European latitudinal gradient were investigated within the EC project European River Margins Study (ERMAS), started in 1993. To achieve this, six mid-sized rivers were selected: Vindel (North-Sweden), Helgeån (South-Sweden), Trent (UK), Adige-Noce (Italy), Garonne (France) and Danube (Romania).

Among the tasks of the project was the evaluation of biodiversity using terrestrial invertebrates. While the whole invertebrate community was studied at the taxonomical level of order or family (Maiolini et al., 1998), carabids were identified to species due to the ecological interest of this group. They are a favourable study subject for their great diversity of habitat preference, for being a taxonomical unit of convenient size, and because they are easily sampled (Talvi, 1995).

The various behavioural, morphological and physiological adaptations of riparian and wetland carabid beetles is reflected in their wide range of specialist habitats (Lott, 1996). Furthermore, most species are generalist predators associated more with habitat structure than with specific food plants. Carabids are generally restricted to a given area and environmental factors determining the composition of the community are vegetation structure, moisture, substrate composition and land use patterns (Talvi, 1995).

In terms of quality assessment studies, the carabids represent one of the ecologically most sensitive groups, with many highly specialised taxa; changes in alluvial ecosystems will affect them much more than other invertebrates (Sustek, 1994).

Lastly, identification is favoured by the availability of good identification keys (Casale et al., 1982; Freude, 1976; Jeannel, 1941; Lindroth, 1974, 1985; Porta, 1923; Trautner & Geigenmüller, 1987) and taxonomical literature referring to European species (Brandmayr & Seriani, 1981; Ravizza, 1968; Schatzmayer, 1929; Schmidt, 1994).

Study areas

On each river system the sampling was carried out along a series of transects covering the environmental

* Author for correspondence.

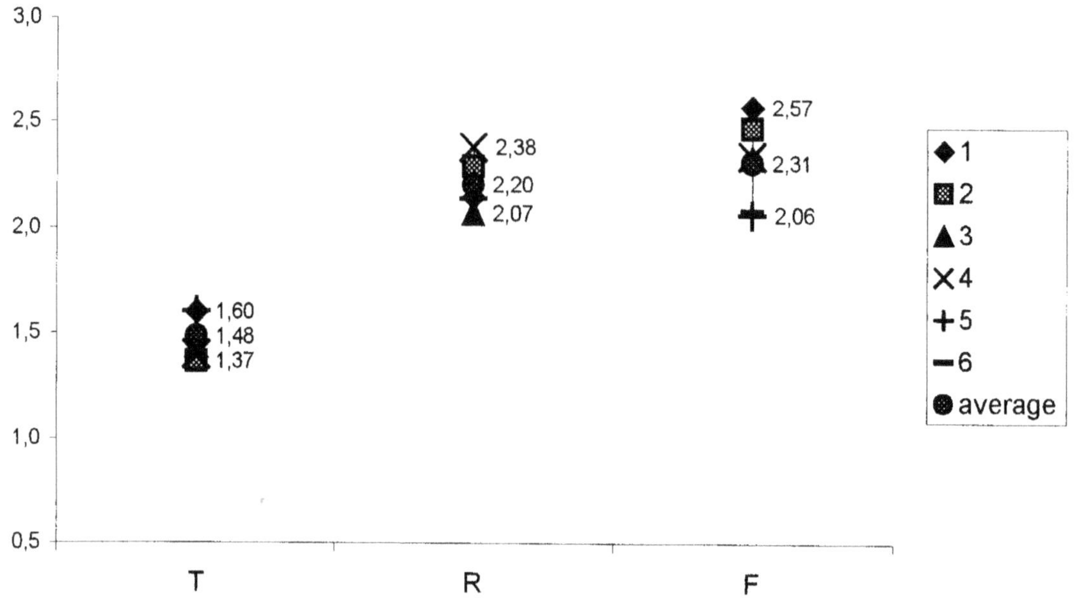

Figure 1. Trent River. Values of Shannon–Wiener's index calculated on total number of carabids for each sampling site.

gradient from the riverbanks to the margin of the floodplain. Three different research sites were selected: frequently flooded (F), near the river margin and submerged for a mean of more than 10 weeks per year; rarely flooded (R), with flooding events occurring from 1 to 10 weeks per year; terrestrial or never flooded (T). Altogether 15 sites, 5 replicates for each of the three site typologies, were selected on each river system.

All sites were classified on the basis of relative elevation from mean water level of the river (Petts & Décamps, 1995).

In this paper we present the results from three areas having different environmental characteristics: ecotones from the rivers Trent (UK), Adige and Noce (Italy), and Garonne (France). Latitude ranges from 52° (Trent) to 44° (Garonne).

All rivers are more or less regulated and their floodplains are flooded to some extent every year. The Trent River is rain fed and has high flows in February, prior to the growing season (from March to late October). River regulation is low in the upper part, but the downstream section is channelised and impounded by weirs and sluices (Petts et al., 1992). Water quality is fair. T sites are characterised by the presence of *Populus* sp. and *Ulmus* sp., while in R and F the dominant species is *Salix fragilis*.

The Italian study sites are located partly on the Adige River and partly on a fully vegetated island in its tributary, the Noce River. Both rivers are strongly regulated for hydroelectric power, and flood events are generally limited to one or two a year. The Adige River basin, however, presents a summer flood dominated system having a strong glacial/snowmelt influence. Water quality is fair. Site characterisation: sampling stations T4, T5, R2, R4, R5 are along the Adige River, located in an old poplar – elder-dominated humid residual forest, 2–3 m above mean water level. The other 10 sites are located upstream (5 km) within a fully vegetated island in the tributary Noce River. This island is characterised by coarse gravel/fluvial boulders with only a thin layer of humic soil.

The Garonne River is rain dominated with high flows in spring. The river is regulated for power generation and irrigation with dams and quays.

Site characterisation: study sites are situated in a forested floodplain corridor in four separate areas 10, 20, 30 and 40 km downstream of Toulouse, respectively. Poplar dominates in the T sites, with *Rubus* sp. and Gramineae as ground species. The tree cover is a mixture of poplar and willow in the R sites and willow in the F ones, associated in both to *Rubus* sp., *Urtica dioica* and *Hedera helix*.

Materials and methods

In each of the 15 sites, nine pitfall or Barber traps were set up and partially filled with a 50% solution of glycol ethylene as a preservative. Eight traps were arranged

Table 1. List of carabid species referred to the sampling sites of the Trent (UK), the Adige–Noce (I) and the Garonne (F) rivers

Species	UK	I	F
Carabus granulatus Linné, 1758	433	260	
Carabus monilis Fabricius, 1792	2		
Carabus nemoralis O.F. Müller, 1764	232		
Carabus auratus Linné, 1761			129
Carabus cancellatus Illiger, 1798			1
Carabus convexus Fabricius, 1775		10	
Carabus coriaceus Linné, 1758		67	
Carabus purpurascens Fabricius, 1787			26
Carabus violaceus Linné, 1758	1	2	
Cychrus angustatus Hoppe, 1825		1	
Cychrus caraboides (Linné, 1758)		19	
Leistus spinibarbis (Fabricius, 1775)	1		
Leistus sardous Baudi, 1883			57
Leistus ferrugineus (Linné, 1758)	5	3	
Nebria brevicollis (Fabricius, 1792)	119	4	281
Notiophilus palustris (Duftschmid, 1812)		4	
Notiophilus rufipes Curtis, 1829			3
Notiophilus biguttatus (Fabricius, 1779)	3		
Elaphrus aureus P. Müller, 1821		107	44
Elaphrus cupreus Duftschmid, 1812	1		
Loricera pilicornis (Fabricius, 1775)	148		
Clivina fossor (Linné, 1758)	34		
Clivina collaris (Herbst, 1784)	16	6	110
Asaphidion flavipes (Linné, 1761)	9	71	103
Metallina lampros (Herbst, 1784)	2	1	
Metallina properans (Stephens, 1828)			6
Phyla obtusum Audinet-Serville, 1821	2		
Notaphus semipunctatus (Donovan, 1806)			2
Notaphus dentellus (Thunberg, 1787)	92		2
Leja gilvipes Sturm, 1825	2		
Philochthus aeneum Germar, 1824	58		
Philochthus biguttatus (Fabricius, 1779)	90		14
Philochthus guttula (Fabricius, 1792)	29		
Bembidion quadrimaculatum (Linné, 1761)	2		3
Ocydromus dalmatinus (Dejean, 1831)			1
Ocydromus tetracolus (Say, 1823)	70	2	630
Synechostictus elongatus (Dejean, 1831)			1
Ocys harpaloides (Serville, 1821)	10		
Epaphius secalis (Paykull, 1790)	53	5	
Lasiotrechus discus (Fabricius, 1801)		1	
Trechus quadristriatus (Schrank, 1781)	13		9
Trechus obtusus Erichson, 1837			4
Trechus subnotatus Dejean, 1831	1		
Patrobus atrorufus (Ström, 1768)	243	6	
Platynus assimilis (Paykull, 1790)	1250	183	788
Parachus albipes (Fabricius, 1796)	1117	1	516
Oxypselaphus obscurus (Herbst, 1784)	2568		59
Anchomenus dorsalis (Pontoppidan, 1763)	129	1	2818
Agonum afrum (Duftschmid, 1812)	151		
Agonum gr. *viduum*		72	1261

Table 1. Continued

Species	UK	I	F
Agonum marginatum (Linné, 1758)	1		
Agonum viridicupreum (Goeze, 1777)			1
Agonum muelleri (Herbst, 1784)			226
Europhilus fuliginosus (Panzer, 1809)	227		
Europhilus micans (Nicolai, 1822)	938		160
Platyderus ruficollis (Marsham, 1802)			34
Platyderus rufus (Duftschmid, 1812)		11	
Synuchus vivalis (Illiger, 1798)	1	1	
Calathus micropterus (Duftschmid, 1812)	1		
Calathus piceus (Marsham, 1802)	8		1
Calathus erratus (Sahlberg, 1827)		54	
Calathus fuscipes (Goeze, 1777)	21		
Dolichus halensis (Schaller, 1783)		1	
Laemostenus terricola (Herbst, 1784)			3
Astigis salzmanni (Germar, 1824)			6
Platysma nigrum (Schaller, 1783)	59	642	
Platysma melanarium (Illiger, 1798)	390	1201	
Platysma minor (Gyllenhal, 1827)	21		
Platysma nigrita (Paykull, 1790)	157	20	21
Platysma anthracinum (Illiger, 1798)			359
Steropus madidus (Fabricius, 1775)	318		
Steropus madidus var. aquitanus Barthe, 1920			196
Steropus madidus var. validus Dejean, 1828			3
Pterostichus funestes Csiky, 1930		64	
Stomis pumicatus (Panzer, 1796)	6	3	1
Argutor vernalis (Panzer, 1796)	17		22
Phonias maritimus n. sp.			30
Phonias strenuus (Panzer, 1797)	1119	8	
Bothriopterus oblungopunctatus (Fabricius, 1787)		7	
Omaseus aterrimus (Herbst, 1784)			4
Poecilus cupreus (Linné, 1758)			206
Poecilus versicolor (Sturm, 1824)		1	
Poecilus lepidus (Leske, 1758)		6	
Tanythrix marginepunctatus (Dejean, 1831)		1	
Abax ater (Villers, 1789)	185	1710	
Abax exaratus (Dejean, 1828)		2	
Amara plebeja (Gyllenhal, 1810)	1		
Amara aenea (Degeer, 1774)	1		11
Amara anthobia Villa & Villa, 1833			1
Amara convexior Stephens, 1828		10	
Amara familiaris (Duftschmid, 1812)	4	2	
Amara ovata (Fabricius, 1792)	1	6	21
Amara similata (Gyllenhal, 1810)		1	37
Amara spreta Dejean, 1831		9	
Amara aulica (Panzer, 1797)		1	
Anisodactylus binotatus (Fabricius, 1787)		14	17
Diachromus germanus (Linné, 1758)			74
Parophonus maculicornis (Duftschmid, 1812)			5
Ophonus rupicola (Sturm, 1818)		1	

Continued on p. 176

Table 1. Continued

Species	UK	I	F
Ophonus rufibarbis (Fabricius, 1792)	2		
Pseudophonus rufipes (Degeer, 1774)	42	102	19
Harpalus affinis (Schrank, 1781)			1
Harpalus punctipennis Mulsant, 1852			1
Harpalus distinguendus (Duftschmid, 1812)			18
Harpalus dimidiatus (Rossi, 1790)			7
Harpalus rubripes (Duftschmid, 1812)		1	
Harpalus atratus Latreille, 1804		17	
Harpalus luteicornis (Duftschmid, 1812)		4	
Harpalus latus (Linné, 1758)		1	
Harpalus tardus (Panzer, 1797)		16	17
Stenolophus teutonus (Schrank, 1781)			8
Stenolophus mixtus (Herbst, 1784)			2
Bradycellus caucasicus (Chaudoir, 1846)		4	
Acupalpus meridianus (Linné, 1761)	1		
Badister bullatus (Schrank, 1798)	6	1	
Panageus bipustulatus (Fabricius, 1775)			2
Panageus cruxmajor (Linné, 1758)			6
Chlaenius spoliatus (Rossi, 1790)			1
Chlaenius velutinus (Duftschmid, 1812)			1
Chlaeniellus vestitus (Paykull, 1790)		1	88
Chlaeniellus nigricornis (Fabricius, 1787)			5
Chlaeniellus nitidulus (Schrank, 1781)		5	91
Chlaeniellus tristis (Schaller, 1783)			1
Callistus lunatus (Fabricius, 1775)			2
Oodes helopioides (Fabricius, 1792)		6	
Demetrias atricapillus (Linné, 1758)	1		2
Philorhizus melanocephalus (Dejean, 1825)	1		
Syntomus obscuroguttatus (Duftschmid, 1812)			4
Microlestes sp. Schmidt-Goebel, 1846			1
Brachinus ganglbaueri Apfelbeck, 1904			6
Brachinus glabratus Latreille & Dejean, 1824			2
Brachinus scolopeta (Fabricius, 1792)			18

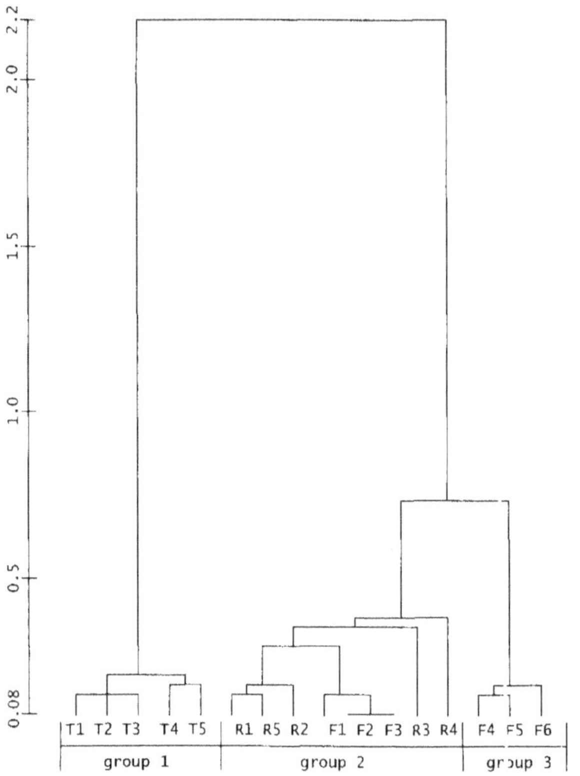

Figure 2. Trent River. Hierarchical classification of sampling sites based on the carabid species database. Three main groups are evidenced.

to form a circle of about 5 m in diameter and the 9th was placed in the middle of the circle. Every fourteen days the traps were collected by site.

This method has several advantages: the survey period can be rather long and is therefore less affected by short-term weather conditions, and traps are more comparable than other techniques such as hand collecting (Lott & Eyre, 1996). Furthermore, sampling by pitfall traps depends less on the skill of the operators (Lott, 1996).

Carabids in particular are efficiently sampled by traps and long-term monitoring can yield interesting results in terms of community fluctuations (Luff, 1996).

The experiment started in March 1993 in Italy, in April in England and France; it ended in late September in Italy and England, and late November in France according to local climate conditions. All samples were sent to the Museum of Natural Sciences of Trento for identification and data processing. A total of more than 400 000 invertebrates were identified to order or family, and carabids were determined to species level. The nomenclature used is according to Vigna Taglianti (1993). The results were computed to form a common database.

Data on the carabid community is used here for biodiversity trends and classification analysis. Hierarchical classification was performed using SYNTAX 4 (Podani, 1990) with Wishart's similarity ratio as distance function and β flexible as clustering methods ($\beta = -0.25$). Shannon–Wiener's index ($\log_n$) has been used owing to its effectiveness and wide diffusion in the literature, allowing comparison with other studies.

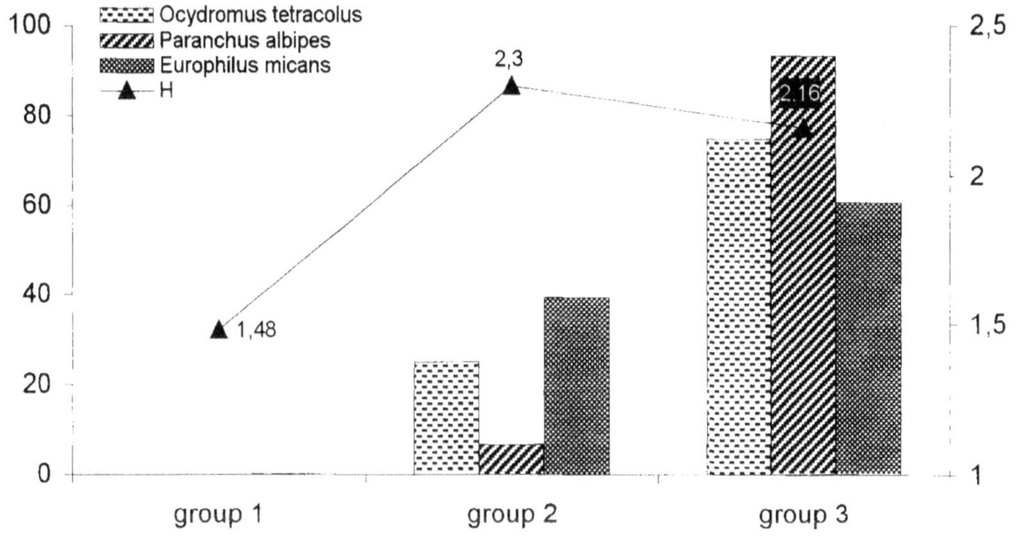

Figure 3. Trent River. Relative abundance of three hygrophilous species calculated for the groups clustered by hierarchical analysis. Shannon–Wiener's index values (H) for the same groups show a clear increase from terrestrial to ecotonal sites.

Results and discussion

In all 23 785 individuals were collected and identified in the sampling areas on the Garonne (F), Adige-Noce (I) and Trent (UK); they belong to 132 species and are distributed as follows:

The complete list of carabid species is presented in Table 1.

Based on the carabid data, the Trent River displays the clearest separation between ecotonal and terrestrial community assemblages. Shannon–Wiener's index (Figure 1) and the dendrogram (Figure 2) based on the Similarity ratio both clearly show this distinction.

Values of Shannon–Wiener's index range from 1.37 to 1.60 in the terrestrial sites and from 2.07 to 2.57 in the rarely and frequently flooded ecotone. The role of riparian areas as hot spots with strongly enhanced biodiversity (Pilon, 1991) is clearly indicated.

Only three species account for 87.8% of the carabid community in the T sites, hence the low values of Shannon–Wiener's index. The dominant species is *Steropus madidus*, considered as eurytopic and occurring in Britain mainly in open country and cultivated areas (Lindroth, 1985).

In the R and F sites, six and five species, respectively, represent 80.8 and 72.5%. The dominant species is *Oxypselaphus obscurus*, a hygrophilous component living in damp and shaded habitats among moss and alder and ash litter (Lindroth, 1985).

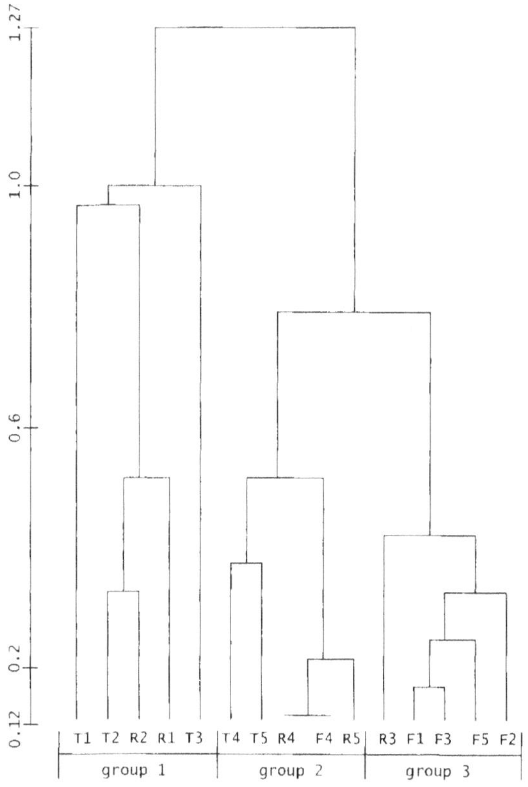

Figure 4. Garonne River. Values of Shannon–Wiener's index calculated on total number of carabids for each sampling site.

The first division separates all T sites from the riparian ones (R and F), while the second clustering separates three stations (F4, F5, F6) from the other ecotonal stations. In order to better understand this

Number of species:	F = 70	I = 56	UK = 58
Number of exclusive species:	F = 42	I = 25	UK = 24
Number of common species:	F-I = 18	F-UK = 21	I-UK = 24
	F-I - GB = 11		
Number of individuals:	F = 8610	I = 4760	UK = 10415

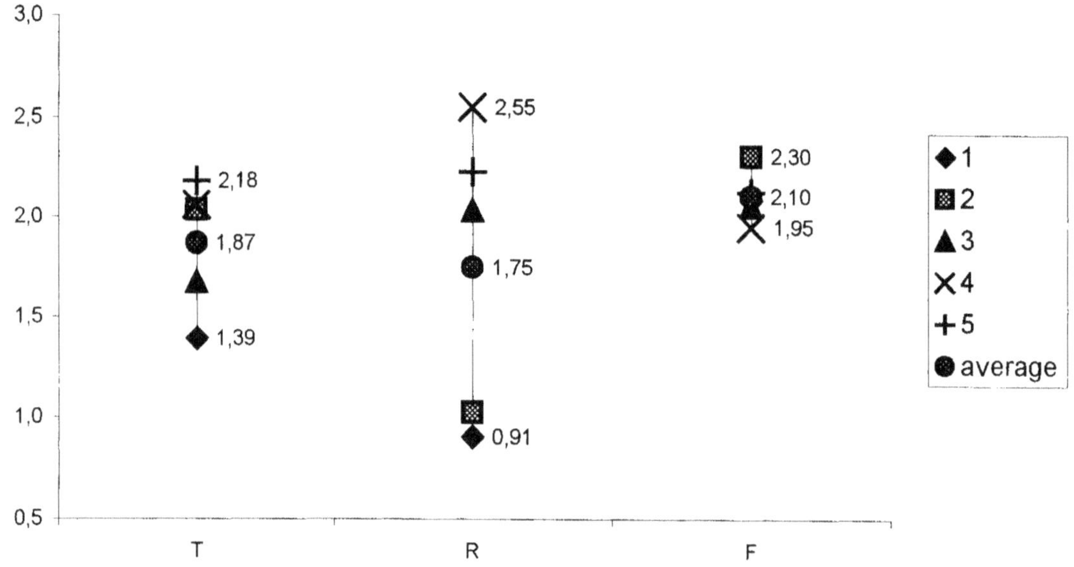

Figure 5. Garonne River. Hierarchical classification of sampling sites based on the carabid species database.

separation the relative abundance of three hygrophilous species was analysed (Figure 3):

Ocydromus tetracolus, an eurytopic species, usually occurring on moist, clayey soils, often along eutrophic waters (Lindroth, 1985). The species is rather indifferent to vegetational cover and the activity of imago is mainly diurnal with a peak of activity in spring, when breeding occurs, and a second peak in late summer (Drioli, 1997).

Paranchus albipes, a shore-dwelling species, found on eutrophic as well as oligotrophic lakes and riverbanks. The adults usually live near the water-edge, in open, almost barren soils, but also in open country with scattered vegetation. The imago is strictly nocturnal and breeding occurs mainly in autumn, although part of the population breeds in spring (Lindroth, 1985).

Europhilus micans, also common along the shores of eutrophic lakes and rivers, particularly in muddy-clay soils shaded by *Salix* or *Alnus*. It is a spring breeder and images hibernate under bark (Lindroth, 1985).

The first two species are absent from the first group of terrestrial sites, while *Europhilus micans* is very rare (0.1%). Their relative presence increases within the second group (25.1, 6.6 and 39.2%, respectively) but most of the individuals (74.9, 93.4 and 60.7%) were found in the third. These results justify the separation of the ecotonal stations in two groups, with F4, F5 and F6 being considered moister or more 'frequently flooded' than the remaining F stations.

Shannon–Wiener's index calculated for clustered groups rather than site typology shows lower values (1.48) for the first group formed by all the terrestrial sites, while the ecotonal groups exhibit higher values.

The major group comprising all rarely flooded sites and the three less moist, frequently flooded ones (F1, F2, F3) has a mean value of 2.30, while the more moist stations (F4, F5, F6) have slightly lower ones (2.16).

The Garonne dendrogram (Figure 4) separates three main groups that do not completely correspond to the T, R and F categories. Note that the Garonne floodplain is very wide and the terrestrial sites quite distant from the river margin. Owing to their relative low elevation, the T stations were occasionally flooded during the sampling period.

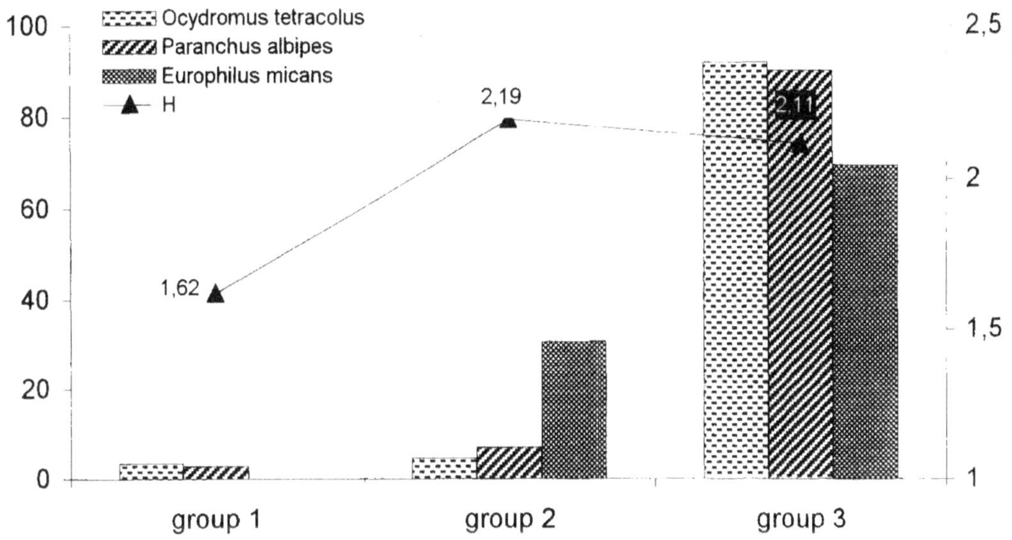

Figure 6. Garonne River. Relative abundance of three hygrophilous species calculated for the groups clustered by hierarchical analysis. Shannon–Wiener's index values (*H*) for the same groups show a clear increase from terrestrial to ecotonal sites.

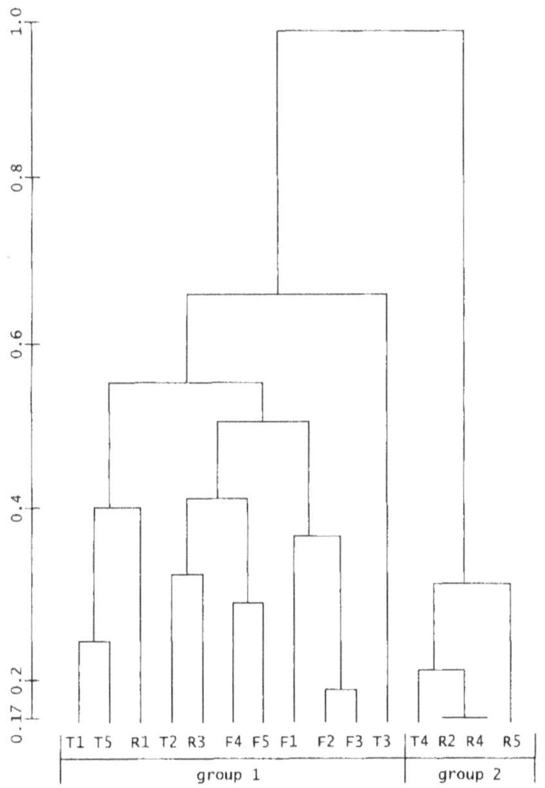

Figure 7. Adige–Noce Rivers. Hierarchical classification of sampling sites based on the carabid species database. Group 1 clusters stations on the Noce River and group 2 the Adige ones.

The riparian canopy reflects this condition; in fact, these areas defined as terrestrial are also dominated by hygrophilous species (*Populus* sp. and *Salix* sp.).

Shannon–Wiener's index values based on each station (Figure 5) do not outline a clear difference between terrestrial and ecotonal sites, as opposed to the situation in the Trent River. Furthermore the mean value of T sites is slightly higher than that of the R sites. The range of values is very wide for the T and R sites, and more restricted around the mean value in the F sites.

The dominant species in all sites (T, R, F) is *Anchomenus dorsalis*, the least hygrophilous of the *Agonum* group, frequently abundant on gravelly and clayey soils (Lindroth, 1985).

The separation suggested by the dendrogram (Figure 4) is justified by the distribution of carabid species. Two main groups are well separated here as well, followed by a second division within the second grouping. We can thus consider three groups: the first with mainly terrestrial sites; the second with a mix of terrestrial and rarely flooded and one frequently flooded; the last formed mainly by F stations.

Analysing the distribution of the same species used for the Trent River, we found that also in this case two of the hygrophilous species have a low relative abundance and one is absent in the first group (Figure 6).

The second group comprises 30.6% *Europhilus micans*, 4.6% *Ocydromus tetracolus* and 7% *Paranchus albipes*. In the third group, the three above species yield most of the collected individuals: 69.4% *Europhilus micans*, 91.9% *Ocydromus tetracolus* and 90.1% *Paranchus albipes*. A trend from humid habi-

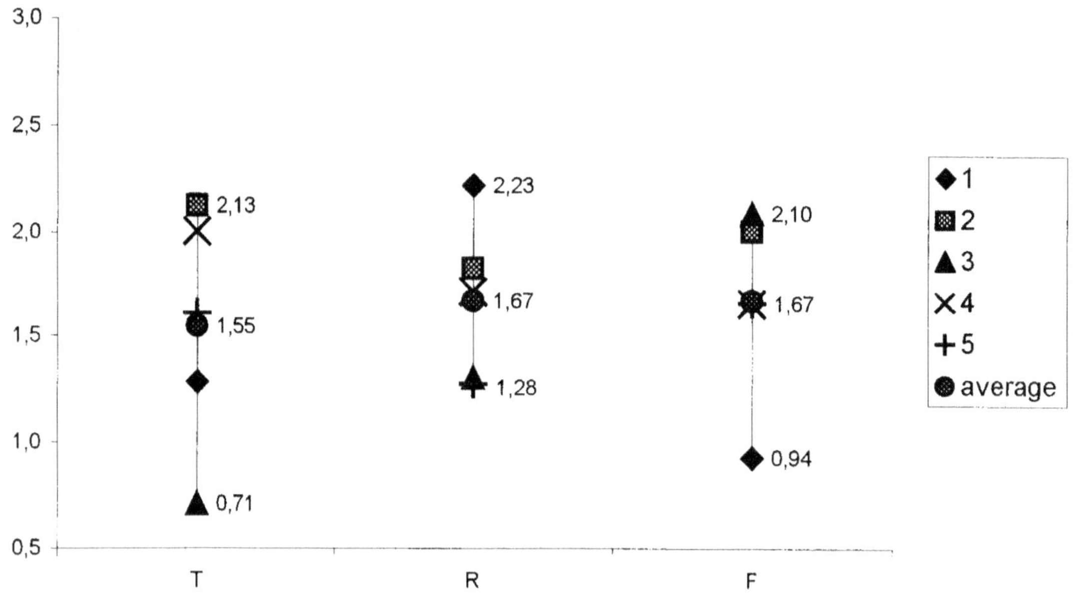

Figure 8. Adige-Noce Rivers. Values of Shannon–Wiener's index calculated on total number of carabids for each sampling site.

tats to drier ones is evident if we consider the abundance of these three hygrophilous species.

Average values of Shannon–Wiener's index (Figure 5) revealed no evident difference between ecotonal and terrestrial sites as was instead the case for the Trent River. Mean values calculated according to the group of sites as clustered by the dendrogram (Figure 6) show an increasing diversity from terrestrial to riparian areas.

In conclusion the Garonne sites show no clear distinction between ecotonal and terrestrial sites; nevertheless, a gradient from moist to drier sites is evident based on the relative abundance of some species and on the diversity indices.

The Italian study sites are the most disturbed by human impacts, both in terms of hydrological regime (Noce) and artificial banking (Adige). The interpretation of the dendrogram (Figure 7) is complex. The first separation clusters all but one of the Adige sites. The second grouping contains the Noce stations and one from the Adige River. Local geographical features apparently influence the ecological conditions due to the flooding regime of the ecotone. This is very probably due to the unnatural hydrographic situation and to the existence of artificial banks that disconnect the riparian area from the river.

Shannon–Wiener's index based on site typology (Figure 8) has mean values of 1.38 for T, 1.81 for R and 1.71 for F sites. These values are lower than for the other two river systems, but the ecotone habitat still exhibits a relatively higher diversity. The same index calculated for homogeneous groups added no further information.

Abax ater is dominant in the T (44.6%) and F sites (49.6%) and co-dominant in the R ones (27.6%). This species is mesophilous and common in forests. No evident distribution of hygrophilous species was evident and they were often found associated with xerophilous species such as *Calathus erratus*.

Our results correspond with findings from other similar studies (Sustek, 1994) where a 1-m drop in the groundwater level (associated with the absence of regular flooding events in riparian areas), changed the carabid community from an hygrophilous biocenosis towards a mesohygrophilous one.

Conclusions

Classification analysis of the carabid data suggests that this insect family is a good indicator to discriminate terrestrial versus riparian habitats when the later are in good ecological condition. The flooding regime of the floodplain appears to be a major determinant of the community structure of ground beetles by maintaining a long-term stability of the gradient from open water to terrestrial environments through marsh, fen and carr (Kangas, 1990). Channelisation may severely modify the species composition when it involves removal of bankside vegetation and changes in substrate composition due to alteration of the sediment/erosion process (Lott, 1996).

The continuum variation of environmental conditions from the shoreline through the ecotone to terrestrial habitats is clearly reflected in the relative abundance of some widely distributed hygrophilous species. Most of these species are adapted to survive short-term disturbance from flooding events and are very sensitive to habitat changes induced by altered flooding regimes (Lott, 1996).

Acknowledgements

We wish to thank M. Cristina Del Prete, Barbara Del Prete, Elisabetta Grigolli and Serenella Marchetti for help in the field at the Italian sites and in sorting all the samples. Eric Chauvet, Gilles Pinay, Aimé Pech and Corinne Calvo for organising and conducting field work on the Garonne. John Taylor and Helen Ruffel for field work on the Trent River and Malcolm Greenwood and Geoffrey Petts for organisation and fruitful discussions. Special thanks are due to John Saddler and Paolo Bonavita for taxonomical assistance. The European Council funded the ERMAS project (Contract EV 5 VCT 929100).

References

Brandmayr, P., M. Seriani, 1981. Schede ecologico-biogeografiche su Coleotteri Carabidi: 1. *Clivina*, *Platynidius*, *Platynus* (Coleoptera, Carabidae). Gortania-Atti Mus. Friul. St. Nat. 2: 195–208.

Casale, A., M. Sturani & A. Vigna Taglianti, 1982. Coleoptera Carabidae. 1 Introduzione, Paussinae, Carabinae. Fauna d'Italia, 18. Calderini, Bologna, 499 pp.

Drioli, G., 1997. Tipi e tempi di sviluppo dei coleotteri geoadefagi presenti sul basso carso triestino. Tip. Adriatica, Trieste, 125 pp.

Freude, H., K. W. Harde & C. A. Lohse, 1976. Die Käfer Mitteleuropas, 2. Goecke & Evers, Krefeld, 300 pp.

Jeannel, R., 1941/42. Coléoptères Carabiques, 1–2. Faune de France 39–40. P. Lechevalier et fils, Paris, 1173 pp.

Kangas, P. C., 1990. Long-term development of forested wetlands. In: Lugo, A. E., M. M. Brinson & S. Brown (eds), Forested Wetlands. Ecosystems of the World 15. Elsevier, Amsterdam: 25–51.

Lindroth, C. H., 1974. Handbooks for the identification of British Insects, 4: 2. Coleoptera Carabidae. Royal Entomological Society of London, London, 148 pp.

Lindroth, C. H., 1985. The Carabidae (Coleoptera) of Fennoscandia and Denmark. Fauna Entomologica Scandinavica 15: 1–2. E. J. Brill/Scandinavian Science Press, Leiden-Copenhagen, 227 pp.

Lott, D. A., 1996. Beetles by rivers and the conservation of riparian and floodplain habitats. In: Eyre, M. D. (ed.), Environmental Monitoring, Surveillance and Conservation Using Invertebrates. EMS Publications, Newcastle upon Tyne: 31–46.

Lott, D. A. & M. D. Eyre, 1996. Invertebrate Sampling Methods. In: Eyre, M. D., (ed.), Environmental Monitoring, Surveillance and Conservation Using Invertebrates. EMS Publications, Newcastle upon Tyne: 9–13.

Luff, M. L., 1996. Environmental assessments using ground beetles (Carabidae) and pitfall traps. In: Eyre, M. D. (ed.), Environmental Monitoring, Surveillance and Conservation Using Invertebrates. EMS Publications, Newcastle upon Tyne: 41–47.

Maiolini, B., A. Franceschini & A. Boscaini, 1998. The role of invertebrate communites as indicators of environmental characteristics of European river margins. In: Joyce, C. B. & P. M. Wade (eds), European Wet Grassland. Biodiversity, Management and Restoration. Wiley, Chichester: 151–162.

Petts, G. E. & H. Decamps, 1995. European river margins as indicators of global change. Unpublished European Union Report.

Petts, G. E., A. R. G. Large, M. T. Greenwood & M. A. Bickerton, 1992. Floodplain assessment for restoration and conservation: linking hydrogeomorphology and ecology. In: Carling, P. & G. E. Petts (eds), Lowland Floodplain Rivers: Geomorphological Perspectives. Wiley, Chichester: 217–234.

Pilon, N., R. Sciaky & C. Violani 1991. La carabidofauna di un biotopo ripario del corso lombardo del Po. Mem. Soc. ent. Ital. 70: 59–77.

Podani, J., 1990. SYN-TAX 4. Computer Programs for Data Analysis in Ecology and Systematics. United Nations Industrial Development Organization. International Centre for Science and High Technology, Trieste.

Porta, A., 1923. Fauna Coleopterum Italica, 1: Adephaga. Ed. Stab. Tipogr. Piacentino, Piacenza, 285 pp.

Ravizza, C., 1968. I *Bembidion* popolanti i greti del basso corso aostano della Dora Baltea. Studi sui *Bembidion* 3° Contributo (Coleoptera, Carabidae). Boll. Soc. ent. Ital. 98: 42–65.

Schatzmayer, A., 1929. I *Pterostichus* italiani. Mem. Soc. ent. Ital. 8: 145–339.

Schmidt, J., 1994. Revision der mit *Agonum* (s.str.) *viduum* (Panzer, 1797) verwandten Arten (Coleoptera, Carabidae). Beitr. Ent. 44: 3–51.

Sustek, Z., 1994. Impact of water management on a Carabid community (Insecta, Coleoptera) in a Central European floodplain forest. Quad. Staz. Ecol. Civ. Mus. St. nat. Ferrara 6: 293–313.

Talvi, T., 1995. Carabid beetle assemblages (Coleoptera) in a wooded meadow and in the adjacent habitats on the Saaremaa Island, Estonia. Ent. Fenn. 6: 169–175.

Trautner, J., K. Geigenmüller, 1987. Tiger beetles, Ground beetles. Illustrated key to the Cicindelidae and Carabidae of Europe. J. Margraf, Aichtal, 488 pp.

Vigna Taglianti, A., 1993. Coleoptera Archostemata, Adephaga 1 (Carabidae). In: Minelli, A., S. Ruffo & S. La Posta (ed.), Checklist delle specie della fauna italiana, 44. Calderini, Bologna.

Hydrobiologia **422/423**: 183–195, 2000.
M. Jungwirth, S. Muhar & S. Schmutz (eds), Assessing the Ecological Integrity of Running Waters.

Secondary production of benthic communities at the habitat scale as a tool to assess ecological integrity in mountain streams

Andrea Buffagni & Ester Comin
Water Research Institute – C.N.R., Via della Mornera 25, I-20047 Brugherio (MI), Italy
E-mail: buffagni@server-mi.irsa.rm.cnr.it; ab1@mail.disat.unimib.it

Key words: benthic community, secondary production, habitat scale, mountain river, ecological integrity, feeding groups, mayfly

Abstract

Secondary production of the benthic community was estimated in the four functional habitats identified in the pristine Pioverna stream (Northern Italy): riffle, pool, transition and bedrock habitats. The instantaneous growth, removal-summation and size-frequency methods were used to estimate production. Twelve taxa reached appreciable densities and presumably accounted for most of total benthic production. Three of them belong to the Plecoptera, seven to the Ephemeroptera and one each to Diptera and Trichoptera. The riffle habitat had the highest secondary production, while bedrock habitat had the lowest. Pool and transition habitats showed intermediate production values. Shredders and predators were the major component of production in pools, with a low contribution of gathering collectors. In riffles, this last category and scrapers had the highest production, with shredders, predators and filtering collectors showing decreasing production values. In general terms, shredders, gathering collectors and scrapers contributed most to total production. The rank-production curves for the habitats are reported. In riffles, a proportionally high evenness of the most productive taxa was noted. In contrast, in the pool habitat a steep curve was recognised. The potential of comparing rank/production curves and changes in reference species production between unaltered and impacted stream habitats and sites is discussed as a means to estimate ecological integrity. For site comparisons of secondary production in the alpine area, the riffle habitat is suggested as the most adequate, as it shows a lower dominance and more taxa contributing to overall production. The calculation of secondary production for indicator and dominant taxa in the different habitats should constitute an effective way to describe some functional aspects of the community to assess the ecological integrity of mountain streams.

Introduction

In general terms, ecological integrity can be intended as the sum of biological and physical elements and processes (Covich et al., 1998; Karr, 1993). Biological elements depend on biological and physical processes for their maintenance, and the degradation of ecological integrity results in global biotic impoverishment. Measures of ecological integrity – as regards biological attributes – should be based on population, community and ecosystem responses to disturbance (Covich et al., op. cit.); they often involve structural measures such as abundance, species richness and modifications of community composition. Functional measures are usually employed at the ecosystem level: e.g., energy flow, nutrient cycling, primary and secondary production. Production is one of the major paths of energy flow through ecosystems (Waters, 1977), and even modest rates of secondary production could be linked to important organic matter processing and nutrient cycling within ecosystems (Fisher & Gray, 1983). In addition, it is a valuable response variable at population, community and ecosystem levels and the most comprehensive representation of success for any population (Benke, 1993). Furthermore, production estimates of invertebrate taxa are a powerful means to assess man-induced stress in aquatic environments in that they are a quantitative measure of ecosystem function (Benke, op. cit.); such estimates could be a key tool in defining ecological integrity. Nevertheless, to the authors' knowledge, few studies regarding benthic invertebrates have dealt with the potential of production for measuring biological or ecological integrity, apart from the impact of organic pollution (e.g., Zelinka et al., 1977). Furthermore, very few of them were performed at the habitat scale (e.g., Smock et al., 1985). It has been argued (Harper et al., 1992) that conserving the river habitats themselves is a cost-effective surrogate for conserving the invertebrate community, and Harper et al. (1998a) demonstrated that the re-establishment of functional habitats is linked with geomorphological and ecological changes. Despite the widespread interest in the ecosystem processes, most of functional (and production) changes at the habitat scale due to river management practices and anthropogenic stress are still to be understood (e.g., Harper et al., 1998b).

The aim of this study is to highlight some aspects of secondary production estimates for aquatic invertebrates that could be used to assess ecological integrity in mountain rivers. Attention will be paid to the relative production of the most representative and abundant taxa, also according to trophic groups and different benthic habitats. In particular, we focus upon the use of the habitat scale in secondary production investigations.

Study site

The Pioverna is a typical mountain stream of the Italian Alps, located in Valsassina (Pasturo, Lecco, North Italy) with a drainage basin of 131.6 km^2. It flows between 418 and 2554 m a.s.l. and its waters run for about 20 km before running into the Como Lake, near Bellano. The collecting site is a pristine river stretch about 200 m long and 3 m wide at base flow, at 830 m a.s.l., close to the city of Pasturo. The minimum discharge was observed in April 1997 (0.011 m^{3}s^{-1}), the maximum in July 1997 (0.241 m^{3}s^{-1}), with an average discharge of 0.076 m^{3}s^{-1}.

The vegetation of the study site, which is only partially shaded, is typical of the southern slope of the Alps, with deciduous trees. *Alnus glutinosa*, *Quercus* spp., *Salix* spp., *Fraxinus excelsior*, *Corylus avellana*, *Acer pseudoplatanus* and *Robinia pseudoacacia* occur along the stream banks and in the steep adjacent areas. The sampling site had almost entirely stony or rocky substratum, with water velocity varying from 0 to about 2 ms^{-1}; maximum water depth was about 1 m. No aquatic macrophytes are present. Water temperature ranged from 3.5°C in December to 15.6°C in August, with an annual average of 7.8°C. The Pioverna stream is characterised by high calcium and magnesium bicarbonate concentration and exhibited average annual conductivity and pH of 469 mS/cm (at 20°C) and 7.7, respectively.

The studied reach showed very good water quality. On the date of sampling, the biological quality was evaluated by means of the Extended Biotic Index (Woodiwiss, 1964) modified by Ghetti (1986) for use in Italian streams. The values assumed by this biotic index were found to be constantly over 10 (Quality Class I), indicating non-polluted water or an environment unaltered in any detectable way.

Materials and methods

Benthic samples: A total of 115 Surber samples (area, 0.15 m^2; mesh size, 0.45 mm) were collected monthly from April 1996 to September 1997. On average three replicates were taken for each of the different habitats sampled. In total, about 17 000 larvae (belonging to 63 different taxa) were collected, sorted out in the field and fixed in 80% ethanol. Subsequently the individuals were identified in the laboratory to the following taxonomic levels: species (Ephemeroptera, Hirudinea, Mollusca, Isopoda and Diptera Athericidae), genus (Plecoptera, Odonata and Gammaridae) and family (Diptera, Trichoptera, Coleoptera, Oligochaeta and Turbellaria).

For each sampling area, the following variables were measured: flow velocity, depth, pH and conductivity. Furthermore, substratum composition (estimated visually as percentage composition) and bed roughness were calculated. Moreover, discharge was determined and instantaneous, minimum and maximum water temperatures were recorded every 2 weeks.

Estimation of secondary production: Densities and standing stocks were estimated for the most abundant taxa present at the study site. To allow life cycle descriptions and production appraisal, we measured body length of the selected taxa from the anterior edge of labrum to the posterior edge of the last abdominal segment (Welton et al., 1982), to the nearest 0.25 mm under a microscope.

To estimate production (dry weight) we used three methods: removal–summation (Boysen-Jensen, 1919; Brooker & Morris, 1978; Waters, 1979), instantaneous growth (Allen, 1949; Ricker, 1946; Waters, 1966) and Hynes–Hamilton (Hynes, 1961; Hamilton, 1969). In the first method, the mortality between successive samples, taking into account the size of organisms during the period of loss, is calculated in terms of weight or biomass. The sum of such observed mortalities over the entire life cycle is equivalent to the total production of the cohort (Waters & Crawford, 1973). If the species is univoltine, the cohort production is equal to annual production. In the instantaneous growth method, for a given interval of time, production is calculated as the product of the instantaneous rate of growth (G) and the mean standing crop in weight (B) during the time interval. The formula used is: $P = G \cdot B$ (Allen, 1949; Ricker, 1946). For a given period of time, G is calculated as the natural logarithm of the ratio of the mean weight at the end of the period to the mean weight at the beginning of the period and B is computed as the average of the standing crops at the beginning and end of the period (Waters & Crawford, op. cit.). The Hynes's method differs from the others in that it divides a population into size classes rather than in sequential sampling periods. It is based upon the calculation of an 'average cohort', or the mean size distribution of animals collected over a year, and production is estimated as the sum of losses between successive size classes rather than between successive sampling dates (Waters, 1979). This method does not require the discrimination of single cohorts and can thus be used whenever the life cycle cannot be deciphered or when it is not possible to identify taxa to the species level. For the application of this method it is necessary to know the cohort production interval (CPI), that is the average period required for growth, from hatching to the attainment of maximum larval size (Georgian & Wallace,1983). For estimating the annual production of Hydropsychidae (Trichoptera) and *Protonemura* (Plecoptera), with no identifiable cohorts, we used this method only (asterisks in Figure 2). In all the other cases, the values resulted from averaging the production values obtained with the three methods. Taxa were assigned to functional feeding groups according to Ghetti (1986). The nymphs of the *Baetis alpinus* group (i.e., *B. alpinus* and *B. melanonyx*) were attributed to the scrapers category.

Production of benthic invertebrates was estimated in functional habitats (habitats with a real ecological meaning: Harper et al., 1992) previously identified in the Pioverna stream (Crosa & Buffagni, 1996) by means of the TWINSPAN analysis. These were defined by the composition and abundance of the benthic community. 'Riffle', 'pool', 'transition' and 'bedrock' habitat types were identified. The four habitats differ from each other mainly with regard to water velocity (pool and transition vs. riffle and bedrock), depth (pool vs. the other habitats) and substratum features: granulometry (bedrock vs. the others) and substratum roughness (pool and bedrock vs. transition and riffle). The term 'transition' was used to label one habitat because it shows somewhat intermediate attributes between pools and riffles, not being properly identified with a 'run' habitat (Crosa & Buffagni, op. cit.). The transition habitat is highly dynamic, as it often occurs in stream areas where hydraulic features are expected to change even after a small discharge variation (e.g., lateral runs and marginal, still water, areas).

Results

In Figure 1, the percentage of occurrence of the collected taxa is reported – at the family or order level – for the four functional habitats. The abundance of a group of 12 taxa reached appreciable values (from 50% of total invertebrates in pools to 79% in bedrock habitats) and presumably accounted for most of total benthic production. Three of these taxa belong to the Plecoptera (*Amphinemura*, *Protonemura* and *Leuctra*), seven to the Ephemeroptera (*Alainites muticus* (Linné), *Baetis alpinus* (Pictet), *B. melanonyx* (Pictet), *B. rhodani* (Pictet), *Ecdyonurus helveticus* Eaton, *Rhithrogena* spp, *Habroleptoides auberti* (Biancheri)), and one each to Diptera (*Ibisia marginata* (Fabricius)) and Trichoptera (Hydropsychidae). Other taxa showed a relatively high density in particular habitats only. For example, the Chironomidae were common in pools and bedrock, the Helodidae in riffle and transition habitats, the Hydracarina in pools and the Oligochaeta in transition areas. These taxa were not included in the secondary production estimate since, for calculating their production, methodological difficulties could emerge, leading to unreliability of results. In many – and relevant – cases, the taxa's percentage of production was quite different from the equivalent percentage of numbers. For example, the Athericidae showed production percentages of 31 and 47 – in transition habitats and pools, respectively – against abundance percentages of 22 and 37. The Heptageniidae displayed, in riffles, 27% of production versus 17% of numbers and, in the transition habitat, 3 versus 20%. In this latter habitat, the Nemouridae contributed 33% to total production, while having only 16% of the individuals. In contrast, in the same habitat, the Leuctridae exhibited 4 vs. 8%, and, in pools, 23% of production against 40% of individuals.

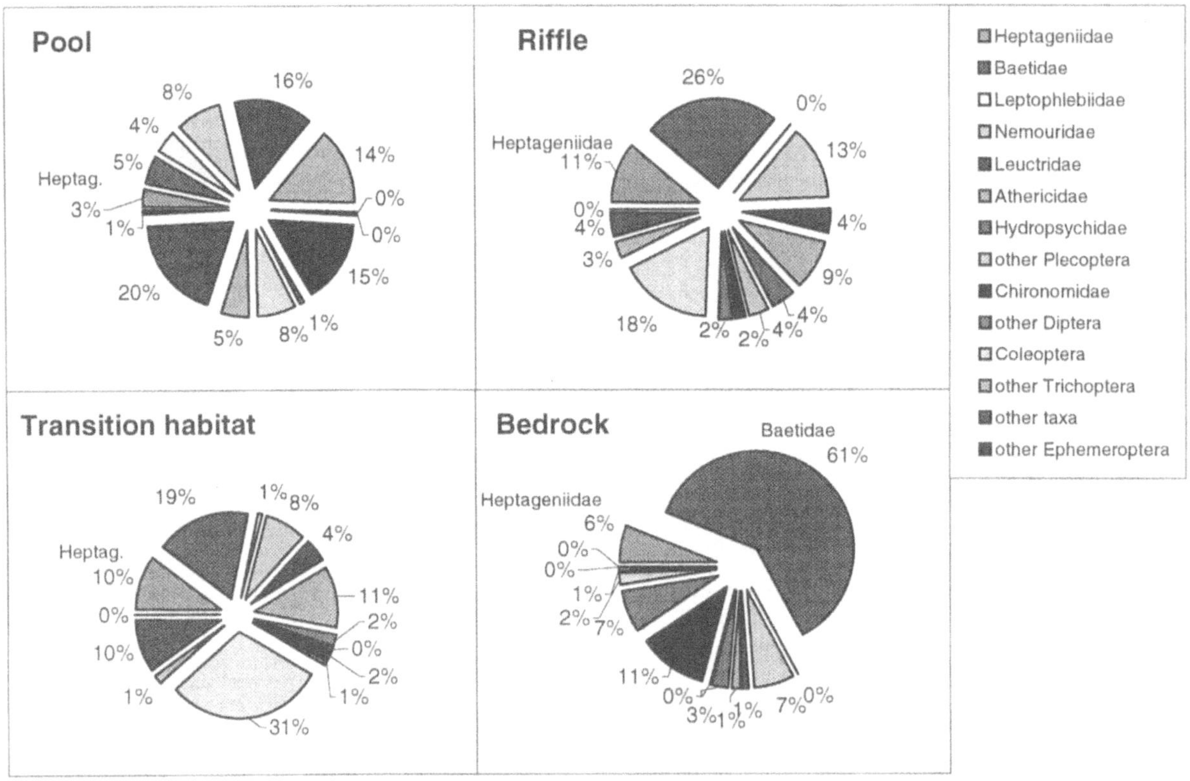

Figure 1. Relative abundance of the different families or orders of macroinvertebrates in each of the four Pioverna stream: pool, riffle, transition and bedrock. Families are reported clockwise from Heptageniidae to 'other Ephemeroptera', following the taxa list.

As regards the life cycle traits necessary to properly estimate production with removal–summation and instantaneous growth methods, the studied taxa showed different numbers of cohorts in one year. While *H. auberti* developed in one synchronous winter generation, all the other taxa exhibited two or three cohorts in 1 year. *E. helveticus*, *R. semicolorata*, *B. rhodani*, *B. alpinus*, *Leuctra* and *Amphinemura* presented three cohorts which could be followed (being bivoltine or multivoltine). These taxa and *I. marginata* were present year-round, while the remaining ones were found seasonally with two cohorts. A detailed presentation of the life cycle and production values of each taxon is beyond the aims of the present paper and will be the subject of a specific study (Buffagni, in prep.). The three methods used to estimate secondary production displayed high correlations. The Spearman's rho correlation coefficient was over 0.7 between each pair of methods and for any group of taxa. High correlations were present for the mayflies and *I. marginata* – identified at the species level – as well as for *Amphinemura* and *Leucra*, especially between the Hynes–Hamilton and removal–summation methods. For these two latter taxa and methods, rho reached the highest value (0.931). This suggests that the non-specific identification – which could lead to inaccuracy in life cycle interpretation and production calculation for the stoneflies – did not obviously affect the production estimates.

The 'riffle' habitat had the highest secondary production, while 'bedrock' habitat had the lowest (Figure 2a). Pool and transition habitats showed intermediate production values. The total production of the community – for riffle, pool and transition habitats, respectively – was about 1.85, 0.66 and 0.65 g m^{-2} $year^{-1}$. Baetid and heptageniid mayflies, along with the dipteran Athericidae and plecopteran Nemouridae, made the greatest contribution to total production (Figure 2b–d). In the bedrock habitat, production was high only for the two mayflies *B. melanonyx* and *B. alpinus* (a total of 0.23 g m^{-2} $year^{-1}$). *I. marginata* exhibited a relatively high production in the three remaining habitats, while the other taxa showed more variability between habitats. For example, the heptageniid production was quite high in the riffle habitat, but close to zero in the other ones. Secondary consumers should be kept separate from primary consumers; in the analysis of habitat production, however, they were considered together, as the only obvious predators are the Athericidae which are consequently easily distinguishable from the bulk of detritivores and herbivores.

Figure 3 a–c reports the rank–production curves for the three habitats pool, riffle and transition. A larger number of taxa (11) contributed to most of the riffle production than to pool production (five), with the transition habitat (eight) being intermediate. In the pool habitat a steep curve is recognisable, with *I. marginata*, *Amphinemura* and *Leuctra* giving the highest productions. In the riffle habitat two *Baetis* species, two Heptageniidae taxa, *Protonemura* and *I. marginata* had production values between 0.2 and 0.3 g m^{-2} $year^{-1}$ (left side of Figure 3b). The remaining taxa displayed lower production (less than 0.1). In the transition habitat, three taxa (*I. marginata*, *Amphinemura* and *B. rhodani*) had a similar production of about 0.18 while all the other taxa had values less than 0.03 g m^{-2} $year^{-1}$. The rank–production curve for the site is reported in Figure 3d. The overall secondary production is here calculated on the basis of the integrated samples from all habitats, assuming that they were present with equal proportions. The resulting curve is, to some extent, similar to the riffle one, even if more uniform. In contrast, the rank ordination of the taxa is different. For example, *B. alpinus* shows at the site scale a much lower production and higher rank than in the riffle habitat, and the other riffle reference taxa change here their relative position along the production gradient.

Figure 4 shows the production values of the four habitats by macroinvertebrate feeding groups. Shredders and predators were the major component of production in pools, with a low contribution by gathering collectors. Anyway, since the Chironomidae production was not calculated, maybe the overall value for gathering collectors in this functional habitat is somewhat underestimated. In riffles, this latter category and scrapers had the highest production, with shredders, predators and filtering collectors showing decreasing production values. In transition habitats, shredders, predators and gathering collectors accounted for the highest proportions of benthic production, while filtering collectors and scrapers, even if present, made only a modest contribution. In general terms, shredders, gathering collectors and scrapers made the greatest contributions to total production. Scrapers and predator production proportions gradually increased and decreased, respectively, along an approximate velocity gradient (i.e. from pool to bedrock habitat).

Discussion and conclusion

Macroinvertebrate assemblages exhibited distinct production patterns among habitat types. This confirms the results of previous studies concerning benthic community production with respect to habitat type (e.g., Smock et al., 1985). In the Pioverna stream, the riffle habitat had the highest secondary production and bedrock habitat had the lowest. Antithetically, Wohl et al. (1995), in the southern Appalachians, found that cobble–riffle habitat had the lowest invertebrate biomass and productivity while the bedrock outcrops had the highest. Nevertheless, the latter habitat was characterised by the presence of standing crops of moss – absent from the Pioverna bedrock habitat – which were a suitable substratum for gathering collectors and filterers and which can explain this seeming discrepancy. In fact, aquatic macrophytes can represent a good habitat for taxa-rich and abundant macroinvertebrate communities (e.g., Gregg & Rose, 1985). For example, Grubaugh et al. (1997), estimating production in cobble and pebble/gravel habitats with and without macrophytes, noted that large standing crops of aquatic plant supported much greater production than plant-free areas. Comparing invertebrate communities of pool and riffles in North American low order rivers, Brown & Brussock (1991) found that riffle habitats contained more numbers, biomass and faunal richness than pools. The same occurred in the Pioverna stream (Buffagni, unpublished data) and corresponds to lower secondary production in pools than in riffles. In this respect, Brown & Brussock (1991) concluded that pool-adapted taxa are more likely to occur in different types of rivers where suitable refugia are available within the pool habitat substratum and flow reversals are less frequent. Hence, the observed lower production in the pool habitat of the Pioverna stream can be linked also to the unstable substrata in these areas, which are susceptible to scouting during spates (Grubaugh et al., 1997). Brown & Brussock (1991) also found – in agreement with the pattern observed in the Pioverna stream – that areas with intermittent flow (i.e., similar to the transition habitat) had lower densities and diversity than permanently submerged habitats (i.e., riffle habitat). The overall production estimated for the Pioverna stream is quite

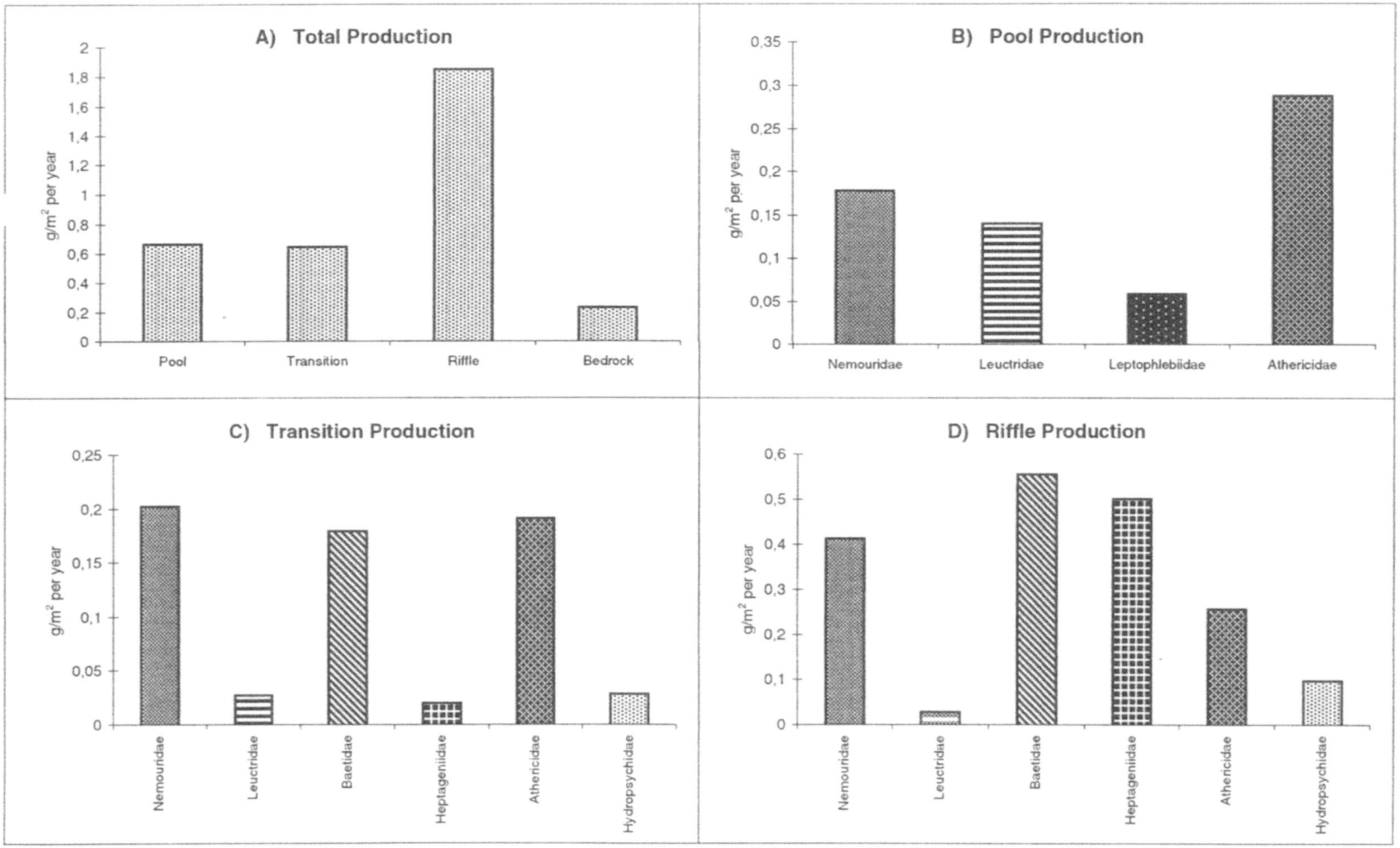

Figure 2. Production values of the whole community in the four functional habitats of the Pioverna stream (A) and of the different families of macroinvertebrates in each of the three more productive habitats: pool (B), transition (C) and riffle (D).

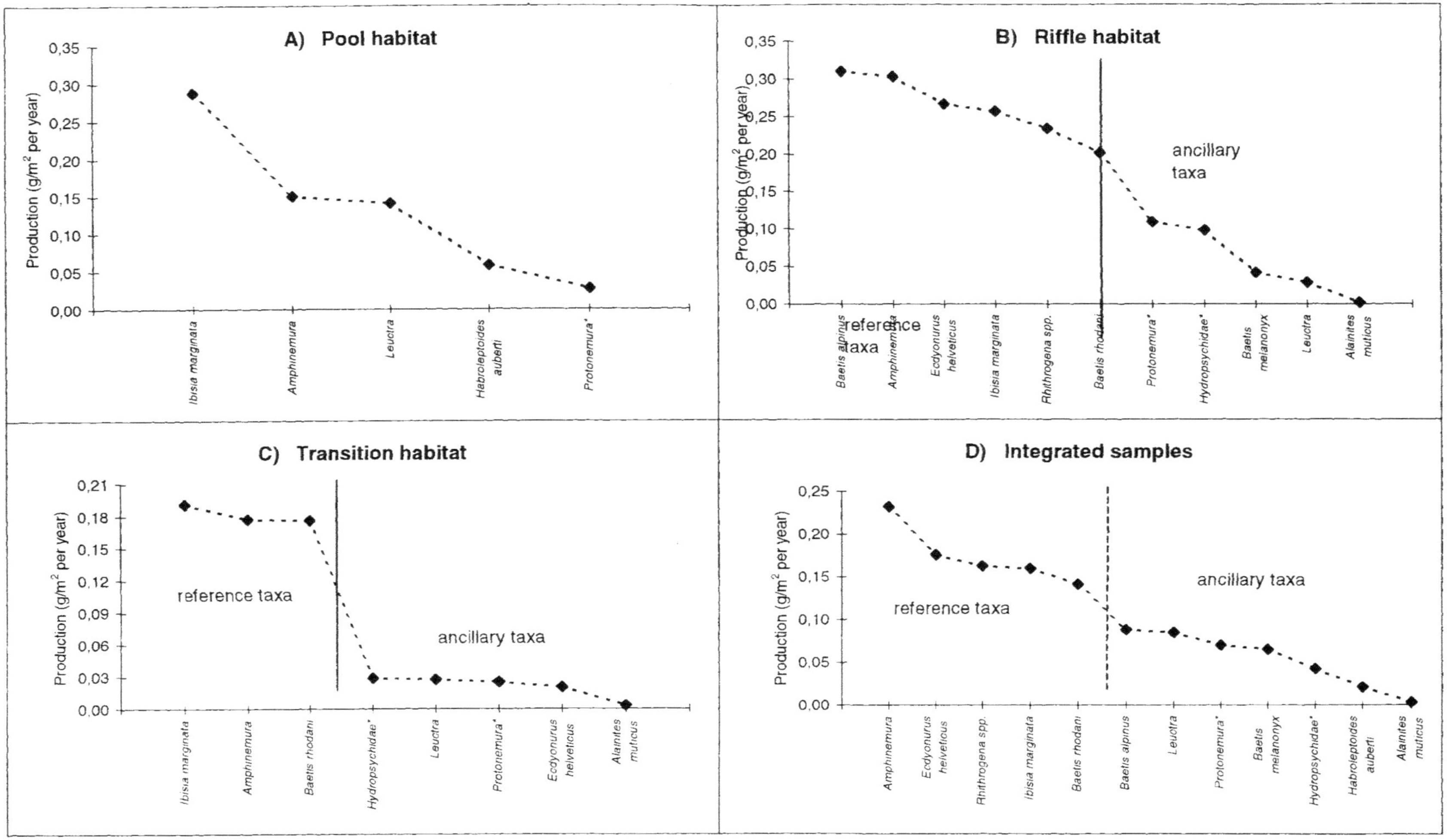

Figure 3. Rank–production curves for the three most productive functional habitats in the Pioverna stream, pool (A), riffle (B) and transition habitat (C), and the same from integrated samples from all the habitats.

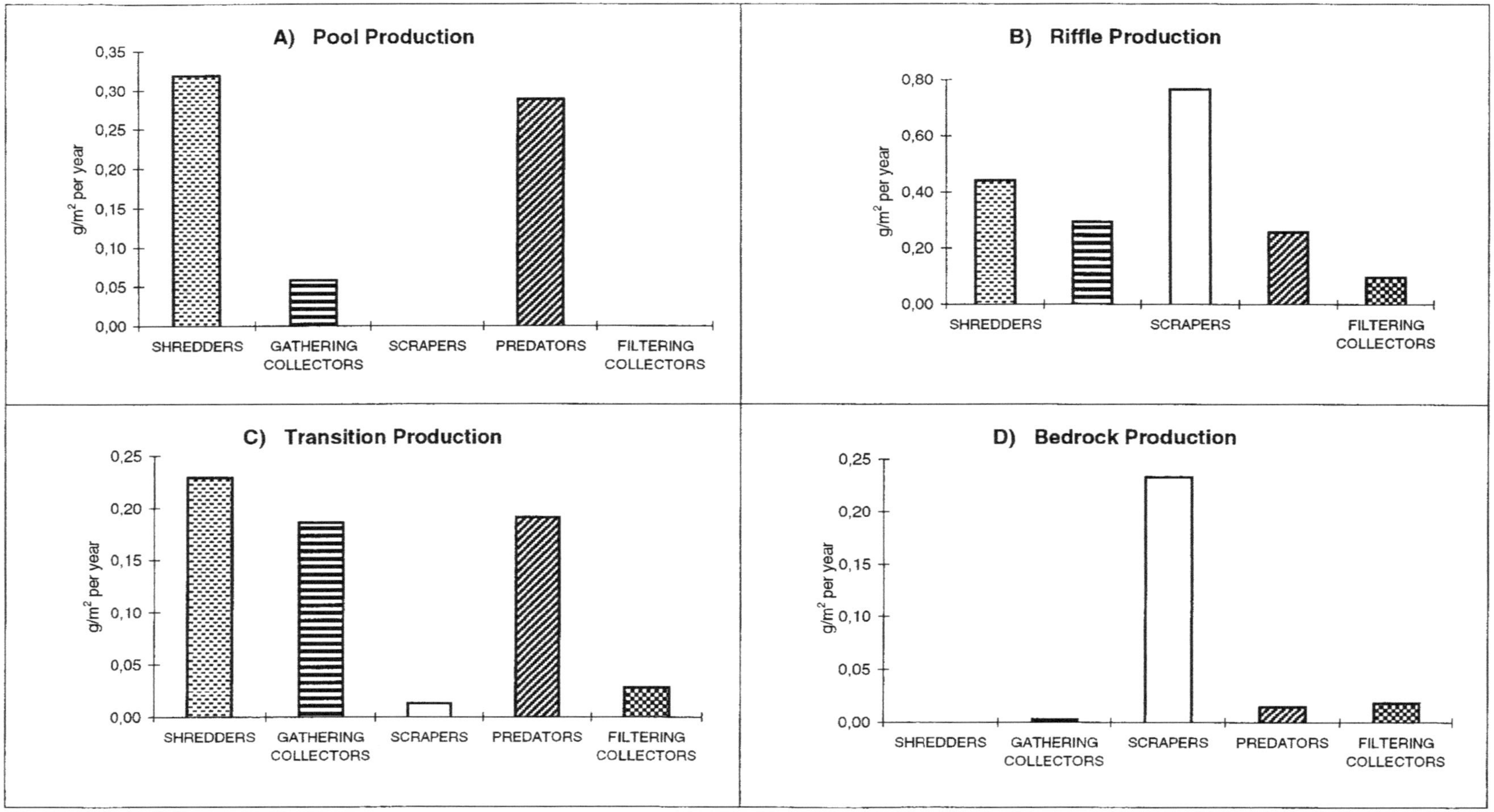

Figure 4. Production values by feeding groups of macroinvertebrates in the four functional habitats of the Pioverna stream: pool (A), riffle (B), transition (C) and bedrock (D) habitat.

low compared to values usually reported for rivers, and this is due to low densities of macroinvertebrates. Explanations can be sought in the frequent spates – and the presumably limited deposition of organic allochthonous detritus – which occurred during the rainy 1996 in the Pioverna stream, and in the paucity of the hyporheic component to the rocky sampling area: both probably affected overall densities (Dole-Olivier et al., 1997). Moreover, the study site is located in a narrow valley with little light and therefore low primary productivity.

According to Cao et al. (1997), it is desirable to split community variation in different forms to better assess man-induced impacts. In the same way, it could be worthwhile to split up the production data of the benthic community. Cao et al. (op. cit.) listed various types of change in species abundance as a result of organic pollution, assigning species as gained, lost or persisting. In the riffle habitat of the pristine Pioverna stream, about 10 taxa contributed with higher production values. A portion of these organisms are expected to modify their production in altered environmental conditions due to human activities (Frutiger, 1985). The production of some of those species will increase while that of others will decrease in the case of pollution, because of different sensitivity and response to stress. Taxa like the Hydropsychidae, *B. rhodani* and to a lesser extent *B. alpinus* would probably had increased production in many cases (e.g., Alba Tercedor et al., 1991; Janeva, 1979; Mackay & Waters, 1986; Zelinka et al., 1977), while the production of Heptageniidae *E. helveticus* and *Rhithrogena* and the Athericidae *I. marginata* would decrease (e.g., Bauernfeind et al., 1995; Leps et al., 1989; Malmqvist, 1996; Rivosecchi, 1984).

The rank–production curve in the riffle habitat reveals a proportionally high evenness (i.e., the relative taxa production is similar) of the most productive taxa ('reference' taxa: *B. alpinus*, *Amphinemura*, etc.; highest portion of the curve) and a quite evident separation from the less productive taxa ('ancillary' taxa: *Protonemura*, Hydropsychidae, etc.; lowest portion of the curve). Note that production of *Protonemura* and Hydropsychidae was estimated with the size–frequency method only, which usually gives higher values than the other methods used here (e.g., Iversen & Dall, 1989; Waters & Crawford, 1973). If studying production in impacted watercourses of the Pioverna stream type, once the species association has been defined, it would therefore be possible to evaluate which taxa, following environmental alteration, move from the dominant to the ancillary category or vice versa, disappear or – for a new taxon – enter the major contributors to the macroinvertebrate production.

Comparing curves from different habitats of the Pioverna stream, it is notable that in the transition habitat (often characterised by unstable flow conditions, low primary productivity and by sharing most species with the riffle habitat (Buffagni & Crosa, unpublished)), three out of six riffle reference taxa show high production; these are *I. marginata*, *B. rhodani* and *Amphinemura*, which can be considered habitat generalists. *I. marginata* also lives within the substratum and is less affected by flow variations. In this case, dissimilarities between the two habitats are presumably due to the naturally differing physical features (e.g., lower water velocity in the transition habitat), although analogous changes in the list of reference species after habitat degradation or water pollution events are expected. In the event of a temporal shift from riffle to transition habitat due to environmental alteration (e.g., flow regulation), three taxa would drop out of the reference category and none would move in from the ancillary category. Furthermore, *Rhithrogena* and *B. melanonyx* would no longer be found in the community.

Examining the rank/production curve and comparing its shape between unaltered stream areas (i.e., with high ecological integrity) and impacted areas should demonstrate notable changes (e.g., Odum, 1988). In a curve closer to the broken-stick model (Wittaker, 1965), as seen in the riffle habitat of the Pioverna stream, the taxa's production trend would change, moving to higher dominance curves (i.e., typical of logarithmic and geometric series). In fact, a benthic community dominated by one or few taxa is often indicative of environmental stress (Plafkin et al., 1989) or low water resource quality (Kerans & Karr, 1994). The dominance of the community can be approximated by the percent contribution of the most productive taxon (Grubaugh & Wallace, 1995) and, where reference communities do exist, indices of community similarity can measure the loss of benthic species as well as changes in taxonomic composition, taxon abundance and production (Plafkin et al., op. cit.). Some indices enable more emphasis to be placed on selected species; they could perhaps be advantageously applied to production data as presented here, e.g., the Pinkham and Pearson Community Similarity Index (Pinkham & Pearson, 1976), as already recommended in biomonitoring protocols (Plafkin et al., op. cit.).

Zelinka et al. (1977) estimated macroinvertebrate production, at the site scale, in three stations along the Bítyska brook, a trout stream in the Czech Republic, characterised by different levels of saprobity: oligosaprobity (reference site), betamesosaprobity and polysaprobity. The unpolluted site showed the lowest dominance (and highest number of reference taxa) and the most strongly impacted one shows high dominance and very few species being productive. The moderately polluted station supported a community with intermediate production attributes (Zelinka et al., op. cit.). Grubaugh & Wallace (1995) estimated production at one site in the southeastern US before and after biotic integrity improved (after a 35-year period), reporting a shift in the dominant taxon from simuliids to hydropsychids. The same authors stated that this shift in dominance indicated an improvement in biotic integrity of the stream (Grubaugh & Wallace, op. cit.). At the site scale, evident gradients of river integrity therefore effect trends in community production relationships and dynamics. Examining the curve of secondary production estimated in the present paper from the integrated samples from all the habitats reveals a similarity with the riffle curve. As the riffle contributed most to total production, it is logical that the shape of the integrated sample curve resembles the riffle one. The rank-production curve from the integrated sample seems a rough surrogate, being reshaped by the production of taxa present in pool, transition and bedrock habitats. The derived, overall picture fails in describing a single habitat (i.e., riffle) reference situation and shows a taxa sequence that is not representative of any habitat, furthermore smoothing down inter-taxa production differences that are apparent within the single habitats. The community structure assessed – merging all the samples – therefore probably describes a situation of unsatisfactory ecological meaning, combining four distinct macroinvertebrate assemblages. The approach of studying secondary production by splitting up river stretches into the major macroinvertebrate functional habitats has been present in the literature for many years (e.g., Benke et al., 1984). In this regard, Smock et al. (1985) argued that it is important to adjust production according to habitat abundance in order to properly examine total stream productivity; this was confirmed by Lugthart & Wallace (1992), who measured the overall production weighting the proportional availability of the habitats. To estimate the production of a river stretch, a preliminary listing of the instream habitats is therefore required to obtain a representative sample, whether taking into account the habitats in equal proportion or according to their relative availability. Furthermore, an environmental impact could act on specific habitat types or on functional habitat occurrence (e.g., Kemp et al., 1999); this may lead to failure when assessing ecological impact using production estimates performed at the site scale. Besides, the occurrence and frequency of instream habitats reflects the main geomorphological processes in the river channel, allowing further assessment of anthropomorphic impact (Harper et al., 1998). Thus, calculating production at the functional habitat scale, rather than solely at the site scale, could represent a thorough and attainable approach to ecological integrity investigations. In addition, the identification and study of distinct aquatic habitats has an intuitive value, since habitats link the impacts on the natural environment and its inhabitants (Harper et al., 1995). Habitats also have a notable potential in river management because they can be easily recognised, whereas most taxa and ecological functions cannot (Harper et al., op. cit.). Harper et al. (1992) emphasised that within the set of habitats in which a taxon can potentially be present there is a smaller set which must be represented for the taxon to succeed. Assessing production at the habitat scale allows to list which habitat offers the best conditions for a particular species to develop; this has direct relevance in biodiversity conservation strategies to production estimates.

When trying to estimate river integrity (other than comparing absolute production values, which – in Alpine streams – can greatly vary among different years due to natural fluctuations), intertaxon and interhabitat relationships could be juxtaposed between reference and impacted sites. Furthermore, in habitats where a few species usually exploit the food resource – in polluted as well as in unaltered sites – production can attain high values (Grubaugh et al., 1997), without any clear link with ecological integrity. For site comparisons of secondary production in the Alpine area, the riffle habitat therefore seems to be the most adequate, as is implicitly assumed in other areas (Krueger & Waters, 1983), because it shows a lower dominance and more taxa contributing to overall production.

As the collected benthic invertebrates were represented by more than one trophic level, analysis can be done separating taxa according to their trophic role (Lugthart & Wallace, 1992). No studies have been conducted in Italian streams regarding the feeding habits and food types selected by invertebrates, and explicit investigations (i.e., gut analysis) should be carried out to allow a truly representative attribution to functional feeding groups. For instance, some of the most productive taxa found in the Pioverna stream, such as the Heptageniidae (i.e., *Ecdyonurus helveticus* and *Rhithrogena*), seem to be able to feed both on scrapers and gatherers, depending on habitat type, food availability and nymphal size (e.g., McShafferty & McCafferty, 1986); the *Baetis* species can easily vary their feeding group too. Notwithstanding these uncertainties on group attribution, in the study of the relative importance of trophic roles, production as a functional measure of ecological integrity seems potentially informative (Benke, 1993). According to longitudinal trend predictions of the River Continuum Concept (Vannote et al., 1980) and to previous studies (Grubaugh et al., 1997), the shredders', gathering collectors' (and here scrapers') contribution to production were the greatest in the low-order Pioverna stream. Similarly to previous works (Grubaugh & Wallace, 1995 for *Baetis* spp.), the *Baetis* species of the *alpinus* group – dominant in the bedrock and riffle habitats – were classified into the scrapers category, leading to a high production percentage of this feeding group. Very small quantities of fine organic material, compared with aufwuchs, were available in the bedrock habitat, supporting this attribution. As far as the proportions of production in the four sampled habitats are concerned, notable differences were recorded among feeding groups. While shredder organisms were a major and filtering collectors a minor component in all of them, gathering collectors, scrapers and predators showed a clear variability among habitats. The relative contributions to annual production by functional groups mostly confirm results from other studies (e.g., Wohl et al., 1995): major differences concern the low contributions of gathering collectors in pools and filter-feeders in bedrock habitats. The absence of the Chironomidae production percentage, not calculated here, from the gathering collectors group can in part explain this discrepancy in pools. In general terms, the proportion of scrapers and predators within single habitats seems suitable to be used as an indicator of biotic integrity in Alpine streams, as stated for other geographic areas (Kerans & Karr, 1994). Apart from classifying taxa into specific functional feeding groups, in order to measure the integrity of the benthic community and habitat quality, Frutiger (1985) suggested using the ratio between production by primary and secondary consumers. This ratio is expected not to exceed 15 in undisturbed Alpine rivers – in the Pioverna stream it was below 8 – and to increase in impacted sites (Frutiger, op. cit.). The application of equivalent ratios at the habitat scale seems promising since it could limit estimation inaccuracy induced by inter-habitat variability of production. The *P/B* ratio, too, can be suitably used at the habitat scale to evaluate shifts of the invertebrate assemblages to smaller taxa with shorter generation time (Lugthart & Wallace, 1992) or to assess the habitat-specific variation of population structure for single indicator taxa after disturbance.

Production studies therefore seem to represent an important tool for examining the effect of disturbance in lotic ecosystems (Benke, 1984). Lugthart & Wallace (1992) demonstrated that production estimates can better assess the impact of disturbance than abundance and biomass, two critera more frequently employed in river integrity evaluations. These authors (Lugthart & Wallace, op. cit.) proved that, in mountain Appalachian streams, measures of abundance did not reflect the importance of shredders and predators, while biomass calculation underestimated the role of small-sized gathering collectors. In general terms, abundance and biomass measures tend to under- or overestimate the roles of taxa in river ecosystems (e.g., Benke et al., 1984; Wohl et al., 1995). Production is potentially a very useful measure as it combines effects on population and community function; it therefore synthesises information about a variety of perturbations. Even if not adequate for rapid and 'one-sample' bioassessment, secondary production estimates can be an effective tool for more detailed studies. These would include regularly monitored river systems or areas of declared social or naturalistic interest. Moreover, they could be fruitfully carried out in long-term programs aimed at identifying changes in ecosystem processes on a basin-wide scale. In addition, in many cases, the product of the mean biomass and the annual *P/B* ratio of each taxon could be used to estimate the annual production (Lapachin & Neveu, 1980; Mortensen & Simonsen, 1983; Waters, 1977). This would reduce time, depress costs and simplify estimates, giving secondary production higher monitoring effectiveness.

As confirmed by the present research, benthic secondary production may vary greatly between the different in-channel habitats (Smock et al., 1985). For this reason, calculating an overall mean value for secondary production could be inadequate to evaluate specific functional aspects of ecosystem integrity. Instead, secondary production should be estimated for each of the most important constituent habitats. In the present study, calculating secondary production for dominant and indicator taxa in the different habitats was an effective way to describe reference conditions against which the ecological integrity of mountain streams can be measured.

Acknowledgements

This study was funded (70%) by the Fondazione Lombardia per l'Ambiente (FLA), Milan. We thank Dr Giuseppe Crosa (Environmental Sciences Department, University of Milan), Dr David Harper and Dr Joanna Kemp (Ecology Unit, Biology Dept., University of Leicester) for habitat concept debate. We also thank Filippo Angelotti and Stefania Costanzo for field and laboratory work and the anonymous reviewers who improved the paper.

References

Alba Tercedor, J., C. Zamora Munoz, A. Sàncez Ortega & I. Guisasola, 1991. Mayflies and stoneflies from the Rio Monachil (Sierra Nevada, Spain) (Ephemeroptera and Plecoptera). In: Alba Tercedor, J. & J. Sanchez Ortega (eds), Overview and Strategies of Ephemeroptera and Plecoptera Sandhill-Crane Press: 529–538.

Allen, K. R., 1949. Some aspects of the production and cropping of fresh waters. Trans. r. Soc. N. Z. 77: 222–228.

Bauernfeind, E., P. Weichselbaumer & O. Moog, 1995. Ephemeroptera. In: Moog, O. (ed.), Fauna Aquatica Austriaca, Teil 3, Lief. 1/1995, BMLF., Wasserkataster, Wien: 1–17.

Benke, A. C., 1984. Secondary production of aquatic insects. In: Resh V. H. & D. M. Rosemberg (eds), Ecology of Aquatic Insects. Praeger Publishers, New York: 289–322.

Benke, A. C., 1993. Concepts and patterns of invertebrate production in running waters. Verh. int. Ver. Limnol. 25: 15–38.

Benke, A. C., T. C. Van Arsdall, D. M. Gillespie & F. K. Parrish, 1984. Invertebrates productivity in a subtropical blackwater river: the importance of habitat and life history. Ecol. Monogr. 54: 25–63.

Boysen-Jensen, P., 1919. Valuation of the Limfjord. 1. Rep. Dan. Biol. Sta. 26: 3–44.

Brooker, M. P. & D. L. Morris, 1978. Production of two species of Ephemeroptera (*Ephemerella ignita* Poda and *Rhithrogena semicolorata* Curtis) in the upper reaches of the R. Wye, Wales. Verh. int. Ver. Limnol. 20: 2600–2604.

Brown, A. V. & P. P. Brussock, 1991. Comparison of benthic invertebrates between riffles and pools. Hydrobiologia 220: 99–108.

Cao, Y., A. W. Bark & W. P. Williams, 1997. Analysing benthic macroinvertebrate community changes along a pollution gradient: a framework for the development of biotic indices. Wat. Res. 31: 884–892.

Covich, A. P., W. H. Clements, K. D. Fausch, J. D. Stednick, J. Wilkins-Wells & S. R. Abt, 1998. Ecological integrity and western water management: a Colorado perspective. Department of the Interior, US Geological Survey. Colorado Water Resources Research Institute: Grant Number 14-08-0001-G2008/3, project 12.

Crosa, G. & A. Buffagni, 1996. L'habitat idraulico quale elemento per la gestione degli ambienti fluviali. S.It.E. Atti. 17: 581–583

Dole-Olivier, M. J., P. Marmonier & J. L. Beffy, 1997. Response of invertebrates to lotic disturbance: is the hyporheic zone a patchy refugium? Freshwat. Biol. 37: 257–276.

Fisher, S. G. & L. J. Gray, 1983. Secondary production and organic matter processing by collector macroinvertebrates in a desert stream. Ecology 64: 1217–1224.

Frutiger, G., 1985. The Production Quotient PQ: a new approach for quality determination of slightly to moderately polluted running waters. Arch. Hydrobiol. 104: 513–526.

Ghetti, P. F., 1986. I macroinvertebrati nell'analisi di qualità dei corsi d'acqua. Provincia autonoma di Trento, Trento, 105 pp.

Georgian, T. & J. B. Wallace, 1983. Seasonal production dynamics in a guild of periphyton-grazing insects in a Southern Appalachian stream. Ecology 64: 1236–1248.

Gregg, W.W. & F. L. Rose, 1985. Influences of aquatic macrophytes on invertebrate community structure, guild structure, and microdistribution in streams. Hydrobiologia 128: 45–56.

Grubaugh, J. W. & J. B. Wallace, 1995. Functional structure and production of the benthic community in a Piedmont river: 1956–1957 and 1991–1992. Limnol. Oceanogr. 40: 490–501.

Grubaugh, J. W., J. B. Wallace & E. S. Houston, 1997. Production of benthic macroinvertebrate communities along a southern Appalachian river continuum. Freshwat. Biol. 37: 581–596.

Hamilton, A. L., 1969. On estimating annual production. Limnol. Oceanogr. 14: 771–782.

Harper, D. M., C. D. Smith & P. J. Barham, 1992. Habitats as the building blocks for river conservation assessment. In: Boon, P. J., P. Calow & Petts G. E. (eds), River Conservation and Management. John Wiley, Chichester: 311–319.

Harper, D. M., C. D. Smith, P. J. Barham & R. Howell, 1995. The ecological basis for the management of the natural river environment. In: Harper, D. M. & Ferguson A. J. D. (eds), The Ecological Basis for River Management. John Wiley & Sons Ltd, Chichester: 219–238.

Harper, D. M., M. Ebrahimnezhad & F. Climent, 1998a. Artificial riffles in river rehabilitation: setting the goals and measuring the successes. Aquat. Conserv. Mar. Freshwat. Ecosyst. 8: 5–16.

Harper, D. M., C. D. Smith, J. L. Kemp & G. A. Crosa, 1998b. The use of 'functional habitats' in the conservation, management and rehabilitation of rivers. In: Bretschko, G. & J. Halesic (eds), Advances in River Bottom Ecology. Backhuys Publishers, Leiden, The Netherlands: 315–326.

Hynes, H. B. N., 1961. The invertebrate fauna of a Welsh mountain stream. Arch. Hydrobiol. 57: 344–388.

Iversen, T. M. & P. Dall, 1989. The effect of growth pattern, sampling interval and number of size classes on benthic invertebrate production estimated by the size-frequency method. Freshwat. Biol. 22: 323–331.

Janeva, I., 1979. Einige Vertreter der Gattung Baetis (Ephemeroptera) als Limnosaprobe Bioindicatoren. In Pasternak & Sowa (eds), Proceedings of the Second International Conference on Ephemeroptera, Krakow: 139–143.

Karr, J., 1993. Defining and assessing ecological integrity: beyond water quality. Envir. Toxicolo. Chem. 12: 1521–1531.

Kemp, J. L., D. M. Harper & G. A. Crosa, 1999. Use of 'functional habitats' to link ecology with morphology and hydrology in river rehabilitation. Aquat. Conserv. Mar. Freshwat. Ecosyst. 9: 159–178.

Kerans, B. L. & J. R. Karr, 1994. A benthic index of biotic integrity (B-IBI) for rivers of the Tennessee Valley. Ecolo. Appli. 4: 768–785.

Krueger, C. C. & T. F. Waters, 1983. Annual production of macroinvertebrates in three streams of different water quality. Ecology 64: 840–850.

Lapachin, L. & A. Neveu, 1980. The production of benthic invertebrates: comparison of different methods. Acta Ecolo. Ecol. Gener. 1: 307–322.

Leps, J., T. Soldàn & V. Landa, 1989. Multivariate analysis of compositional changes in communities of Ephemeroptera (Insecta) in the Labe basin, Czechoslovakia – A comparison of methods. Coenoses 4: 29–37.

Lugthart, G. J. & J. B. Wallace, 1992. Effects of disturbance on benthic functional structure and production in mountain streams. J. N. am. Benthol. Soc. 11: 138–164.

Mackay, R. J. & T. F. Waters, 1986. Effects of small impoundments on hydropsychid caddisfly production in Valley Creek, Minnesota. Ecology 67: 1680–1686.

Malmqvist, B., 1996. Bäckbromsen, *Atherix ibis* (Diptera: Athericidae), i Sverige: en rödlistad flugas utbredning och status. Ent. Tidskr. 117: 23–28.

McShafferty, D. & W. P. McCafferty, 1986. Feeding behaviour of Stenonema interpunctatum (Ephemeroptera: Heptageniidae). J. N. am. Benthol. Soc. 5: 200–210.

Mortensen, E. & J. L. Simonsen, 1983. Production estimates of the benthic invertebrate community in a small Danish stream. Hydrobiologia 102: 155–162.

Odum, E. P., 1988. Basi di ecologia. Piccin ed., Padova, 544 pp.

Pinkham, C. F. A. & J. B. Pearson, 1976. Applications of a new coefficient of similarity to pollution surveys. J. Wat. Pollut. Contr. Fed. 48: 717–723.

Plafkin, J. L., M. T. Barbour, K. D. Porter, S. K. Gross & R. M. Hughes, 1989. Rapid bioassessment protocols for use in streams and rivers: benthic macroinvertebrates and fish. EPA/444/4-89-001, Office of Water (WH-553), US Environmental Protection Agency.

Ricker, W. E., 1946. Production and utilization of fish populations. Ecol. Monogr. 16: 373–391.

Rivosecchi, L., 1984. Guide per il riconoscimento delle specie animali delle acque interne italiane. 28.

Ditteri (Diptera). C.N.R., S. Ruffo (ed.), Verona, 177 pp.

Smock, L. A., E. Gilinsky & D. L. Stoneburner, 1985. Macroinvertebrate production in a southeastern United States stream. Ecology 66: 1491–1503.

Vannote, R. L., G. W. Minshall, K. W. Cummins, J. R. Sedell & C. E. Cushing, 1980. The river continuum concept. Can. J. Fish. aquat. Sci. 37: 130–137.

Waters, T. F., 1966. Production rate, population density and drift of a stream invertebrates. Ecology 47: 595–604.

Waters, T. F., 1977. Secondary production in inland waters. Adv. Ecol. res. 10: 91–164.

Waters, T. F., 1979. Influence of benthos life history upon the estimation of secondary production. J. Fish. Res. Bd Can. 336: 1425–1430.

Waters, T. F. & G.W. Crawford, 1973. Annual production of a stream mayfly population: a comparison of methods. Limnol. Oceanogr. 18: 286–296.

Welton, J. S., M. Ladle & J.A.B. Bass, 1982. Growth and production of five species of Ephemeroptera larvae from an experimental recirculating stream. Freshwat. Biol. 12:103–122.

Whittaker, R. H., 1965. Dominance and diversity in land plant communities. Science 147: 250–260.

Wohl, D. L., J. B. Wallace & J. L. Meyer, 1995. Benthic macroinvertebrate community structure, function and production with respect to habitat type, reach and drainage basin in the southern Appalachians (U.S.A.). Freshwat. Biol. 34: 447–464.

Woodiwiss, F. S., 1964. The biological system of stream classification used by the Trent River Board. Chem. Indust. 14: 443–447.

Zelinka, M. & Working Team, 1977. Production conditions of the polluted trout brook. Folia Fac. Sci. Nat. Univ. Purkynianae Brunensis XVIII. Biol. 60: 1–105.

Hydrobiologia **422/423:** 197–208, 2000.
M. Jungwirth, S. Muhar & S. Schmutz (eds), Assessing the Ecological Integrity of Running Waters.

Performance of different biotic indices and sampling methods in assessing water quality in the lowland stretch of the Tiber River

Angelo G. Solimini, Paolo Gulia, Monica Monfrinotti & Gianmaria Carchini
Department of Biology, University of Rome 'Tor Vergata', Via della Ricerca Scientifica, 00133 Rome, Italy
E-mail: carchini@uniroma2.it

Key words: Biomonitoring, macrobenthos, chronological clustering, artificial substrates

Abstract

The lowland stretches of rivers are usually the ones where a macroinvertebrate species collection is more difficult to obtain because of the depth and turbidity of the water. Since the choice of the index and of the specific sampling technique is critical for the final result, there is a strong interest into setting up reliable methods for assessing the biological water quality. The aim of this paper is to elucidate the performances of four biotic indices and the influence of four sampling methods on the water quality assessment of a deep and turbid stretch of a river. We collected macroinvertebrates seasonally at 10 sampling sites along the final stretch of the Tiber River using four sampling methods (including hand net, wood and plastic artificial substrates). At the same time, chemical, physical, and bacteriological variables were recorded monthly along the same river stretch. Considering those variables, sampling sites were grouped into three clusters with different water quality, using the chronological clustering method. The same result was obtained from macroinvertebrate data using the same method. We then examined the ability of four biotic indices widely used in Italy to discriminate between the three clusters. The indices are: BMWP' (British Monitoring Working Party, modified for Spain), ASPT' (Average Score Per Taxon, modified for Spain), EBI (Extended Biotic Index) and IBE (Indice Biotico Esteso). BMWP' and ASPT' showed the best results. BMWP' and ASPT' scores – but not those of EBI and IBE – were sensitive to the different sampling methods utilised. We concluded that in this type of river stretch, BMWP' and ASPT' could be more useful than EBI and IBE although the relative efficiency of the chosen sampling method should be taken into account.

Introduction

Macroinvertebrates is the group of organisms most frequently used in biomonitoring studies on rivers (see Rosenberg & Resh, 1993). One of the possible and most popular approaches is the use of taxonomic lists to calculate a 'biotic index' which summarises in a single number the information provided by the community structure. According to De Pauw et al. (1992), more than 30 biotic indices have been proposed in Europe. They differ mainly in the taxonomic level required for determination and in the way in which the final score is calculated. Because the different indices are related to the conditions of the area for which a particular biotic index has been set up, the assumptions on which they are based should be considered critically (Norris & Georges, 1993). However, the importance of choosing the right biotic index for assessing the water quality in a given river has often been underrated.

In the Mediterranean area the landscape is mainly mountainous, and plains are scarce and are near the coast where many large cities are also placed. Consequently, many final zones of rivers, or potamal(sensu Illies & Botosaneanu, 1963), are associated with large human population concentrations. City sewage outlets, along with intensive agriculture runoff and factory outlets, lower the river water quality making effective methods of water quality biomonitoring a major practical issue. Unfortunately, in the potamal zone of rivers, popular sampling methods such as the Surber net or the Hesse's sampler are impossible to use because of the depth and slowness of the water. Furthermore, the dredge does not allow effective sampling on river vegetation and on river bottoms with coarse sand or even bigger granulometry. This results in an underestimation of the taxa richness and lowers the score of the biotic index (Battegazzore et al., 1995; Modde & Drewes, 1990). In such potamal zones several handbooks on the application of biotic index (Ghetti, 1986, 1997; Hellawell, 1986) therefore suggest using artificial substrata. However, no standard sampling methods have been proposed, and the choice of devices has often been left to the individual researcher. Nevertheless if and how the sampling method influences the final score of a biotic index has not been investigated.

Finally, in biomonitoring studies, not enough attention has been paid to examining methods for statistical analysis of the data stemming from the application of the indices. In particular, the lack of within site replicates appears to be the weak point of previous studies (Norris, 1995). Valdecasas & Baltanas (1990) suggested the use of Jacknife and Bootstrap procedures to statistically estimate the differences between single biotic index values, but complex iterative re-sampling procedures have rarely been applied in biomonitoring studies (Norris & Georges, 1993). A simpler, potentially useful approach is to form clusters of samples, so that samples included in each group can be regarded as replicates and compared by some statistical tests (but see Norris & Georges, 1993, for warnings about this approach).

Therefore, the aims of this work are (1) to compare the performance of different biotic indices in assessing the water quality in the last stretch of a large Mediterranean river and (2) to evaluate the sensitivity of these biotic indices to the use of different sampling methods, including artificial substrates. To achieve the first objective, sampling sites were grouped applying the chronological clustering technique (Legendre et al., 1985) to two different data sets which include physical, chemical and biological variables (see Methods). The chronological clustering is a constrained hierarchical clustering technique separating, either in a temporal or spatial sequence, groups of contiguous samples in which the homogeneity is maximized. A randomization test gives a measure of the significance of the group identification. This statistical technique has been applied to date to study the succession of a zooplankton community (Legendre et al., 1985) but, to our knowledge, not to biomonitoring data.

Study area

The study area was the last 50 km of the Tiber River. In this stretch the Tiber flows through the city of Rome (about 3 million inhabitants) and is altered by extensive regulation works, which include gabions and brick walls on the banks. Consequently, the overall physical features of the river such as river width, current velocity, type of sediments and bank slope are similar everywhere. Riparian vegetation is dominated by trees (mainly *Populus*) and the aquatic vegetation, which is not very abundant, by *Phragmites*, *Typha* and *Myriophyllum*. A total of 10 sampling sites, indicated as S1–S10 (Figure 1), were established. S1 (at 18 m above sea level) is upstream from Rome and S10 is located at the summer limit of seawater mixing (Pagnotta et al., 1989). In the studied stretch the Tiber receives, from upstream to downstream, small streams which drain agricultural lands and villages, the North Rome Sewage Plant runoff, the East Rome Sewage Plant runoffs (treating the effluents of about 1 million people), the heavily polluted tributary Aniene River and the South Rome Sewage Plant runoff (treating the effluents of about 1.5 million people). Although the city runoffs are adequately processed, the water quality of the effluents is poor. Furthermore, the sewage of about half a million people is not linked to the sewage water system and flows directly into the Tiber, especially into the last stretch of real urban tract. In our study area, the physical, chemical, and bacteriological variables of Tiber water were recorded by the city water authority (ACEA, Table 1) at 13 sites, indicated as T1–T13 (Figure 1). In the following text, we indicate these data as 'abiotic', even if they also include information on a part of the river biota (bacteria abundance).

Methods

Macroinvertebrate samplings

Macroinvertebrates were collected seasonally at each site from May 1993 to March 1994, using four sampling devices. (1) A hand net (mouth, 50×40 cm; net, 21 mesh per cm) was used on all the different microhabitats we could reach from a boat (bottom, detritus, submerged macrophytes etc.) for an overall time of 3 min at each site. (2) An artificial substrate made of a 20×20×50-cm faggot of brush wood (2 kg fresh weight) wrapped in a plastic net with hexagonal mesh of 2 cm. The faggot was kept floating by a buoy. (3) An identical faggot was kept on the river bottom by a weight. These faggots imitated a traditional fishing device used for eels in the Tiber, which

Table 1. Mean, minimum and maximum values of 19 monthly samples of chemical, physical and bacteriological variables along the studied stretch. Data were collected and processed by ACEA as descripted in IRSA (1976) Abbrevations are as following: Temp, water temperature; F. coli, faecal coliforms; T. coli, total coliforms; F. strep., faecal streptococcus; MBAS, anionic surface-actives

Site		Temp (°C)	pH (u pH)	DO (mg l^{-1})	BOD5 (n mg l^{-1})	COD (n mg l^{-1})	F. coli (n mg l^{-1})	T. coli (n mg l^{-1})	F. strep. (n mg l^{-1})	N-NH_4 (mg l^{-1})	N-NO_3 (mg l^{-1})	N-NO_2 (mg l^{-1})	Org-N (mg l^{-1})	Tot-N (mg l^{-1})	P-PO_4 (mg l^{-1})	Tot-P (mg l^{-1})	MBAS (mg l^{-1})
T1	*Mean*	15.6	7.4	8.9	2.7	13.7	1759	5445	274	0.41	2.39	0.08	0.69	1.11	0.03	0.15	<50
	Minimum	8.7	7.1	6.8	1.0	4.4	33	130	0	0.04	1.19	0.04	0.00	0.00	0.01	0.04	<50
	Maximum	23.6	7.7	10.4	6.0	28.8	3800	22 000	1400	1.62	4.05	0.21	2.79	2.88	0.05	0.98	<50
T2	*Mean*	16.2	7.6	8.7	2.2	13.8	8066	26 826	645	0.41	2.46	0.08	0.57	0.99	0.04	0.09	52
	Minimum	7.3	7.3	7.3	1.0	4.8	160	3300	79	0.08	0.68	0.04	0.03	0.54	0.01	0.03	<50
	Maximum	23.8	8.0	10.5	5.4	32.0	79 000	130 000	3600	0.90	4.32	0.17	1.04	1.71	0.08	0.14	79
T3	*Mean*	16.4	7.6	8.5	2.3	12.9	8868	28 574	855	0.49	2.71	0.09	0.47	1.01	0.04	0.11	<50
	Minimum	7.7	7.2	6.9	1.2	4.8	800	3600	13	0.00	1.72	0.04	0.00	0.37	0.01	0.03	<50
	Maximum	24.1	7.9	10.3	5.3	27.3	79 000	172 000	3300	1.00	4.32	0.19	0.90	1.80	0.08	0.19	<50
T4	*Mean*	16.5	7.6	8.6	2.3	11.9	12 200	38 568	1325	0.57	2.62	0.09	0.62	1.22	0.05	0.11	<50
	Minimum	8.0	7.2	7.1	1.4	5.3	1200	3600	120	0.00	1.71	0.04	0.00	0.00	0.01	0.03	<50
	Maximum	25.2	7.9	10.7	3.9	25.5	79 000	120 000	4900	1.65	4.30	0.20	1.76	2.98	0.09	0.22	<50
T5	*Mean*	16.4	7.5	7.4	3.2	15.4	84 789	273 947	7182	0.78	2.30	0.09	0.74	1.52	0.06	0.14	69
	Minimum	8.1	7.0	5.5	1.6	6.3	200	9000	360	0.19	0.31	0.04	0.03	0.64	0.02	0.05	<50
	Maximum	23.4	7.8	9.3	4.7	29.3	700 000	800 000	33 000	2.15	4.00	0.19	1.68	2.83	0.11	0.21	207
T6	*Mean*	17.1	7.5	6.9	3.2	17.1	49 305	200 453	8250	0.82	2.37	0.10	1.14	2.02	0.06	0.15	63
	Minimum	8.0	7.2	5.0	1.6	2.9	3600	3600	200	0.00	1.59	0.04	0.00	1.04	0.03	0.05	<50
	Maximum	24.5	7.8	9.4	5.4	28.0	230 000	490 000	34 000	1.78	4.21	0.20	5.64	6.41	0.08	0.20	208
T7	*Mean*	16.3	7.6	6.8	4.6	14.0	61 889	254 421	7414	0.85	2.45	0.12	0.75	1.56	0.07	0.16	54
	Minimum	8.1	7.2	4.1	1.5	4.9	3300	23 000	200	0.17	1.73	0.04	0.00	0.00	0.04	0.10	<50
	Maximum	23.5	7.9	8.5	8.8	31.7	330 000	600 000	33 000	2.29	3.50	0.27	1.79	2.72	0.11	0.27	89
T8	*Mean*	16.4	7.5	6.7	4.7	16.7	46 105	314 947	16 145	0.88	2.46	0.12	0.86	1.73	0.07	0.18	55
	minimum	8.0	7.1	4.1	2.0	6.8	9000	9000	300	0.21	1.72	0.04	0.03	0.80	0.04	0.15	<50
	Maximum	23.8	7.8	8.6	9.6	46.3	130 000	1 100 000	172 000	1.76	3.50	0.30	2.04	3.00	0.12	0.25	127
T9	*Mean*	16.5	7.5	6.4	4.3	14.3	67 895	363 737	10 736	0.92	2.44	0.12	0.90	1.83	0.08	0.18	71
	Minimum	8.2	7.2	3.9	1.8	5.9	9000	23 000	360	0.27	1.71	0.04	0.06	1.07	0.04	0.11	<50
	Maximum	23.8	7.8	8.4	13.2	25.5	230 000	1 100 000	33 000	1.55	3.50	0.29	1.63	2.84	0.10	0.54	220
T10	*Mean*	16.5	7.5	6.4	5.9	18.0	98 800	485 684	14 312	1.08	2.44	0.13	0.95	2.03	0.10	0.22	85
	Minimum	8.5	7.2	4.1	2.3	7.8	9000	33 000	360	0.16	1.71	0.04	0.14	1.07	0.06	0.13	<50
	Maximum	23.5	7.8	8.1	9.3	36.6	240 000	1 100 000	49 000	1.87	3.50	0.36	1.88	3.13	0.16	0.46	420
T11	*Mean*	16.7	7.5	6.2	4.8	16.8	80 437	483 211	19 470	1.23	2.41	0.13	0.86	2.08	0.10	0.21	76
	Minimum	8.6	7.2	4.1	1.4	7.0	15 000	79 000	93	0.17	1.69	0.04	0.12	1.09	0.06	0.13	<50
	Maximum	23.8	7.7	8.0	15.6	30.2	240 000	1 400 000	49 000	3.19	3.50	0.35	1.68	3.36	0.14	0.37	270
T12	*Mean*	16.8	7.6	6.0	4.1	13.3	134 647	728 947	10 106	1.13	2.39	0.14	0.83	1.95	0.10	0.21	91
	Minimum	8.6	7.3	3.3	1.7	0.0	7300	79 000	320	0.15	1.68	0.05	0.18	0.80	0.05	0.12	<50
	Maximum	24.0	7.8	8.5	6.5	25.4	1 100 000	4 600 000	33 000	2.18	3.50	0.34	1.91	3.40	0.14	0.58	536
T13	*Mean*	17.0	7.5	5.6	3.9	8.6	109 000	700 053	9492	1.26	2.39	0.14	0.84	1.98	0.11	0.22	73
	Minimum	9.2	7.3	3.5	1.4	0.0	8000	43 000	360	0.23	1.69	0.05	0.00	0.00	0.06	0.11	<50
	Maximum	24.3	7.7	8.0	7.6	27.0	330 000	2 400 000	32 000	2.18	3.68	0.31	1.91	3.58	0.15	0.72	122

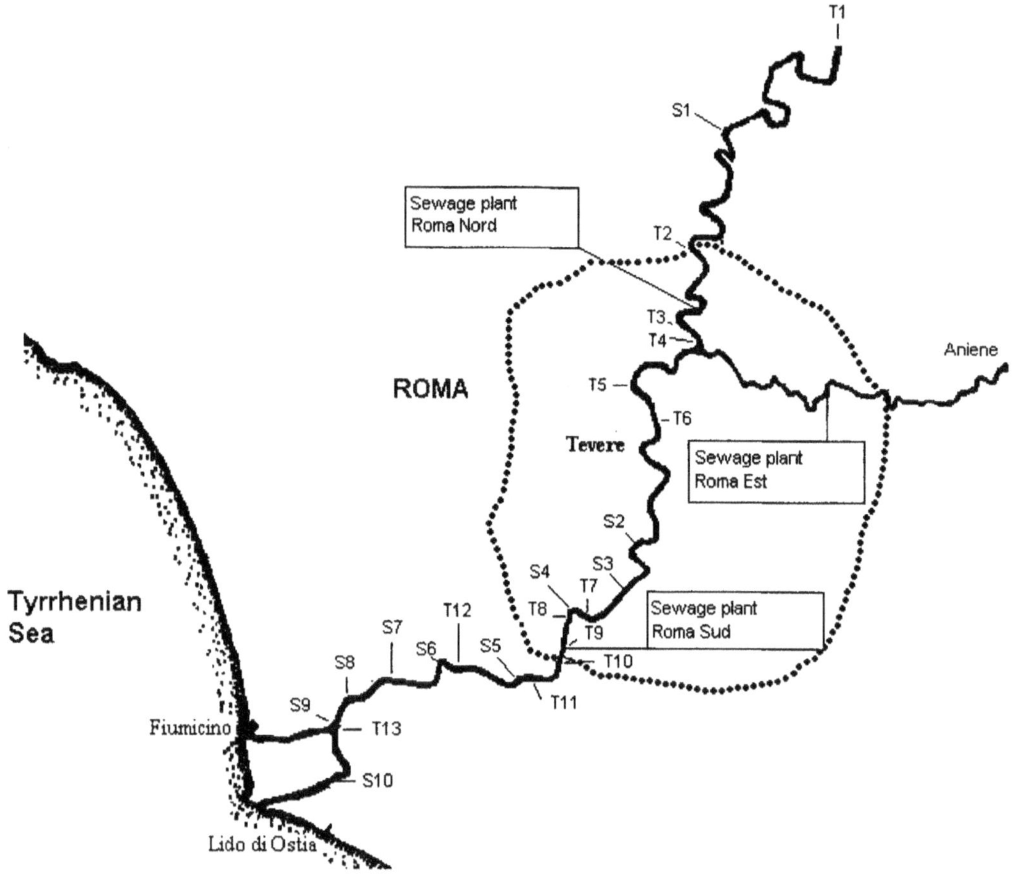

Figure 1. Map of the last tract of the river Tiber showing the sampling sites for chemical, physical and bacteriological variables (T1–T 3) and for macroinvertebrates (S1–S10). Dotted line bounds the urban area of Rome.

we observed to be very rich in macroinvertebrates. (4) A bunch of plastic branches similar to natural branches (produced to replace the latter in broom-making). The bunch was roughly conical, 60 cm long and 30 cm large, with a total surface of 67.5 dm^2. This device is similar to *Myriophyllum*, almost the only submerged macrophyte in the studied stretch of the Tiber. It was kept just under the water surface by a buoy. Following published observations about the time required to stabilize a macroinvertebrate community on artificial substrates (De Pauw et al., 1986; Meier et al., 1979), all artificial substrates were left in the water for more than two months before the sampling dates (including the first date). The artificial substrates were seasonally dipped out with a hand net (the same used for direct sampling), and carefully searched for macroinvertebrates, which were preserved in 75% ethyl alcohol, sorted, and determined in the laboratory using literature keys. The original sampling design was to collect a total of 160 samples (4 dates × 10 sites × 4 methods). Unfortunately we lost all the artificial substrata at S1 site and at some others. This resulted in a total of 128 samples collected during the year of study, 39 with the hand net and 89 with artificial substrates (31 with the wood floating substrate, 33 with the wood bottom substrate and 25 with the plastic substrate).

Biotic indices

The biotic indices compared in this study were those most frequently used for Italian rivers. These include: (1) the Extended Biotic Index (EBI) (Woodiwiss, 1978) as modified by Ghetti (1986); (2) the Indice Biotico Esteso (IBE) (Ghetti, 1997), which is a further adaptation for Italian rivers of the EBI; (3) the BMWP'; and (4) the ASPT', which are the British Monitoring Working Party (Hellawell, 1978) and the related Average Score Per Taxon (Armitage et al., 1983) adapted for Spain by Alba-Tercedor & Sanchez-Ortega (1988). The EBI and IBE have similar tables for establishing the quality score, which is dependent on the presence of particular taxa and on the number of systematic units. The latter are assessed by determining of the macroinvertebrate taxa at various systematic levels. The EBI requires a more detailed determination level (genus for most taxa) than the IBE (family for most taxa). According to Ghetti (1986; 1997), the scores of both indices can be grouped into the following 'quality classes': score >9 = class 1 (the best), scores 8 or 9 = class 2, scores 6 or 7 = class 3, scores 4 or 5 = class 4, scores <4 = class 5 (the worst). The BMWP and ASPT, require the same identification level (family or higher), and a quality score is awarded to each taxon. The BMWP quality score is the sum, while the ASPT quality score is the average of the taxon's scores. The modified BMWP' and ASPT' differ from the original ones only by the addition of some taxa which are absent in Great Britain but present in both Spain and Italy. According to Alba-Tercedor & Sanchez-Ortega (1988) and Rico et al. (1992), BMWP' scores can also be grouped into 'quality classes' as follows: score >101 = class 1 (the best), scores between 61 and 101 = class 2, scores between 36 and 60 = class 3, scores between 35 and 15 = class 4, scores <15 = class 5 (the worst).

Data analysis

In a first step we assessed the water quality along the studied tract by both the abiotic variables (Table 1) and the biotic indices.

Then, in order to identify homogeneous groups of sites, we applied chronological clustering (Legendre et al., 1985) to two distance matrices obtained from the abiotic data and the macroinvertebrate community composition. The first matrix was calculated using the Euclidean distance after centering and scaling each variable. The second matrix was calculated from a matrix of presence–absence data using the Sokal & Michener (1958) algorithm. Computations were made using the *R* Package (Legendre & Vaudor, 1991). To obtain a graphical description of inter-site similarities, we used the multi dimensional scaling (MDS) technique.

In order to compare biotic index performances in discriminating between different site groups, the biotic index scores were processed with a non-parametric

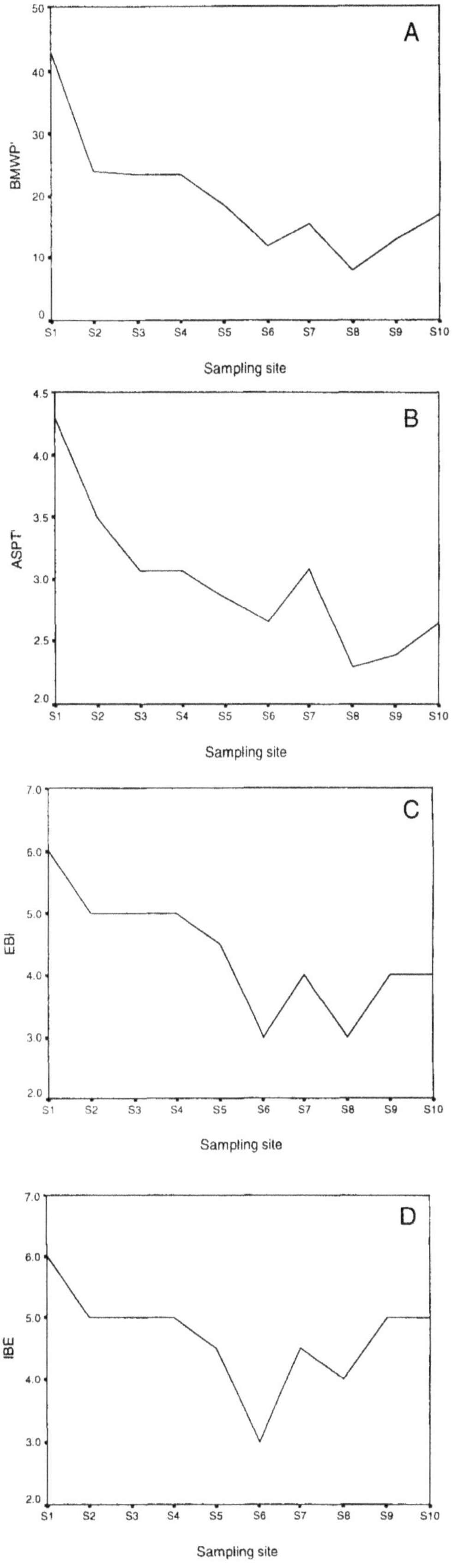

Figure 2. Trends of the biotic indices along the studied tract of the river Tiber. The values of each site are the median of four seasonal scores. (A) BMWP' (British Monitoring Working Party, modified for Spain); (B) ASPT' (Average Score Per Taxon, modified for Spain); (C) EBI (Extended Biotic Index); (D) IBE (Indice Biotico Esteso).

rank F-test (Neter et al., 1996), considering time (sampling date) and space (site group) as variation sources. Subsequent multiple comparisons were carried out by the Bonferroni procedure (Neter et al., 1996).

In order to check the sensitivity of the biotic index scores to different sampling methods, we also used a non-parametric rank F-test, considering time (sampling date) and type of device (sampling method) as variation sources. Again the Bonferroni procedure was used for multiple comparisons.

The use of a non-parametric test to analyse biotic index data can be justified because these biotic indices do not define a unit of measure. Therefore, the scores should be considered as ordinal rather than cardinal numbers, and rank-based statistical methods should be preferred. Time was considered in the tests in order to check for the presence of an interaction between time and groups or devices.

Results

Assessment of water quality

Table 1 reports the average values of the abiotic data recorded monthly from 1993 to 1994 at the 13 sampling sites (T1–T13) by the city water authority (ACEA). For many variables the values are typical of water affected by organic pollution and a previous study showed significant differences in all variable values except $N\text{-}NO_3$ (Solimini et al., 1997) between sites T1, T9 and T13 (each one is included in each one of the site groups pointed out below).

Figure 2 shows the median values of the four biotic indices at all macrobenthic sampling sites (S1–S10). Biotic index scores were calculated from the hand net data only, because at S1 the artificial substrates were missing, possibly due to vandalism by local fishermen. S1 had the highest scores for all the indices, but is included only in the third quality class of BMWP', EBI and IBE. Downstream the water quality worsens, reaching the lowest class at S6 and at some sites further downstream than S6.

Sampling site clustering

The chronological clustering from abiotic data, a matrix of 13 sites × 16 variables (means of 19 monthly values, Table 1), yielded three site groups: G1 including T1–T4; G2 including T5–T9; and G3 including T10–T13 sites (Figure 3). The randomization test significantly separated these three groups ($p = 0.008$ for all comparisons). The study area map (Figure 1) shows that between T4 and T5 the Tiber receives the heavily polluted Aniene River as well as with the outlet of the East Rome Sewage Plant. The outlet of the biggest sewage plant (South Rome Sewage Plant) enters between T9 and T10. The three groups therefore appear to be separated by the bigger pollution inflows.

The same analysis was applied to the matrix of presence–absence data (a matrix of 10 macrobenthic sampling sites × 41 taxa from four pooled seasonal hand net samples, Table 2). The chronological clustering again revealed three groups of sites, separated by the same pollution inputs (Figure 4): G1 contains only S1, G2 contains S2–S4, and G3 contains S5–S10 sampling sites. The randomization test gave a probability $P = 0.25$ between G1and G2 and a probability $P = 0.005$ between G2 and G3. The first result was because of the low sensitivity of the chronological clustering when one of the clusters consists of one element only (Legendre et al., 1985).

The MDS ordination gave a good representation (having low stress values) of the same three groups from both abiotic and macrobenthic data.

Comparison between biotic index performances

Table 3 reports the results of applying the non-parametric rank F-test to the values of the four biotic indices (data from hand net only). This yielded the same grouping as in the previous analysis: group G1 including the S1 site only, group G2 including the S2–S4 sites, and group G3 including the remaining six sites. The results showed that time is not an influencing factor on the biotic index values, either alone or when interacting with space, but space (i.e. groups) appears to influence the values of all indices. Nevertheless, looking at multiple comparisons (Table 3), the BMWP' and ASPT' were able to discriminate among all three site groups, while the EBI and IBE failed in comparing the adjacent groups and were only able to detect the difference between G1 and G3 groups. The EBI had a better performance, almost reaching significance in the G2 versus G3 match.

Sensitivity of the biotic indices to different sampling methods

Table 4 reports the results of applying the non-parametric rank F-test to biotic index values assessed by four different sampling methods. The S1 site was

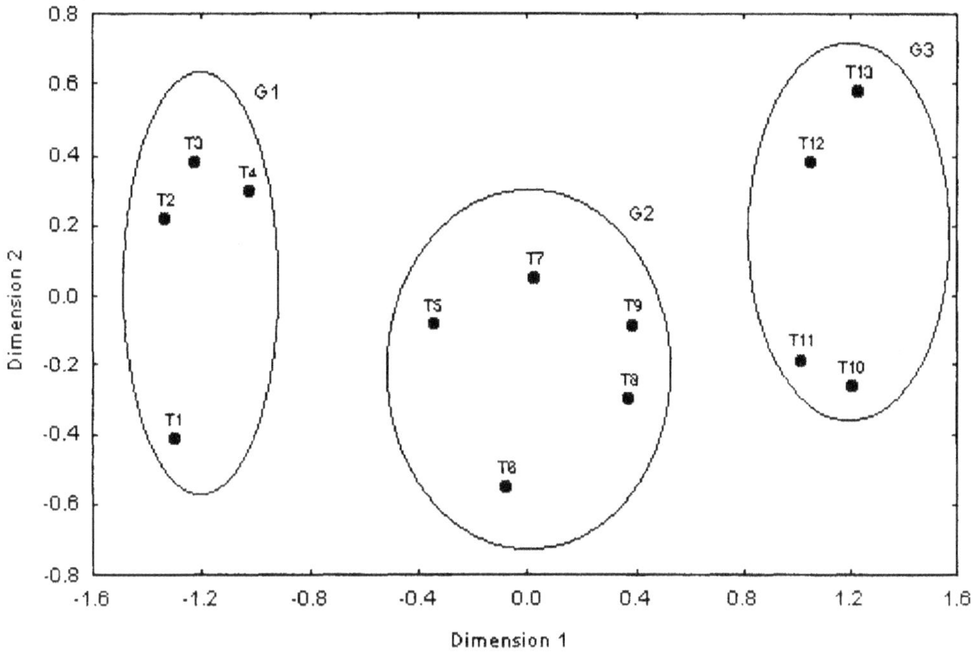

Figure 3. MDS ordination of abiotic variables sampling sites (T1–T13). Stress value after 900 iterations was 0.02. Ellipses enclose the groups of sites (G1, G2, G3) resulting by the chronological clustering analysis (see text).

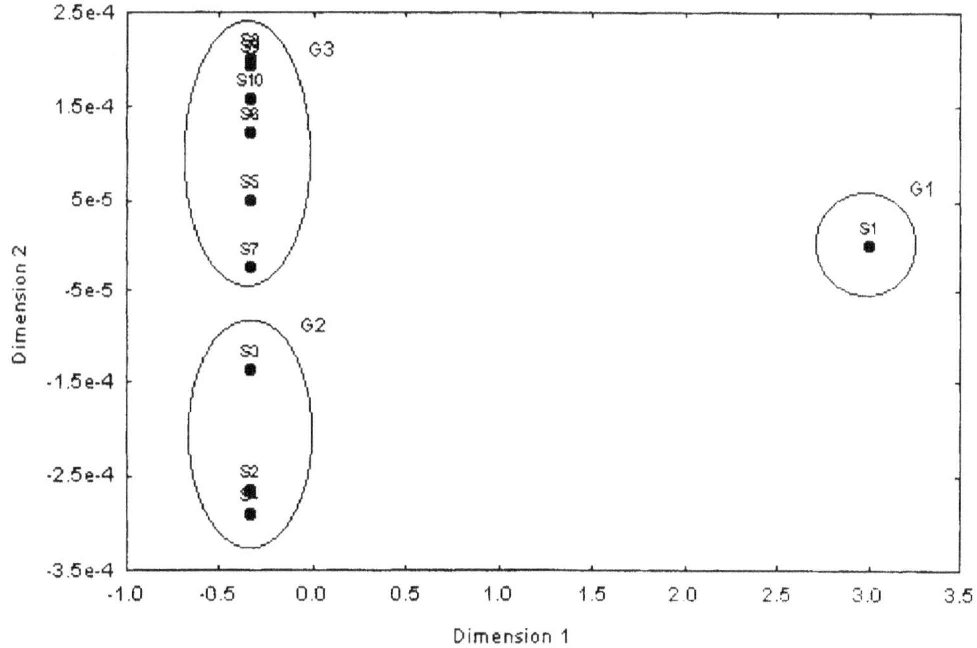

Figure 4. MDS ordination of macroinvertebrate sampling sites (S1–S10). Stress value after 900 iterations was 0.00004. Ellipses enclose the groups of sites (G1, G2, G3) resulting by the chronological clustering analysis (see text).

Table 2. Presence of macroinvertebrate taxa (at the best possible determination level) at 10 sites along the Tiber River. Data from hand net samplings, four seasonal samples pooled

Taxon	Sites									
	S1	S2	S3	S4	S5	S6	S7	S8	S9	S10
Dugesia sp	1	1	0	0	0	0	1	0	0	1
Tubifex sp	1	1	1	1	1	1	1	1	1	1
Limnodrilus sp	1	1	1	1	1	1	1	1	1	1
Psammoryctides sp	0	0	1	0	1	0	0	0	0	1
Naididae	0	1	0	1	1	0	1	0	1	1
Enchytraeidae	0	0	0	1	0	0	0	0	0	0
Lumbricidae	1	0	0	0	0	0	0	0	0	0
Haplotaxidae	1	0	0	0	0	0	0	0	0	0
Dina lineata	1	0	0	0	0	0	0	0	0	0
Erpobdella testacea	0	0	1	0	1	1	0	1	0	0
Helobdella stagnalis	0	1	0	0	0	0	0	0	0	0
Piscicola geometra	0	0	1	0	0	0	0	0	1	0
Bithynia tentaculata	1	1	1	1	1	1	1	1	1	1
Physa acuta	1	1	1	1	1	1	1	1	1	1
Potamopyrgus antipodarum	1	0	0	0	0	0	0	0	0	0
Theodoxus fluviatilis	1	1	0	0	1	0	0	0	0	0
Pisidium sp	1	0	0	0	0	1	0	0	1	0
Sphaerium sp	1	0	0	0	0	0	0	0	0	0
Asellus aquaticus	1	0	0	1	0	0	0	0	0	1
Proasellus coxalis	1	1	1	1	1	1	1	1	1	1
Palaemonetes antennarius	1	0	0	0	0	0	1	1	1	1
Atyaephyra desmaresti	0	1	0	1	0	0	0	0	0	0
Echinogammarus veneris	1	1	1	1	1	0	1	0	1	1
Caenis sp	1	0	0	0	0	0	0	0	0	0
Cercion lindeni	1	1	1	1	0	1	1	0	0	0
Erythromma viridulum	1	0	1	1	1	1	1	0	0	0
Ischnura elegans	1	1	1	1	0	0	0	0	0	0
Platycnemis pennipes	1	0	0	0	0	0	0	0	0	0
Coenagrium puella	1	0	0	0	0	0	0	0	0	0
Anax sp	1	0	0	0	0	0	0	0	0	0
Corixa sp	1	0	0	0	0	1	0	0	0	1
Micronecta sp	1	1	0	0	0	0	0	1	1	1
Aphelocheirus aestivalis	0	0	0	0	0	0	0	0	0	1
Sisyridae	0	0	1	0	0	0	0	0	0	0
Haliplidae	1	0	0	0	0	0	0	0	0	0
Polycentropodidae	1	0	0	0	0	0	0	0	0	0
Ceratopogonidae	0	1	0	1	0	0	0	0	0	0
Chironomidae	1	1	1	1	1	1	1	1	1	1
Culicidae	0	0	0	0	1	0	0	0	0	0
Psycodidae	1	0	1	1	0	0	0	0	0	0
Tipulidae	0	1	0	0	0	0	0	0	0	0

Table 3. Summary of non parametric rank *F*-test table of studied biotic indices. Sources of variation are the date and the groups (G1, G2 and G3) of sampling sites. Multiple comparisons were carried out by the Bonferroni procedure. Scores of all indices differ between group 1 (better water quality) and group 3 (worst water quality). Only BMWP' and ASPT' scores differ also between group 2 (intermediate water quality) and group 3 (see text)

Index	Source	df	rank *F*	*P*	Probability of multiple comparisons		
					G1–G2	G1–G3	G2–G3
IBE	Model	10	2.39	0.034			
	Date	3	0.70	0.559			
	Group	2	6.71	0.004	0.131	0.007	0.204
	Date × Group	5	1.48	0.227			
	Error	28					
EBI	Model	10	2.30	0.04			
	Date	3	0.27	0.85			
	Group	2	8.86	0.001	0.102	0.002	0.057
	Date × Group	5	0.843	0.53			
	Error	28					
BMWP'	Model	10	3.73	0.003			
	Date	3	0.17	0.913			
	Group	2	15.8	0.000	0.029	0.000	0.004
	Date × Group	5	0.73	0.608			
	Error	28					
ASPT'	Model	10	3.41	0.005			
	Date	3	0.59	0.621			
	Group	2	14.3	0.000	0.028	0.000	0.000
	Date × Group	5	0.92	0.480			
	Error	28					

not considered in this analysis because artificial substrates were lacking. Time was not an influencing factor in this analysis either, except for the BMWP' values, and no interaction existed between time and sampling methods, although sampling methods appeared to influence the values of BMWP' and ASPT'. The multiple comparisons reveal that this result was because the hand net differed from most other methods (Table 4). The table of mean ranks by sampling methods (Table 5), shows that the artificial substrates yielded higher scores than the hand net. In contrast, the EBI and IBE values did not appear to be significantly influenced by sampling methods (Table 4), which gave almost the same mean scores (Table 5).

Discussion

The aim of dividing a stretch of river into clusters of similar sites appears to have been reached using chronological clustering. We obtained the same three groups using both the biotic and the abiotic data, and they were in very good accordance with what is known about the location of urban pollution sources of the Tiber here. Other sources of disturbance to the river environment, which could cause discordance between biotic and abiotic assessment of the water quality (for an example of discord between these types of data, see Fontoura & De Pauw, 1991), certainly exist due to regulation works, but seem to have the same effects all along the studied stretch. The scores of all biotic indices and the values of physical, chemical, and bacteriological variables pointed out the progressive

Table 4. Summary of non-parametric rank *F*-test table of studied biotic indices. Sources of variation are the date and the different sampling methods (a, hand net; b, wood floating substrate; c, wood bottom substrate; d, plastic substrate). Multiple comparisons were carried out by the Bonferroni procedure. The scores of BMWP' and ASPT' are sensitive to the method used to collect macroinvertebrates (see text)

Index	Source	df	*F*	*P*	Probability of multiple comparisons					
					a–b	a–c	a–d	b–c	b–d	c–d
IBE	Model	15	1.26	0.238						
	Date	3	0.78	0.506						
	Method	3	0.45	0.719						
	Date×Method	9	1.76	0.085						
	Error	109								
EBI	Model	15	1.264	0.238						
	Date	3	0.783	0.566						
	Method	3	0.448	0.719						
	Date×Method	9	1.757	0.085						
	Error	109								
BMWP'	Model	15	2.23	0.009						
	Date	3	4.26	0.007						
	Method	3	4.13	0.008	0.020	0.019	0.568	1.000	1.000	1.000
	Date×Method	9	1.00	0.442						
	Error	109								
ASPT'	Model	15	3.70	0.000						
	Date	3	1.67	0.177						
	Method	3	11.75	0.000	0.000	0.160	0.001	0.006	1.000	0.464
	Date×Method	9	1.67	0.104						
	Error	109								

deterioration of water quality from upstream to downstream of the city of Rome and, therefore, between subsequent clusters of sites.

The comparison between biotic indices showed that those two (BMWP' and ASPT') based on a less precise determination level performed better in discriminating between site groups with different water quality. This is quite surprising, although it is in accordance with the results of Rico et al. (1992) from a different environment: 'short, shallow and turbulent' rivers in the Basque Country. These authors, comparing several biotic indices with the Chandler Biotic Score (CBS), which they assumed to be the best, found BMWP' to be better than both the EBI (Woodiwiss, 1978) and the Trent Biotic Index or TBI (Woodiwiss, 1964, the ancestor of EBI and IBE), despite the fact that the latter requires determination to the species level. In our study IBE appeared to be slightly worse than EBI, in accordance with a less precise determination level.

Time (season) did not appear to play an important role in determining the water quality level in our study for all indices except BMWP'. In general, reduced fluctuation in time is a desirable feature of a biotic index, which should result from the length of the life cycles of the considered taxa. However, Murphy (1978) reported seasonal variations of TBI, while, more recently, Zamora-Munoz et al. (1995) reported that BMWP' (but not ASPT') is not significantly correlated with season (assessed as temperature variations) in unpolluted sites: both indices show significant correlation with seasons in polluted sites. The absence of a time effect in our study is probably because the pollution inputs in the studied stretch of the Tiber come from a city and thus are hardly affected by seasons. Furthermore, the obvious seasonal variations of the river flow, which may change the dilution rate

Table 5. Mean ranks of the biotic index scores for the different sampling methods (a, hand net; b, wood floating substrate; c, wood bottom substrate; d, plastic substrate)

Sampling method	Biotic Index			
	BMWP'	ASPT'	EBI	IBE
a	47	41	67	58
b	72	85	61	65
c	72	58	60	65
d	63	74	66	66

of the pollutants, are altered by several dams; their discharges are determined by electricity power demands thus are not linked to seasons. Further research is necessary in order to elucidate this point, which was not among the goals of our study. Regarding the effect of the sampling method on biotic index scores, the BMWP' and ASPT' proved to be sensitive to the use of different sampling methods. The differences were mainly between the use of the hand net and the use of artificial samplers, the former giving the worst scores. This could be due to the features of the last stretch of the Tiber, where the depth is usually 10 m, turbidity is high and vegetation is very scarce. All these conditions render the use of the hand net unproductive. In contrast, there were fewer differences among the scores stemming from the different artificial substrates. De Pauw et al. (1986) also found that the nature of the substrates was not a crucial factor for macroinvertebrate colonization, although they used a different type of artificial substratum. In the Tiber, we observed that the substrates were quickly covered by a thin layer of detritus, so that the type of the substrate material, at least for some months, was probably not important for macroinvertebrate colonization.

Conclusion

Chronological clustering, although to our knowledge not yet applied in river biomonitoring studies, proved to be effective in grouping sampling sites with homogeneous macrobenthic fauna. Several multidimensional ordination methods were applied for this purpose in other studies (Norris & Georges, 1993); their use together with the chronological clustering is quite interesting because it adds probability levels to the groups' indentification which would be lacking if ordination was used alone. The better performances of BMWP' and ASPT' suggest the need to more frequently use these two indices in biomonitoring studies in Italy, until now mainly based on EBI and IBE. This can be done without additional macroinvertebrate sampling and determination efforts. However, BMWP' and ASPT' indices should be adapted to the Italian river fauna because the score list does not presently include certain macroinvertebrate families. Moreover, in the potamal zone of the rivers, the choice of a particular sampling method appears to be crucial for biomonitoring studies, and artificial substrates should be more widely used.

Acknowledgements

Dr L. Romano and Dr A. Borgioli (ACEA) provided chemical, physical and bacteriological data. We thank Mr A. Bergamini and C. Bergamini, professional fishermen of the Tiber, for their invaluable help during field work. The manuscript was greatly improved by two anonymous referees' comments. This study was partially supported by a MURST grant to G. Carchini (Program: Popolamento animale del Mediterraneo occidentale).

References

Alba-Tercedor, J. & A. Sanchez-Ortega, 1988. Un metodo rapido y simple para evaluar la calidad biologica de las aguas corrientes basado en el de Hellawell (1978). Limnetica 4: 51–56.

Armitage, P. D., D. Moss, J. F. Wright & M. T. Furse, 1983. The performance of a new biological water quality score system based on macroinvertebrates over a wide range of unpolluted running-water sites. Wat. Res. 17: 333–347.

Battegazzore, M., A. Guzzini, R. Pagnotta & R. Marchetti, 1995. The importance of investigatory and analytical techniques in biological water-quality investigations. In Harper, D. M. & A. J. D. Ferguson (eds), The Ecological Basis for River Management. John Wiley & Sons, New York, USA: 193–209.

De Pauw, N., D. Roels & A. P. Fontoura, 1986. Use of artificial substrates for standardized sampling of macroinvertebrates in the assessment of water quality by the Belgian biotic Index. Hydrobiologia 133: 237–258.

De Pauw, N., P. F. Ghetti, P. Manzini & R. Spaggiari, 1992. Biological assessment methods for running water. In Newman, P. J., M. A. Piavaux & R. A. Sweeting (eds), River Water Quality Ecological Assessment and Control. Commission of the European Communities, Bruxelles, Belgium: 217–249.

Fontoura, A. P. & N. De Pauw, 1991. Macroinvertebrate community structure and impact assessment of dams and impounding reservoir in the Cavado river basin (Northern Portugal). Verh. int. Verein. Limnol. 24: 1353–1359.

Ghetti, P. F., 1986. I macroinvertebrati nell'analisi dei corsi d'acqua. Indice Biotico: E.B.I. modificato. Manuale di applicazione. Provincia Autonoma di Trento, Trento, Italy, 111 pp.

Ghetti, P. F., 1997. I macroinvertebrati nel controllo della qualità di ambienti di acque correnti. Indice Biotico Esteso (I.B.E.). Manuale di applicazione. Provincia Autonoma di Trento, Trento, Italy, 222 pp.

Hellawell, J. M., 1978. Biological Surveillance of Rivers. Water Research Center, Stevenage, UK, 322 pp.

Hellawell, J. M., 1986. Biological Indicators of Freshwater Pollution and Environmental Management. Elsevier Applied Science, London, UK, 546 pp.

Illies, J. & L. Botosaneanu, 1963. Problémes et méthodes de la classification et de la zonation écologique des aux courantes, considerée surtout du point de vue faunistique. Mitt. Int. Verein. Theor. Angew. Limnol. 12, 57 pp.

IRSA (Instituto di Ricerca sulle Acque del C.N.R.), 1976. Metodi analitici per le acque. Quad. Ist. Ric. Acque 11, C.N.R., Roma, 130 pp.

Legendre, P. & A. Vaudor, 1991. The R Package: Multidimensional analysis, spatial analysis. Département de Sciences Biologiques, Université de Montréal, 142 pp.

Legendre, P., S. Dallot & L. Legendre, 1985. Succession of species within a community: chronological clustering, with application to marine and freshwater zooplankton. Am. Nat. 125: 257–288.

Meier, P. J., D. L. Penrose & L. Polak, 1979. The rate of colonization by macro-invertebrates on artificial substrate samplers. Freshwat. Biol. 9: 381–392.

Modde, T. & H. G. Drewes, 1990. Comparison of biotic index values for invertebrate collections from natural and artificial substrates. Freshwat. Biol. 23: 171–180.

Murphy, P. N., 1978. The temporal variability in biotic indices. Envir. Pollut. 17: 227–236.

Neter, J., M. H. Kutner, C. J. Nachtsheim & W. Wasserman, 1996. Applied Linear Statistical Models. Irwin, Chicago, 1408 pp.

Norris, R. H., 1995. Biological monitoring: the dilemma of data analysis. J. N. am. Benthol. Soc. 14: 440–450.

Norris, R. H. & A. Georges, 1993. Analysis and interpretation of benthic macroinvertebrate surveys. In Rosenberg, D. M. & V. H. Resh (eds), Freshwater Biomonitoring and Benthic Macroinvertebrates. Chapman & Hall, New York: 234–286.

Pagnotta, R., C. M. Blundo, T. La Noce, M. Pettine & A. Puddu, 1989. Nutrient remobilization processes at the Tiber River mouth (Italy). Hydrobiologia 176/177: 297–306.

Rico, E., A. Rallo, M. A. Sevillano & M. L. Arretxe, 1992. Comparisons of several biological indices based on river macroinvertebrate benthic community for assessment of running water quality. Ann. Limnol. 25: 147–156.

Rosenberg D. M. & Resh V. H., 1993. Introduction to freshwater biomonitoring and benthic macroinvertebrates. In Rosenberg, D. M. & V. H. Resh (eds.), Freshwater Biomonitoring and Benthic Macroinvertebrates. Chapman & Hall, New York: 1–9.

Sokal, R. R. & C. D. Michener, 1958. A statistical method for evaluating systematic relationships. Univ. Kansas Sci. Bull. [illegible]: 2–18.

Solimini, A. G., G. A. Tarallo & G. Carchini, 1997. Life history and species composition of the damselfly assemblage along the urban tract of a river in central Italy. Hydrobiologia 356: 21–32

Valdecasas, A. G. & A. Baltanas, 1990. Jacknife and bootstrap estimation of biological index of water quality. Wat. Res. 24: 1279–1283.

Woodiwiss, F. S., 1964. The biological system of stream classification used by the Trent River board. Chem. Ind. 83: 443–447.

Woodiwiss, F. S., 1978. Comparative study of biological–ecological water quality assessment methods. Second practical demonstration. Summary report. Commission of the European Communities.

Zamora-Munoz, C., C. Sainz-Cantero, A. Sanchez-Ortega & J. Alba-Tercedor, 1995. Are biological indices BMWP' and ASPT' and their significance regarding water quality seasonally dependent? Factors explaining their variation. Wat. Res. 29: 285–290.

Hydrobiologia **422/423:** 209–223, 2000.
M. Jungwirth, S. Muhar & S. Schmutz (eds), Assessing the Ecological Integrity of Running Waters.

Urbanization gradients in streams of Anchorage, Alaska: a comparison of multivariate and multimetric approaches to classification

A. M. Milner[1,2] & M. W. Oswood[2]
[1]*School of Geography and Environmental Science, University of Birmingham, Edbgaston, Birmingham B15 2TT, U.K.*
[2]*Institute of Arctic Biology, University of Alaska, Fairbanks, AK 99775, U.S.A.*

Key words: stream classification, bioassessment, similarity, multimetric, multivariate analyses

Abstract

The classification of stream sites into groups of differing habitat quality is central to the evaluation of water quality. In bioassessment approaches using macroinvertebrates, debate exists as to the merit of multivariate approaches compared to multimetric techniques for site classification. We analyze similarity coefficients between pairs of sites within classification groupings, compared to pairs of sites between groupings, as a test of the strength of different approaches to site classification. Streams within the Municipality of Anchorage in Alaska were used to investigate the effects of marked urbanization gradients on benthic communities without the influence of other disturbances in their watersheds (e.g. agriculture or timber harvest). In 1989, macroinvertebrates were collected at 27 stream sites in 13 different streams and then sites were classified into groups of apparent habitat quality in one of four ways; (1) an *a-priori* classification of the sites based on land use characteristics and known inputs of contaminants, (2) a TWINSPAN classification using abundances of taxa, (3) a multimetric approach using 5 metrics combined into a single index which was then trisected into three groups, and (4) a TWINSPAN classification using the metric data. Analyses of faunal similarities between and within habitat quality groups showed that the TWINSPAN classification using numerical data produced the classification with the highest overall strength, while the multimetric approaches resulted in the lowest strength of classification. However, all classification approaches identified essentially the same suite of sites in the 'poor' group, suggesting considerable robustness among bioassessment approaches in identifying sites most in need of monitoring and remediation.

Introduction

Using bioasssessment approaches to classify habitat quality of stream sites

Within the last decade, there has been a shift in approaches to assessing the health and integrity of aquatic ecosystems, from monitoring of chemical and physical variables to an organism-based approach that addresses the complexity of ecological interactions (Karr, 1999). The advantages of using the biota as indicators are substantial; communities of organisms integrate the effects of episodic or chronic stressors, acting as 'living dataloggers'.

Several major taxonomic groups of freshwater organisms have been used in bioassessment of running waters: diatoms, macrophytes, protozoa, meiofauna, macroinvertebrates and fish (Norris & Thoms, 1999). The most widely used bioassessment systems for running waters analyze the community composition of benthic macroinvertebrates. Benthic communities at sites that are impaired by activities of humans (or their domestic animals) are presumed to show predictable changes in composition in response to changes in water quality, i.e. the common anthropogenic stressors to streams (water temperature changes, toxins, oxygen-depleting organic matter, sediments). Predictable responses of benthic macroinvertebrates to reduced water quality include decreased taxonomic diversity and an increase in the relative abundance of taxa less sensitive to perturbation at the expense of sensitive taxa (Rosenberg & Resh, 1993). However, there is no 'tricorder' that provides unambiguous and easily understood interpretations of the welter of raw data (taxa abundances) provided as the typical first step of bioassessment. Bioassessment approaches have evolved from a simple search for indicator organisms, whose presence or absence may be indicative of impaired water quality, to complex systems potentially capable of providing deep insights into the health of freshwater ecosystems (Norris & Thoms, 1999).

Recently, two major evolutionary clades of stream bioassessment have emerged. One major branch, the multivariate approach, uses the various techniques of multivariate statistics to simultaneously consider all of the information in a taxa × site data matrix (sometimes including information about environmental characteristics of sites). The goal is to find patterns among sites (potentially indicative of differences in habitat quality) in the community composition of benthic macroinvertebrates. In the U.K., the multivariate approach has found expression in RIVPACS, a multivariate model that predicts the taxa expected in a community at a site from a number of key physicochemical variables (Moss et al., 1987). Comparison of the expected with the observed invertebrate community allocates sites into categories of water quality impairment when similarity to the expected community is less than 90%. Another major branch of (macroinvertebrate-based) bioassessment is the multimetric approach (Barbour et al., 1995). From the taxa × site data matrix, metrics are calculated, each metric potentially capturing key information (e.g. taxonomic diversity, exposure of the community to stressors) about the community. Metrics can be combined (hence '*multi*metric') into combinations that are potentially more predictive than individual metrics because metrics can be chosen to extract different aspects of community structure.

Divergence of the multivariate and multimetric methods is due, at least in part, to geographic isolation of their development; the multivariate method is centered in the U.K. but is being investigated for use in other European countries and Australia (Smith et al., 1999), while the multimetric approach has dispersed across the state and federal agencies responsible for water quality in the U.S. Both multivariate and multimetric approaches have inherent advantages and disadvantages and (somewhat unusually for freshwater ecology) each approach has its following of strong adherents. Whether one or the other approach is universally the better at assessing all types of anthropogenic effects on running waters in all regions and whether hybridization between the approaches is possible has not been adequately explored.

The fundamental goal of bioassessment is to group stream sites into constellations that are meaningful (reflecting real ecological patterns rather than artifacts of the analytical technique) and useful (in helping resource managers ameliorate human impacts to running waters). The multivariate and multimetric approaches have spawned many techniques for grouping stream sites into classes of water (or habitat) quality (e.g. 'good'; 'fair'; 'poor'). In the multivariate approach, one or more of a large number of multivariate statistical procedures (e.g. cluster analysis) can be used to produce groupings of sites of similar community structure. An advantage of the multivariate approach is that benthic data (the taxa × sites matrix) are inherently multivariate (many taxa at each site) and so techniques (multivariate) designed for such data presumably allow maximum extraction of information (Norris, 1995). Further, multivariate techniques "let the data speak for themselves" in that the techniques presumably find natural groupings of sites. However, there are countervailing disadvantages to the multivariate approach. The various multivariate techniques have underlying assumptions that may not be met by the data set (James & McCulloch, 1990). Choosing among the very large number of multivariate techniques (and the permutations of techniques within a technique – e.g. clustering algorithms and distance/similarity measures of cluster analysis) at best requires a sound knowledge of the often complex underlying mathematics; at worst, the patterns found may reflect the choices of mathematical techniques as much as natural divisions of sites. The multimetric approach is conceptually much simpler. The metrics calculated from the data matrix are computationally simple and intuitive; e.g. % of the benthic fauna composed of EPT taxa – taxa generally less tolerant of stressors than many non-EPT (Diptera and non-insect macroinvertebrate) taxa. However, reduction of the taxa × sites data matrix to a series of such simple metrics inevitably involves loss of information (Norris & Norris, 1995); the question remains whether such radical 'transformations' produce optimal groupings of sites. Further questions arise regarding how best to combine the separate metrics (producing a single multimetric measure of the health of a stream site). Simply obtaining the sum (or average) of several metrics seems to have utility based upon the ability of such multimetrics to discriminate between least and most impaired stream sites (e.g. Fore et al., 1996). However, simply adding together metrics assumes that the metrics are additive and should be weighted equally in contribution to the multimetric score, assumptions that seem in need of testing (Norris, 1995).

Finally, a common procedure for producing groups of sites in the multimetric approach is to simply divide the range of the multimetric value (or use the 95th percentile as the highest value of the range) into equal portions (e.g. quadrisection yielding four habitat quality classes; trisection yielding three classes). Such 'imposed' divisions among sites are computationally simple and easy for water resource managers to understand and use (compared to the often hard-to-understand but potentially 'natural' divisions produced by multivariate analyses), but whether such imposed divisions produce the best groupings of sites has not been examined in depth.

Evaluating the goodness of the habitat quality ratings produced by bioassessment techniques is not easy. If one could find watersheds in which the true habitat quality of stream sites was unequivocally known, it would be a simple matter to try any number of bioassessment techniques and compare the habitat quality ratings of each site with the true habitat quality of sites (and likewise compare group memberships of sites (e.g. 'good'; 'fair'; 'poor') produced by various bioassessment techniques with the 'known-to-be true' group memberships). Alas, it is hard to imagine a watershed in which the true habitat quality of stream sites could be known. Perhaps an otherwise pristine system having a single point source of a contaminant whose toxicological effects are well understood would suffice. In nearly all real world situations (e.g. stream sites in watersheds with urbanization, mining, forest harvest), the human effects on streams are an array of physical and chemical perturbations whose effects play out in complex biotically-mediated (competition, predation) shifts in community structure. In nearly all cases, we must assess the goodness of the site classifications produced by different bioassessment techniques obliquely.

Using a steep 'urbanization gradient' in Alaska to assess stream bioassessment approaches

In Alaska, community structure of macroinvertebrates is better documented than for most other groups of organisms except fish, although collection records are not extensive and are often in close proximity to road corridors where access is easier (Oswood, 1989). Taxonomically the macroinvertebrate fauna in Alaskan streams is not well understood beyond genera level, particularly in the case of Diptera, which may make up to 50% by number of the organisms found in the benthic community (Oswood, 1989). In addition, the number of EPT (Ephermeroptera, Plecoptera, Trichoptera) genera in an undisturbed stream rarely exceeds 20 and decreases with increasing latitude. For example, in small streams on the Alaskan North Slope this number can be less than 10 (Hershey et al., 1997). One important question for the utilization of macroinvertebrates for bioassessment in Alaska is "Is there enough diversity within the sensitive EPT taxa to detect water quality impairment?"

Alaska has vast areas of wilderness with no roads and very low population densities and thus many streams and rivers are as 'near pristine' as possible. The extent of agriculture and livestock grazing is small given the size of the state: the major anthropogenic disturbances to watersheds are urbanization, timber harvest and placer mining. The major areas of urbanization are Juneau (population = 28 000) in southeast Alaska, Anchorage (238 000) in southcentral Alaska and Fairbanks (56 000) in interior Alaska. Our study area, within the Municipality of Anchorage, contains streams that flow over a short distance (< 20 km) from the mountains to the sea and do not have the complication of disturbances in their upper watersheds. Thus, urbanization influences on streams occur over very short distances (minimizing biogeographic differences in faunas), making this high latitude region a good place to test the utility of various bioassessment tools. Habitats range from pristine to low density suburban land use to urban habits resembling urban areas in much more developed regions to the south.

In this paper, we compare the site classifications produced by different bioassessment approaches using stream sites in the steep 'urbanization gradient' of Anchorage, Alaska. We obtained quantitative samples of benthic macroinvertebrates from riffle habitats. We examine four approaches to classification of stream sites: 1. *a-priori* site classification based upon land use into three groups; 2. use of a multivariate technique, TWINSPAN, to produce a classification of stream sites into three groups; 3. calculation of a multimetric index of habitat quality of stream sites, with classes of stream sites generated by trisection of the range of the multimetric index values, and (4) a hybrid multivariate/multimetric technique – calculation of rapid bioassessment metrics for each site, which are then used as input variables to a multivariate classification technique (TWINSPAN). Norris (1995) suggests that if several metrics are used they should lend themselves well to multivariate analyses as they are multiple measures of water quality. The 'goodness' of the classifications produced by these four approaches is compared via a simple graphical technique ("mean similarity dendrograms"; Van Sickle, 1997) that visually compares the relative isolation and compactness of groups of sites produced by different classifications.

Study area

The Municipality of Anchorage covers an area of 5180 ha (greater than the state of New Jersey) lying within the 'Cook Inlet' ecoregion (Gallant et al., 1995). Streams have their headwaters in the Chugach Mountains and then some streams flow through the urbanized areas of Anchorage (where over half the population of Alaska resides) in their lower reaches. There are no disturbances (e.g. agriculture or timber harvest) in the headwaters that would likely impair water quality. Hence, this allows a study of the direct effects of urbanization on stream communities. Thirteen streams (Figure 1) were selected for study in 1989 after a pilot study in 1988 had indicated substantial downstream changes in benthic community structure at some of the more urbanized sites (Milner & Oswood, unpublished data).

Some of these sites were potentially reference sites (well outside the metropolitan area in regions with very low population density) meeting the criteria of Omernik (1995), who considers that reference sites represent 'least-disturbed conditions' and are located in old-growth forests, inaccessible to livestock and possess a buffer of natural riparian vegetation. Other sites, while within the metropolitan area, were either near their headwaters, upstream of most potential point and nonpoint sources or were on streams with sizable greenbelts. These stream sites would often represent 'best available' reference sites in regions where population densities are higher and where areas between cities are semi-urban. Finally, some of our most downstream sites in urban Anchorage were clearly subject to point and nonpoint disturbances.

Methods and materials

Sampling and laboratory processing

Macroinvertebrates were collected at a total of 27 stations in the 13 streams using five replicate Surber samples with a 353 μm mesh net at two different time periods in 1989; one set was collected in early May and the other early September. This mesh size may not retain first or second instars of certain taxa, although finer mesh sizes require disproportionate time to sort these early instar (and often unidentifiable) organisms from the sediment and organic matter. Samples were only collected in riffle areas where water depths were less than 30 cm and velocities greater than 40 cm s^{-1}. Parsons & Norris (1996) support the 'riffle only' protocol as it reduces inter-habitat variations and thus comparisons are more easily made between equivalent environmental units. Riffles are certainly one of the most dominant habitat types in most Alaskan rivers. Samples were not pooled so that it was possible to use ANOVA to test for significant difference between metrics at given sites.

Streams that traversed areas of potential habitat impairment were sampled at a minimum of two stations; one station in each stream was established above likely areas of impairment to act as a potential control for other stations on the stream. However, a suitable control station, upstream of potential sources of water quality degradation, could not be located on Fish Creek. Invertebrates were identified to genera for the EPT orders and to families for all other taxa, except oligochaetes (at class). The choice of taxonomic level represents a compromise between the desire for increased information from species-level identification with the cost in time and required expertise to achieve it (Resh & Jackson, 1993). The time taken to key certain families (e.g. Chironomidae) to genera and species would be excessive and would require expert taxonomic assistance (although species-level identifications are essential to detect species extirpations and assess similar biodiversity issues). Even though lack of species-level information can decrease sensitivity and reduce the ability of a study to detect subtle changes (Resh & McElravy, 1993) caused by some stressors, species- and family-level identifications have produced similar patterns in changing community structure along the profile of a large French river (Bournard et al., 1996). All organisms in a sample were enumerated. A total of 30 macroinvertebrate taxa were collected at the 27 stream sites in 1989. Although we sampled in both spring and autumn (times of highest taxonomic diversity in Alaskan streams), substantial spates occurred prior to the autumn sampling and so we report here only data from spring (late April/early May).

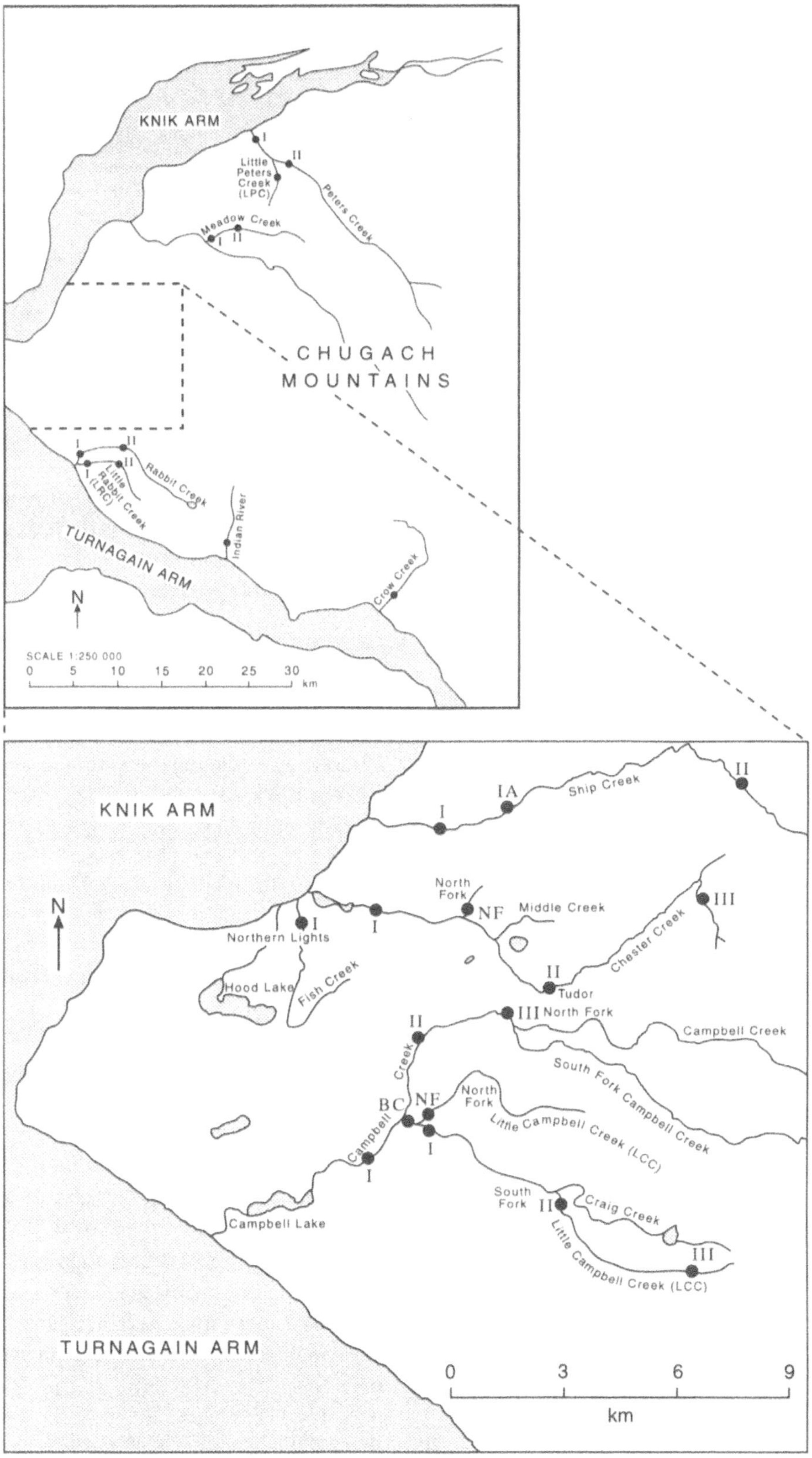

Figure 1. Map of the Municipality of Anchorage showing the 27 study sites in 13 streams.

Data analysis

Our goal was to compare four approaches to the classification of stream sites into groups (categories) potentially indicative of habitat quality.

Approach 1, a-priori classification of sites based upon land use

We used site location in relation to potential local and upstream disturbances to classify stream sites into one of three categories: 1. 'good' = sites outside of urban Anchorage, with intact riparian vegetation, and in watersheds with very low population density and no substantial resource use; 2. 'urban reference sites' in urban Anchorage but either in areas upstream of potential urban effects or in streams with substantial greenbelts buffering urban effects; 3. 'urban sites' in urban Anchorage and potentially affected by both local disturbances and cumulative effects of both point and nonpoint upstream disturbances.

Approach 2, TWINSPAN classification of stream sites based upon quantitative data (organism abundances)

We used a TWINSPAN (Hill, 1979) classification of the 30 macroinvertebrate taxa collected at the 27 stream sites in 1989 to produce groups of stream sites representing possible classes of habitat quality. TWINSPAN is a widely used multivariate tool for producing classifications of sites based upon the abundances of taxa at each site; additionally, each group defined by TWINSPAN is characterized by 'differential taxa' that appear to be useful discriminators of group membership. Zamora-Munoz & Alba-Tercedor (1996) and Wright et al. (1984) provide examples of use of TWINSPAN in bioassessment (using macroinvertebrate communities) of running waters. In our TWINSPAN analyses, rare taxa were downweighted using the 'pseudo-species' (see Yongman et al., 1995) facility with abundance cut levels of 10, 20, 50 and 100. Taxa had to be present at three sites to be included in the TWINSPAN classification.

Approach 3, Multimetric classification of stream sites into 3 categories of habitat quality

We calculated five biotic metrics that were used in the original EPA RBP III protocol (Plafkin et al., 1989) as these metrics are still arguably the most defensible for showing gradients to urbanization effects as summarized below:

1. the number of families within the class Insecta;
2. the number of genera in the orders Ephemeroptera, Trichoptera and Plecoptera (EPT);
3. the number of EPT individuals in a sample divided by the number of individuals of all the taxa expressed as a percentage (percent EPT);
4. the percent that the most abundant taxon contributes to the total organisms in the sample (percent dominant taxon);
5. Family Biotic Index, calculated as the total number of individuals in each family, multiplied by the score (relative tolerance to organic pollution; Hilsenoff, 1988) for that family. These values are summed across the total number of families collected. The FBI is then calculated by dividing by the total number of individuals in the sample. A value of 0 indicates a family extremely sensitive to water quality degradation, while a value of 10 indicates a family very tolerant of water quality degradation. Oligochaeta were not included in Hilsenhoff's original index, but were given a tolerance rating of 8 because they are known to be generally tolerant of poor water quality conditions.

Values for the five individual metrics were amalgamated into a single multimetric value (Barbour et al., 1995; Gerritsen, 1995) by allocating scores. A value of 1 (strong deviation from reference condition), 3 (moderate deviation) or 5 (similar or deviating slightly from reference condition) was assigned using a three-level ordinal scale as suggested by Fore et al. (1996) by trisecting the range of metric values (using the 95th percentile as the highest attainable value). Scores for each metric were then summed to produce a single multimetric value for each site which was then allocated into one of three classes by trisecting the maximum score attainable (i.e. 25) The groupings were then: 'good' (reference or close to reference), 'fair' (slight to moderate impairment) and 'poor' (significant impairment).

Approach 4, Classification of stream sites by TWINSPAN analysis, based upon scores on biotic metrics: a hybrid multimetric/multivariate approach

We again used TWINSPAN classification of the 27 stream sites but in this approach the scores for the five metrics outlined above were used in the analysis rather than taxa abundance. TWINSPAN default values were applied.

Van Sickle (1997) has proposed a conceptually and computationally simple technique for graphically comparing the goodness of different classifications of sites into groups. In brief, one first calculates a matrix of similarity values (Jaccard coefficients, based upon presence/absence of taxa at sites) for all possible combinations of sites. Jaccard coefficients are calculated as

$$s_{ij} = C_{ij}/(C_{ij} + U_i + U_j),$$

where C_{ij} is the number of species in common at sites i and j, U_i is the number of taxa unique to site i, and U_j is the number of taxa unique to site j. The Jaccard index ranges from 0 (no taxa in common) to 1 (all taxa present at both sites). The within-group and average between-group similarities are then compared graphically as a 'mean similarity dendrogram.' The best ('strongest') classifications produce groups in which the groups are 'isolated' from one another (the between-group similarities are small) and 'compact' (the within-group similarities are high and similar among groups). We present mean similarity dendrograms for each of the four approaches to site classification as a means of comparing the strength of these approaches.

Van Sickle (1997) also suggests calculating an M statistic, where $M = \bar{B}/\bar{W}$ where $\bar{B}$ is the mean of all the between-class similarities and $\bar{W}$ is the mean of all the within-class similarities. If no classification structure was apparent, then between- and within-class similarities should be similar (i.e. M = close to 1), but the stronger the classification the lower the value of M.

One-way analysis of variance was undertaken to determine whether there were significant differences in the metric scores for group classifications using the *a-priori* classification and the trisection approaches. $p < 0.05$ was taken as significant and < 0.01 as highly significant.

Results

The multivariate TWINSPAN classifications of the 27 stream sites using the 30 taxa for the spring of 1989 and the multimetric scores both separated three distinct groups at division 2 (Table 1). Further divsions were not meaningful, separating groups with only one stream site. The *a-priori* groupings and groupings from the trisection of the multimetric score are also provided in Table 1.

The four tables of Jaccard coefficients for each group with the mean similarities between and within groups are provided in Tables 2–5. For each of the classification approaches, a dendrogram was constructed with branches for each class joined at a node that is equal to the mean Jaccard index for between-group similarity (Figure 2, see p. 221). Dendrogram branches were not of equal length and the longest branch was 'good' followed by the 'fair' group. The lowest differentiation of within-group similarity and between-group similarity was the 'poor' group. M values varied from 0.50 for the TWINSPAN classification using numerical data to 0.75 for the TWINSPAN classification approach using metrics.

Analysis of variance showed that two of the metrics (percent dominant taxa and percent EPT) were significantly different when comparing the 'fair' *a-priori* group with the 'good' group, but four of the five (exception was percent domiant taxa) were highly significant in comparing the 'poor' group with 'good'. Using the multimetric trisection approach, all of the metrics, except percent EPT for 'fair', showed significant differences when comparing 'fair' or 'poor' with 'good' groups (Table 6).

Discussion

TWINSPAN classifications produced three distinct groups at the second division stage. Our *a-priori* classifications also assigned sites to one of three groups. Multimetric values were trisected (not quadrisected as sometimes recomended; Barbour et al., 1995). Therefore, each of our classification procedures assigned sites to one of three classes of habitat quality (loosely, 'reference/good', 'fair', 'poor'), facilitating comparisons among techniques (Table 1). We can concoct a simple measure of the correlation or congruency of assignments of sites to classes between techniques by calculating the % of the sites (27) assigned to the same classes by two techniques (Table 7). Assignments of sites to classes were very nearly congruent between the *a-priori* and TWINSPAN (using taxa) approaches (85% agreement), suggesting that the TWINSPAN approach tracked the apparent 'urbanization gradient' closely. However, four sites considered 'fair' in the *a-priori* assignment were ranked as 'good' by the TWINSPAN (using taxa) approach, suggesting that detailed bioassessment can reveal sites having better than expected water quality, as well as identify biologically troubled sites. In contrast, congruence of assignments to classes between the mutimetric approach and any of the other three approaches was weaker (41–52% agreement), indicating that the mutimetric approach produced a substantially different classification of sites. This difference might result

Table 1. Grouping of the 27 stream sites into three categories by four classification approaches. Groups represent provisional classes of habitat quality: group 1 = 'good', group 2 = 'fair' or 'urban reference sites', and group 3 = 'poor' or 'urban site'

Stream site	*a-priori* grouping	TWINSPAN using taxa	Multimetric trisection	TWINSPAN using metrics
Crow Creek	1	1	2	1
Campbell I	2	2	1	2
Campbell II	2	2	1	2
Campbell III	2	2	1	1
Chester NF	3	3	3	3
Chester I	3	3	3	3
Chester II	3	3	3	3
Chester III	2	2	1	2
Fire Creek	1	1	1	2
Fish Creek	3	3	3	3
Indian River	1	1	1	1
LCC BC	3	3	3	3
LCC I	3	3	2	1
LCC II	2	1	2	1
LCC III	2	1	1	1
LPC	1	1	1	2
LRC I	1	1	1	1
LRC II	1	1	2	1
Meadow I	1	1	1	2
Meadow II	1	1	2	1
Peters I	1	1	2	1
Peters II	1	1	2	1
Rabbit I	2	1	1	1
Rabbit II	2	1	2	1
Ship I	3	3	3	3
Ship IA	2	2	2	1
Ship II	2	2	1	2

from the loss of information inherent in the reduction of the taxa × site data matrix to metric values or from the trisection of metric values (see Norris & Georges, 1993; Norris & Norris, 1995) compared to the 'natural' classification of sites into classes produced by land use (*a-priori*) considerations or by multivariate (TWINSPAN) analyses.

It is possible that the sites allocated to the 'fair' group by the TWINSPAN (taxa × site matrix) analysis are comparable in many ratio-based metric values, but possess different abundance and community composition due to possible nutrient enrichment within the urbanized areas of Anchorage. Many streams at these latitudes are oligotrophic and some enrichment could influence community composition without impairing water quality significantly. This may account for the difference between 'good' and 'fair' classifications with respect to TWINSPAN (taxa × site matrix) and the multimetric technique. Disagreement among techniques in assignment of sites to classes (of habitat quality) involved mainly placements of sites into 'good' vs. 'fair' categories. All techniques identified essentially the same sites in the 'poor' class. From a management viewpoint this is heartening; it suggests that very different analytical techniques agree in identification of sites most in need of further investigation. However, discrimination of 'good' from 'fair' sites is less certain, suggesting that better identification of sites at risk (analogous to early diagnosis of a serious but treatable disease) may be a valuable goal for research in bioassessment.

Table 2. Jaccard Similarities between sites grouped by the *a-priori* approach

		Crow Creek	Fire	Indian	LPC	LRC I	LRC II	Meadow I	Meadow II	Peters I	Peters II	Rabbit I	Rabbit II	LCC II	LCC III	Campbell I	Campbell II	Campbell III	Chester III	Ship IA	Ship II	Chester NF	Chester I	Chester II	Fish	LCC BC	LCC I	Ship I
Good	Crow Creek	1																										
	Fire	0.33	1																									
	Indian	0.38	0.29	1																								
	LPC	0.57	0.40	0.64	1																							
	LRC I	0.67	0.46	0.57	0.85	1																						
	LRC II	0.63	0.47	0.61	0.65	0.77	1																					
	Meadow I	0.56	0.43	0.63	0.75	0.78	0.75	1																				
	Meadow II	0.47	0.43	0.63	0.67	0.68	0.65	0.88	1																			
	Peters I	0.73	0.47	0.53	0.65	0.77	0.86	0.65	0.56	1																		
	Peters II	0.53	0.33	0.53	0.57	0.67	0.63	0.65	0.65	0.53	1																	
Fair	Rabbit I	0.60	0.42	0.67	0.77	0.71	0.68	0.80	0.79	0.60	0.68	1																
	Rabbit II	0.55	0.44	0.55	0.65	0.67	0.63	0.74	0.74	0.55	0.55	0.76	1															
	LCC II	0.35	0.32	0.36	0.48	0.48	0.35	0.45	0.45	0.29	0.35	0.44	0.46	1														
	LCC III	0.45	0.29	0.46	0.64	0.57	0.45	0.48	0.41	0.38	0.38	0.52	0.55	0.67	1													
	Campbell I	0.24	0.27	0.38	0.32	0.36	0.30	0.27	0.27	0.30	0.44	0.33	0.24	0.35	0.32	1												
	Campbell II	0.30	0.28	0.57	0.42	0.42	0.36	0.33	0.33	0.36	0.36	0.44	0.30	0.48	0.44	0.77	1											
	Campbell III	0.35	0.26	0.58	0.42	0.41	0.42	0.38	0.38	0.42	0.50	0.50	0.33	0.33	0.30	0.69	0.72	1										
	Chester III	0.24	0.27	0.38	0.32	0.36	0.30	0.27	0.27	0.30	0.24	0.28	0.29	0.35	0.32	0.53	0.58	0.50	1									
	Ship IA	0.26	0.30	0.42	0.29	0.33	0.41	0.30	0.37	0.41	0.33	0.36	0.32	0.14	0.17	0.41	0.40	0.56	0.41	1								
	Ship II	0.30	0.33	0.65	0.61	0.48	0.43	0.46	0.52	0.36	0.36	0.57	0.46	0.41	0.50	0.43	0.55	0.55	0.43	0.47	1							
Poor	Chester NF	0.07	0.06	0.13	0.10	0.12	0.07	0.13	0.13	0.07	0.15	0.11	0.11	0.14	0.13	0.15	0.12	0.14	0.15	0.08	0.12	1						
	Chester I	0.13	0.18	0.17	0.14	0.16	0.13	0.18	0.18	0.13	0.20	0.14	0.15	0.27	0.24	0.29	0.22	0.19	0.29	0.14	0.16	0.40	1					
	Chester II	0.14	0.23	0.33	0.28	0.26	0.19	0.23	0.23	0.36	0.19	0.24	0.20	0.37	0.27	0.47	0.53	0.53	0.67	0.35	0.38	0.17	0.31	1				
	Fish	0.20	0.11	0.24	0.19	0.22	0.20	0.25	0.25	0.20	0.20	0.20	0.21	0.19	0.17	0.13	0.16	0.27	0.20	0.23	0.22	0.40	0.25	0.31	1			
	LCC BC	0.19	0.24	0.22	0.18	0.21	0.19	0.24	0.24	0.19	0.19	0.19	0.20	0.18	0.16	0.12	0.15	0.25	0.27	0.21	0.21	0.33	0.38	0.29	0.57	1		
	LCC I	0.29	0.45	0.25	0.31	0.35	0.29	0.32	0.38	0.29	0.23	0.32	0.33	0.40	0.25	0.23	0.29	0.27	0.50	0.25	0.35	0.14	0.27	0.44	0.36	0.43	1	
	Ship I	0.28	0.14	0.53	0.36	0.29	0.28	0.32	0.32	0.28	0.21	0.38	0.33	0.26	0.30	0.28	0.42	0.50	0.44	0.40	0.59	0.20	0.15	0.38	0.36	0.33	0.26	1

Mean similarities

	Good	Fair	Poor
Good	0.60		
Fair	0.43	0.44	
Poor	0.22	0.26	0.32

Table 3. Jaccard Similarities between sites grouped by TWINSPAN classification using a taxa × site matrix

		Crow Creek	Fire	Indian	LPC	LRC I	LRC II	Meadow I	Meadow II	Peters I	Peters II	Rabbit I	Rabbit II	LCC II	LCC III	Campbell I	Campbell II	Campbell III	Chester III	Ship IA	Ship II	Chester NF	Chester I	Chester II	Fish	LCC BC	LCC I	Ship I
Good	Crow Creek	1																										
	Fire	0.33	1																									
	Indian	0.38	0.29	1																								
	LCC II	0.35	0.32	0.36	1																							
	LCC III	0.45	0.29	0.46	0.6	1																						
	LPC	0.57	0.40	0.64	0.48	0.64	1																					
	LRC I	0.67	0.46	0.57	0.48	0.57	0.85	1																				
	LRC II	0.63	0.47	0.61	0.35	0.45	0.65	0.77	1																			
	Meadow I	0.56	0.43	0.63	0.45	0.48	0.75	0.78	0.75	1																		
	Meadow II	0.47	0.43	0.63	0.45	0.41	0.67	0.68	0.65	0.88	1																	
	Peters I	0.73	0.47	0.53	0.29	0.38	0.65	0.77	0.86	0.65	0.56	1																
	Peters II	0.53	0.33	0.53	0.35	0.38	0.57	0.67	0.63	0.65	0.65	0.53	1															
	Rabbit I	0.60	0.42	0.67	0.44	0.52	0.77	0.71	0.68	0.80	0.79	0.60	0.68	1														
	Rabbit II	0.55	0.44	0.55	0.46	0.55	0.65	0.67	0.63	0.74	0.74	0.55	0.55	0.76	1													
Fair	Campbell I	0.24	0.27	0.38	0.35	0.32	0.32	0.36	0.30	0.27	0.27	0.30	0.44	0.33	0.24	1												
	Campbell II	0.30	0.28	0.57	0.48	0.44	0.42	0.42	0.36	0.33	0.33	0.36	0.36	0.44	0.30	0.77	1											
	Campbell III	0.35	0.26	0.58	0.33	0.30	0.42	0.41	0.42	0.38	0.38	0.42	0.50	0.50	0.33	0.69	0.72	1										
	Chester III	0.24	0.27	0.38	0.35	0.32	0.32	0.36	0.30	0.27	0.27	0.30	0.24	0.28	0.29	0.53	0.58	0.50	1									
	Ship IA	0.26	0.30	0.42	0.14	0.17	0.29	0.33	0.41	0.30	0.37	0.41	0.33	0.36	0.32	0.41	0.40	0.56	0.41	1								
	Ship II	0.30	0.33	0.65	0.41	0.50	0.61	0.48	0.43	0.46	0.52	0.36	0.36	0.57	0.46	0.43	0.55	0.55	0.43	0.47	1							
Poor	Chester NF	0.07	0.06	0.13	0.14	0.13	0.10	0.12	0.07	0.13	0.13	0.07	0.15	0.11	0.11	0.15	0.12	0.14	0.15	0.08	0.12	1						
	Chester I	0.13	0.18	0.17	0.27	0.24	0.14	0.16	0.13	0.18	0.18	0.13	0.20	0.14	0.15	0.29	0.22	0.19	0.29	0.14	0.16	0.40	1					
	Chester II	0.14	0.23	0.33	0.37	0.27	0.28	0.26	0.19	0.23	0.23	0.36	0.19	0.24	0.20	0.47	0.53	0.53	0.67	0.35	0.38	0.17	0.31	1				
	Fish	0.20	0.11	0.24	0.19	0.17	0.19	0.22	0.20	0.25	0.25	0.20	0.20	0.20	0.21	0.13	0.16	0.27	0.20	0.23	0.22	0.40	0.25	0.31	1			
	LCC BC	0.19	0.24	0.22	0.18	0.16	0.18	0.21	0.19	0.24	0.24	0.19	0.19	0.19	0.20	0.12	0.15	0.25	0.27	0.21	0.21	0.33	0.38	0.29	0.57	1		
	LCC I	0.29	0.45	0.25	0.40	0.25	0.31	0.35	0.29	0.32	0.38	0.29	0.23	0.32	0.33	0.23	0.29	0.27	0.50	0.25	0.35	0.14	0.27	0.44	0.36	0.43	1	
	Ship I	0.28	0.14	0.53	0.26	0.30	0.36	0.29	0.28	0.32	0.32	0.28	0.21	0.38	0.33	0.28	0.42	0.50	0.44	0.40	0.59	0.20	0.15	0.38	0.36	0.33	0.26	1

Mean similarities

	Good	Fair	Poor
Good	0.57		
Fair	0.36	0.53	
Poor	0.22	0.28	0.32

Table 4. Jaccard Similarities between sites grouped by the multimetric approach

		Campbell I	Campbell II	Campbell III	Chester III	Fire	Indian	LCC III	LPC	LRC I	Meadow I	Rabbit I	Ship II	Crow Creek	LCC I	LCC II	LRC II	Meadow II	Peters I	Peters II	Rabbit II	Ship IA	Chester NF	Chester I	Chester II	LCC BC	Fish	Ship I
Good	Campbell I	1																										
	Campbell II	0.77	1																									
	Campbell III	0.69	0.72	1																								
	Chester III	0.53	0.58	0.50	1																							
	Fire	0.27	0.28	0.26	0.27	1																						
	Indian	0.38	0.57	0.58	0.38	0.29	1																					
	LCC III	0.32	0.44	0.30	0.32	0.29	0.46	1																				
	LPC	0.32	0.42	0.42	0.32	0.40	0.64	0.64	1																			
	LRC I	0.36	0.42	0.41	0.36	0.46	0.57	0.57	0.85	1																		
	Meadow I	0.27	0.33	0.38	0.27	0.43	0.63	0.48	0.75	0.78	1																	
	Rabbit I	0.33	0.44	0.50	0.28	0.42	0.67	0.52	0.77	0.71	0.80	1																
	Ship II	0.43	0.55	0.55	0.43	0.33	0.65	0.50	0.61	0.48	0.46	0.57	1															
Fair	Crow Creek	0.24	0.30	0.35	0.24	0.33	0.38	0.45	0.57	0.67	0.56	0.60	0.30	1														
	LCC I	0.23	0.29	0.27	0.50	0.45	0.25	0.25	0.31	0.35	0.32	0.32	0.35	0.29	1													
	LCC II	0.35	0.48	0.33	0.35	0.32	0.36	0.67	0.48	0.48	0.45	0.44	0.41	0.35	0.40	1												
	LRC II	0.30	0.36	0.42	0.30	0.47	0.61	0.45	0.65	0.77	0.75	0.68	0.43	0.63	0.29	0.35	1											
	Meadow II	0.27	0.33	0.38	0.27	0.43	0.63	0.41	0.67	0.68	0.88	0.79	0.52	0.47	0.38	0.45	0.65	1										
	Peters I	0.30	0.36	0.42	0.30	0.47	0.53	0.38	0.65	0.77	0.65	0.60	0.36	0.73	0.29	0.29	0.86	0.56	1									
	Peters II	0.44	0.36	0.50	0.24	0.33	0.53	0.38	0.57	0.67	0.65	0.68	0.36	0.53	0.23	0.35	0.63	0.65	0.53	1								
	Rabbit II	0.24	0.30	0.33	0.29	0.44	0.55	0.55	0.65	0.67	0.74	0.76	0.46	0.55	0.33	0.46	0.63	0.74	0.55	0.55	1							
	Ship IA	0.41	0.40	0.56	0.41	0.30	0.42	0.17	0.29	0.33	0.30	0.36	0.47	0.26	0.25	0.14	0.41	0.37	0.41	0.33	0.32	1						
Poor	Chester NF	0.15	0.12	0.14	0.15	0.06	0.13	0.13	0.10	0.12	0.13	0.11	0.12	0.07	0.14	0.14	0.07	0.13	0.07	0.15	0.11	0.08	1					
	Chester I	0.29	0.22	0.19	0.29	0.18	0.17	0.24	0.14	0.16	0.18	0.14	0.16	0.13	0.27	0.27	0.13	0.18	0.13	0.20	0.15	0.14	0.40	1				
	Chester II	0.47	0.53	0.53	0.67	0.23	0.33	0.27	0.28	0.26	0.23	0.24	0.38	0.14	0.44	0.37	0.19	0.23	0.36	0.19	0.20	0.35	0.17	0.31	1			
	LCC BC	0.12	0.15	0.25	0.27	0.24	0.22	0.16	0.18	0.21	0.24	0.19	0.21	0.19	0.43	0.18	0.19	0.24	0.19	0.19	0.20	0.21	0.33	0.38	0.29	1		
	Fish	0.13	0.16	0.27	0.20	0.11	0.24	0.17	0.19	0.22	0.25	0.20	0.22	0.20	0.36	0.19	0.20	0.25	0.20	0.20	0.21	0.23	0.40	0.25	0.31	0.57	1	
	Ship I	0.28	0.42	0.50	0.44	0.14	0.53	0.30	0.36	0.29	0.32	0.38	0.59	0.28	0.26	0.26	0.28	0.32	0.28	0.21	0.33	0.40	0.20	0.15	0.38	0.33	0.36	1

Mean similarities

	Good	Fair	Poor
Good	0.48		
Fair	0.45	0.45	
Poor	0.24	0.22	0.32

Table 5. Jaccard Similarities between sites grouped by TWINSPAN classification using multimetric values

		Crow Creek	Campbell III	Indian	LCC I	LCC II	LCC III	LRC I	LRC II	Rabbit I	Rabbit II	Ship IA	Peters I	Peters II	Meadow II	Campbell I	Campbell II	Chester III	Meadow I	Fire	LPC	Ship II	Chester NF	Chester I	Chester II	LCC BC	Fish	Ship I
Good	Crow Creek	1																										
	Campbell III	0.35	1																									
	Indian	0.38	0.58	1																								
	LCC I	0.29	0.27	0.25	1																							
	LCC II	0.35	0.33	0.36	0.40	1																						
	LCC III	0.45	0.30	0.46	0.25	0.67	1																					
	LRC I	0.67	0.41	0.57	0.35	0.48	0.57	1																				
	LRC II	0.63	0.42	0.61	0.29	0.35	0.45	0.77	1																			
	Rabbit I	0.60	0.50	0.67	0.32	0.44	0.52	0.71	0.68	1																		
	Rabbit II	0.55	0.33	0.55	0.33	0.46	0.55	0.67	0.63	0.76	1																	
	Ship IA	0.26	0.56	0.42	0.25	0.14	0.17	0.33	0.41	0.36	0.32	1																
	Peters I	0.73	0.42	0.53	0.29	0.29	0.38	0.77	0.86	0.60	0.55	0.41	1															
	Peters II	0.53	0.50	0.53	0.23	0.35	0.38	0.67	0.63	0.68	0.55	0.33	0.53	1														
	Meadow II	0.47	0.38	0.63	0.38	0.45	0.41	0.68	0.65	0.79	0.74	0.37	0.56	0.65	1													
Fair	Campbell I	0.24	0.69	0.38	0.23	0.35	0.32	0.36	0.30	0.33	0.24	0.41	0.30	0.44	0.27	1												
	Campbell II	0.30	0.72	0.57	0.29	0.48	0.44	0.42	0.36	0.44	0.30	0.40	0.36	0.36	0.33	0.77	1											
	Chester III	0.24	0.50	0.38	0.50	0.35	0.32	0.36	0.30	0.28	0.29	0.41	0.30	0.24	0.27	0.53	0.58	1										
	Meadow I	0.56	0.38	0.63	0.32	0.45	0.48	0.78	0.75	0.80	0.74	0.30	0.65	0.65	0.88	0.27	0.33	0.27	1									
	Fire	0.33	0.26	0.29	0.45	0.32	0.29	0.46	0.47	0.42	0.44	0.30	0.47	0.33	0.43	0.27	0.28	0.27	0.43	1								
	LPC	0.57	0.42	0.64	0.31	0.48	0.64	0.85	0.65	0.77	0.65	0.29	0.65	0.57	0.67	0.32	0.42	0.32	0.75	0.40	1							
	Ship II	0.30	0.55	0.65	0.35	0.41	0.50	0.48	0.43	0.57	0.46	0.47	0.36	0.36	0.52	0.43	0.55	0.43	0.46	0.33	0.61	1						
Poor	Chester NF	0.07	0.14	0.13	0.14	0.14	0.13	0.12	0.07	0.11	0.11	0.08	0.07	0.15	0.13	0.15	0.12	0.15	0.13	0.06	0.10	0.12	1					
	Chester I	0.13	0.19	0.17	0.27	0.27	0.24	0.16	0.13	0.14	0.15	0.14	0.13	0.20	0.18	0.29	0.22	0.29	0.18	0.18	0.14	0.16	0.40	1				
	Chester II	0.14	0.53	0.33	0.44	0.37	0.27	0.26	0.19	0.24	0.20	0.35	0.36	0.19	0.23	0.47	0.53	0.67	0.23	0.23	0.28	0.38	0.17	0.31	1			
	LCC BC	0.19	0.25	0.22	0.43	0.18	0.16	0.21	0.19	0.19	0.20	0.21	0.19	0.19	0.24	0.12	0.15	0.27	0.24	0.24	0.18	0.21	0.33	0.38	0.29	1		
	Fish	0.20	0.27	0.24	0.36	0.19	0.17	0.22	0.20	0.20	0.21	0.23	0.20	0.20	0.25	0.13	0.16	0.20	0.25	0.11	0.19	0.22	0.40	0.25	0.31	0.57	1	
	Ship I	0.28	0.50	0.53	0.26	0.26	0.30	0.29	0.28	0.38	0.33	0.40	0.28	0.21	0.32	0.28	0.42	0.44	0.32	0.14	0.36	0.59	0.20	0.15	0.38	0.33	0.36	1

Mean similarities

	Good	Fair	Poor
Good	0.48		
Fair	0.45	0.43	
Poor	0.23	0.25	0.32

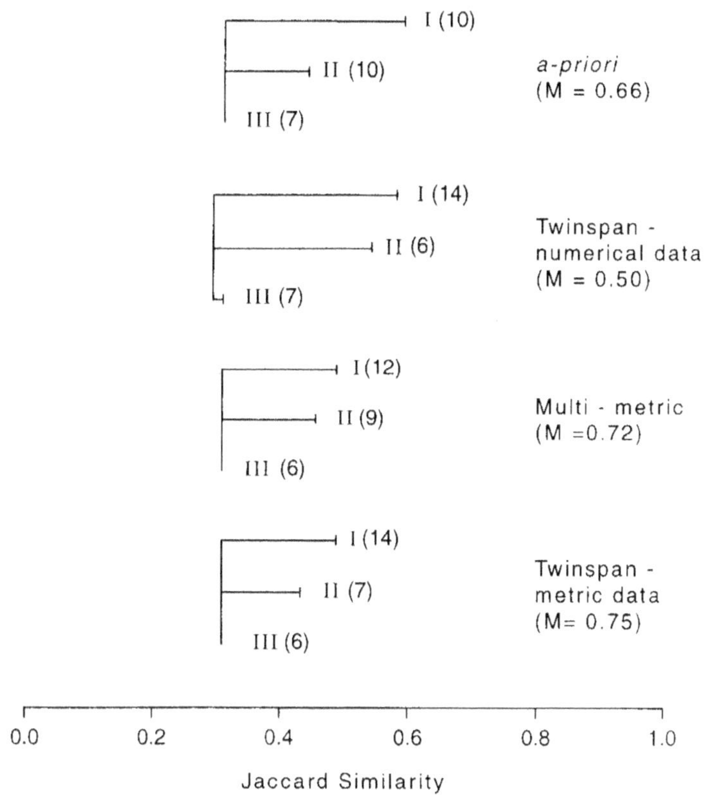

Figure 2. Mean similarity dendrogram for invertebrate assemblages in three groups as assigned by four different classification approaches. (I = 'good', II = 'fair', III = 'poor'. See text for calculation of M.)

Table 6. One-way analyis of variance to test for signficance of five metrics between groups differentiated by the *a-priori* approach and the multimetric approach (compared to 'good' streams). $p < 0.05$ = significant*, $p < 0.01$ = highly significant**

Metric	'fair' *a-priori*	'poor' *a-priori*	'fair' (multimetric)	'poor' (multimetric)
No. families	1.0	<0.01**	<0.01**	<0.01**
No. EPT genera	0.11	<0.01**	<0.05*	<0.01**
% EPT	<0.05*	0.24	<0.23	<0.05*
% Dom. taxa	<0.05*	<0.01**	<0.01**	<0.01**
FBI	0.86	<0.01**	<0.01**	<0.01**

Table 7. Congruence of classifications of sites into categories among classification techniques. Values are percentages of sites placed in same classes ('good', 'fair', 'poor' – see Table 1)

a-priori	–			
TWINSPAN – numerical data	85	–		
Multimetric	52	52	–	
TWINSPAN – metric data	63	78	41	–
	a-priori	TWINSPAN numerical data	Multimetric	TWINSPAN metric data

Mean similarity diagrams (Figure 2) allow comparison of the strength of different classifications. A 'strong' classification is indicated by groups that are well isolated from one another (i.e. the between-group similarities are low and the within-group similarities are high), indicated by long branches arising from a low average between site (Van Sickle, 1997). In our analyses, the branches were markedly disimilar in length, with 'good' having the longest branches for the four techniques of classification. For example, with TWINSPAN classification using the entire site × taxa matrix, site pairs in the 'good' classification shared a mean of 57% of their taxa, whereas with the multimetric trisection, classification sites within the 'good' group shared an average of only 48% of their taxa. These values compared to a mean similarity near 30% for site pairs in different groups.

In marked contrast, 'poor' sites as designated by all approaches showed the least similarity and most variation in their taxa composition, with some site pairs sharing less than 20% of their taxa. Sites identified by TWINSPAN (using the site by taxa matrix) or by multimetric trisection as falling in the 'poor' group shared an average of only 32% of their taxa. Although these data would indicate that the 'poor' group overall showed the least compactness and isolation, this could be a result of the low overall taxa diversity within this group. This group contained the most impaired sites, some with few or no EPT genera and taxa richness generally less than six. Where diversity is low, Jaccard's coefficient values may vary significantly when one or two taxa are different between sites. Overall, however, this group had the most consistent membership of sites by all classification approaches, and analysis of variance indicated that most metrics were significantly different when compared to the reference sites. This seemingly paradoxical situation indicates that such 'poor' habitat quality sites, while different among themselves in taxonomic composition, were sufficiently differentiated from 'fair' and 'good sites' to be identifiable by all techniques.

The M value (ratio of average between-site taxonomic similarity to average within-site similarity) of 0.50 was lowest for the TWINSPAN classification using the taxa × site matrix, indicating a more distinct compactness and isolation of groups than the other three methods. The classification using the multimetric data had the highest M values (0.72–0.75), indicating less compactness and isolation of the groups.

A question of importance is whether taxonomic similarity between groups is directly relevant to bioassessment classifications. Does it matter, for example, that three genera of stoneflies have been replaced in the macroinvertebrate community by three different genera? Such faunal substitutions may trigger a difference in TWINSPAN classifications, but not necessarily a change in metric values such as the number of EPT genera or percent EPT. We argue that faunal similarity provides a useful first approximation of functional similarity (because, to some extent, each taxon has a unique role in ecosystem processes); accordingly, as faunas diverge in taxonomic similarity, then community function must differ as well, hopefully reflecting habitat or water quality changes. For example, there could be a replacement of three taxa of shredder caddisflies by three taxa of collector-grazers caddisflies; this may reflect degradation of riparian function and loss of leaf litter inputs.

Our conclusions – that a multivariate technique (TWINSPAN – using the taxa × site data matrix) produced the strongest classification and produced site designations that most closely matched the apparent urbanization gradient of Anchorage, Alaska – are tempered by several constraints. First, all techniques, including the multimetric approach, successfully identified the stream sites in urbanized (and apparently lower habitat quality) areas. This suggests that biomonitoring using macroinvertebrates may provide a relatively robust (to differences in analytical technique) means of identifying sites likely in need of monitoring or remediation. Second, we have approached the classification of stream sites into groups potentially indicative of water or habitat quality as a problem in applied biogeography. Our measure of the relative strength of different classifications is faunal similarity within and among groups of sites. Certainly other measures of the goodness of site classifications are necessary (e.g. congruence of site classifications with physical and chemical characteristics of sites).

Gerritsen (1995) suggests that reference sites must be selected independently of biological knowledge of the sites. While this is necessary as a first step, we showed in these analyses that *a-priori* classification based on land use criteria was weaker than using the macroinvertebrate data in a site by taxa matrix with both impaired and clean sites included. TWINSPAN classification of sites as applied here does not provide a mechanism of placing new sites into a habitat quality class, but it does provide a way to differentiate 'good' (reference) sites that have the most similarity in their invertebrate taxa structure, as suggested by Resh et al. (1995).

Finally, our analyses obviously apply only to an urbanization gradient in a high latitude (taiga forest) biome. Studies of other biomes and of other perturbations (e.g. logging, mining, agriculture and grazing) are needed to test the utility of different bioassessment protocols. Nonetheless, such comparative studies are essential if we are to converge on a 'best tool set' for bioassessment of running waters.

Acknowledgements

We would like to thank Mark Little of the Water Quality Section, Municipality of Anchorage for providing the spark to apply biomonitoring to streams of this area and his enthusiastic support for the project. This project was funded by the Municipality of Anchorage. Maureen Milner provided the maps. MWO gratefully acknowledges the National Science Foundation (NSF-LTER: DEB-950594) for support during analysis and writing. Constructive comments by Dr P. J. Boon and an anonymous reviewer substantially improved this paper.

References

Barbour, M. T., J. B. Stribling & J. Karr, 1995. Multimetric approach for establishing biocriteria and measuring biological condition. In Davis, W. S. & T. P. Simon (eds), Biological Assessment and Criteria; Tools for Water Resource Planning and Decision Making. Lewis Publishers, Boca Raton, FL: 63–77.

Bournard, M., B. Cellot, P. Richoux & A. Berrahou, 1996. Macroinvertebrate community structure and environmental characteristics along a large river: congruity of patterns for identification to species or family. J. n. am. Benthol. Soc. 15: 232–253.

Fore, L. S., J. R. Karr & R. W. Wisseman, 1996. Assessing invertebrate responses to human activities: evaluating alternative approaches. J. n. am. Benthol. Soc. 15: 212–231.

Gallant, A. L., E. F. Binnan, J. M. Omernik & M. B. Shasby, 1995. Ecoregions of Alaska. U.S. Geological Survey Professional Paper 1567, Washington, D.C.: 73 pp + map.

Gerritsen, J., 1995. Additive biological indices for resource management. J. n. am. Benthol. Soc. 14: 451–457.

Hershey, A. E., W. B. Bowden, L. A. Deegan, J. E. Hobbie, B. J. Peterson, G. W. Kipphut, G. W. Kling, M. A. Lock, R. W. Merritt, M. C. Miller, J. R. Vestal & J. A. Schuldt, 1997. The Kuparuk River: a long-term study of biological and chemical processes in an arctic river. In Milner, A. M. & M. W. Oswood (eds), Freshwaters of Alaska: Ecological Syntheses. Springer-Verlag, New York: 107–129.

Hill, M. O., 1979. TWINSPAN – a FORTRAN program for arranging multivariate data in an ordered two-way table by classification of the individuals and the attributes. Ecology & Systematics, Cornell University, Ithaca, NY.

Hilsenoff, W. L., 1988. Rapid field assessment of organic pollution with a family level biotic index. J. n. am. Benthol. Soc. 7: 65–68.

James, F. C. & C. E. McCulloch, 1990. Multivariate analysis in ecology and systematics: pancea or Pandora's box? Ann. Rev.

Jongman, R. H. G., C. J. F. ter Braak & O. F. R. Van Tongeren (eds), 1995. Data Analysis in Community and Landscape Ecology. Cambridge University Press, New York.

Karr, J. R., 1999. Defining and measuring river health. Freshwat. Biol. 41: 221–234

Moss, D., M. T. Furse, J. F. Wright & P. D. Armitage, 1987. The prediction of the macroinvertebrate fauna of unpolluted running water sites in Great Britain using environmental data. Freshwat. Biol. 17: 41–52.

Norris, R. H., 1995. Biological monitoring: the dilemma of data analysis. J. n. am. Benthol. Soc. 15: 440–450.

Norris, R. H. & A. Georges, 1993. Analysis and interpretation of benthic invertebrate surveys. In Rosenberg, D. M. & V. H. Resh (eds), Freshwater Biomonitoring and Benthic Macroinvertebrates. Chapman and Hall, New York: 234–286.

Norris, R. H. & K. R. Norris, 1995. The need for biological assessment of water quality: Australian perspective. Aust. J. Ecol. 20: 1–6.

Norris, R. H. & M. C. Thoms, 1999. What is river health? Freshwat. Biol. 41: 197–210.

Omernik, J. M., 1995. Ecoregions: a spatial framework for environmental management. In Davis, W. S. & T. P. Simon (eds), Biological Assessment and Criteria; Tools for Water Resource Planning and Decision Making. Lewis Publishers, Boca Raton, FL: 49–62.

Oswood, M. W., 1989. Community structure of benthic invertebrates in interior Alaskan (USA) streams and rivers. Hydrobiologia 172: 97–110.

Parsons, M. & R. H. Norris, 1996. The effect of habitat-specific sampling on biological assessment of water quality using a predictive model. Freshwat. Biol. 36: 419–434.

Resh, V. H. & E. P. McElravy, 1993. Contemporary quantitative approaches to biomonitoring using benthic macroinvertebrates. In Rosenberg, D. M. & V. H. Resh (eds), Freshwater Biomonitoring and Benthic Macroinvertebrates. Chapman and Hall, New York: 159–194.

Resh, V. H. & J. K. Jackson, 1993. Rapid assessment approaches to biomonitoring using benthic macroinvertebrates. In Rosenberg, D. M. & V. H. Resh (eds), Freshwater Biomonitoring and Benthic Macroinvertebrates. Chapman and Hall, New York: 195–233.

Resh, V. H., R. H. Norris & M. T. Barbour, 1995. Design and implementation of rapid bioassessment approaches for water resource monitoring using benthic macroinvertebrates. Aust. J. Ecol. 20: 108–121.

Rosenberg, D. M. & V. H. Resh (eds), 1993. Freshwater Biomonitoring and Benthic Macroinvertebrates. Chapman and Hall, New York.

Smith, M. J., W. R. Kay, D. H. D. Edward, P. J. Papas, K. S. J. Richardson, J. C. Simpson, A. M. Pinder, D. J. Cale, P. H. J. Horwitz, J. A. Davis, F. H. Yung, R. H. Norris & S. A. Halse, 1999. AuRivAS: using macroinvertebrates to assess ecological condition of rivers in Western Australia. Freshwat. Biol. 41: 269–282.

Van Sickle, J., 1997. Using mean similarity dendrograms to evaluate classifications. J. agric. Biol. Envir. Stats. 2: 370–388.

Wright, J. F., D. Moss, P. D. Armitage & M. T. Furse, 1984. A preliminary classification of running-water sites in Great Britain based on macro-invertebrate species and the prediction of community type using environmental data. Freshwat. Biol. 14: 221–256.

Zamora-Munoz, C. & J. Alba-Tercedor, 1996. Bioassessment of organically polluted Spanish rivers, using a biotic index and multivariate methods. J. n. am. Benthol. Soc. 15: 332–352.

Hydrobiologia **422/423**: 225–232, 2000.
M. Jungwirth, S. Muhar & S. Schmutz (eds), Assessing the Ecological Integrity of Running Waters.

Macroinvertebrate functional feeding group methods in ecological assessment

C. Rawer-Jost, J. Böhmer, J. Blank & H. Rahmann
Institute for Zoology, University of Hohenheim, Garbenstraße 30, D-70593 Stuttgart, Germany
E-mail: rawer@uni-hohenheim.de

Key words: benthic macroinvertebrates, functional feeding groups, disturbance, metrics, index, Germany

Abstract

Six functional feeding group (FFG) indices applied in ecological assessment were tested for their accuracy in detecting impairment in 12 cases of disturbance on 11 streams in south-western Germany. Three metrics – percentage of predators, percentage of shredders and Rhithron-Ernährungstypen-Index RETI – performed well in small highland streams up to five m width. The FFG metrics were compared to the taxonomic measures 'percentage of Chironomidae' and 'percentage of ephemeropteran, plecopteran and trichopteran taxa' (EPT) used the U.S.A., and to a metric based on locomotion type (percentage of sessile taxa). As the implementation of such metrics implies the use of ecological data tables, some aspects of the pro and contra of European ecological lists published in recent years were critically assessed.

Introduction

The distribution of macroinvertebrate functional feeding groups in running waters is supposed to reflect process-level aquatic ecosystem attributes. Specialized feeders, such as shredders and scrapers, are presumed to be more sensitive to perturbation, while generalists, such as gatherers and filterers, are more tolerant to pollution that might alter the availability of certain food (Barbour et al., 1996). Therefore, in recent years, functional feeding group measures have been used in biomonitoring approaches such as the "Benthic Index of Biotic Integrity" (Kerans & Karr, 1994), the "Florida Stream Condition Index" (Barbour et al., 1996), and the "Rapid Bioassessment Protocols" (Barbour et al., 1997). As European biomonitoring also aims at an assessment of ecological integrity, the use of the functional feeding group concept could increase the knowledge on the complex dynamic processes of running water ecosystems (Moog, 1995). Nevertheless, examinations of the accuracy of those measures mostly have had discouraging results (Resh, 1994; Resh et al., 1995) and meanwhile they are regarded as not especially useful as components of a multimetric index (Karr, 1999). As part of a project aiming at the development of a multimetric index for use in Germany, the functional feeding group measures were tested for their accuracy in detecting disturbance and compared with other ecological guild measures using locomotion type or taxonomic categories similar to those also used in the Rapid Bioassessment Protocols (Barbour et al., 1997). European ecological data tables published in recent years ("Fauna Aquatica Austriaca", Moog, 1995; and "Ökologische Typisierung der aquatischen Makrofauna" "ÖTM", Bayerisches Landesamt für Wasserwirtschaft, 1996) contain autecological information such as feeding type, preferred habitat, preferred current and locomotion form for several thousand macroinvertebrate species. The pro and contra of using these data in calculating the metrics was critically assessed.

Methods

Study sites

We used our own data sets from 11 highland streams in four regions of Baden-Württemberg, southern Germany. Two study sites have been compared within each stream: an undisturbed reference site and one (in

Table 1. Stream data, disturbance type and distance of impaired site from reference site

Stream code	Region geology	Stream length	Ø Stream width	Disturbance type	Distance from reference site
Bärenbach BÄ	Suebian Alb Jurassic	2.5 km	1.5 m	landfill / heavy metals	0.8 km downstream
Fils FI	Suebian Alb Jurassic	45 km	4 m	1. defective purification plant 2. heavy metals deposit	1.3 km downstream 21.2 km downstream
Ziegelbach ZI	Suebian Alb Jurassic	3.5 km	1.5 m	chicken farm waste water	0.7 km downstream
Buchenbach BU	Rems Valley	8 km	2 m	fish farm	0.5 km downstream
Brettach BR	Hohenlohe Keuper	29 km	5 m	residual flow	2 km downstream
Kressbach KR	Hohenlohe Keuper	6 km	0.5 m	influx of untreated domestic waste water	1 km downstream
Laibach LA	Hohenlohe Keuper	4 km	1 m	influx of untreated domestic waste water	0.8 km downstream
Mühlbach MÜ	Hohenlohe Keuper	12 km	3 m	increased chloride	2 km downstream
Seckach SE	Hohenlohe Keuper	25 km	4 m	influx of chemical industry purified waste water	0.05 km downstream
Kaltenbach KA	Black Forest Sandstone	5 km	1.5 m	acidification: pH of headwater about 5	reference is a second headwater to Kaltenbach, pH $\approx$ 7
Vorderer Seebach VS	Black Forest Sandstone	3 km	2 m	acidification: pH of headwater about 5–6	reference is a section 0.45 km downstream, pH $\approx$ 7

Table 2. Tested measures

Measure	Author(s)	Hypothesized effect of impact
Ratio shredders to total collectors	Merritt & Cummins, 1996	< 0.25
Ratio Filtering Collectors to Gathering Collectors	Merritt & Cummins, 1996	< 0.5
Rhithron-Ernährungs-Typen-Index RETI	Schweder, 1990	decline
% scrapers	Cummins, 1988	decline
% shredders	Cummins, 1988	decline
% predators	Kerans & Karr, 1994	increase
% Chironomidae	Barbour et al., 1997	increase
% Ephemeroptera+Plecoptera + Trichoptera (EPT)	MACS, 1996	decline
% sessile taxa	Bayr. Landesamt f. Wasserwirtschaft, 1996	increase

Table 3. Significance of differences between reference and impaired sites (n = 14) using the Wilcoxon Matched Pairs Test for nine tested metrics. *SH/TC = ratio Shredders to Total Collectors; FC/GC = ratio Filtering Collectors to Gathering Collectors; RETI= Rhithron-Ernährungstypen-Index; n.s. = not statistically significant*

Tested metric	z	Significance level P
SH/TC	3.17	<0.01
FC/GC	3.29	<0.001
% scrapers	0.78	n.s.
RETI	2.98	<0.01
% shredders	3.04	<0.01
% predators	2.42	<0.05
% EPT	0.03	n.s.
% Chironomidae	2.35	<0.05
% sessile taxa	3.30	<0.001

one case two) site(s) influenced by a known disturbance (Table 1).

The reference sites had to be structurally intact, with saprobity index levels ranging from 'not polluted' to 'least polluted'. They are located in the headwaters except the two acidic Black Forest brooks and the Brettach River. For the latter we chose a reference section 2 km upstream of the residual flow stretch as a site with comparable structural conditions. The reference site for acidic Kaltenbach is a near-parallel neutral headwater; the comparison site for Vorderer Seebach is located 0.5 km downstream, at a stream section with neutral pH due to a change in geology.

The disturbances can be loosely sorted into five groups:

1. In five cases, problems are caused by an increase of organic and inorganic nutrients; due to a defective purification plant on Fils River or to villages with no purification of waste water at all as in Kressbach and Laibach; to a fish farm in the upper reaches of Buchenbach; or due to a chicken farm situated immediately on the banks of Ziegelbach.
2. On Brettach River, a diversion channel powers a mill, resulting in a very low discharge in the original streambed.
3. Bärenbach is influenced by a near-by landfill: cadmium concentrations five times as high as the threshold value of 0.2 mg Cd kg $^{-1}$ sediment permitted by law were detected. On the second site of Fils River several metalworking plants upstream caused heavy metal deposits in the sediment.
4. Mühlbach is influenced by brine from nearby saltwater springs, which results in total chloride concentrations of 650 mg l $^{-1}$.
5. The purification plant of a factory for transformer insulation compounds of cellulose and synthetic resins releases its wastewater directly into Seckach River, one of the few remaining direct industrial point sources in Baden-Württemberg.
6. Two streams are acidic, caused by acid rain and catchments with extremely low buffering capacity.

The macroinvertebrate data sets originate from a series of qualitative benthic samples collected in these streams from spring 1997 through fall 1998. Samples were collected by hand-picking animals for 30 min from different kinds of substratum, proportionately to the habitats present in the river section, followed by kick-sampling at five randomly chosen section of the streambed. Macroinvertebrates were identified in the laboratory, generally to species, in some groups to family level (Oligochaeta and Diptera). For the preliminary results presented in this paper, we used only the data sets of spring 1998 to exclude seasonal influences.

Tested indices

We selected six functional feeding type indices (Table 2), one of them of European origin (Rhithron-Ernährungstypen-Index RETI; Schweder, 1990), the other five metrics used in rapid bioassessment in the U.S.A. The metrics '% scrapers' and '% shredders' have been proposed by Cummins (1988) and tested by Resh (1994) and Resh et al. (1995). The two ratios were suggested by Merritt & Cummins (1996) as indicators of stream ecosystem attributes, while '% predators' is used in the current B-IBI (Karr, 1999) and the Rapid Bioassessment Protocols RBP (Barbour et al., 1997). For comparison, we chose the metrics '% EPT' and '% Chironomidae' also used in North American multimetric indices, as well as the 'percentage of sessile taxa' (Table 2).

The ratios 'Filtering to Gathering Collectors' and 'Shredders to Total Collectors' as well as the Rhithron-Ernährungstypen-Index RETI yield threshold values and are supposed to be > 0.5, > 0.25 and at least > 0.5, respectively, in undisturbed rhithral benthic communities. RETI was supposed to have detected impairment when dropping below 0.1 from the value of the reference site. The percentage metrics were judged to have discovered impact when decreasing or increasing at least ten percent.

Functional feeding group and locomotion type assignments

The use of functional feeding type metrics involves the use of information on feeding habits of benthic taxa. In North America, such information was mostly derived from the ecological data tables by Merritt & Cummins (1984) or the field guide designed by Cummins & Wilzbach (1985). In 1995 and 1996, respectively, ecological data tables for central European taxa have been published in Austria (Moog, 1995) and Germany (Bayr. Landesamt für Wasserwirtschaft, 1996). For this study, macroinvertebrates were assigned to functional feeding groups and to the locomotion type 'sessile' according to the German 'Ökologische Typisierung der aquatischen Makrofauna'.

Results

Index testing

The nine metrics (Table 2) were tested on 12 cases of disturbance in 11 streams (Table 1). Each was compared to an undisturbed reference site. Significance of differences between reference and disturbed sites were tested using the Wilcoxon Matched Pairs Test (Table 3). Of the six functional feeding group metrics tested, three detected impairment in up to four cases, whereas three indices revealed disturbance in 8–11 cases out of 12. The three other percentages showed a reaction in the expected direction in seven (% EPT), 10 (% Chironomidae) and 12 (% sessile taxa) cases, respectively (Tables 4 and 6).

The proportion of 'Shredders to Total Collectors' is expected to be greater than 0.25 within a normal shredder summer or fall association linked to a functioning riparian system (Merritt & Cummins, 1996). It was less than 0.25 at four impaired sites: the residual flow section, two of the organically polluted sites and in acidic Vorderer Seebach.

The ratio 'Filtering Collectors to Gathering Collectors' is supposed to be less than 0.5 as a result of increased light from loss of riparian zone, increased sedimentation or heavy metals, nutrient loading or toxins absorbed to the suspended load (Merritt & Cummins, 1996). Using our database, either both sites (reference and impairment) reached values less than 0.5 (seven cases) or the value for the impacted site was > 0.5 and at the same time higher than that for the reference site (five cases).

In four cases, the proportion of scrapers (impacted sites versus reference sites) decreased as expected: by 35% in the Brettach residual flow, by 50% in the organically polluted Kressbach, while the two acidic stretches showed the largest decrease with 67 and 69%. Rhithron-Ernährungstypen-Index (RETI) values for the reference sites were at least 0.7, except the reference stream for acidic Kaltenbach (RETI = 0.59). RETI decreased by at least 0.2 in eight cases. Within five of these, RETI values for the impacted sites were well below the postulated 0.5 threshold value for an undisturbed rhithral benthic community. In the residual flow stretch on Brettach River and the heavy metal influence on Fils River, RETI values were slightly over the threshold (0.54 and 0.56, respectively), while the site situated below the fish farm on Buchenbach still reached 0.63. Schweder (1990) recommended the use of the scrapers or shredders proportions as a more appropriate index in cases where RETI values just slightly exceed the limit. In all of the latter cases, shredders were drastically reduced. The additional calculation of '% shredders' detected the high chloride concentrations in Mühlbach and the chemical industry point source on Seckach, which RETI did not.

The percentage of shredders clearly declined by at least 14%, at most 100%, in 10 cases of disturbance. The metric missed one case of acidification and the chicken farm waste water. The measure '% predators' increased in all cases but one, where the benthic community was so much depleted that *Chironomus thummi* was virtually the only species remaining. The increase was low in acidic Kaltenbach (14.2%), since predatory stoneflies were also abundant at the reference site, and in the Brettach residual flow (28.6%). In all other cases, it increased by at least 50%.

The percentage of ephemeropteran, plecopteran and trichopteran taxa (EPT) declined in seven of the 12 cases, but less than the required 10% at cadmium-stressed Bärenbach. In six cases the decline was 18% or more.

The metric 'percentage of Chironomidae', used also in the Rapid Bioassessment Protocols, increased in all cases, but only very slightly with the chicken farm wastewater (0.6%) and at acidic Kaltenbach (3%).

The proportion of sessile taxa at the impaired sites also increased in every case. With the exception of Laibach (36%), the increase was drastic, amounting to 200% or more. In 'Ökologische Typi-

Table 4. Index values. Values for the ratios Shredders to Total Collectors and Filtering Collectors to Gathering Collectors are total values for the impaired sites. Values for Rhithron-Ernährungstypen-Index are absolute differences (minus or plus for the impaired sites); values for the percentage metrics are increase or decrease in percent. *SH/TC = ratio Shredders to Total Collectors; FC/GC = ratio Filtering Collectors to Gathering Collectors; RETI= Rhithron-Ernährungstypen-Index. * = Value for FC/GC < 0.5 for both sites*

Stream code	SH/TC	FC/GC	RETI	% scrapers	% shredders	% predators	% EPT	% Chironomidae	% sessile taxa
BÄ	2.32	0.08*	−0.01	+12	−14	+200	−6	+0.6	+200
FI 1	0.22	0.16	−0.41	+23	−80	+333	+69	+1885	+670
FI 2	0.81	0.73	−0.20	+38	−49	+233	–	+79	–
ZI	1.12	1.00	−0.03	+12	+3	+50	−37	+83	+135
BU	0.37	0.06*	−0.20	+113	−78	+50	+307	+3	+10713
BR	0.18	0.46*	−0.25	−35	−60	+29	−21	+1270	+662
KR	0.70	0.10*	−0.32	−50	−38	−57	−100	+1185	+6433
LA	0.00	0.57	−0.40	+44	−100	+125	−70	+90	+36
MÜ	1.05	0.25*	+0.02	+129	−53	+50	+245	+1692	+1190
SE	0.28	0.08*	−0.08	±0	−68	+212	+0.5	+555	+527
KA	0.48	2.83	−0.20	−69	+9	+14	−27	+760	+693
VS	0.18	1.23	−0.35	−67	−82	+143	−18	+325	+263

Table 5. Comparison of functional feeding group assignments according to 'Fauna Aquatica Austriaca' (FAA: Moog, 1995) and 'Ökologische Typisierung der aquatischen Makrofauna' (ÖTM; Bayr. Landesamt für Wasserwirtschaft, 1996). *SH = shredder; SC = scraper/grazer; GC = gathering collector; FC = filtering collector; PR = predator*

Species	SH		SC		GC		FC		PR	
	FAA	ÖTM	FAA	ÖTM	FAA	ÖTM	FAA	ÖTM	FAA	ÖTM
Asellus aquaticus, Crustacaea	3	7	3	1	4	2	–	–	–	–
Gammarus pulex, Crustacea	6	7	1	1	2	2	–	–	1	–
Hydropsyche siltalai, Trichoptera	–	–	2	1	–	–	5	7	3	2
Odontocerum albicorne, Trichoptera	3	3	3	2	–	3	–	–	4	2
Sericostoma personatum, Trichoptera	9	7	–	1	–	1	–	–	1	1

sierung der aquatischen Makrofauna' (Bayerisches Landesamt für Wasserwirtschaft, 1996) the sessile category includes taxa like the families Tubificidae (10 points out of 10), Simuliidae (seven points) and Chironomidae (four points), Mollusca like *Sphaerium corneum* (five points), but also Trichopteran taxa, e.g. several members of the family Hydropsychidae (four to seven points) or *Odontocerum albicorne* (two points).

Functional feeding group assignments

In Table 5, we compared FFG designations of five abundant species, based on Fauna Aquatica Austriaca (FAA) (Moog, 1995) and on Ökologische Typisierung der aquatischen Makrofauna (ÖTM) (Bayerisches Landesamt für Wasserwirtschaft, 1996). Depending on the source of information, the lists differ in their classifications, resulting in different scores for the calculation of indices.

Discussion

The proportion of 'Shredders to Total Collectors', expected to be > 0.25 within a normal shredder summer or fall macroinvertebrate association, was less than that only in four cases. As we used a spring sample for testing, the validity of this metric remains open. The ratio 'Filtering Collectors to Gathering Collectors', supposed to be less than 0.5 with an impairment, either reached values less than 0.5 for both sites, or the value for the impacted site was > 0.5, while the reference sites showed lower values than the impaired

Table 6. Descriptive statistics for tested metrics (medians and quartiles) (n = 14). *SH/TC = ratio Shredders to Total Collectors; FC/GC = ratio Filtering Collectors to Gathering Collectors; RETI= Rhithron-Ernährungstypen-Index. R = reference site; I = impaired site*

Metric	Site	Median	Minimum	Maximum	Lower quartile	Upper quartile
SH/TC	R	2.34	0.70	3.40	1.16	2.80
	I	0.42	0.00	2.50	0.20	1.09
FC/GC	R	0.07	0.00	0.42	0.02	0.10
	I	0.24	0.02	2.91	0.09	0.73
RETI	R	0.75	0.58	0.80	0.72	0.77
	I	0.58	0.29	0.78	0.38	0.64
% scrapers	R	0.15	0.10	0.61	0.13	0.26
	I	0.19	0.05	0.50	0.13	0.38
% shredders	R	0.56	0.15	0.66	0.29	0.60
	I	0.18	0.00	0.58	0.07	0.36
% predators	R	0.04	0.01	0.21	0.02	0.07
	I	0.07	0.02	0.43	0.04	0.14
% EPT	R	0.23	0.85	0.79	0.09	0.65
	I	0.42	0.00	0.82	0.16	0.62
% Chironomidae	R	0.15	0.00	0.12	0.00	0.04
	I	0.50	0.00	0.96	0.01	0.35
% sessile taxa	R	0.01	0.00	0.06	0.00	0.03
	I	0.16	0.00	0.44	0.03	0.28

sites. Both ratios yielded high ranges, and it was therefore impossible to define threshold values or interpret our results in more detail. Pending further investigation, these ratios do not seem to be appropriate for ecological assessment in southern Germany.

The Rhithron-Ernährungstypen-Index RETI, often successfully used in southern Germany (Bayrisches Landesamt für Wasserwirtschaft, 1996) missed only the four cases less frequently detected by all the feeding type metrics. Nevertheless, the proportion of shredders, suggested by Schweder (1990) as an alternative in doubtful cases, detected impairment in two of the streams RETI missed.

The proportion of scrapers (impacted sites versus reference sites) decreased in four cases, among them the two acidic streams. Studies of an accidental acid spill in a Californian stream revealed that the percentage of scrapers was the only functional measure to indicate change (Resh & Jackson, 1993; Resh, 1994) and judged as a suitable component of an ideal protocol for monitoring acid and thermal impacts in California coastal streams (Resh & Jackson, 1993). For both cases of acidification, the decrease in scrapers was similar (67 and 69%), while the values for all other metrics were rather variable (Table 4). The scraper community may be especially sensitive to this kind of impairment.

The percentage of predators was the most successful functional feeding group metric. Except for one obvious case, it was the only FFG index to indicate change in all categories of impairment and seems to be a rather reliable metric.

Experience gained by using functional feeding group indices as components of several multimetric indices in the U.S.A. lead to the conclusion that they might not be as useful in bioassessment as expected. In the current B-IBI, only one out of the ten measures used is a feeding group (Karr, 1999). While they are still regarded as metrics providing information not readily obtainable from taxonomic metrics (Barbour et al., 1996), functional feeding group measures deliver highly variable responses for a variety of ecoregions. Percentage of predators, still retained in the B-IBI, was considered uninformative and, therefore, eliminated from the Florida Stream Condition Index SCI, as was% scrapers (Barbour et al., 1996).

The number or percentage of Ephemeroptera, Trichoptera and Plecoptera taxa, used singly or combined as EPT taxa, have been tested and considered effective throughout the U.S.A. (DeShon, 1995; Barbour et al., 1996, 1997; Karr, 1999). The metric 'percentage of Chironomidae' is also suggested by the Rapid Bioassessment Protocols (Barbour et al., 1997), and a 'behavioural' measure such as 'percentage of clingers', proposed by Karr (1999), has been found useful in low-gradient as well as high-gradient streams (M. Barbour, pers. com.). While '% EPT' was somewhat less successful in detecting impairment than several of the FFG metrics, '% Chironomidae' increased drastically in 10 out of 12 cases with our data. The behavioural type 'clinger' is not a classification used in Ökologische Typisierung der aquatischen Makrofauna (Bayerisches Landesamt für Wasserwirtschaft, 1996), but the category 'sessile taxa', calculated for our data, showed a clear increase at all the impaired sites.

In summary, the two latter metrics performed as well or better than the three most successful FFG measures and should be clearly included in a future multimetric index as well.

The proper choice of a reference site may be a crucial point in the implementation of "guild" metrics. Our reference sites were located in the same stream as the disturbed site and shared, therefore, the same running water type. However, in studies using multimetric approaches in the U.S.A., references of necessity were based on the ecoregion level, resulting in higher variation and consequently a lower ability to clearly separate disturbed from reference sites. Our direct approach might therefore add to the accuracy of FFG measures in detecting disturbance.

Functional feeding group assignments

The question of the assignment of a functional feeding group may be a crucial one in using FFG metrics. Cummins (1988) criticised the interchangeable use of functional group designations by trophic levels (herbivores, detritivores, carnivores and omnivores, or specialists *versus* generalists) and by the means of food acquisition, emphasising that the latter, based on mouthpart morphology and behaviour, should be preferred. Published functional feeding group classifications do not always indicate obligate resource utilisation (Mihuc & Minshall, 1995). In contrast to the assumption that the scraper and shredder group contains specialized feeders, Mihuc (1997) stated that the scraper group is composed of taxa occupying the trophic roles of herbivores, detritivores and generalists, and shredders contain both generalists and detritivores. The practice of grouping taxa into a single trophic category based on FFG assignment may not be an appropriate use of this concept (Hawkins & MacMahon, 1989). One macroinvertebrate individual can use different strategies either at one time or during different life stages. European ecological lists try to take these problems into account by distributing ten points for each species rather than assigning it to a single feeding type.

Although a 'weighted' list may be more accurate than assignment to a single feeding type, one of the problems is the splitting of one taxon into several FFG entities. For instance, 500 specimens of *Gammarus pulex* at a given study site split into 300 shredders, 50 scrapers/grazers , 100 gathering collectors and 50 predators, when using the ÖTM, although they do not exist as real animals. In contrast, assigning the same specimens of *Gammarus pulex* to feeding groups according to FAA yields no predators because the distribution of points diverges, giving emphasis to different feeding groups. These disparities may result in considerable variances in the calculation of metrics. We calculated the percentages of scrapers and of shredders for our data according to ÖTM (Bayrisches Landesamt für Wasserwirtschaft, 1996) or FAA (Moog, 1995), respectively. The percentage of scrapers differed between lists from 6 to 27% at almost 30% of the study sites, and the percentage of shredders varied from 5 to 54% at over 50% of the sites. This illustrates that accurate autecological information is still lacking (Mihuc, 1997); also despite a wealth of examples emphasising the usefulness of the guild concept, guild theory does little to elucidate ecosystem structure and functioning (Hawkins & MacMahon, 1989). The use of functional feeding group metrics still might be promising, but much work has to be done in autecological research in order to improve and standardise FFG assignments.

References

Barbour, M. T., J. Gerritsen, B. D. Snyder & J. B. Stribling, 1997. Revision to Rapid Bioassessment Protocols for use in streams and rivers. Periphyton. benthic macroinvertebrates and fish. Http://www.epa.gov/owowwtr1/monitoring/AWPD/RBP/html.

Barbour, M. T., J. Gerritsen, G. E. Griffith, R. Frydenborg, E. McCarron, J. S. White & M. L. Bastian, 1996. A framework for biological criteria for Florida streams using benthic macroinvertebrates. J. n. am. Benthol. Soc. 15: 185–211.

Bayerisches Landesamt für Wasserwirtschaft (ed.), 1996. Ökologische Typisierung der Aquatischen Makrofauna. Informationsberichte 4/96. München: 543 pp.

Cummins, K. W., 1988. Rapid bioassessment using functional analysis of running water invertebrates. In Simon T. P., L. L. Holst & L. J. Shepard (eds), Proceedings of the First National Workshop on Biological Criteria. EPA-905/9-89/003. U.S. Environmental Protection Agency, Chicago: 49–54.

Cummins, K. W. & M. A. Wilzbach, 1985. Field procedures for analysis of functional feeding groups of stream macroinvertebrates. Contribution 1611, Appalachian Environmental Laboratory, University of Maryland, Frostburg: 18 pp.

DeShon, J. E., 1995. Development and applicability of the Invertebrate Community Index. In Davis W. S. & T. P. Simon (eds), Biological Assessment and Criteria: Tools for Water Resource Planning and Decision-making. Lewis Publishers, Boca Raton, FL: 415 pp.

Hawkins, C. P. & J. A. MacMahon, 1989. Guilds: The multiple meanings of a concept. Ann. Rev. Ent. 34: 423–451.

Karr, J. R., 1999. Defining and measuring river health. Freshwat. Biol. In press.

Kerans, B. L. & J. R. Karr, 1994. A benthic index of biotic integrity (B-IBI) for rivers of the Tennessee Valley. Ecol. Appl. 4: 768–785.

MACS (Mid-Atlantic Coastal Streams Workgroup), 1996. Standard operating procedures and technical basis: Macroinvertebrate collection and habitat assessment for low-gradient nontidal streams. Delaware Department of Natural Resources and Environmental Conservation, Dover, DE.

Merritt, R. W. & K. W. Cummins (eds), 1984. An Introduction to the Aquatic Insects of North America. 2nd edn. Kendall/Hunt Publishing Co., Dubuque, IA: 862 pp.

Merritt, R. W. & K. W. Cummins, 1996. Trophic relations of macroinvertebrates. In Hauer, F. R. & G. A. Lamberti (eds), Methods in Stream Ecology. Academic Press Inc., San Diego: 453–474.

Mihuc, T. B., 1997. The functional trophic role of lotic primary consumers: generalist versus specialist strategies. Freshwat. Biol. 37: 455–462.

Mihuc, T. B. & G. W. Minshall, 1995. Trophic generalists vs. trophic specialists: implications for food web dynamics in post-fire streams. Ecology 76: 2361–2372.

Moog, O. (ed.), 1995. Fauna Aquatica Austriaca. Wasserwirtschaftskataster, Bundesministerium für Land- und Fortswirtschaft. Wien, Loseblattsammlung.

Resh, V. H., 1994. Variability, accuracy, and taxonomic cost of rapid assessment approaches in benthic macroinvertebrate biomonitoring. Boll. Zool. 61: 375–383.

Resh, V. H. & J. K. Jackson, 1993. Rapid assessment approaches to biomonitoring using benthic macroinvertebrates. In Rosenberg, D. M. & V. H. Resh (eds), Freshwater Biomonitoring and Benthic Macroinvertebrates. Chapman & Hall, New York: 195–223.

Resh, V. H., R. H. Norris & M. T. Barbour, 1995. Design and implementation of rapid assessment approaches for water resource monitoring using benthic macroinvertebrates. Aust. J. Ecol. 20: 108–121.

Schweder, H., 1990. Neue Indizes für die Bewertung des ökologischen Zustandes von Fließgewässern, abgeleitet aus der Makroinvertebraten-Ernährungstypologie. In Friedrich, G. & J. Lacombe (eds), Ökologische Bewertung von Fließgewässern. Limnologie aktuell 3. G. Fischer Verlag, Stuttgart: 353–377.

Hydrobiologia **422/423:** 233–243, 2000.
M. Jungwirth, S. Muhar & S. Schmutz (eds), Assessing the Ecological Integrity of Running Waters.

The statistical power of selected indicator metrics using macroinvertebrates for assessing acidification and eutrophication of running waters

Leonard Sandin & Richard K. Johnson
Department of Environmental Assessment, Swedish University of Agricultural Sciences, Box 7050, S-750 07 Uppsala, Sweden

Key words: biological indices, macroinvertebrates, acidification, eutrophication, statistical power, type II error

Abstract

Ten benthic macroinvertebrate indicator metrics commonly used in the assessment of running waters (taxa richness, total density, number of EPT taxa, Shannon's index, Simpson's index, ASPT, DFI, and three acidity indices) were evaluated for effect size, standardized effect size, statistical power and the number of samples needed to detect an impact of either eutrophication or acidification. The eutrophication and acidity indices were also evaluated for type I and type II errors. Indices were evaluated using the coefficient of variation of data taken from a spatial dataset of some 700 sites distributed across Sweden, and a temporal dataset of five streams sampled over a time period of 6–11 years. Richness measures (taxa richness and number of EPT taxa) and the two eutrophication indices (ASPT and DFI) had the highest standardized effect sizes as well as a high statistical power and therefore needed fewer samples to detect changes caused by perturbation. Conversely, total density of individuals was the least informative metric with the lowest standardized effect size and the highest spatial, temporal and sample variability.

Introduction

Running waters are among the most important natural resources, but also among the most threatened on earth. Eutrophication, i.e. the disposal of organic matter into aquatic systems, is one of the oldest and most fully documented forms of pollution. In northwestern Europe and eastern North America, acidification is also a serious environmental problem (Johnson et al., 1993). It is therefore very important to have adequate tools to assess and monitor the water quality and biota of streams and rivers. Since the beginning of the 20th century, a large number of biological indicator metrics have been developed for fresh waters (e.g., De Pauw & Vanhooren, 1983; Johnson, 1995; Metcalfe, 1989) and, indeed, biological indices are among the most commonly used methods for biological monitoring of rivers (Norris & Georges, 1993). Moreover, the use of biological indices often influences policy decisions regarding surface water quality in Europe, and they are used as tools for water management and to monitor the effects of pollution control measures (Metcalfe, 1989).

In biomonitoring, the use of indicators often entails comparing an observed index or score with an expected (reference) value. The ability to detect change using biological indicator metrics is constrained and confounded, however, by a number of factors of which the lack of well-addressed questions and weakness in study design are among the most important (Johnson, 1995). The use of biological indicators in biomonitoring involves the test of a null hypothesis (i.e., no change in indicator value or score). The power of the study design is affected by a number of factors, the most important being: (i) the α level selected; (ii) statistical assumptions; (iii) sample size and variability; and (iv) the effect to be detected (Lipsey, 1990). Type I error or α is the probability of rejecting the null hypothesis when it is in fact true; the β or type II error is the probability to fail to reject the null hypothesis when in fact it is false; and statistical power $(1-\beta)$ is the capacity of a test to correctly reject the null hypothesis (Underwood, 1993). Commonly, the investigator reports some level of α or probability of type I error in biomonitoring studies (Johnson, 1995);

traditionally, inferential statistical decision making has been preoccupied with type I errors, when in fact an incorrect conclusion of no impact (i.e., type II error or β) might be considered more serious because environmental degradation is occurring but not being detected (Mapstone, 1995).

In this study, we evaluate 10 biological indicator metrics commonly used in the assessment of running waters. Two richness measures (taxa richness and the number of EPT taxa), two community indices (Shannon's index and Simpson's index), an enumeration index (total density), and five biotic indices (i.e., two eutrophication indices, ASPT and DFI, and three acidity indices) were evaluated. The aim of this study was to examine how well the different indices performed in detecting anthropogenic impact, and to determine the number of independent samples (either spatial or temporal) needed to detect change with a known degree of certainty (i.e., statistical power).

Materials and Methods

Spatial data

Data were obtained from the National Stream Survey (Wilander et al., 1998), where 694 stratified and randomly selected sites were sampled in autumn 1995 for benthic macroinvertebrates using a handnet (0.5-mm mesh size) and kick sampling in riffles according to Swedish standard (SIS, 1994). Incorrectly sampled (n = 38), limed (n = 99), and sites moderately affected by acidification and eutrophication ($5.5 < \text{pH} < 6.5$ and a total phosphorous content of $15 < \text{TP} < 50\ \mu\text{g l}^{-1}$, n = 227) were removed. The remaining streams were divided into those not significantly affected by anthropogenic stress (i.e., here referred to as 'spatial' reference sites and defined as having a pH $\geq$ 6.5 and a total phosphorous content $\leq 15\ \mu\text{g l}^{-1}$, n = 246) and sites affected by acidification (pH $\leq$ 5.5, n = 8) or eutrophication (total phosphorous content $\geq 50\ \mu\text{g l}^{-1}$, n = 78). Most sites affected by eutrophication and all acidified sites were situated in the southern parts of the country, and therefore equal sized groups of reference sites situated close to the perturbed streams were used in calculating the effect size and standardized effect size (Figure 1).

Figure 1. Streams sampled for benthic macroinvertebrates: acidified streams (n=8), eutrophicated streams (n=78), non-affected streams, i.e., spatial references (n=246); some of the non-affected streams were also used as reference for acidified streams (n=8) and eutrophicated streams (n=78), and temporal references (n=5).

Neither altitude nor stream velocity differed between the acidified and acidification reference sites (P > 0.05). The eutrophicated sites, on the other hand, were found at a lower altitude (P < 0.001) and were more slow flowing (P < 0.001) than the eutrophication reference sites (Table 1). To determine if indicator values for the eutrophication reference sites could be considered as references for eutrophicated sites, the eutrophication reference dataset was compared between two groups; one group with the lower quartile of sites found between 1 and 80 m.a.s.l. and the other group consisting of the upper quartile of sites found between 230 and 465 m.a.s.l. No significant differences in indicator values were found for taxa richness, total density, number of EPT taxa, Shannon's index, Simpson's index, and ASPT (P > 0.05). For

Table 1. Selected chemical and physical characteristics (mean ± 1 SE) of reference (*R*) (n=246), eutrophicated (E) (n=78) eutrophication references (ER) (n=78) acidified (A) (n=8) and acidification references (AR) (n=8) streams

	R	E	ER	A	AR
Altitude[a]	314 ± 11	38 ± 6	160 ± 13	117 ± 16	138 ± 31
Stream velocity[b]	2.2 ± 0.1	1.2 ± 0.1	2.0 ± 0.1	2.4 ± 0.3	2.1 ± 0.2
pH	7.03 ± 0.02	7.14 ± 0.06	6.88 ± 0.03	4.89 ± 0.13	6.81 ± 0.07
Total nitrogen ($\mu g\ l^{-1}$)	432 ± 26	3353 ± 330	703 ± 62	864 ± 135	887 ± 228
Total phosphorous ($\mu g\ l^{-1}$)	8.8 ± 0.2	116.6 ± 11.2	10.8 ± 0.3	18.6 ± 2.9	11.3 ± 0.8
Alkalinity (meq l^{-1})	0.33 ± 0.04	2.02 ± 0.19	0.36 ± 0.08	−0.03 ± 0.01	0.25 ± 0.07
TOC (mg l^{-1})	5.9 ± 0.3	10.6 ± 0.6	9.1 ± 0.5	18.2 ± 1.0	9.2 ± 1.3

[a]Meters above sea level. [b]See text.

DFI, lower indicator values were observed at higher altitudes ($P < 0.05$).

Temporal data

Five small forested streams were sampled for macroinvertebrates using a handnet (0.5-mm mesh size) and kick sampling in riffles as described above. Five replicate samples were taken once a year (in the period April to July) from these 'temporal' reference streams over a period of 6–11 years (1986–1996). Chemical measures were taken once a month during this period. The streams ranged from acidified (median pH 4.3) to well buffered (median pH 6.8), and all streams were relatively nutrient poor (median total phosphorous ranged from 8 to 15 $\mu g\ l^{-1}$). The streams were also situated along a latitudinal gradient, from the nemoral ecoregion of southern Sweden to the boreal regions in the north (Figure 1).

Taxa were identified to a predetermined taxonomic resolution (generally to species or species group, with the exception of chironomids and oligochaetes). Stream velocity of streams sampled in the national survey (spatial dataset) was classified as: (0) 0 m s^{-1}; (1) 0–0.2 m s^{-1}; (2) 0.2–0.7 m s^{-1}; (3) >0.7 m s^{-1}. A complete list of the taxonomic resolution and the chemical methods used in the national survey is found in Wilander et al. (1998).

Biological indicator metrics

Ten biological indices commonly used in biomonitoring programs (Johnson et al., 1993) were evaluated for spatial (within and among stream) and temporal (among-year) variability, as well as statistical power. The metrics used to assess general ecological quality were taxa richness, richness of Ephemeroptera, Plecoptera and Trichoptera (EPT) taxa (Lenat, 1988), and total density. For acidification three indices were evaluated: in this paper referred to as acidity index I (Raddum et al., 1988), acidity index II (Degerman et al., 1994) and acidity index III (Bækken & Aanes, 1995). For eutrophication, two indices were assessed: Average Score Per Taxon (ASPT) (Armitage et al., 1983) and the Danish Fauna Index (DFI) (Kirkegaard et al., 1992). Lastly, two diversity indices, Shannon's index (Shannon, 1948) and Simpson's diversity index (Simpson, 1949) were evaluated. Type I and type II errors were also evaluated for the three acidification indices and the two eutrophication indices since these indices have defined scores where a sampled site is deemed as affected according to the index (Table 2).

Table 2. The score of the biological indices was divided into three categories: NA, non-affected; MA, moderately affected; and A, affected according to the index

Index	NA	MA	A
ASPT	≥6	5	≤4
DFI	1.5	2	≥2.5
Acidity index I	1	0.5	≤0.25
Acidity index II	5	4	≤3
Acidity index III	1	2	≥3

All three acidity indices are similar in that site classification is based on the most sensitive individual taxon found at a site (e.g., the recording of one individual from the non-affected class is all that is required to classify the site as such). For example, in acidity index I if a taxon that only tolerates pH > 5.5 is found at a site, then the site is given the score of 1, if no taxa with scores of 1 are found, but a taxon that tolerates pH > 5.0 is present, then the site is scored as 0.5 and so on. Acidity index I developed for the western parts of Norway ranges from 0 (highly acidified) to 1 (no acidification stress) and a list of 43 taxa is used in the classification of sites. Acidity index II is an alteration of acidity index I modified for Sweden. The scoring system of this index ranges from 1 (highly acidified), where all taxa found tolerate pH < 4.0, to a score of 5 (non-acidified), where at least one taxon that does not tolerate pH < 5.5 is found. The index includes some 140 taxa in the taxa list. Acidity index III, developed to assess acidification in Norway, ranges from 1 (no acidification), where at least one taxon that does not tolerate pH < 5.5 is found, to 4 (highly acidified), where only taxa that tolerate pH < 4.7 are found. Thirty-three taxa are included in this index. The ASPT indicator metric was developed to assess the water quality of rivers in the UK; each family found at a site is given a score between 1 and 10, where pollution-tolerant families (e.g., Oligochaeta) are given the lowest, and sensitive families (e.g., Siphlonuridae) the highest score. The ASPT index is then calculated as the total score divided by the number of scoring taxa. The DFI is similar to the Trent Biotic Index (Woodiwiss, 1964) and was developed to evaluate pollution of running waters in Denmark. Values range from 1 to 4, with 1 representing the best ecological quality. The Shannon diversity index was calculated as

$$H' = -\sum_{i=1}^{s} \frac{n_i}{N} \log_2 \frac{n_i}{N}, \qquad (1)$$

where n_i is the abundance of species i, N is the total number of individuals and s is the number of species.

Simpson's diversity index was calculated as

$$D_s = 1 - \frac{\sum_{i=1}^{s} n_i(n_i - 1)}{n(N-1)}, \qquad (2)$$

where n_i, N and s are interpreted as above.

Indicator variability

Indicator metric variability was compared across spatial and temporal scales. Spatial variability was calculated as

$$\mathrm{CV}_{\mathrm{spatial}} = \frac{\text{std metric}}{\text{metric}}, \qquad (3)$$

using the 246 spatial reference sites. Sites that lacked indicator taxa for acidity indices were not included in the calculations (i.e., one site for acidity index I and 3 sites for acidity index III).

Temporal variability was calculated as

$$\mathrm{CV}_{\mathrm{temporal}} = \frac{\sum_{n=1}^{5} \mathrm{CV}_{(m)n}}{n}, \qquad (4)$$

where n is the number of streams (5) and m is the number of years (6–11). For each sampling occasion the mean of the five kick samples were used in the calculations.

Among-sample variability was calculated as

$$\mathrm{CV}_{\mathrm{Sample}} = \frac{\sum_{n=1}^{5} \sum_{s=1}^{11} \mathrm{CV}_{\mathrm{kick}}}{ns}, \qquad (5)$$

where n is the number of streams and s is the number of years. $\mathrm{CV}_{\mathrm{kick}}$ is the coefficient of variation of the five kick samples taken on each sampling occasion.

Effect size

To evaluate the effect size for the different metrics the spatial stream data were divided into affected and non-affected sites as described above. Assuming that the macroinvertebrate communities of eutrophicated and acidified streams were similar to the communities of non-impacted streams before impact, then the difference between indicator metrics of impacted and non-impacted streams provides a measure of the magnitude of impact or effect size (δ) from eutrophication and acidification, respectively (Johnson, 1998). The temporal references provide an estimate of among-year variability of the indicator metrics. To calculate the expected effect size, the mean index values of the 78 eutrophicated sites (with a total phosphorous content $\geq 50\ \mu g\ l^{-1}$) were compared to the mean index value of the same number of non-impacted sites (eutrophication references, with a total phosphorous content $\leq 15\ \mu g\ l^{-1}$) situated geographically close to each other. Similarly, for acidification, the mean index values for sites with pH ≤ 5.5 ($n = 8$) were compared with the mean index value of the same number of non-impacted (pH ≥ 5.5, $n = 8$) adjacent sites (acidification references) (Table 3).

Effect size was calculated as

$$\text{Effect size}(\delta) = \delta_{\mathrm{B}} - \delta_{\mathrm{A}}, \qquad (6)$$

where δ_{B} and δ_{A} are the mean values of indicator values of non-impacted (B, before period, i.e., reference conditions) and impacted (A, after period, i.e., eutrophicated or acidified conditions) streams; in the

Table 3. Indicator values (mean ± 1 SE) for 10 biological indicator metrics commonly used in the assessment of running waters. Mean indicator values for reference condition (R) (n=246) were calculated using streams with a total phosphorous level $\leq 15\ \mu g\ l^{-1}$ and a pH $\geq$ 6.5. Eutrophicated streams (E) (n=78) with a total phosphorous level $\geq 50 \mu g\ l^{-1}$, eutrophication references (ER) (n=78) with a total phosphorous level $\leq 15 \mu g\ l^{-1}$, acidified streams (A) (n=8) with a pH $\leq$ 5.5 and acidification references (AR) (n=8) with a pH $\geq$ 6.5

Indicator metric	R	E	ER	A	AR
Taxa richness	21.9 ± 0.5	15.9 ± 0.7	25.1 ± 0.9	17.5 ± 2.9	25.5 ± 3.0
Total density	1311 ± 95	2051 ± 305	1759 ± 196	935 ± 359	1433 ± 385
EPT	13.5 ± 0.4	4.9 ± 0.4	14.5 ± 0.7	9.3 ± 1.9	15.4 ± 2.0
Shannon's index	2.90 ± 0.05	2.27 ± 0.08	3.06 ± 0.09	2.44 ± 0.28	3.20 ± 0.21
Simpson's index	0.76 ± 0.01	0.67 ± 0.02	0.78 ± 0.02	0.70 ± 0.07	0.82 ± 0.03
ASPT	6.41 ± 0.72	4.71 ± 0.08	6.28 ± 0.07	–	–
DFI	1.50 ± 0.59	2.45 ± 0.52	5.97 ± 0.14	–	–
Acidity index I[a]	0.95 ± 0.01	–	–	0.44 ± 0.13	0.78 ± 0.15
Acidity index II	4.78 ± 0.04	–	–	3.25 ± 0.37	4.25 ± 0.41
Acidity index III[b]	1.10 ± 0.03	–	–	2.50 ± 0.46	1.38 ± 0.38

[a]245 reference sites.
[b]243 reference sites used in calculations.

comparisons of effect sizes and standardized effect sizes, log-transformed indicator values were used.

To obtain a measure of confidence or precision of the effect size, standardized effect sizes were calculated. Here the temporal indicator variability is taken into account and the effect size is expressed in standard deviation units:

$$\text{Standardized effect size} = |\delta_B - \delta_A|/(2S_\delta), \quad (7)$$

where S_δ is the standard deviation of deltas calculated in the same way as $CV_{temporal}$ above. The temporal standard deviation of deltas is doubled since a double-sided test is assumed. Indicator metrics with a large standardized effect size (i.e., relatively large effect size and small variability) will yield more powerful assessments of impact than metrics with a small standardized effect size (Osenberg et al., 1994).

Type I and II errors

For the three acidity indices and the two eutrophication indices, the type I or α and type II or β error were calculated using the spatial reference sites. Individual spatial reference sites were scored according to the various indices and then divided into three categories: (i) no detection of impact, (ii) small or intermediate effects, or (iii) eutrophicated or acidified according to the biological index (Table 2). The type I error was then calculated as the number of spatial reference sites deemed as affected by the biological index divided by the total number of spatial reference sites. Similarly, the frequency of type II errors was calculated as the number of impacted sites (i.e., acidified or eutrophicated) that were deemed as non-impacted by the biological index divided by the total number of impacted sites.

Statistical power

Statistical power (1–β), or the probability of rejecting the null hypothesis when the hypothesis is in fact false and should be rejected (Zar, 1996), was calculated for the 10 metrics. The indicator effect size (δ) used was the difference between the impacted and non-impacted sites as described above. Statistical power was estimated using a two-tailed t-test design with α = 0.05 and n = 5 where the test statistic ϕ was calculated according to Zar (1996) as

$$\phi = \sqrt{\frac{n\delta^2}{4s^2}}, \quad (8)$$

where n is the number of samples, δ is the effect size (log-transformed) and s^2 is the mean variance of the samples, calculated in the same way as CV_{sample}. The statistical power was obtained by observing where ϕ met the power curve for the appropriate number of samples. The number of independent samples needed to detect a change as large as the effect size for each index with a statistical power of 95% and α = 0.05 was also calculated assuming either a t-test or linear regression. The number of samples needed assuming a one-tailed t-test was solved iteratively according to Zar (1996) by changing the sample size until the desired statistical power was attained and using the sample variance calculated in the same way as CV_{sample}. The number of samples needed using linear regression was solved using the algorithm of Gerrodette (1987; 1991):

$$r^2 n(n-1)(n+1) = 12\mathrm{CV}_{\mathrm{temporal}}{}^2, \times (Z_\alpha + Z_\beta)^2 x\{1 + r(n-1)(1 + r/6(2n-1))\}, \quad (9)$$

where n is the number of samples (years) needed, r is the rate of change of the indicator metric assuming a response time of 10 years (i.e., $r = (1 - (\delta_A/\delta_B)) / (q - 1)$ where δ_A is the mean index value of the perturbed sites, δ_B is mean index value of the unperturbed sites and q = 10 years), $CV_{temporal}$ is the temporal indicator metric variability, and α and β are the probabilities of type I and type II errors, respectively; they were both set to 0.05.

Results

Indicator variability

Indicator metric variability was generally higher for the among-stream than for temporal (among year) and within-site (among replicate) data (Figure 2). For the three indices of general ecological quality, total density had the highest CV of all indicator metrics tested (114%), whereas taxa richness (37%) and number of EPT taxa (44%) had a much lower variability for the 246 spatial reference sites. The acidity indices had spatial CVs between 12% (acidity index II) and 50% (acidity index III). The eutrophication indices had spatial CVs of 11% for ASPT and 39% for the DFI, and the diversity indices had CVs of 26% (Shannon's index) and 20% (Simpson's index). For the temporal reference sites, total density again had the highest CV (67%), but acidity index I also exhibited relatively high among-year variability (CV = 46%). ASPT and DFI, on the other hand, had relatively low among-year variability, 7 and 4%, respectively. All other indicator metrics ranged between 16% for Simpson's diversity index to 29% for Shannon's index. Within-site or among-sample variability was again highest (42%) for total density, and similarly two of the acidity indices had relatively high variability (32% for acidity index I and 31% for acidity index III). The remaining indices had values between 9% for the DFI and 26% for number of EPT taxa.

Table 4. Type I and II errors for selected indicators of acidification and eutrophication. NA/A, non-affected sites (according to chemistry, n=246, pH $\geq$ 6.5 and total phosphorous $\leq$ 15 μg l^{-1}) classified as affected by the biological index. A/NA, affected sites (according to chemistry, pH $\leq$ 5.5 and total phosphorous $\geq$ 50 μg l^{-1}) classified as non-affected by the biological index. Type I, the percentage of sites where the biological index detects an impact, but the site is not perturbed (according to chemistry) and type II, the percentage of sites where the site is perturbed (according to chemistry) but the biological index cannot detect it

Index	NA/A	A/NA	Type I (%)	Type II (%)
ASPT	13	5	5.3	6.4
DFI	33	7	13.4	9.0
Acidity index I[a]	63	2	25.7	25.0
Acidity index II	33	0	13.4	0.0
Acidity index III[b]	56	2	23.1	25.0

[a] 245 reference sites.
[b] 243 reference sites used in calculations.

Effect size

The highest effect size for eutrophication was found for total number of EPT taxa (0.46), whereas taxa richness (0.19) and the two eutrophication indices DFI (0.14) and ASPT (0.11) were somewhat lower (Fig. 3a). For acidification, total density had the highest absolute effect size (0.26), and both taxa richness and acidity index III had effect sizes of 0.17, and the remaining two acidity indices had effect sizes $\leq$ 0.10.

Another order of ranking of indicator metrics was observed when metrics were compared using standardized effect sizes (Figure 3b). For eutrophication, DFI had a relatively high standardized effect size (4.8) due to relatively low temporal variability. The number of EPT taxa (3.2), ASPT (2.2) and taxa richness (1.0) also had relatively high standardized effect sizes. Total density, on the other hand, had a very low value (0.16) due to its relatively high temporal variability. For acidification, the number of EPT taxa had the highest standardized effect size (1.5), whereas the three acidity indices had values between 1.2 (acidity index III) and 0.85 (acidity index II), and, again, total density had the lowest standardized effect size (0.45).

Type I and II errors

For the acidity indices, acidity index II had the lowest type I (13%) and II (0%) errors, whereas acidity index I had both the highest type I (26%) and II (25%) errors (Table 4). Type I and II errors were lower for

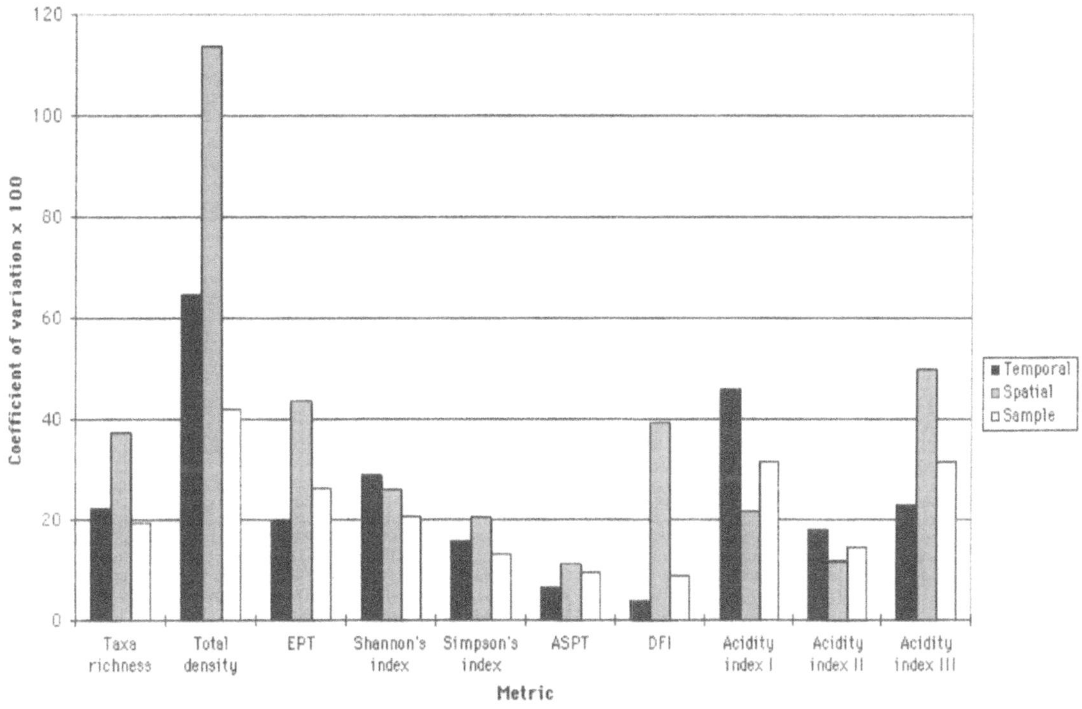

Figure 2. Spatial, temporal and sample variability (coefficient of variation, CV) of 10 biological indicator metrics commonly used in the assessment of running waters. Spatial variability calculated from 246 sites classified as non-affected by acidification or eutrophication. Temporal variability and sample variability calculated from five time-series reference streams sampled between 6 and 11 years. Temporal variability was calculated using the mean of five kick samples at each sampling location, whereas the sample variability was calculated using all the individual samples.

Table 5. Statistical power estimates for 10 biological indicator metrics commonly used in the ecological assessment of running waters. Power ($1-\beta$) was calculated using $\alpha = 0.05$, $n = 5$, mean among sample variability (s^2) calculated from time-series data and an effect size (δ) calculated as the difference in indicator value between affected (eutrophicated and acidified, respectively) and non-affected sites. Sample units needed to detect a change as large as the effect size, assuming either a *t*-test (N_{95}) or linear regression (R_{95}) with a 10-year response time, were also calculated. Both calculations were done assuming a one-tailed test, a statistical power of 95% and a 5% probability of a type I error. For the *t*-test sample variability, and for the linear regression, temporal variability was used in the calculations

	Power Eutrophicated	Power Acidified	N_{95} Eutrophicated	N_{95} Acidified	R_{95} Eutrophicated	R_{95} Acidified
Taxa richness	> 0.99	> 0.99	<3	3	20	22
Total density	> 0.99	> 0.99	4	<3	78	55
EPT	> 0.99	> 0.99	<3	<3	13	17
Shannon's index	0.67	0.62	9	10	31	33
Simpson's index	0.40	0.45	17	17	27	27
ASPT	0.95	–	5	–	10	–
DFI	0.97	–	5	–	4	–
Acidity index I[a]	–	0.93	–	6	–	37
Acidity index II	–	0.72	–	9	–	22
Acidity index III[b]	–	0.35	–	20	–	13

[a]245 reference sites.
[b]243 reference sites used in calculations

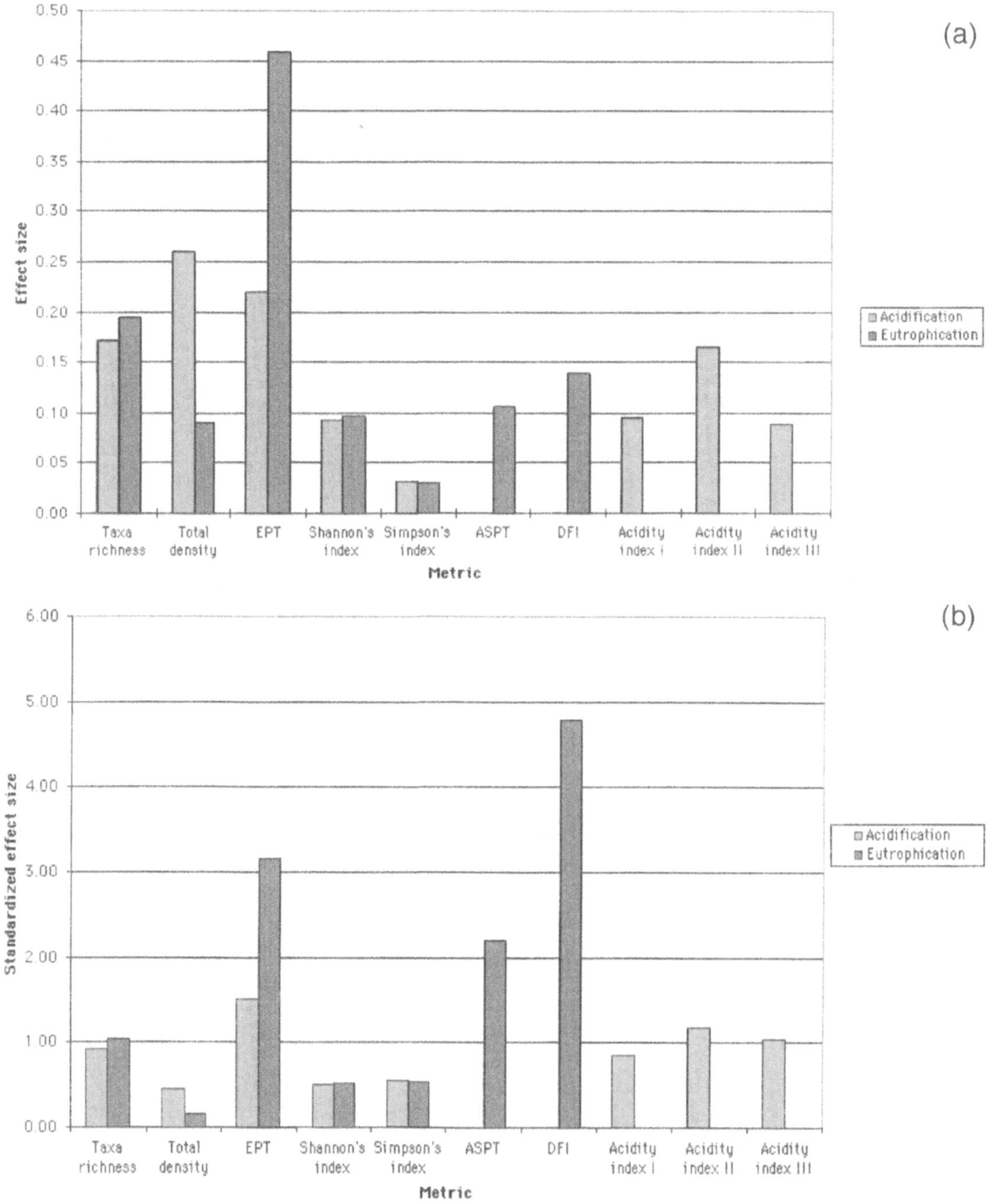

Figure 3. Mean absolute effect size (a) and standardized effect size (b) for indicator metrics. Effect sizes calculated for eutrophicated streams (n=78) and acidified streams (n=8) using the same number of reference streams as affected streams in the calculations.

ASPT, 5.3% and 6.4%, respectively, compared to the DFI (13% and 9%) (Table 4).

Statistical power

The statistical power to detect a difference between the eutrophicated and non-impacted sites was > 0.99 for taxa richness, total density, and number of EPT taxa (Table 5). ASPT and DFI also had relatively high power ≥ 0.95. For acidification effects, taxa richness, total density, and number of EPT taxa had a power > 0.99, while the acidification indices gave power estimates between 0.35 (acidity index III) and 0.93 (acidity index I). The diversity indices had the lowest statistical power to detect eutrophication and also among the lowest to detect acidification. The number of samples needed to detect an effect of either eutrophication or acidification were in almost all cases higher using linear regression than a *t*-test. For a *t*-test, the number of samples needed was lowest for taxa richness, total density and number of EPT taxa (≤ 4 samples). ASPT, DFI and acidity index I required < 7, while the remaining indices needed between nine and 20 samples. Using linear regression, the number of independent samples (e.g., years) needed to detect an impact of eutrophication was ≥10 for all of the indices tested with the exception of DFI, which required four samples. For acidification, acidity index III required the fewest samples (13), compared with 17 samples for EPT. Similar to absolute and standardized effect sizes, total density required the highest number of samples to detect change for both eutrophication (78) and acidification (55).

Discussion

The power of any test of environmental impact is simultaneously constrained by several factors, for example the variability of the data, the magnitude of the impact in question, and the number of independent sampling events (Osenberg et al., 1994). Moreover, the spatial and temporal variability of indicator metrics often confound interpretation and data analysis of environmental impact assessment due to the inverse relationship between natural variability and detection of impact (Johnson, 1998). A robust sampling design to detect environmental impact should have a known probability of detecting a predicted change of a specified magnitude. Without this knowledge the monitoring program may have little or no chance of detecting anything but catastrophic impacts (Bernstein & Zalinski, 1983). Statistical power can and should also be used to optimize sampling effort. Although habitat stratification of sampling effort is a well-accepted method for reducing sample variance (Resh & Jackson, 1993), the simultaneous effect of both spatial and temporal variability has been largely ignored in impact assessment studies by aquatic ecologists (Resh & Rosenberg, 1989).

In this study, the spatial (among-site), temporal (among-year) and sample (among-sample) variability were calculated as the coefficient of variation (CV); the magnitude of the impact was measured as the effect size or, taking into account the natural variability, the standardized effect size. The CV differed markedly among the different indices. In general the temporal and sample variability were lower than the spatial variability, probably because the five temporally sampled streams were more homogeneous regarding local-scale factors than the spatial reference streams. The temporal reference sites were also sampled only once a year and a more intensive sampling scheme would probably reveal a larger (i.e., within-year) variability. Total density had the highest temporal, sample, and spatial CVs. This is consistent with other studies where enumeration and functional measures have been found to be more variable than richness measures (Barbour et al., 1992; Resh & Jackson, 1993), and where a larger spatial than temporal variability has been found (Johnson, 1998). The lowest variability was generally found for the two indices specifically designed to detect the effects of eutrophication or organic pollution, together with Simpson's diversity index and the acidification index developed for Sweden (acidity index II). Results from comparisons of effect sizes and standardized effect sizes were similar; i.e., the index with the largest effect size (total density) was also the most variable and therefore had a low standardized effect size. On the other hand, ASPT and DFI had relatively low effect sizes, but these metrics also had relatively low natural variability, resulting in relatively high standardized effect sizes. DFI was found to be significantly (negatively) correlated with altitude, hence estimates of expected effect size and statistical power were overestimated. Our findings support those of earlier studies in that it is not only important how large an effect size a metric has, but it is also very important how the metric is influenced by the natural variability. If the human impact is small compared to natural variability it will be difficult to detect any change with any degree of confidence (Osenberg et al., 1994).

Taking into account the natural variability of the selected metrics, standardized effect sizes were higher when assessing eutrophication than acidification effects, with the two eutrophication indices and number of EPT taxa as the most promising for monitoring of eutrophication or organic enrichment. These three indices all include some aspect of taxa richness either directly (EPT taxa) or indirectly (ASPT and DFI), which may partly explain their somewhat better performance. Armitage et al. (1983) showed, for example, that ASPT was less sensitive to sampling effort and seasonal change than BMWP (ASPT = BMWP/number of scoring taxa). In contrast, the common denominator for all three acidity indices is that they are solely based on the most pH-sensitive specimen (taxon) found at a site. The acidity indices may therefore have a high power to detect improvement of habitat quality, since the colonization of the site by one sensitive taxon will improve the score of that site (if that taxon is found when that site is sampled). On the other hand, they may have low power to detect impoverishment of a site, since all individuals of all sensitive taxa have to disappear before the site is deemed as affected, an important consideration in biomonitoring studies. For metrics like these acidification indices, careful consideration should be given to the number and type of taxa included in the taxa list. The type II errors were zero for acidity index II developed for Sweden, whereas type II errors were 25% for both of the Norwegian acidification indices. Acidity index II also had the lowest type I error. This might be an effect of the different organisms used and/or different physical and geographical conditions between the two countries. Furthermore, the frequency of type I and II errors may be partly due to the taxonomic resolution used. Lower taxonomic resolution (e.g., family level) should result in lower natural variability and hence lower frequencies of type I errors and higher standardized effect size. For example, the metrics used to detect acidification are based on a taxonomic identification to the species or genus level, whereas for ASPT identification is to the family level; similarly, for DFI, identification is mainly to family level. It has been shown, at least in marine pollution studies, that very little information is lost by using a higher taxonomic level and one may more easily detect a community response to human perturbations above the noise of natural variability (Warwick, 1993). For freshwater systems, however, the use of species-versus family-level identification in biomonitoring and assessment is still debated. In a review by Resh & McElravy (1993) more than half of the papers examined emphasized the importance of species-level identification because (i) the species is seen as the basic biological unit, (ii) it is assumed to have the highest information content, and (iii) species-level information increases sensitivity and, without it, subtle changes may be difficult to detect. On the other hand, authors who recommended an identification to any level above species did so for mainly two reasons: (i) the unavailability of keys and descriptions of species; and (ii) the cost/time to obtain species level identifications.

Even though most studies use a fixed type I error or α value, the consequences of type II errors may be much more costly than type I errors (Fairweather, 1991; Peterman, 1990). The incorrect conclusion of 'no impact' might also mean that serious environmental degradation occurs before the real impact is noticed (Mapstone, 1995). Statistical power ($1 - \beta$) is the capacity of a test to correctly reject the null hypothesis (Underwood, 1993). Several of the metrics tested here had a statistical power > 0.99 both for acidification and eutrophication, and the number of samples needed to detect a change as large as the effect size was ≤ 4 for taxa richness, total density and number of EPT taxa when using a *t*-test and a BACI statistical design. For most of the other indices ≥ 6 samples were needed to detect change. This finding is distressing, since often only three to five samples are taken in stream benthic macroinvertebrate studies (Resh & McElravy, 1993). One should recall that the effect size to detect change was quite large in this study since the intermediate disturbed sites, i.e., those sites with a total phosphorous content between 15 and 50 g l^{-1} and a pH between 5.5 and 6.5, had been removed from the calculations. To detect a change as large as the effect size using regression and an assumed change over a 10-year time period (i.e., slope), most of the metrics tested required a minimum of 20 sample units (e.g., years) and for total density up to 78 sampling occasions was estimated. Again, this finding is alarming, as according to the review by Resh & McElravy (1993) most impact assessments (80%) lasted less than one year.

Some of the metrics evaluated in this study appear to be good candidates for environmental assessment studies, e.g., taxa richness and the number of EPT taxa. The use of richness measures in impact studies is supported by Rapport (1991), who demonstrated that ecosystems affected by stress often show: (i) a reduced taxa richness, with the disappearance of sensitive taxa, (ii) a predominance of pollution-sensitive taxa, and (iii) more monospecific communities. In a study by Resh (1994), richness measures were also among the best indicators to detect impact. In environmental assessment studies it is very important to know what effect size and variability one can expect from a metric so that the money spent on an impact study is used in a cost-effective way.

Acknowledgements

Thanks to the many people who assisted in the sampling and processing of the large number of samples collected in the 1995 national stream survey. In particular, we would like to thank Lars Eriksson, who was responsible for macroinvertebrate taxonomy. Also, the comments and suggestions given by two anonymous referees were appreciated.

References

Armitage, P. D., D. Moss, J. F. Wright & M. T. Furse, 1983. The performance of a new biological water quality score system based on macroinvertebrates over a wide range of unpolluted running-water sites. Wat. Res. 17: 333–347.

Barbour, M. T., J. L. Plafkin, B. P. Bradley, C. G. Graves & R. W. Wisseman, 1992. Evaluation of EPA's Rapid Bioassessment benthic metrics: metric redundancy and variability among reference stream sites. Envir. Toxicol. Chem. 11: 437–449.

Bernstein, B. B. & J. Zalinski, 1983. An optimum sampling design and power tests for environmental biologists. J. Envir. Mgmt. 16: 35–43.

Bækken, T. & K. J. Aanes, 1995. Use of macroinvertebrates to classify water quality. Report 2 A. Acidification, Norsk Institutt for Vannforskning, Oslo, Norway.

De Pauw, N. & G. Vanhooren, 1983. Method for biological quality assessment of watercourses in Belgium. Hydrobiologia 100: 153–168.

Degerman, E., B. Fernholm & P-E. Lingdell, 1994. Bottenfauna och fisk i sjöar och vattendrag. Utbredning i Sverige. SNV 4345, Naturvårdsverket, Solna, Sweden.

Fairweather, P. G., 1991. Statistical power and design requirements for environmental monitoring. Aust. J. Mar. Freshwat. Res. 42: 555–567.

Gerrodette, T., 1987. A power analysis for detecting trends. Ecology 68: 1364–1372.

Gerrodette, T., 1991. Models for power of detecting trends. A reply to Link and Hatfield. Ecology 72: 1889–1892.

Johnson, R. K., 1995. The indicator concept in freshwater monitoring. In Cranston, P. (ed.), Cironomids: From Genes to Ecosystems. Proceedings of the 12th International Chironomid Symposium, CSIRO Publications, Canberra.

Johnson, R. K., 1998. Spatiotemporal variability of temperate lake macroinvertebrate communities: detection of impact. Ecol. appl. 8: 61–70.

Johnson, R. K., T. Wiederholm & D. M. Rosenberg, 1993. Freshwater biomonitoring using individual organisms, populations, and species assemblages of benthic macroinvertebrates. In Rosenberg, D. M. & V. H. Resh (eds), Freshwater Biomonitoring and Benthic Macroinvertebrates. Chapman & Hall, New York:

Kirkegaard, J., P. Wiberg-Larsen, J. Jensen, T. Moth Iversen & E. Mortensen, 1992. Biologisk bedømelse af vandløbskvalitet. Teknisk anvisning fra Danmarks Miljøundersøgelser no. 5, Miljøministeriet, Copenhagen, Denmark.

Lenat, D. R., 1988. Water quality assessment of streams using a qualitative collection method for benthic macroinvertebrates. J. N. am. Benthol. Soc. 7: 222–233.

Lipsey, M. W., 1990. Design Sensitivity: Statistical Power for Experimental Research. Sage Publications, London: 207 pp.

Mapstone, B. D., 1995. Scalable decision rules for environmental impact studies: effect size, type I, and type II errors. Ecol. appl. 5: 401–410.

Metcalfe, J. L., 1989. Biological Water Quality assessment of running waters based on macroinvertebrate communities: history and present status in Europe. Envir. Pollut. 60: 101–139.

Norris, R. H. & A. Georges, 1993. Analysis and interpretation of benthic macroinvertebrate surveys. In Rosenberg, D. M. & V. H. Resh (eds), Freshwater Biomonitoring and Benthic Macroinvertebrates. Chapman & Hall, New York: 488 pp.

Osenberg, C. W., R. J. Schmitt, S. J. Holbrook, K. E. Abu-Saba & A. R. Flegal, 1994. Detection of environmental impacts: natural variability, effect size, and power analysis. Ecol. appl. 4: 16–30.

Peterman, R. M., 1990. Statistical power analysis can improve fisheries research and management. Can. J. Fish. aquat. Sci. 47: 2–15.

Raddum, G. G., A. Fjellheim & T. Hesthagen, 1988. Monitoring of acidification by the use of aquatic organisms. Verh. int. Ver. Limnol. 23: 2291–2297.

Rapport, J. D., 1991. Evaluating ecosystem health. J. Aquat. Ecosyst. Health 1: 15–24.

Resh, V. H., 1994. Variability, accuracy, and taxonomic approaches in benthic macroinvertebrate biomonitoring. Boll. zool. ital. 61: 375–383.

Resh, V. H. & J. K. Jackson, 1993. Rapid assessment approaches to biomonitoring using benthic macroinvertebrates. In Rosenberg, D. M. & V. H. Resh (eds), Freshwater Biomonitoring and Benthic Macroinvertebrates. Chapman & Hall, New York: 488 pp.

Resh, V. H. & E. P. McElravy, 1993. Contemporary quantitative approaches to biomonitoring using benthic macroinvertebrates. In Rosenberg, D. M. & V. H. Resh (eds), Freshwater Biomonitoring and Benthic Macroinvertebrates. Chapman & Hall, New York: 488 pp.

Resh, V. H. & D. M. Rosenberg, 1989. Spatio-temporal variability and the study of aquatic insects. Can. Ent. 121: 941–963.

Shannon, D. E., 1948. A mathematical theory of communication. Bell Syst. Tech. J. 27: 379–423, 623–656.

Simpson, E. H., 1949. Measurement of diversity. Nature 163: 688.

SIS, 1994. Water quality – methods for biological sampling – guidance on handnet sampling of aquatic benthic macroinvertebrates. SS-EN 27 828.

Underwood, A. J., 1993. The mechanics of spatially replicated sampling programmes to detect environmental impacts in a variable world. Aust. J. Ecol. 18: 99–116.

Warwick, R. M., 1993. Environmental impact studies on marine communities: Pragmatical considerations. Aust. J. Ecol. 18: 63–80.

Wilander, A., R. K. Johnson, W. Goedkoop & L. Lundin, 1998. Riksinventering 1995. En synoptisk studie av vattenkemi och bottenfauna i svenska sjöar och vattendrag. SNV 4813, Statens Naturvårdsverk, Stockholm, Sweden.

Woodiwiss, F. S., 1964. The biological system of stream classification used by the Trent River Board. Chem. Ind. 14: 443–447.

Zar, J. H., 1996. Biostatistical Analysis. Prentice Hall, New Jersey: 662 pp.

Hydrobiologia **422/423:** 245–256, 2000.
M. Jungwirth, S. Muhar & S. Schmutz (eds), *Assessing the Ecological Integrity of Running Waters.*

Evaluating physical habitat integrity in relation to the biological potential of streams

Charles F. Rabeni
Missouri Cooperative Fish and Wildlife Research Unit, Biological Resources Division, U.S. Geological Survey, 302 Natural Resources Building, University of Missouri, Columbia, MO 65211-7240, U.S.A.

Key words: invertebrates, scale, bioassessment, watershed, ecoregion, microhabitat

Abstract

Physical habitat integrity, while a necessary condition for ecological integrity, is not well defined and rarely examined in relation to the biological potential of a stream. The relations of physical habitat to the biota need to be quantified so as to better establish reference conditions and to document those physical habitat alterations that actually impact the biota. I evaluated the relation of physical habitat to benthic invertebrate communities in Missouri, U.S.A., streams – at both reference quality and habitat-altered sites. Six common within-stream habitats were delineated from each of 45 streams of reference quality in three ecoregions. Distinctive assemblages of invertebrates were associated with each habitat type. A spatial hierarchy of the influence of habitat was evident. At the largest scale, physical attributes unique to ecoregions were more influential than local conditions in structuring invertebrate communities. Within an ecoregion, however, small-scale habitat conditions were more of an influence than stream location because communities collected from the same habitat type at different streams were more similar than those collected from different habitat types at the same stream. I used this information to develop reference habitat conditions for streams of each ecoregion. I then evaluated invertebrate communities from 20 streams with varying habitat conditions to model the variation and sensitivity of biological metrics in order to assess how the biota responded to habitat changes. Biological responses to habitat alterations were readily documented by taking into account both the presence or absence of particular habitats and the quality of each habitat type.

Introduction

Evaluating the condition of waterbodies by examining the resident fauna has been well established in both theory and practice for several decades (Davis, 1995). The Saprobien system developed around the turn of the 20th century (Kolkwitz & Marsson, 1908) listed indicator organisms associated with different zones of decomposition below sewage inflows and effectively evaluated organic degradation and downstream recovery. Biologists have since worked to refine this approach and to incorporate it into management and regulatory strategies for the improvement of 'polluted' waters. In the U.S.A., the 1972 Amendments to the Federal Water Pollution Control Act (often called the Clean Water Act) incorporated the concept of biological integrity; this had potentially profound effects on water management because it changed the perspective from one concentrated on what enters a stream or river to the effect of anthropogenic activities on aquatic life. The emphasis in regulatory agencies was supposed to shift from examining point-source problems with well-defined causes to examining nonpoint source problems with multifactor causes. Unfortunately, the goal of assessment of biological integrity has not been achieved because most bioassessment programs still concentrate on measuring the response of organisms to physicochemical aspects of water quality while giving only lip service to measures of other aspects of integrity (Karr, 1991).

While there is debate on exactly what constitutes ecological integrity, it is generally agreed that the physical habitat surrounding an organism – for example water velocity, water depth, substrate conditions and cover conditions – is particularly important to the normal functioning of an aquatic system. Additionally, other factors considered to be important to integrity (water quality, water quantity and energy sources) are at least indirectly influenced by physical conditions within, along and in the watershed of a stream. Therefore, it is important to recognize how physical habitat conditions are related to aquatic life, at what spatial scale these effects operate and how to measure biotic response to habitat alteration.

A widely used approach to evaluating biological quality of surface waters in the U.S.A. involves the establishment of reference conditions. They describe characteristics of waterbodies least impaired by anthropogenic activities and are used to define attainable habitat and biological conditions. Reference biotic conditions are the standard by which impairment is judged. Often a subjective habitat evaluation is conducted during the biological evaluation, but usually without any real understanding of the relations between habitat variables and biota. An approach is needed that is explicitly geared to assessing habitat conditions that, when altered, directly reduce biotic potential. Research is required to determine how and under what conditions habitat alterations control the distribution and abundance of the biota, and the information should be used to develop criteria for habitat conditions at reference sites.

Physical habitat

Physical habitat can be defined in the classical sense as a place to live – specifically the small-scale area that constitutes an organism's daily environment. To make the definition more functionally useful to bioassessment, it has been broadened to be thought of as a discrete area within a stream with unique characteristics of substrate, depth and velocity. Important considerations for bioassessment include how these habitats are differentially used by the various taxa, how they change due to anthropogenic activities and how they are sampled. Choosing the number and types of habitats, also commonly termed habitat units and habitat patches, is not very well standardized. For example, the U.S. Geological Survey's NAWQA program has one of the more elaborate systems and employs a three-tiered hierarchy of geomorphic channel units, channel boundaries and channel features totaling 54 habitat types (Cuffney et al., 1993). At the other extreme, one of the U.S. Environmental Protection Agency's (EPA) Rapid Bioassessment protocols recommends sampling just 'the most productive habitat' (Plafkin et al., 1989). We in Missouri, U.S.A., use habitat types corresponding to riffle, pool, vegetation beds, snags, rootmats, boulder areas and leafpacks. The Florida Department of Environmental Protection uses snags, aquatic vegetation, leafpacks, leaf mats, rocky outcrops, roots, gravel/rock/cobble, undercut banks, sand and mud/silt (Barbour et al., 1996). For Australian streams, Parsons & Norris (1996) used riffle, edge, pool rocks and macrophytes. Some agencies, e.g. the Ohio EPA, avoid the habitat question by using artificial substrates for their sampling method (Ohio EPA, 1987).

Bioassessment programs must be concerned with physical features – and their processes – that occur at larger spatial scales than the habitat, but which have the potential to maintain, alter, restructure or eliminate habitats and affect their invertebrate communities. These controlling factors may operate at multiple spatial scales and often relate to condition of channel morphology, riparian integrity and watershed land use (Wiley et al., 1997). Habitats where invertebrates reside are influenced both qualitatively and quantitatively by anthropogenic activities at larger scales. Quantitative changes occur when channelization, gravel extraction, riparian tree harvest or streamside grazing by livestock alter the relative abundance of habitats and perhaps eliminate some altogether. Alternately, a habitat may not be eliminated but may become less optimal for the biota, such as a riffle becoming embedded, pools becoming more shallow or vegetation beds less dense. Both types of habitat changes contribute to ecological integrity that should be recognized and assessed. The protocols of most major assessment programs measure some of these controlling factors, usually in close proximity to the biological sampling site, but because of our incomplete understanding of the ecological role of many of the measured variables, approaches vary widely as to what and how to evaluate.

The objectives of this paper are to explain why local within-stream habitat conditions need to be considered in habitat assessment programs, to demonstrate the feasibilty of assessing both the quality and quantity of reference habitat conditions using common biotic metrics and to emphasize assessment at multiple scales.

Methods

Information for this paper was obtained during a study to develop biological criteria for the state of Missouri, U.S.A. (Rabeni et al., 1997). Two data sets are used. In 1993, 44 reference quality streams, minimally impacted, of comparable size and distributed throughout the state, were sampled for benthic invertebrates. In 1995, we collected invertebrates from a set of 10 reference-quality streams paired with a set of 10 streams with some degree of physical habitat alteration, but little or no water quality problems. Sampling was conducted within a 3-week period each year. All sampling locations were in wadable 4th or 5th order streams. The basic sampling location was the reach (sensu Frissell et al., 1986), which is a length of stream about 20× the mean bank width, averaging about 450 m in length, and representative (i.e. possessing typical morphology) of the entire stream segment.

All available standard habitat types were sampled at each reach. Standard habitats were visually differentiated and included areas of water flow over coarse substrate (cs flow), water flow over fine substrate (fs flow), snags, rootmats, nonflowing water over fine deposition (nonflow), vegetation beds and terrestrial leafpacks. All habitats except snags were sampled using a kick net (Wildlife Supply Company, Saginaw, Michigan) with a mesh size of 500 μm. Snags were sampled by shaking and scraping invertebrates into a 500-μm Nitex mesh bag. We maximized the spatial distribution of collections for each habitat type by making a composite sample from at least six different areas within the reach. Invertebrates from each habitat were kept separate. A 200-invertebrate subsample was enumerated from each habitat (Plafkin et al., 1989 as modified by Canton, 1991). This allowed calculation of reach metrics by using data from single or multiple habitats, and allowed comparisons of sites by single, multiple or habitats in common data sets. The metrics chosen were: Total taxa = total number of taxa; EPT = the sum of all Ephemeroptera, Plecoptera and Trichoptera taxa; Biotic index = Hilsenhoff biotic index modified for local fauna; two diversity indices – Shannon's and Simpson's; % Dominant taxa = the proportion of total number of individuals comprised of the most abundant taxon; Family = total number of invertebrate families (Barbour et al., 1998). Field collections and laboratory analyses followed standard protocols (Rabeni et al., 1997) based upon EPA protocols (Plafkin et al., 1989).

Habitat evaluation

The habitat evaluation procedure used in this study was developed for the U.S. Environmental Protection Agency (Plafkin et al., 1989) and is in use by numerous U.S. state and federal agencies. It derives a single numeric value by totaling the scores from a number of physical features which are separated into three categories (Barbour & Stribling, 1991). Primary parameters characterize stream microhabitats and probably have the most direct influence on the biota-substrate conditions, flow conditions and shading. Secondary parameters relate to channel morphology, including estimates of channel alteration, bottom scouring and deposition and riffle-width/bend-width ratios. Tertiary parameters are concerned with riparian and bank structure upstream from the sampling site, which are estimates of bank stability, vegetative stability, stream-side cover and riparian vegetative zone width.

It is evident that this procedure measures conditions within habitats (primary factors) as well as physical features at larger spatial scales thought to influence habitat (secondary and tertiary factors). While this procedure has not been tested for its relevance to the biological potential of a stream, and though many of the variables measured are intercorrelated, it appears to give an adequate index of overall site quality, at least as humans subjectively perceive quality. Its use might be better understood if this and other habitat evaluation procedures were termed 'physical feature evaluations' so as to reserve the term habitat to its more restrictive definition.

Analyses

Boxplots were used to display relations of metrics in comparing reference to habitat-altered streams because they provide a visual representation of several important features of a data set, namely central tendency, the median value and the variation of the data set at the 10th, 25th, 75th and 90th percentiles, as well as the skewness of the data and any outliers of data points. We scored the difference between two boxplots (comparing reference to test streams) after Barbour et al. (1996) from 0 = no detectable difference to 3 = maximum difference, where 3 points were awarded if there was no overlap in interquartile ranges, 2 points if there was some overlap in interquartile range but both medians were outside the interquartile range overlap and 1 point if there was moderate overlap of interquartile ranges but at least one median was outside the interquartile range overlap. No points were awarded for extensive overlap of interquartile ranges or when both medians were within the overlap. We also compared the data sets of reference streams to those of impaired streams for each metric using a paired t-test.

Detrended Correspondence Analysis (DCA) was used to visualize relations among communities from different habitat types within an ecoregion. DCA was chosen because this indirect technique performs an ordination on just the community data, allowing an expression of pure community gradients (Okland, 1996). Sites, represented by points on the ordination, are distributed along both axes based upon their similarity of taxa occurrence (using relative abundances of taxa), with more similar sites occurring closer together. Equal distances within the ordination correspond to equal differences in taxa composition (Gauch, 1982). By coding sites from each habitat type, the patterns of communities from one habitat type can be compared to all others.

Results

Evaluating the natural condition at different scales

Relation of habitat scores to metrics on a statewide basis

Habitat assessment in an ecoregion context provides a conceptual framework for defining biological potential (Hughes et al., 1990). Habitat variables that are believed to control aquatic biota are a consequence of terrestrial features of geology, geomorphology and vegetation that also form the basis of regionalization efforts (Bailey, 1995; Omernik, 1995). An ecoregion assessment of habitat allows refinements of protocols so that they are low in variation, are pertinent to important causal variables that may be unique to a particular ecoregion and reflect the range of habitat conditions likely to affect the biota.

The state of Missouri has three generally accepted major natural areas: a Prairie region in the north (glaciated) and west central, an Ozark region in the south, and a Mississippi Alluvial plain or lowland in the southeast 'bootheel' (Figure 1). The 44 reference-quality streams were selected to proportionally represent all three ecoregions.

The characteristic fauna of the streams from each region often yielded significantly different bioassessment metric values (Figure 2) between regions. Ozark streams ranked best, lowland streams the worst, and Prairie streams intermediate. This was paralleled by

Table 1. Correlation coefficient p values between metric score and habitat score

	Habitat score		
	Statewide ($N = 41$)	Prairie ecoregion ($N = 12$)	Ozark ecoregion ($N = 25$)
Total taxa	<0.00	0.962	0.216
Family	<0.00	0.536	0.260
EPT	<0.00	0.222	0.123
% Dominant taxa	<0.00	0.357	0.412
Biotic index	<0.00	0.433	0.012
Shannon diversity index	<0.00	0.661	0.197
Simpson diversity index	<0.00	0.682	0.305

associations between habitat scores and many of the invertebrate bioassessment indices (Figure 3). The strength of the relations were all $p < 0.001$ suggesting a strong habitat influence on a statewide basis.

Within each ecoregion, however, there was only a single significant correlation between any metric and its habitat score (Table 1). This further suggests that over a large spatial scale, habitat conditions at these reference streams were likely to be one of the factors influencing invertebrate distributions. In contrast, at a smaller scale, i.e. ecoregion, habitat conditions in reference streams vary little and show few associations with invertebrate community indices. Thus, within any single ecoregion (Figure 3) the reference invertebrate communities were located in streams with a narrow range of habitat scores. Reference sites with consistent habitat characteristics and low variation are important in developing criteria for habitat conditions of reference streams on a regional basis, and for evaluating effects of habitat alteration. This low variation among reference habitat conditions is one of two important considerations – the other being sensitivity to change – in developing a useful protocol.

There appears to be a habitat potential for streams in each ecoregion in Missouri that influences the potential of the invertebrate community (Figure 3). Spatial scales smaller than the ecoregion deserve consideration in bioassessment. Longitudinal changes of invertebrate communities along stream systems (Illies & Botosaneau, 1963; Vannote et al., 1981; Statzner et al., 1988) require considering site location within a watershed and using similarly sized sites for comparison. Within streams of comparable size and flow the selection of habitat types and invertebrates collec-

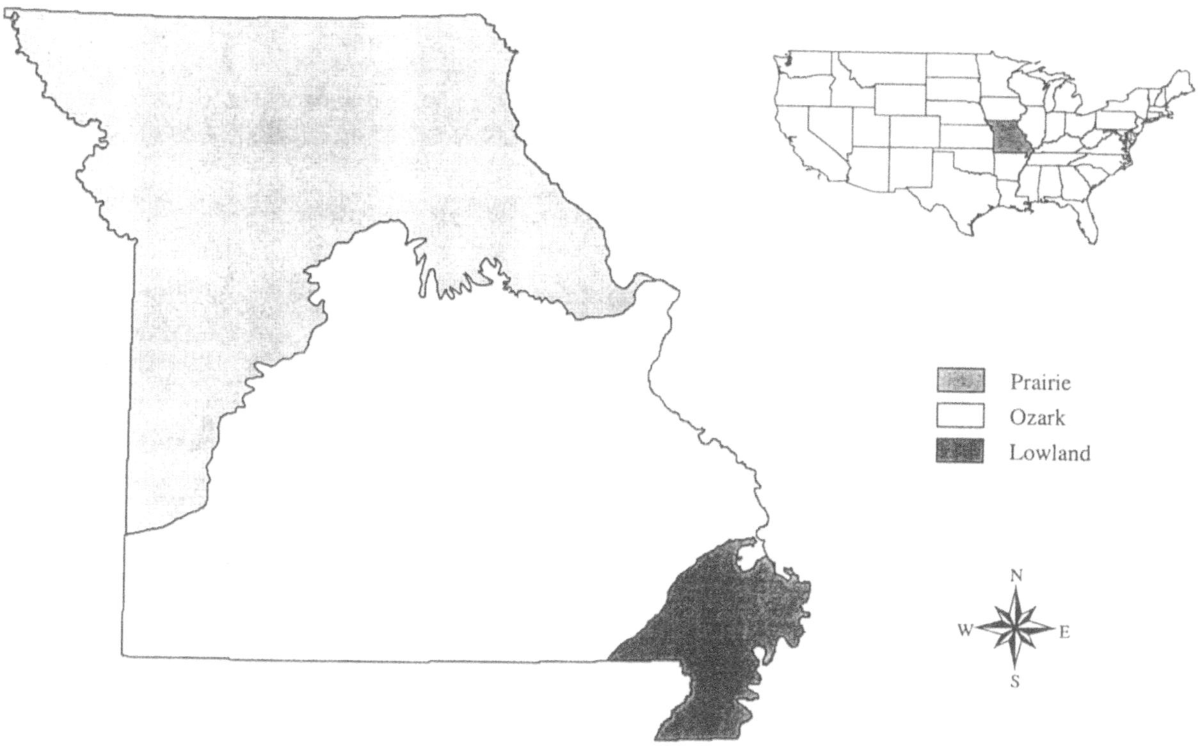

Figure 1. The state of Missouri, U.S.A., delineating the three major ecoregions.

tion methods will strongly influence whether habitat alterations can be detected. It is useful to examine how invertebrate communities are distributed among those common in-stream habitats that are often sampled.

Evaluation of habitat-specific invertebrate communities

The similarities among invertebrate communities from different habitats within a region can be visualized from the DCA ordination of reference sites (Figure 4). In the Ozark ecoregion, most communities grouped by habitat type and each tended to be separate from communities of other habitats. The faunas from vegetation and snags were most alike, while the faunas of flow over coarse substrate (cs flow) and nonflow were quite distinct. In Prairie streams, communities from different habitats were also distinct, especially cs flow and nonflow. These analyses show the importance of habitat-specific sampling; within the same ecoregion, communities from the same habitats at different sites were more similar than communities from different habitats at the same site – a situation noted by others (Brussock & Brown, 1991; Parsons & Norris, 1996).

Seven commonly-used bioassessment metrics were calculated for individual habitat types from the reference streams of the Prairie ecoregion (Figure 5). The variation of values from different habitats was substantial and in many cases statistically significant for each metric. Additionally, there was a lack of consistency in the ranking of scores of each habitat among metrics, with none of the metrics for a particular habitat type consistently scoring high or low. Sampling protocols need to consider that metric scores at a stream bioassessment site vary widely depending upon the habitat evaluated. Two philosophies regarding sampling a site for bioassessment purposes are prevalent. For example, the U.S. EPA Rapid Bioassessment protocol (Plafkin et al., 1989) recommends single habitat sampling to limit the effect of interhabitat variation on assessment, while the British Institute of Freshwater Ecology's RIVPACS (River Invertebrate Prediction and Classification Scheme; Wright et al., 1984) recommends collecting from all major habitats and then compositing the sample. Multihabitat sampling is sometimes favored because it is believed that communities from different habitats are differentially affected by impairment and that a single-habitat analysis

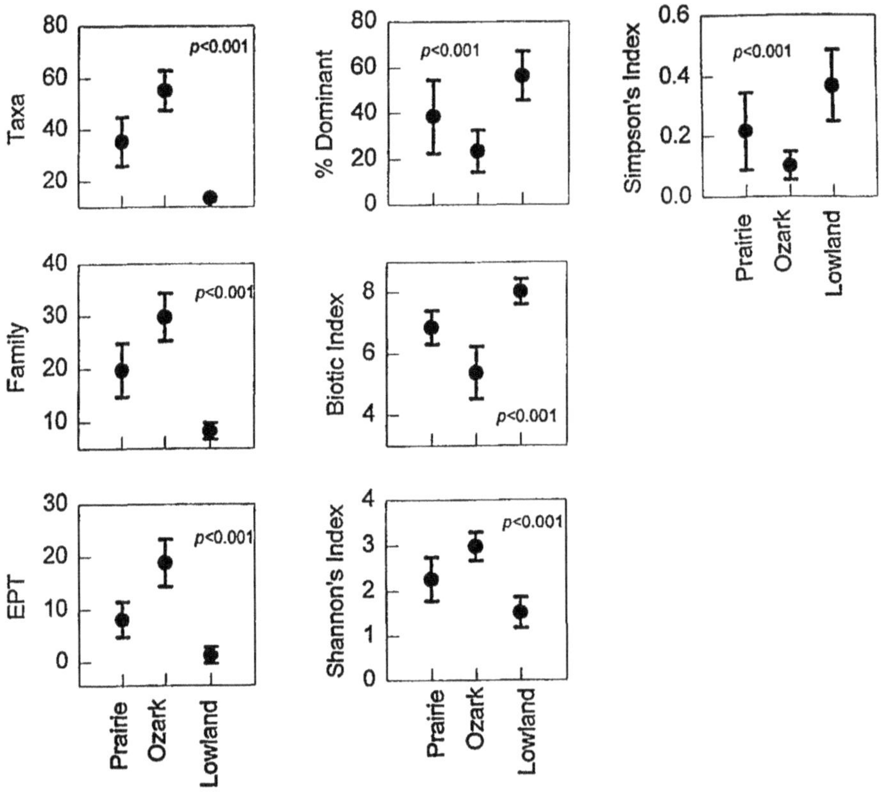

Figure 2. Among-region comparison of common bioassessment metrics calculated from reference stream invertebrate collections (nean and standard deviations).

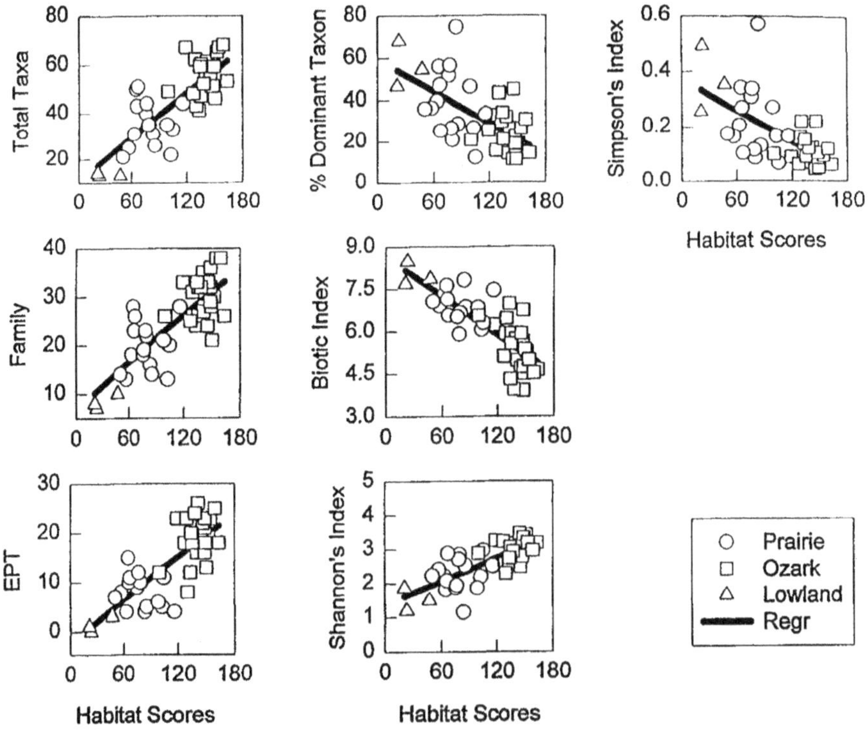

Figure 3. The relation of habitat score to individual bioassessment metrics for reference streams.

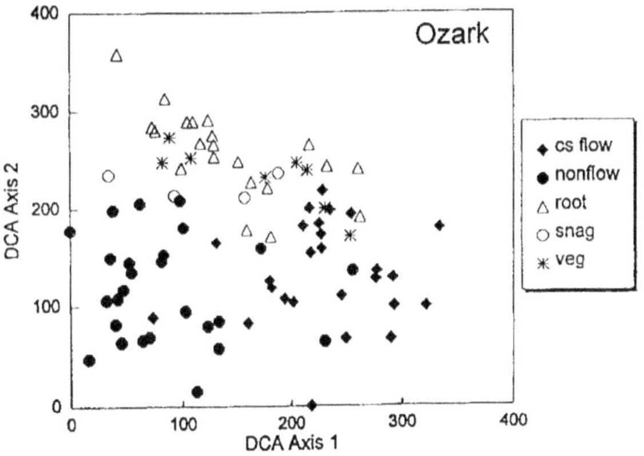

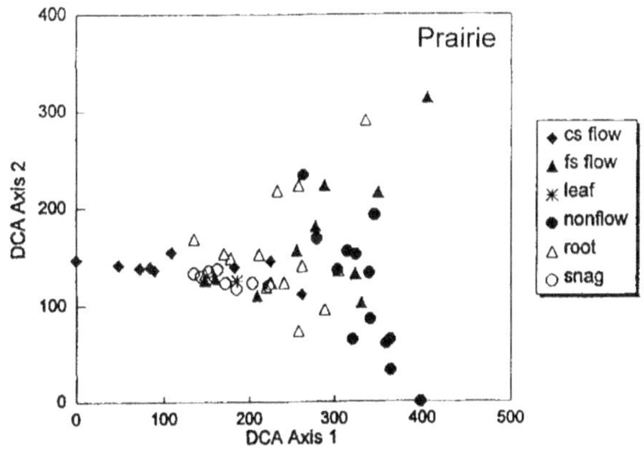

Figure 4. DCA ordination of invertebrate communities from different habitat types in reference streams from the Ozark and Prairie ecoregions; csfl = flow over coarse substrate, fsfl = flow over fine substrate, leafp = leafpack, nonfl = nonflow over fine substrate, root = rootmats from riparian trees, snag = trees fallen into water.

may miss these effects; the single-habitat school, in turn, regards multiple sampling as redundant, variable and a waste of resources (Parson & Norris, 1996).

Results of this study (Figures 4 and 5), showing how each habitat within a reach possesses a characteristic fauna, imply that a sample from a single habitat may produce a nonrepresentative fauna. For example, stream invertebrates are noted for their ability to find refugia, and sampling a single habitat may sample the only remaining livable place in the stream. Even though that habitat may have 'good integrity,' it could falsely characterize the entire stream. While multihabitat analyses potentially yield a better overall understanding of the entire stream community, Figures 4 and 5 clearly show that a metric calculated from several habitats can increase variation substantially and thus make distinctions among reference and test situations more difficult to detect.

Table 2. Habitat characteristics of reference and habitat-altered streams

	N	$\bar{x}$-habitat score	$\bar{x}$ no. of habitat types sampled
Reference	10	119	3.2
Habitat-altered	10	82	2.2

Evaluating habitat impairment

Qualitative evaluation

Measuring the quality of individual habitats (i.e. the capacity to support a rich, diverse fauna) should be part of any bioassessment. I have had the most success examining habitat quality using invertebrates from the flow over coarse substrate or 'riffle' habitat. Habitat alterations likely to affect the biota include siltation, changes in depths and flow conditions and temperature alterations. As an example, I present the 1995 data set in which metrics from a series of reference-quality sites are compared with a series of habitat-altered sites whose habitat scores averaged 70% of the reference site. Every metric showed some level of discrimination between reference-habitat sites and habitat-altered sites (Column A, Figure 6). The diversity index responded most to habitat changes and, along with % Dominant taxon, showed maximum differences. The remainder of the metrics showed low but detectable differences.

Quantitative evaluation (effects of habitat simplification)

Habitat-altered streams in Missouri often have riparian vegetation removed, which eliminates rootmat habitat, changes channel morphology, flow patterns and substrate conditions and destroys vegetation beds. Such sites tend to have fewer distinct habitats available for sampling (Table 2). When I compared metrics from reference versus habitat-altered streams (habitat score averaging 70% of reference), the level of discrimination shown by each metric depended upon the number and type of habitats sampled. The discriminatory power of the metrics in a Box and Whisker analysis increased when all available habitats were used (Column C, Figure 6) over that when habitats in common were

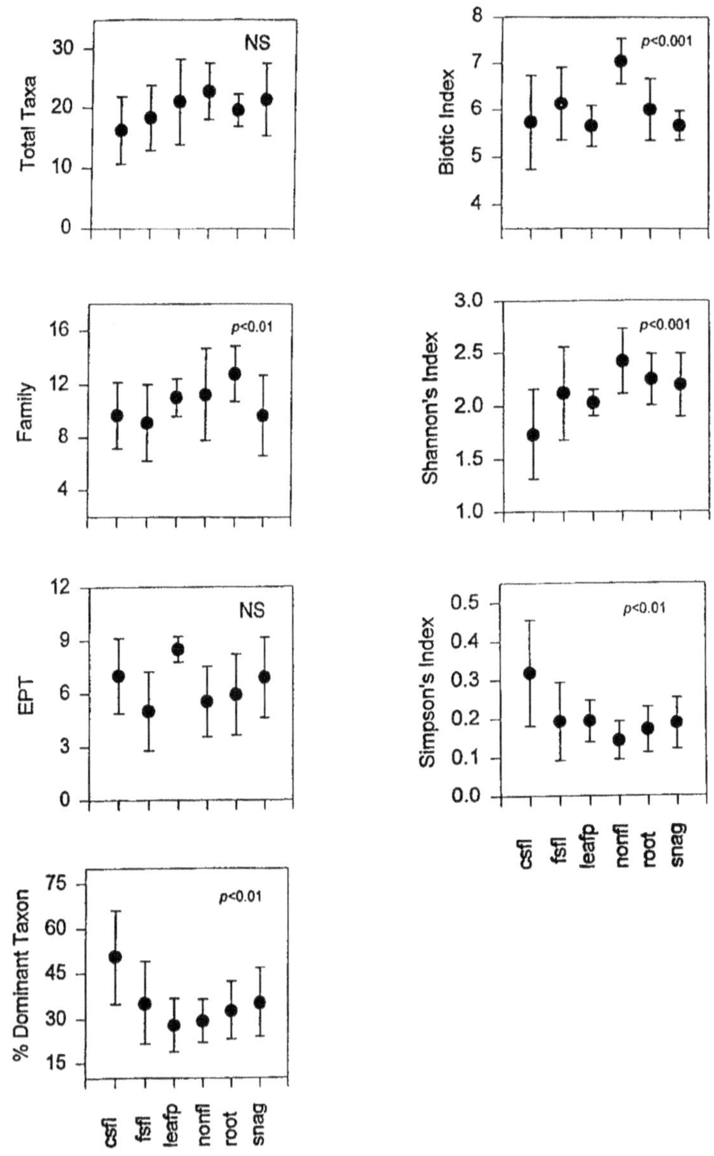

Figure 5. Mean metric values (±SD) for invertebrate communities collected from different habitat types in prairie reference streams; csfl = flow over coarse substrate, fsfl = flow over fine substrate, leafp = leafpack, nonfl = nonflow over fine substrate, root = rootmats from riparian trees, snag = trees fallen into water.

compared. Total taxa, Family and Shannon's diversity index show excellent discrimination (value = 3), while EPT, Biotic index, Simpson's Index, and % Dominant taxon all show a moderate (value = 2) discriminatory power. Compare this to the analysis of habitats in common (Column B, Figure 6) where five metrics showed no discriminatory power and the two that did were only moderately successful. I conclude that if habitat simplification is a departure from reference attainable conditions, all available habitats should be taken into account and habitat alteration can be measured with some degree of precision.

Thus, effects of habitat alteration on invertebrate communities can be evaluated by modifying existing protocols, but will probably require both a single habitat evaluation (to detect changes in habitat quality within specific habitats) and an evaluation using all habitats (to detect changes in habitat quantity – i.e. the number of total habitats which is a measure of overall site quality).

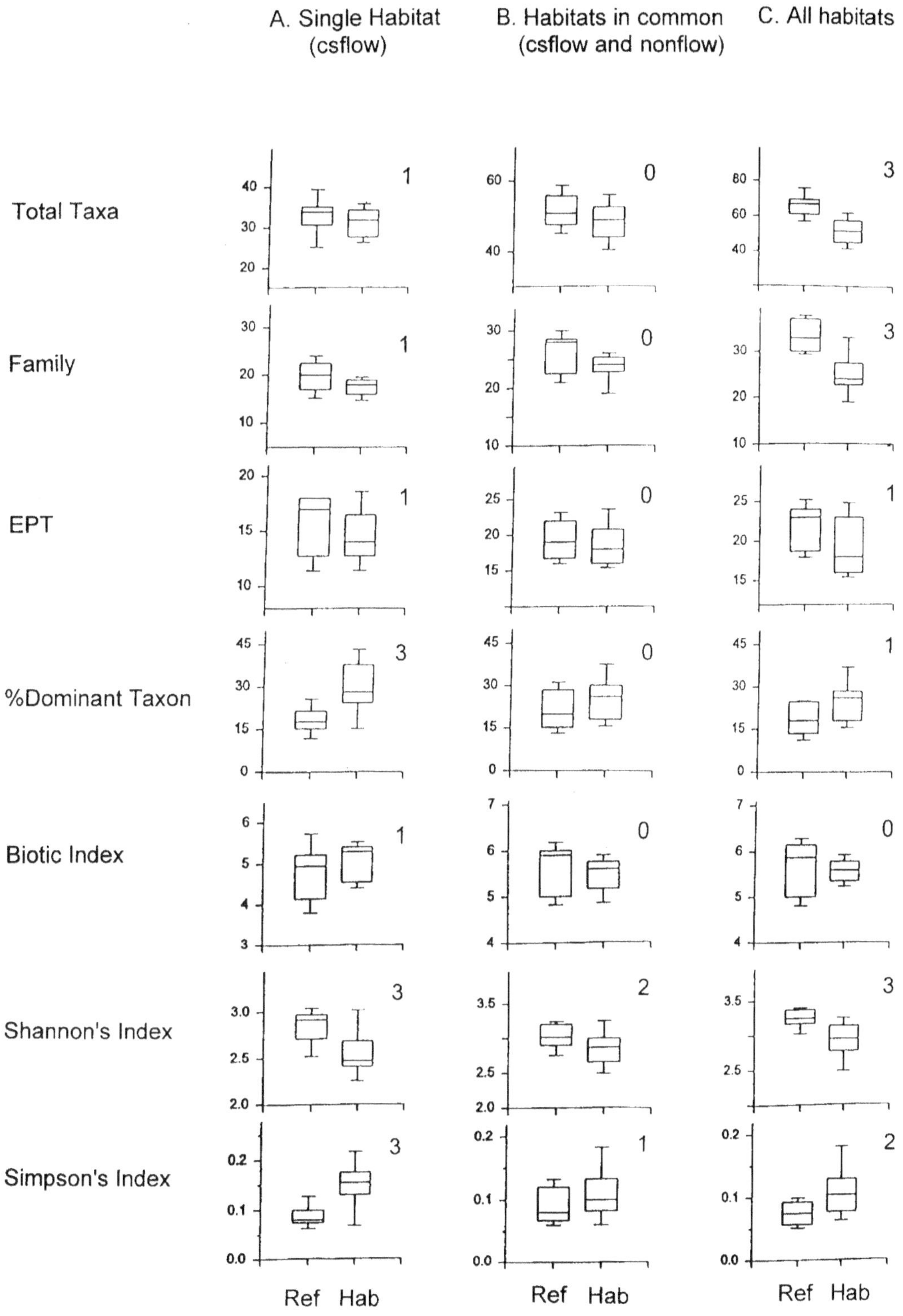

Figure 6. Discriminatory power of metrics for reference (Ref; $N = 10$) and habitat-altered (Hab; $N = 10$) streams, Ozark ecoregion 1995. Each column represents comparisons calculated using different combinations of habitats: Column A = single habitat, B = habitats in common, and C = all available habitats.

Discussion

Preserving unmodified physical habitat is basic to maintaining diverse functional aquatic communities (Rankin, 1995), yet its evaluation and integration into assessment programs has been minimal (Karr, 1991). Much basic knowledge of how populations and communities relate to their unmodified habitat and to habitat changes is unavailable, and even the most elegant protocols for habitat assessment do not offer insights into the relation of biological capacity and habitat condition. The contention of the U.S. EPA (Rapid Bioassessment Protocols; Barbour et al., 1998) that the relation between habitat quality and biological condition can be predicted is perhaps true but has yet to be supported by much evidence. Barbour & Stribling (1991) conceptualized a sigmoid curve as a probable relation between habitat quality and biological condition. This model needs to be validated and then calibrated with data from different situations in order to better determine what changes in habitat quality actually affect biological condition. This has important practical implications for rehabilitating or preserving streams. Effective management strategies will require knowing whether there is a straight line relation (e.g. Figure 3), a sigmoid curve with a plateau as Barbour & Stribling (1991) suggest or some other curve. Our results (Figure 6), using only slight modification of the Barbour & Stribling (1991) habitat evaluation protocol, indicate that a situation where habitat quality is about 70% reference quality probably will not support a reference-level biological community. Unfortunately, our data set is not extensive enough to extrapolate on either side of this point.

Refining habitat evaluations to better reflect the biological potential for invertebrates will not be easy, mainly because it is difficult to use recent and site-specific physical evaluations to quantify the effects of past or remote conditions. Habitat for an assemblage of invertebrates in a stream reach may be defined as the local features of that site that constitute their daily environment. An important consideration in evaluating ecological integrity is that while invertebrates respond to local conditions, some of those conditions originate far from the stream. Some locally expressed elements of the habitat are reflections of broader scale watershed influences and include water temperature, nutrients, dissolved oxygen, flow variation and flow regime. Habitat assessment for bioassessment purposes is traditionally done at the site where the biological sampling occurs – the 'representative reach' of a few hundred meters in length. Recent landscape-level and multiple spatial scale studies are showing predictable responses of biota to large-scale patterns of geology, hydrology and land use with sometimes contradictory relations with site-scale physical variables (Fausch et al., 1994; Rabeni & Sowa, 1996; Roth et al., 1996; Wang et al., 1997; Wiley et al., 1997). This spatially explicit nature of biota-habitat relations makes it difficult to define and quantify habitat quality.

Many habitat alterations are slow acting, cumulative and show effects on a temporal scale not easily measured. For example, open range grazing on the flood plain of Missouri's Ozark streams, a practice abandoned several decades ago, is thought to be the chief cause of the graveling in of many streams occurring today (Jacobson, 1995).

Furthermore, nonpoint source impairment is rarely a single-factor phenomenon. Cumulative alterations, of which physical habitat is only one, makes determination of causes difficult. Yet, factoring out the relative contribution of individual stressors to a decrease in ecological integrity is certainly worthwhile. Simply documenting that the biotic community is degraded or in the process of degrading is akin to a doctor's diagnosing that the patient has a high fever. Not knowing the underlying causes of the biotic condition does not portend well for rehabilitating, restoring or preserving streams.

Recommendations

Habitat evaluation models that are currently used in conjunction with invertebrate monitoring need to be validated as to their relation to the biota, and calibrated for ecoregion and stream size. Fisheries biologists have developed a wide array of habitat models for particular sport fish populations and their efforts may be instructive to those involved in the bioassessment of ecological integrity. A review of fish models (Rabeni, 1992) found that many produce an index of habitat quality that is not amenable to biological testing. Many fish models integrate into their evaluation both habitat characteristics – i.e. local within-stream factors – with larger-scale physical attributes, thus confusing cause and effect. Models that do successfully relate habitat features to the biota, usually multiple regression models, are invariably situation specific and not transferable to other geographic regions (Fausch et al., 1988).

Surveys of regional reference streams will determine habitats common to reaches of all streams and, therefore, important. Bioassessment should include qualitative and quantitative aspects of habitat. Habitats should be common to all reference streams and be present in sufficient amount to be ecologically meaningful. Such habitats should also have standard definitions and minimum standards of measurement (Rankin, 1995).

This study indicated that common metrics can identify a change in the biota presumably due to qualitative changes in habitat patches. More effort is needed to examine different types of habitats, determine the variability of the biological condition in reference conditions and determine how the biota within each habitat responds to physical alteration at larger spatial scales.

Metrics that are specifically sensitive to habitat alterations need to be identified. While biological responses may not be unique to habitat change, some indices appear more sensitive than others. In this example, and in our unpublished studies, diversity indices appear particularly sensitive to habitat change – especially when evaluating habitat quality. This is not particularly surprising as much habitat degradation is merely habitat simplification and the positive relation between habitat diversity and biotic diversity is well known (e.g. Gorman & Karr, 1978).

I recommend sampling all available habitats that are present in some minimum amount, analyzing them separately and treating them as if all were equal in abundance because this will increase the significance of important, but areal-limited, habitats and downplay the overriding dominance of widely occurring habitats. Also, data from individual habitats can be combined for many typical comparisons. Sampling habitats in proportion to their abundance and compositing for analysis implies that each habitat will be represented proportionally to its area in the final index or metric – but this is not the case unless densities of organisms are equal at each habitat. A habitat with a very small relative area could dominate a metric if it had high densities of organisms. Furthermore, I doubt our abilities to accurately sample in proportion to habitat availability unless considerable and probably time-prohibitive quantitative measurements are made beforehand.

The integration of comprehensive habitat analyses into bioassessment programs would certainly add some costs and efforts, but the returns in useful information would be substantial. Understanding biota-physical habitat linkages will allow more sensitive detection of any cause of a change in ecological integrity, and if the cause is habitat alteration, much more effective preservation or rehabilitation strategies are possible.

Acknowledgements

I thank project personnel Randy Sarver, Greg Wallace and Ning Wang and appreciate the suggestions of Kathy Doisy. This is a contribution from the Missouri Cooperative Fish and Wildlife Research Unit (Biological Resources Division, U.S. Geological Survey; Missouri Department of Conservation; University of Missouri; and Wildlife Management Institute cooperating).

References

Bailey, R. G., 1995. Description of the ecoregions of the United States (2nd edn). Miscellaneous publication No. 1391, Map Scale 1:7,500,000 U.S. Department of Agriculture, Forest Service.

Barbour, M. T. & J. B. Stribling, 1991. Use of habitat assessment in evaluating the biological integrity of stream communities. In U.S. EPA Biological criteria: research and regulation. EPA-440/5-91-005, Washington, D.C.

Barbour, M. T., J. Gerritsen & J. S. White, 1996. Development of the Stream Condition Index (SCI) for Florida. Tetra Tech Inc., Owings Mills, MD.

Barbour, M. T., J. Gerritsen, B. D. Snyder & J. B. Stribling, 1998. Rapid bioassessment protocols for use in streams and wadeable rivers: periphyton, benthic macroinvertebrates and fish (2nd edn). EPA/841/B/98-010 U.S. Environmental Protection Agency Office of Water, Washington, D.C.

Brussock, P. P. & A. V. Brown, 1991. Riffle-pool geomorphology disrupts longitudinal patterns of stream benthos. Hydrobiologia 220: 109–117.

Canton, L. W., 1991. Improved subsampling methods for the EPA 'Rapid bioassessment' benthic protocols. Bull n. am. Benthol. Soc. 8: 317–319.

Cuffney, T. F., M. E. Gurtz & M. R. Meadow, 1993. Methods for collecting benthic invertebrate samples as part of the National Water Quality Assessment Program. U.S. Geological Survey Open File Report 93-406, Raleigh, NC.

Davis, W. S., 1995. Biological assessment and criteria: building on the past. In Davis, W. S. & T. P. Simon (eds), Biological Assessment and Criteria. Lewis Publishers, Boca Raton, FL: 15–29.

Fausch, K. D., C. L. Hawkes & M. B. Parsons, 1988. Models that predict standing crop of stream fish from habitat variables. USDA, Forest Service Gen. Tech. Rep. WO-213: 2950–85.

Fausch, K. D., S. Nakano & K. Ishigaki, 1994. Distribution of two congeneric charrs in streams of Hokkaido Island, Japan: considering multiple factors across scales. Oecologia 100: 1–12.

Frissell, C. A., W. J. Liss, C. E. Warren & M. D. Hurley, 1986. A hierarchical framework for stream habitat classification: viewing streams in a watershed context. Envir. Manage. 10: 199–214.

Gauch, H. G., 1982. Multivariate analysis in community ecology. Cambridge University Press.

Gorman, O. T. & J. R. Karr, 1978. Habitat structure and stream fish communities. Ecology 59: 507–515.

Hughes, R. M., T. R. Whittier, C. M. Rohm & D. P. Larsen, 1990. A regional framework for establishing recovery criteria. Envir. Manage. 14: 673–683.

Illies, J. & L. Botosaneau, 1963. Problème et méthodes de la classification et de la zonation écologique des eaux courantes considerées surtout du point de vue faunistique. Int. Ver. theor. angew. Limnol. 12: 1–57.

Jacobson, R. B., 1995. Spatial controls on patterns of land-use induced stream disturbance at the drainage-basin scale – an example from gravel-bed streams of the Ozark Plateaus, Missouri. In Costa, J. E., A. J. Miller, K. W. Potter & P. R. Wilcock (eds), AGU Geophysical Monograph 89, The Wolman Volume: 219–239.

Karr, J. R., 1991. Biological integrity: a long-neglected aspect of water resource management. Ecol. Appl. 1: 66–84.

Kolkwitz, R. & M. Marsson, 1908. Principles for the bioassessment of waterbodies according to their flora and fauna. Mitt. Der Klg. Pruf. fur Wass. Abw. Berlin-Dahlem 1: 33–72.

Ohio Environmental Protection Agency, 1987. Biological criteria for the protection of aquatic life: Volume II. Users manual for biological field assessment of Ohio surface waters. Ohio EPA, Columbus, OH.

Okland, R. H., 1996. Are ordination and constrained ordination alternative or complementary strategies in general ecological studies? J. Vegetation Sci. 7: 289–292.

Omernik, J. M., 1995. Ecoregions: a spatial framework for environmental management. In David, W. S. & T. P. Simon (eds), Biological Assessment and Criteria: Tools for Water Resource Planning and Decision Making. CRC Press, FL.

Parsons, M. & R. H. Norris, 1996. The effect of habitat-specific sampling on biological assessment of water quality using a predictive model. Freshwat. Biol. 36: 419–434.

Plafkin, J. L., M. T. Barbour, K. D. Porter, S. K. Gross & R. M. Hughes, 1989. Rapid Bioassessment Protocols for use in streams and rivers: benthic macroinvertebrates and fish. EPA/444/4-89-001, Office of Water Regulations and Standards, U.S. Environmental Protection Agency, Washington, D.C.

Rabeni, C. F., 1992. Habitat evaluation in a watershed context. In Stroud, R. H. (ed.), Fisheries Management and Watershed Development. Proceedings, Fisheries Management: Dealing with development in the watershed. Northeastern Division, American Fisheries Society Symposium 13, Bethesda, MD: 57–67.

Rabeni, C. F. & S. P. Sowa, 1996. Integrating biological realism into habitat restoration and conservation strategies for small streams. Can. J. Fish. Aquat. Sci. 53 (Suppl. 1): 252–259.

Rabeni, C. F., R. J. Sarver, N. Wang, G. S. Wallace, M. Weiland & J. T. Peterson, 1997. Development of regionally based biological criteria for streams of Missouri. Report to Missouri Department of Natural Resources from Missouri Cooperative Fish and Wildlife Research Unit, University of Missouri, Columbia, MO.

Rankin, E. T., 1995. Habitat indices in water resource quality assessment. In Davis, W. S. & T. P. Simon (eds), Biological Assessment and Criteria. Lewis Publishers, Boca Raton, FL: 181–214.

Roth, N. E., J. D. Allan & D. L. Erickson, 1996. Landscape influences on stream biotic integrity assessed at multiple spatial scales. Landscape Ecol. 11: 141–156.

Statzner, B., J. A. Gose & V. H. Resh, 1988. Hydraulic stream ecology: observed patterns and potential applications. J. n. am Benthol. Soc. 7: 307–360.

Vannote, R. L., G. W. Minshall, K. W. Cummins, J. R. Sedell & C. E. Cushing, 1981. The river continuum concept. Can. J. Fish. Aquat. Sci. 37: 130–137.

Wang, L., J. Lyons, P. Kanehl & R. Gatti, 1997. Influences of watershed land use on habitat quality and biotic integrity in Wisconsin streams. Fisheries 22(6): 6–12.

Wiley, M. J., S. L. Kohler & P. W. Seelbach, 1997. Reconciling landscape and local views of aquatic communities: lessons from Michigan trout streams. Freshwat. Biol. 37: 133–148.

Wright, J. F., D. Moss, P. D. Armitage & M. T. Furse, 1984. A preliminary classification of running-water sites in Great Britain based on macro-invertebrate species and the prediction of community type using environmental data. Freshwat. Biol. 14: 221–256.

Hydrobiologia **422/423:** 257–270, 2000.
M. Jungwirth, S. Muhar & S. Schmutz (eds), Assessing the Ecological Integrity of Running Waters.

Application of an adapted index of biotic integrity to rivers of Lithuania

V. Kesminas & T. Virbickas
Institute of Ecology, Akademijos 2, 2600 Vilnius, Lithuania

Key words: fish, biotest, Index of Biotic Integrity

Abstract

The species composition, community structure, trophic composition, abundance and biomass were investigated in 60 Lithuanian rivers. Leaning on the results of investigations, four main types of river fish communities (fish communities of the brooks, streams, middle-sized and large rivers) differing in species number and composition, trophic composition, density and biomass were singled out. Referring to the characteristic features of fish communities, the Index of Biotic Integrity (Karr et al., 1986) was modified and adapted to the conditions of Lithuania. In order to test biointegrity, nine metrics in three categories (species composition, abundance and biomass; trophic composition and fish condition) were selected. The metrics of this biotest assess attributes that are assumed to be correlated with biotic integrity. These metrics, by the sum of their ratings, characterize the underlying biotic integrity of the sampling site. Six integrity classes (excellent, good, fair, poor, very poor and no fish) of river fish communities were distinguished. Individual tests were adapted to brooks (up to 10 km in length), streams (up to 50 km and riverhead of bigger rivers), middle-sized rivers (up to 200 km and the middle reaches of the large rivers) and large rivers (over 200 km and the lower reaches of the middle-sized rivers). While brooks and streams mainly show intact conditions (class 1 and 2), middle-sized and large rivers are significantrly more impacted (class 2 to 3).

Introduction

Biointegrity is the ability to support and maintain a balanced, integrated, adaptive community of organisms having a species composition, diversity and functioning organization comparable to that of the natural habitat of the region. Aquatic systems possessing biotic integrity can withstand or rapidly recover from most perturbations imposed by natural environmental processes and survive many major disruptions induced by humans. Aquatic systems that lack integrity are often already degraded and further natural or human-induced events only accelerate the degradation. In brief, biotic integrity is a feature of aquatic ecosystems whose composition, structure and function have not been adversely impaired by human activities (Karr et al., 1986).

Many groups of organisms have been used as indicators to estimate environmental quality. Algae and benthic invertebrates are typical species in biological monitoring (Patrick, 1975; Resh & Unzicker, 1975). The structure and successions of fish communities also reflect the ecological conditions of the water body (Virbickas, 1988). The test of biotic integrity of running waters using fishes as biological monitors was originated by Karr et al. (1986). However, rivers of different regions, as well as their fish communities, are distinctive. Therefore, leaning on multiple-year data on Lithuanian river fish communities, we have slightly modified and adapted Karr's Index of Biotic Integrity to the conditions of Lithuania.

Materials and methods

We have used the research data from 60 Lithuanian rivers of various size (a total of 101 river sites) investigated between 1981 and 1995. The research was usually carried out in summer (June–August).

In each river site, we investigated the mean channel width (m), depth (m), water velocity (m s^{-1}), water temperature (°C), dissolved oxygen concentration (mg l^{-1}), as well as habitat structure (bottom structure, presence of rapids, pits and coves, % of bottom area overgrown with submerged vegetation) were recorded. Data on water quality (concentration of total nitrogen

(N), mg l^{-1}; phosphates (P), mg l^{-1}; BOD_5, mg l^{-1}), were taken from the reports of the Ministry of Environment of Lithuania (Lietuvos upių vandens kokybės 1993 m. metraštis, 1994).

Fish were caught by electrofishing (pulsed current 600–900 V, frequency 20–80 Hz, impulse duration 2–12 ms^{-1}) in river sections not less than 150 m in length (usually more than 200 m). Three successive electrofishings were carried out at intervals of 45 min. The theoretical density of individuals for each fish species was calculated by the Zippin method of three removals, modified by Junge & Libosvarsky (1965). Fish density and biomass were extrapolated to values for one hectare (ind ha^{-1} and kg ha^{-1}). The specimens from successive electrofishings in a sample were separated according to species. For each individual, length (L cm) and weight (Q, g) were measured; scale, otolith or gill cover samples were taken for age determination.

The assignment of rivers to types was carried out according to the similarity of fish communities, by the method of Sorensen (1948). The dendrograms of similarity of fish communities were produced by the polythenial method of connecting clusters (Diuran & Odel, 1977).

Results

Fish species composition and the number of species in the community mainly depend on the river size, biotope structure and level of nutrient enrichment. The highest diversity – up to 50 fish species – was recorded in the Nemunas River (Zhukov, 1965). Other rivers of Lithuania typically contained up to 40 fish species (Virbickas & Maniukas, 1972). According to our research and published data, at present 47 fish and lamprey species belonging to 16 families inhabit Lithuanian rivers. Forty-five fish species are found in the large rivers, 38 in the middle-sized rivers, 23 in the streams and eight in the brooks. Of the total number of fish species, rheophilic forms amount to 45%, limnophilic 36% and benthic 19%. Many of fish communities include all these ecological groups of fishes, especially those of large rivers (Table 1a).

Since rivers differ by their size, yield, water velocity, temperature and, most importantly, they change from upstream to downstream, the influence of various factors on different fish species and communities is diverse. Taking into consideration the wide spectrum of alteration of species composition and applying mathematical approaches, we created the dendrogram of river ichthyofauna resemblance (according to Chekhanovskij-Sörensen, L°(Ri ∩ Rj). In the dendrogram, Lithuanian rivers divide into four major types according to the resemblance of communities: brooks – up to 10 km in length, streams – up to 50 km (and headwaters of bigger rivers), middle-sized rivers – up to 200 km (and the middle reaches of the large rivers) and large rivers – over 200 km (and the lower reaches of the middle-sized rivers) (see Figure 1). According to the mean water temperature in summer, Lithuanian rivers divide conditionally into cold-water (<17 °C) and warm-water (>17 °C) ones. Thus, depending on thermal regime, each river type was subdivided into subtypes with different dominant and indicator fish species; these can be further subdivided into variants according to character of flow regulation (see Figure 2) (Kesminas et al., 1996).

In the brooks and headwaters, 1–5 fish species are present in most cases (on the whole 8 fish species were established): the warm-water brooks support species belonging to borealic and arctic faunistic complexes, while the cold-water ones support fish species belonging to foothill borealic and arctic faunistic complexes. Depending on thermal regime of the water, the main brook fish community includes three-spined stickleback, brown trout, bullhead, brook lamprey and the facultative species: minnow, stone loach, gudgeon. Some brooks in Lithuania are inhabited solely by three-spined stickleback, brown trout or even juvenile sea trout. On the whole, fish production in the brooks is low, many of brooks are regulated.

The small rivers (streams) and headwaters provide a habitat for about 10–15 fish species (overall, 24 species are present here). According to the species composition, two types of streams were distinguished: cold-water streams, where brown trout (and in places sea trout) and other rheophilic fish species prevail, and warm-water streams are not inhabited by brown trout. Besides fish species characteristic for brooks, most stream fish communities includes dace, gudgeon, grayling, roach, pike, perch and burbot. In more nutrient-enriched streams, the prevalence of typical rheophilic species decreases, while that of eurybiontic species (roach, pike, bleak) increases. Gudgeon, dace and perch are common in many of the biotopes. In dammed streams or those linking the lakes, limnophilic species such as carp, crucian carp, bream and silver bream often occur in the communities. These species are not abundant and merely supplement the facultative part of the community. Migratory species –

Table 1a. List of fish species reported from Lithuanian rivers (scientific and common names)

	Fish families and species		Status	River type specific status			
	Scientific names:	Common names:		large rivers	middle-sized rivers	streams	brooks
	Petromyzonidae						
1	*Petromyzon marinus* (L.)	Sea lamprey	+	+			
2	*Lampetra fluviatilis* (L.)	River lamprey	++	++	++	+	
3	*Lampetra planeri* (Bloch.)	Brook lamprey	++	+	+	++	++
	Acipenseridae						
4	*Acipenser sturio* (L.)	Atlantic sturgeon	extinct				
5	*Acipenser ruthenus* (L.)	Sterlet	introduced, extinct				
	Clupeidae						
6	*Alosa fallax* (Leceped)	Twaite shad	+	+			
	Salmonidae						
7	*Salmo salar* (L.)	Atlantic salmon	+	+	+		
8	*Salmo trutta trutta* (L.)	Sea-trout	+	+	+	+	+
9	*Salmo trutta fario* (L.)	River trout	++	+	++	++	+
10	*Oncorhynchus mykiss* (Gib.)	Rainbow trout	+		+		
	Coregoninae						
11	*Coregonus lavaretus* (L.)	European whitefish	+				
12	*Coregonus peled* (Gmelin)	Peled whitefish	introduced in lakes, in the rivers casual				
	Thymallidae						
13	*Thymallus thymallus* (L.)	European grayling	+	+	+	+	
	Osmeridae						
14	*Osmerus eperlanus* (L.)	European smelt	+++	+++			
	Esocidae						
15	*Esox lucius* (L.)	Pike	++	++	++	+	
	Anguillidae						
16	*Anguilla anguilla* (L.)	European eel	+	+	+	+	
	Cyprinidae						
17	*Rutilus rutilus* (L.)	Roach	+++	+++	+++	++	
18	*Leuciscus cephalus* (L.)	Chub	++	++	++	+	
19	*Leuciscus leuciscus* (L.)	Dace	+	+	+	+	
20	*Leuciscus idus* (L.)	Ide	++	+	+	+	
21	*Phoxinus phoxinus* (L.)	Minnow	++	++	++	++	+
22	*Scardinius erythrophthalmus* (L.)	Rudd	+	+	+	+	
23	*Aspius aspius* (L.)	Asp	+	++	+		
24	*Leucaspius delineatus* (Heck.)	Sunbleak	+	+	+		
25	*Tinca tinca* (L.)	Tench	+	+	+		
26	*Chondrostoma nasus* (L.)	Undermouth	+	+	+		
27	*Gobio gobio* (L.)	Gudgeon	++	++	++	++	+
28	*Barbus barbus* (L.)	Barbel	+	+	+		
29	*Alburnus alburnus* (L.)	Bleak	+++	+++	+++	++	
30	*Alburnoides bipunctatus* (Bloch.)	Riffle minnow	+++	+++	++	+	
31	*Blicca bjoerkna* (L.)	Silver bream	++	++	+		
32	*Abramis brama* (L.)	Bream	++	++	+		
33	*Vimba vimba* (L.)	Vimba	+	+	+	+	
34	*Pelecus cultratus* (L.)	Chekhon	+	+			

Continued on p. 260

Table 1a. contd.

35	*Rhodeus sericeus amarus* (Bloch.)	Bitterling	++	++	+		
36	*Carassius carassius* (L.)	Crucian carp	+	+			
37	*Carassius auratus gibelio* (Bloch.)	Gibel	+	+	+		
38	*Cyprinus carpio* (L.)	Carp	+	+	+		
	Cobitidae						
39	*Noemacheilus barbatulus*	Stone loach	+++	+++	+++	++	++
40	*Cobitis taenia* (L.)	Spined loach	++	++	++	++	
41	*Misgurnus fossilis* (L.)	Pond loach	+	+	+		
	Siluridae						
42	*Silurus glanis* (L.)	European wels	+	+			
	Gadidae						
43	*Lota lota* (L.)	Burbot	+	+	+	+	
	Percidae						
44	*Gymnocephalus cernuus* (L.)	Ruff	+	+	+		
45	*Perca fluviatilis* (L.)	Perch	++	++	++	+	
46	*Stizostedion lucioperca*	Pike-perch	+	+			
	Cottidae						
47	*Cottus gobio* (L.)	Bullhead	++	++	++	++	++
	Pleuronectidae						
48	*Platichthys flesus* (Duncker)	Flounder	+	+			
	Gasterosteidae						
49	*Pungitius pungitius* (L.)	Nine-spined stickleback	+	+	+		
50	*Gasterosteus aculeatus* (L.)	Three-spined stickleback	+++	+++	+++	+++	++
		Total	**47**	**45**	**38**	**23**	**8**

Note: +++ - abundant; ++ - intermediate; + - rare.

Table 1b. Indicator species in the order of their sensitivity in different river types

River type	Brooks		Streams		Middle-sized rivers		Large rivers	
	gudgeon	↓	bullhead, minnow	↓	minnow, vimba, gudgeon	↓	barbel, asp, vimba	↓
Warm-water	stone loach	↓	riffle minnow, gudgeon	↓	riffle minnow, chub, dace	↓	chub, ide, bream	↓
	stickleback	↓	roach, pike	↓	roach, pike, perch	↓	roach, pike, tench	↓
	trout	↓	river trout, bullhead	↓	sea-trout, river trout, bullhead	↓	salmon, sea-trout, river trout	↓
Cold-water	bullhead	↓	minnow, gudgeon	↓	grayling, dace, gudgeon	↓	chub, dace, vimba	↓
	stone loach	↓	roach, pike	↓	roach, pike, perch	↓	roach, pike, bleak	↓

vimba and Atlantic salmon – are rare in the streams. Vimba is sometimes found in the warm-water streams, Atlantic salmon in deeper cold-water streams (Lakaja, Pershokshna). In the Jura, Minija, Zheimena and Bartuva river basins, juvenile brown trout and sea trout predominate.

The middle-sized rivers up to 100 km in length contain 20–25 fish species, whereas the rivers up to 200 km (and the middle reaches of big rivers) bear up to 25–30 species. In most cases, the fish community is formed of 9–16 species. The presence of different ecological conditions predetermines a diverse ichthyofauna and complex community structure. The ichthyofauna of many rivers of this type is compound, i.e. composed of fishes belonging to various ecological groups (rheophilic, limnophilic, ground fishes). Fish community structure in the middle-sized rivers changes from upstream to downstream. The following common features indicate that fish communities of the brooks and those of the headwaters of streams, middle-sized and large rivers are similarly structures: in all of them, stenobiontic rheophilic fish species usually prevail. In the middle and lower reaches of bigger rivers, the influence of succession leads to the gradual elimination of stenobiontic species and the formation of compound fish communities. Besides fish species characteristic for brooks and streams, most middle-sized rivers includes chub, bleak, riffle minnow, spined loach, ide, asp, bream, vimba and migratory salmonids.

Fish communities of large rivers are complex and consist of many components. They are formed of migratory, rheophilic and limnophilic fish species. Large rivers are characterized by a succession of fish communities of brooks, streams, middle-sized rivers and ultimately large-river ichthyofaunas further downstream. Therefore, the number of fish species from upstream to downstream increases too. The communities mostly consist of 9–19 species; roach, pike, gudgeon, perch and some smaller species usually prevail. The increase in community size is accompanied by an increase in the main body of the community: besides the fish species mentioned, chub, bleak, bream and vimba occur. Fish communities of the Nemunas and Neris rivers are especially abundant: on average 24–26 fish species are found here. However, such multi-component communities are not stable because they depend on the seasonal physical features of the water (duration of freshets, etc.) as well as on the prevalence of different migratory (Atlantic salmon, sea trout, vimba) and semi-migratory (European smelt, chekhon) species.

The influence of river size, river length, width, and total area of river basin on the number of fish species also can be illustrated by the following equations:

$$y = e(ai + bi \cdot xi);\ p < 0.005,$$

where:

a) xi = river length (km), ai = 0.793; bi = 0.774;

b) xi = ln river width (m), ai = −0.493; bi = 0.942;

c) xi = ln basin area (ha), ai = −0.469; bi = 0.459.

Thus, fish communities of the rivers of various types differ in characteristic species and the sensitivity of the latter to environmental alterations is different also. In the course of regressive succession, fish species are eliminated sequentially. The main indicator species characteristic for the different riverine fish community types are listed in the order of their sensitivity to human impacts in Table 1b (derived from Kesminas, 1992).

Fish density and biomass in Lithuanian rivers are also related to the river size (r = 0.5264 and r = 0.4518; p < 0.0001); the mean fish density and biomass were therefore calculated separately for each type.

Depending on river size, fish biomass varies from 2 to 250 kg/ha, and the density reaches up to ~ 15 600 ind /ha, being the lowest in brooks and the highest in large rivers (Figure 3). Fish density and biomass in the warm-water streams, middle-sized and large rivers are higher than those in the cold-water ones, while in the brooks, on the contrary, fish density and biomass are higher in the cold-water ones.

According to their feeding habit, Lithuanian riverine fishes are divided into five groups: bentophagous – bream, silver bream, barbel, gudgeon, vimba, carp, stone loach, spined loach, pond loach, ruff, minnow, three-spined stickleback, nine-spined stickleback, grayling; ichthyophagous – pike, perch, pike-perch, wels, burbot, asp, sea trout, Atlantic salmon, brown trout; euryphagous – roach, dace, chub, ide, crucian carp, tench; phytophagous – rudd, bitterling, undermouth; planktivorous – bleak, riffle minnow, sunbleak. Since feeding habits change with age, species are divided into trophic groups according to the character of nutrition of mature individuals.

Bentophagous fishes dominate about 45% of the total number of species clearly predominate. Predators amount to ~ 22%, euryphagous 17%, phytophagous and planktivorous 8%. The percentage of fishes

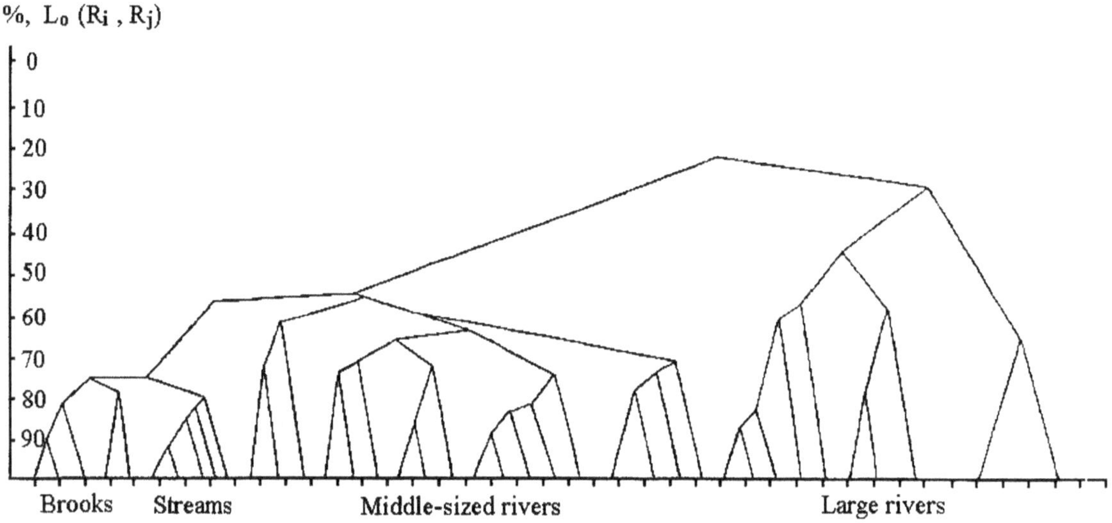

Figure 1. Dendrogram of river types according to the similarity of fish communities.

The brook fish community	warm-water		— 1 subtype
	cold-water		— 2 subtype
The stream fish community	warm-water	unregulated	— 3 subtype
		regulated	— 4 subtype
	cold-water	unregulated	— 5 subtype
		regulated	— 6 subtype
The middle-sized river fish community	warm-water	unregulated	— 7 subtype
		regulated	— 8 subtype
	cold-water	unregulated	— 9 subtype
		regulated	— 10 subtype
The large river fish community	warm-water	unregulated	— 11 subtype
		regulated	— 12 subtype
	cold-water	unregulated	— 13 subtype
		regulated	— 14 subtype

Figure 2. Types of Lithuanian river fish communities.

Table 2. Integrity scheme for brooks

Category	Metric	Scoring criteria		Estimation ratings	Integrity class	Sum of ratings	Attributes
		Warm-water brooks	Cold-water brooks				
Species composition, abundance and biomass	1. Total number of fish species	> 2	> 4	10	Excellent	62–56	Comparable to the best situation without human disturbance. All expected species for fish community of river type are present. Fish communities of this type usually include up to 5 fish species, the main body of fish community is formed of 1–2 species (in the cold-water brooks – stone loach, three-spined stickleback, and in the warm-water ones – river trout, bullhead). Trophic composition is well balanced. Fish density reaches approximately 1000 ind/ha, and biomass – 10 kg/ha.
		2	4–2	6			
		1	1	3			
	2. Indicator species	Gudgeon	River trout	10			
		Stone loach	Bullhead	6			
		Stickleback	Stone loach	3			
		—	—	0			
	3. Fish abundance (ind/ha)	> 800	> 1300	10	Good	52–48	Species richness is less than expected due to the loss of the most intolerant forms (indicator species). Some species are present with less than optimal abundance, trophic composition shows some signs of stress.
		800–500	1300–1000	6			
		< 500	< 1000	3			
	4. Fish biomass (kg/ha)	> 4	> 9	10			
		4–3	9–6	6			
		< 3	< 6	3			
Trophic composition	5. Predators	present	present	6	Fair	40–34	Species composition still more poor, abundance of intolerant species is considerably lower, highly skewed trophic structure, number of euryphagous species increases, total fish density is goes down.
		absent	absent	0			
	6. Euryphagous species	absent	absent	6			
		present	present	0			
	7. Proportion of individuals of bentophagous fish species (%) to the total number of fishes	> 50	> 60	6	Poor	24–19	Species richness and abundance continues to decrease. There are no clear dominants, predators are solitary only, individuals with disease, anomalies and hybrids are found.
		50–40	60–40	4			
		< 40	< 40	2			
Fish condition	8. The presence of individuals with disease and anomalies	absent	absent	2	Very poor	6–12	Fish density is very low, individuals with disease and anomalies are common.
		present	present	0			
	9. The presence of individuals as hybrids	absent	absent	2	No fish	–	
		present	present	0			

Table 3. Integrity scheme for streams

Category	Metric	Scoring criteria		Estimation ratings	Integrity class	Sum of ratings	Attributes
		Warm-water streams	Cold-water streams				
Species composition, abundance and biomass	1. Total number of fish species	> 8	> 8	10	Excellent	62–56	Comparable to the best situation without human disturbance. Fish communities of this type usually include up to 10 fish species. The main body of fish community is formed of 3–4 species. In the main bodies of warm-water streams, the roach, pike and gudgeon predominate. In those of cold-water streams, the trout, bullhead, minnow prevail. Trophic composition is well balanced, dominated by bentophagous species. There are quite a lot of predatory species and only a few euryphagous ones. Fish density reaches up to 1500–2000 ind/ha, biomass – 35–40 kg/ha.
		8–6	8–6	6			
		< 6	< 6	3			
	2. Indicator species	Bullhead, minnow	River trout, bullhead	10			
		Riffle minnow, gudgeon,	Minnow, gudgeon	6			
		Roach, pike	Roach, pike	3			
		—	—	0			
	3. Fish abundance (ind/ha)	> 2000	> 1500	10	Good	52–46	Species richness is less than expected due to the loss of the most intolerant forms (indicator species). Some species are present with less than optimal abundance, trophic composition shows some signs of stress.
		2000–1000	1500–1000	6			
		< 1000	< 1000	3			
	4. Fish biomass (kg/ha)	> 40	> 35	10			
		40–20	35–20	6			
		< 20	< 20	3			
Trophic composition	5. Proportion of individuals of predatory fish species (%) to the total number of fishes	> 16	> 14	6	Fair	40–33	Species composition still more poor, considerably lower the abundance of intolerant species, highly skewed trophic structure (number of euryphagous species increases, while of others – decreases), total fish density goes down.
		16–10	14–8	4			
		< 10	< 8	2			
	6. Proportion of individuals of euryphagous fish species (%) to the total number of fishes	< 7	< 7	6			
		7–14	7–14	4			
		> 14	> 14	2			
	7. Proportion of individuals of bentophagous fish species (%) to the total number of fishes	> 50	> 60	6	Poor	27–20	Species richness and abundance continues to decrease. There are no clear dominants, further decreases abundance of predatory species, individuals with disease, anomalies and hybrids are found.
		50–40	60–40	4			
		< 40	< 40	2			
Fish condition	8. The presence of individuals with disease and anomalies	absent	absent	2	Very poor	19–14	Fish density is very low, individuals with disease and anomalies are common.
		present	present	0			
	9. The presence of individuals as hybrids	absent	absent	2	No fish	–	
		present	present	0			

Table 4. Integrity scheme for middle-sized rivers

Category	Metric	Scoring criteria: Warm-water rivers	Scoring criteria: Cold-water rivers	Estimation ratings	Integrity class	Sum of ratings	Attributes
Species composition, abundance and biomass	1. Total number of fish species	> 12 12–8 < 8	> 11 11–8 < 8	10 6 3	Excellent	62–56	Comparable to the best situation without human disturbance. All expected species for fish community of river type are present. Fish communities of this type usually include more than 10 fish species. The main body of fish community is formed of 5–7 species. In fish communities of warm-water rivers the roach, pike, perch, gudgeon, minnow, dace predominate. Migratory fish species are rare. In cold-water river fish communities the stenobiontic rheophilic local and migratory species prevail: river trout, bullhead, grayling, sea-trout, dace, etc. Fish density reaches up to 2500–3000 ind/ha, biomass – 40–60 kg/ha. Trophic composition is well balanced.
	2. Indicator species	Minnow, vimba, gudgeon Riffle minnow, chub, dace Roach, pike, perch —	Sea-trout, river trout, bullhead, Grayling, dace, gudgeon Roach, pike, perch —	10 6 3 0			
	3. Fish abundance (ind/ha)	> 3000 3000–2000 < 2000	> 2500 2500–1500 < 1500	10 6 3	Good	52–46	Species richness is less than expected due to the loss of the most intolerant forms (indicator species). Some species are present with less than optimal abundance, trophic composition shows some signs of stress.
	4. Fish biomass (kg/ha)	> 60 60–40 < 40	> 40 40–25 < 25	10 6 3			
Trophic composition	5. Proportion of individuals of predatory fish species (%) to the total number of fishes	> 14 14–8 < 8	> 14 14–8 < 8	6 4 2	Fair	40–33	Species composition still more poor, intolerant species entirely vanishes, skewed trophic structure, number of euryphagous species increases, while that of predators – decreases.
	6. Proportion of individuals of euryphagous fish species (%) to the total number of fishes	< 10 10–20 > 20	< 10 10–20 > 20	6 4 2			
	7. Proportion of individuals of bentophagous fish species (%) to the total number of fishes	> 50 50–40 < 40	> 60 60–45 < 45	6 4 2	Poor	27–20	Species richness still more poor, density of predatory species is very low, individuals with disease, anomalies and hybrids are found.
Fish condition	8. The presence of individuals with disease and anomalies	absent present	absent present	2 0	Very poor	19–14	Fish density is very low, individuals with disease and anomalies are common.
	9. The presence of individuals as hybrids	absent present	absent present	2 0	No fish	–	

Table 5. Integrity scheme for large rivers

Category	Metric	Scoring criteria: Warm-water rivers	Scoring criteria: Cold-water rivers	Estimation ratings	Integrity class	Sum of ratings	Attributes
Species composition, abundance and biomass	1. Total number of fish species	> 15 15–11 < 11	> 13 13–10 < 10	10 6 3	Excellent	62–56	Comparable to the best situation without human disturbance. All expected species for fish community of river type are present. Fish communities of this type usually include 13–17 fish species. In the main bodies of warm-water river fish community the bream, asp, bleak, chub, etc. predominate. In those of cold-water rivers the dace, river trout, gudgeon,and also migratory species – salmon, sea-trout, vimba prevail. Trophic composition is well balanced, dominated by bentophagous species, abundance of predators is relatively high, euryphagous species are less abundant. Fish density reaches approximately 4000–5000 ind/ha, biomass – 70–80 kg/ha.
	2. Indicator species	Barbel, asp, vimba Chub, ide, bream Roach, pike, tench —	Salmon, sea-trout, river trout Chub, dace, vimba Roach, pike, bleak —	10 6 3 0			
	3. Fish abundance (ind/ha)	> 4500 4500–3000 < 3000	> 4000 4000-3000 < 3000	10 6 3	Good	52–46	Species richness is less than expected due to the loss of intolerant forms (indicator species). Some species are present with less than optimal abundance, trophic composition shows some signs of stress.
	4. Fish biomass (kg/ha)	> 70 70–50 > 50	< 50 50–40 < 40	10 6 3			
Trophic composition	5. Proportion of individuals of predatory fish species (%) to the total number of fishes	> 14 14–8 < 8	> 20 20–12 < 12	6 4 2	Fair	40–33	Species composition still more poor, total fish density goes down, intolerant species entirely vanishes, skewed trophic structure, number of euryphagous species increases, while that of predators – decreases.
	6. Proportion of individuals of euryphagous fish species (%) to the total number of fishes	< 15 25–15 > 25	< 15 25–15 > 25	6 4 2			
	7. Proportion of individuals of bentophagous fish species (%) to the total number of fishes	> 40 40–30 < 30	> 40 40–30 < 30	6 4 2	Poor	27–20	Species diversity and abundance still more poor, density of predatory species is very low, individuals with disease, anomalies and hybrids are found.
Fish condition	8. The presence of individuals with disease and anomalies	absent present	absent present	2 0	Very poor	19–14	Fish density is very low, individuals with disease and anomalies are common.
	9. The presence of individuals as hybrids	absent present	absent present	2 0	No fish	–	

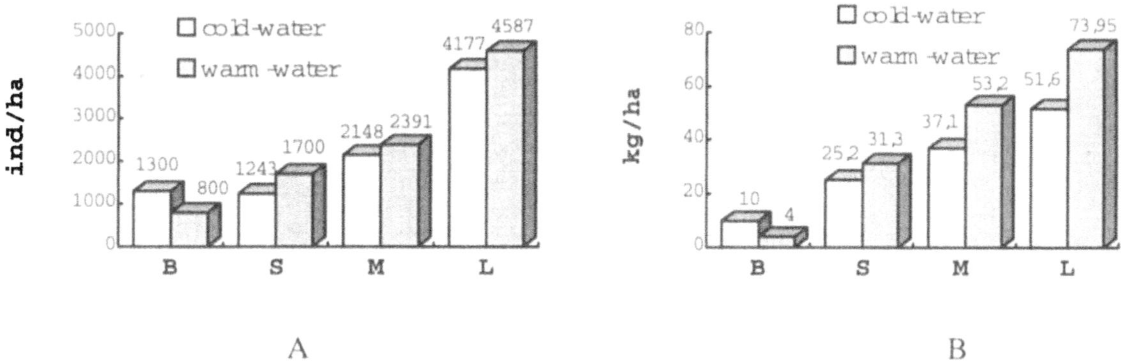

Figure 3. The mean fish density (A) and biomass (B) in rivers of different types; B = brooks, S = streams, M = middle-sized rivers, L = large rivers.

belonging to different trophic groups varies in the different river types.

Alterations in water quality or other habitat conditions change fish communities due to fluctuating food resources, the varying resistance of various species to different pollutants and changes in habitat structure. We found that only the biomass of euryphagous fish species correlates positively with concentration of the biogenous elements (especially phosphorus, r=0.868; p<0.0005) and river thermal regime (r=0.567; p<0.0001), while biomasses of other trophic groups change indifferently or negatively.

Leaning on this set of characteristic features of fish communities, we have modified and adapted Karr's index of biotic integrity to Lithuanian rivers. For testing biointegrity, nine metrics in three categories (species composition, abundance and biomass, trophic composition, and fish condition) were selected. The combined ratings of these metrics characterize the underlying biotic integrity (excellent, good, fair, poor, very poor and no fish) of the sampling site.

Species composition, abundance and biomass

This category is the most important because it characterizes the main quantitative and qualitative metrics of the fish community. Those metrics change depending on the river type and level of biointegrity.

Total number of fish species

Number of species in the community depends on the river type and state of the fish community – while biocoenosis degrades, species diversity decreases.

Indicator species

The presence or absence of indicator fish species refers to the state of the community, because they are sensitive to many ecological parameters: water quality, habitat degradation, siltation, pollution, etc. In the tests of biointegrity of various river fish communities, the indicator species are listed in the order of their sensitivity to human impacts.

Fish abundance

In the different river types, fish density (ind/ha) ranges within characteristic limits. It decreases with degradation of fish community.

Fish biomass

(in kg/ha). It also has characteristic limits for every river type, decreases with degradation of the fish community.

Trophic composition

Altered water quality or other habitat conditions change fish communities due to fluctuating food resources. Species are divided into trophic groups according to the feeding habits of mature individuals.

Predators

Include species feeding mainly on fish. Some possess compound nutrition (for example brown trout). Predatory species occurrence is one of the attributes of trophically balanced fish communities. In brooks, due to overall low diversity, their presence or absence in the communities are used as the criterion.

Table 6. The state of fish communities in Lithuanian rivers

River type	Subtype	Name	Sum of ratings[a]	Class of integrity	No. of sites investigated
Brooks	Cold-water	1. Beržupis	62	Excellent	1
		2. Karklė	62	Excellent	1
		3. Strūzda	62	Excellent	1
		4. Avirė	60	Excellent	1
		5. Pasgrinda	58	Excellent	1
		6. Bilsinyčia	48	Good	2
		7. Samė	48	Good	1
	Warm-water	8. Bagdonas	52	Good	1
		9. Eglynupis	41	Fair	1
		10. Kanalas	38	Fair	1
		11. Bevardis	31	Fair	1
Streams	Cold-water	12. Bražuolė	52	Good	2
		13. Judrė	51	Good	1
		14. Ratnyčia	51	Good	2
		15. Lomena	49	Good	2
		16. Skroblus	48	Good	2
		17. Gauja	44	Good	2
		18. Kena	44	Good	1
		19. Taurija	44	Good	1
		20. Verseka	43	Good-Fair	2
		21. Derežna	40	Fair	3
		22. Graumena	39	Fair	1
	Warm-water	23. Kertuša	56	Excellent	1
		24. Vašuoka	51	Good	1
		25. Viešinta	47	Good	2
Middle-sized rivers	Cold-water	26. Šalpė	49	Good	4
		27. Vilnia	43	Good-Fair	8
		28. Virinta	39	Fair	3
		29. Verknė	33	Fair	3
	Warm-water	30. Dirvinta	43	Good-Fair	8
		31. Lėvuo	40	Fair	3
		32. Virvytė	36	Fair	4
Large rivers	Cold-water	33. Merkys	33	Fair	7
	Warm-water	34. Šventoji	49	Good-Fair	3

[a] If more than 1 station was tested in the river, the mean of sum of ratings was calculated.

Euryphagous fishes

Increased abundance of euryphagous fish species indicates degraded trophic conditions. Specific food resources become even less accessible for trophically specialized species and, as a result, the abundance of euryphagous species increases.

Bentophagous species

Predominating trophic group in fish communities of all the rivers types. The drop in diversity here is related to poverty of trophic resources due to degradation.

Fish condition

It is being assessed visually:

The presence of individuals with disease and anomalies

Habitat degradation reduces fish resistance to parasites and diseases (ligula, saprolegnia, cancer, wounds, curvature of the spine, etc.).

The presence of individuals as hybrids

Environmental degradation can lead to increased frequency of hybridization as a result of a decline in spawning places and a decrease of fish abundance in populations.

Discussion

Four main river types are distinguished in Lithuania. The structure of their fish communities differs in a range of attributes. For this reason, the individual test of biointegrity was made up for each river type: brooks, streams, middle-sized rivers and large rivers (see Tables 2, 3, 4 and 5). Rivers of each type are subdivided into cold-water and warm-water ones, which differ in fish community structure and abundance.

Reference conditions, to which actual conditions of the rivers of each type can be compared, were established as follows. Since the water quality in most investigated sites was good (BOD_5: <3.5 mg l^{-1},N: <0.75 mg l^{-1} , P: < 0.05 mg l^{-1}) and river morphology was natural, we assumed that the average values of different metrics (number of fish species, abundance, biomass, trophic composition in %) was characteristic for fish communities of medium level of integrity. Therefore, medium ratings were assigned in the biotest for these values. Correspondingly, for average maximal values the maximal ratings were assigned and for average minimal values the minimal ratings were assigned. These values differ in the various river types.Indicator fish species in the order of their sensitivity are given in Table 1b, derived from the scheme of the fish community structure in Lithuanian rivers (Kesminas, 1992).

Fishes must be divided according to species, the total number of species must be established, and indicator species need to be defined. For example, if the brown trout and bullhead are found in a sample from a cold-water stream, then the rating of metric 2 is 10 (see Table 3); if at least one of the species mentioned is absent in a sample, but minnow and gudgeon are present, then the rating is 6; if those species are absent too, but roach and pike are present, the rating is 3; if all of them are absent, the rating is 0. Subsequently, fish density (ind /ha) and biomass (kg/ha) are estimated (metrics 3 and 4).

The next step is to determine the distribution of fish species according to feeding habit. The ascription of Lithuanian river fish species to a specific trophic group is given in the text.

The presence of fishes with disease and anomalies, and the presence of hybrids, is being assessed visually. Usually such individuals are rare, they are more frequent in heavily disturbed rivers. After the ratings are assigned to all the metrics, the total score of ratings (it can range from 0 to 62) indicates the class of integrity of the sample site under investigation.

The method was tested in 34 different Lithuanian rivers and the results are given in Table 6: accordingly, the integrity of fish communities is excellent in most cold-water brooks. This was expected because most are under protection in Lithuania as the spawning places of salmonids (brown trout, sea trout). Moreover, the load of anthropogenic activies in such brooks is low, with most flowing through woodlands.

The majority of warm-water streams flow through open agricultural landscapes and are consequently heavily affected by human activities (channelization, pollution), yielding poor fish communities. That is indicated by the biotest: the integrity of fish communities of warm-water brooks is in most cases fair.

The integrity of fish communities in most cold-water streams is good. The majority, as in the case of the cold-water brooks, are under protection as potential salmonid spawning sites. However, the impact of human activities on them is much greater (sporfishing, dams). Their basins are more extensive, increasing the probability of pollution.

A similar situation is observed in the warm-water streams. Although they are less protected, more resistant eurybiontic fish species are characteristic for fish communities here. Therefore, in spite of considerable anthropogenic load, fish communities of warm-water streams exhibit rather high productivity.

Fish communities of the middle-sized and large rivers of Lithuania suffer the greatest anthropogenic load. Although the integrity of fish communities in the various river segments changes from 'excellent' (mostly upstream) to 'fair', the mean sum of ratings indicates the integrity of fish communities of such rivers as being 'fair'.

References

Diuran, B. & P. Odel, 1977. Klasternyj analiz. In Statistika. Moskva: 1–127.

Junge, Ch. O. & J. Libosvarsky, 1965. Effects of size selectivity on population estimates based on successive removals with electrofishing gears. Zool. Listy 14: 171–178.

Karr, J. R., K. D. Fausch, P. L. Angermeier, P. R. Yant & I. J. Schlosser, 1986. Assessing biological integrity in running waters: a method and its rationale. Illinois Natural History Survey Special Publication 5. 28 pp.

Kesminas, V., 1992. Struktura i dinamika rybnogo naseleniya rek Litvy. Institut Ekologii AN Litvy, Vilnius. 175 pp.

Kesminas, V., T. Virbickas & J. Virbickas, 1996. Apie buvž išteklius Lietuvos upėse. Žuvininkystė Lietuvoje 2: 229–236.

Lietuvos upių vandens kokybės, 1993 m. metraštis, 1994. Lietuvos Respublikos Aplinkos Apsaugos ministerija, Vilnius. 49 pp.

Patrick, R., 1975. Stream communities. In Cody, M. L. & J. M. Diamond (eds), Ecology and Evolution of Communities. Harvard University Press, Cambridge, Massachusetts: 445–459.

Resh, V. & J. D. Unzicker, 1975. Water quality monitoring and aquatic organisms: the importance of species identifications. J. Wat. Pollut. Cont. Fed. 47: 9–19.

Sorensen, T., 1948. A method of establishing groups of equal amplitude in plant sociology based on similarity of species. Copenhagen. 34 pp.

Virbickas, J., 1988. Struktura i razvitiye ichtiocenozov ozer Litvy. Acta Hydrobiologica Lituanica 8: 74–92.

Virbickas, J. & I. Maniukas, 1972. Fauna ryb vnutrennykh vodoemov Litvy i metody po ee preobrazovaniyu. In Voprosy razvedeniya ryb i rakoobraznykh v vodoemakh Litvy. Vilnius: 7–34.

Zhukov, P. I., 1965. Ryby basejna Nemana. Minsk. 414 pp.

Hydrobiologia **422/423:** 271–278, 2000.
M. Jungwirth, S. Muhar & S. Schmutz (eds), Assessing the Ecological Integrity of Running Waters.

Fish as indicators for the assessment of the ecological integrity of large rivers

F. Schiemer
Department of Limnology, Institute of Ecology and Conservation Biology, University of Vienna, Althanstraße 14, A-1090 Vienna, Austria

Key words: biodiversity, bioenergetics, bioindicators, disturbance, fish, floodplains, ecological integrity, monitoring, restoration, river engineering

Abstract

Fish communities in large rivers are characterized by a high diversity, which reflects the structural diversity and habitat richness of inshore zones and connected floodplains. The connectivity of the different habitat elements in a broad spatio-temporal context, i.e., at various scales from catchment to microhabitat – and their nestedness – define the fitness of fish species both on the individual level (e.g., growth performances) and population level (i.e., population structure, mortality, etc.). Relevant spatial scales can be the whole river course in the case of some anadromous migrators or the availability of complementary microhabitat elements, e.g., during the early ontogeny of a species. The significance of connectivity at various scales from whole river to local reach have to be evaluated based on the requirements, reaction norms, and ecological flexibility of individual species. Integrity has to be evaluated in several respects:

(1) with regard to population genetics over extensive biogeographic areas and in long time scales;
(2) with regard to supplementary habitats in the course of the life cycle of individual species with ontogenetic habitat shifts and specific requirements during the reproductive phase;
(3) with regard to longitudinal and lateral transport and exchange processes determining local habitat conditions and the food supply for fish.

Due to these dependencies the status and condition of the fish fauna is a critical sensor of integrity at different scales and thus a good monitoring tool especially with regard to river engineering. The structure of the fish population – in particular the relative abundance of specifically adapted floodplain species – provides information on the overall conditions. For a more detailed evaluation the local distribution pattern of the fish fauna, population structure, seasonality, and growth performances of the early life history stages can indicate the quality and extent of connectedness at a finer scale and can be used to define management and engineering criteria.

Introduction

Over the past 10 years several concepts for understanding functional processes and biodiversity patterns of large rivers have been developed emphasizing their four-dimensional nature and hierarchical structure with regard to spatial and temporal properties. It is well understood that the key factors of floodplain rivers are the hydrological connectivity between river and floodplain and – on a longer time scale – the flood-controlled geomorphologic processes. These hydrologically driven disturbance regimes lead to river- and reach-specific levels of patch dynamics as well as to the characteristic ecotonal and mosaic structure and successional patterns. Ecological integrity of floodplain rivers is mainly defined by the extent of the interphase between geomorphology and hydrology, resulting in a reach-specific level of habitat diversity and habitat characteristics (Bayley, 1995; Gregory et al., 1991; Junk et al., 1989; Naiman et al., 1988; Petts & Amoros, 1996; Schiemer & Zalewski, 1992; Ward, 1998; Ward & Stanford, 1995a).

River engineering has disrupted ecological integrity with regard to the longitudinal continuity, the lateral interactions with the bordering riparian zones, and the structural subunits within the instream channel. This has led to deficiencies in functional processes, in habitat diversity, and in the diversity of characteristic biota. These disruptions have to be viewed from a catchment scale to the local microhabitat conditions (Frissell et al., 1986; Noss, 1990; Townsend, 1996; Ward, 1998; Ward & Stanford, 1995b). Existing concepts of river ecology provide a useful framework to develop restauration and management schemes (Karr, 1991, 1994; Schiemer et al., 1999; Sparks, 1995; Stanford et al., 1996).

The term 'ecological integrity' is valuable in this context since it focuses on a system-orientated view. However, the term is used in at least two different ways: firstly as a monitoring index – analogously to the 'saprobic index' – for assessing deterioration due to river engineering and secondly – in a more conceptual and analytical approach – towards a deeper understanding of the functioning of river systems and their biodiversity patterns.

In both meanings it is necessary that abiotic parameters be calibrated to the requirements of characteristic floodplain animal and plant species. Fish play a specific role as indicators since a broad spectrum of abiotic variables of different spatio-temporal scales are linked to the habitat requirements of particular species and their ontogenetic stages (see Jungwirth et al., 2000). With more knowledge about these manifold requirements their significance as bioindicators will further increase.

Elements of ecological integrity in large rivers and their relevance for fish

The term 'ecological integrity' provides a good focus for river ecology and river conservation even though it is not very clearly defined. In the Oxford Concise Dictionary integrity is described as 'wholeness, entirety and soundness', whereas the Austrian Dictionary emphasizes 'the undeteriorated state of a system'. We can combine these two meanings to arrive at a useful definition as 'functional entity with regard to its original state'. This is in line with the 'EC Water Framework Directive'.

Ecological integrity can be described by its physical and abiotic correlates. Ultimately, however, these have to be set in relation to the biological requirements of characteristic species, depending on the level of ecological insight necessary for a particular goal of assessment (e.g., nationwide monitoring of river conditions in contrast to a restoration scheme of a particular river segment).

Ecological integrity comprises several elements which are schematically outlined in a set of diagrams (Figure 1 a–c). The graph is based on the overall situation in large European rivers like the Danube or Rhine after river regulation, which started in the 19th century and resulted in channelized and stabilized river channels along with narrow floodplains and inundation zones. Different aquatic and semiaquatic habitat types can be distinguished: (a) the main channel and its tributaries; (b) open, connected sidearms (parapotamon); (c) disconnected larger sidearms with strong ongoing terrestrification processes and a strong development of aquatic vegetation (plesiopotamon); (d) strongly disconnected and fragmented waterbodies and wetlands (palaeopotamon).

Integrity has a hydrological meaning (Figure 1a) which refers to hydrological pulses via surfacewater connections and through the aquifer. Integrity in this sense means (a) connected waterbodies as habitat, (b) water supply to disconnected waterbodies within the floodplain and – on a longer time scale – (c) the geomorphologic processes within the river–floodplain relief and thus the sustained existence of certain habitat elements. Integrity is here defined as geomorphologic dynamics and habitat rejuvenation by erosive, scouring floods and sedimentation processes.

A second meaning of integrity concerns the flux of nutrients and living and dead organic matter (Figure 1b). This aspect is significant because it essentially defines habitat quality with regard to trophic conditions for fish. The number within the arrows indicates exchange processes at different scales:

(1) at a catchment-macroscale referring to the continuum of ecosystem interactions along the river corridor;
(2) at a reach-mesoscale referring to exchange processes between river and riparian zone;
(3) at a microscale referring to exchange processes of inshore storage zones within the river and between river and connected sidearms at rising and sinking water levels.

Lateral exchange processes are especially relevant during periods when floods open up new food resources. The extent of interaction between river and floodplains depends on the interphase of the geomorphology of the floodplain and the river stage. It is apparent that the more gradual the ecotones and

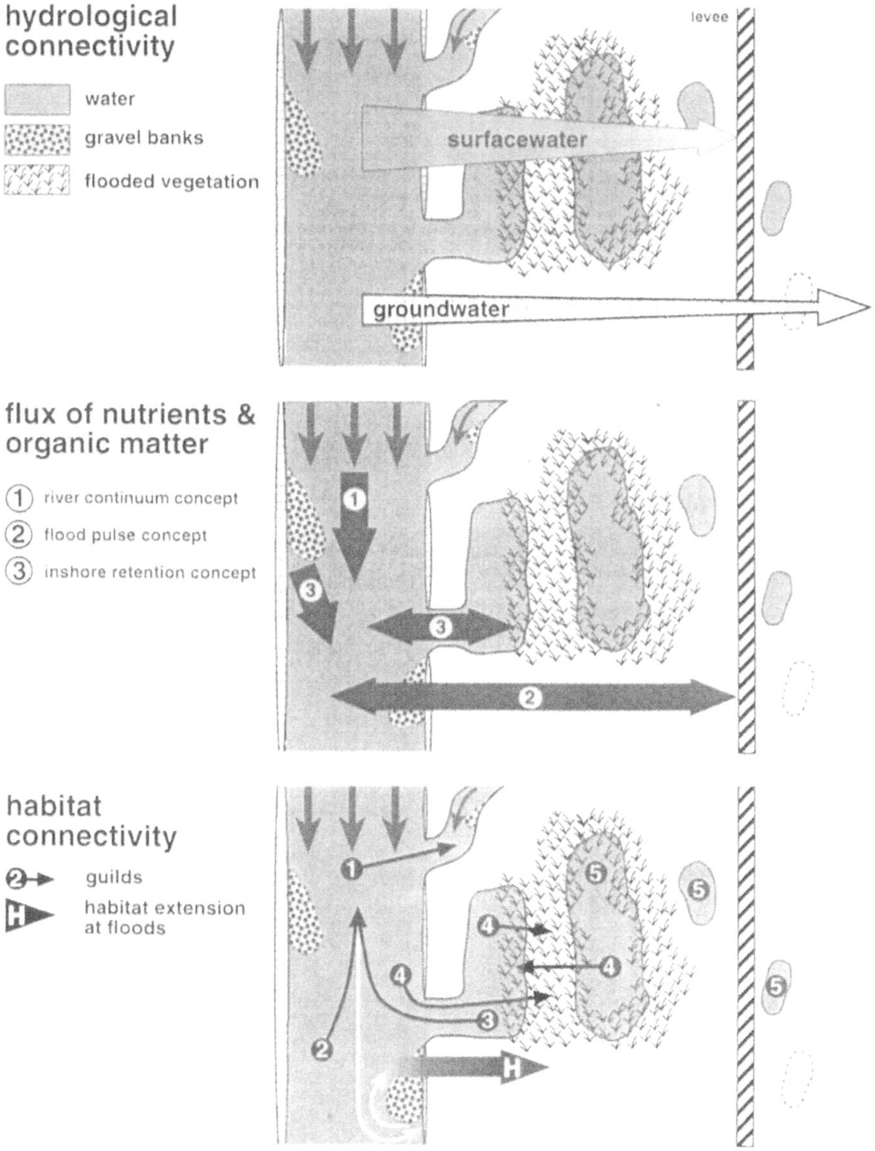

Figure 1. Elements of ecological integrity of floodplain rivers. Different habitat types are distinguished based on the situation of regulated European rivers. Integrity is defined with regard to: (a) hydrological connectivity; (b) flux of nutrients and organic matter; (c) habitat connectivity for fish. For explanations see text.

transition zones the higher the exchange processes at changing water levels.

A third element of integrity refers to the habitat availability and habitat connectivity for fish (Figure 1c). The graph distinguishes five guilds according to fish life cycle ranges, preferred adult habitat (circles), and spawning and nursery sites (black arrows). It addresses the meaning of integrity with regard to:

(a) ontogenetic niche shifts and the required connectivity of complementary habitat elements (especially nurseries) in the river–floodplain system, the population exchange between source and sink areas either by active migration or passive larval drift is a main determinant for population sustainance;

(b) the requirements for supplementary and seasonal habitats (e.g., seasonal change in habitat use);

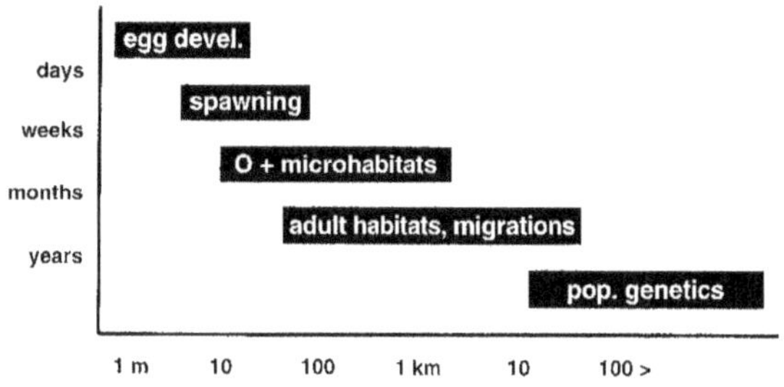

Figure 2. Significant spatio-temporal scales of the nase, *Chondrostoma nasus.*

(c) the connectivity and availability of refugia – which are particularly relevant in large river systems with continuously changing river stage and high stochasticity.

It is apparent that ecological integrity has an entirely different meaning for the various guilds. Large anadromous migrators like sturgeons require integrity at a catchment scale. Stagnotopic species like *Umbra krameri* or *Misgurnus fossilis* are exclusively found in strongly fragmented and vegetated pools in the wetlands. Integrity for such 'disconnectance specialists' means the long-term existence of such habitat types which are dependent on groundwater exchange and geomorphologic processes leading to their renewal and recreation at a centennial scale. These species belong to a strongly endangered guild, which indicates that much larger wetland areas with this habitat type existed prior to river regulation.

The most endangered guild of large river systems is, however, the riverine species (see below). In Figure 2 the required integrity at different spatio-temporal scales is exemplified for a well-studied European representative of this guild, the nase (*Chondrostoma nasus*).

Significance of integrity for fish: from descriptive indicators towards an analytical understanding

Fish are good indicators for ecological conditions of floodplain rivers because the various guilds integrate over their life cycle a wide range of riverine conditions from properties of bed sediments for egg development to longitudinal integrity for spawning migrations (Copp, 1989; Gaudin, in press; Persat et al., 1995; Schiemer et al., 1991). The occurrence, population dynamics, and growth pattern of individual species, along with their contribution to community structure, react sensitively to the connectedness of supplementary and complementary habitats (see Schlosser, 1995). They are also sensitive to the specific habitat condition with regard to temperature, current velocity profiles, and food availability, which are direct or indirect expressions of the ecological integrity of river systems. Several overall trends between hydrological conditions and the diversity, population dynamics, and production of fish have been proposed (Figure 3, see Schiemer et al., in press). Junk & Welcomme (1990) discussed models of fish production depending on the hydrological regime of river systems.

The community structure of fish is frequently monitored to describe river conditions (Schiemer & Zalewski, 1992). An example is given in Figure 4. It compares the assemblage structure of the free-flowing Austrian Danube with that of an impoundment 5 years after the dam was closed (Schiemer & Waidbacher, 1992). In the free-flowing river the proportion of the rheophilic guild is high. Note that most of these elements are on the Red List of endangered taxa. Five years after the dam was closed the composition had

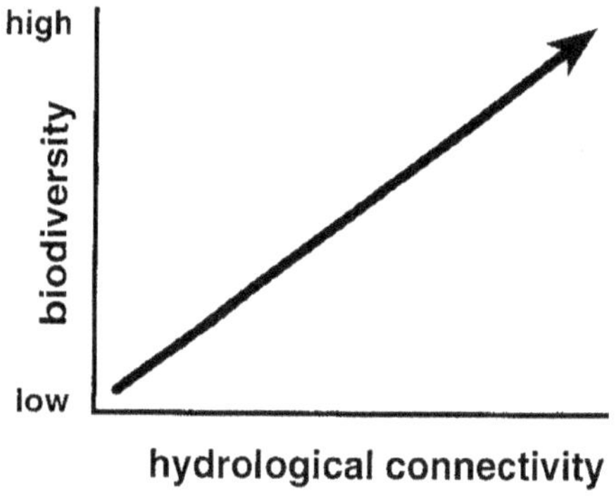

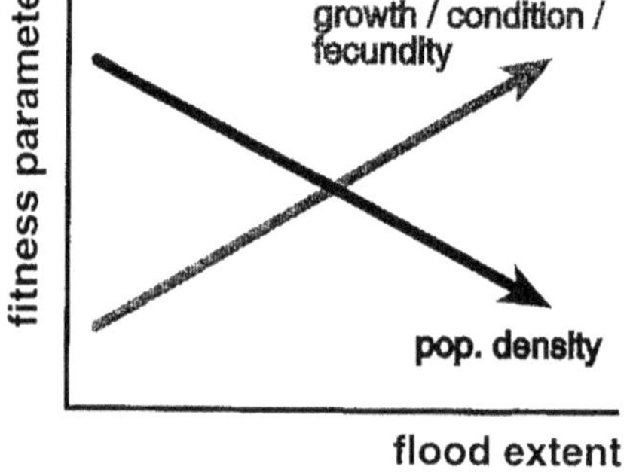

Figure 3. (a) Overall trends between hydrological connectivity within a floodplain and the biodiversity of fish; (b) overall trends between the duration and extent of floods and its effect on population size as well as growth, condition, and fecundity.

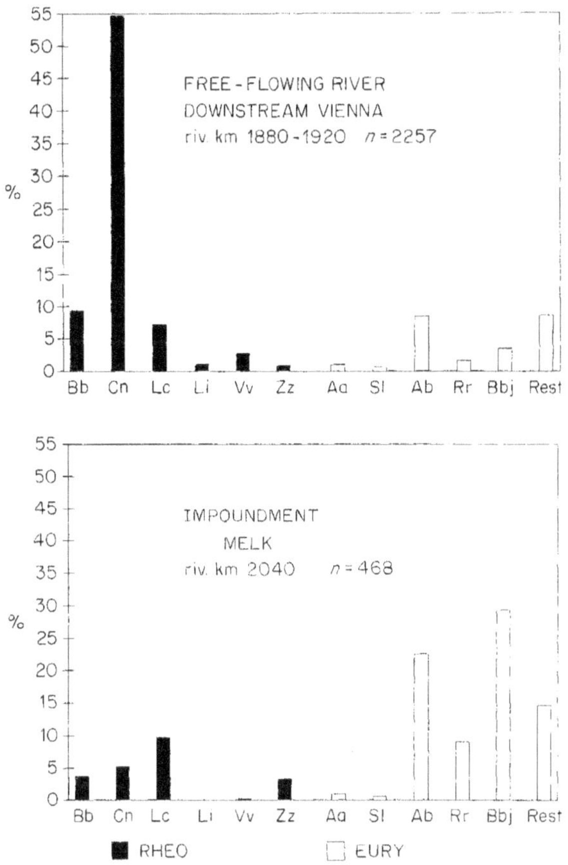

Figure 4. Assemblage structure of fish (percentage composition) in the free-flowing section of the Austrian Danube compared to an impoundment (Melk) 5 years after the dam was closed. Catches obtained by standard electrofishing techniques in the inshore zone. Black columns, rheotopic species; white columns, eurytopic species. Bb, *Barbus barbus*; Cn, *Chondrostoma nasus*; Lc, *Leuciscus cephalus*; Li, *L. idus*; Vv, *Vimba vimba*; Zz, *Zingel zingel*; Aa, *Alburnus alburnus*; Sl, *Stizostedion lucioperca*; Ab, *Abramis brama*; Rr, *Rutilus rutilus*; Bbj, *Blicca bjoerkna*; Rest, other species (From Schiemer & Waidbacher, 1992).

drastically changed. The graph indicates the decline of riverine species and the increased proportion of eurytopic forms.

Although such assessments are indicative and valuable, they give no immediate cues on the ultimate causes for the observed changes. For a detailed explanation the reaction norm (Schiemer & Wieser, 1992) of individual species has to be analysed by a combination of field and experimental studies. The value of fish as bioindicators will increase the further such an analytical understanding is developed. The type of research required is exemplified by studies on the rheophilic guild of the Danube. In a series of papers we have shown (Schiemer & Spindler, 1989; Schiemer & Waidbacher, 1992; Schiemer et al., 1991) that the early life history phase is the critical stage due to its narrow requirements with regard to current velocity, temperature, and food availability. The match or mismatch between these requirements and environmental conditions is decisive for recruitment and year-class strength. The main cause for the decline of rheophilic species in regulated and dammed rivers is the deficiency of riverine inshore structures – they form the spawning and nursery habitats of this guild. Richly structured inshore zones are rare in regulated rivers like the Rhine or Danube. For example, a very limited number of gravel bars, which function as nursery areas, exist in the free-flowing 45-km long stretch of the National Park downstream of Vienna. As an index of ecological quality for such stretches we found that the complexity of the shoreline (its littoral development per 100 m linear distance) is well correlated with the number of species in the 0+ stage, the occurrence of endangered taxa, and community organisation (Schiemer et al., 1991; Wintersberger, 1996). Inshore structure apparently combines and integrates several elements required for successful recruitment at a micro- and mesoscale, namely (a) the proximity and connectance of spawning sides and nurseries, (b) the availability of supplementary microhabitats during early ontogeny, (c) the quality of microhabitats with regard to current velocity, temperature, and food availability, and (d) the availability of refugia in the case of spates.

Ecological quality, however, is not merely dependent on the structural properties but also on the intercept between geomorphology and hydrology. Water level fluctuations lead to a continuous shift of microhabitat position and determine the availability, connectance, and quality of microhabits and refugia. In Figure 5 the relationship of inshore configuration and water level is analysed for different types of shorelines of the 'Alluvial Zone National Park' area downstream of Vienna. The inserts schematically show three shoreline-scenarios. The black dots (insert c) refer to linear ripraps, which are uninhabitable for most 0+ fish. The broken line (insert a) is hypothetical, indicating the conditions prior to regulation. The best present-day conditions are larger-sized gravel bars, which have been formed after regulation over the last 100 years (insert b). The shoreline configuration index increases with water level up to a river stage where the laterally constructed side dams are reached. Such high water levels represent a structural discontinuity, a disruption of habitat availability, and disturbances for 0+

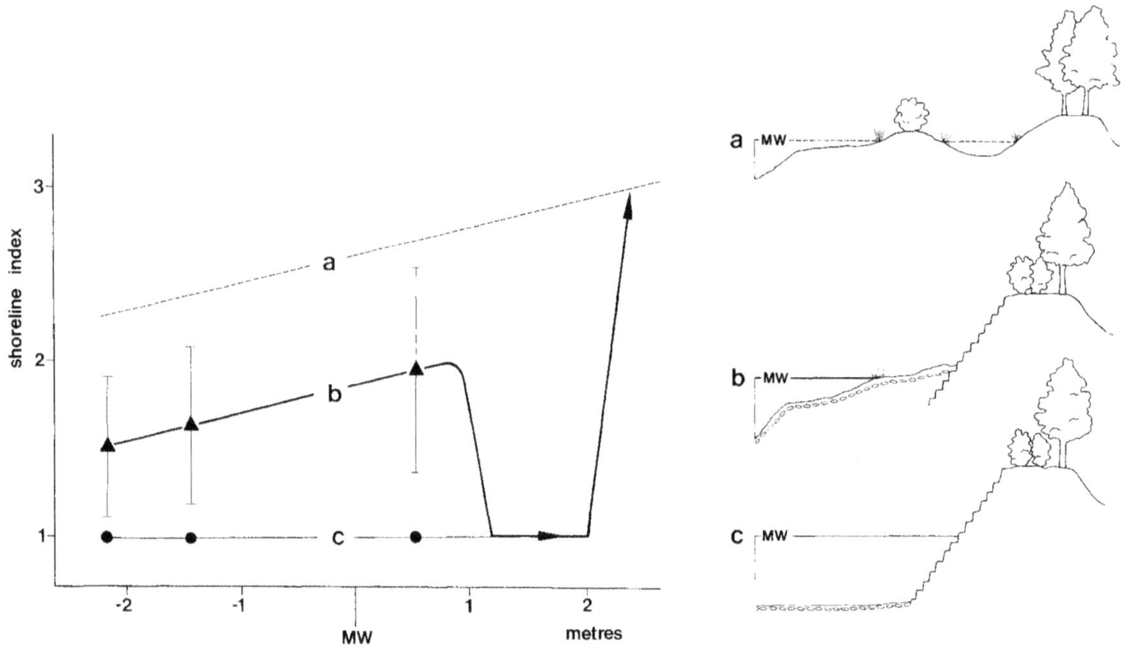

Figure 5. Relationship between inshore configuration and water level (related to the mean water level, MW) in the free-flowing section of the Austrian Danube downstream of Vienna. Arrows and vertical bars refer to the 18 gravel bar zones which represent the best nursery conditions. Black dots, linear rip-raps; broken line, conditions prior to restoration.

fish. They lead to population losses if refugia are not available and appropriately connected.

Disturbances are influenced by three conditions (Figure 6): (a) water level fluctuations, (b) the structural properties of the inshore zone, and (c) the size of young fish. Population losses will depend on the extent to which the effects of water level fluctuation are buffered by structural heterogeneity of the shoreline. Body size determines critical current velocities and the escape range of fish.

An important approach to understanding environmental thresholds and limitations is to analyse of the energy budget of an individual fish over the course of its early life history development (Schiemer et al., in press). In order to illustrate this point the insert in Figure 6 compares experimental data for the early life history stages of a characteristic rheophilic species, *Chondrostoma nasus*, at optimal temperature (20°C) and ad libitum food supply for larvae (5 mg fresh weight) and juveniles (50 mg fresh weight). Daily production (*P*, in form of body growth) and respiration (*R*) are given in percent of energy consumed (*C*). Assimilation (*A*) is assumed to be 70% of consumption. The graph shows the growing scope for additional energy expenditures with fish size. In a continuously shifting, stochastic environment, the costs of station maintenance are apparently high and such additional costs or lower consumption will constrain growth and

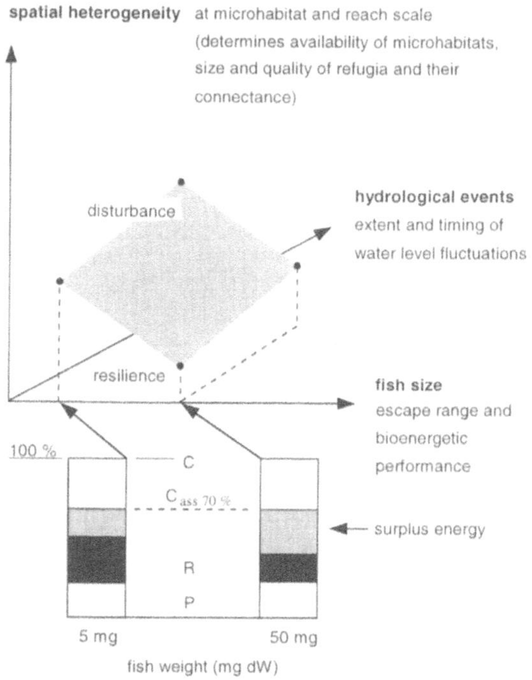

Figure 6. A schematic definition of 'disturbance' for riverine fish populations dependent on three main axes: (a) spatial heterogeneity, (b) hydrological events, and (c) fish size. The histogram on the bioenergetic budget for two size classes of 0+ *Chondrostoma nasus* is based on experimental data (see text). It demonstrates the increasing scope for escape reaction with size.

Table 1. The significance of various elements of 'ecological integrity' of floodplain rivers for community structure, population dynamics, and performance in fish.

biotic indicators ↓ / integrity parameters →	habitat diversity	connectance of complementary habitat elements	refugia	physiographic and trophic habitat condition
community structure				
biodiversity	++	+		
assemblage structure	+	++	++	
endangeredness	+	++	++	
0+ assemblage		++	++	++
population dynamics				
year class strength		+	++	
recruitment/mortality rates		+	++	+
population exchange processes (e.g. migration, drift)		+		
bioenergetics				
growth				++
condition				++
egg size/egg quality				++

extend the duration of the critical period. With increasing body size the energetics become less constrained and the escape range is enhanced. A detailed account of this experimental study and a discussion on the significance of a bioenergetic approach is given in Schiemer et al. (in press).

Experimental data of this kind on the performance of fish in relation to temperature conditions, food availability, and current velocity throughout early ontogeny will provide insight into the causes behind the endangerment of individual species.

Derived evaluation criteria and conclusions

The development of restoration and management concepts for floodplain rivers requires biological indicators (Schiemer, 1994; Schiemer et al., 1999). In order to identify the practical value of fish for the assessment of ecological integrity and their significance as a monitoring tool, three questions have to be addressed:

(1) Which elements of ecological integrity are indicated by fish?
(2) How precise is the indication?
(3) How efficient /costly (in terms of time and personnel) is the application?

The matrix (Table 1) relates various elements of integrity and their significance for the community organisation, population dynamics, and bioenergetic performances of fish. It is quite apparent that the structure of the assemblage, the presence or absence of individual species of fish, and their state of endangeredness provide a broad overview and first overall indication of the situation. A comparison with the original situation ('reference standard', 'Leitbild') and its characteristic fish assemblages, which for many river types in Europe is relatively well assessed, reveals deficiencies. For a more detailed evaluation the local distribution pattern of the fish fauna, its population structure, seasonality of occurrence, and growth performances – especially of the early life history stages – must be assessed. These parameters can indicate the quality and extent of connectivity at a finer scale and can be used to define management and engineering criteria.

In conclusion: there is a need to develop management criteria and restoration strategies for large river systems in order to restore ecological properties and functions. There is a need for appropriate indicators to document and analyse environmental deficiencies. The goal is to develop flexible, broadly applicable and lean tools to monitor and manage large river systems. Fish are the most significant single indicator group.

In order to further strengthen their value as indicators, however, research is required on a broad range of target species. Fitness has to be addressed both at the level of population dynamics and the ecophysiology of critical stages.

References

Bayley, P. B., 1995. Understanding large river-floodplain ecosystems. BioScience 45: 153–158.

Copp, G. H., 1989. The habitat diversity and fish reproductive function of floodplain ecosystems. Envir. Biol. Fishes 26: 1–26.

Frissell, C. A., W. J. Liss, C. E. Warren & M. D. Hurley, 1986. A hierarchical framework for stream habitat classification: viewing streams in a watershed context. Envir. Mgmt. 10: 199–214.

Gaudin, P., in press. Habitat shifts in juvenile riverine fishes. Arch. Hydrobiol.

Gregory, S. V., F. J. Sweanson, W. A. McKee & K.W. Cummins, 1991. An ecosystem perspective of the riparian zones. BioScience 41: 540–551.

Jungwirth, M., S. Muhar & S. Schmutz, 2000. Fundamentals of fish ecological integrity and their relation to the extended serial discontinuity concept. Hydrobiologia 422/423: 85–97.

Junk, W. J. & R. L. Welcomme, 1990. Floodplains. In Patten B. C. (ed.), Wetlands and Shallow Continental Water Bodies. SPB Academic Publishing, Vol. 1: 491–524.

Junk, W. J., P. B. Bayley & R. E. Sparks, 1989. The flood-pulse concept in river floodplain systems. Can. spec. Publ. Fish. aquat. Sci. 106: 110–127.

Karr, J. R., 1991. Biological integrity: a long neglected aspect of water resource management. Ecol. appl. 1: 66–84.

Karr, J. R., 1994. Landscapes and management for ecological integrity. In Kim, K. C. & R. D. Weaver (eds), Biodiversity and Landscapes. Cambridge University Press, New York: 229–251.

Naiman, R. J., H. Décamps, J. Pastor & C. A. Johnston, 1988. The potential importance of boundaries to fluvial ecosystems. J. north. am. Benthol. Soc. 7: 289–306.

Noss, R. F., 1990. Indicators for monitoring biodiversity: a hierarchical approach. Conserv. Biol. 4: 355–364.

Persat, H., J.-M. Olivier & J. P. Bravard, 1995. Stream and riparian management of large braided Mid-European rivers, and consequences for fish. In Armantrout, N. B. (ed.), Condition of the World's Aquatic Habitats: Proc. of the World Fisheries Congress, Theme 1. Oxford & IBH Publishing Co, New Delhi: 139–169.

Petts, G. E. & C. Amoros (eds), 1996. Fluvial Hydrosystems. Chapman and Hall, London.

Schiemer, F., 1994. Monitoring of floodplains: Limnological indicators. In Aubrecht, G., G. Dick & C. Pentrice (eds), Monitoring of Ecological Change in Wetlands of Middle Europe. Stapfia 31: 95–107.

Schiemer, F., T. Spindler, H. Wintersberger, A. Schneider & A. Chovanec, 1991. Fish fry associations: important indicators for the ecological status of large rivers. Verh. int. Ver. Limnol. 24: 2497–2500.

Schiemer, F. & H. Waidbacher, 1992. Strategies for conservation of the Danubian fish fauna. In Boon, P. J., P. Calow & G. E. Petts (eds), River Conservation and Management. John Wiley, Chichester: 364–384.

Schiemer, F. & W. Wieser, 1992. Epilogue: food and feeding, ecomorphology, energy assimilation and conversion in cyprinids. Envir. Biol. Fishes 33: 223–227.

Schiemer, F. & M. Zalewski, 1992. The importance of riparian ecotones for diversity and productivity of riverine fish communities. Neth. J. Zool. 42: 323–335.

Schiemer, F. & T. Spindler, 1989. Endangered fish species of the Danube River in Austria. Regul. Rivers Res. Mgmt. 4: 397–407.

Schiemer, F., C. Baumgartner & K. Tockner, 1999. Restoration of floodplain rivers: The 'Danube Restoration Project'. Regul. Rivers Res. Mgmt. 115: 231–244.

Schiemer, F., H. Keckeis, G. Winkler, L. Flore & W. Reckendorfer, in press. Large rivers: ecotonal structure and hydrological properties and their relevance for the fish fauna. In Zalewski, M., F. Schiemer & J. Thorpe (eds), The Role of Ecotones for Freshwater Fish. Unesco MaB.

Schlosser, I. J., 1995. Critical landscape attributes that influence fish populations in headwater streams. Hydrobiologia 303: 71–81.

Sparks, R. E., 1995. Need for ecosystem management of large rivers and their floodplains. BioScience 45: 168–182.

Stanford, J. A., J. V. Ward, W. J. Liss, C. A. Frissell, R. W. Williams, J. A. Lichatowich & C. C. Coutant, 1996. A general protocol for restoration of regulated Rivers. Regul. Rivers Res. Mgmt. 12: 391–413.

Townsend, C. R., 1996. Concepts in river ecology: pattern and process in the catchment hierarchy. Arch. Hydrobiol. Suppl. 113: 3–21.

Ward, J. V., 1989. The four-dimensional nature of lotic ecosystems. J. n. am. Benthol. Soc. 8: 2–8.

Ward, J. V., 1998. Riverine landscapes: biodiversity patterns, disturbance regimes, and aquatic conservation. Biol. Conserv. 83: 269–278.

Ward, J. V. & J. A. Stanford, 1995a. The serial discontinuity concept: Extending the model to floodplain rivers. Regul. Rivers Res. Mgmt. 10: 159–168.

Ward, J. V. & J. A. Stanford, 1995b. Ecological connectivity in alluvial river ecosystems and its disruption by flow regulation. Regul. Rivers Res. Mgmt. 11: 105–119.

Ward, J. V., K. Tockner & F. Schiemer, 1999. Biodiversity of floodplain river ecosystems: ecotones and connectivity. Regul. Rivers Res. Mgmt. 15: 125–139.

Wintersberger, H., 1996. Spatial resource and species assemblages of larval and juvenile fish. Arch. Hydrobiol./ Suppl. 115: 29–44.

Hydrobiologia **422/423:** 279–289, 2000.
M. Jungwirth, S. Muhar & S. Schmutz (eds), Assessing the Ecological Integrity of Running Waters.

A multi-level concept for fish-based, river-type-specific assessment of ecological integrity

S. Schmutz, M. Kaufmann, B. Vogel, M. Jungwirth & S. Muhar
Dept. Hydrobiology, Fisheries and Aquaculture, Univ. of Agricultural Sciences Vienna, Max Emanuel-Strasse 17, A-1180 Vienna
E-mail: schmutz@mail.boku.ac.at

Key words: water quality, assessment criteria, biological hierarchy, reference conditions, MuLFA, Austria, Europe

Abstract

We propose a **mul**ti-level concept for **f**ish-based **a**ssessment (MuLFA) of the ecological integrity of running waters. This concept is designed for large-scale monitoring programmes such as required for the proposed Water Framework Directive of the EU. Out of five different biological organisation levels (fauna, community, guild, population and individual), we propose seven criteria: River-type-specific species, species with self-sustaining populations, fish region, number of guilds, guild composition, population size and population age structure. The principle of the MuLFA is based on assessing the deviation from undisturbed reference conditions. Reference conditions have to be compiled for every distinct river type using historical fish and abiotic data, present river-type-specific reference sites and reference models. The final assessment procedure is done by comparing the assessment reach with the reference conditions using a 5-tiered normative scheme and assigning that reach to the level of highest coincidence. The benefit of the MuLFA is its potential for consistent sensitivity to low- and high-dose human alterations, and due to its general character, its adaptability to all river types.

Introduction

Associating the health of fish stocks with water quality conditions has a comparatively long tradition. Nonetheless, standardised fish-based methods to assess the ecological integrity (EI) of running waters have been available only since the 1980s. Various single-species-oriented methods were subsequently developed for salmonids, among them the HQI (Habitat Quality Index, Binns & Eiserman, 1979), the HSI (Habitat Suitability Model, Raleigh et al., 1984; Wesche et al., 1987) and other models (Bowlby & Roff, 1986; Scarnecchia & Bergersen, 1987; Fausch et al., 1988; Jowett, 1992; Barnard & Wyatt, 1995) based on fish and habitat relationships, and the community oriented IBI (Index of Biotic Integrity, Karr, 1981) for non-salmonid rivers. The transferability of these methods to other catchments or ecoregions without considerable adaptations is limited (Angermeier & Karr, 1986; Miller et al., 1988). In Europe, the development of new and the adaptation of existing fish-based methods is still in the early stage (Oberdorff & Hughes, 1992; Oberdorff & Porcher, 1994; Hughes & Oberdorff, 1998). However, these methods are rapidly gaining importance, not least because the new EU water policy specifies fishes as one of the four biotic elements which are decisive for assessing EI (Water Framework Directive, WFD).

The principle behind the new EU approach is based on assessing the deviation of the present status from the undisturbed river-type-specific conditions. The assessment procedure is carried out by assigning the present status to five predefined EI-levels ranging from 'high' to 'bad' quality. While the general outline of the new method is already embodied in the Directive, detailed information on the definitive assessment approach is still entirely lacking. The purpose of this paper is to develop a new concept for a fish-related EI-assessment method based on existing methods and embedded in a theoretical framework. The new method should enable a large-scale, nationwide EI-assessment in accordance with the proposed WFD by simultaneously considering the uniqueness of different river types. Special emphasis is placed on the selection of assessment criteria, as well as on the characterisation of river-type-specific reference conditions.

Fishes as indicators for the ecological integrity of running waters

While the chemical/physical water quality of running waters has been considerably improved in many industrial countries over the last decades, the hydrology, morphology and connectivity situation continues to deteriorate. Both in theoretical river concepts (Amoros & Roux, 1988; Junk et al., 1989; Ward & Stanford, 1995; Jungwirth, 1998) and in current research and management activities (e.g. Persat et al., 1995; Dieplinger et al., 1996; Tockner & Schiemer, 1997; Muhar et al., 1998a), the significance of connectivity has been recognised as one of the crucial elements of running water integrity.

Due to their complex requirements, fishes are sensitive indicators for habitat quality at various spatial scales (Bayley & Li, 1996), and as migratory organisms fishes are the ideal indicator for habitat connectivity conditions (Peter, 1998; Schmutz & Jungwirth, 1999; Jungwirth et al., 2000). At the micro-scale, fishes use clearly defined, life-stage-specific habitats. Shifts between these distinct habitats take place via short-, medium- or long-distance movements/migrations (e.g. drift of larvae, spawning runs). Therefore, the spatial/temporal availability of connectivity between these habitats plays an important role for the life cycle of riverine fishes. Movement/migration takes place in all spatial dimensions: vertical in the bed sediments (e.g. larval development in salmonids), longitudinal (e.g. spawning) and lateral (e.g. reproduction, refuge). In addition, at the temporal scale, the longevity of fishes helps to 'register' environmental alterations across long time periods. As top predators, they subsume the trophic conditions across the food chain; as members of a specific trophic guild, they also provide detailed information on the respective trophic level. Finally, fishery and sport fishing has a long tradition in which fishes have been used as indicators for water quality. These phenomena clearly distinguish fishes from other aquatic organisms and underline their significance as essential indicators to assess the EI of running waters.

A disadvantage of using fishes as indicators arises whenever fishery-caused alterations such as species transfer, continuous stocking, overfishing, etc. make it more difficult to distinguish between fishery-caused and other human alterations.

Assessment concept

Assessment criteria

The selection of adequate criteria takes place in the field of tension between ecological considerations that guarantee a sound assessment approach and practical limitations caused by methodological and financial constraints. The resulting selection procedure always represents a compromise between these contradictory objectives, adjusted to the state-of-the-art ecological knowledge and the socio-economic value of the issue (Moog & Chovanec, 2000).

The concept of the proposed assessment method is based on the hierarchical organisation of the biota (Odum, 1971; Allen & Starr, 1982; O'Neill et al., 1986) and the linkages of the various organisation levels to temporal/spatial scales (Frissell et al., 1986; Bayley & Li, 1996; Habersack, 2000) as shown in Figure 1a. The underlying theoretical principle implies that higher levels (e.g. fauna/zoogeographic area) – compared to lower levels (individual/microhabitat) – are more persistent, more stable, relevant for larger areas and affect lower levels more than vice versa (top-down effects). Consequently, the more human activities alter an ecosystem, the higher the levels affected. Since no single level of organisation is fundamental, the hierarchical concept suggests that EI should be monitored at multiple levels of organisation covering multiple spatial and temporal scales (Noss, 1990; Hughes & Noss, 1992). Therefore, a set of assessment criteria is selected from different hierarchical levels in order to guarantee that various intensities of human alterations can be detected (Figure 1b).

The faunal composition of a given reach is the result of its zoo- and physiogeographic characteristics; it remains fairly constant within a specific river type of an ecoregion and persists over comparatively long time scales, compared to criteria of lower hierarchical levels. Along the river continuum, environmental gradients form a sequence of different fish regions inhabited by distinct communities. These coenoses are composed of characteristic sets of guilds that, in turn, consist of several species. Each population of a species can be characterised by its size (biomass or density). The description of the population age structure requires individual information such as the body length (Figure 1).

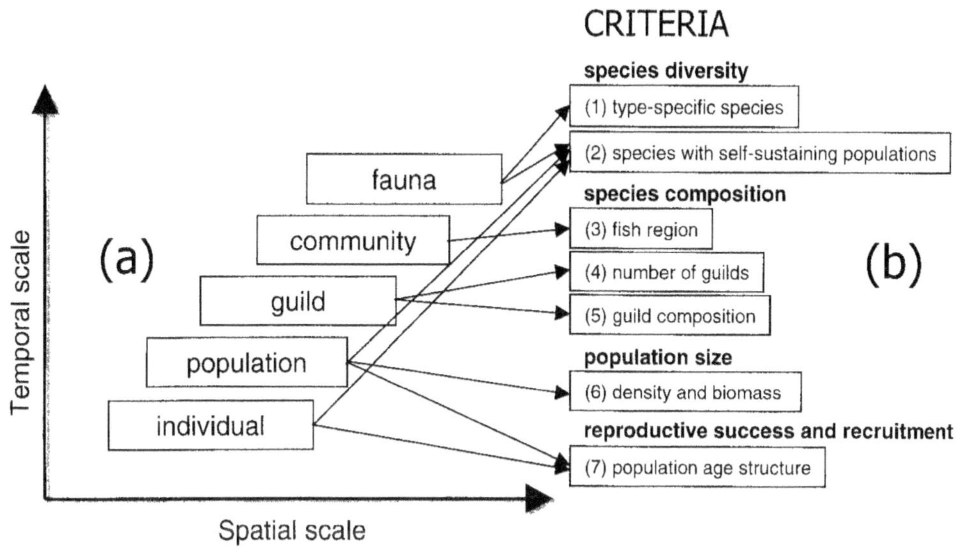

Figure 1. Hierarchical organisation of biota, relationships to temporal/spatial scales (a), and derived assessment criteria (b).

Out of the five biological organisation levels (fauna, community, guild, population and individual), we selected seven criteria: River-type-specific species, species with self-sustaining populations, fish region, number of guilds, guild composition, population size and population age structure.

Species diversity

Number of river-type-specific species (RTS-species) This criterion reflects the indigenous fish fauna naturally occurring in the specific type of a river, excluding species not native in a given area (e.g. country) and not autochthonous in the specific river.

Non-RTS-species are not considered in the assessment method. Compared to the total number of species, the RTS-species criterion remains sensitive in situations where diversity is still high, but native species are replaced by non-native species. Numerous examples prove that negative effects on the autochthonous fish fauna by non-RTS-species, in particular by non-native species, are possible (Stauffer, 1984; Welcomme, 1984); they are not, however, obligatory. In addition, the proof and the quantification of negative effects would need a considerable effort not affordable in a routine method. The occurrence of alien species is, therefore, not an adequate assessment criterion. At any rate, potential negative effects of non-autochthonous species are indirectly considered in the assessment method because they are reflected in other criteria if they become decisive.

Number of species with self-sustaining populations (SSP-species) This criterion reflects the type-specific fauna (RTS-species) composed of species meeting the following minimum criteria: the species are self-reproducing, which is proved by the occurrence of juvenile fishes (that must not be stocked) and maintain at least a minimum population size. According to the 50/500-rule of Franklin (1980), a minimum population size should comprise at least 50, or better 500, individuals capable of reproduction in order to guarantee sufficient genetic variability. This rule merely represents a tentative clue to guarantee the genetic viability of a population and values can vary according to the available information for a given species and situation.

The SSP-criterion excludes RTS-species that fail to fulfil any requirement for successful reproduction and whose population size is below a level guaranteeing mid- or long-term survival. For example, species that have experienced a major stock decline below this threshold, and will probably become extinct in near future, are omitted.

Species composition

Fish region The riverine fish fauna is characterised by a predictable sequence of distinct communities along the river course. According to the fish zonation concept (Thienemann, 1925) and the biocoenotic region concept (Illies & Botosaneanu, 1963), fish regions are classified and named after the dominating key-species, which are associated with other specific species of that region:

1. Epirhithral – upper trout region
2. Metarhithral – lower trout region
3. Hyporhithral – grayling region
4. Epipotamal – barbel region
5. Metapotamal – bream region
6. Hypopotamal – brackish water region

Human alterations often result in a shift of fish regions to upper or lower regions. For example, channelisation may cause a so-called 'rhithralisation effect', a shift to rhithral species; vice-versa, impoundments may lead to a 'potamalisation effect' (Jungwirth et al., 1995).

In contrast to the diversity-criteria, the fish region-criteria not only consider the occurrence of single species or groups of species, but also reflect the coenotic composition of the fish fauna. Fish region is a more robust criterion because it is less affected by insufficient reference data such as incomplete historical species lists and it is more indicative for severe human impacts; a shift of fish region only takes place if human activities lead to a complete alteration of the overall character of the river type.

Guild composition Species of a guild take over the same function within ecosystem, show similar strategies of resource utilisation and have developed similar life-forms with respect to habitat use, temperature preference, reproductive strategy, feeding habit, migratory behaviour, etc. (Kryzhanovsky, 1948). Therefore, the loss of a guild is much more significant for the EI than the equivalent loss of single species. Guilds have an advantage over species because they can also be used if our knowledge on species occurrence is incomplete.

Although a huge amount of information on the autecology of European fish species already exists, appropriate classification schemes that cover all major ecological requirements are still missing. For example, current classifications with respect to reproduction (Balon, 1975, 1981) or feeding habits (Michel & Oberdorff, 1995) only provide information for some European fish species. In the case of the Danube fish fauna, for example, the only available guild scheme is a nearly complete list of species classified according to their main habitat requirements (rheophilous (-A,-B), stagnophilous, eurytopic; Schiemer & Waidbacher, 1992). With respect to migratory behaviour, the available information allows only very coarse differentiations, such as the distinction into diadromous and potamodromous species or short-, mid- and long-distance migrants. Similarly, the classification based on temperature preferences is only feasible on a very general level (e.g. cold, cool and warm water species) due to incomplete knowledge.

Since standard guild classifications are largely missing, individual guild schemes must be established for specific river types (see 'Discussion'). Once the guilds are defined, their composition is assessed twofold. First, minor changes in the relative abundance of guilds are used to demonstrate slight human impacts and, second, the absence of certain guilds in connection with a complete alteration of the overall guild composition indicates major impacts.

Guild composition is a commonly used criterion for biotic indices. For example, the trophic guild composition plays a major role within the Index of Biotic Integrity (Karr, 1981). The advantage of the guild concept is its suitability for different zoogeographic regions and its applicability for situations with incomplete species knowledge. However, guilds that naturally consist of few species (e.g. stagnophilous species in Europe) tend to cause the same problems as single species.

Population size

Fish population size reflects human-induced alterations before these impacts become limiting for the existence of the species. These low-dose alterations can be detected in both directions, either in an increase (e.g. caused by eutrophication) or, more commonly, in a decrease of population size.

The population size of a species should be characterised by quantitative measures and estimates of the density and biomass per area or river length. Whenever quantitative estimates are prevented by methodological or financial constraints, semi-quantitative sampling methods (e.g. CPUE) are recommended.

Population size criteria are commonly used to assess water quality. In particular, many fish-habitat models have been developed based on quantitative population estimates to assess and predict water quality in salmonid rivers (e.g. Binns & Eiserman, 1979; Fausch et al., 1988).

Population age structure

Reproduction plays a crucial role within the life history of fish, and the larval and juvenile life stages often are more sensitive than adults. Human impacts can, therefore, be detected by analysing the reproductive success and the recruitment before they manifest themselves in criteria at higher organisational levels.

As the direct measurement of reproductive success is very labour- and cost-intensive, indirect proof of reproduction is used by analysing length-frequency-plots of the population age structure.

Characterisation of reference conditions

The principle of our method is based on assessing the deviation from clearly defined, river-type-specific reference conditions ('Leitbild', Muhar et al., 1995). Reference conditions equate to undisturbed EI defined as the maintenance of all internal and external processes and attributes interacting with the environment in such a way that the biotic community corresponds to the natural state of the relevant aquatic habitat (Adamicka et al., 1992).

In principle, the river type of the reference site must be comparable to that of the assessment reach. The type of a river is characterised by general abiotic and biotic criteria and features such as stream order, gradient, river morphology, flow regime, altitude, fish region and ecoregion (Muhar et al., 1993).

Various methods of data acquisition are used to characterise reference conditions (Hughes, 1995; Muhar et al., 1995). The 'Leitbild' is compiled with historical abiotic data, historical fish data, data of reference sites and reference models (Figure 2).

Historical abiotic data

Old maps and historical reports provide useful information to identify and characterise undisturbed conditions at the river-type scale. In particular, the description of the pristine river morphology and hydrology helps to reconstruct the main habitat types and their dynamics. Historical maps and hydrographs provide data about original channel type (constrained, braided, meandering) and floodplain extension. Based on such information, the historical evidence of significant mainstem and floodplain habitat types can be verified. In Austria, this approach is being successfully applied for habitat integrity assessment (Muhar et al., 1993, 1998b, 2000), as well as for restoration concepts (Muhar et al., 1995, 1996, 1998a).

Historical fish data

Fishes are the only group of freshwater organisms for which substantial historical information is available. This is based on their use as an essential human food resource. Even for the riverine fish fauna of Europe, commercial data about fish catches, leasing contracts, fish sales, fishmarkets and biological data on the distribution of species is accessible. Historical descriptions go back several centuries, but relevant information is documented mainly between 1800 and 1950. A preliminary Austria-wide literature survey revealed useful historical data for 13 rivers (Haidvogl & Waidbacher, 1997). While the historical occurrence of most of the original species is documented, quantitative data about species composition and population size are very rare and vague. In general, historical data mainly focus on species of commercial interest, whereby altered nomenclature, various synonyms and unclear taxonomy can complicate matters.

Historical data should refer to a time frame in which riverine ecosystems were still in their natural state (early to mid-19th century). In Europe, major human impacts were caused by systematic channel regulation measures since the early to mid-20th century followed by the construction of large hydropower stations and chains of dams after 1950. Prior to this time frame, European cultural landscapes were subject to land cover alterations, but the fundamental functions of most river systems remained intact and system-inherent processes could still run their natural course (Muhar et al., 2000). In respect to the fish fauna, human impacts occasionally go back to the Middle Ages. For example, the drastic decline of anadromous Acipenseridae in the Austrian Danube, caused by overfishing, already took place in the 16th century (Jungwirth, 1984). Therefore, the reliability of historical data has to be checked in every single case.

Reference sites

The simplest approach is to use field data of reference sites along the same river. Field data are usually missing for the situation before the assessment reach was affected; this forces a switch to up- or downstream sites. It should be noted, however, that the river-type of reference sites located up- or downstream may be different due to natural river zonation. If adequate reference sites are not available in the river under study, an alternative is to switch to other rivers within the same ecoregion with comparable river type.

Reference models

As reference sites are generally rare in Europe and other parts of the world, all available information should be used to establish reference models. Fish data in context with physiographical and habitat-related data of undisturbed and near-natural sites provide useful information to generate models valid across various catchments or within an entire ecoregion (Oberdorff et al., in print). Prerequisites of such models are

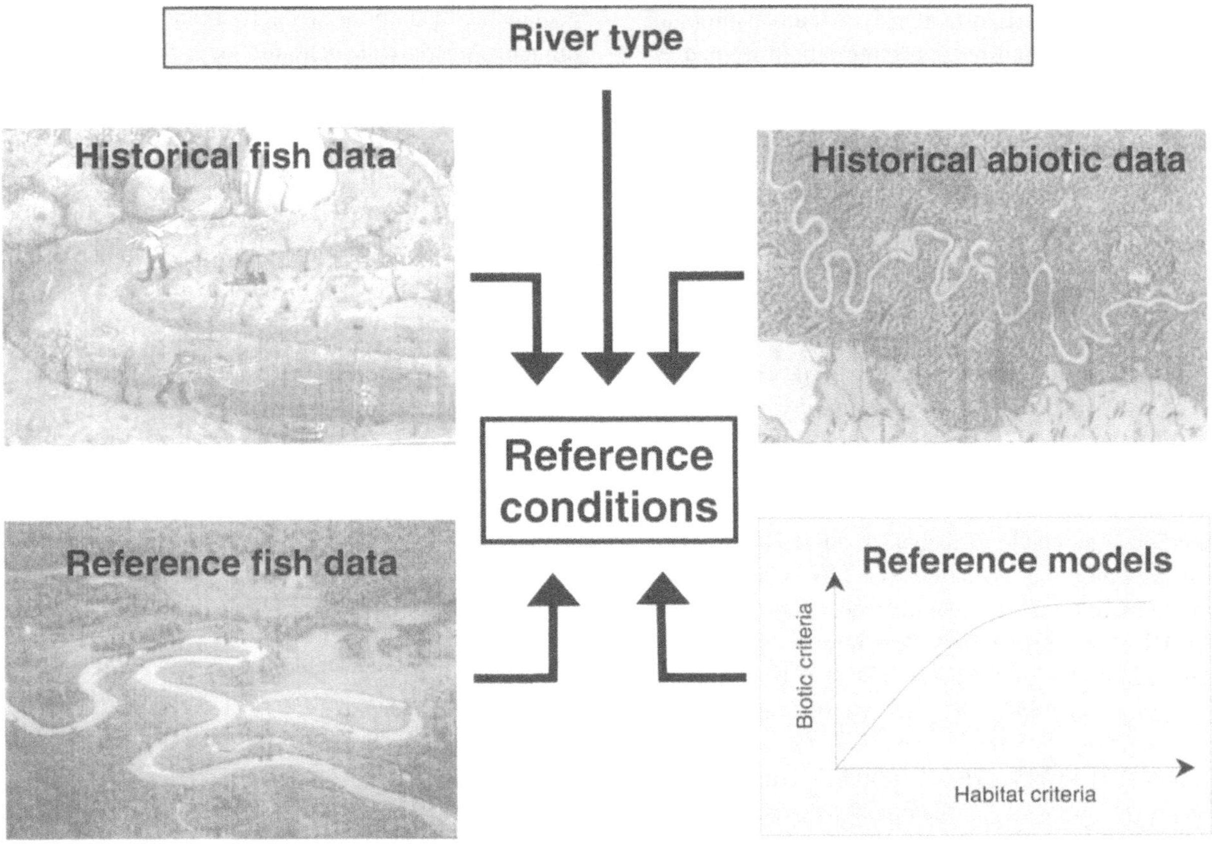

Figure 2. Integrative approach to characterising river-type-specific reference conditions using historical fish data, historical abiotic data, present reference site data and reference models.

the availability of adequate monitoring programmes, standardised sampling protocols and proper data management.

Integrative approach

Each method of reference characterisation has its advantages and shortcomings. The application of only one method does not yield a fully satisfactory view of the undisturbed situation. Therefore, compiling all available information into an integrated picture is the best strategy for a sound description of the reference conditions (Figure 2).

The first step is to identify and characterise the original river type of the assessment reach. In severely impacted rivers, this cannot be determined merely by analysing the present situation. In such cases, a precondition for any further step is to identify the original river type by historical data as shown above. Once the specific river type is known, type-specific information can be gathered by sampling reference sites in other rivers of the same type, by historical surveys and by employing type-specific models. Historical fish data reveal preliminary information on the originally occurring fish species. Sometimes an estimate of the species composition can be derived. State-of-the-art knowledge about the ecological requirements of potentially occurring species, combined with reference models, allow adequate assignment of additional species to the various habitat types indicated by historical surveys. The data of sampled reference sites complete the view of the reference conditions. In addition, samples of low- and medium-impaired assessment reaches can reveal supplementary information which should be included if necessary (Figure 2). The definition of guild composition, population size and population age structure will be mainly based on present reference data and models. The final result of this integrative approach is a characterisation of reference conditions for every criterion, at the species level, for example, the number and identity of originally occurring species (Figure 3).

Figure 3a shows examples of species criteria for some larger Danube tributaries (data from Jungwirth et al., 1989, 1990; Kaufmann et al., 1991; Zauner, 1993, 1994). In the case of the Drau, Inn and Salzach, historical records widely coincide with the list of originally occurring species, compiled according the integrative approach. At the Mur and Thaya, however, only little more than half of the original fauna is historically documented, underlining the deficiency of historical data. At the Thaya, even a recent survey revealed 25% more species than shown by historical data (Figure 3a, b).

Figure 3b clearly demonstrates the negative effects of human activities using RTS- and SSP-criteria. The near-natural Thaya shows the highest proportion of presently occurring species compared to the original fauna. The straightened rivers Mur, Drau and Salzach continue to maintain nearly 50% of the originally occurring species, while the Inn contains less than 10% of the original fauna due to channelisation and hydropeaking impacts.

Spatial assessment unit

The spatial assessment unit – the unit of a river where the method is applied – is predefined by the proposed assessment criteria. As all of the criteria refer to a river-type-specific fish community, the assessment unit has to reflect all attributes characteristic of a specific river type. Therefore, the smallest possible unit is a subset of a river type comprising all types of water bodies (main channel, floodplain) and habitats (runs, pools, glides, coves, side arms, oxbows, etc.) characteristic of the specific river type. Furthermore, the assessment unit should be large enough to theoretically hold a minimum population size of RTS-species.

Normative classification scheme of EI-levels and assessment procedure

Table 1 presents a 5-tiered scheme to assess the EI-status based on the above-described assessment criteria. The number of EI-levels, their gradation and the naming of the levels of the classification scheme correspond to the normative definitions of ecological status developed within the WFD.

The status of the 'high' EI-level totally or nearly totally corresponds to the undisturbed reference conditions. 'Good' EI-level is characterised by significant but still slight deviations from the undisturbed conditions. At this level, some RTS-species and several SSP-species are missing and both guild composition and population size are slightly altered; fish zonation and the number of occurring guilds, however, remain unaltered. At 'fair' EI-level, all criteria show substantial deviations from the undisturbed conditions. Fish regions are shifted and single guilds are missing. 'Poor' EI-level is characterised by strong, and 'bad' by extreme deviations from the undisturbed conditions.

The assessment method uses five levels of EI. A differentiation into more levels would not coincide with the accuracy of the method and is therefore not justifiable. In addition, practical considerations, such as compatibility with the WFD schema, factor into this gradation. The final assessment is a quite simple procedure and is done by comparing the present situation of the assessment reach with the undisturbed reference conditions (Table 1). The described criteria of the present situation are assigned to that predefined water quality level in Table 1 showing the highest coincidence. According to the outlined principles of the assessment concept and criteria selection, higher weight is given to the species- than to the population-criteria. The final weighting of the criteria is individually defined for every river type (see 'Discussion').

Discussion

Modern river ecology concepts perceive running waters as a unity of river channel, riparian area and interrelating floodplains (e.g. Amoros & Roux, 1988; Junk et al., 1989). Since river/floodplain-systems consist of a great variety of different habitat types and corresponding fish coenoses, adequate EI-assessment methods must be developed such that every individual river type with its unique coenotic characteristic can be handled. Our assessment concept is designed for large-scale EI-assessment monitoring programmes covering a variety of different catchments and ecoregions. It must, therefore, be applicable to every river type.

Under these circumstances, a single method or even a single set of guilds cannot cover the entire range of potential coenotic structures in different river types. This calls, on the one hand, for a general concept applicable to every river type and, on the other hand, for a specific assessment approach developed for a particular river type. These requirements are incorporated by our method insofar as the proposed criteria correspond to a very general level applicable to all potential situations. The final specification of the criteria, the degree of differentiation and the relative significance

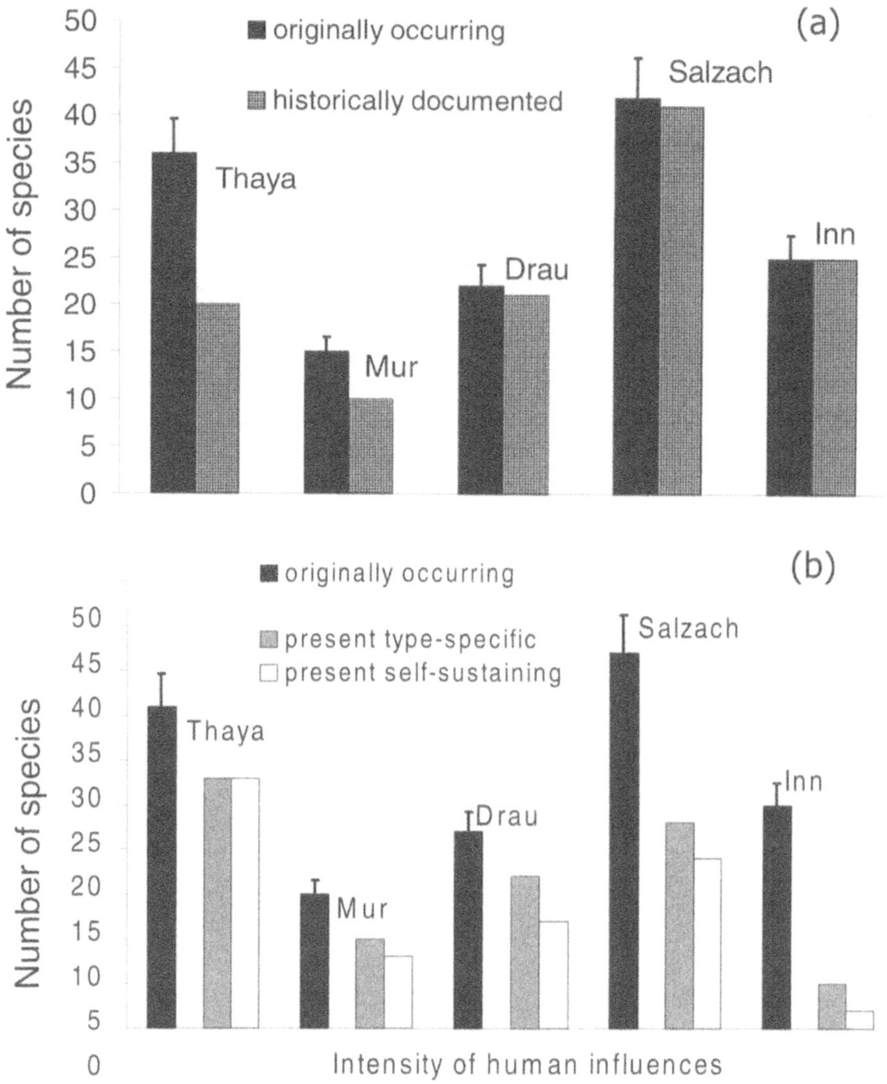

Figure 3. Comparison of originally occurring and historically documented species (a) and the species decline in relation to human influences (b) at five major Danube tributaries (data from Jungwirth et al., 1989, 1990; Kaufmann et al., 1991; Zauner, 1993, 1994).

Table 1. Five-tiered normative classification scheme for fish-based assessment of ecological integrity

Criteria	Ecological integrity levels				
	1 high	2 good	3 fair	4 poor	5 bad
(1) Type-specific species	none or nearly none missing	some species missing	several species missing	many species missing	most species missing
(2) Self-sustaining species	none or some missing	several species missing	many species missing	most species missing	nearly all species missing
(3) Fish region	no shift	no shift	shift	shift	shift
(4) Number of guilds	no guild missing	no guild missing	single guilds missing	many guilds missing	most guilds missing
(5) Guild composition	no alteration	slight alteration	substantial alteration	complete alteration	complete alteration
(6) Biomass and density	no or nearly no changes	slight changes	substantial changes	heavy changes	extremely changed
(7) Population age structure	no or nearly no changes	slight changes	substantial changes	heavy changes	extremely changed

of criteria derived from different organisation levels have to be specified in detail for every river type.

Due to the hierarchical organisation of biological systems and the derived MuLFA-approach, species diversity criteria, for example, should generally receive more attention than population criteria (Figure 1). The weighting of the criteria, however, will be different in an epirhithralic stream compared to a large potamalic river/floodplain-system. In the first case, even under natural conditions fish species diversity is very low and both diversity and species composition are, thus, not sensitive criteria. On the other hand, quantitative information on the population size and age structure of each species is attainable at acceptable costs and without methodological sampling problems. In the river/floodplain-system, species diversity is usually high: the sensitivity of the diversity and species composition criteria are, therefore, much higher and quantitative information becomes less important. In moderately diverse fish faunas, a combination of qualitative and quantitative attributes – e.g. the SSP-species criteria – will yield the highest sensitivity.

Future efforts should aim at standardised assessment approaches for defined river types. A prerequisite for that is a fish-based river classification; a set of original river types is needed for which individually defined assessment approaches could be developed. In Europe, such a classification could be based on the ecoregions defined by Illies (1967) and on a refinement of the fish region concept (Thienemann, 1925; Illies & Botosaneanu, 1963) in accordance with modern classification approaches (Naiman et al., 1992). As long as such a classification is not available, the method must be adapted to the specific river type in every single assessment case.

In situations where the autecology of species and the coenotic structure and organisation are well known, assessment methods can be developed at a very detailed level. Here, species composition criteria could be defined as relative distribution of guilds, families or even single species. In other cases, such as in larger river/floodplain-systems like the Danube, the available fish-coenotic knowledge – not least due to methodological limitations of field sampling – is less developed (Schiemer & Waidbacher, 1992; Schiemer et al., 1994). This reduced possibility of coenotic characterisation and differentiation lowers the attainable precision level of the assessment method.

The river-type-specific approach first requires a characterisation of the type-specific reference conditions for each criterion. The relative deviation is then assessed from this baseline. We use undisturbed and not 'best available' or 'best attainable' conditions as the reference: they represent an absolute benchmark that is unaffected by regional differences in degradation and is independent of socio-economic importance or of the status of public awareness toward running water ecosystems. The 'high quality' level is equivalent to the reference conditions, yet also includes minor alterations related to the difficulties in accurately assessing reference conditions. Our experience shows that selecting and using reference sites without verifying their naturalness with historical data often leads to a misjudgement of actual status. Using only 'best available' sites results in inaccurate reference descriptions in areas such as Central Europe, where rivers are exhaustively degraded.

A promising approach is the development of reference models. Fish data of undisturbed and near-natural sites covering a wide range of river types, combined with physiographical and habitat-related attributes, provide the basis for models valid for larger areas. Initial attempts have been undertaken in Europe by several authors (Milner et al., 1985, 1993, 1995; Baran et al., 1995; Barnard & Wyatt, 1995), but models for Central Europe and the Danube basin are completely lacking.

The precision of an assessment method depends on the weakest link, i.e. the low exactness in characterising the reference conditions. Since many steps within the assessment procedure are based on professional judgement, we prefer a categorical gradation over a numerical algorithm. Single numerical indices attributed to the criteria would feign an accuracy not achieved in such a method. Extensive sampling efforts to characterise the present situation and highly sophisticated assessment algorithms do not improve assessment accuracy unless a comparably high standard is achieved for the reference conditions.

Conclusions

The type-specific approach is a compromise between the intention to consider the individuality of running waters, the inability to reconstruct the original characteristic of degraded rivers in detail and cost-efficiency. Consequently the precision of the method is limited. For the large-scale monitoring programmes specified for the WFD, this degree of exactness is sufficient. A higher precision is probably neither impartable nor politically executable. Nonetheless, other objectives such as environmental impact assessments or evaluations of restoration efforts require a more accurate method.

The advantage of the MuLFA-approach is its potential of consistent sensitivity to low- and high-dose human alterations. Moreover, its general character makes it adaptable to all river types. Successful application relies on the availability of accurate river-type-specific reference characterisations, detailed methodological approaches for specific river types, as well as on the provision of more sophisticated and more complete guild classifications. Above all, more emphasis should be placed on migratory guilds in order to underline the indicative value of fishes for the connectivity conditions.

Acknowledgements

We would like to thank G. Zauner and J. Eberstaller for fruitful discussions and comments, and M. Stachowitsch for improving the English.

References

Adamicka, P., G. Bretschko, E. Danecker, J. Hinteregger, G. Imhof, M. Jungwirth, M. Leichtfried, O. Moog, C. Moritz, G. Müller, R. Pechlaner, E. Pipp, E. Polzer & N. Schulz, 1992. Zur Gewährleistung, Beeinträchtigung und Beurteilung der ökologischen Funktionsfähigkeit von Gewässern. Österreichs Fischerei 45: 120–121.

Allen, T. F. H. & T. B. Starr, 1982. Hierarchy: Perspectives for ecological complexity. University of Chicago Press, Chicago. 310 pp.

Amoros, C. & A. L. Roux, 1988. Interaction between water bodies within the flood plains of large rivers: Function and development of connectivity. In Schreiber, K. F. (ed.), Connectivity in Landscape Ecology. Münstersche Geographische Arbeiten: 125–130.

Angermeier, P. L. & J. R. Karr, 1986. Applying an index of biotic integrity based on stream-fish communities: considerations in sampling and interpretation. N. am. J. Fish. Manage. 6: 418–429.

Balon, E. K., 1975. Reproductive guilds of fishes: a proposal and definition. J. Fish. Res. Bd Can. 32: 821–864.

Balon, E. K., 1981. Additions and amendments to the classification of reproductive styles in fishes. Envir. Biol. Fishes 6: 377–389.

Baran, P., M. Delacoste, G. Poizat, J. M. Lascaux, S. Lek & A. Belaud, 1995. Multi-scales approach of the relationships between brown trout (*Salmo trutta* L.) populations and habitat features in the Central Pyrenees. Bull. fr. Peche Piscic. 337/338/339: 399–406.

Barnard, S. & R. J. Wyatt, 1995. An analysis of predictive models for stream salmonid population. Bull. fr. Peche Piscic. 337/338/339: 365–373.

Bayley, P. B. & H. W. Li, 1996. Riverine Fishes. In Petts, G. & P. Calow (eds), River Biota. Blackwell Science, Oxford: 92–122.

Binns, N. A. & F. M. Eiserman, 1979. Quantification of fluvial trout habitat in Wyoming. Trans. am. Fish. Soc. 108: 215–228.

Bowlby, J. N. & J. C. Roff, 1986. Trout biomass and habitat relationships in southern Ontario streams. Trans. am. Fish. Soc. 115: 503–514.

Dieplinger, K., S. Leberl & G. Zauner, 1996. Biotope management on the Austrian Danube. Arch. Hydrobiol. Suppl. 113, Large Rivers 10: 219–228.

Fausch, K. D., C. L. Hawkes & M. G. Parsons, 1988. Models that predict standing crop of stream fish from habitat variables: 1950–85. U.S. Forest Service General Technical Report PNW-GTR-213, Portland, OR. 52 pp.

Franklin, I. R., 1980. Evolutionary change in small populations. In Soulé, M. E. & B. A. Wilcox (eds), An Evolutionary-Ecological Perspecitve. Sinauer, Sunderland, Mass.: 135–149.

Frissell, C. A., W. J. Liss, C. E. Warren & M. D. Hurley, 1986. A hierarchical framework for stream habitat classification: viewing streams in a watershed context. Envir. Manage. 10: 199–214.

Habersack, H.M., 2000. The river scaling concept (RSC): a basis for ecological assessments. Hydrobiologia 422/423: 49–60.

Haidvogl, G. & H. Waidbacher, 1997. Ehemalige Fischfauna an ausgewählten österreichischen Fließgewässern. Abt. Hydrobiologie, BOKU, Wien. 85 pp.

Hughes, R. M., 1995. Defining acceptable biological status by comparing with reference conditions. In Davies, W. S. & T. P Simon (eds), Biological Assessment and Criteria; Tools for Water Resource Planning and Decision Making. Lewis Publishers, Boca Raton, FL: 31–47.

Hughes, R. M. & R. F. Noss, 1992. Biological diversity and biological integrity: current concerns for lakes and streams. Fisheries 17: 11–19.

Hughes, R. M. & T. Oberdorff, 1998. Application of IBI concepts and metrics to waters outside the United States. In Simon, T. P. (ed.), Assessment Approaches for Estimating Biological Integrity using Fish Assemblages. Lewis Press, Boca Raton, FL: 79–83.

Illies, J., 1967. Limnofauna Europaea. Gustav Fischer Verlag, Stuttgart. 474 pp.

Illies, J. & L. Botosaneanu, 1963. Problèmes et méthodes de la classification et de la zonation écologique des eaux courantes considerées surtout du point de vue faunistique. Int. Verh Limnol. 12: 1–57.

Jowett, I. G., 1992. Models of the abundance of large brown trout in New Zealand rivers. N. am. J. Fish. Manage. 12: 417–432.

Jungwirth, M., 1984. Die fischereilichen Verhältnisse in Laufstauen alpiner Flüsse, aufgezeigt am Beispiel der österreichischen Donau. Österreichische Wasserwirtschaft 36: 104–111.

Jungwirth, M., O. Moog, S. Schmutz & H. Wiesbauer, 1990. Gutachten Raum- und Umweltverträglichkeitsprüfung der Kraftwerksprojekte Obere Drau I Sachsenburg – Spittal – Mauthbrücken, Teilgutachten 4. Ökologie des aquatischen Lebensraumes, Hydrobiologie, BOKU, Wien.

Jungwirth, M., S. Schmutz & H. Waidbacher, 1989. Fischökologische Fallstudie Inn. Fischerei-Revierausschuß Innsbruck Stadt u. Land, Hall. 93 pp.

Jungwirth, M., S. Muhar & S. Schmutz, 1995. The effects of recreated instream and ecotone structures on the fish fauna of an epipotamal river. Hydrobiologia 303: 195–206.

Jungwirth, M., 1998. River Continuum and Fish Migration – Going Beyond the Longitudinal River Corridor in Understanding Ecological Integrity. In Jungwirth, M., S. Schmutz & S. Weiss (eds), Fish Migration and Fish Bypasses. Fishing News Books, Oxford: 19–32.

Jungwirth M., S. Muhar & S. Schmutz, 2000. Fundamentals of fish ecological integrity and their relation to the extended serial discontinuity concept. Hydrobiologia 422/423: 85–97.

Junk, W. J., P. B. Bayley & R. E. Sparks, 1989. The flood pulse concept in river-floodplain systems. Int. Riv. Symp. 106: 110–127.

Karr, J. R., 1981. Assessment of biotic integrity using fish communities. Fisheries 6: 21–27.

Kaufmann, T., S. Muhar, J. Raderbauer, O. Rathschüler, S. Schmutz, H. Waidbacher & G. Zauner, 1991. Fischökologische Studie Mur. Stadl bis Gratkorn, Hydrobiologie, BOKU, Wien. 104 pp.

Kryzhanovsky, S. G., 1948. Ecological groups of fishes and principles of their development. Naucho-Issled 27: 1–114.

Michel, P. & T. Oberdorff, 1995. Feeding habits of fourteen European freshwater fish species. Cybium 19: 5–46.

Miller, D. L., P. M. Leonard, R. M. Hughes, J. R. Karr, P. B. Moyle, L. H. Schrader, B. A. Thompson, R. A. Daniels, K. D. Fausch, G. A. Fitzhugh, J. R. Gammon, D. B. Halliwell, P. L. Angermeier & D. J. Orth, 1988. Regional applications of an index of biotic integrity for use in water resource management. Fisheries 13: 12–20.

Milner, N. J., R. J. Hemsworth & B. E. Jones, 1985. Habitat evaluation as a fisheries management tool. J. Fish. Biol. 27: 85–108.

Milner, N., R. Wyatt & M. Scott, 1993. Variability in the distribution and abundance of stream salmonids, and the associated use of habitat models. J. Fish. Biol. 43: 103–119.

Milner, N. J., R. J. Wyatt, S. Barnard & M. D. Scott, 1995. Variance structuring in stream salmonid populations, effects of geographical scale and implications for habitat models. Bull. fr. Peche Piscic. 337/338/339: 387–398.

Moog, O. & A. Chovanec, 2000. Assessing the ecological integrity of rivers: walking the line among ecological, political and administrative interests. Hydrobiologia 422/423: 99–109.

Muhar, S., 1996. Habitat improvement of Austrian rivers with regard to different scales. Reg. Riv. Res. Manage. 12: 471–482.

Muhar, S., A. Muhar, S. Schmutz, R. Wimmer, H. Wiesbauer, B. Hozang, G. Imhof & P. Tschemernig, 1993. Ausweisung naturnaher Fließgewässerabschnitte in Österreich, Vorstudie. Blaue Reihe BM Umwelt, Jugend u. Familie, Wien. 175 pp.

Muhar, S., S. Schmutz & M. Jungwirth, 1995. River restoration concepts – goals and perspective. Hydrobiologia 303: 183–194.

Muhar, S., M. Jungwirth, J. Eberstaller, C. Moritz, K. Michor, G. Egger & J. Petutschnig, 1998a. Der Zustand der Oberen Drau - Wertigkeiten und Defizite aquatischer und semi-/terrestrischer Zönosen. Österr. Wasser- und Abfallwirtschaft 50: 25–31.

Muhar, S., M. Kainz & M. Schwarz, 1998b. Ausweisung flußtypspezifisch erhaltener Fließgewässerabschnitte in Österreich. Teil 2. BMLF, Wien. 177 pp.

Muhar, S., M. Schwarz, S. Schmutz & M. Jungwirth, 2000. Identification of rivers with high and good habitat quality: methodological approach and applications in Austria. Hydrobiologia 422/423: 343–358.

Naiman, R. J., D. G. Lonzarich, T. J. Beechie & S. C. Ralph, 1992. General principles of classification and the assessment of conservation potential in rivers. In Boon, P. J., P. Calow & G. E. Petts (eds), River Conservation and Management. Wiley, Chichester: 95–125.

Noss, R. F., 1990. Indicators for monitoring biodiversity: a hierarchical approach. Conserv. Biol. 4: 355–364.

Oberdorff, T. & R. M. Hughes, 1992. Modification of an index of biotic integrity based on fish assemblages to characterize rivers of the Seine Basin, France. Hydrobiologia 228: 117–130.

Oberdorff, T. & J. P. Porcher, 1994. An index of biotic integrity to assess biological impacts of salmonid farm effluents on receiving waters. Aquaculture 119: 219–235.

Oberdorff, T., D. Pont, B. Hugueny, J. P. Porcher & D. Chessel (in press). A probabilistic model characterizing riverine fish communities of French rivers: a framework for the adaptation of a fish-based index. J. appl. Ecol.

Odum, E. P., 1971. Fundamentals of Ecology. Saunders, Philadelphia. 574 pp.

O'Neill, R. V., D. L. DeAngelis, J. B. Waide & T. F. H. Allen, 1986. A Hierarchical Concept of Ecosystems. Princeton University Press, Princeton, NJ. 253 pp.

Persat, H., J. M. Olivier & J. P. Bravard, 1995. Stream and Riparian Management of Large Braided Mid-European Rivers, and Consequences for Fish. World Fisheries Congress, New-Dehli, Oxford & IBH Publishing Co. Theme 1: 139–169.

Peter, A., 1998. Interruption of the river continuum by barriers and consequences for migratory fish. In Jungwirth, M., S. Schmutz & S. Weiss (eds), Fish Migration and Fish Bypasses. Fishing News Books, Oxford: 99–112.

Raleigh, R. F., L. D. Zuckerman & P. C. Nelson, 1984. Habitat suitability index models and instream flow suitability curves: brown trout. U.S. Fish and Wildlife Services Biological Services Program, FWS/OBS-82/10.71.

Scarnecchia, D. L. & E. P. Bergersen, 1987. Trout production and standing crop in Colorado's small streams, as related to environmental features. N. am. J. Fish. Manag. 7: 315–330.

Schiemer, F., M. Jungwirth & G. Imhof, 1994. Die Fische der Donau- Gefährdung und Schutz. Styria Medien Service, Graz.

Schiemer, F. & H. Waidbacher, 1992. Strategies for conservation of a Danubian fish fauna. In Boon, P. J., P. Calow & G. E. Petts (eds), River Conservation and Management. Wiley, Chichester: 363–382.

Schmutz, S. & M. Jungwirth, 1999. Fish as indicators of large river connectivity: the Danube and its tributaries. Arch. Hydrobiol. Suppl. 115/3: 329–348.

Stauffer, J. R., 1984. Colonization theory relative to introduced populations. In Courtenay, W. R. & J. R. Stauffer (eds), Distribution, Biology and Management of Exotic Fishes. Hopkins Univ. Press, Baltimore: 8–21.

Thienemann, A., 1926. Die Binnengewässer Mitteleuropas. Die Binnengewässer, Band I. Schweizerbart'sche Verlagsbuchhandlung, Stuttgart.

Tockner, K. & F. Schiemer, 1997. Ecological aspects of the restoration strategy for a river-floodplain system on the Danube River in Austria. Glob. Ecol. Biogeo. Let. 6: 321–329.

Ward, J. V. & J. A. Stanford, 1995. Ecological connectivity in alluvial river ecosystems and its disruption by flow regulation. Reg. Riv. 11: 105–119.

Welcomme, R. L., 1984. International transfers of inland fish species. In Courtenay, W. R. & J. R. Stauffer (eds), Distribution, Biology, and Management of Exotic Fishes. Hopkins Univ. Press, Baltimore: 22–40.

Wesche, T. A., C. M. Goertler & W. A. Hubert, 1987. Modified habitat suitability index model for brown trout in southeastern Wyoming. N. am. J. Fish. Manage. 7: 232–237.

Zauner, G., 1993. Fischökologische Studie untere Thaya. Hydrobiologie, BOKU, Wien.

Zauner, G., 1994. Fischökologische Untersuchung, Teiluntersuchung 1.4.2 im Rahmen der Gesamtuntersuchung Salzach. Österreichisches Institut für Raumplanung, Wien. 135 pp.

Hydrobiologia **422/423:** 291–304, 2000.
M. Jungwirth, S. Muhar & S. Schmutz (eds), Assessing the Ecological Integrity of Running Waters.

Variation in a Great River Index of Biotic Integrity over a 20-year period

James R. Gammon[1] & Thomas P. Simon[2]
[1]*DePauw University, Department of Biological Science, Greencastle, IN 46135, U.S.A.*
[2]*U.S. Environmental Protection Agency, Water Division, 77 West Jackson Blvd., WW-16J, Chicago, IL 60604, U.S.A.*

Key words: biological integrity, IBI, fish assemblage, fish community structure and function, land use, gear effects, recruitment

Abstract

The Index of Biotic Integrity (IBI) has proven to be an important assessment tool for evaluating the resource quality of aquatic ecosystems. We used a modified IBI for Great Rivers based on collections from the Wabash River in west-central Indiana. We evaluated natural, inherent variation in natural phenomena, gear effects and land use modification over a 20-year period. Sixty-three stations distributed through 260 km of river were sampled 2–3 times per year to evaluate fish community structure and function. Natural changes in variation included temporal variation, hydrologic cycle and recruitment. Our results suggest that removing gizzard shad, whose population levels fluctuate widely, from percentage metrics enhanced assessment and exposed subtle affects. IBI scores consistently declined in a downstream direction. IBI profiles based on numbers were almost always higher than those using biomass, but patterns corresponded well for both metric systems. Combining biomass catches by seining and electrofishing did not substantially change the observed patterns compared to numeric electrofishing catches. The large number of small species taken by seining overwhelmed the subtle patterns observed at least impacted stations, but greater departures in quality were evident at impaired downstream locations. Floods decimated fish communities throughout the river. Fish moved out of stressed areas during severe droughts and into areas near clean tributaries. Certain structured attributes of Great River fish communities showed statistically significant changes based on ecoregion scales. Total number of species, number of sensitive species, number of centrarchid species and percentage of simple lithophils declined between the Corn Belt Plain and Interior River Lowland ecoregions, but water quality factors might also be influential. These changes in metric attributes lowered IBI scores by 6–10 cumulative points.

Introduction

The index of biotic integrity (IBI) was originally developed for small midwestern streams and rivers (Karr, 1981; Karr et al., 1986) and has become a family of multi-metric indices that are regionally adapted and calibrated (Miller et al., 1988; Simon & Lyons, 1995; Barbour et al., 1995). The IBI has proven to be responsive to a wide variety of disturbances that affect fish assemblage stability and function (Karr, 1981; Karr et al., 1986; Davis & Simon, 1995; Simon, 1998). It is reliable for estimating differences in assemblage structure when the IBI values differ by 4–8 points.

The metrics and application of the IBI for Great River ecosystems are in their early developmental stages (Simon & Emery, 1995; Simon & Sanders, 1998; Reash, 1998; Emery et al., 1998). Fish community structure and function have been used widely for a variety of resource types including streams (Karr et al., 1986), inland lakes (Jennings et al., 1998), wetlands (Simon, 1998; Simon & Stewart, 1998) and Great Lake nearshore (Minns et al., 1994; Thoma, 1998) to develop biological reference condition and assessment conditions. Great Rivers are distinctive in that they are few in number and comprise the largest component of the resource (Vannote et al., 1980). In addition, most of our largest rivers are disproportionately degraded (Karr et al., 1985).

Few investigators have a database as extensive as that available for the Wabash River (Gammon, 1998).

The effects of natural variation were evaluated to determine the changes in IBI score and various changes in assemblage structure and function. Our research topics include an analysis of inherent natural variation, including temporal variation, hydrologic cycle and recruitment, and differences in sampling gear. Our study to evaluate the contributions of different kinds of variation and the effects on the final IBI score and patterns observed uses the Index of Well Being (I_{wb}) (Gammon, 1976, 1980).

Year-to-year variations in population size may result from fluctuations in year class strength (Nikolskii, 1969). Some researchers believe that variation in year class strength is primarily biologically based on availability of food during the transition from the yolk stage to external feeding, a difficult investigative task in nature or laboratory (Braum, 1967). Another view is that abiotic factors such as temperature, weather, river flow, thickness of snow cover, etc. are the primary determinants of survival. An alternative, broader approach extends survival factors to include specific adaptations of the fish larvae themselves and even to the nutritional state of their parents.

The long-term studies on which this contribution is based have permitted us to:

(a) relate changes in the fish communities to natural events and man-induced factors,
(b) identify problem sections of the river,
(c) evaluate ecological changes associated with operational modifications by industry, and
(d) clarify ecological interactions among the major biotic components of the ecosystem.

The substantial population changes which have been observed have, however, provided little direct information about basic reproductive processes or early life history of major species.

Material and methods

Study area

The Wabash River is the largest northern tributary of the Ohio River and drains 65.6% of Indiana and 35% of Illinois in the midwestern United States. Simon (1992) and Simon & Lyons (1995) defined Great Rivers as those that possess a unique fauna comprised of the large river faunal group species (Pflieger, 1971) and that have drainage areas of at least 3226 km^2. The Wabash River becomes a Great River in its lower two-thirds where the mean slope is 0.123 m km^{-1} (Gammon, 1998). A single dam is located at rkm 662 (RM 411), after which the river flows freely to the Ohio River.

Land use in the Wabash River watershed is 55.2% agricultural cropland with an additional 8.2% being pasture or grassland. Forest or woodland accounts for only 13.5% of the basin's land use (Gammon, 1998). The Wabash River drains two principal ecoregions, the Eastern Corn Belt Plain and the Interior River Lowland (Omernik, 1987). Within the study area, differences in bedrock geology, potential natural vegetation and soil type are the most prominent natural differences from adjacent ecoregions. Loamy, high lime till plains in the northern section are characterized by loamy, high lime, late-Wisconsinan glacial till, soils which are mostly Alfisols and potential climax forests of oak-sugar maple and beech. The glaciated Wabash lowlands constitute the southern section with bedrock of Pennsylvanian shale, sandstone and coal. Mollisol soils bordering the river and potential climax vegetation consisting of oak-hickory, beech and scattered prairies.

Collection methods

We usually conducted 2–3 passes of electrofishing at 63 locations between June and October 1967–1997 between Delphi and Merom, Indiana, a distance of 274 km (169 mi). Weather and flow conditions sometimes necessitated modifying collections since high discharge negatively influenced sampling efficiency. The river was divided into 12 reaches to assist in assessment. Each collection consisted of a two or three person crew using a boat-mounted generator and a Smith-Root Type IV electroshocker with a safety switch foot pedal; this unit produced 600 V pulsed, DC output capable of 6–8 A. Bow-mounted circular stainless steel anodes were suspended at the water-surface by booms and included long copper-woven cathodes along the gunwales. A single netter was positioned on the bow of the boat and used a long-handled dip net with 6.25 mm^2 mesh to dip all fish seen. Each station was fished downstream for 500 m along a single shoreline. Gammon (1983) indicated that species richness asymptotes at distances of about 500 m on the Wabash River. We recorded the sampling time as the time that the power was applied to the electrodes; this typically averaged 600–1000 s depending on river discharge. During 1977 and 1978, a supplemental seining component was added to the collection techniques (Rogellin, 1979). A total of 25 sites were seined during mid-July from Delphi to 10 km upstream of Terre Haute. A 3.6×1.2 m seine with 6.25 mm^2 mesh was used to collect fish. Two 10 m seine hauls were made at each site in a downstream direction. In September 1993, 14 sites were sampled once by a different protocol.

All captured fish were placed into a holding tank and were processed after the entire station was sampled. Each fish was identified to species, measured and weighed, and inspected each individual specimen for deformities, eroded fins, lesions and tumors (DELT). We preserved representative voucher specimens and small species such as minnows, madtoms and darters in the field and returned to the laboratory for processing. We identified fish in the laboratory using keys from Gerking (1955), Smith (1973) and Trautman (1981). The special collections made in September 1993 included all fish longer than 25 mm. The regular summer collecting series excluded young-of-year prior to calculations in order to normalize annual data.

Statistics

Statistical analyses were performed using a two-way analysis of variance (ANOVA) to determine differences between gear types, seasons, hydrologic cycle, and recuitment (SAS, 1985). Trend lines were generated by means of locally weighted sum of squares (Lowess) plots (TriMetrics, Inc., 1994).

Results

Reference conditions

Simon & Emery (1995) developed an IBI for Great Rivers based on fish assemblage information from the Ohio River. We modified this index and developed biological expectations for the middle Wabash River fish assemblage, exploring two forms of each metric, one based on numbers of fish and the other based on biomass (Table 1). These expectations were developed with the assumption that 'least impacted' conditions would emerge from the cumulative data set without having to choose reference sites (Simon, 1998a). Large floodplain river reference conditions must consider species composition variation by season (Tejerino-Garro et al., 1998), response to habitat variation (Gelwick & Matthews, 1990; Edds, 1992) and longitudinal zonation (Ibarra & Stewart, 1989).

IBI metrics were modified to best reflect the Wabash River data set following Angermeier & Karr (1986). Metrics were calibrated following Fausch et al. (1984). We followed Goldstein et al. (1994) and Niemela et al. (1998) in substituting the percentage of trophic metrics as individuals to percentages based on biomass.

The development of the composite Index of Well-Being (Iwb) was motivated by the need to simplify data. During the late 1960s, it was necessary to produce a collecting protocol for large rivers and methods for analyzing the resulting data in order to assess the environmental impact of electric-generating stations (Gammon, 1971, 1973, 1976). The environmental criteria which were being established at that time were to establish physical and chemical criteria which would provide protection for resident aquatic biota. Zalewski et al. (1990) found that fish assemblage dynamics were regulated by abiotic factors that controlled assemblage structure.

The electric power industry sought and received from regulatory authorities an exemption from thermal limits if a healthy population of aquatic organisms lived in heated waters produced by electric-generating plants. To achieve this exemption it had to be shown – through the so-called 316(a) demonstration – that the aquatic communities immediately downriver from electric generating plants were both diverse and numerous: the biotic integrity of the community had to be maintained despite excursions of temperature beyond the established criteria.

We examined several quantitative expressions of biotic diversity, including the total number of species and the Shannon–Weiner or Shannon–Weaver index of diversity, for their utility in evaluating the impact of electric power plants. Quantitative expressions of abundance were also examined for their usefulness, including numeric abundance and aggregate weight or biomass. All of these community parameters exhibited the same basic pattern through space; therefore, two indices of diversity and two suitably weighted indices of abundance were combined into a single composite Index of Well-Being or I_{wb} which would reflect both the diversity and abundance of the fish community:

$$I_{wb} = 0.5 \ln \text{No/km} + 0.5 \ln \text{Kg/km} + \text{Div}_{(no)} + \text{Div}_{(wt)}.$$

Two advantages of the composite index are 1. simplification and 2. reduced variability compared to individual components. Figure 1 shows the variation of I_{wb} from 1973 through 1997 over 265 km of river subdivided into 12 Reaches.

Table 1. Metrics used to compute Index of Biotic Integrity (IBI) scores for Wabash River sites

Metric category	Metric attribute
Species richness and composition	
	1.[a] (a) and (b) Total number of species
	2.[b] (a) Number of round-bodied sucker species
	(b) RBS species as a% of total species
	3.[c] (a) Number of centrarchid species
	(b) Centrarchid species as a % of total species
	4.[d] (a) % Great River individuals (no.)
	(b) % Great River individuals (biomass)
Species tolerance	
	5.[e] (a) Number of sensitive species
	(b) Number of sensitive species as % of total species
	6.[f] (a) % tolerant individuals (no.)
	(b) % tolerant individuals (biomass)
Trophic composition	
	7.[g] (a) % macrovorous individuals (no.)
	(b) % macrovorous individuals (biomass)
	8.[h] (a) % insectivorous individuals (no.)
	(b) % insectivorous individuals (biomass)
	9.[i] (a) % omnivorous individuals (no.)
	(b) % omnivorous individuals (biomass)
Reproductive guild	
	10.[j] (a) % simple lithophils (no.)
	(b) % simple lithophils (biomass)
Fish abundance	
	11.[k] (a) Catch per unit effort – (no./km)
	(b) Catch per unit effort – (kg/km)
Fish condition	
	12.[l] (a) and (b) Percentage DELT anomalies (deformities, fin erosion, lesions, tumors)

[a] Included exotic and non-indigenous species, but excluded hybrids and subspecies
[b] Members of the sucker family Catostomidae including *Moxostoma, Hypentelium, Minytrema* and *Cycleptus*. These species are considered benthic, invertivore specialists that are sensitive to various human-caused disturbances.
[c] Members of the family Centrarchidae, including the black basses (*Micropterus*) and sunfish (*Lepomis, Pomoxis* and *Ambloplites*. This group is responsive to pool habitat and structural complexity.
[d] This group includes multiple families which are sensitive to a variety of perturbations. The list should be limited to 10% of the fauna.
[e] A pollution-tolerant group from various families that increase in relative abundance in degraded habitats.
[f] These species are great river specialists which decline with the loss of associated habitats.
[g] Various species which ingest a considerable amount of plant (25%) and animal (25%) material.
[h] Species from multiple families which consume aquatic macroinvertebrates, zooplankton and crustaceans.
[i] Species which are primarily piscivores or consume crayfish and molluscs.
[j] Species which exhibit simple spawning behavior over gravel.
[k] Numbers of individuals and biomass collected per unit effort (time or distance).
[l] Percentage of individuals in poor health, excluding parasitism and protozoan infections.

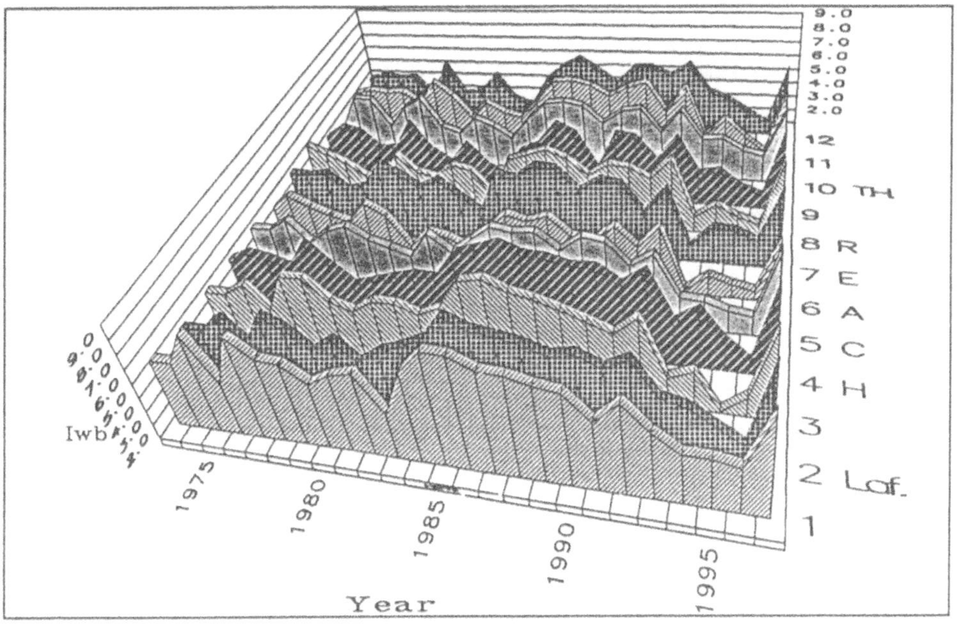

Figure 1. Mean composite index of well-being (I_{wb}) values for 12 reaches of the middle Wabash River 1973–1997.

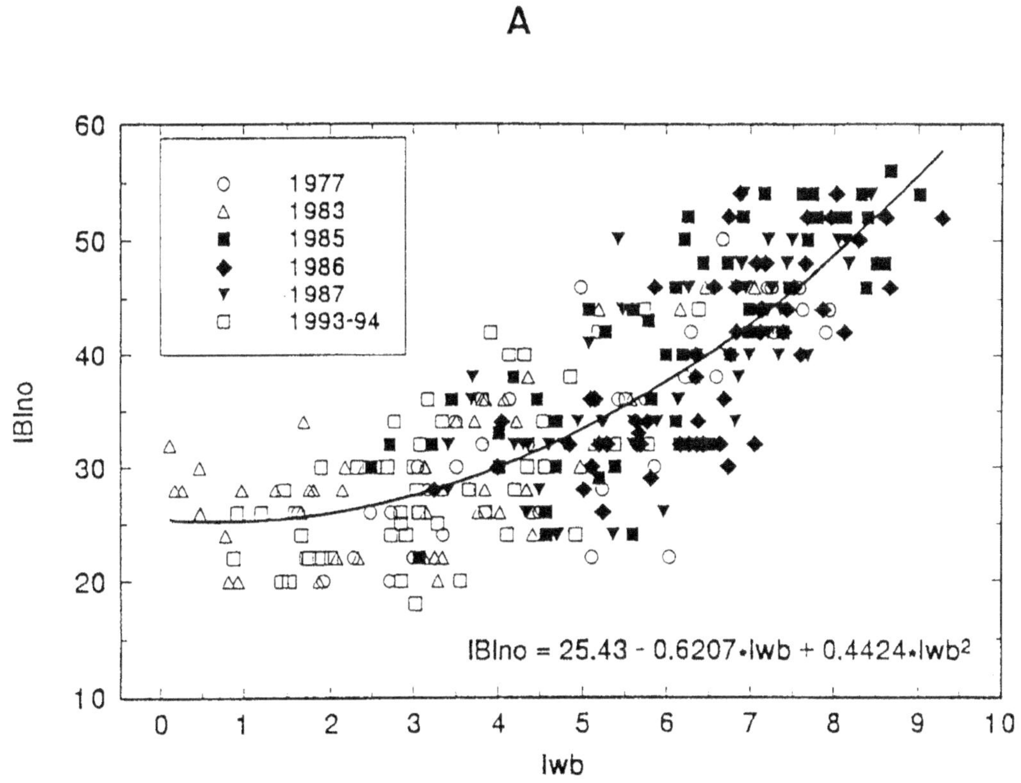

Figure 2a. (A) Curvilinear regression of raw IBI scores on I_{wb} values. (B) Linear regression of modified IBI scores on I_{wb} values.

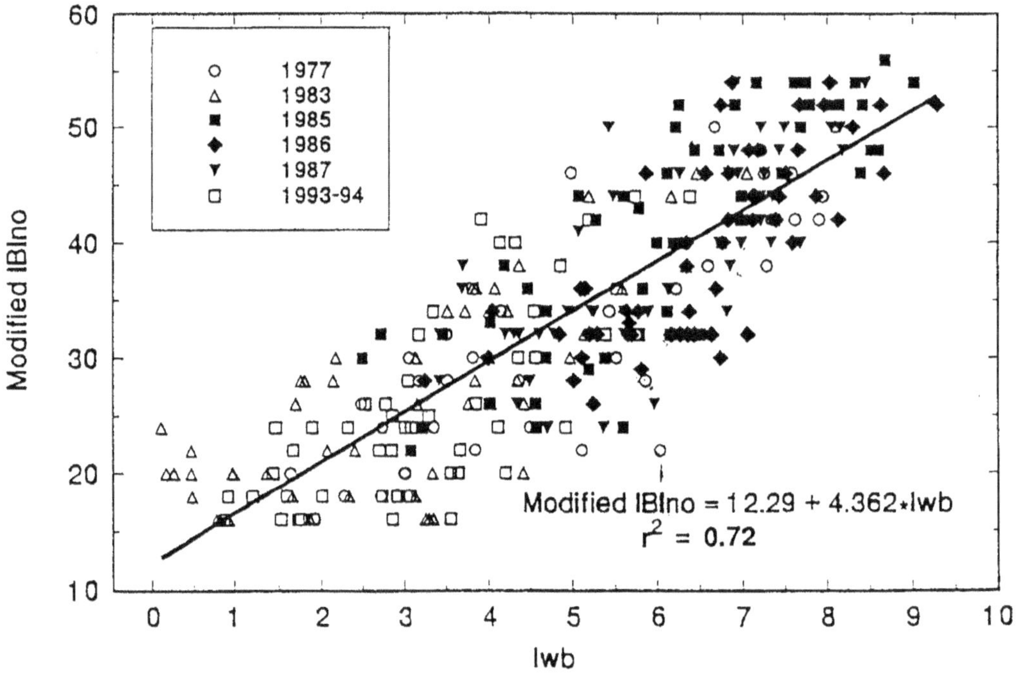

Figure 2b. (A) Curvilinear regression of raw IBI scores on I_{wb} values. (B) Linear regression of modified IBI scores on I_{wb} values

Only the catch data from 1977, 1983, 1985, 1986, 1987 and 1993–4 were used to establish the relationship between the IBI and I_{wb} using numbers of fish. A regression of IBI on I_{wb} is shown in Figure 2A. Modifications were required to achieve linearity when catch rates were very low with few species as was the situation at many sites during 1983 and 1993–94. The relationship is approximately linear for IBI values greater than about 25 and I_{wb} values of about 3.0, but when I_{wb} values are lower than 3.0 the relationship becomes distinctly curvilinear. Catches with low IBI scores were skewed primarily because of high values for the Great River Species and Macrovore metrics.

Trophic dynamics of Great River fish assemblages are determined by changes in habitat, eutrophication, and naturally possess fewer top level carnivores. These top level carnivores are capable of affecting indirect relationships among prey items that may cause cascading trophic level relationships. Johnson & Dropkin (1993) found that the diet composition varied over a diel period, while Godinho et al. (1999) found diet in a single species to vary depending on environmental conditions. Turner & Mittlebach (1990) found that predator species indirectly affected trophic dynamics between planktivores and plankton.

Low catch rates of only one or two species can often make a difference in IBI scoring since these rare species are often difficult to collect. Pusey et al. (1995) described low catch rates in two northern Australian rivers, while Niemela et al. (1998) found that alternative descriptive metrics were necessary to characterize the depauperate fish fauna of the Red River of the North. Therefore, when catch rates were less than 10 fish/km we decided to score the Great River species and Macrovore metrics as '1'. Modification of the IBI based on numeric data is shown in Figure 2B. The equation (Modified IBI=12.29+4.362 I_{wb}; r=0.72) indicates satisfactory correspondence of the lowest IBI and I_{wb} values, IBI=12 and I_{wb}=0.

IBI profiles based on numbers were almost always higher than biomass calculations (Figure 3). However, both profiles for any given year were very similar, with corresponding peaks and valleys. Most of the following analyses are based on numeric data alone.

Natural variation

Effects of fish species attributes

Changes in IBI score can be attributed to either temporal variability or changes in dominance of select species. Species that are schooling species, r-strategists, or tend to clump in the environment should be examined to determine whether they should be removed from analysis. Our results show that removal of gizzard shad (*Dorosoma cepedianum*) and possibly emerald shiner (*Notropis atherinoides*) would be an

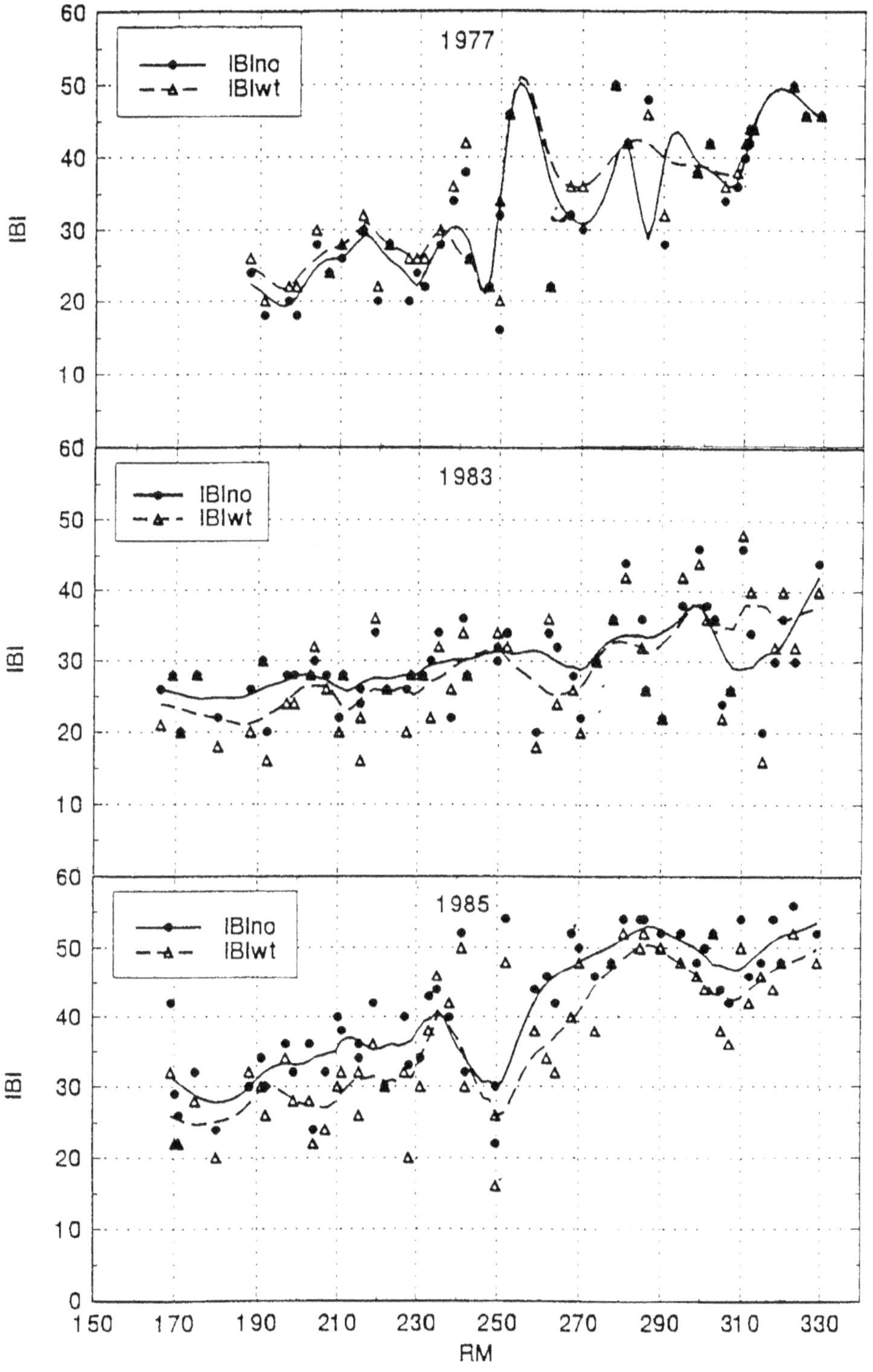

Figure 3. Index of Biotic Integrity (IBI) profiles based on numbers and biomass for 1977, 1983 and 1985.

important consideration in exposing more subtle affects. Simon & Emery (1995) and Thoma (1998) likewise recommended removal of schooling species. Gizzard shad was usually abundant and often comprised approximately half of the total catch, but became scarce in the late 1980s as a result of predator pressure. This variability tended to obscure trends in IBI over time. In order to improve trend assessment, we have removed select species from the IBI calculations in order to expose subtle patterns.

Hydrologic cycles

The Wabash River experienced droughts in 1976–77, 1988 and 1991, and prolonged periods of high water during the summers of 1981, 1992, 1993 and 1998. Electrofishing collections during drought years were characterized by low diversity and abundance, but not all river sections were affected to the same degree (Figure 1). The 1983 and 1991 droughts were most severe as indicated by the depression of fish communities throughout all reaches. The 1976–77 and 1988 droughts did not influence communities in the uppermost reach, but depressed them further downriver.

Stations located near clean tributaries served as refugia during periods of stress, as was described during an episode of lower dissolved oxygen in 1977 (Gammon & Reidy, 1981). However, most stations remote from tributaries exhibited altered communities with fewer species overall, reduced numbers of round-bodied sucker and sensitive species, and relative increases in percent omnivores and tolerant species (Figure 4). The primary omnivore, the carp (*Cyprinus carpio*), was no more common in the affected segment than further up- or downriver, but it was relatively more abundant because of the scarcity of more sensitive species. The gizzard shad, however, was much more common in the affected segment than elsewhere. This phenomenon, which was noticed previously, may be caused by a local increase in phytoplankton and benthic algae, the result of nutrient additions from the Lafayette-West Lafayette area. The overall stress in this segment may also involve a change in the diurnal pattern of dissolved oxygen. Channel morphology might also be a factor in that drought conditions should expose more of the bottom sediments to light. This critical section of river clearly requires more research. A more restricted, but somewhat comparable problem area downriver from the Cayuga Electric Generating Station was described by Parke & Gammon (1986).

Floods, compared to droughts, profoundly affected fish communities throughout the Wabash River. The 1993 flood devastated most species populations and strongly depressed values of both I_{wb} and IBI through 1994 (Figure 1). Post-flood catch rates were less than half of the 1992 rates for many species as young year-classes were virtually eliminated from the populations. The 1998 summer floods were even longer in duration and a similar depression of the fish communities throughout the river is expected.

Chronic human-induced variation

As shown previously, IBI scores consistently declined in a downstream direction. The 1977 profile is similar to most years from 1973 to 1982, with high values in the upper reach trending toward values lower by 25 IBI points in the lower reach (Figure 5). The sharp declines at RM 310 and 210 represent effects from Lafayette and Terre Haute, Indiana, industrial complexes, the largest small cities along the Wabash River. Improvements in waste treatment at both cities resulted in increases in the I_{wb} 'sag curves' downriver (Gammon, 1998) An electric generating station (EGS) at RM 250 also produced a shorter dip because thermally sensitive species avoided the heated downstream segment during the summer months (Gammon, 1971).

IBI scores were low during 1983 for unknown reasons, but reproductive success was spectacular. Summer discharge rates were uniform and moderate. The biochemical oxygen demand (BOD) was reduced to about half of its 1960s value, the legacy of improved waste treatment at communities throughout the river basin during the 1970s. Furthermore, 1983 was the year of PIK (payment-in-kind), when farmers in the U.S. were paid not to grow crops. In Indiana, this lead potentially to a 25% reduction in agricultural loadings to the river.

From 1984 through 1987, populations of most fish species expanded greatly, except for gizzard shad (*Dorosoma cepedianum*) and carp (*Cyprinus carpio*). Predator species quadrupled in abundance, as did many other species. Diversity increased as some species' populations extended their ranges into previously vacant sections of the river, both up- and downriver. Despite the increased recruitment, the average size of many species increased. These improvements in the fish community are reflected in the elevated IBI profiles for the period 1985–87 (Figure 5). IBI levels gradually declined for several years thereafter.

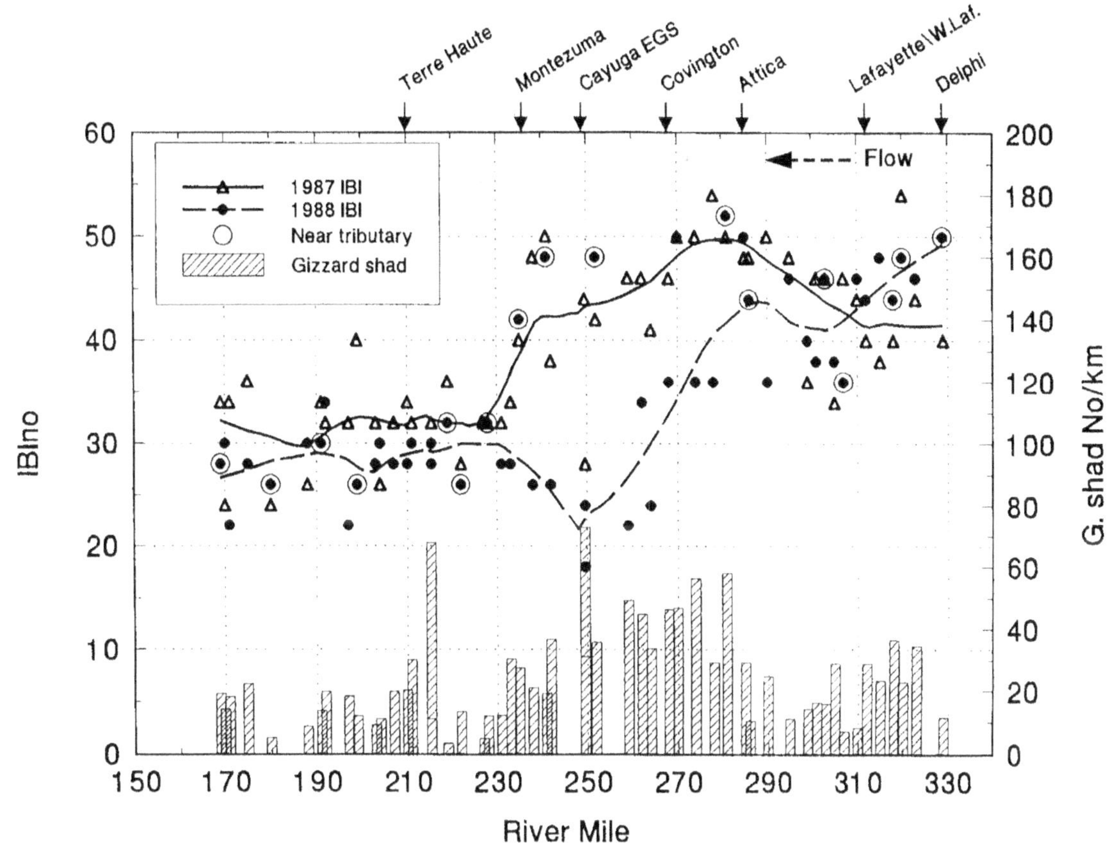

Figure 4. LOWESS plots of Index of Biotic Integrity (IBI) profiles based on numbers for 1977, 1983, 1985, 1986, 1987 and 1993–94.

The effects of different electrofishing protocols

IBI calculations based on intense electrofishing collections in mid-September 1993 at 14 stations included juveniles, y-o-y, and adult fish following the sampling protocol of Simon & Emery (1995). Two series of IBI values were calculated from this data set: (a) all species and sizes above 25 cm and (b) the catch data from which all small species such as darters and small minnow species were removed prior to calculation. The number of species was reduced 38% (from 68 to 42) and numbers were decreased 52% (from 3964 to 2084). Removing these small species prior to IBI calculation changed IBI profiles only slightly (Figure 6), except at the three lower-most sites. However, these differences were statistically significant (t=2.811, df=12, p=0.016).

Larval specimens less than 25 mm TL were excluded from the September, 1993 IBI analysis. However, species spawned early in the year could have exceeded 25 mm TL by late fall. If these juveniles are included in the IBI calculations, the scores might be inflated, thereby invalidating comparisons with collections taken earlier. A contrary argument for their inclusion, however, is that a juvenile walleye that is not sexually reproducing still functions as a predator and should be included.

In calculations of I_{wb} based on regular, annual collections, the y-o-y are routinely excluded because of the impossibility of predetermining the time of sampling in a free-flowing river such as the Wabash when sampling was sometimes initiated in late May and in other years not until late-August. It is believed that comparisons would otherwise be invalid for long-term monitoring evaluations.

Figure 6 also compares the results of the special September 1993 IBI study to that of a single regular electrofishing catch in late summer 1993 following the flood combined with multiple collections in 1994. The September 1993 IBI profile is more than 14 units higher than the regular 1993–94 IBI profile, but there is a good correspondence of peaks and valleys except at RM 252.

Differences of IBI profiles based on two different protocols were statistically significant (mean

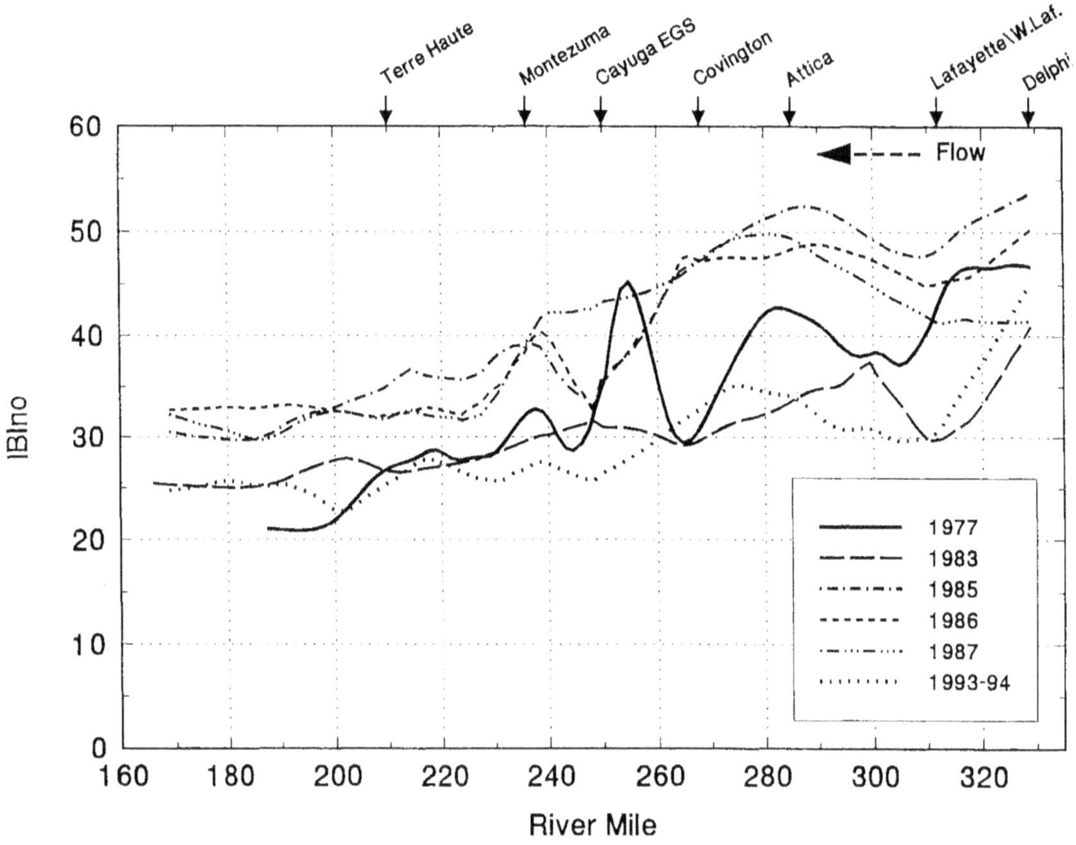

Figure 5. Changes in the Index of Biotic Integrity (IBI) profile during the 1988 drought compared to the 1987 profile.

diff.=14.182, SD=4.895, range=8–25) (ANOVA F=16.458, p=0.021).

Construction at a bridge located less than 1 km upriver from the collecting site at RM 252 produced high turbidity and sedimentation downstream during summer, 1993, causing fish to avoid the area and reducing the IBI at that site far below the expected normal. The river discharge had fallen and bridge construction activity ceased by September resulting in dramatic improvement in catches (Figure 6).

Effects of combining seining and electrofishing collections

Combined seining and electrofishing samples from 1977 were compared to electrofishing-only data from the closest site and evaluated by the Great River metrics based on trophic numbers and biomass percentages. Twenty-five sites were seined in mid-July 1977. Most seined species were adult cyprinids and some percids together with juveniles of larger resident species such as *Carpiodes carpio*.

The combined biomass catches produced IBI spatial patterns (using LOWESS) similar to those based on electrofishing alone using either numbers or biomass (Figure 7). However, the combined numeric catches produced an IBI profile which differed significantly from the others except in the extreme upper section, where environmental quality was better. The large numbers of fish (mean=214/sample) and additional species taken by seine basically overwhelm the smaller catches and lower number of species captured by electrofishing.

Electrofishing biomass and seining data combined generally enhanced IBI score by less than 8 IBI points, however, comparisons based on numbers were significantly different (t-test=1.652, df=23, p=0.112). Differences in IBI scores based on biomass between electrofishing and combined catches were statistically significantly different (F=2.026, df=8, p=0.109) in the lower river.

IBI scores based on trophic percentage metrics, separate analyses for number and biomass, were compared to combined gear, include a mean of seining number and weight. Electrofishing results based on number were not statistically different from catches based on biomass, however, the mean electrofish-

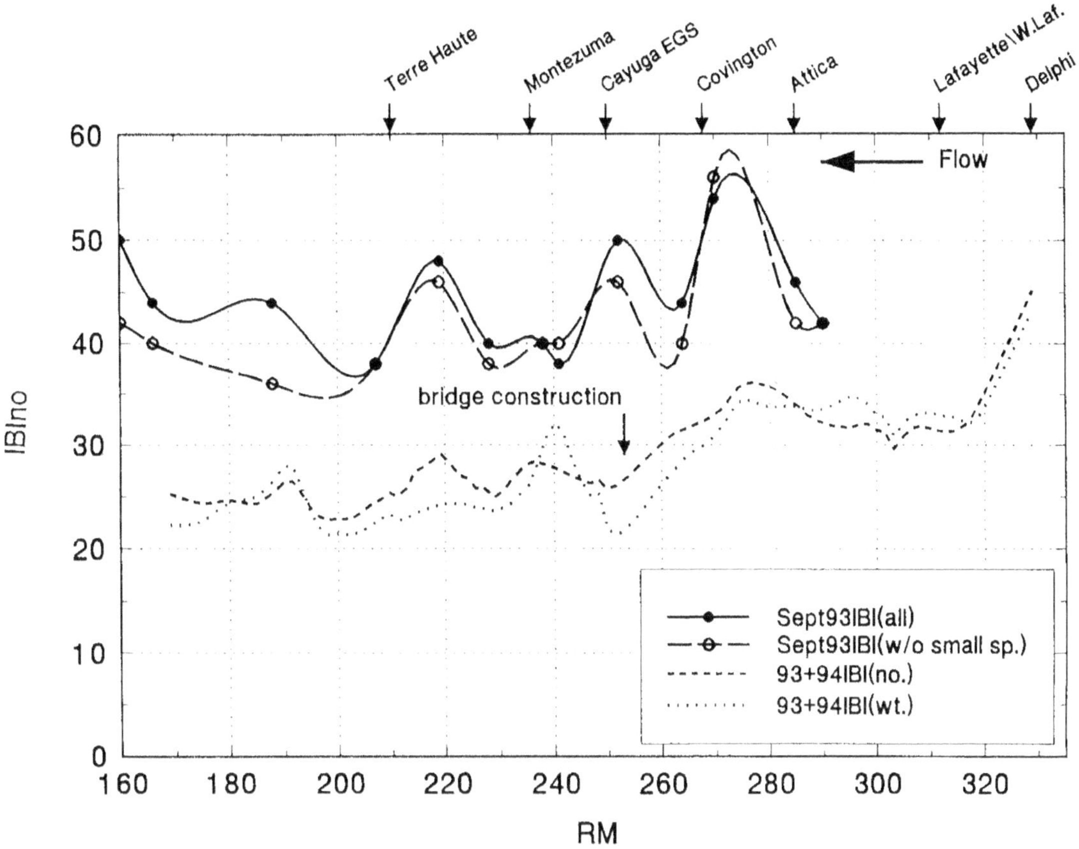

Figure 6. Index of Biotic Integrity (IBI) profiles obtained by two different electrofishing protocols. The September 1993 profiles compare the effect of excluding small species of fish such as darters and minnows, but include young-of-year for larger species.

ing results combined with seining was significantly different (t-test=2.36, df=22.7, p=0.027) from either electrofishing method in the lower river. Combined gear achieved significantly higher IBI scores (x=40.6, SD=6.0) that were not statistically different between the least impacted and the impacted lower river (F=0.761, df=8, p=0.641).

Our observations are similar to Simon & Emery (1995) and Thoma (1998), where degraded areas in large systems showed wider fluctuations in IBI scores when smaller species are included in the IBI score. Smaller species such as minnows, darters and madtoms appear to improve IBI scores for the lower river. However, these short-lived species, in comparison to long-lived suckers, black basses and catfishes, also reduced the clarity of trends in the impaired portions of the river.

Discussion and conclusions

The Wabash River still supports most of the fish species present in 1800, albeit in altered proportions. Sight-feeding fishes are scarce, while those which mostly rely on olfactory and cutaneous senses are probably much more common than formerly. The fish community changed little from 1967 to 1983 and the first signs of positive change came in 1984. Significant population increases for most species followed during the next few years. The average catch rate quadrupled as most fish populations enlarged. Some fishes extended their range into areas of the river in which they did not or could not live previously. The density of carp, however, remained unchanged and the gizzard shad population actually declined sharply because of increased predator pressure. The average size of many species increased, possibly because of greater longevity and/or faster growth.

The river's environment has been changed most by the development of a fertilizer-fueled agriculture and an expanding population which together produce sustained algal blooms during warm summer months. Improved waste treatment at communities and industries reduced the mean annual BOD to 4.5–5.0 mgl^{-1}

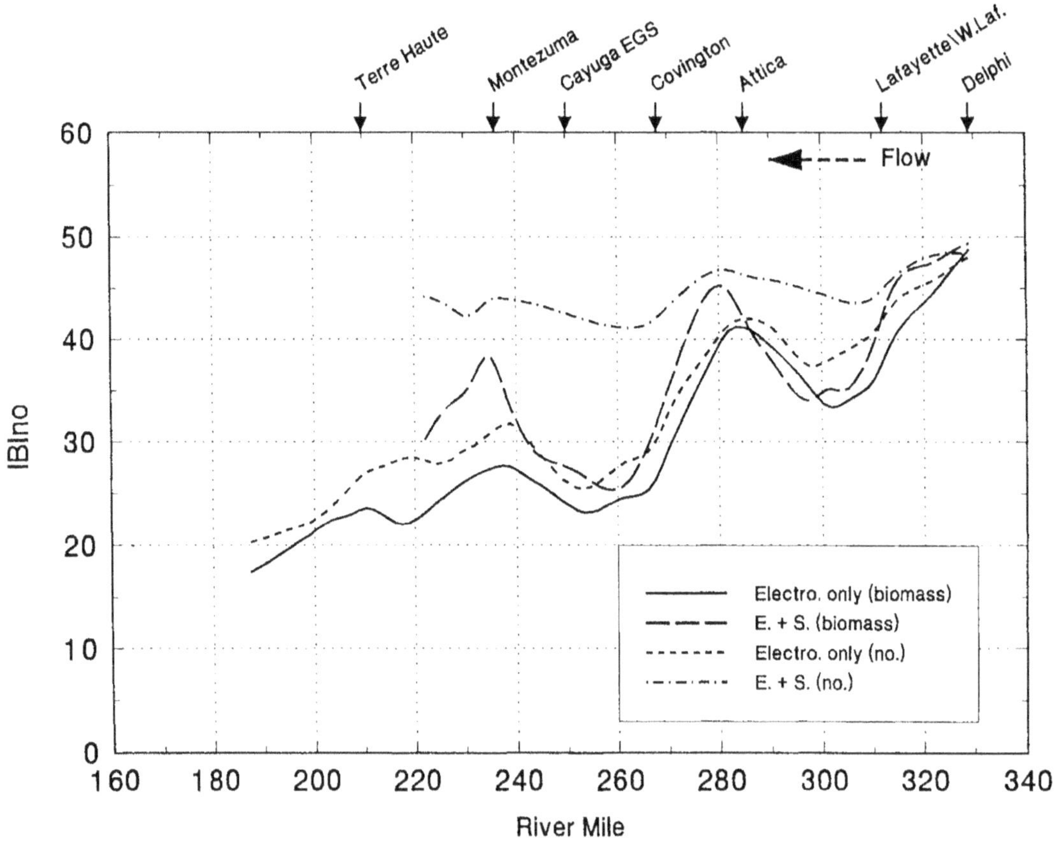

Figure 7. Index of Biotic Integrity (IBI) profiles for 1977 data based on electrofishing only and combined electrofishing and seining samples.

in the 1960s to 2.5–3.0 mgl^{-1} by the 1980s. Multivariate statistical analysis of catch data and patterns of I_{wb} and IBI indicate that negative local impacts have decreased during the past 10 years at Terre Haute and Lafayette, Indiana.

However, BOD is lower overall only during cooler periods when phytoplankton is less abundant. During summer phytoplankton density reduces secchi transparency to about 35 cm in the upper reaches and to 20–25 cm in the lower reaches. Maximum summer phytoplankton density approaches 120 000 cells/ml, chlorophyll *a* nearly 250 *u*g/l, pheophytin *a* 6.5–51 *u*g/l, gross primary productivity (GPP) ranging from 70 to 850 mg $C/m^2/h$, and net primary productivity 90–784 mg $C/m^2/h$ (Bridges et al., 1986). A dissolved oxygen deficit (DOD), estimated using a version of the DIURNAL computer model (HydroQual, Inc., 1984), is produced with low-flow DOD values of 2.0–2.5 mg Ol^{-1} upper reaches increasing to about 4.0 mg Ol^{-1} downriver from Terre Haute, Indiana.

Phytoplankton respiration accounted for 50–60% of the DOD in the upper reaches and about 70% downriver from Terre Haute. Carbonaceous BOD (CBOD) accounted for only about 10% of the DOD in the upper reaches and over 15% downriver from Terre Haute. Sediment oxygen demand (SOD) was the third largest oxygen sink overall and was locally important occasionally at certain lentic sites during extremely low flows.

The quality of the fish community as measured by I_{wb} and IBI is, therefore, negatively correlated with several physiochemical parameters. Floods and droughts superimpose altered environmental conditions within this 'normal' low-flow regime. The decimation of several year classes of fish by the 1993 flood 'reset' the fish community to lowered populations consisting of larger adults. One species to benefit from floods was the freshwater drum (*Aplodinotus grunniens*) with its planktonic eggs. Frequent periods of high water during mid-summer in the 1990s continue to obscure stages of recovery. Droughts caused local reorganization of the fish communities which require more detailed research to explain.

Integrated community indices such as IBI and I_{wb} are useful for monitoring changes in the fish community within the highly changable environment of large rivers, but the collecting methodology and metrics both require further development. Our results suggest that gizzard shad, whose population levels fluctuate widely, should be removed from percentage metrics in order to enhance assessment and expose subtle affects. IBI profiles based on numbers were almost always higher than those using biomass, but patterns corresponded well for both numeric and biomass metric systems. Adding a seining component to electrofishing sampling enlarges the species coverage of the overall fish communities, but does not seriously alter basic IBI patterns produced by electrofishing alone when using biomass. However, a much less desirable approach was the use of numbers of fish from seining and electrofishing in calculating IBI scores.

Certain structured attributes of Great River fish communities showed statistically significant changes based on ecoregion scales. Total number of species, number of sensitive species, number of centrarchid species and percentage of simple lithophils declined between Corn Belt Plain and Interior River Lowland ecoregions, but other factors might also be influential.

Acknowledgements

We appreciate the assistance of many students and colleagues that collected, crunched, and entered information. We appreciate the constructive discussions with R. M. Hughes, C. O. Yoder, E. B. Emery and R. E. Sanders who helped us frame part of this study. The opinions expressed do not necessarily represent those of the U.S. Environmental Protection Agency. No official endorsement by that agency should be inferred. Sustained support of the overall monitoring effort was provided by Cinergy Corp. and Eli Lilly and Company.

References

Angermeier, P. L. & J. R. Karr, 1986. Applying an index of biotic integrity based on stream-fish communities: considerations in sampling and interpretation. N. am. J. Fish. Manage. 6: 418–429.

Barbour, M. T., J. B. Stribling & J. R. Karr, 1995. Multimetric approach for establishing biocriteria and measuring biological condition. In Davis, W. S. & T. P. Simon (eds), Biological Assessment and Criteria: Tools for Water Resource Planning and Decision Making. Lewis Publishers, Boca Raton, FL: 63–77.

Braum, E., 1967. The survival of fish larvae in reference to their feeding behaviour and the food supply. In Gerking, S. D. (ed.), The Biological Basis of Freshwater Fish Production. John Wiley & Sons, Inc., New York: 113–131.

Bridges, C. L., H. L. BonHomme, G. R. Bright, D. E. Clark, J. K. Ray, J. L. Rud & J. L. Winters, 1986. Summer primary productivity and associated ata for four Indiana rivers. Proceedings of the Indiana Academy of Science 96 (1986): 309–314.

Davis, W. S. & T. P. Simon (eds), 1995. Biological Assessment and Criteria: Tools for Water Resource Planning and Decision Making. Lewis Press, Boca Raton, FL: 415 pp.

Edds, D. R., 1993. Fish assemblage structure and environmental correlates in Nepals Gandaki River. Copeia 1993(1): 48–60.

Emery, E. B., T. P. Simon & R. Ovies, 1998. Influence of the family Catostomidae on the metrics developed for a Great Rivers index of biotic integrity. In Simon, T. P. (ed.), Assessing the Sustainability and Biological Integrity of Water Resources using Fish Communities. CRC Press, Boca Raton, FL: 203–224.

Fausch, K. D., J. R. Karr & P. R. Yant, 1984. Regional appliation of an index of biotic integrity based on stream fish communities. Trans. am. Fish. Soc. 113: 39–55.

Gammon, J. R., 1971. The response of fish populations in the Wabash River to heated effluents. Proc. 3rd National Symposium on Radioecology: 513–523.

Gammon, J. R., 1973. The effect of thermal inputs on the populations of fish and macroinvertebrates in the Wabash River. Technical Report No. 32. Purdue University Water Resources Research Center: 105 pp.

Gammon, J. R., 1976. The fish populations of the middle 340 km of the Middle Wabash River. Water Resources Research Center, Purdue University. Technical Report No. 86: 73 pp.

Gammon, J. R., 1980. The use of community parameters derived from electrofishing catches of river fish as indicators of environmental quality. Proceedings Water Quality Trade-offs, Washington, D.C., EPA 905/9-80-009: 335–363.

Gammon, J. R., 1983. Changes in the fish community of the Wabash River following power plant start-up: projected and observed. In Bishop, W. E., R. D. Cardwell & B. B. Heidolph (eds), Aquatic Toxicology and Hazard Assessment: Sixth Symposium. ASTM STP 802, Philadelphia, PA: 350–366.

Gammon, J. R., 1998. The Wabash River Ecosystem. Indiana University Press, Bloomington: 250 pp.

Gammon, J. R. & J. M. Reidy, 1981. The role of tributaries during an episode of low dissolved oxygen in the Wabash River. Am. Fish. Soc. Warmwater Streams Symposium: 396–407.

Gelwick, F. P. & W. J. Matthews, 1990. Temporal and spatial patterns in littoral zone fish assemblages of a reservoir (Lake Texoma, Oklahoma-Texas, U.S.A.). Envir. Biol. Fishes 27: 107–120.

Gerking, S. D., 1955. Keys to the fishes of Indiana. Inv. Ind. Lakes and Streams 4: 49–86.

Godinho, F. N., M. T. Ferreira & R. V. Cortes, 1999. The environmental basis of diet variation in pumpkinseed sunfish, Lepomis gibbosus, and largemouth bass, Micropterus salmoides, along an Iberian river basin. Envir. Biol. Fishes 50: 105–115.

Goldstein, R. M., T. P. Simon, P. A. Bailey, M. Ell, K. Schmidt & J. W. Emblom, 1994. Proposed metrics for the index of biotic integrity for the streams of the Red River of the North basin. Proc. N. Dak. Wat. Qual. Symp., March 30–31, Fargo, ND.

HydroQual, Inc., 1984. Dissolved oxygen analysis of the Wabash River. A Report for Eli Lilly and Company, Indianapolis, Indiana.

Ibarra, M. & D. J. Stewart, 1989. Longitudinal zonation of sandy beach fishes in the Napo River basin, eastern Ecuador. Copeia 1989(2): 364–381.

Jennings, M. J., J. Lyons, E. E. Emmons, G. R. Hatzenbeler, M. Bozek, T. D. Simonson, T. D. Beard, Jr. & D. Fago, 1998. Toward the development of an index of biotic integrity for inland lakes in Wisconsin. In Simon, T. P. (ed.), Assessing the Sustainability and Biological Integrity of Water Resources using Fish Communities. CRC Press, Boca Raton, FL: 541–562.

Johnson, J. H. & D. S. Dropkin, 1993. Diel variation in diet composition of a riverine fish community. Hydrobiologia 271: 149–158.

Karr, J. R., 1981. Assessment of biotic integrity using fish communities. Fisheries 6: 21–27.

Karr, J. R., L. A. Toth & D. R. Dudley, 1985. Fish communities of midwestern rivers: a history of degradation. BioScience 35: 90–95.

Karr, J. R., K. D. Fausch, P. L. Angermeier, P. R. Yant & I. J. Schlosser, 1986. Assessing Biological Integrity in Running Waters: A Method and its Rationale. Ill. Nat. Hist. Surv. Spec. Publ. 5. Champaign: 23 pp.

Miller, D. L., P. M. Leonard, R. M. Hughes, J. R. Karr, P. B. Moyle, L. H. Schrader, B. A. Thompson, R. A. Daniels, K. D. Fausch, G. A. Fitzhugh, J. R. Gammon, D. B. Halliwell, P. L. Angermeier & D. J. Orth, 1988. Regional applications of an index of biotic integrity for use in water resource management. Fisheries 13(5): 12–20.

Minns, C. K, V. W. Cairns, R. G. Randall & J. E. Moore, 1994. An index of biotic integrity (IBI) for fish assemblages in the littoral zone of Great Lakes' areas of concern. Can. J. Fish. aquat. Sci. 51: 1804–1822.

Niemela, S., E. Pearson, T. P. Simon, R. M. Goldstein & P. A. Bailey, 1998. Development of an index of biotic integrity for the species-depauperate Lake Agassiz Plain Ecoregion, North Dakota and Minnesota. In Simon, T. P. (ed.), Assessing the Sustainability and Biological Integrity of Water Resources using Fish Communities. CRC Press, Boca Raton, FL: 339–366.

Nikolskii, G. V., 1969. Theory of Fish Population Dynamics. Oliver & Boyd, Edinburgh: 323 pp.

Omernik, J. M., 1987. Ecoregions of the conterminous United States. Ann. Ass. am. Geogr. 73: 133–136.

Parke, N. J. & J. R. Gammon, 1986. An investigation of phytoplankton sedimentation in the middle Wabash River. Proceedings of the Indiana Academy of Science 95: 279–288.

Pflieger, W. L., 1971. A Distributional Study of Missouri Fishes. Occ. Pap. Mus. Nat. Hist. Univ. Kansas, 20: 225–570.

Pusey, B. J., A. H. Arthington & M. G. Read, 1995. Species richness and spatial variation in fish assemblage structure in two rivers of the West Tropics of northern Queensland, Australia. Envir. Biol. Fish 42: 181–199.

Reash, R. J., 1998. Considerations for characterizing midwestern large-river habitats. In Simon, T. P. (ed.), Assessing the Sustainability and Biological Integrity of Water Resources using Fish Communities. CRC Press, Boca Raton, FL: 463–473.

Rogellin, E. M., 1979. The role of seining in the analysis of the middle Wabash River. M.A. thesis, DePauw University, Greencastle, Indiana.

Simon, T. P., 1992. Development of biological criteria for large rivers with emphasis on an assessment of the White River drainage, Indiana. U.S. Environ. Protect. Agency, Chicago, IL. EPA905/R-92-006.

Simon, T. P., 1998. Modification of an index of biotic integrity and development of reference condition expectations for dunal palustrine wetland fish communities along the southern shore of Lake Michigan. Aquat. Ecosys. Health & Manage. 1: 57–70.

Simon, T. P. & E. B. Emery, 1995. Modification and assessment of an index of biotic integrity to quantify water resource quality in Great Rivers. Reg. Rivers: Res. Manage. 11: 283–298.

Simon, T. P. & J. Lyons, 1995. Application of the index of biotic integrity to evaluate water resources integrity in freshwater ecosystems. In Davis, W. S. & T. P. Simon (eds), Biological Assessment and Criteria: Tools for Water Resource Planning and Decision Making. Lewis Press, Boca Raton, FL: 245–262.

Simon, T. P. & P. M. Stewart, 1998. Application of an index of biotic integrity for dunal, palustrine wetlands: emphasis on assessment of nonpoint source landfill effects on the Grand Calumet Lagoons. Aquat. Ecosys. Health & Manage 1: 71–82.

Simon, T. P. & R. E. Sanders, 1998. Applying an index of biotic integrity based on Great River fish communities: considerations in sampling and interpretation. In Simon, T. P. (ed.), Assessing the Sustainability and Biological Integrity of Water Resources using Fish Communities. CRC Press, Boca Raton, FL: 475–505.

Smith, P. W., 1979. The Fishes of Illinois University of Illinois Press, Champaign: 314 pp.

Statistical Analysis Systems (SAS), 1985. SAS User's Guide: Statistics, Version 5. SAS Institute, Cary, NC: 956 pp.

Tejerino-Garro, F. L., R. Fortin & M. A. Rodriques, 1998. Fish community structure in relation to environmental variation in floodplain lakes of the Araguaia River, Amazon Basin. Envir. Biol. Fishes 51: 399–410.

Thoma, R. F., 1998. Biological monitoring and an index of biotic integrity for Lake Erie's nearshore waters. In Simon, T. P. (ed.), Assessing the Sustainability and Biological Integrity of Water Resources using Fish Communities. CRC Press, Boca Raton, FL: 417–461.

Trautman, M. B., 1981. The Fishes of Ohio. The Ohio State University Press, Columbus: 782 pp.

TriMetrics, Inc., 1994. Axum User's Guide. TriMetrics, Ind., Seattle, WA 98115: 326 pp.

Turner, A. M. & G. G. Mittlebach, 1990. Predator avoidance and community structure: interactions among piscivores, planktivores, and plankton. Ecology 71: 2241–2254.

Vannote, R. L., G. W. Minshall, K. W. Cummins, J. R. Sedell & C. E. Cushing, 1980. The river continuum concept. Can. J. Fish aquat. Sci. 37: 130–137.

Zalewski, M., P. Frankiewicz, M. Pryzybylski, J. Banbura & M. Nowak, 1990. Structure and dynamics of fish communities in temperate rivers in relation to the abiotic-biotic regulatory continuum concept. Polskie Archwum Hydrobiologii 37: 151–176.

Hydrobiologia **422/423:** 305–317, 2000.
M. Jungwirth, S. Muhar & S. Schmutz (eds), Assessing the Ecological Integrity of Running Waters.

Effects of residual flow and habitat fragmentation on distribution and movement of bullhead (*Cottus gobio* L.) in an alpine stream

Sabine Fischer[1] & Helmut Kummer[2]
[1]*Biological Station Lunz, Austrian Academy of Sciences, Lunz/See, Austria*
[2]*Department of Hydrobiology, Fisheries and Aquaculture, University of Agricultural Sciences, Vienna, Austria*

Key words: abundance, multiple capture-recapture method, intermittent flow

Abstract

The present study investigates the population structure and dynamics of *Cottus gobio* in three adjoining, hydrologically very different stream sections that are affected by water abstraction (residual flow). During the one-year investigation from October 1997 to September 1998 the selected stretches were repeatedly sampled by electrofishing. Multiple mark-recapture method was applied to determine fish density along the individual stretches and the individual marking technique (VI-tag) was used to record the migratory activity of individual fish. Stretches with intermittent flow (section A) are temporarily colonised by adult *Cottus gobio* from downstream sites only at higher water levels. These stretches are completely dry for about half of the year and do not enable populations to become established. Small populations persist in those transitional stretches that are intermittently isolated from other surface waters (section B). However, population size is less than one-quarter the size of those in directly adjoining downstream stretches with permanent water flow (section C). In addition, bullheads in section B have considerably poorer conditions. Changes in discharge led to increased migratory activity over the entire study area, whereby the radius of activity of the bullhead was largely (88%) restricted to distances of < 150 m. Elevated discharges mostly triggered upstream longitudinal and lateral dispersion. When the water drops to low water level and large areas in these sections dry out, the bullheads have to withdraw into deep pools. This might be the main reason for the lack of the probably less mobile juvenile stages. The results clearly demonstrate that habitat fragmentation resulting from water abstraction has drastic effects on all life stages of *Cottus gobio* in respect to habitat availability, spatial distribution, movement, fitness and survival. These effects can be seen at an extremely fine scale and are revealed in the significant differences in abundance and condition of *Cottus gobio* between the short (94 m–188 m) adjacent sections. In order to maintain and restore ecological integrity of running waters, the preservation of intact flow conditions is one of the key issues for river management.

Introduction

The intensive exploitation of natural water resources over the last century has drastically changed environmental conditions for aquatic organisms. One of the main impacts in freshwater systems are hydroelectric power plants; they affect running water ecosystems in particular by changing the flow regime and by fragmenting the longitudinal corridor through dam construction, reservoir operation and interbasin diversion (Dynesius & Nilson, 1994; Jungwirth, 1998).

Longitudinal connectivity becomes a central issue when focusing on fish migration. The term 'migration' is often applied to adaptive, long-distance movements that occur predictably in the life cycle of a species (Smith, 1985). The term 'movement' or 'restricted movement' describes habitat shifts within the normally used areas, i.e. within territories and home ranges (Gerking, 1958; Peter, 1998). Negative impacts of artificial barriers on migratory opportunities are well described (Peter, 1998; Waidbacher & Haidvogl, 1998; Williams, 1998), but fragmentation by de-watering of residual flow stretches can also have dramatic effects on local fish fauna (Peter, 1998). Residency of stream fish populations includes migratory flexibility in life history patterns dependent on ontogenetic stage, competition, season, etc. as well as on the ability to react appropriately to environmental changes. Therefore, resident fish must undergo restricted movements and depend on barrier-free streams to maintain their viability. For various small-sized fish species with low jumping or swimming ability (like the bullhead *Cottus gobio*), reductions of local populations as an effect of migratory barriers have been documented (Barandun, 1990; Bless, 1990; Roth & Utzinger, 1993; Stahlberg-Meinhardt, 1994).

The effects of reduced flow caused by water abstraction on the aquatic fauna have been demonstrated for a wide range of different types and sizes of running waters (Jungwirth et al., 1990; Baran et al., 1995; Moog & Jungwirth, 1993). Especially in the Austrian headwaters or in their tributaries, numerous abstractions are already authorised and little water remains for the natural environment; in some cases no permanent surface flow is guaranteed. The minimum flow required to avoid adverse damage to the ecosystem is, therefore, one of the key issues for river management allowing for ecological aspects (Sale & Otto, 1991; Moog & Jungwirth, 1993; Maile et al., 1995).

Our current knowledge on the biology of the bullhead (*Cottus gobio* L.) stems largely from comparatively near-natural running waters (Smyly, 1957; Mills & Mann, 1983; Marconato & Bisazza, 1988; Gaudin & Caillere, 1990; Waterstraat, 1992). Increasingly, however, this species is being recognized as an important indicator of anthropogenic impacts. Barandun (1990), Bless (1990), Roth & Utzinger (1993) and Stahlberg-Meinhardt (1994), focused on the effects of river engineering measures and migratory barriers on bullhead populations. Moreover, Roth & Utzinger (1993) and Hofer & Bucher (1996) discuss the usefulness of the bullhead as an indicator for the chemical status of running waters.

Based on the hypothesis that hydrology and the longitudinal continuum have a major influence on the population structure and dynamics of *Cottus gobio*, the present study investigates three adjoining, hydrologically very different stretches of a stream that is utilized to produce hydropower (residual waters).

Study site

The Oberer Lunzer Seebach (OLS) is a submontane alpine stream of stream order 2 with a nivo/pluvial discharge regime (mean annual flow: 1.10 m^3 s^{-1}). The OLS extends from the Lunzer Obersee (1116 m a.s.l.) to the Lunzer Untersee (608 m a.s.l.) and drains a catchment approximately 20 km^2 in the northeast

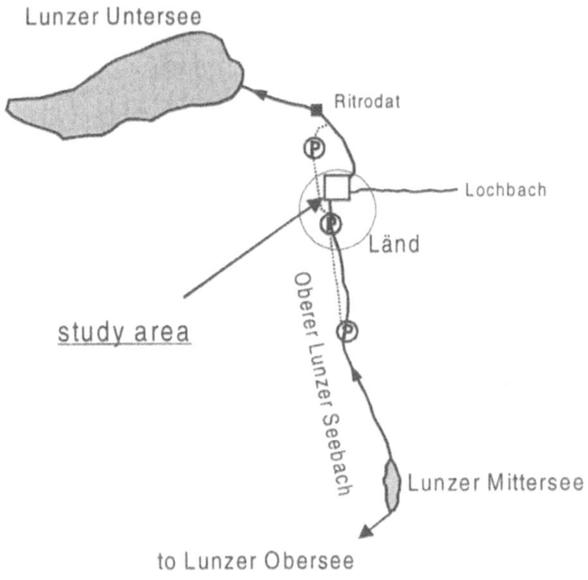

Figure 1. 'Oberer Lunzer Seebach' (OLS) with the study area (□), hydropower stations (P) and water diversion (....).

Limestone Alps. As is typical for karstic regions long stretches are subterranean. A total of three small power stations are located between the Lunzer Mittersee and the Lunzer Untersee and virtually the entire stream is a residual water stretch (Figure 1).

The study stretches are located at the lower end of a valley widening (the so-called 'Länd') in which the stream periodically shows intermittent flow. The OLS is characterised by high natural discharge fluctuations. The extremely low natural discharges at low water (mean annual low flow: 0.31 m^3 s^{-1}) are reduced even further by hydropower production. The present studies were conducted in three hydrologically different river sections, in each of which two stretches were selected (Figure 2). According to the data provided by the power station operators, the residual water volume ranged between 0.05 m^3 s^{-1} and 5 m^3 s^{-1} during our surveys.

Section A – temporarily completely dry

Stretch A1 (100 m^2) and stretch A2 (120 m^2), which are at times entirely desiccated, are located at the lower end of the 'Länd' valley widening. The river bed dries out on the surface several times a year. During these times, the OLS flows several hundred meters solely in the gravel bed. Stretch A1 is a 27 m-long and about 80 cm-deep pool, stretch A2 is a 37 m-long

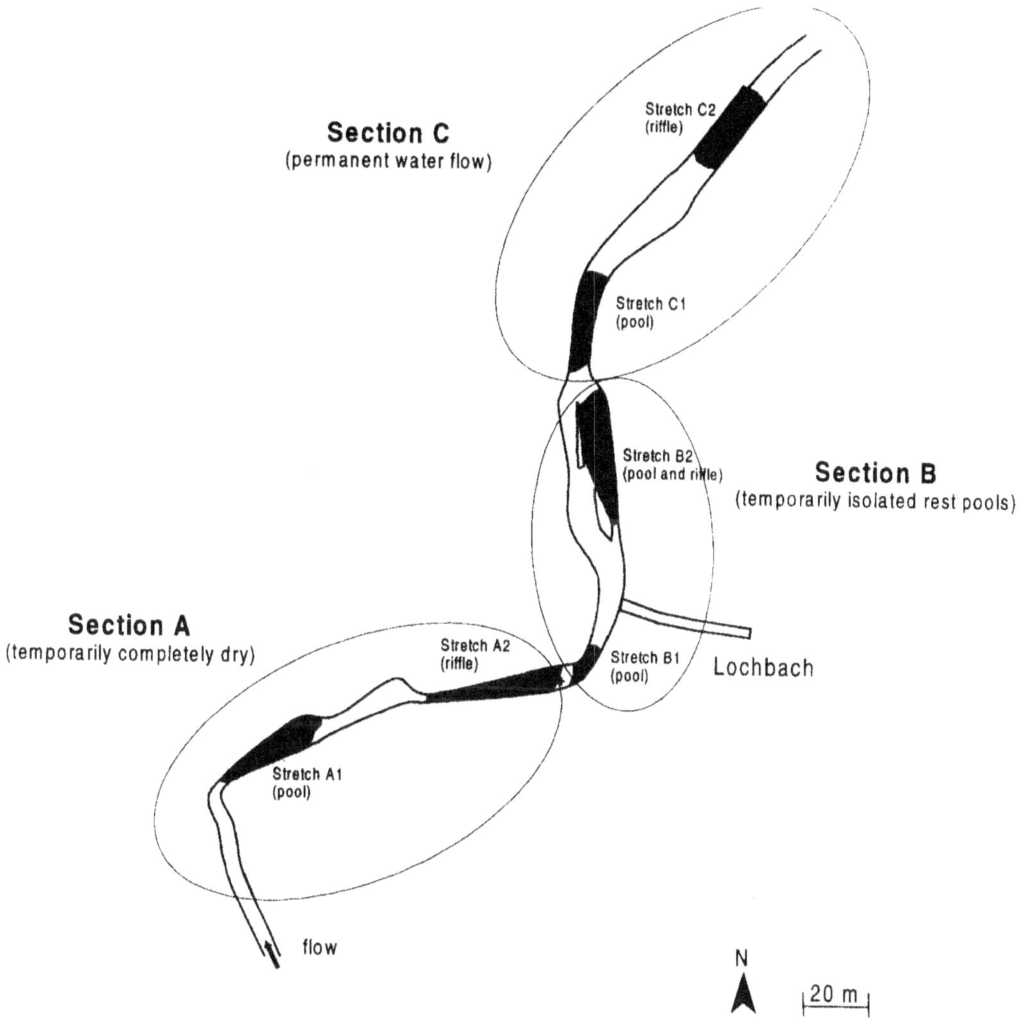

Figure 2. Study area with three different sections (shaded areas) each including two stretches (black areas).

riffle averaging approximately 15 cm in depth at mean water level.

Section B – temporarily isolated pools

This section largely dries out at low water. The surface flow continuum is then interrupted, although isolated pools that are sufficiently deep remain. Stretch B1 is a 13 m-long and about 100 cm-deep pool (52 m^2). Stretch B2 cannot be classified precisely: it is a pool/riffle sequence with a small, 50 cm-deep, upstream pool. At mean water level the average depth of the 37 m-long and 4 m-wide stretch (148 m^2 surface area) is 35 cm (50–20 cm).

Section C – permanent water flow

Stretch C1 (104 m^2) and stretch C2 (88 m^2) bear water year-round and, at least in 1997 and 1998, the surface flow continuum remained uninterrupted downstream to the Lunzer Untersee (Figure 1). Stretch C1 is a 26 m-long and 70 cm-deep pool at mean water level and stretch C2 is a 25 m-long riffle with an average depth of 20 cm at mean water level.

The individual sections are separated from one another by short, cataract-like riffle-pool transitions.

In order to determine the radius of movement of the tagged bullhead in sections A–C, additional 11 areas (A3, B3, B4, C3, C4, C5 in Figure 3 and D1, D2, D3, D4, D5 in Table 2) along a total stream length of 1150 m were sampled 2–3 times each between May and August 1998. The location of the additional sampling areas adjoining sections A, B, C is shown in Figure 3

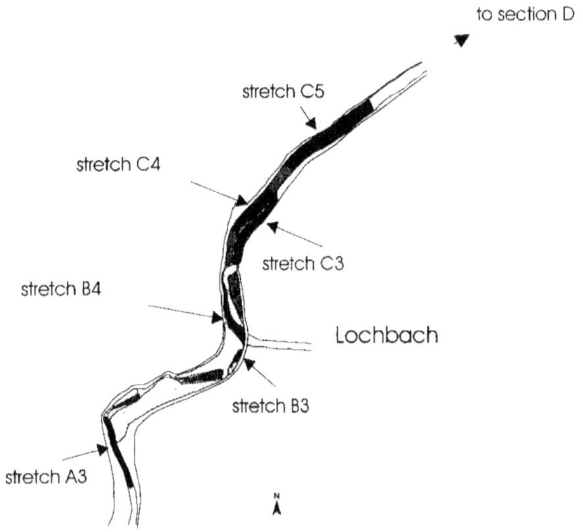

Figure 3. Additional sampling sections in study area (black), situation in March at low water. Dark shaded areas = stretches of sections A, B, C; brightly shaded areas = dry gravel bars.

Table 1. Additional sample-areas downstream of section C; distance and size

Stretch	Distance from C5	Length (m)	Area m^2
D1	110	70	273
D2	270	100	850
D3	370	95	827
D4	465	153	1377
D5	618	150	1350

and describes the situation in March, with discharge being near to low water. The 50 m-long stretch (A3) directly upstream of A1 and the 17 m-long stretch (B3) downstream of B1 are riffle areas which temporarily completely dry out at low water level. Stretch B4 of section B is a temporarily isolated 30 m-long and about 50 cm-deep run downstream of the mouth of the 'Lochbach', which only bears water after heavy rainfall or snowmelt. The additional stretches located in section C (C3: 75 m long; C4: 47 m long; C5: 90 m long) are riffle areas with permanent surface flow.

Sampling was also conducted in five stretches further downstream (section D). These stretches include riffle and pool sequences and their sizes vary from 273 m^2 to 1377 m^2 (Table 1).

Methods

During the one-year investigation from October 1997 to September 1998 the selected stretches were repeatedly sampled by electrofishing with a DC gas-powered backpack unit (1.5 kW). The multiple capture-recapture method was applied to determine fish density along the individual stretches. At each capture period, the individual stretches were intensively sampled several times in March, October, May/June and August (Table 2). The stretches B1, B2, C1 and C2 were fished a total of 16–18 times (Table 2). The temporarily water-bearing stretches (A1 and A2) were entirely desiccated in October and largely dried out in March, with very shallow pools that were not sufficiently deep for bullhead. Sampling of these stretches was, therefore, conducted only once in March and 5 times each in the two following periods (Table 2). The duration of the fishing runs along each stretch were recorded in minutes and the number of captures per unit time (10 min) or per surface area (100 m^2) was calculated. 2–5-day intervals were allowed between the individual runs in order to minimise abnormal, fishing-related behaviour.

The captured bullhead were measured to the nearest mm (total length), weighed to the nearest gram, tagged and subsequently released in the same stretch. The individual marking of the bullhead (Visible Implant Tag; Northwest Technology Inc., Seattle, WA) enabled the migratory activity of individual fish to be recorded.

Sampling of the 11 additional stretches specified in Table 1 and Figure 3 was conducted 2 or 3 times each between May and August 1998. Each single fish was measured to the nearest mm (total length) and checked for VI-tags, yielding information in respect to movement and length-frequency.

Site fidelity is calculated as the percentage of individuals recaptured in the same area compared to total recapture and analysed for single sampling periods (fidelity within a period), for different periods (fidelity between periods) and for all sampling runs (total fidelity). Site fidelity of individuals captured more than two times refers solely to the site at recapture and the one at the capture directly before recapture.

Population sizes were calculated according to Seber (1965), whereby only marked bullhead from a single sampling period were considered.

$$N = \frac{(m+1)(c+1)}{(r+1)} - 1,$$

Table 2. Sampling periods and number of runs in each stretch (str)

Sampling period		Str. 1	Str. 2	Str. 3	Str. 4	Str. 5	Str. 6
October	14.10.–28.10.1997	0	0	4	3	5	4
March	25.03.–30.03.1998	1	1	3	3	3	3
May/June	25.05.–12.06.1998	5	5	5	5	5	5
August	10.08.–21.08.1998	5	5	5	5	5	5

where

- N = calculated population per stretch;
- m = number of marked fish per stretch;
- c = captures per stretch and sampling run;
- r = recapture of tagged bullhead per stretch and run.

The increasing proportion of tagged individuals with each run improves the accuracy of the calculation. As the time interval between the first and last run was a maximum of 15 days, mortality was omitted from the calculations. Observed migrations into and out of the sampling areas, however, were considered (see Figure 8).

In order to represent the nutritional status of the bullhead in the individual stretches and at the individual fishing dates, the condition factor (K') according to Fulton (in Bagenal, 1978) was calculated,

$$K' = (100 * \text{weight(in g)})/(\text{length(in cm)})^b$$

whereby the exponent (b) is determined from the regression of the logarithmically transformed length/weight pairs; log (weight) = log a + b* log (length). The average values of the condition factors along the individual stretches were tested for significant differences by a one-way analysis of variance (ANOVA).

In order to register water flow, rod gages were positioned at all six stretches in late October 1997 and read at least on every sampling date. Data from an automatic gauge station (GANSER water-stage recorder) installed in the RITRODAT-study area of the Biological Station Lunz, which is 1.5 km further downstream, were also available and allow correlation and back calculation of water level in sampling areas.

Results

The discharge regime during the investigation period showed large fluctuations typical for alpine streams in karstic catchment (Figure 4). The stretches with temporary surface waters (1 and 2) fall dry below a water level of 135 cm measured at the automatic gauge station (Figure 4).

This occurred over a total of 78 days between October 1997 and September 1998, with dry periods varying in duration between several days and one month (Table 3).

Site fidelity and migration

A total of 398 bullheads were captured and tagged between October 1997 and September 1998 along the six stretches. 194 of these individuals (48.7%) were recaptured at least once. During the entire investigation period, the proportion of individually marked bullheads recaptured within the same stretch was higher in the temporarily isolated section (stretch 3: 60%; stretch 4: 66.7%) than in the sections with permanent flow (stretch 5: 47.8%; stretch 6: 46.2%). In the stretches that occasionally fell dry, only 6 bullheads could be initially captured and tagged. Three (into stretch 1) and four bullheads (into stretch 2) temporarily immigrated from downstream sections. Within the maximum 15-day sampling periods, only a single bullhead that had immigrated into this section was recorded twice (27.5.98 and 30.5.98). The site fidelity within a sampling period at the downstream stretches (B1, B2, C1 and C2) was on the average very high (76.8%; Table 4).

The migratory activity within a survey period increased with increasing water level (Figure 5). A potential relationship between water level and total migration rate (= number of individuals captured in two different areas * 100 / number of individuals captured at least 2 times within the sampling period), upstream migration and downstream migration was tested with Spearman's Correlation Coefficient (r_s). This revealed a correlation between water level and total migration (r_s= 0.671; $p < 0.05$), but not between

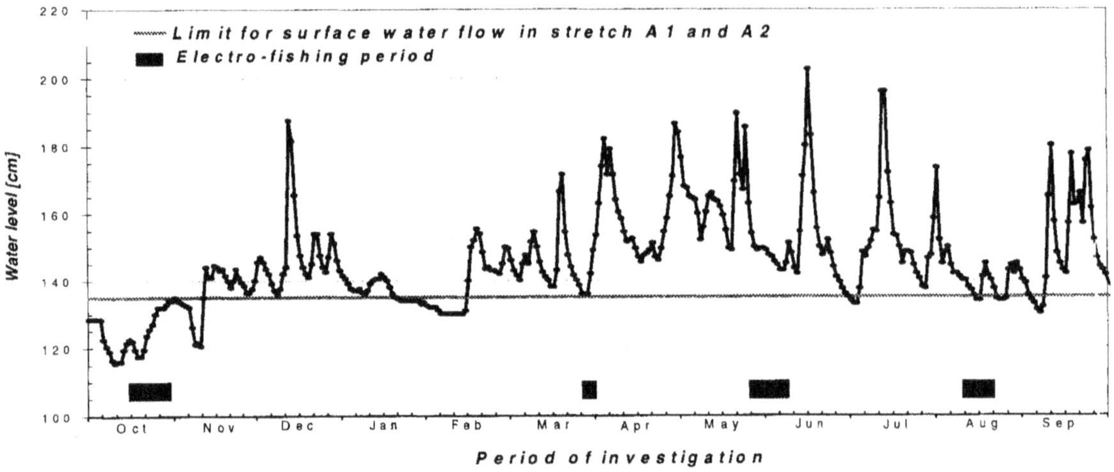

Figure 4. Water level from Oct. 1997 to Sept. 1998, read at gauge station. Sampling periods are represented as bars on the time axis.

Table 3. Duration of dry river bed at study section A (days in each month)

Oct.	Nov.	Dec.	Jan.	Feb.	March	April	May	June	July	Aug.	Sep.
30	9	0	13	12	0	0	0	3	0	8	3

water level and upstream migration ($r_S = 0.429$; $p > 0.2$) or downstream migration ($r_S = 0.265$; $p > 0.5$).

Between May and August 1998 in total 1357 individuals were captured in all 17 stretches (Figure 6). 149 (11%) of these bullheads were tagged individuals and therefore originated from stretches A1+2, B1+2 and C1+2 (Figure 6). As the distribution of the recaptures shows, *Cottus gobio* does not migrate extensively (Figure 6). Most of the tagged fish (75%) were recaptured in the tagging stretches, 13.4% in stretches (B3+4 and C3+4) which are located parallel to the tagging areas and 10.1% in the about 90 m-long stretch (C5) located directly downstream of C2. Although 792 bullheads (=58.4%) were captured in the downstream section D, only 1 fish was recaptured (Figure 6). Above stretch A2, two sampling efforts failed to yield a single bullhead in stretch A3.

Of the 149 recaptured bullhead, 56 individuals (37.6%) were caught in another stretch. The radius of activity of the vast majority (87.5%) of the migrated bullhead was restricted to 150 m (Figure 7). The maximum recorded downstream migration was 330 m (from stretch A1 to C5); the longest upstream migration measured about 250 m (from stretch D1 to C2).

Population survey

On the one hand, population surveys based on 3–5 fishing runs within one sampling period guarantees accurate population size estimates. On the other hand, tagged bullhead migrated into and out of the stretches. The proportion of tagged individuals that remained in the stretches (Figure 8) shows a highly significant negative correlation to the interval between capture and recapture ($r_S = 0.964$, $p < 0.001$). The increasing number of tagged bullhead (M) with each run was therefore corrected by using the regression equation (M^*) indicated in Figure 8:

$$M^* = -19.975\mathrm{Ln}(M) + 104.36.$$

No population estimates could be made in section A, which repeatedly fell dry. Only 10 individuals each were captured in both stretches of this section and solely on the May/June date at higher water levels.

The population density in the section B whose surface water was occasionally isolated is significantly lower than in the permanently water-bearing section C (Figure 9). Multiple comparisons (Bonferroni Post Hoc test of ANOVA) revealed highly significant differences ($p < 0.001$) between stretch C1 and the two stretches in section B, a significant difference ($p < 0.01$) between the two riffle stretches (C2 and B2)

Table 4. Number of captured and recaptured bullhead in the individual stretches

	Capture	Recapture within period		Recapture between period		Total recapture	
	ind.	%	fidelity %	%	fidelity %	%	fidelity %
Stretch 1	6	0.0	0.0	0.0	0.0	0.0	0.0
Stretch 2	6	0.0	0.0	0.0	0.0	0.0	0.0
Stretch 3	31	38.7	69.2	29.0	60.0	64.5	60.0
Stretch 4	51	49.0	81.8	15.7	66.7	58.8	66.7
Stretch 5	187	23.5	75.4	28.9	47.6	49.2	47.8
Stretch 6	117	26.5	77.1	19.7	29.6	44.4	46.2
Total	**398**	**28.1**	**76.8**	**23.6**	**46.5**	**48.7**	**51.6**

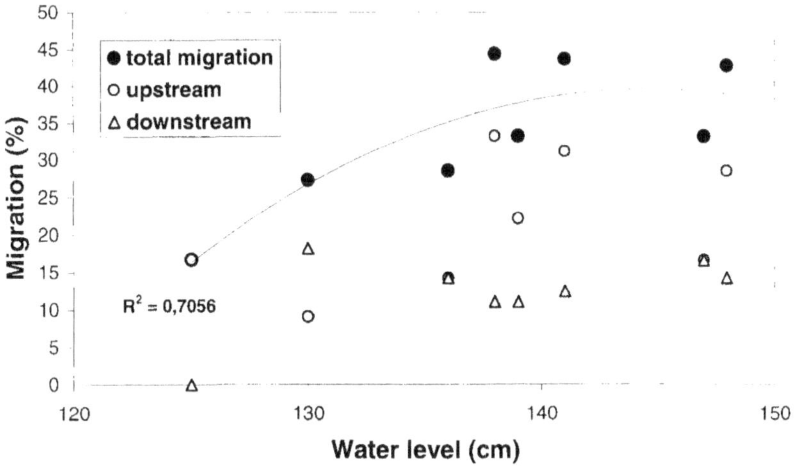

Figure 5. Migration rate of the recaptured bullhead within one sampling period in relation to water level (R^2 = coefficient of determination).

and a significant difference ($p < 0.05$) between C2 and B1 (Table 5). A comparison of the population density within a section (Bonferroni Post Hoc test of ANOVA) revealed no significant difference between the stretches (Table 5).

Catch per unit effort data of the different stretches (per 100 m^2) support the results of population estimates using multiple mark-and-recapture-method. The average of total capture per time or area is lower in temporarily isolated sections (stretches 3 and 4) than in permanent water-bearing sections (stretches 5 and 6); however, there is high variability in catch between the different sampling periods (Table 6).

Nutritional status

The temporarily isolated stretches (3 and 4) not only showed distinctly lower population densities than the permanently water-bearing sections (5 and 6), but the nutritional status (condition factor) of the fish captured in 3 and 4 was on average significantly (ANOVA, $p < 0.001$) poorer (Figure 10).

Discussion

The population density calculated for the residual stretches in the OLS ranges from 17 ind./100 m^2 (stretch 4) to 94 ind./100 m^2 (stretch 5) and reflects the numbers reported in the literature (4 to 260 ind./100 m^2) (Orsag & Zelinka, 1974; Welton et al., 1983; Daniels, 1987; Waterstraat, 1992; Stahlberg-Meinhardt, 1994). Based on a literature review, Stahlberg-Meinhardt (1994) specifies the average density of *Cottus gobio* in suitable habitats of small streams in Europe to be in the range of 50–100 ind./100 m^2 (ind. > 50 mm total length). These

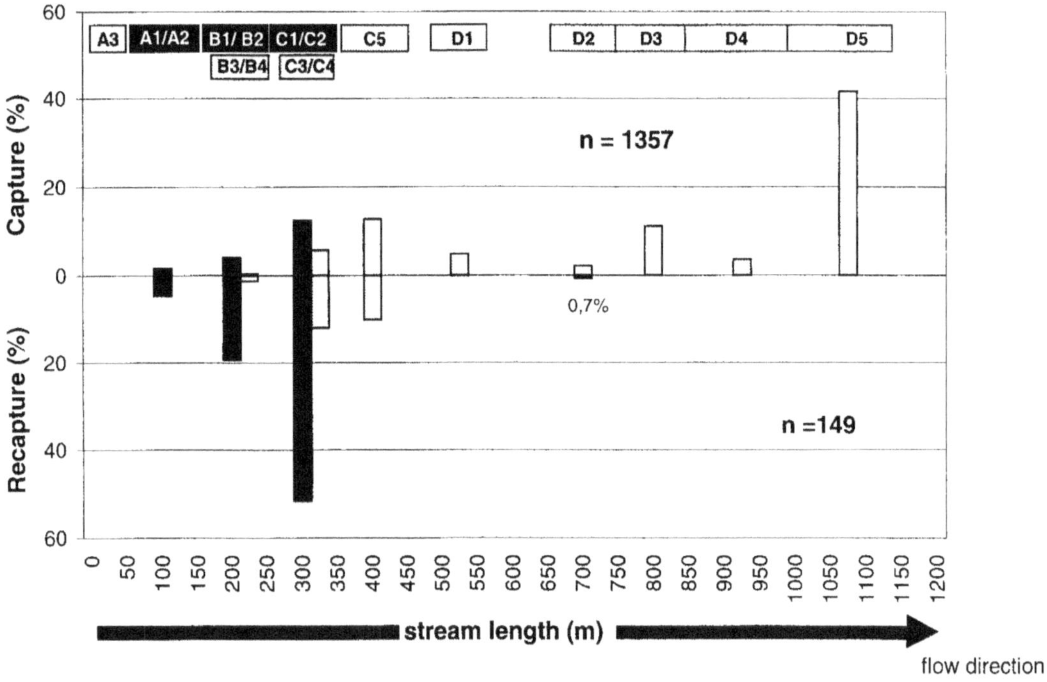

Figure 6. Proportion of captured and recaptured bullhead in the individual sections between May and August 1998. Bullhead were tagged in sections A, B and C.

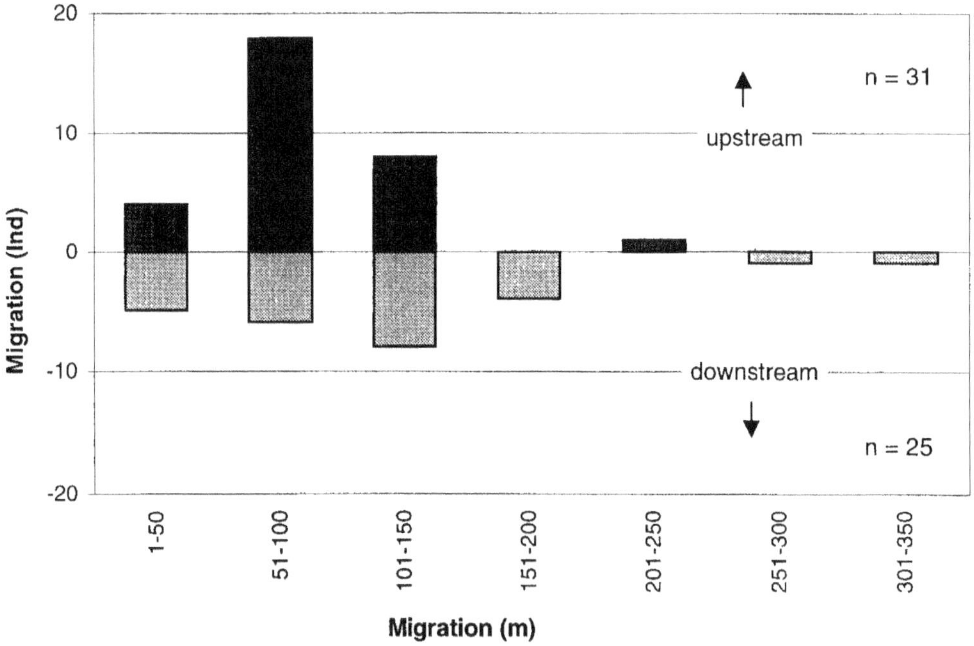

Figure 7. Number and distance of the tagged bullhead that migrated either up- or downstream from May to August 1998.

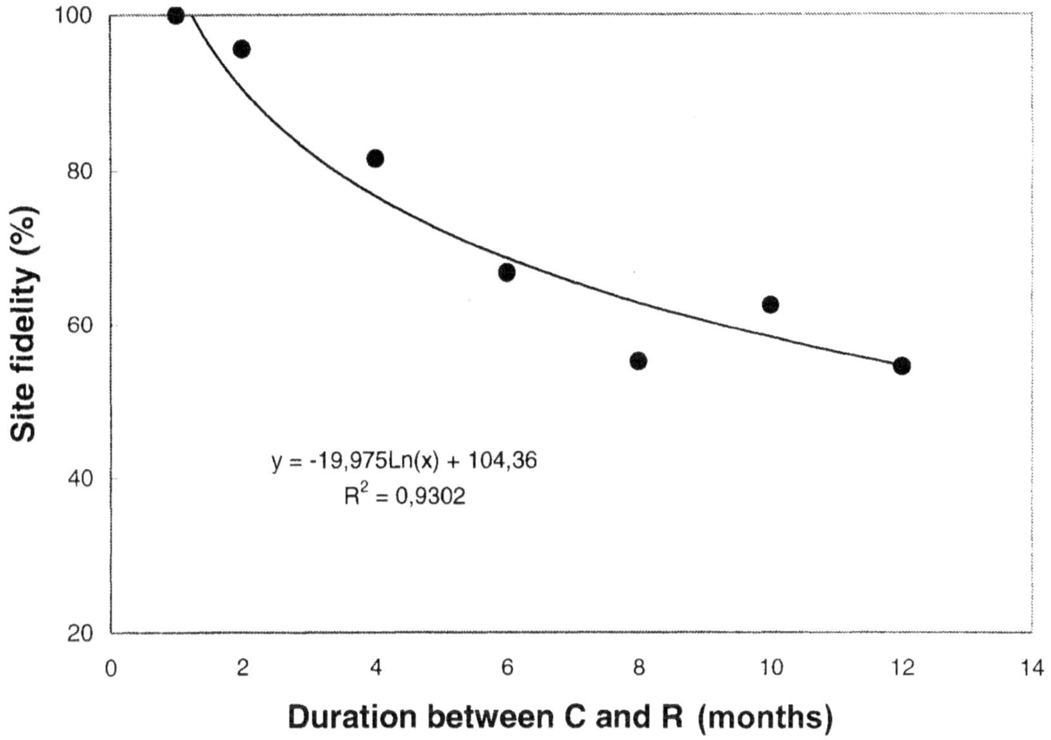

Figure 8. Correlation between interval from capture (C) and recapture (R) and site fidelity of tagged bullhead. Immigrated individuals were not treated as recaptures.

values generally ignore or under-represent juvenile stages. Mills & Mann (1983), who included one-year-old bullhead, reported densities of up to 2100 ind./100 m^2, while Andreasson (1971) reported between 1200 and 2100 ind./100 m^2. The highest recorded densities – 7500 ind./100 m^2 – stem from a karstic river in England (Mills & Mann, 1987).

Those stretches that only temporarily bore water were inhabited exclusively by adult *Cottus gobio* from downstream sites and then only briefly and at high water levels. According to the lowest water level at which individuals were captured in this section, *Cottus gobio* could theoretically use this area on 169 days (46%) in the year 1998. However, the short-term increase of water level within the sampling periods in March and August indicates that adequate discharge for at least 1 week is necessary to trigger immigration into this section from downstream. The correlation between discharge and migratory activity has been described by a number of authors (e.g. McCleave, 1964; Korolev, 1991), whereby elevated discharges mostly trigger upstream longitudinal migrations. At any rate, the brief periods with sufficient flow prevented a population from establishing itself. When the water drops, most of the bullhead again migrate downstream.

Small populations can establish themselves in those transitional stretches that are intermittently isolated from other surface waters (section B); however, these populations are less than one-quarter the size of those in directly adjoining downstream stretches with permanent water flow.

Based on the site fidelity of many bullhead in the intermittently isolated areas and in those areas with permanent water flow over the entire investigation period (46–66.7%), one can assume that the fish also reproduce there. Nonetheless, no juvenile stages were recorded. Comparing the lengths of those bullhead caught in August with those from sites further downstream indicates that the absence of juvenile stages is definitely not attributable solely to methodological difficulties in surveying small bullhead by electrofishing. The investigated stretches clearly did not contain naturally structured populations (Figure 11).

Freeze-Core samples from stretches further downstream show that the interstitial space of the gravel bed is extensively inhabited by juvenile stages (up to 63 mm body length) (Adamicka, 1987; Fischer, 1998). All of the bullheads caught in the study area (97%) were larger than the individuals captured in the cores. It can therefore be excluded that the bullheads recor-

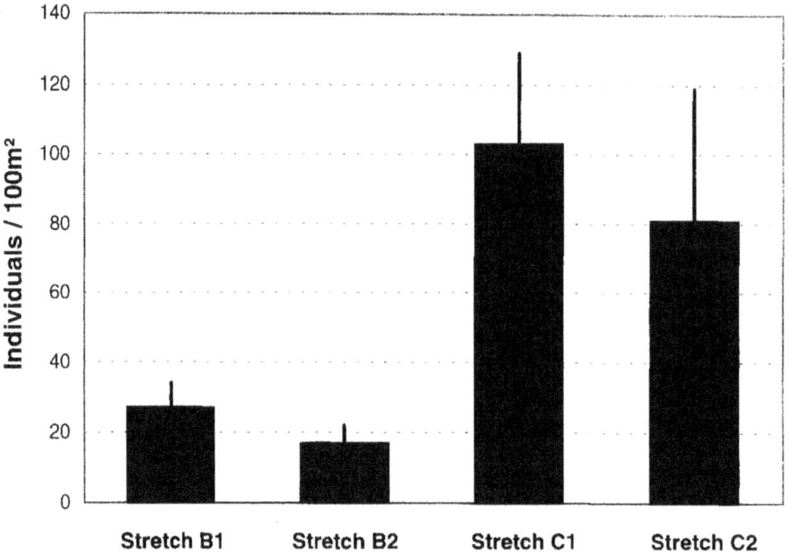

Figure 9. Average bullhead population (95% Confidence) in the individual stretches.

ded in the study area migrate into the interstitial at low water.

Not only do the transitional stretches (B1 and B2) exhibit considerably smaller populations than the stretches (C1 and C) that are permanently water bearing and are connected to downstream sites by surface waters, but the bullheads also have a significantly lower nutritional status. These sections apparently have poorer food resources because the approximately 500-m-long river section above the transitional stretches repeatedly dries out in summer (during the growth phase of *Cottus gobio*). If not completely dry the wetted surface at least shrinks to isolated small pools. Extensive investigations would be necessary to shed more light on this topic.

Changes in the discharge led to increased migratory activity over the entire study area, whereby the radius of activity of the bullheads was generally restricted to distances of <150 m. Elevated discharges mostly trigger upstream longitudinal migrations, but are also responsible for lateral dispersion in which the fish take advantage of the more extensively wetted river bed. At lower water levels the bullheads retreat into the deep pools. The correlation between discharge and migratory activity has been described by a number of authors (McCleave, 1964; Korolev, 1991; Stahlberg-Meinhardt, 1994). Furthermore, the study by McCleave (1964) also revealed a distinctly higher proportion of upstream *versus* downstream migrations (by 23%). According to the interpretation of Stahlberg-Meinhart (1994), this upstream movement is a function of dispersion dynamics. Bless (1990,

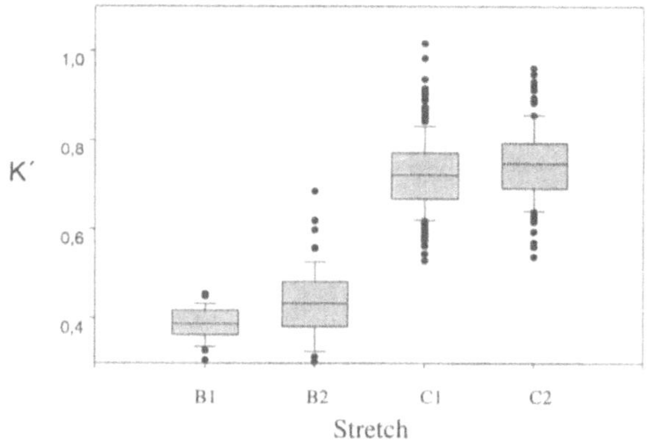

Figure 10. Condition factor (K') of the bullheads in the different stretches (95% Confidence).

1997) provides an additional explanation based on the compensation of losses due to drift of juveniles. In the investigated stretches, the distinctly higher upstream movement was not correlated with a conspicuous population increase: in addition to mortality, emigration and drift upstream movement probably also regulates population size.

In general *Cottus gobio* turns out to be a species with remarkable adaptive radiation in order to survive in a wide range of environmental conditions. As a characteristic species of the salmonid region, bullheads depend on migration opportunities within the running water systems. Sectional isolation of bullhead populations or subpopulations due to cross structures is well described in several studies (Barandun, 1990; Bless, 1990; Roth & Utzinger, 1993; Stahlberg-

Table 5. Bonferroni Post Hoc test of ANOVA, probability of error (p) not significant (p >0.05), significant (p <0.05), very significant (p <0.01), highly significant (p <0.001)

Multiple comparisons						
					95% Confidence interval	
(I) STRETCH	(J) STRETCH	Mean difference (I–J)	Std. error	Sig.	Lower bound	Upper bound
B1	B2	10.0394	16.121	1	−35.8559	55.9348
	C1	−75.8685*	16.121	0.000	−121.7638	−29.9732
	C2	−54.5774*	16.687	0.018	−102.0836	−7.0712
B2	B1	−10.0394	16.121	1	−55.9348	35.8559
	C1	−85.9080*	16.121	0.000	−131.8033	−40.0126
	C2	−64.6169*	16.687	0.004	−112.1230	−17.1107
C1	B1	75.8685*	16.121	0.000	29.9732	121.7638
	B2	85.9080*	16.121	0.000	40.0126	131.8033
	C2	21.2911	16.687	1	−26.2151	68.7973
C2	B1	54.5774*	16.687	0.018	7.0712	102.0836
	B2	64.6169*	16.687	0.004	17.1107	112.1230
	C1	−21.2911	16.687	1	−68.7973	26.2151

*The mean difference is significant at the 0.05 level.

Table 6. Catch per unit effort data (per 10 min and per 100 m^2) – Results in different stretches and sampling periods

	Stretch 1		Stretch 2		Stretch 3		Stretch 4		Stretch 5		Stretch 6	
	100 m^2	10 min	100 m^2	10 min	100 m^2	10 min	100 m^2	10 min	10 min	100 m^2	10 min	100 m^2
October	0	0	0	0	9.14	2.51	3.38	3.03	21.15	8.38	15.62	6.77
March	0	0	0	0	2.89	0.95	1.01	0.79	35.60	10.83	6.29	2.75
May/June	1.25	0.96	1.39	1.67	6.92	3.47	4.46	4.60	6.15	3.63	4.57	3.22
August	0	0	0	0	8.65	5.33	9.05	7.54	25.77	11.09	19.66	9.41
Total	0.50	0.38	0.57	0.68	7.44	3.37	5.32	4.76	19.62	8.03	12.73	6.21
Total (n)	10		10		58		120		310		160	

Meinhardt, 1994). Habitat analyses indicate that the minimum depth of patches selected by bullheads is 10 cm (Roussel & Bardonnet, 1996; Fischer, 1998). Residual water stretches with very shallow or completely dry pool/riffle transitions allow only very restricted movements. Therefore the stretches with temporary flow are colonised by adult *Cottus gobio* from downstream sites only temporarily and at higher water levels. The short duration of water flow does not enable populations to become established.

'Obere Lunzer Seebach' displays a very flashy hydrograph because of its karstic catchment. The 5-year average of spate occurrence is 15.9 $m^3\ s^{-1}$ (Bretschko & Moog, 1990). More than 50% of these spates occur in the summer months during and after rain storms, and show very high rates of discharge increase (max.

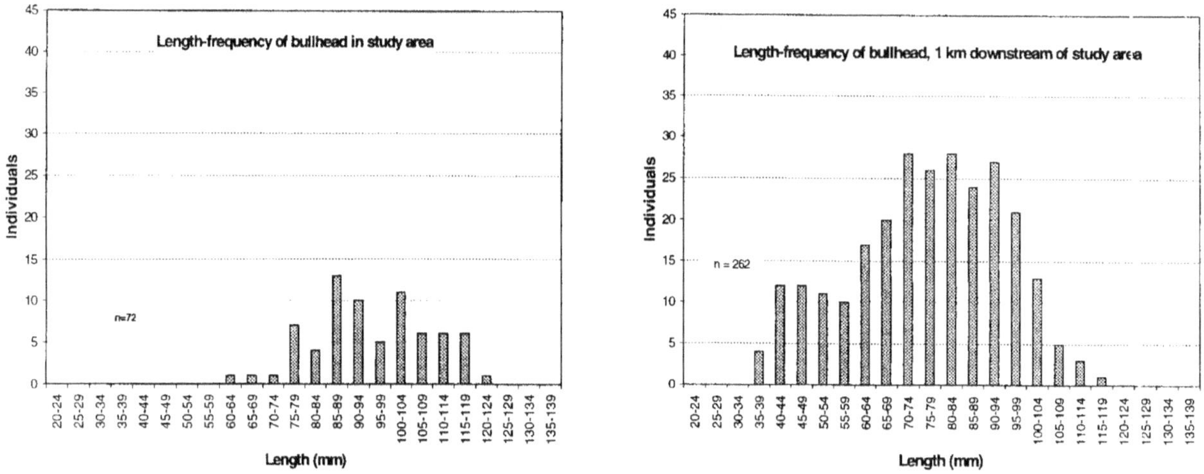

Figure 11. Comparison of length–frequency distributions in the study area and the stretches located 1 km downstream (August survey).

25 cm per h). Discharge between the 'spate phase' is often extremely low and even further reduced by water abstraction. Such extreme and rapid changes in flow conditions lead to sustainable reductions in bullhead population density by mortality and downstream migration but also in zoobenthic biomass (Späh & Beisenherz, 1986; Bretschko & Moog, 1990).

Restocking experiments in two small rhithral streams in Germany indicate a high dispersion rate of bullheads (Späh & Beisenherz, 1986), but recolonisation from downstream areas to compensate losses (Bless, 1990, 1997) is only possible in periods with high water level. Field and laboratory experiments (Bless, 1981, 1983) illustrate that bullhead larvae normally show downstream drift behaviour before the yolksac supply is exhausted and tend to migrate upstream when the swim ability allows movement against the flow direction. The disruption of the longitudinal corridor during long periods of the year certainly handicaps drift and compensatory upstream migration of all life stages. This might be one important reason for the lack of juveniles in the two upstream sections, but does not really explain their absence in the permanently water-bearing section C. At any rate, the residual water stretches, which are characterised by high fluctuations in discharge and extreme low water levels, are only a moderately suitable habitat for adult fish but uninhabitable for the probably much less mobile juvenile stages.

During low water periods large areas in section B regularly dry out and the wetted surface shrinks to isolated pools. Long dry periods often decimate zoobenthic biomass (Bretschko & Moog, 1990) and reduce prey availability for fish. The significantly lower bullhead population and its considerably poorer condition in the intermittently isolated transitional stretches (section B) as opposed to the directly adjoining downstream stretches with permanent water flow (section C) clearly demonstrate the importance of maintaining the river continuum for *Cottus gobio*.

Improved understanding of the linkage between animal behaviour and physical environment allow more sensitive detection of changes in ecological integrity. In order to document and assess biodiversity and ecological integrity of the native fish fauna in running waters, intensive surveys regarding stream habitat quality and migratory possibilities are necessary. The results of the present study clearly demonstrate that the fragmentation of the longitudinal stream corridor in the investigated residual flow stretches at low water have drastic effects on all life stages of the local bullhead population in respect to habitat and prey availability, movement, lifetime fitness, and survival. The investigated stretches clearly did not contain naturally structured populations.

Acknowledgements

Thanks to the colleagues of the biological station in Lunz am See, especially to the head of the station Gernot Bretschko, for his scientific and financial assistance. We also acknowledge Michael Stachowitsch for translating the manuscript.

References

Adamicka, P., 1987. Nahrungsuntersuchungen an der Koppe (*Cottus gobio* L.) im Gebiet von Lunz. Österr. Fischerei 40: 8–10.

Andreasson, S., 1971. Feeding habits of a sculpin (*Cottus gobio* L. Pisces) population. Rep. Inst. Freshwat. Res. Drottningholm 51: 5–30.

Bagenal, T., 1978. Methods for assessment of fish production in fresh waters, 3rd edn. Blackwell Scientific Publications, Oxford, London, Edinburgh, Melbourne: 365 pp.

Baran, P., M. Delacoste, F. Dauba, J. M. Lascaux, A. Belaud & S. Lek, 1995. Effects of reduced flow on brown trout (*Salmo trutta* L.) populations downstream dams in French Pyrenees. Regulated Rivers: Research & Management 10: 347–361.

Barandun, J., 1990. Auswirkungen von Ausbreitungsbarrieren auf das Vorkommen von Groppen (*Cottus gobio*) – Anregungen für den Artenschutz. Nat. u. Landsch. 65: 66–68.

Bless, R., 1981. Untersuchungen zum Einfluß von gewässerbaulichen Maßnahmen auf die Fischfauna in Mittelgebirgsbächen. Nat. u. Landsch. 56: 243–252.

Bless, R., 1983. Untersuchungen zur Substratpräferenz der Groppe, Cottus gobio LINNAEUS. Senckenbergiana biol. 63: 161–165.

Bless, R., 1990. Die Bedeutung von gewässerbaulichen Hindernissen im Raum-Zeit-System der Groppe (*Cottus gobio* L.). Nat. u. Landsch. 65: 581–585.

Bless, R., 1997. Möglichkeiten und Grenzen der Zustandserfassung und Bewertung bei Populationen von Fischarten der FFH-Richtlinie. Nat. u. Landsch. 72: 496–498.

Bretschko, G. & O. Moog, 1990. Downstream Effects of Intermittent Power Generation. Wat. Sci. Tech. 22: 127–135.

Daniels, R. A., 1987. Comparative life histories and microhabitat use in three sympatric sculpins (Cottidae: *Cottus*) in northeastern California. Envir. Biol. Fishes 19: 93–110.

Dynesius, M. & C. Nilsson, 1994. Fragmentation and flow regulation of river systems in the northern third of the world. Science 266: 753–762.

Fischer, S., 1998. Verteilung und Wanderverhalten der Mühlkoppe (*Cottus gobio* L.) in einem astatischen Bachabschnitt. Biologische Station Lunz, Universität Salzburg: 71 pp.

Gaudin, P. & L. Caillere, 1990. Microdistribution of *Cottus gobio* L. and fry of *Salmo trutta* L. in a first order stream. Pol. Arch. Hydrobiol. 37: 81–93.

Gerking, S. D., 1958. The restricted movement of fish populations. Biol. Rev. 34: 221–242.

Hofer, R. & F. Bucher, 1996. Die Koppe (*Cottus gobio* L.) als Indikator für Umweltbelastungen. Fischökologie 10: 47–62.

Jungwirth, M., 1998. River continuum and fish migration – going beyond the longitudinal river corridor in understanding ecological integrity. In Jungwirth, M., S. Schmutz & S. Weiss (eds), Fish Migration and Fish Bypasses. Fishing News Books, Oxford: 19–32.

Jungwirth, M., O. Moog & S. Schmutz, 1990. Auswirkungen der Veränderungen des Abflußregimes auf die Fisch-und Benthosfauna anhand von Fallbeispielen. Abt. für Hydrobiologie und Fischereiwirtschaft, Universität für Bodenkultur, Wien.

Korolev, V. V., 1991. The common sculpin, *Cottus gobio*, of the Upper Pechora River Basin. J. Ichthyol. 31: 36–41.

Maile, W., T. Heilmair & T. Strobl, 1995. Relationships between hydraulic and biological parameters in residual flow reaches as a basis for stipulating residual discharge levels. International Journal on Hydropower and Dams, Barcelona: 137–147.

Marconato, A. & A. Bisazza, 1988. Mate choice, egg cannibalism and reproductive success in the river bullhead, *Cottus gobio* L. J. Fish Biol. 33: 905–916.

McCleave, J. D., 1964. Movement and population of the mottled sculpin (*Cottus bairdi* Girard) in a small Montana stream. Copeia 3: 506–513.

Mills, C. A. & R. H. K. Mann, 1983. The bullhead *Cottus gobio*, a versatile and successful fish. Rep. Freshwat. Biol. Ass. 51: 76–88.

Moog, O. & M. Jungwirth, 1993. Berücksichtigung ökologischer Gesichtspunkte bei der Wasserkraftnutzung durch Ausleitungskraftwerke – Kurzfassung. Österreichische Wasserwirtschaft Jg. 45, Heft 7/8.

Orság, L. & M. Zelinka, 1974. Zur Nahrung der Arten *Cottus poecilopus* Heck. und *Cottus gobio* L. Zool. Listy 23: 185–196.

Peter, A., 1998. Interruption of the river continuum by barrriers and the consequences for migratory fish. In M. Jungwirth, S. Schmutz & S. Weiss, Fish Migration and Fish Bypasses. Fishing News Books, Oxford: 99–112.

Roth, C. & J. Utzinger, 1993. Ökologie der Groppe (*Cottus gobio* L.) und deren Eignung als Indikatorfisch für den chemischen und morphologischen Zustand eines Fließgewässersystems. Diplomarbeit, Abt. für Umweltnaturwissenschaften XB, ETH Zürich (unveröffentlicht): 138 pp.

Roussel, J. M. & A. Bardonnet, 1996. Changements d'habitat de la truite (*Salmo trutta*) et du chabot (*Cottus gobio*) au cours du nycthémère. Approches multivariées a différentes échelles spatiales. Cybium 20: 43–53.

Sale, M. J. & R. G. Otto, 1991. Improving the assessment of instream flow needs for fish populations. Waterpower 91, 1: 76–84.

Seber, G. A. F., 1965. A note on the multiple-recapture census. Biometrika 52: 249–259.

Smith, R. J. F., 1985. The control of fish migration. Springer Verlag, Berlin.

Smyly, W. J. P., 1957. The life-history of the bullhead or miller's thumb (*Cottus gobio* L.). Proc. zool. Soc. Lond. 128: 431–453.

Späh, H. & W. Beisenherz, 1986. Wiederbesiedlung von Forellenbächen. Erfolgreiche Versuche mit Groppen. Mitteilungen LÖLF 3: 28–34.

Stahlberg-Meinhardt, S., 1994. Verteilung, Habitatansprüche und Bewegungen von Mühlkoppe (*Cottus gobio* L.) und Bachforelle (*Salmo trutta* L.) in zwei unterschiedlich anthropogen beeinflußten Fließgewässern im Vorharz. Dissertation, Zoologisches Institut der Technischen Universität Braunschweig (unveröffentlicht): 197 pp.

Waidbacher, H. & G. Haidvogl, 1998. Fish migration and fish passage facilities in the Danube: past and present. In Jungwirth, M., S. Schmutz & S. Weiss (eds), Fish Migration and Fish Bypasses. Fishing News Books, Oxford: 85–98.

Waterstraat, A., 1992. Populationsökologische Untersuchungen an *Cottus gobio* L. und anderen Fischarten aus zwei Flachlandbächen Norddeutschlands. Limnologica 22: 137–149.

Welton, J. S., C. A. Mills & F. L. Rendle, 1983. Food and habitat partitioning in two small benthic fishes, *Noemacheilus barbatulus* L. and *Cottus gobio* L. Arch. Hydrobiol. 97: 434–454.

Williams, J. G., 1998. Fish passage in the Columbia River, USA and its tributaries: problems and solution. In Jungwirth, M., S. Schmutz & S. Weiss (eds), Fish Migration and Fish Bypasses. Fishing News Books, Oxford: 180–191.

Hydrobiologia **422/423**: 319–330, 2000.
M. Jungwirth, S. Muhar & S. Schmutz (eds), Assessing the Ecological Integrity of Running Waters.

Habitat condition and fish assemblage structure in a coastal mediterranean stream (Yarqon, Israel) receiving domestic effluent

Sarig Gafny[1], Menachem Goren[2] & Avital Gasith[1]
[1]*Institute for Nature Conservation Research and* [2]*Department of Zoology; George S. Wise Faculty of Life Sciences, Tel-Aviv University, Ramat-Aviv 68878, Israel*

Key words: urban stream, Multi Dimensional Scaling, stream pollution, BOD

Abstract

The Yarqon stream is the largest urban stream in Israel. In the past 40 y it was severely impacted by human action. Fish kills occur along the polluted sections of the stream almost regularly following winter floods and when poor-quality effluent is discharged into the stream. These events attest to the serious ecological state of the Yarqon stream. We studied the interaction between habitat conditions and fish assemblage structure for four consecutive years. Comparison of certain water quality variables among selected sites along the upper and central stream sections of the Yarqon indicated that the two stream sections differ significantly in electric conductivity and organic load (BOD) but not in temperature and dissolved oxygen. The average species richness and fish abundance in the upper, relatively undisturbed section is significantly higher than in the downstream, upper reach of the central section where domestic effluents enter the stream. In contrast, fish biomass is usually higher in the central stream section. Organic load was the only factor among the four water quality variables measured that significantly correlates with variables of the fish assemblage. A curve fit analysis between organic load and fish species richness and abundance suggests a threshold at ca. 10 mg l^{-1} BOD, above which most fish species totally avoid polluted habitats. Separate Multi Dimensional Scaling (MDS) analyses of water quality variables and fish assemblages clearly separated between the sites in the upper and the central stream sections. When the result of the fish MDS analysis was graphically superimposed on the result of the water quality MDS analysis, we found a close overlap between the two, suggesting a strong association between water quality and the fish assemblage structure in the Yarqon stream.

Introduction

Streams and rivers are unique aquatic systems characterized by an unidirectional flow of water. Water and materials from the drainage basin cascade down along the geomorphologic gradient and determine water quality and habitat features. The riparian vegetation that grows along the banks limits the light that enters the stream, provides coarse particulate organic matter and habitats for wildlife and acts as a buffer zone against anthropogenic influences (Naiman & Bilby, 1998). In mediterranean-type streams, flow presents a high seasonal variability, with high discharge during flood events in winter and minimum or no flow in summer. These seasonal changes in the hydrological regime and its associated water quality conditions provide a structural framework for mediterranean-stream ecosystems. In turn, these factors may be strongly influenced by human activity yielding detrimental ecological effects (Gasith & Resh, 1999).

During the 20th century, river and stream pollution and misuse have decimated species richness (Gore, 1985; Haslam, 1990). The study of restoration and rehabilitation of streams and rivers is growing in importance (for a review of case studies see: Sparks, 1995; Kauffman, 1997), and the understanding of the complex processes involved in the maintenance of healthy, functioning river and stream ecosystems is improving (Herricks & Osborne, 1985; Allen, 1995).

In water-stressed regions the competition for water between human needs for water and that of nature may result in severely diminished stream flow. This human impact is often exacerbated by water pollution (e.g. Gasith, 1992; Gasith & Resh, 1999). The degraded status of the coastal streams of Israel is an example of the failure to recognize and control the compounded effect of water diversion and pollution (Gasith, 1992). Despite the continued deterioration of water quality and the intensification of water diversion from streams in Israel, since the early 1970s (Goren, 1974) no new information on the status of the fish populations in the coastal streams of Israel has been published until the present study on the Yarqon stream was initiated (Gasith et al., 1998).

The Yarqon stream is the largest urban stream in Israel. In the past 40 years the stream has been severely impacted by human action. Fish kills occur along the polluted sections of the stream almost regularly following winter floods and when poor quality effluent is discharged into the stream (e.g. when the operation of waste-water treatment fails). These events attest to the serious ecological state of the Yarqon stream.

Presently, efforts are being made to rehabilitate the Yarqon and other coastal streams in Israel. The aim of the present study is to assess the ecological health of the Yarqon stream and to identify the rehabilitation measures needed to mitigate the adverse human impact. For that purpose we used fish assemblage structure as an indicator of habitat condition. The working hypothesis is that habitat conditions influence fish species richness, abundance and size distribution. Ecosystem perturbation such as enrichment with organic matter is expected to reduce species richness, enhance or inhibit fish growth relative to fish in unpreturbed sites (Angermeier & Karr, 1994). Comparing of the structure of the fish assemblages in less perturbed sites with those in polluted sites is therefore expected to reveal the extent of the pollution impact along the Yarqon stream.

Methods

The sites studied

The Yarqon stream is the southernmost perennial stream in the Coastal Plain of Israel. This lowland stream (gradient 0.06%), meanders along 29 km through the Tel-Aviv metropolitan region and drains into the Mediterranean. The Yarqon drainage basin covers an area of approximately 1800 km^2, with an average annual rainfall of about 600 mm. The stream's complex environment is composed of at least three sub-systems: (a) the channel and its aquatic life; (b) the riparian zone (mostly modified by human activity); and (c) the built area, the center of human activities. Prior to 1950, the Yarqon stream carried over 200 million cubic meters (MCM) of freshwater annually discharged from its main springs (Rosh Ha'Ayin); the discharge resulting from run-off varied between less than one MCM in drought years and up to 300 MCM in rainy years (Ben-Zvi et al., 1995). Diversion of the Yarqon's main springs in the mid-1950s for domestic uses has reduced the base flow to less than 1% of the average annual discharge (about 1.5 MCM). If not for the discharge of effluents (about 4.5 MCM yearly), this former 'Yarqon River', second in discharge after the Jordan River, would have become an intermittent stream in drought years (Gasith, 1992).

Based on the degree of perturbation and water quality the Yarqon stream can be divided into three segments: (a) an upper 7.5 km relatively unperturbed section of relatively high water quality (Gasith et al 1998). This unpolluted section ends at the confluence of the Qane tributary and the Yarqon stream, where municipal effluents from a waste-water treatment plant (Kefar-Sava–Hod-Hasharon) enter the Yarqon stream; (b) the central section of about 17.5 km that is severely impacted by pollution, mostly of municipal effluents (entering via the Qane tributary and from the Ramat-Hasharon oxidation ponds); (c) the lowermost 4 km section which is under tidal influence and therefore is variably saline. Our study was limited to the freshwater, upper and central sections of the Yarqon stream.

Sampling and analyses

Analysis of habitat conditions and fish assemblage structure was conducted on a monthly basis from June 1995 to August 1997. We included in the analysis data obtained in a preliminary study (Oct. 1994–June 1995; Gasith et al., 1998).

The sampling sites along the upper and central sections of the Yarqon stream are shown in Figure 1. Samples were taken monthly in three or four sites in the upper section (Sts. 2a, 2b, 3a and 3b), two sites in the upper reach of the central section (Sts. 3bc and 3c) and one in the lowermost reach (St. 6b) of the central section. Stream morphometry (width ca. 8 m) was similar at all sites. Sites 2a and 3a are above small rocky dams and represent lentic (pool) habitats (depth=1–2 m), whereas sites 2b, 3b, 3bc, 3c and 6b are below rocky dams and represent lotic (riffle or shallow run) habitats (depth=0.5–1.5 m). The small dams allow upstream migration of some species all year round (except in extremely dry summers) and winter upstream migration of the others.

Temperature, dissolved oxygen and conductivity were measured with a YSI model 85 Oxygen-Temperature-Conductivity meter, turbidity was measured with a Hack turbidity meter (model 2100A) and organic load (BOD) was measured following the standard procedure (APHA, 1995). Fish samples were obtained by a non-selective electrofishing technique using a shore-operated electroshocker (EFKO model FEG 6000) producing 350 Volts (17 Amp), non pulse DC. Each site was sampled for 10 min. The fish were sampled without block nets but attempts were made to collect all fish encountered in the sampled segment. The fish were sorted, identified, measured (total length to the nearest 1 mm) and weighed (wet weight to the nearest 0.1 g).

Analysis of variance using Systat for Windows (1992) and Statistica (Statsoft Inc., 1995) was applied to test the difference in water quality and fish assemblage variables among sites (statistical significance of $p<0.05$), and the ARIMA model (Diggle, 1990) was applied to the time series data of each variable to correct for autocorrelation and white noise.

The overall among-site similarity in the measured variables was calculated using monotonic Multi Dimensional Scaling analysis (MDS, Bernstein et al., 1988) performed separately on the water quality (Euclidean distance) and the fish assemblage data (both Euclidean distance and Bray-Curtis similarity measure). The output of the Euclidean distance fish assemblage MDS was then graphically overlaid on the output of the water quality to detect similar patterns in the two analyses.

Results

Water quality conditions

Mean values (±SD) of pH, chemical oxygen demand (COD), suspended solids (SS), ammonia (NH_3–N) and turbidity from sites at the upper and central sections of the Yarqon stream are presented in Table 1. Suspended solids, pH and turbidity did not vary significantly among the upper and central stream section ($t=0.13$, $t=1.67$ and $t=0.49$ respectively, $p>0.1$), whereas NH_3–N was significantly higher at the central section ($t=6.8$ $p<0.001$).

Similar ranges, median and average values of water temperature were recorded at the different sites along the Yarqon stream throughout the study period (Figure 2a). In most sites the temperature ranged from 12 °C in winter to 26 °C in summer. The highest variations in temperature were recorded in the lowermost reach of the central section (St. 6b), where the summer temperature reached 28 °C.

The electric conductivity varied significantly among sites ($F=5.25$; $df=6$, 92; $P<0.001$). It also varied temporally, with minimum values recorded in winter (800–1000 μMohs cm^{-1} at 20 °C) and maxima in summer (1300–1700 μMohs cm^{-1}). Similar conductivity was recorded in the different sites along the upper section (Sts. 2a, 2b, 3a, 3b; $F=0.415$; $df=40$; $P>>0.1$), where it ranged from a minimum of 800 μmhos cm^{-1} at 20 °C in winter to a maximum of 1520 in summer (annual mean ca. 1060). The conductivity in the central section (Sts. 3c, 6b) was slightly higher than in the upper section (annual mean 1170; Figure 2b). The highest range and average values of conductivity were recorded in the lowermost reach of the central section (St. 6b; annual mean 1315).

The daytime dissolved oxygen concentration varied among the sites studied, although the difference was not statistically significant ($F=1.33$; $df=118$; $P>>0.1$). It was usually 30–40% below saturation in the upper section and varied from a minimum of 2 to a maximum of 11 mg l^{-1} (20 and 110% saturation, respectively) in the central section (Figure 2c). The lowest oxygen concentrations were recorded below the inflow of the Qane tributary (St. 3c) during a 3-month period (Feb.–April) after a flood event in 1996. Below cascades formed by dams the dissolved oxygen concentration was slightly but significantly elevated ($t=3.26$ $P<0.01$) compared to that recorded above the dams (e.g. St. 2b compared with St. 2a and St. 3b compared with St. 3a, Figure 2c).

In the upper stream section (Sts. 2a, 2b, 3a) and in the lowermost reach of the central section (St. 6b) the dissolved oxygen exhibited seasonal dynamics, being higher in winter than in summer. In the upper reach of the central section (Sts. 3bc, 3c) oxygen concentration varied with no apparent seasonal pattern. Following operation of an upgraded waste-water treatment plant (May, 1996) the dissolved oxygen concentration in this reach slightly but significantly increased ($t=2.26$, $P<0.05$).

The organic load (expressed in terms of BOD) in the upper stream section was low (< 3 mg l^{-1}), but it abruptly increased by at least 2–3 fold in the upper reach of the central stream section below the inflow of the Qane tributary (Sts. 3bc, 3c; Figure 2d). Highest

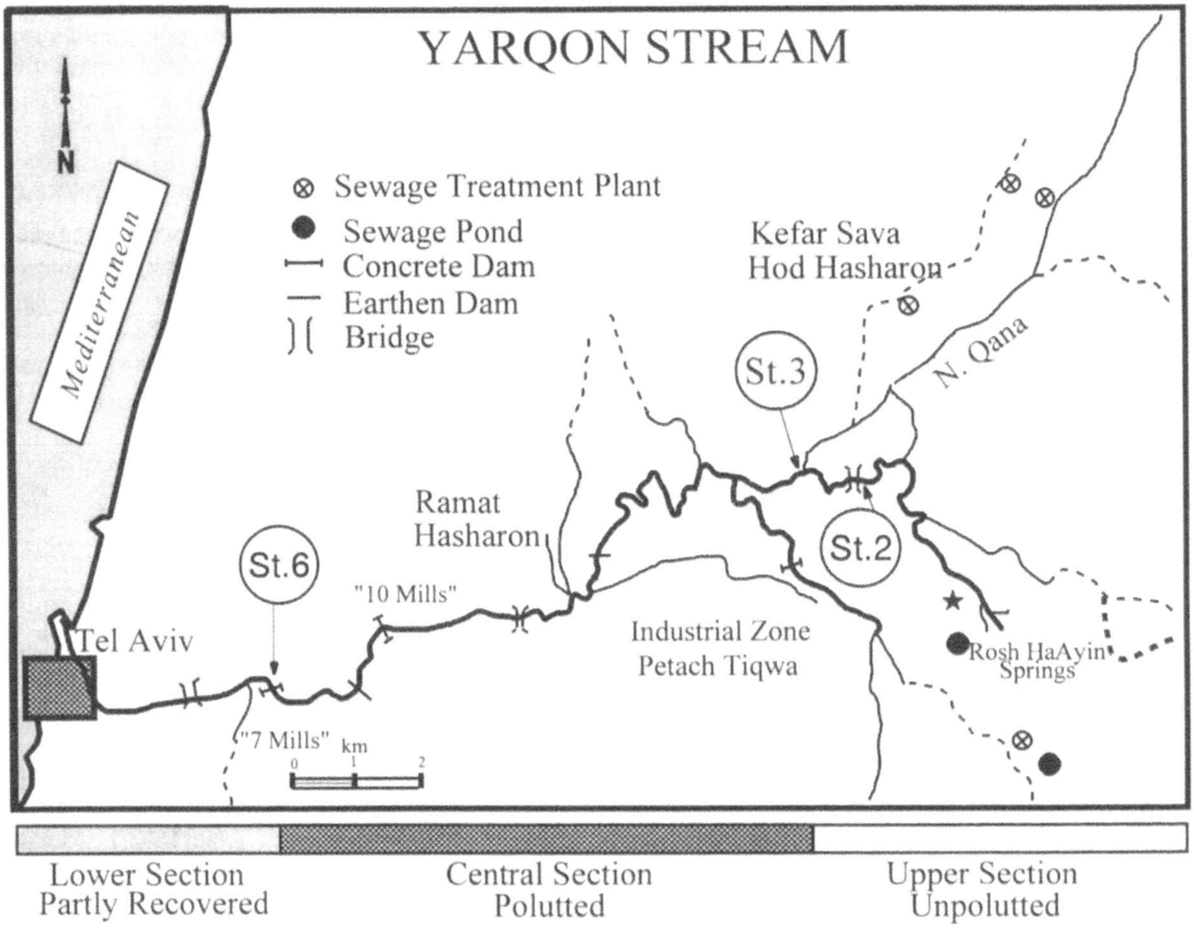

Figure 1. Sampling sites along the Yarqon stream. Upper section – lentic habitats: sites 2a and 3a; lotic habitats: sites 2b and 3b. Central section -- lotic habitats: sites 3bc, 3c and 6b.

concentrations of organic matter (>50 mg l^{-1} BOD) were measured in this reach in winter, following flood events. After putting an upgraded waste-water treatment plant into operation, the BOD loading to the Yarqon stream via the Qane tributary declined. This decline was particularly apparent when the effect of the watershed was minimized (by excluding the winter values from the calculation of the average BOD concentration) and the average BOD concentration before (55.8±19.8 mg l^{-1}) and after (10.7±2.9 mg l^{-1}) the new waste-water treatment plant went into operation were compared.

The MDS analysis of the measured water quality variables clearly separated between the sites in the upper (2a, 2b and 3a, 3b) and the central section (3bc, 3c, 6b) of the Yarqon stream (Figure 3). The lowermost site of the central section (St. 6b) was relatively far apart (Euclidean distance) from the sites in the upper reach of this section (St. 3bc, 3b), suggesting that it differed in water quality from the latter sites. As shown in Figure 3, sites in the upper stream section that differed in flow but were not far apart (< 10 m; i.e. Sts. 2a and 2b and Sts. 3a and 3b) were separated by a small Euclidean distance. On the other hand, a relatively large Euclidean distance separated sites that had similar flow characteristics (i.e. lentic conditions in Sts. 2a and 3a, and lotic conditions in Sts. 2b and 3b) but were relatively distant (ca. 1.5 km).

Fish species richness, abundance and biomass

The total number of fish and their respective relative occurrence in the selected study sites is summarized in Table 2. Nine species of fish were recorded in the Yarqon stream in the present study, including

Table 1. Average values (±SD) of selected water quality variables from sites at the upper and central sections of the Yarqon stream (1995–1997)

	Upper section		Central section	
	2a, 2b	3a, 3b	3bc, 3c	6b
pH*	7.8±0.06	7.7±0.09	7.8±0.15	8.4±0.15
COD (mg l^{-1})*	11.2± 4.3	11.7±4.4	126.5±115	43.7±18.5
SS (mg l^{-1})*	33.4±10	38.3±28.4	86.1±69.7	18.4±10.4
NH_3 -N (mg l^{-1})*	0.2±0.4	0.2±0.4	12.1±5.6	11.3±5.1
Turbidity (NTU)	17.8±9.6	18.6±12.9	21.6±11.8	16.5±6.7

*Data courtesy of the Yarqon River Authority.

two exotic species (*Xiphophorus* sp. and *Gambusia affinis*). Two native species – *Aphanius mento* (Cyprinodontidae) and the cichlid *Oreochromis niloticus* – that were recorded in the Yarqon stream until the 1970s (Goren, 1974) have so far been absent from our samples. In 1994–95, the cichlid *Sarotherodon galilaeus*, the cyprinid *Acanthobrama telavivensis*, and the sword-fish, *Xiphophorus* sp. were the only species exclusively found in the upper section of the Yarqon (Gasith et al 1998). In the subsequent two years the cichlid *S. galilaeus* was absent from our samples; the other two species were consistently found in the upper stream section but were occasionally also found in smaller numbers in the upper reach of the central stream section (Sts. 3bc, 3c) when higher water quality conditions prevailed here. The carp *Cyprinus carpio* and the catfish *Clarias gariepinus* were found both in the upper and central stream sections but were more abundant in the central section. Consistent with the preliminary study (Gasith et al 1998), the cichlid *Tilapia zillii*, the mosquito-fish *G. affinis* and the eel *Anguilla anguilla* were found in the upper stream section but occasionally also in the lowermost reach of the central section (St. 6b). *G. affinis* was more abundant in the upper section, whereas *T. zillii* and *A. anguilla* were more abundant in the lowermost reach of the central section. In the last 2 y of the study, *G. affinis* and *T. zillii* were occasionally found also in the upper reach of the central section (Sts. 3bc, 3c) when higher water quality conditions prevailed here. As found in our preliminary study, the mullet *Mugil cephalus* was the only species that was consistently recorded only in the lowermost reach of the central section (St. 6b), about 4 km upstream of the mouth of the stream.

Species richness was highest in summer, fall and early winter and lowest in mid-winter and in spring (Figure 4). Species richness per standard fishing effort (10 min) varied from 0 to 6 in the upper section and from 0 to 5 and from 2 to 6 in the upper and lowermost reaches of the central section, respectively. The median richness value was lowest in the upper reach of the central section and highest at the lowermost reach of this section (Figure 5a). A significant negative correlation was found between organic load and fish species richness ($S^{0.5}$=2.1+0.1 ln (BOD)^2, r^2=0.76, F=108.2, p>0.001). Until August 1996 the average species richness in the upper section (Sts. 2a, 2b, 3a) was significantly higher than in the upper reach (St. 3bc, St. 3c) of the central section (paired t=5.58, p<<0.001). During late summer and fall (August 1996–February 1997) species richness in these two stream segments was similar (t=1.58, p>0.1). At this time young individuals of *T. zillii*, adult and young *G. affinis* and of *Xiphophorus* sp., and a few adult specimens of *A. telavivensis* were found along with carp in the upper reach of the central section (St. 3c). Species richness here declined after winter floods carried contaminants from the watershed into this section via the Qane tributary and since February 1997 it was significantly lower than in the upper section (t=3.97, p=0.0074).

Fish abundance (CPUE) varied seasonally: it was highest in summer and fall and lowest in winter. The lowest and highest abundance was recorded in the upper sites (Sts. 3bc and 3c) and lowermost reach (St. 6b) of the central section, respectively (Figure 5b). A significant negative correlation was found between organic load and fish abundance (CPUE^{-1}=0.02+0.01 (BOD)^2, r^2=0.92, F=319.8, p>>0.0001). Most remarkable was the abrupt decline in fish abundance between two adjacent sites (< 100 m apart) representing the lowermost reach of the upper stream section (St. 3a) and the uppermost reach of the central section (Sts. 3bc, 3c; Figure 5b). This also suggests a threshold at ca. 10 mg l^{-1} BOD, above which the fish avoid the polluted habitats. Improved water quality in

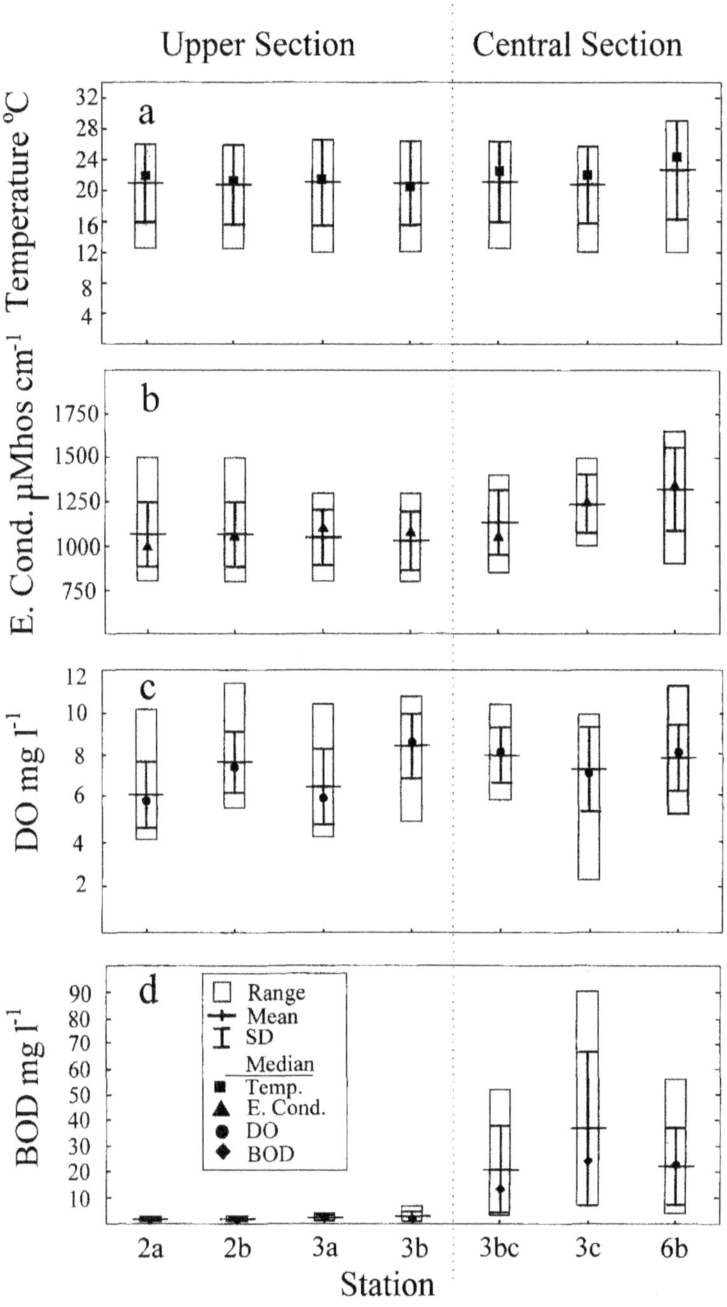

Figure 2. Range (min–max), median and mean (±SD) of (a) water temperature; (b) conductivity; (c) dissolved oxygen; (d) organic load – BOD in selected sites along the Yarqon stream during 1994–1997.

the upper reach of the central section after the wastewater plant went into operation (May 1996) increased the fish abundance here (starting from July 1996). However, the fish disappeared from this stream reach in February 1997 after winter flood events (Figure 6).

Fish biomass was highest but also varied the most in the central section (Figure 5a). The largest range of biomass was recorded in the upper reach of the central stream section (St. 3c), where fish richness and abundance were the lowest (Figure 5a,b,c). It corresponded to the greater frequency of occurrence of larger fish here (Figure 7). The difference in size distribution of the fishes in the upper and central stream sections was best demonstrated by the size distribution of the cichlid *T. zillii*, which is currently the most common fish in the Yarqon (Table 2). Relatively small *T. zillii* (31–40

Table 2. Total number (*n*) and relative frequency of occurrence (%, in italics) of the fishes in different sites and habitat types (len= lentic; lot= lotic) of the upper and central sections of the Yarqon stream (1994–1997)

Site	*n*	Upper section (%)				Central section (%)		
		2a	2b	3a	3b	3bc	3c	6b
Habitat type		len	lot	len	lot	lot	lot	lot
Taxon								
CICHLIDAE								
T. zillii	8896	*8.9*	*12.2*	*16.2*	*12.5*	*4.3*	*3.5*	*42.4*
S. galilaeus	51	*29.2*	*62.5*	*8.3*	*0*	*0*	*0*	*0*
CYPRINIDAE								
C. carpio	671	*6.3*	*0.3*	*1.8*	*1.3*	*5.2*	*4.6*	*80.4*
A. telavivensis	1045	*13.2*	*26.7*	*25.8*	*29.4*	*4.7*	*0.2*	*0*
CLARIIDAE								
C. gariepinus	12	*8.3*	*0*	*0*	*0*	*0*	*16.7*	*75*
ANGUILLIDAE								
A. anguilla	199	*5.5*	*4.0*	*7.5*	*16.6*	*7.5*	*0*	*58.8*
POECILLIDAE								
G. affinis	3920	*18.4*	*29.4*	*25.3*	*18.1*	*2.3*	*3.4*	*3.1*
Xiphophorus sp.	354	*13.8*	*8.8*	*37.3*	*23.4*	*9.3*	*7.3*	*0*
MUGILIDAE								
M. cephalus	893	*0*	*0*	*0*	*0*	*0*	*0*	*100*
Total *n*	16041	1769	2589	2874	2253	604	505	5447

mm TL) prevailed in the upper section and their size distribution was skewed toward small individuals (Figure 7). Small numbers of relatively large *T. zillii* (>150 mm) were recorded in the upper reach of the central section and smaller *T. zillii* (71–80 mm) prevailed in the lowermost, partially recovered site which was the only location where the fish size was normally distributed (Figure 7). Likewise, the carps collected in the upper reach of the central section (120<TL<340 mm) were much larger than those collected in the upper stream section (TL <140 mm).

An MDS analysis of the fish data separated the assemblages found in the upper section (Sts. 2a, 2b, 3a, 3b) from those of the central reaches (Sts. 3bc, 3c, 6b). Within the upper section higher similarity (separated by a small Euclidean distance) was found for the fish inhabiting similar habitat types (lentic – Sts. 2a, 3a; lotic – Sts. 2b, 3b) than for those from adjacent sites that differed in habitat nature (St. 2a compared with St. 2b and St. 3a compared with St. 3b; Figure 3). The fish assemblage in the lowermost reach of the central section (St. 6b) differed (separated by a relatively large Euclidean distance) from those in the upper reach of the same stream section (Sts. 3bc, 3c) even though all these sites were lotic.

Discussion

An underlying assumption in water resource management is that altered stream habitats contain altered biological communities (Karr et al., 1985; Scott & Hall, 1997). In recent years, the use of biota to assess the anthropogenic impacts on the integrity of streams is attaining greater importance (e.g. Miller et al., 1988; Rosenberg & Resh, 1993; Gerritsen, 1995). Fish community structure is one of the options suggested for assessing biotic integrity in rivers and streams in different parts of the world (e.g. Karr,

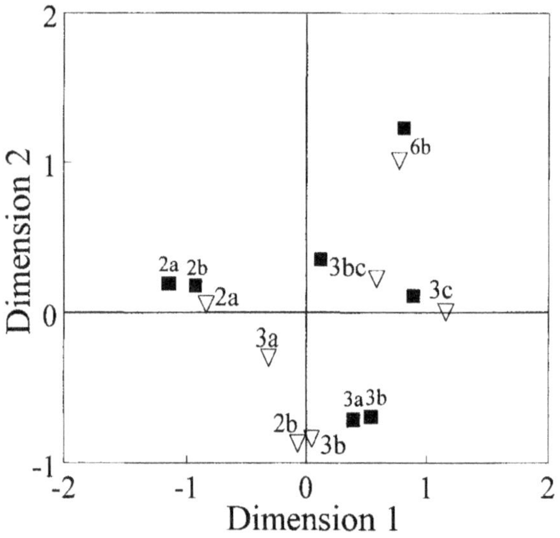

Figure 3. Multi-Dimensional Scaling analysis of fish (clear triangles) and water quality (black squares) data collected during 1994–1997 in selected study sites along the Yarqon stream.

1981; Edds, 1993, Lyons et al., 1995; Paller et al., 1996). We followed the latter approach and examined the relationship between habitat condition and fish assemblage structure in a coastal, mediterranean-type stream that has been stressed by water diversion and pollution for the past 50 y (Gasith, 1992).

Similar to other coastal streams in Israel the Yarqon stream underwent a drastic reduction in stream flow since the mid-1950s; unlike other streams, however, a small upper stream section (ca. 25% of the stream) remained relatively pollution free. Today it is the only perennial freshwater stream section in the central coastal plain that can sustain indigenous plant and animal populations. We examined to what extent the upper stream section differs in water quality and fish populations from the central section, which is perturbed by pollution discharged into Yarqon stream and how site-specific habitat attributes influence fish assemblage structure.

Comparison of certain water quality variables among selected sites along the upper and central stream sections of the Yarqon indicated that the two stream sections differ significantly in conductivity and organic load (BOD) but not in temperature and dissolved oxygen. This difference is most apparent at the confluence of the Yarqon stream and the polluted Qana tributary, where water quality quickly deteriorates from unpolluted (site 3a) to slightly enriched (3b) to partly polluted (3bc) to heavily polluted (3c) within less than 100 m. The higher BOD values in the central stream section are a result of the effluent discharged into the Yarqon via the Qane tributary. The lower load of BOD in the lowermost reach of the central section relative to that in the upper reach indicates partial recovery of the water quality along the central section by instream 'self-purification' processes. In contrast, the conductivity in the lowermost reach of the central section is highest, probably as a result of a tidal effect which mixes freshwater with sea water. Since the temperature of the water entering the Yarqon stream from all sources is similar and the canopy cover of the riparian vegetation along most of the stream channel is restricted, we expected no significant difference in temperature among the different sites. In light of the large and significant difference in the organic load between the two stream sections the insignificant difference in the average oxygen regime between these sections was somewhat surprising. This may in part be attributed to the low oxygen saturation of the water in the upper, unpolluted stream section; this masks the combined effect of higher respiration and higher photosynthetic activity (consequence of higher organic load and nutrient concentration, respectively) in the central stream section. The reason for the relatively low oxygen saturation throughout the upper stream section remains unclear. It may reflect the effect of reduced flow and thus reduced aeration, combined with high natural turbidity (10–27 NTU, Gasith et al., 1998), which limits photosynthetic activity (submersed vegetation is indeed uncommon). The assumption of reduced aeration is supported by the higher oxygen concentration in the more vigorously aerated water at cascades formed by dams in the upper section. However, this difference is not strong enough to distinguish between lentic and lotic sites by the MDS analysis, which does separate the sites in the upper stream section based on their geographic location. In contrast, the MDS analysis of the fish assemblage data separates the sites in this stream section based on habitat type (lentic, lotic). It is, therefore, possible that small differences in oxygen regime or other flow-related habitat attributes influence the segregation of the fish among habitats.

Among the water quality variables examined, BOD best delineates the upper, clean from the central, polluted freshwater sections of the Yarqon stream. The close overlap of site segregation based on independent MDS analysis of the water quality and that of the fish assemblage data suggests a close relationship between the two. A similar overlap was also reported by Matthews et al. (1992), who used several other

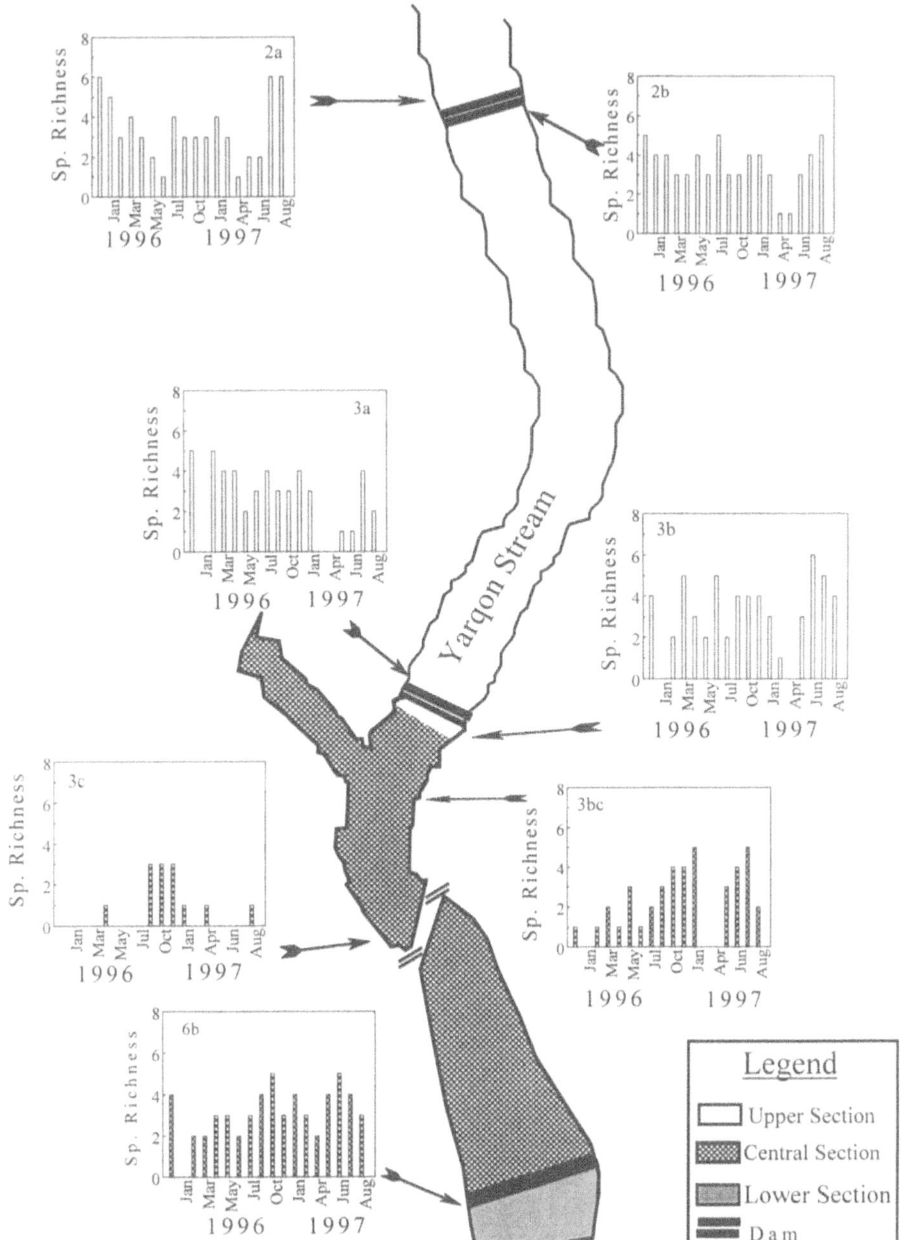

Figure 4. Patterns of annual change of fish species richness in the upper (Sts. 2a, 2b, 3a, 3b) and lower (Sts. 3bc, 3c, 6b) stream sections of the Yarqon during 1996–97.

multivariate analysis methods and suggested high concordance between water quality and fish assemblage regardless of the specific multivariate analysis method.

The association between water quality and fish assemblage structure is most apparent in the highly polluted upper reach of the central section. The curve fit analysis between organic load and species richness or abundance suggests a threshold at ca. 10 mg l^{-1} BOD, above which most fish species totally avoid polluted habitats (Gafny et al., 1997). Markedly higher average than median BOD values in the central stream section suggest that the frequency of occurrence of extremely high organic load is relatively low. However,

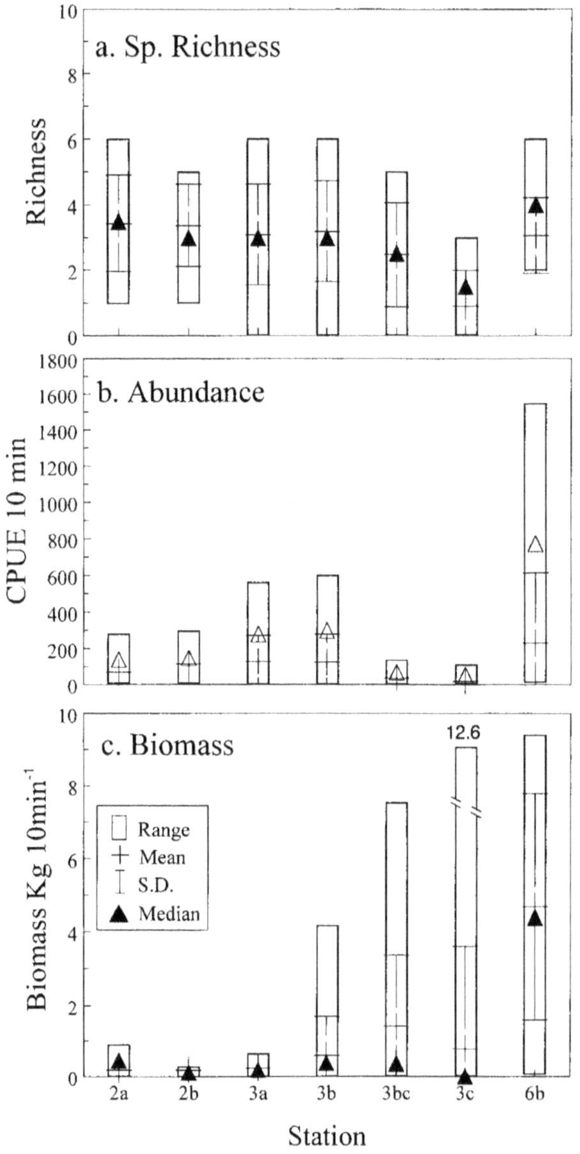

Figure 5. Range (min-max), median and mean (±SD) of: (a) fish species richness; (b) abundance; and (c) biomass in selected study sites along the Yarqon stream during 1994–1997.

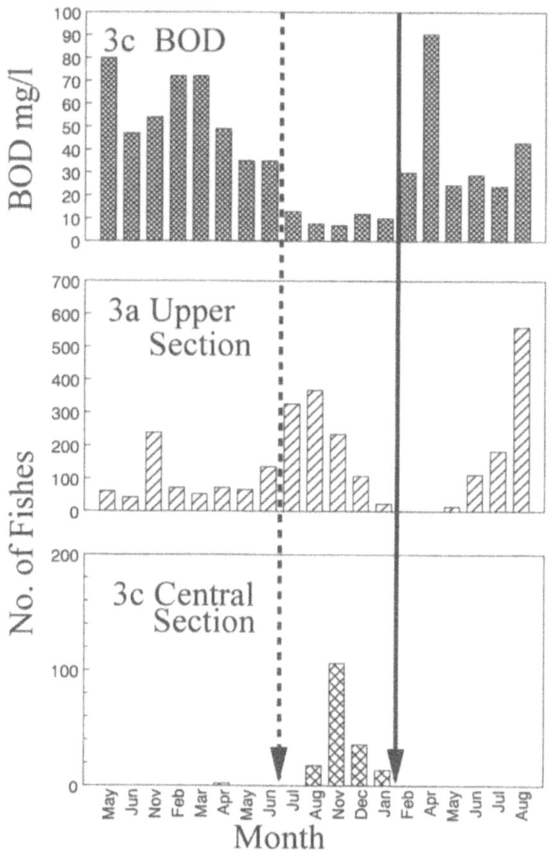

Figure 6. Comparison of the annual changes in organic load (BOD) and fish abundance at a selected site at the upper section (St. 3a) and in an adjacent site (<50 m apart) in the upper reach of the central section (St. 3c) before and after the start-up of operation of an upgraded waste-water treatment plant (dashed line) and before and after a major flood event (solid line).

it is the extreme conditions that set the limits for the biota.

Our study suggests that organic pollution may also affect the fish assemblage when the level of pollution is lower than the above threshold, and fish can survive. Such effects are noticeable on the fish size structure of most species currently inhabiting the Yarqon stream (Gafny et al., 1997). The size distribution of these species is skewed toward small individuals in the unpolluted sites, whereas it is skewed toward large individuals in even mildly polluted sites (see differences in size structure between St. 3a and 3b; Figure 7). A similar effect was reported in other studies that described an older age structure (Adams et al., 1992) or an increase in condition factor (Munkittrich et al., 1991) or lipid levels (Servos et al., 1994) of fish collected in polluted sites compared with unpolluted ones. Mild pollution levels (BOD 5–10 mg l^{-1}) can also affect fish distribution, as certain fish species (e.g. *S. galilaeus*) avoid even mildly polluted habitat.

Deteriorating water quality is among the major factors contributing to the disappearance of some species in many streams and rivers (Karr et al., 1985). We, therefore, suggest that the pollution effect is reflected both at the community and population level of certain fish. At the community level, pollution reduces both species diversity and abundance. Only certain species and only few individuals can survive the pollution and, thus, polluted habitats are poorer in fish. However,

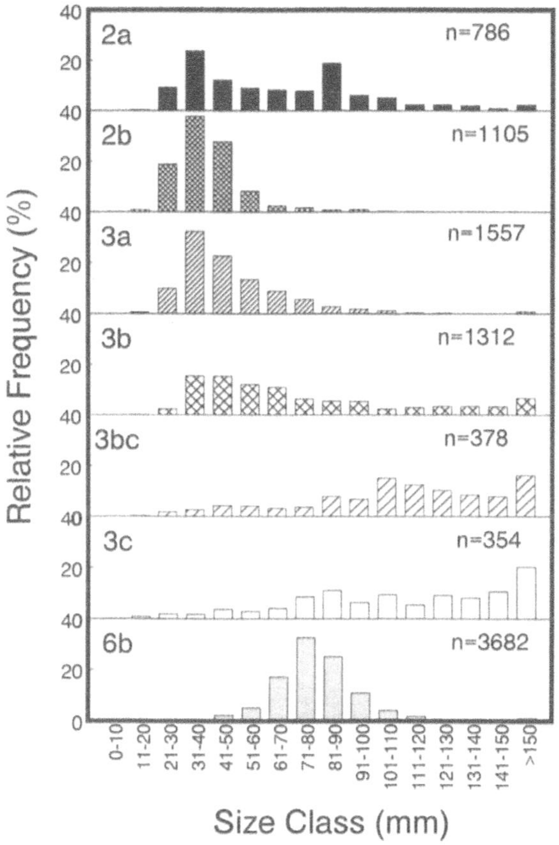

Figure 7. Size class distribution (total length) of *Tilapia zillii* (Cichlidae) in selected study sites in the upper and lower stream sections of the Yarqon stream (1994–1997).

the difference in species richness between unpolluted and mildly polluted habitats is less apparent, mostly because of seasonal changes in the water quality.

Conclusions

1. Both the water quality and the biotic variables of the fish community structure clearly divide the freshwater segment of the Yarqon stream into the upper unpolluted section (7.5 km) and the central, polluted stream section (17.5 km). 2. A negative correlation exists between the organic load (BOD level) and fish species richness and abundance. 3. Certain fish species such as the endemic cyprinid *A. telavivensis*, the cichlid *S. galilaeus* and the exotic sword-fish *Xiphophorus* sp. seem to be good bioindicators for high water quality conditions. 4. Floods are typical events in mediterranean-type streams. They play an important role in flushing and cleaning the stream channel. Under existing conditions of pollution sources in the drainage basin of the Yarqon, the first flood events are associated with degradation of the water quality and subsequent adverse impact on the fish community. 5. Diversion of the stream sources severely reduces base flow during summer. Habitats of runs and riffles are extremely rare and are limited to a few meters on top of and below dams. The significance of the flow regime, i.e. lentic (standing or slowly flowing water) versus lotic (running water), is thus difficult to assess; however, preliminary observations suggest that lotic habitats sustain a higher species richness and abundance of fish than lentic habitats. 6. Water quality conditions are a major barrier for the distribution of sensitive fish species in the Yarqon stream. Improvement of water quality to less than 10 mg l^{-1} BOD is required for at least partial recovery of most of the fish populations in the Yarqon stream.

Acknowledgment

The study was supported by grants from the Porter Super-Center for Ecological and Environmental Studies, the Israel Ministry of the Environment, the Ministry of Science and the Belpher Foundation. We wish to thank Merav Bing for technical assistance and the graduate students of the Institute for Nature Conservation Research for helping in field sampling. We are in debt to the Yarqon River Authority and its director, David Pargament, for logistic support throughout the study and to Yonatan Raz and Philip Rubinzaft for their invaluable assistance in sample collection.

References

Adams, M. S., W. D. Crumby, M. S. Greeley Jr., L. R. Shugart & C. F. Saylor, 1992. Response of fish populations and communities to pulp mill effluents: a holistic assessment. Ecotox. Envir. Safety 24: 347–360.

Allen, D. J., 1995. Stream ecology: structure and function of running waters. Chapman & Hall, London: 388 pp.

Angermeier, P. L. & J. R. Karr, 1994. Biological integrity *versus* biological diversity as policy directives: protecting biotic resources. BioScience 44: 690–697.

APHA, 1995. Standard methods for the examination of water and wastewater. American Public Health Association: 1268 pp.

Ben-Zvi, A., L. Gvaskneeht & B. Azmon, 1995. The hydrology of the Yarqon. In Pargament, D. (ed.), The Yarqon. The Yarqon Stream Authority, Ramat-Gan, Israel: 38–44.

Bernstein, I. H., C .P. Garbin & G. K. Teng, 1988. Applied multivariate analysis. Springer-Verlag, New York: 508 pp.

Diggle, P. J., 1990. Time Series: a Biostatistical Introduction. Clarendon Press, Oxford: 257 pp.

Edds, D. R., 1993. Fish assemblage structure and environmental correlates in Nepal's Gandaki river. Copeia 1993: 48–60.

Gafny, S., M. Goren & A. Gasith, 1997. Effect of habitat conditions on fish biological response in the Yarqon stream. Final project report 123–6. submitted to the Israeli Ministry of the Environment. 68 pp. (In Hebrew).

Gasith, A., 1992. Conservation and management of the coastal streams of Israel: An assessment of stream status and prospect for rehabilitation. In Boon, P. J., P. Calow & G. E. Petts (eds), River Conservation and Management. John Wiley & Sons Ltd., New York: 51–64.

Gasith A. & V. H. Resh, 1999. Streams in mediterranan climates: abiotic and biotic responses to predictable seasonal events. Ann. Rev. Ecol. Syst. 30: 51–81.

Gasith, A., M. Bing, Y. Raz & M. Goren, 1998. Fish community parameters as indicators of habitat conditions: The case of the Yarqon, a lowland, polluted stream in a semi-arid region (Israel). Verh. Int. Ver. Limnol. 26: 1023–1026.

Gerritsen, J., 1995. Additive biological indices for resource management. J. n. am. Benthol. Soc. 14: 451–457.

Gore, J. A., 1985. The restoration of rivers and streams, theories and experience. Butterworth Pub., Boston, MA: 280 pp.

Goren, M., 1974. The freshwater fishes of Israel. Israel. J. Zool. 23: 67–118.

Haslam, S. M., 1990. River Pollution: an Ecological Perspective. John Wiley & Sons, Chichester, GB: 253 pp.

Herricks, E. E. & L. L. Osborne, 1985. Water quality restoration and protection in streams and rivers. In Gore, J. A. (ed.), The Restoration of Rivers and Streams, Theories and Experience. Butterworth Pub., Boston, MA: 1–20.

Karr, J. R., 1981. Assessment of biotic integrity using fish communities. Fisheries 6: 21–27.

Karr, J. R., L. A. Toth & D. R. Dudley, 1985. Fish communities of midwestern rivers: a history of degradation. BioScience 35: 90–95.

Kauffman, J. B., R. L. Beschta, N. Otting & D. Lytjen, 1997. An ecological perspective of riparian and stream restoration in the western United States. Fisheries 22: 12–24.

Lyons, J., S. Navarro-Perez, P. A. Cochran, A. C. Sartana & M. Gusman-Arroyo, 1995. Index of biotic integrity based on fish assemblages for the conservation of streams and rivers in West-Central Mexico. Conservation Biology 9: 569–584.

Matthews, W. J., D. J. Hough & H. W. Robison, 1992. Similarities in fish distribution and water quality patterns in streams of Arkansas: congruence of multivariate analyses. Copeia 1992: 296–305.

Miller, K. L. & 13 co-authors, 1988. Regional application of an index of biotic integrity for use in water resource management. Fisheries 13: 12–38.

Munkittrick, K. R., C. B. Port, G. J. Van Der Kraak, I. R. Smith & D. A. Rokosh, 1991. Impact of bleached kraft mill effluent on population characteristics, liver MFD activity and serum steroid levels of a Lake Superior white sucker (*Catostomus ommersoni*) population. Can J. Fish. aquat. Sci. 48: 1371–1380.

Naiman, R. J. & Bilby, 1998. River Ecology and Management: Lessons from the Pacific Coastal Ecoregion. Springer-Verlag, New York: 705 pp.

Paller, M. H., M. J. M. Reichert & J. M. Dean, 1996. Use of fish communities to assess environmental impacts in South Carolina coastal plain streams. Trans. am. Fish. Soc. 125: 633–644.

Rosenberg, D. M. & V. H. Resh, 1993. Freshwater Biomonitoring and Benthic Macroinvertebrates. Chapman and Hall New York: 488 pp.

Scott, M. C. & L. W. Hall, 1997. Fish assemblages as indicators for environmental degradation in Maryland coastal plain streams. Trans. am. Fish. Soc. 126: 349–360.

Servos, M. R., S. Huestis, D. W. Whittle, G. J. Van Der Kraak & K. R. Munkittrick, 1994. Survey of receiving water environmental impacts associated with discharge from pulp mills. III. Polychlorinated dioxins and furans in muscle an liver of white sucker (*Catostomus ommersoni*), Envir. Toxicol. Chem. 13: 1103–1115.

Sparks, R. E., 1995. Need for ecosystem management of large rivers and their floodplains. BioScience 45: 168–178.

Statsoft, Inc., 1995. Statistics for Windows 5.0. Tulsa, OK.

SYSTAT (for Windows), 1992. Statistics version, 5th edn. Systat Inc., Evanston, IL: 750 pp.

Hydrobiologia **422/423:** 331–342, 2000.
M. Jungwirth, S. Muhar & S. Schmutz (eds), Assessing the Ecological Integrity of Running Waters.

Benthic invertebrate and fish communities as indicators of morphological integrity in the Enz River (south-west Germany)

W. Jansen, J. Böhmer, B. Kappus, T. Beiter, B. Breitinger & C. Hock
Institute for Zoology, University of Hohenheim, Stuttgart, Germany

Key words: river morphology, impoundment, habitat use, functional feeding groups, *Oligoneuriella rhenana*, bioindication

Abstract

Between 1994 and 1995 we sampled invertebrate and fish communities of the lower Enz River, a second order tributary of the Rhine River (total length, 108 km, mean annual discharge, 16.6–23.9 m^3 s^{-1}), in the vicinity of man-made weirs, their associated fishways, and more natural river sections. We focused on the ecological consequences of these man-made alterations to river morphology and flow conditions on the presence/absence and relative densities of invertebrate feeding groups, including the effects on fish and benthos habitat preference types. We further attempted to identify species useful as indicators of morphological integrity, particularly of physical disturbance of lotic habitats typical for the Enz River. We found pronounced differences in species composition and abundance, and in the contribution of feeding groups and habitat preference types, between disturbed and more natural sites. For example, the benthic community characteristics of two bypass fishways were quite similar to each other, but differed from those found in the respective adjacent river sections, suggesting a local 'rhithralisation' effect of fishways that mimic artificial streams. Among the 109 invertebrate and the 28 fish taxa identified, both the presence/absence and the abundance of some species or functional groups were useful indicators of the morphological integrity of typical riverine habitats. The mayfly *Oligoneuriella rhenana*, for example, occurred almost exclusively in riffle sections, and its density correlated positively with turbulent, high velocity currents and the presence of larger rocks.

Introduction

In most countries lotic ecosystems have been profoundly altered by human impacts over the last 200 years. Water pollution (Katz & Gauffin, 1933), channelization (Fruget, 1992; Steiger et al., 1998), and reduced longitudinal connectivity (Beamish & Northcote, 1989) are among the most notable of these impacts. Recent efforts in pollution control in many industrialized regions (e.g., LfU, 1992) have meant that water quality often no longer limits the distribution of lotic macroinvertebrates and fish. Despite this positive development, improvements to stream morphology have been less successful. Even today, the construction of a fish pass is often considered sufficient to maintain or to restore longitudinal connectivity for aquatic organisms in streams altered by human dams or weirs. The deleterious effects of instream barriers and channelization on the distribution and abundance of fish and macroinvertebrates are well known (Jungwirth et al., 1993; Li et al., 1987; Pehofer & Sossau, 1995), and substantial work has been done evaluating fishway performance (Schmutz et al., 1995; Schwalme et al., 1985). However, relatively few studies simultaneously document the effects of instream barriers on fish and macrozoobenthos, and data on the use of fishways by small fish species and invertebrates are largely lacking. Furthermore, assessing the ecological integrity of stream ecosystems requires determining useful indicators of the morphological quality of lotic habitats in order to complement those already established as water quality indicators in Europe. Drawing from recent concepts of river ecology, ecological integrity views running waters as complex, four-dimensional systems (Jungwirth, 1998). By focusing on the longitudinal, and to a lesser extent, lateral and vertical dimensions (including conditions of sediment, shoreline development and water velocity), we address a subset referred to as morphological integrity. In this context we quantified the distribution and abundance of invertebrates and fish with their respective functional feeding groups and habitat preference types in more natural and physically disturbed sites over a 60-km long section of a mid-sized river. We particularly focused on locations upstream and downstream of larger weirs and inside their associated fishways. Finally, we related the presence and abundance of certain species to physical habitat variables in order to select potential indicator species.

Description of locations studied

This study was carried out at three locations on the lower Enz River, a tributary of the Neckar River, which in turn flows into the Rhine River west of Heidelberg (Figure 1). The Enz is a carbonate-rich stream 108 km long with a mean annual discharge of 16.6 $m^3\ s^{-1}$ at the first sampling location near the city of Pforzheim, 50 km downstream from the source. Location II (Oberriexingen, 18.5 $m^3\ s^{-1}$ discharge) and location III (Besigheim, 23.9 $m^3\ s^{-1}$ discharge) were situated 29 and 56 km, respectively, further downstream (Figure 1). General morphology was similar at all three locations, where the Enz River can be characterized as epipotamal. According to the German rating system ('Saprobien Index'), the Enz is moderately polluted within the study area (Class II; LfU, 1992) and electrical conductivity during the study period was between 329 and 505 $\mu S\ cm^{-1}$. At all three locations, weirs of hydroelectric plants and their associated structures have altered river morphology to different degrees, and at locations I and II have created impounded sections of approximately 1 km length. Fishways of different designs have been installed at the three weirs. The fishway at Pforzheim is a 75 m long, 2 m wide concrete bypass filled with coarse gravel and regularly placed large boulders (Table 1; a more detailed descriptions of all three fishways is given in Jansen et al., 1999). Its upstream end lies more than 4 m above the river bottom and leads abruptly into the almost stagnant impounded section. The fishway at Oberriexingen is an 83 m long, 2–4 m wide artificial stream, connecting upstream with the bank of the impoundment. At Besigheim a classical step and pool fishway with four concrete basins separated by wooden boards with 10 × 10 cm wide openings has been installed. At each location, three to nine sites were chosen, yielding a total of 21 sites (Table 1) for fish and invertebrate sampling. Preliminary results had indicated that the mayfly *Oligoneuriella rhenana* (Imhoff, 1852) is a potential indicator of specific habitat conditions. To test this hypothesis with a larger sample size, two additional sampling sites each were selected for this species between locations I and II, and II and III.

Material and methods

Aquatic macroinvertebrates were captured at 17 sites (Table 1), except *O. rhenana* which was collected from 21 sites. All sites at Pforzheim and Oberriexingen were sampled four times each between May and September 1995, and the three samples at Besigheim were taken in June 1993 and 1994, and in September 1995. Invertebrates were obtained from several standardized kick-samples of bottom substrates with pond nets (0.5 mm mesh), and by hand-sampling of larger rocks, woody debris, and macrophytes. In addition, deeper bottom substrates from the impounded sites were sampled with a metal dredge (0.5×0.3 m opening, 0.3 mm mesh) which was lowered to the sediment from a small boat and twice pulled towards the shore over approximately 20 m distance. All samples were initially sorted in the field, abundances of taxa estimated, and preserved samples taken for later identification in the laboratory. Except all Acari and most Diptera and Oligochaeta, animals were determined at least to genus, and for more than 95% of all individuals, species identity was established.

Fish were collected between August 1994 and September 1995 from 14 sites (Table 1). Except for site P3, for which a single sample was obtained with pond nets when the concrete basin below the weir at Pforzheim was pumped dry on October 1, 1994, fish were collected at all other sites by electrofishing (8 kW, 200–660 V DC electro shocker, EFKO Co., D-88299 Leutkirch). Each site was fished on two to three occasions, including at least one spring and one fall sample. Sites P1.1, P4.2, and O3.3 were not completely fished. Instead, random samples were taken from all distinct habitats within these sites. All fish caught were identified to species and their total length was estimated to the nearest centimetre. Since the capture success of fish <5 cm length was quite variable between sites and was difficult to quantify, all analyses presented in the tables and figures are based on larger fish.

Table 1. Characterization of 21 sites at three locations Pforzheim (P), Oberriexingen (O), Besigheim (B) sampled for fish and macrozoobenthos on the Enz River. Sites sampled only for fish (F) or only for macrozoobenthos (MZB) are indicated. Means (range) for current speeds (C) are from 20–30 measurements at a depth corresponding to half of the actual water depth. Main bottom substrates are listed

Site	Characteristics
P 1.1	1.5 km above weir, 500 m above end of impoundment, 30 m length; river width, 50 m; depth, 0.1–1.0 m; C, 0.44 (0.3–0.6) m s^{-1}; substrate: rocks and coarse gravel; mainly natural shoreline; fished area, 750 m^2.
P 1.2	MZB; end of the impounded river section, 20 m length; river width, 40 m; depth, 0.3–1.5 m; C, 0.25 (0.2–0.4) m s^{-1}; substrate: larger rocks, gravel, sand; shoreline armoring, steep banks, few trees.
P 1.3	MZB; impoundment 500 m upstream of weir, 20 m length, river width, 50 m; depth, 0.5–2.5 m; C, 0.14 (0–0.3) m s^{-1}; substrate: few larger rocks, sand, detritus; shoreline as at P1.2.
P 1.4	MZB; impoundment 60 m upstream of weir, 10 m length; river width: 50 m; depth, 0.5–4.0 m; C, 0.11 (0–0.2) m s^{-1}; shoreline as at P1.2.
P 2.1	downstream part of fishway, 30 m length; width, 2 m; depth, 0.3–0.4 m; 10% slope; C, 0.82 (0.2–1.8) m s^{-1}; concrete lining with large boulders and gravel (2–8 cm; gravel partially washed out).
P 2.2	upstream part of fishway, 45 m length; width, 2 m; depth, 0.4–0.5 m; 2.2% slope; C, 0.51 (0.1–1.1) m s^{-1}; substrate as in P 2.1, no washed-out section; fished area, 90 m^2.
P 3	F; pool below weir; length, 15 m; width, 35 m; depth, 1.3–2.0 m; C, 0.1 (0–0.2) m s^{-1}; complete concrete lining, some sand cover, debris accumulations on water surface; fished area, 525 m^2.
P 4.1	just downstream of fishway entrance; length, 25 m; width, 10 m (river width 40 m); depth: 0.2–0.4 m; C: 0.37 (0.3–0.5) m s^{-1}; substrate: shoreline armoring, mix of rocks, gravel, and sand; fished area, 250 m^2.
P 4.2	350 m below weir, 80 m length; river width, 40–45 m; depth, 0.1–1.3 m; C, 0.5 (0.2–1.4) m s^{-1}; substrate, larger rocks and coarse gravel; mainly natural shoreline with partial tree cover; fished area, 750 m^2.
O 1.1	MZB; 2.5 km above fishway, 1.0 km above impoundment, 20 m length; river width, 25–30 m; depth, 0.1–1.1 m; C, 0.42 (0.2–1.1) m s^{-1}; substrate, rocks and gravel, mainly natural shoreline with mature trees.
O 1.2	MZB; end of the impounded river section, 20 m length; river width, 30 m; depth, 0.3–1.5 m; C, 0.28 (0.2–0.5) m s^{-1}; substrate, larger rocks, gravel, sand, little detritus; mature trees on shoreline.
O 1.3	MZB; impoundment 750 m upstream of weir, 15 m length, river width, 30–35 m; depth, 0.5–2.0 m; C, 0.19 (0–0.4) m s^{-1}; substrate, few larger rocks, gravel, sand, little detritus; mature trees on shoreline.
O 1.4	MZB; impoundment 10 m upstream of weir, 15 m length; depth, 0.5–2.5 m; otherwise as in O1.3.
O 2.1	downstream part of fishway, 45 m length; width, 2–3.5 m; depth, 0.3–0.5 m; C, 0.86 (0.3–1.7) m s^{-1}; slope, 5–9%; substrate, large and small rocks; banks with riprap, wood armoring and rooted willow branches; fished area, 120 m^2.
O 2.2	upstream part of fishway, 35 m length; width, 3–4.5 m; depth, 0.4–0.8 m; C, 0.61 (0.3–1.6) m s^{-1}; slope, 2–5%; substrate, rocks (20–90 cm) and coarse gravel; banks as at O 2.1; fished area: 125 m^2.
O 3.1	just downstream of fishway entrance, 30 m length; width, 6 m; depth, 0.2–0.8 m, gravel bank dropping steeply into channel; C, 0.68 (0.4–1.3) m s^{-1}; substrate, rocks and coarse gravel; fished area, 180 m^2
O 3.2	area below weir; depth, 0.1–0.2 m, one small pool with 0.6 m depth; C, 0.2 (0.1–0.3) m s^{-1}, temporally higher depending on discharge over the weir; substrate, rocks, gravel, sand, debris; fished area, 1500 m^2.
O 3.3	F; 40 m below fishway, length, 80 m; river width, 25–30 m; depth, 0.1–0.5 m (1.2 m in 7×8 m pool); C, 0.41 (0.2–0.7) m s^{-1}; substrate, rocks and gravel, right shore with riprap, mature trees; fished area, 1200 m^2.
B 2	F; pool-and-step fishway, length, 10 m; width, 1 m; depth, 0.5–0.8; C, 0.68 (0.6–0.8) m s^{-1}; substrate, complete concrete lining; fished area, 10 m^2.
B 3.1	F; area around downstream entrance of fishway, length, 15 m; width, 7 m; depth, 0.4–0.6 m; C, 0.1–0.2 (0.5 near fishway entrance) m s^{-1}; substrate, concrete lining, some sand and detritus; fished area, 100 m^2.
B 3.2	area below weir, 15 m length; width, 50 m; depth, 0.2–0.6 m; C, 0.64 (0.3–0.9) m s^{-1} (often very turbulent); substrate, larger and smaller rocks, some coarse gravel; fished area, 700 m^2.

At every site, physical habitat variables (water temperature, depth, current speed) were measured at each fish or invertebrate sampling. Water velocity was obtained from a model MiniAir2 (Schildknecht Co., Herisau, Switzerland) current meter and water temperature was measured electrometrically (Model LF 191, WTW Co., D-82362 Weilheim).

In order to also evaluate fish and invertebrate community structure from a functional perspective, all fish species and most invertebrate taxa identified were categorized into habitat preference types. For fish, this

Table 2. Mean number of individuals (>5 cm) of 22 fish species caught from two to three electrofishings at 14 sites on the Enz River between August 1994 and September 1995. Habitat preferences (HP) with regard to current and substrate conditions (R1, obligatory rheophilic; R2, facultative rheophilic; L, limnophilic; I, indifferent, see text) are indicated

Species	Common name	HP	P 1.1	P 2.1	P 2.2	P 3[a]	P 4.1	P 4.2	O 2.1	O 2.2	O 3.1	O 3.2	O 3.3	B 1.1	B 1.2	B 2
Lampetra planeri (Bloch)	Brook lamprey	R2	3.0	–	–	–	–	1.0	–	–	–	–	–	–	–	–
Anguilla anguilla (L.)	European eel	I	0.5	2.0	1.0	1.0	0.3	4.0	0.3	0.7	0.7	0.5	1.0	–	3.0	–
Abramis brama (L.)	Bream	L	–	–	–	11.0	–	–	–	–	–	–	–	–	–	–
Alburnoides bipunctatus (Bloch)	Spirlin	R1	16.5	1.7	4.0	16.0	13.0	12.5	–	–	1.3	1.0	1.0	–	–	–
Alburnus alburnus (L.)	Bleak	I	–	–	–	–	–	–	–	–	–	–	–	35.0	15.0	4.5
Barbus barbus (L.)	Barbel	R1	25.5	2.3	10.0	24.0	10.7	13.0	23.3	19.3	12.0	38.0	2.0	1.5	11.0	1.5
Gobio gobio (L.)	Gudgeon	R1	4.5	1.0	–	14.0	18.7	5.5	1.0	0.7	4.0	0.5	9.0	26.0	8.0	–
Leuciscus cephalus (L.)	Chub	R2	3.0	–	6.0	53.0	18.3	6.0	1.0	8.0	5.7	8.0	–	9.5	18.0	4.5
Leuciscus leuciscus (L.)	Dace	R1	–	–	–	–	0.3	4.0	–	0.3	–	0.5	–	5.0	1.0	3.0
Leuciscus souffia agassizi (Risso)	Soufie	R1	–	–	–	–	0.3	1.0	–	–	0.3	–	–	–	–	–
Phoxinus phoxinus (L.)	European minnow	R2	16.5	0.3	–	7.0	45.5	5.5	2.7	0.3	7.0	29.0	–	–	–	–
Rutilus rutilus (L.)	Roach	I	–	0.3	1.0	–	–	2.5	–	3.0	0.7	4.0	2.0	40.0	10.5	20.0
Tinca tinca (L.)	Tench	L	–	–	–	4.0	–	–	–	–	–	–	–	–	–	–
Barbatula barbatula (L.)	Stone loach	R2	27.5	6.7	14.0	–	26.3	12.5	30.7	10.7	47.7	55.5	56.0	–	7.0	–
Esox lucius (L.)	Pike	I	–	–	–	–	–	–	–	–	–	2.5	–	0.5	–	–
Thymallus thymallus (L.)	Grayling	R1	9.0	–	–	–	1.7	3.0	–	–	1.3	–	–	–	–	–
Salmo trutta (L.)	Brown trout	R1	3.0	1.0	4.5	7.0	4.0	1.0	3.3	1.7	1.0	–	4.0	–	2.0	0.5
Lepomis gibbosus (L.)	Pumpkinseed	I	–	–	–	–	–	–	–	–	–	–	–	1.5	1.5	0.5
Gymnocephalus cernuus (L.)	Ruffe	I	–	–	–	–	–	–	–	–	–	–	–	4.5	–	–
Perca fluviatilis (L.)	Eurasian perch	I	–	–	–	1.0	–	–	–	–	–	–	–	7.0	–	–
Gasterosteus aculeatus (L.)	3-spined stickleback	I	–	–	–	1.0	–	–	–	–	–	–	–	–	–	–
Cottus gobio (L.)	Bullhead	R2	5.5	19.7	2.5	5.0	8.7	12.5	1.7	1.0	1.7	1.5	4.0	–	–	–
Total		–	115	35	43	144	148	84	64	46	83	141	79	131	77	35

[a] Data are from a single fishing.

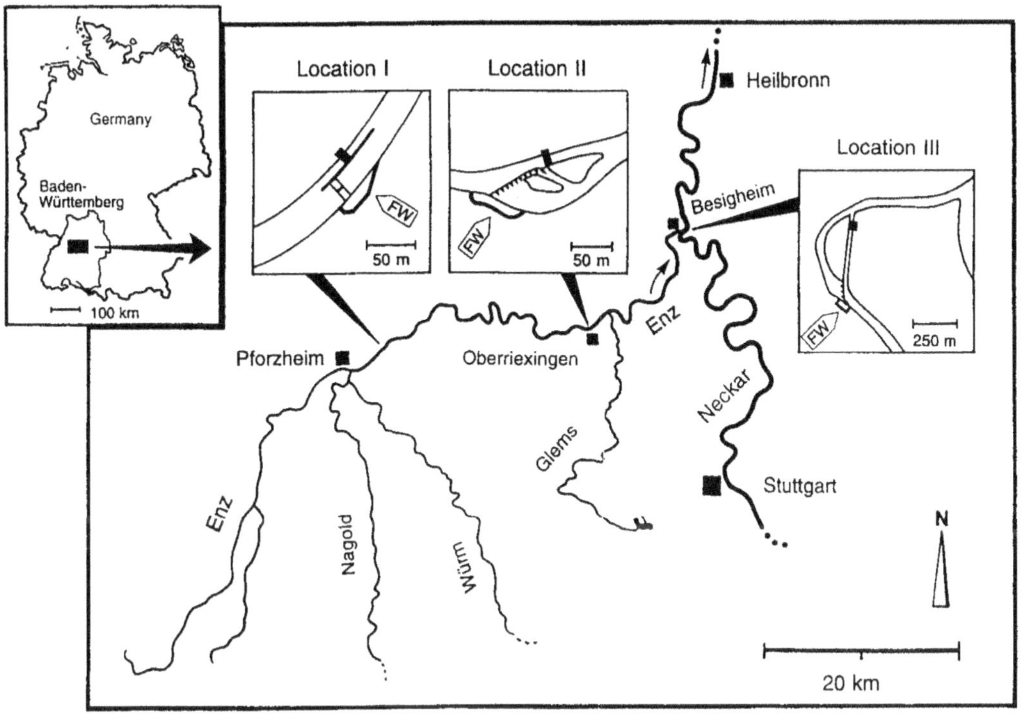

Figure 1. Map of Enz River showing the three sampling locations; FW, fishway.

classification was based on criteria regarding substrate type and current velocities established from a larger study on the Danube River system (Jungwirth, 1984). In our study, this classification was modified to reflect the habitat requirements over the entire range of a species and also to encompass regional aspects relevant to the study area. As a consequence, in addition to limnophilic (LP) and indifferent (ID) species we split the rheophilic group into obligatory rheophilic species (R1) and facultative rheophilic (R2) species (Table 2). For invertebrates, the habitat preference classification of Moog (1995) was adopted. In addition, invertebrates were also categorized into six functional feeding groups (detritivore, collector–filterer, collector–gatherer, scraper, shredder, predator), an approach that was introduced by Cummings (1973). To evaluate relative similarity of fish and invertebrate communities among sampling sites, species were clustered by applying an average linkage method described in Mountford (1962, cited in Spellerberg, 1991) on their Renkonen indices (Renkonen, 1938). This index provides a comparison of the similarity of communities based on both the present/absence of species and their relative abundance.

Results

A total of 28 species of fish and 109 taxa of invertebrates were identified in this study. However, six fish species for which just one individual was caught (*Carassius auratus gibelio, Pseudorasbora parva, Salvelinus fontinalis, Scardinius erythrophthalmus, Stizostedion lucioperca*), or species present at only one site due to recent stocking efforts (*Oncorhynchus mykiss*), were excluded from further analysis.

Fish community

Of the 22 species used in the analysis, eight were present at all three locations, seven occurred at two locations, and seven were found at only one location (Table 2). The barbel (*Barbus barbus*) was present at all 14 sites, and, after the stone loach (*Barbatula barbatula*), was also the most abundant species on the lower Enz River (Table 2). The number of species found at each site ranged from seven to 14. One of the relatively undisturbed sites (P4.2), which also exhibited the highest degree of habitat diversity, yielded the highest number of species, whereas at the similarly dimensioned but structurally more uniform site O3.2 (Table 1), the second lowest number of species was found (Figure 2). For each location, species numbers

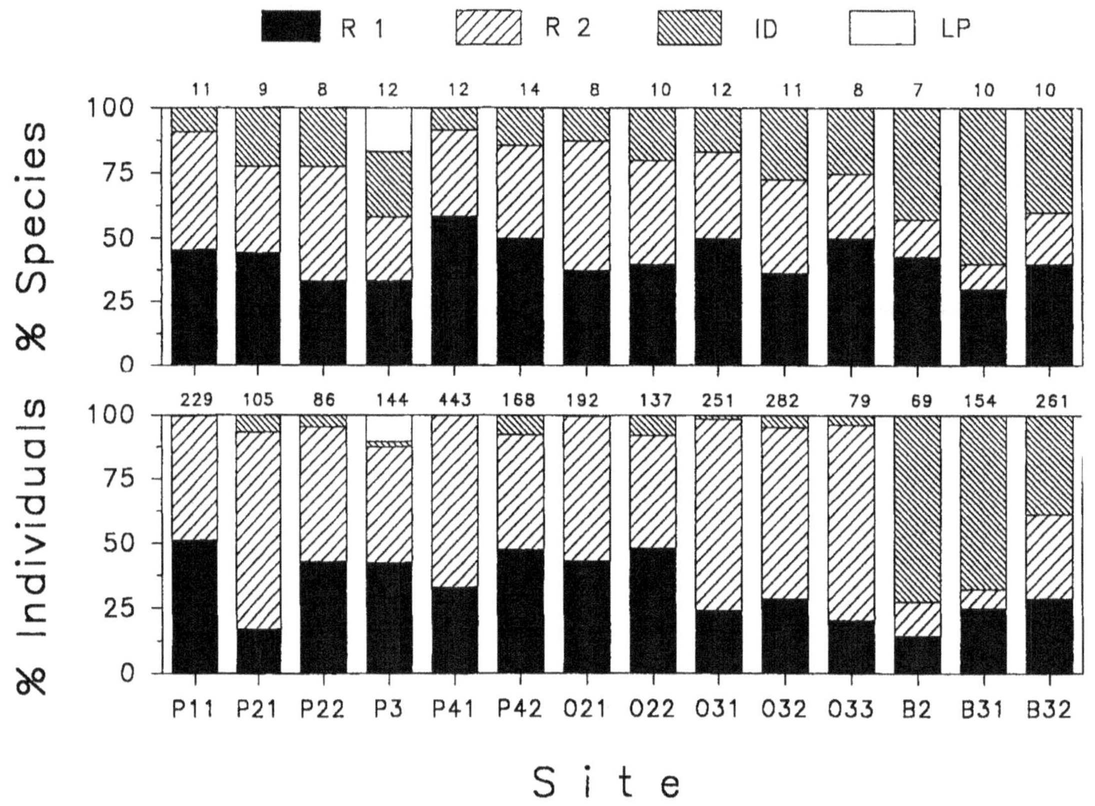

Figure 2. Contribution of fish habitat preference types (R1, obligatory rheophilic; R2, facultative rheophilic; ID, indifferent; LP, limnophilic, see text) to the total number of species (numbers above top panel) and the total number of individuals (numbers above bottom panel) at 14 sites on the Enz River. For the site numbers, periods have been omitted.

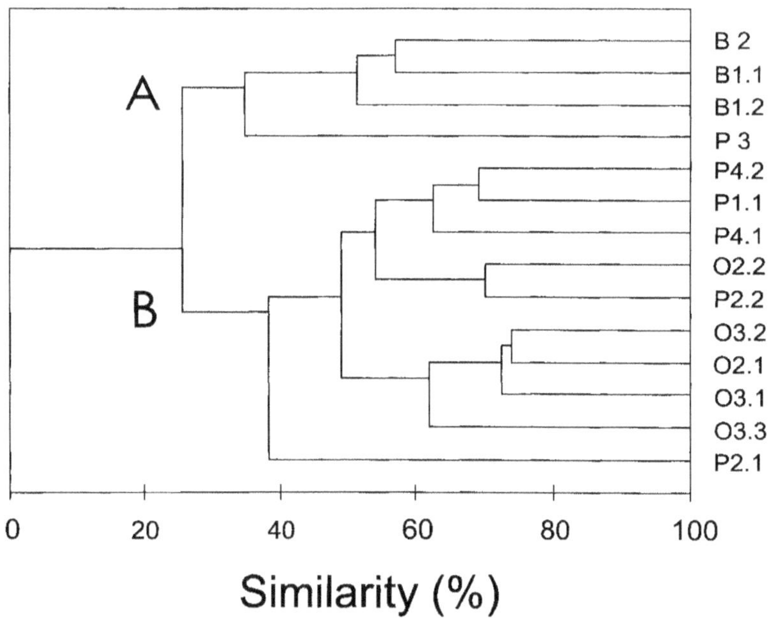

Figure 3. Similarity of fish communities from 14 sites on the Enz River based on average linkage clustering of their Renkonen indices.

inside the fishway were lower than at all other sites (Figure 2).

Tench (*Tinca tinca*) and bream (*Abramis brama*), the two limnophilic species found in this study, where exclusively caught at site P3 (Table 2), a mainly stagnant, deep pool immediately below the weir at Pforzheim. As evidenced by the inclusion in cluster A (Figure 3), the adult fish assemblage at this site was more similar to those at Besigheim 56 km downstream than to the other sites at Pforzheim. Within cluster B, with 70% similarity, the fish communities of the lower gradient, upstream sections of the fishways in Pforzheim (P2.2) and Oberriexingen (O2.2) were relatively closely associated, whereas the fish assemblages of the high gradient, downstream sections showed no such affinity. In fact, the assemblage from the downstream section of the fishway at Pforzheim (P2.1) was most distinct from all sites included in cluster B (Figure 3). This was the only site where the catch predominantly consisted of bullhead (*Cottus gobio*).

The number of rheophilic species was highest at the more natural sites (P1.1 and P4.1) and also at the downstream entrance of the two bypass fishways (P4.1 and O3.1, Figure 2). The relatively high contributions of facultative rheophilic fish at all three sites downstream of the fishway at Oberriexingen (O3) were mainly attributable to the large numbers of stone loach (Table 2). This species was captured in the relatively shallow, slow-flowing waters with sediments dominated by stones of 10–15 cm diameter at the low flow sections. Species indifferent to current conditions and sediment composition such as bleak (*Alburnus alburnus*) and roach (*Rutilus rutilus*) dominated the fish communities at Besigheim, but contributed less than 8% to individual numbers at all sites in Pforzheim and Oberriexingen (Figure 2). These large differences in the abundance of indifferent species contributed to the distinctness of fish communities between locations (cluster A and B in Figure 3).

Macrozoobenthic community

Appoximately 45 000 individuals were identified in this study, and between 16 and 56 taxa of invertebrates were found at each site. Since not all taxa could be assigned to habitat preference types and functional feeding groups, the numbers for taxa and individuals shown in Figures 4 and 5 are somewhat lower than the respective totals for each site. Particularly two of the 'unclassified' taxa, Chironomidae and Acari, contributed between 5–58 and 1–19%, respectively, to the sum of all individuals counted, although values >30% for chironomids were only observed at the impoundments. Except for sites on impounded river sections (P1.2–1.4, O1.2–1.4), the percentage contribution of rheophilic taxa to the community consistently ranged between 50 and 65%, and that of limnophilic taxa was always less than 6%, with the remainder made up by indifferent taxa (Figure 4). For those same lotic sites, the contribution of individuals of rheophilic taxa was much more variable, ranging from 30% at site B1.2 to 96% at site P1.1 (Figure 4). More than 97% of these 'rheophilic individuals' were simulid dipterans, four species of baetid Ephemeroptera, *O. rhenana*, *Ephemerella ignita*, three species of hydropsychid Trichoptera, *Rhyacophila dorsalis* (group), *Athripsodes bilineatus*, and the Coleoptera *Limnius volckmari*.

At the impounded sites, the contribution of rheophilic species to the invertebrate community was dramatically lower than at the lotic sites, falling to less than 5% of the total number of individuals. At the impoundment in Pforzheim we found large (and from site P1.2 to P1.4 increasing) numbers of the corixid *Micronecta minutissima*, which was almost entirely responsible for the substantial contribution of limnophilic species to the respective assemblages (Figure 4). Whereas numerical community composition was dominated by indifferent taxa like *Gammarus* at Oberriexingen, species of this genus were rare at Pforzheim, and this not only at the impounded sites. With regard to functional feeding groups, the lack of amphipods contributed to an almost complete absence of shredders from all Pforzheim sites, whereas this feeding group constituted at least 17% of all individuals found at the other two locations, and completely dominated the communities within the impounded river section at Oberriexingen (Figure 5). At the impoundment in Pforzheim, which featured even lower current speeds and more extensive organic sediments than in Oberriexingen, detritivores made up a large part of the invertebrate assemblage (Figure 5). The distribution of filterers was less clearly defined by location than by specific habitat features. High abundances of filterers were associated with deeper, faster flowing water, as typically found at the more natural sites P1.1 and P4.2, and also inside the fishways (P2.1, P2.2, O2). Sites featuring mainly shallower waters with moderate current speeds (e.g., P4.1, O3.1, O3.2) had relatively few filterers, but more scrapers.

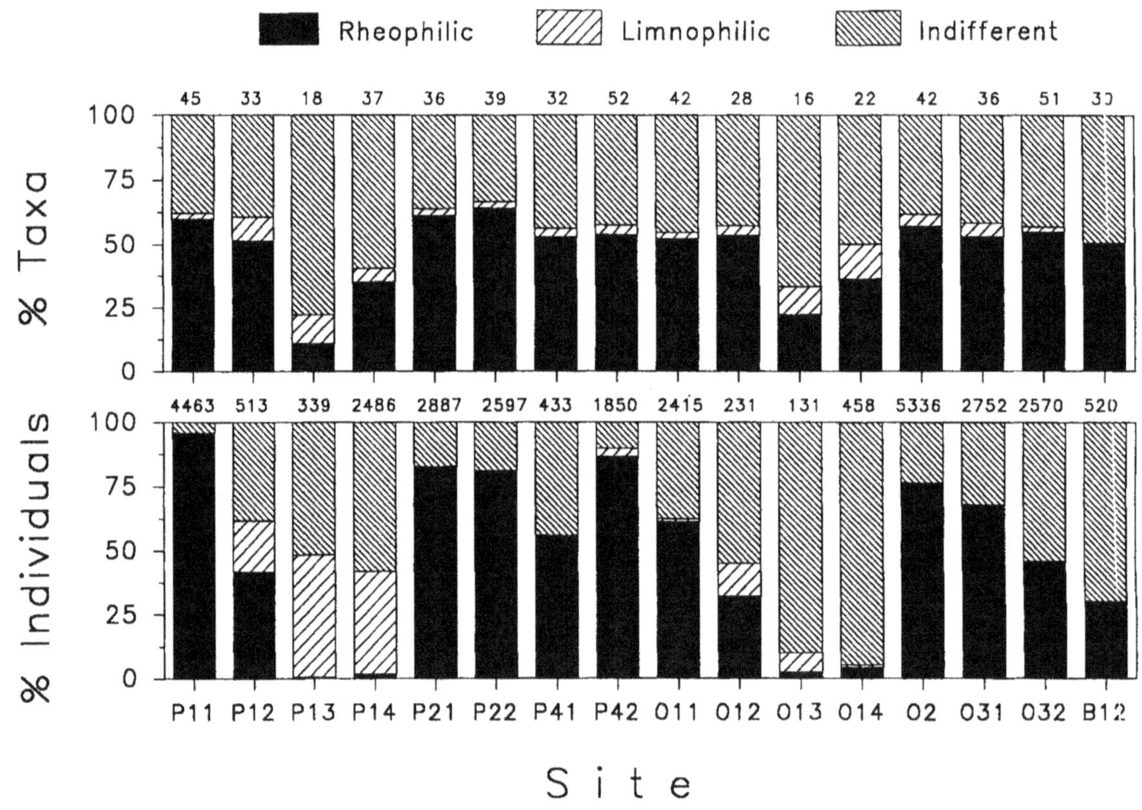

Figure 4. Contribution of invertebrate habitat preference types to the total number of species (numbers above top panel) and the total number of individuals (numbers above bottom panel) at 14 sites on the Enz River. For the site numbers, periods have been omitted.

Based on the Renkonen Index, invertebrate communities separated into three major clusters (A,B,C in Figure 6). The typical impounded sections at Oberriexingen (O1.3 and 1.4) showed only 21% similarity with the remaining assemblages, including those of the impounded sections at Pforzheim. These deeper and slow-flowing sites and the shallow site just downstream of the fishway (cluster C) shared only 26% of the species (weighted by abundance) with the remaining lotic sites from all three locations. Within the large cluster B, all the Oberriexingen sites (except for the fishway) and the one Besigheim site associated more closely. The fishway community (O2) clustered with the more natural sites (P4.2, P1.1) and those inside the fishway (P2.1, 2.2) at Pforzheim.

Distribution of the mayfly Oligoneuriella rhenana

The Ephemeroptera *Oligoneuriella rhenana* showed the most distinct abundance distribution pattern of all invertebrate species identified. This mayfly was found throughout the study range, including four locations not generally sampled for invertebrates. Even when the two bypass fishways were excluded, where even higher larval values were recorded, abundances of *O. rhenana* reached up to 800 animals per sample. The abundance of this mayfly correlated positively with faster (>0.8 m s^{-1}), turbulent currents and the presence of larger (>20 cm diameter) rocks not completely imbedded in the sediment. These conditions were usually associated with sites that were little impacted structurally.

Discussion

Fish community

The inter-site pattern of species richness found in this study confirms the general observation that habitat homogeneity diminishes aquatic species diversity and abundance (Bowlby & Roff, 1986; Jungwirth et al., 1993). The relative low number of species caught inside the fishways can be explained both by the relatively small areas of the fishpasses and by the fact that single electrofishings do not catch species that use

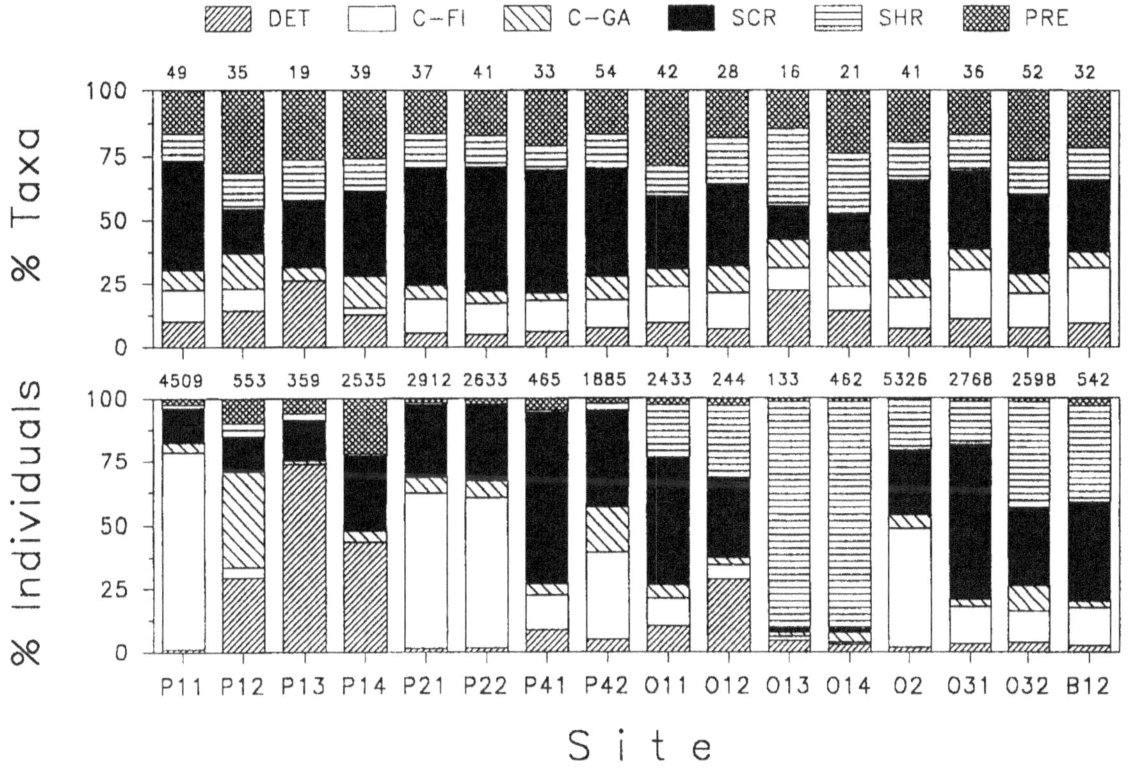

Figure 5. Contribution of engulfing detritivores (DET), collector–filterers (C-FI), collector–gatherers (C-GA), scrapers (SCR), shredders (SHR), and predators (PRE) to the total number of invertebrate taxa (numbers above top panel) and the total number of individuals (numbers above bottom panel) at 16 sites on the Enz River. For the site numbers, periods have been omitted.

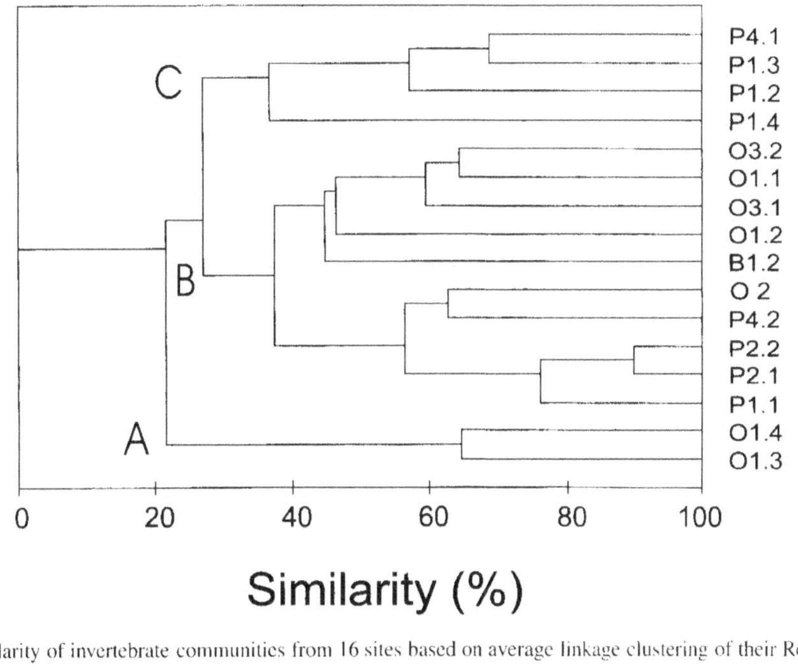

Figure 6. Similarity of invertebrate communities from 16 sites based on average linkage clustering of their Renkonen indices.

the fishways solely to migrate through during relatively short time periods. If longer term trap catches inside the fishways were also considered, fish species numbers for the bypass channels, and particularly for the artificial stream, were similar to the total number found at the respective locations (Jansen et al., 1999). This also indicates that the nature-like bypass allowed passage of most, if not all species present in its vicinity, and confirms observations that this type of fishway is effective for the upstream passage of fish and as habitat for site-specific assemblages (Eberstaller et al., 1998). Based on both these criteria, the concrete bypass was less effective in this respect. For example, the high abundance of bullheads in its downstream section suggests that this species was severely restricted in its ability to swim past a 3-m long section of almost sheer concrete with uniformly high velocities at the upper end of this part of the fishway (Jansen et al., 1999).

In addition to the larger fish, we also caught a school of more than 4000 young-of-year gudgeon (*Gobio gobio*), spirlin (*Alburnoides bipunctatus*), stone loach, and barbel when netting at the concrete pool of site P3 in October (Jansen et al., 1999). Thus, although severely altered, this site illustrates the importance of deep pools for the seasonal habitat use pattern of juveniles of fluvial specialists. This aspect of natural stream habitat diversity is often lost in channelized streams (Jungwirth et al., 1993).

The distinct differences in community composition found between locations must be viewed not only with regard to disturbances in river morphology, but also in the context of longitudinal succession (Sheldon, 1968; Shelford, 1911). This concept refers to an increase in fish species richness and diversity as habitat complexity and environmental stability increases with increasing stream size (Gormann & Karr, 1978). In contrast to those predictions, nominal species richness actually decreased downstream from 18 species at Pforzheim to 14 species at Oberriexingen, and 13 species at Besigheim. However, four of the 18 species at Pforzheim were limnophilic or indifferent taxa found only at the deep concrete pool below the weir, a habitat not naturally found at this location. Even if these fish were subtracted from the total, species numbers do not increase downstream. Habitat complexity was lower at Besigheim because of more pronounced channel regulation. Thus, the longitudinal pattern of species richness in the lower Enz River primarily followed local habitat conditions and provides further evidence that human disturbances of morphological integrity diminish or even obliterate natural community patterns.

The longitudinal succession in species diversity observed in our data reflects not so much species addition (most notably bleak, but also pumpkinseed (*Lepomis gibbosus*) and ruffe (*Gymnocephalus cernuus*)), but species replacement (most notably spirlin, European minnow (*Phoxinus phoxinus*), bullhead, and brook lamprey (*Lampetra planeri*)). At least in small and mid-sized rivers in southern Germany, these two groups of species are typically associated with more downstream and more upstream locations, respectively (Berg et al., 1989). They therefore indicate that despite the lack of pronounced hydrological and morphological differences, the lower Enz River encompasses a transitional gradient.

Overall, the fish distribution patterns observed in this study reinforce the view that community composition is a result of processes operating on both the local (i.e., substrate type) and regional (i.e., position along the continuum) scale. According to Goldstein (1981) both these determinants have to be considered in any type of environmental assessment approach, including that of ecological integrity.

Macrozoobenthic community

Our results show distinct and consistent shifts in macroinvertebrate community composition in relation to primarily human-induced alterations in river morphology. Most notable was the dramatic loss of the otherwise dominating rheophilic taxa within all impounded sections. Despite this common reduction in one habitat preference type, community similarities between the impoundments at the most upstream and the central location were low. This probably reflected the fact that the two impoundments differed sufficiently in sediment type and current regime and thus in two of the principal determinants of invertebrate distribution (Gore & Bryant, 1986; Mauch, 1963) to be inhabited by functionally different communities. One important difference was the almost complete absence of *Gammarus* not only from the impoundments but from all sites at Pforzheim, a phenomenon that has also been observed in other years (Rawer-Jost et al., 1998). The reason for the very low abundance of this otherwise common amphipod is not known, although the considerably lower allochthonous input of leaves at Pforzheim compared to Oberriexingen may be partially responsible. Again, this example cautions, that the assessment of local disturbances to stream morphology must be validated also in light of large-scale distribution patterns of the taxa involved.

Nevertheless, phenomena operating on a more local scale are also important with regard to macroinvertebrate distribution. For example, the distribution pattern of functional feeding groups in the lower Enz River, in particular that of filterers, was clearly site specific rather than location specific. This confirms and extends results by Gore & Bryant (1986), who suggested that for invertebrates, in partial contrast to fish (see above), functional group similarities are correlated more closely to local habitat features than to stream order.

The community similarities found in this study between the natural upstream site and the fishway 30 km downstream suggest that relatively short bypass fishways with constructional features that mimic natural streams apparently do not present distributional barriers. Instead, they seem to reset invertebrate communities to a lower-order stream community. In the lower Enz River this resulted in a local 'rhithralisation' effect.

Conclusions

The data from the Enz River allow a number of conclusions that are of general relevance with regard to the ecological integrity of streams and its assessment. Among the fishpasses investigated, the nature-like fishway is best suited to restore or to improve interrupted migratory pathways of fishes, and thus an important component of river ecological integrity. Indicators suitable to assess further components of morphological integrity, smaller, locally abundant fish species (i.e., bullhead, stone loach) were instrumental in documenting deficits in sediment structure or velocity patterns. For the Enz River, and probably other streams of similar or smaller size in central Europe, these species have to be given greater attention (see also Peter, 1998). Among the invertebrates identified in this study, ***O. rhenana*** must be considered a potential indicator for the morphological integrity of streams. This mayfly is currently expanding into its former river habitats in Germany (Jansen, in press), and its easy identification makes it suitable for large-scale surveys. However, for benthic invertebrates in general, site differences on the Enz River, and thus the degree of morphological deterioration, were much more evident from the relative abundance of taxa or functional groups than from simple taxa composition. This suggests that also semi-quantitative sampling methods and guild analyses should be included in studies that attempt to assess the ecological integrity of running waters.

Acknowledgements

We thank Christine Bratrich, Barbara Danner, Pearl Fok, Ina Maihöfer, and Claudia Rawer-Jost for their assistance during field sampling. Discharge data for the Enz River were obtained from the 'Landesanstalt für Umweltschutz' (LfU) Baden-Württemberg. The 'Regierungspräsidium' Karlsruhe (Dr Kieckhäfer) and Stuttgart (Dipl.-Ing. Wnuck), and the angling clubs of Pforzheim, Niefern, Oberiexingen, and Besigheim granted permission for the electrofishings. Mr Walter from the angling club Niefern and Mr Geist from the angling club Besigheim helped collect fish from the fishways at site I and III, respectively, and provided information on fish stockings. *The original manuscript benefitted from the constructive criticism by two anonymous reviewers.* Financial support was provided by the 'Gewässerdirektion Neckar, Bereich Besigheim'.

References

Beamish, R. J. & T. G. Northcote, 1989. Extinction of a population of anadromous parasitic lamprey, *Lampetra tridentata*, upstream of an impassable dam. Can. J. Fish. aquat. Sci. 46: 420–425.

Berg, R., S. Blank & T. Strubelt, 1989. Fische in Baden-Württemberg. Ministerium für Ländlichen Raum, Ernährung, Landwirtschaft und Forsten Baden-Württemberg, 158 pp.

Bowlby, J. & J. Roff, 1986. Trophic structure in southern Ontario streams. Ecology 67: 1670–1679.

Cummins, K. W., 1973. Trophic relations in aquatic insects. Ann. Rev. Ent. 18: 183–206.

Eberstaller, J., M. Hinterhofer & P. Parasiewicz, 1998. The effectiveness of two nature-like bypass channels in an upland Austrian river. In Jungwirth, M., S. Schmutz & S. Weiss (eds), Fish Migration and Fish Bypasses. Fishing New Books, Oxford: 363–383.

Fruget, J. F., 1992. Ecology of the lower Rhône after 200 years of human influence: a review. Regul. Rivers Res. Mgmt 7: 233–246.

Goldstein, R. M., 1981. Longitudinal succession in impact assessment of river system fish communities. Wat. Res. Bull. 17: 75–81.

Gore, J.A. & R.M. Bryant Jr., 1986. Changes in fish and benthic macroinvertebrate assemblages along the impounded Arkansas River. J. Freshwat. Ecol. 3: 333–345.

Gorman, O. T. & J. R. Karr, 1978. Habitat structure and stream fish communities. Ecology 59: 507–515.

Jansen, W., in press. Bestandssituation der 'Rheinmücke' *Oligoneuriella rhenana* (Oligoneuriidae, Ephemeroptera) in Deutschland. Verh. Westdt. Entom. Tag 1998.

Jansen, W., B. Kappus, J. Böhmer & T. Beiter, 1999. Fish communities and migrations in the vicinity of fishways in a regulated river (Enz, Baden-Württemberg, Germany). Limnologica: 425–435.

Jungwirth, M., 1984. Die fischereilichen Verhältnisse in Laufstauen alpiner Flüsse, aufgezeigt am Beispiel der oesterreichischen Donau. Oesterreichische Wasserwirtschaft 36: 104–111.

Jungwirth, M., 1998. River continuum and fish migration — going beyond the longitudinal river corridor in understanding ecological integrity. In Jungwirth, M., S. Schmutz & S. Weiss (eds), Fish Migration and Fish Bypasses. Fishing New Books, Oxford: 19–32.

Jungwirth, M., O. Moog & S. Muhar, 1993. Effects of river bed restructuring on fish and benthos of a fifth order stream, Melk, Austria. Regul. Rivers Res. Mgmt. 8: 195–204.

Katz, M. & A. R. Gauffin, 1933. The effects of sewage pollution on the stream fish population of a midwestern stream. Trans. am. Fish. Soc. 82: 156–165.

LfU, 1992. Gütezustand der Gewässer in Baden-Württemberg. 7. Zustandsuntersuchungen auf biologisch-ökologischer Grundlage. Ministerium für Umwelt Baden-Württemberg, Wasserwirtschaftsverwaltung Heft 27, Karlsruhe, 64 pp.

Li, H. W., C. B. Schreck, C. E. Bond & E. Rexstad, 1987. Factors influencing changes in fish assemblages of Pacific northwest streams. In Mathews, W. J. & D. C. Heins (eds), Community and Evolutionary Ecology of North American Stream Fishes. Oklahoma University Press, Norman, OK: 193–202.

Mauch, E., 1963. Untersuchungen über das Benthos der deutschen Mosel unter besonderer Berücksichtigung der Wassergüte. Mitt. Zool. Mus. Berlin 39: 3–172.

Moog, O., 1995. Fauna Aquatica Austriaca. Katalog zur autökologischen Einstufung aquatischer Organismen Österreich, 1st edn, Wien, Bundesministerium für Land- und Forstwirtschaft, Loseblattsammlung.

Pehofer, H. E. & C. Sossau, 1995. Ecological changes in an impoundment in a fast-flowing gravelbed river (Innstau Kirchbichl. Tyrol, Austria). Limnologica 25: 109–128.

Peter, A., 1998. Interruption of the river continuum by barriers and the consequences for migratory fish. In Jungwirth, M., S. Schmutz & S. Weiss (eds), Fish Migration and Fish Bypasses. Fishing New Books, Oxford: 99–112.

Rawer-Jost, C., B. Kappus, J. Böhmer, W. Jansen & H. Rahmann, 1998. Upstream movements of benthic macroinvertebrates in two different types of fishways in south-west Germany. Hydrobiologia 391(1/3): 47–61.

Renkonen, O., 1938. Statistisch-ökologische Untersuchungen über die terrestrische Käferwelt der finnischen Bruchmoore. Ann. zool. Soc. Vanamo 6: 1–231.

Schmutz, S., H. Mader & G. Unfer, 1995. Funktionalität von Potamalaufstiegshilfen im Marchfeldkanalsystem. Österreichische Was. Abfallwirtschaft 47: 43–58.

Schwalme, K., W. C. Mackay & D. Lindner, 1985. Suitability of vertical slot and denil fishways for passing north-temperate, nonsalmonid fish. Can. J. Fish. aquat. Sci. 42: 1815–1822.

Sheldon, A. L., 1968. Species diversity and longitudinal succession in stream fishes. Ecology 49: 193–198.

Shelford, V. E., 1911. Ecological succession. I. Stream fishes and the method of physiographic analysis. Biol. Bull. 21: 9–35.

Spellerberg, I. F., 1991. Monitoring Ecological Change. Cambridge University Press, Cambridge: 334 pp.

Steiger, J., M. James & F. Gazelle, 1998. Channelization and consequences on floodplain system function on the Garonne River, SW France. Regul. Rivers Res. Mgmt. 14: 13–23.

Hydrobiologia **422/423**: 343–358, 2000.
M. Jungwirth, S. Muhar & S. Schmutz (eds), Assessing the Ecological Integrity of Running Waters.

Identification of rivers with high and good habitat quality: methodological approach and applications in Austria

Susanne Muhar, Michaela Schwarz, Stefan Schmutz & Mathias Jungwirth
University of Agricultural Sciences Vienna Department of Hydrobiology, Fisheries and Aquaculture, Max-Emanuelstr. 17, A-1180 Vienna, Austria
Tel.: [+43] 1 47654-5210. Fax: [+43] 1 47654-5210. E-mail: smuhar@edv1.boku.ac.at

Key words: river habitat, habitat quality, habitat integrity, ecological integrity, physical environment, habitat assessment, assessment method

Abstract

The investigation of the 52 largest Austrian rivers with catchment areas >500 km^2 (Danube River excluded) provides a national estimate of the ecological status of Austria's rivers and an example of the current status of European alpine rivers. Emphasis is placed on evaluation criteria, such as morphological character, instream structures, longitudinal river corridor, lateral connectivity and hydrological regime compared with original conditions. This assessment and evaluation of nearly 5000 river kilometers identifies the remaining river stretches with high habitat quality as well as those stretches that have been altered by systematic channelisation or hydropower development. Altogether, about 80% (3900 km) of the stretches in Austria's 'larger' rivers are moderately to heavily impacted by human activity and thus no longer correspond to their original channel form and dynamics. Impoundment, water diversion and hydropeaking each impact the hydrology of approximately 20% of the above stretches. Only about 20% of all investigated river stretches can still be classified as having high or good habitat quality and as retaining essentially natural functions. Using natural conditions as a reference, the most severe losses have occurred in reaches that were originally braided or meandering. Analysing the amount of unaltered river stretches according to biocoenotic regions, 37% (523 km) of the originally rhithral (trout) region (1416 km) and 18% (480 km) of the originally hyporhithral (grayling) region (2739 km) remain today in high or good habitat quality. Only a few intact river stretches (totaling 43 km) can be found along formerly potamal river reaches (731 km). These data underline the rarity and, therefore, national importance of the remaining unspoiled aquatic habitats. The results presented here form a basis for water resource management and nature conservation strategies, as well as for restoration programs. In particular, these investigations help to fulfill the requirements of the EU-Water Framework Directive by assisting to delineate and characterise reference sites of different river types.

Introduction

The morphology, longitudinal and lateral connectivity, as well as bed load- and discharge regime of most running waters in Central Europe are severely disturbed. In their general overview of the alpine region, Martinet & Dubost (1992) reported that only 10% of the river reaches can be classified as 'near-natural'. Based on Müller's (1991) vegetation studies, only fragments of the large braided river systems have been preserved. This development is by no means restricted to the densely populated regions of Central Europe. Only 2% of 5.2 million kilometers of U.S.A. rivers were classified as relatively pristine (Benke, 1990). In Australia, a national habitat evaluation of rivers and streams revealed a relatively high degree of impairment through logging, grazing and construction of reservoirs for drinking water and irrigation (Stein et al., 1998). Dynesius & Nilsson (1994) produced a rough estimate of the proportion of rivers that have been degraded through fragmentation or flow regulation in the northern third of the world (North America north of Mexico, Europe and the Republics of the former Soviet Union). Of the 139 largest river systems (defined as rivers with an average annual discharge of >350 m^3 s^{-1}), the hydrological regime of 77% is classified as moderately or strongly altered.

Channel straightening and stabilisation, daming, water diversion and hydropeaking, bedload retention and wastewater inputs often severely alter the aquatic fauna and flora. This is reflected in an increasing loss of species, as well as other negative community level impacts. From an ecological perspective, the remaining river stretches with largely intact morphological and hydrological conditions are, therefore, essential components of water resource management, particularly within heavily impacted systems. Efforts to secure and renew drinking water reserves increasingly focus on such river sections (Stanford & Ward, 1992). These areas also function as refuges for species whose habitats have been changed or destroyed, and provide a source of organisms recolonizing adjoining, formerly impacted habitats that have been improved by remediation or mitigation.

Future programs directed at protecting, restoring or using aquatic habitats will, therefore, require basic nationwide inventories or surveys of intact river ecosystems. The results of such investigations form the basis for management strategies, including identification of priority areas for restoration, conservation, environmental protection and biotope networks. At the same time, these data produce an underlying grid system for detailed future investigations of abiotic and biotic features, exchange processes and ecological functions of near-natural reference stretches (Naiman et al., 1995).

National inventories of river quality can also provide a valuable input into future guidelines for Europe-wide water management policies currently under discussion by identifying the remaining intact river stretches with 'high or good water quality' (EUROPEAN COMMISSION, 1999). Therefore, the objective of this paper is to describe the results of such an inventory for all Austria's rivers with catchments >500 km^2. The Danube, as the largest river in Austria, was excluded from these investigations due to its unique hydrological and morphological dimensions.

Habitat quality assessment methods

A survey of internationally applied assessment methods

In comparison with water quality studies, the large-scale survey and evaluation of physical habitat components is a relatively young discipline. Methods designed to document and evaluate the physical integrity of aquatic habitats over entire catchments were not developed until the 1980s. In German-speaking countries, initial efforts have concentrated on the morphology and structural features of the river bed and riparian zone (LÖLF, 1985; Böttger, 1986; Patzner, 1986; Werth, 1987; Spiegler et al., 1989; Rose, 1990; Bauer, 1992; Konold, 1992; Bohl et al., 1992; LAWA, 1993). Results of such studies were mostly published as maps primarily referencing structural components of the rivers (LFU, 1994). These methods were expanded to assess entire river/floodplain systems, as it was done for example in Germany in the framework of a nationwide investigation on aquatic habitats (Binder & Kraier, 1999). The 'modular concept for comprehensive assessment and evaluation of streams in Switzerland' (Bundi et al., 2000) is another example of an application of large-scale habitat assessment systems, complemented by five biological modules.

In North America, physical habitat assessment has increasingly become a standard element in government monitoring programs. Abe et al. (1992) report that most state water quality agencies in the U.S.A. have initiated investigations to qualitatively address physical habitat integrity. Especially in Ohio, qualitative evaluation of habitat features is already an integrated part of surface water monitoring (Ohio EPA, 1988; Rankin, 1995). The U.S. Environmental Protection Agency's Environmental Monitoring and Assessment Program conducts quantitative habitat assessments at all randomly selected sites, as well as a landscape characterisation of those sites' catchments (Kaufmann & Robinson, 1998; Hughes et al., 2000). Physical habitat inventories of streams are also conducted by the U.S. Forest Service and U.S. Bureau of Land Management.

Similar approaches are registered in the U.K., where the river corridor survey (NRA, 1992) and the subsequent river habitat survey (RHS, Raven et al., 1997, 1999) are being systematically carried out to assess the state of river habitats nationally. SERCON is a further, conservation-oriented evaluation program combining physical, chemical and biological features of rivers and their floodplains (Boon et al., 1997; Boon, 2000).

Method for identifying Austria's running water habitats with high and good quality

The methodological approach to evaluate and identify

intact aquatic habitats was formulated in the framework of a preliminary methodological study (Muhar et al., 1993); it builds on well-known ecomorphological assessment approaches, especially those applied in German-speaking countries, as mentioned above, e.g. by LÖLF (1985), Werth (1987), Spiegler et al. (1989), Konold (1992) and Bohl et al. (1992). It focuses on surveying numerous fundamental elements of ecologically healthy watersheds as, for example, described by Naiman (1992) and Muhar & Jungwirth (1998). In addition to considering established morphological-structural parameters, emphasis is placed on hydrological conditions, as well as their modification through water diversion, hydropeaking or daming; anthropogenic changes related to associated wetlands and, therefore, lateral connectivity are also included. The evaluation also gives consideration to criteria affecting the dynamics of running waters (erosion and sedimentation processes, flooding), as well as to longitudinal connectivity. Finally, the collected data on the individual river habitats are related to the original river type, as promoted both in the legal regulations of Austria's Water Rights Act and in the current version of the EU's Water Framework Directive.

The evaluation and identification of largely intact aquatic habitats is based on the conditions that characterised Austria's rivers and streams prior to systematic channel regulation measures (end of 19th to mid 20th century) and the construction of large hydroelectric stations and chains of dams. In contrast to modifications caused by local flood protection measures, clog drivers, mills, channel adaptation for navigation etc. before 1800, these systematic impacts lead to large-scale alterations of whole river systems. For the major European rivers, Petts (1989) sets the date for the dramatic alterations of their hydrological and geomorphological characteristics at 1850, although anthropogenic changes caused by weirs, improvements for navigation etc. have been significant over a longer timescale.

By setting this date, the current evaluation refers back to a scenario of 'naturally developed' Central European cultural landscapes in which the fundamental functions of most river systems were largely intact and system-inherent processes could still run their natural course. In Austria, a national land survey map was developed in 1806–1850, providing a high-quality physical reference for all rivers.

The first step of the assessment procedure (see Figure 1) involves the characterisation/classification of the river systems to be investigated. To describe different types of rivers, primarily physiographic and hydrological features and one biological component – the original fish regions – are used (compare Table 1). The relevant data are collected and analysed based on

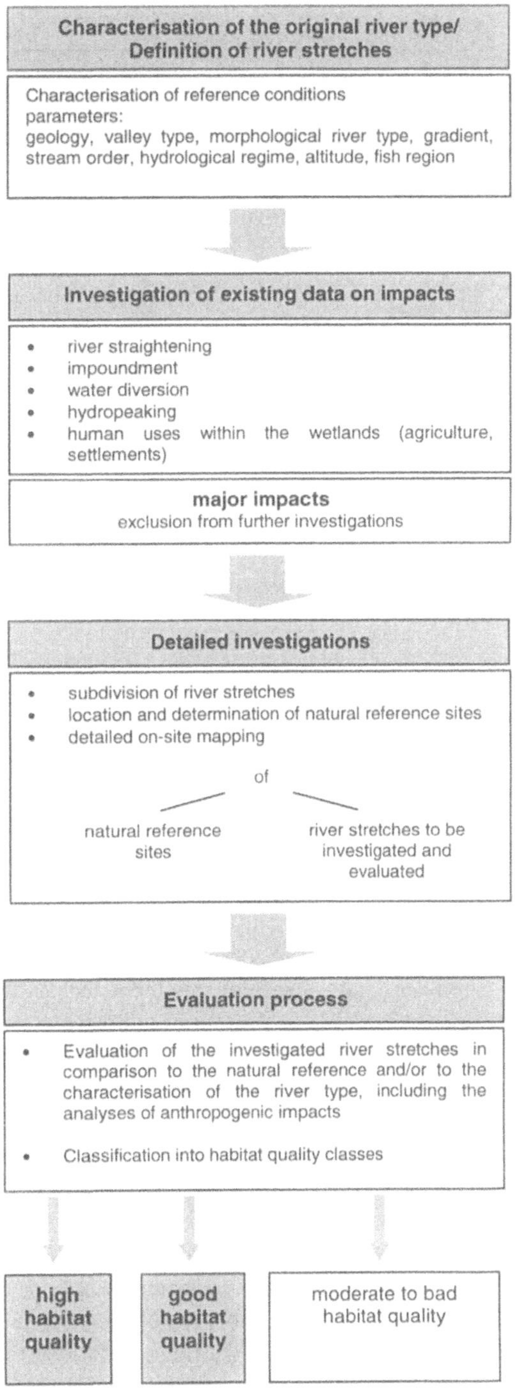

Figure 1. Flow diagram of the assessment process.

Table 1. Attributes for the general characterisation of the investigated river stretches

Characteristics	Source/reference
Geology	Geologic map 1: 100 000
Altitude/vegetation zone	Topographic map (ÖK 1: 50 000), after Ellenberg, 1996
Valley type	Field survey, classification after Wilhelmy, 1990
Gradient	Topographic map (ÖK 1: 50 000)
Morphological river type[a]	Analyses of historic data (classification after Mangelsdorf et al., 1990)
Stream order	Strahler, 1957; Wimmer & Moog, 1994
Mean discharge[b]	Historical hydrographs
Hydrological regime[b]	Pardé, 1947; Steidl, 1991; Mader et al., 1996
Biocoenotic region[a]	Interpretation of existing data (historic and present), after Thienemann, 1925; Illies, 1961

[a]Reference scale: 1st half of the 19th century.
[b]Reference scale: prior to hydrological alterations caused by hydroelectric plants.

available records. These data, in particular geology, morphological river type and gradient, form the basis for the definition of study units. Anthropogenic impacts are used as criterion for a further division into sub-units.

To define the habitat conditions and biocoenoses prior to systematic channel regulation or to disturbances in the discharge regime caused by hydroelectric development, all available historical data are analysed. This includes historical hydrographs (compare also Galat & Lipkin, 1999), fisheries reports (Haidvogl & Waidbacher, 1997; compare also Schmutz et al., 1999) and the Austrian land survey from the early 1800s.

The next stage involves determining which stretches to exclude from further investigation due to major impacts such as systematic channel straightening, impoundment, water diversion and/or hydropeaking. This is done using available data from topographic maps, aerial photographs, water resource management plans and regulation project documents.

This is followed by detailed field investigations of natural reference sites (corresponding to the above-defined pre-industrial reference conditions) as well as of all remaining river stretches to be investigated. Table 2 indicates the assessment criteria used.

During field mapping (supplemented by data research), all these listed assessment parameters are recorded by assigning them to frequency classes (dominant, frequent, rare, absent) or by verbal description. Considering the purpose of the study and the working scale a further quantitative differentiation has not been regarded as practical.

In addition, recognisable impacts in the river and its surroundings are registered (reinforcements, dams, gravel mining, agricultural land use in the floodplain area, etc.). A specially developed data sheet provides a framework enabling standardised documentation of the fieldwork.

Subsequently, the collected data on the individual river stretches are related to the general characteristics of the original river type and to the detailed data of existing reference sites either within the river system itself or in neighbouring systems with a comparable character. Finally, the stretches are assigned to the following habitat classes:

High habitat quality
Conforms totally to the reference conditions in terms of morphology, river-bed dynamics, floodplain conditions and longitudinal connectivity. No influence on the discharge regime through impoundment, water diversion, or hydropeaking occurs (see Table 3).

As the reference level is set very high, very few rivers of the cultural landscape are able to meet these requirements. Therefore, a second quality class – *good habitat quality* – is defined to identify those river stretches that still retain their overall character as they are not altered through industrialised development (e.g. river engineering or hydroelectric stations). Only slight deviations from the type-specific natural state, primarily due to land use practises (agriculture, silviculture) within the floodplains, occur (see Table 3).

The final assessment process is based on professional interpretation and judgement of differences in habitat characteristics between reference conditions and investigated river stretches.

Moderately to heavily impacted
As the main goal of this study was to document the remaining rivers with intact running water habitats,

Table 2. Attributes for detailed field investigations

Criteria	Assessment parameters	Data type
River morphology	• gradient	numerical
	• width	size classes (3)
	• river bed morphology	
	– pool	
	– riffle	
	– bar	frequency classes (4)
	– island	
	– cut off bank	
	– slip off bank	
Instream structures	• structure type	
	– single blocks	
	– zones of lateral erosion	
	– fine sediment bar	
	– coarse sediment bar	frequency classes (4)
	– woody debris	
	– root wads	
	– macrophytes	
	– overhanging vegetation	
Substrate	• mineralic substrate type	
	– silt	
	– sand	
	– fine gravel	
	– medium gravel	frequency classes (4)
	– coarse gravel	
	– cobble	
	– boulder	
	• organic substrate type	
	– coarse organic matter	
	– xylal	
	– fine organic matter	
	– saprobel	
	– phytal	
Hydrology/hydraulics	• low/mean/high flow	numerical
	• flow pattern	
	– fall	
	– rapid (cataract)	frequency classes (4)
	– rippled	
	– smooth	
	– no flow	
Riparian vegetation	• width/structure	frequency classes (4)
	• linear extension	frequency classes (4)
	• dominant plant species	species
Floodplain conditions	• extent of wetland-vegetation	verbal description
	• dominant plant species	species
	• types of water bodies	verbal description
Lateral connectivity	• groundwater table	numerical
	• flooding	frequency classes (2)
	• inundation area	numerical
Longitudinal corridor	• free migration/artificial barriers	verbal description/numerical

Table 3. Habitat conditions of river stretches with high and good habitat quality

	High habitat quality	Good habitat quality
Overall river character	Its inherent functions and processes **reflect the original river type**. The aquatic, semiaquatic and terrestrial habitat characteristics of this type are abundant.	Its inherent functions and processes **reflect the original river type**. The aquatic, semiaquatic and terrestrial habitat characteristics of this type are present, but distinctly reduced.
Morphology Instream structures Hydraulics Substrate	Morphology is **unchanged**. Features, structures and hydrology **correspond** to the river type. Substrate classes are all present in the appropriate proportions. **Shifts in the relative contributions of the fractions do not alter dominance relationships**.	Morphology is **largely unchanged**. Features, structures and hydrology principally **correspond** to the river type. Substrate classes are all present in the appropriate proportions. **Shifts in the relative proportions of the fractions are possible**.
Riparian and floodplain vegetation Processes at the floodplain level	Extent, size-class structure and species composition of the riparian and floodplain vegetation show **no or only minimal human impact**. Key site-specific conditions such as flooding, typical seasonal water levels and their amplitudes, the development of pioneer locations, etc. are **maintained**. Water bodies of the floodplains **are present in their typical diversity and abundance**; they are **periodically** subject to natural processes such as erosion, sedimentation, and new formation.	Extent, size-class structure and species composition of the riparian and backwater vegetation show **some human impact**. Key site-specific conditions such as flooding, typical seasonal water levels and their amplitudes, the development of pioneer locations, etc. are **predominantly maintained**. Although the typical diversity and abundance of water bodies of the floodplains are **reduced**, they are **at least episodically** subject to natural processes such as erosion, sedimentation, and new formation.
Connectivity	Dynamic and exchange processes in the longitudinal, lateral and vertical dimension are **not** or only **minimally restricted** by anthropogenic impacts.	Dynamic and exchange processes in the longitudinal, lateral and vertical dimension are **partially** restricted by anthropogenic impacts.
Discharge regime	Hydrological and hydraulic conditions are not or only extremely minimally influenced by water diversion and/or hydropeaking.	Hydrological and hydraulic conditions are not or only extremely minimally influenced by water diversion and/or hydropeaking.

no further differentiation has been made for rivers which do not fall under one of the above-mentioned categories.

Results

Characteristics of Austria's larger running waters

The 52 investigated rivers with a total length of 4915 km are distributed relatively homogeneously over Austria's different landscape types. Included are the larger rivers of the Bohemian Massif, those along the eastern margin of the Alps, as well as the main alpine river systems.

The size and distribution of the rivers are clearly reflected in stream order and average annual discharge (mean annual flow prior to major hydroelectric plant construction). Stream order classes range from 3 to 8, but stream order 5 and 6 make up 60% of all investigated systems (Figure 2). Although the focus of the study was to cover the largest rivers of Austria, the investigated river stretches represent a wide range of mean annual flow values. This is due to the fact that the upper sections (beginning with stream order 3) of all investigated rivers have also been included (Figure 3).

Channel morphology broadly integrates a range of features such as 'adjacent wetlands' (including different types of waterbodies within the floodplains, flood-

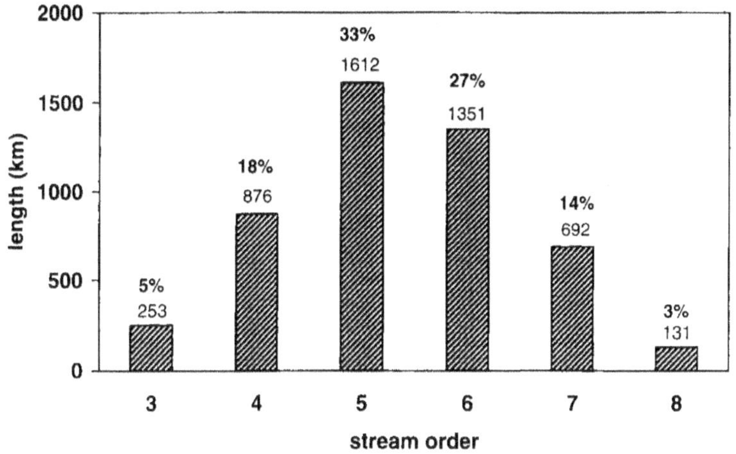

Figure 2. Frequency histogram of the total number of stream kilometers in eight stream order classes for 52 Austrian rivers.

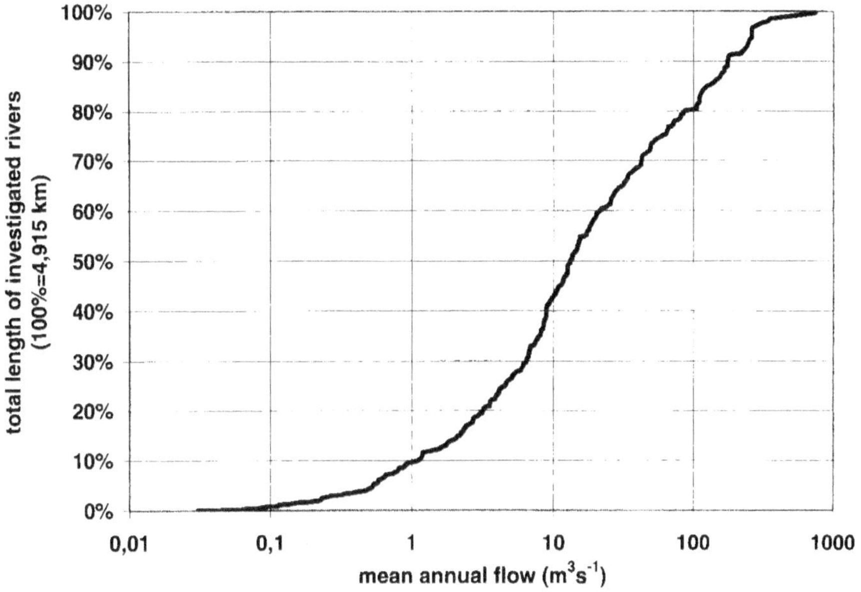

Figure 3. Cumulative distribution of mean annual flow *versus* total length of 52 investigated Austrian rivers (data pertain to the situation prior to power plant construction).

plain vegetation; see also Allen, 1977), discharge-/bedload-dynamics, longitudinal and transversal connectivity (Ward & Stanford, 1995), river bed morphology, substrate and structural conditions. Based on the national land survey (1805–1860, Figure 4), the present Austria-wide investigations show that braided (including the transitional 'winding' form) and meandering channels were formerly typical for the larger river systems. Originally, these two channel types each represented about 25% of all investigated systems (compare also Figure 8, definition of morphological characters after Mangelsdorf et al., 1990).

Constrained river stretches were and still are the typical channel form in alpine-montane regions. They occur frequently in upper reaches, as well as in side tributaries of the large longitudinal valleys. Typical, extensively braided sections were most common in the broad alluvial valley bottoms of the larger inner alpine rivers. The relatively high proportion of formerly meandering rivers can be explained by the extensive basin landscapes in the study area. Nevertheless, both braiding, as well as meandering channel types, could also be found in the formerly glaciated bottom of high alpine valleys: this situation is very typical of the longitudinal profile of many alpine rivers.

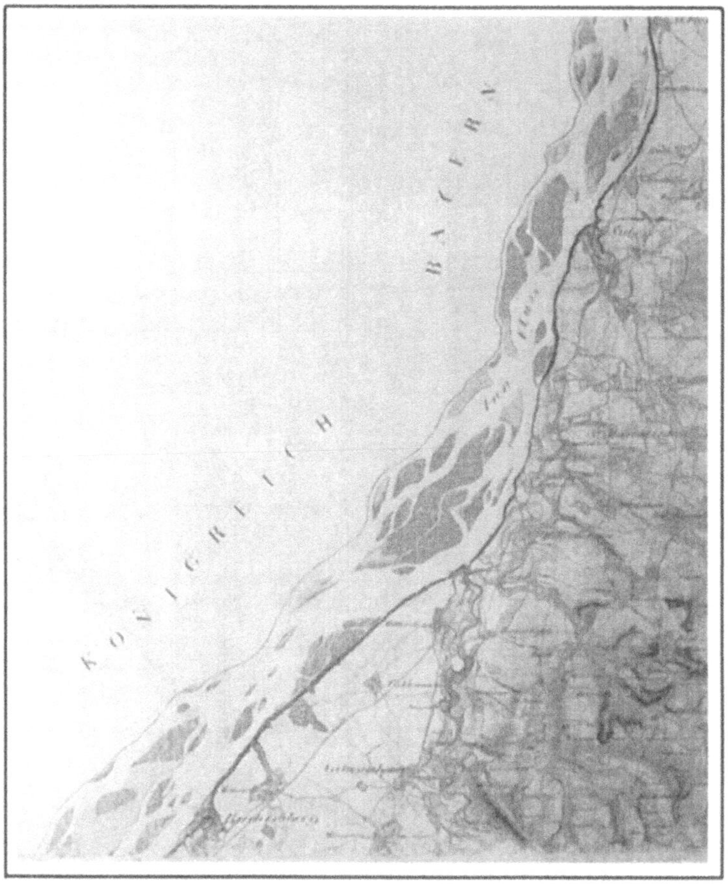

Figure 4. Example of the national land survey (1805–1860) showing the morphological character of a braiding river system (Inn River, border area Austria/Germany) prior to impacts caused by channelisation and impoundment.

Regarding the original fish communities, approximately 84% of the river stretches (approximately 3300 km; based on the situation prior to systematic channelisation and hydroelectric development) can be characterised as rhithral or dominated by rhithral faunal elements (*Epirhithral* to *Hyporhithral/Epipotamal*; upper trout region to grayling region and transition, definition of 'fish regions' after Illies, 1961). Of this total, the pure grayling region (Hyporhithral) made up the greatest proportion (1000 km). Typically, in Austria's alpine rivers the grayling regions corresponded frequently to braided reaches, mentioned above.

Some 750 km, or 15% of the total, were originally potamal in nature (*Epipotamal* to *Metapotamal*; barbel to bream region) and can typically be found in the large basin landscapes in eastern Austria bordering the Pannonian Basin.

Anthropogenically altered rivers

The mapping results and evaluation of the 52 rivers reflect the current status of the investigated running water habitats (Figure 5): Nearly 80% (3869 km) correspond to the category '*moderately to heavily impacted*' and can, therefore no longer be defined as reflecting the original morphological and hydrological character.

Based on their physiographic/hydrologic characteristics, virtually all river systems that formerly influenced their valleys and floodplains through flooding, as well as erosion and sedimentation processes, have been altered by channel straightening and/or stabilisation measures.

The dynamic processes within such river systems have been largely inhibited by flood prevention measures and hydroelectric plants. Intensive land use along valley floors, which often extends directly up to the

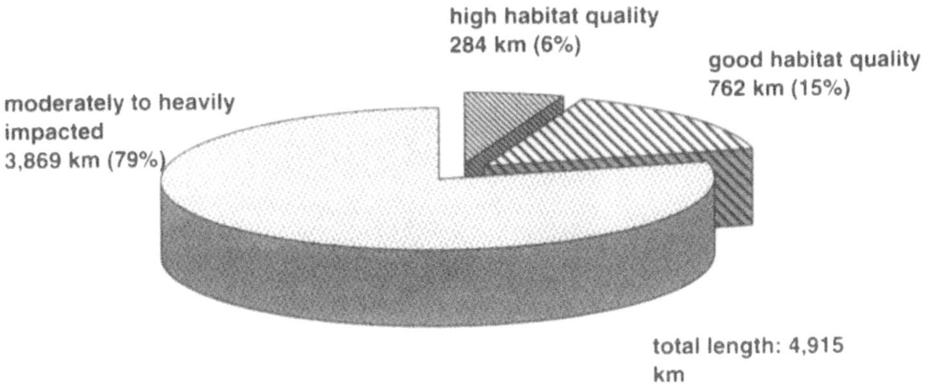

Figure 5. Percentage and total number of stream kilometers in three habitat quality classes for 52 Austrian rivers.

river banks, continues to successively remove the remaining shore structures and constrict the last remnants of former wetlands. Changes in the morphological river type can be used to roughly quantify those human interventions that have altered channel morphology, as well as degraded and minimized former backwater areas, as well as floodplain vegetation. Meandering stretches have been reduced from a total of 1317 km to 242 km or 18% of their former length. The braided and winding channel systems which were once found throughout Austria are preserved in mere fragments (compare also Figure 8).

Surveys of Austria's wetland vegetation (Lazowski, 1997) confirm these results: the overwhelming majority of near-natural riverine forests has been replaced by agricultural or residential areas and is now restricted to isolated sites. The large-scale damage caused by river regulation has another visible repercussion: a wide range of highly diverse aquatic systems is lost when riverine wetlands shrink. According to Gepp et al. (1985), a mere 1/10 of Austria's original backwaters are preserved today. Of these, only 1/3 are considered to have relatively intact hydrologic dynamics.

Running waters, such as the constrained channel rivers in V-shaped valleys, which 'take up' less surrounding land, tend to be more heavily impacted hydrologically/hydraulically. Such impacts in the studied stretches were further differentiated into alterations related to impoundment, water diversion and hydropeaking (Figure 6).

The riverine character of 16% of Austria's 'larger' rivers has been altered by impoundments. Additionally, 19% contain diversions, whereby the original bed is either dry or residual flow is far less than the average monthly low-water discharge. Hydropeaking-related phenomena through reservoir systems occur in 14% of the investigated stretches. Many of these stretches with altered hydrology are additionally impaired through river straightening or stabilisation measures.

Overlaying these data with the results of the morphological field investigations delineates all those river stretches which remain morphologically intact but hydrologically altered either by water diversion or hydropeaking and which, therefore, cannot be put into the high or good habitat quality class. Impoundment, which totally restricts fluvial dynamics of rivers and, therefore, has a low restoration potential, is omitted from this analysis. The results delineate 380 km or 8% of the total river length as being of '*potentially* high or good habitat quality', comprising all morphologically unaltered river stretches with residual flow or alteration through hydropeaking.

River stretches with high and good habitat quality

The spatial distribution of river stretches with high and good habitat quality is shown in Figure 7a. Approximately 6% (284 km) of all investigated river kilometers are classified as having high habitat quality (see 'Methods for identifying Austria's running water habitats with high and good quality' and Figure 5). This quality class contains a total of 61 river stretches divided among 25 rivers, with an average length of 4.7 km. About 30% of the total length of rivers with high habitat quality are longer than 10 km (Figure 7b). Additionally, 206 river stretches with a total length of 762 km (15%) represent good habitat quality, out of which 23% are longer than 10 km (Figure 7c). The average length of these stretches (3.7 km) is very short, also.

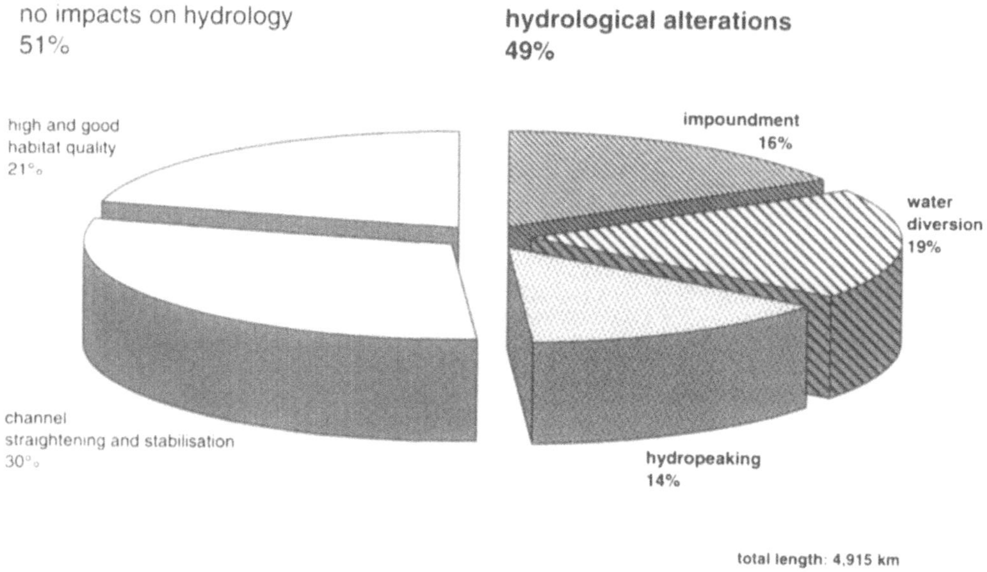

Figure 6. Proportion of Austrian river stretches with hydrological alterations caused by impoundment, water diversion and hydropeaking *versus* river stretches with no impacts on hydrology.

The evaluation of morphological character clearly reveals the rareness of intact river stretches of specific river types (Figure 8). Today, only few meandering stretches (with a mere 16 km having high habitat quality) reflect largely undisturbed habitat conditions both in the immediate river area and in the associated wetlands. Along with unimpaired, braiding river sections (10 km), these stretches belong to those rare areas where the river and the adjoining wetlands continue to form an intact functional unit: at least episodically, river dynamics still shape the entire system. In these running waters, lateral connectivity is a critical habitat feature that is also essential for healthy fish assemblages (Jungwirth et al., 2000).

Constrained rivers belong to the less degraded morphological river type, with a total of 41% of the stretches evaluated having high (234 km) and good (202 km) habitat quality.

The fish-based assessment highlights the tremendous loss of intact rivers originally corresponding to the potamal and hyporhithral region (Figure 9). Of the stretches with rhithral coenoses, 18% (480 km) have remained morphologically and hydrologically largely undisturbed, providing intact aquatic habitats for the hyporhithral fish fauna. These rivers formerly dominated alpine landscapes in Central Europe. Originally 2730 river kilometers (55%) of Austria's larger rivers belonged to this region. Thus, healthy river habitats of the typical, dominating biocoenotic regions have been quantitatively diminished. Unimpaired rivers with a potamal fauna are currently represented by very few river stretches (total only 43 km).

However, these data show the current proportion of river stretches with high or good habitat quality, respectively, the loss of intact stretches within the different fish regions only in a simplified way. We are currently unable to quantify the degree to which the above impacts have also led to shifts in the fish assemblages. Nonetheless, numerous examples demonstrate that altered hydrological and morphological conditions in running waters can reduce the number of fish species, shift the dominance relationships of these species, as well as reduce density and biomass (Karr et al., 1987; Schmutz, 1996). River straightening tends to trigger so-called 'rhithralisation' effects in which the habitat features of a river are reinforced or altered in the direction of rhithral characteristics. Fish-ecological investigations clearly document this phenomenon (Zauner, 1993; Zauner & Schmutz, 1994). Impoundments tend to change the respective river stretches in the sense of a 'potamalisation', whereby the often steady low water temperatures of alpine rivers stand in contrast to the extremely slowed currents and high fine-sediment fractions in impoundments.

The above results are cartographically represented in Figures 10a, b. These maps, however, only differentiate between rhithral and potamal river stretches. Figure 10a underlines the former extensive range of

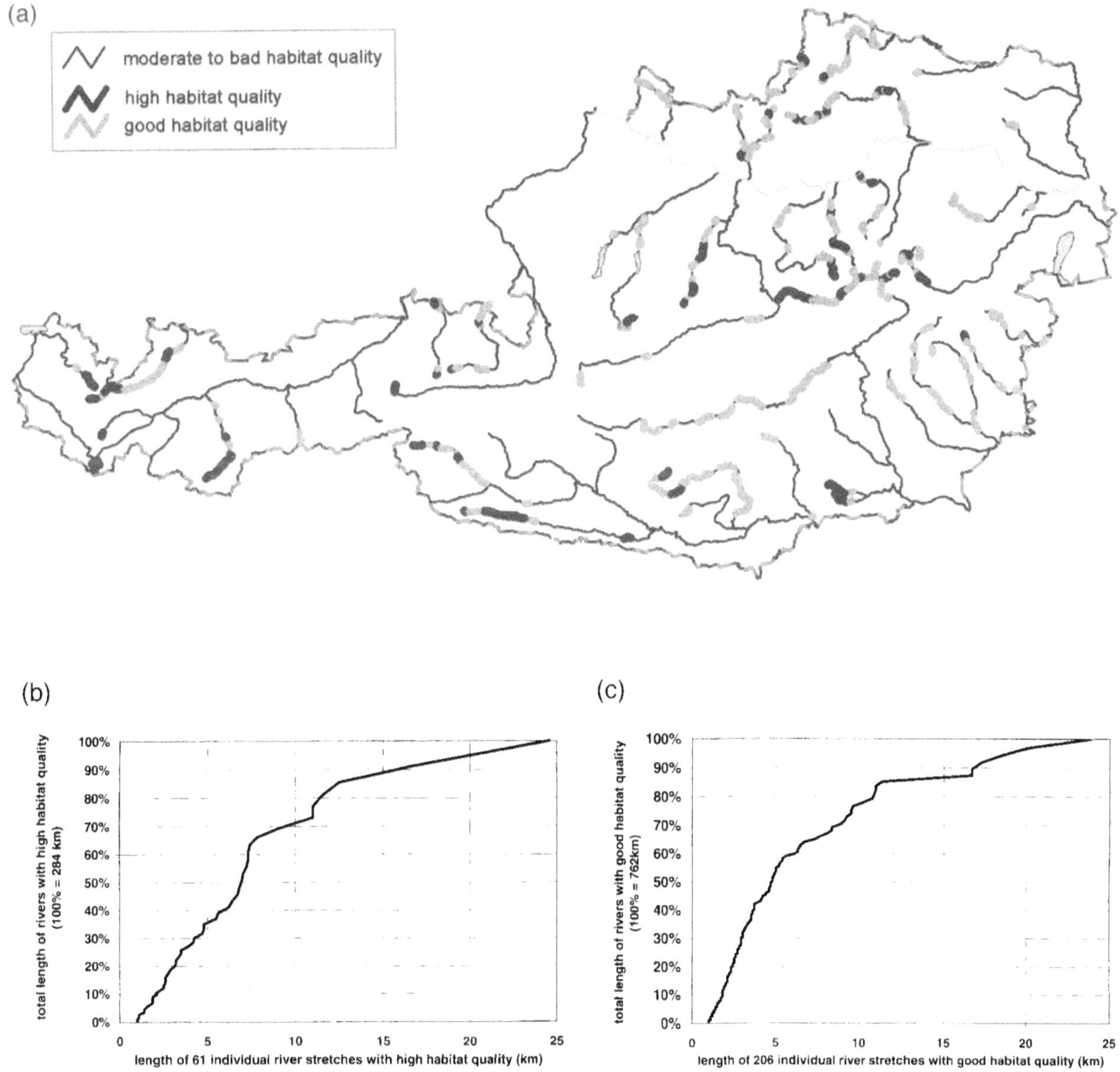

Figure 7. (a) Spatial distribution of river stretches with high and good habitat quality for 52 Austrian rivers. (b) Cumulative distribution of length of individual river stretches with high habitat quality versus total length of 52 investigated Austrian rivers. (c) Cumulative distribution of length of individual river stretches with good habitat quality versus total length of 52 investigated Austrian rivers.

rhithral stretches (thick grey lines) throughout the Alpine regions of Austria, including the upper reaches of virtually all investigated rivers. Almost all rhithral rivers continue to at least exhibit fragmentary riverine habitats that are morphologically and hydrologically intact; these are often concentrated in the upper reaches (thick black lines). Nonetheless, only a few rivers still show more extensive rhithral stretches corresponding to the river type. On the other hand, potamal fish assemblages (see Figure 10b) are naturally more or less restricted to the running waters of the Tertiary hills and basins to the East (thick grey lines). Remnants with high or good habitat quality were found in only six rivers (thick black lines).

Summary and outlook

The results presented here are part of a comprehensive study on the status of all rivers within Austria's boundaries that have a catchment area >500 km^2 (excluding the Danube).

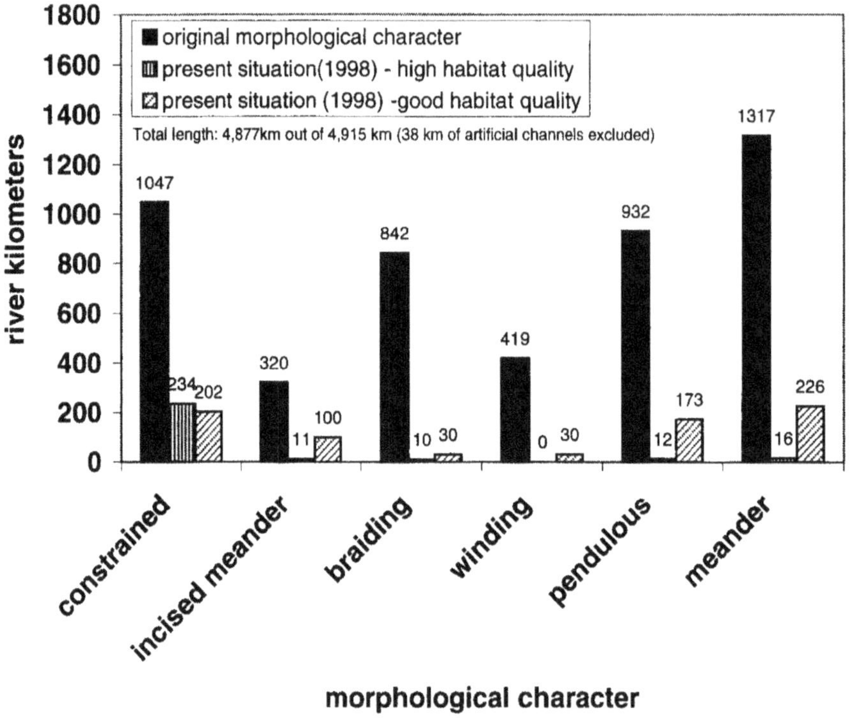

Figure 8. Frequency histogram of the total number of stream kilometers in six morphological character-classes for 52 Austrian rivers. The black bars symbolise the original morphological character (reference scale: 1st half of the 19th century), both adjoining hatched bars the proportion of stretches in the respective class, that are still of high or good habitat quality.

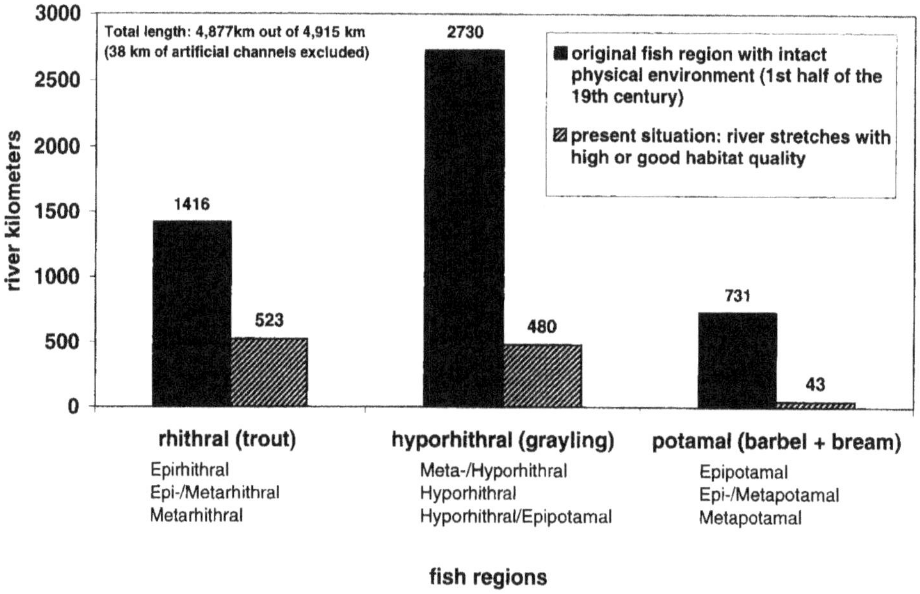

Figure 9. Proportion of remaining river stretches with high and good habitat quality of the rhithral region, hyporhithral region, and potamal region and their relation to the original fish region (reference scale: 1st half of the 19th century).

The combined data identify the stretches that still reflect their original morphological and hydrological character. Of the 52 rivers and 4915 km studied, 284 km (ca. 6%) were evaluated as river stretches with high and 764 km (ca. 15%) as river stretches with good habitat quality. But 3867 km (ca. 79%) no longer reflect the original character. Of these, 381 km (ca. 8%) are largely intact morphologically

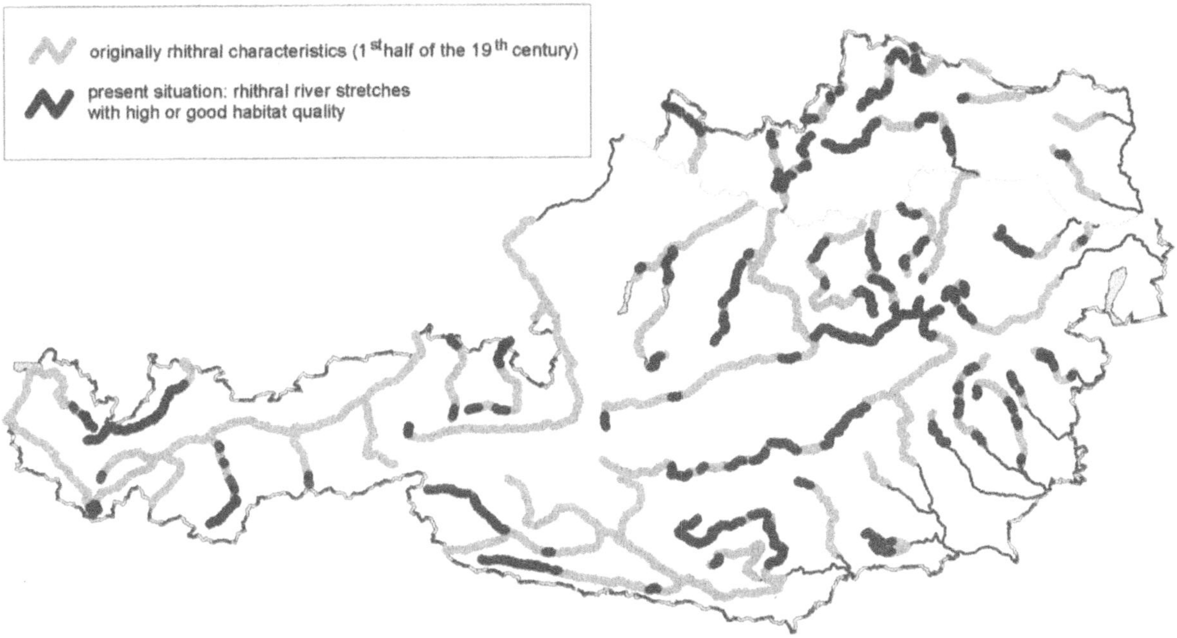

Figure 10a. (a) Loss of intact river stretches with rhithral characteristics. The thick gray lines show the spatial distribution of the original rhithral fish regions with intact physical environments (reference scale: 1st half of the 19th century); the thick black lines indicate all rhithral river stretches with high and good habitat quality (compare Figure 10b). (b) Loss of intact river stretches with potamal characteristics. The thick gray lines show the spatial distribution of the original potamal fish regions with intact physical environments (reference scale: 1st half of the 19th century); the thick black lines indicate all potamal river stretches with high and good habitat quality (compare Figure 10a).

yet have altered discharge regimes through water diversion and/or hydropeaking. The results point to an urgent need for action on a number of levels in future river management schemes.

Absolute priority must be given to protecting river stretches with high habitat quality. While current law mandates the preservation of such aquatic systems, practical experience shows that legal regulations are often insufficient. Future strategies must go beyond the 'classical' measures foreseen under nature conservation laws and must increasingly integrate aquatic conservation concerns into regional planning and stream management concepts.

The river status documented here reveal a critical need for measures designed to improve the ecological integrity of impaired rivers. From the geomorphological perspective, stretches changed by straightening and bank stabilisation have a high regeneration potential, provided that their bedload balances have not been severely changed and that dynamic processes remain or can be restored. Priority should be given to avoiding any measures that would further reduce the ecological integrity of these stretches/reaches and to strengthening efforts to incorporate such systems into ongoing stream management schemes or restoration programs. The long-term goal is the systematic, nationwide improvement of hydromorphological conditions, which play a key role in regenerating running waters and their biocoenoses.

Yet, the restoration potential of those stretches impacted by hydroelectric stations, is in most cases relatively low, especially where chains of power stations/impoundments block the development of original dynamics. This holds true for the current situation in Europe, whereas in the U.S.A. several dams have been removed in order to improve the ecological status of rivers. The prognosis for those river stretches affected by hydropeaking or water diversion is better, both from the legal standpoint and based on the potential success for improving ecological integrity using state-of-the-art technology. The strategy here would mainly be to minimise hydrological impacts (dampen hydropeaking through compensation reservoirs and/or coordinated reservoir operations, increasing residual water flow, etc.; see Moog, 1993 and Moog et al., 1993).

The results presented here conform fully with ongoing activities and agreements in the realm of aquatic

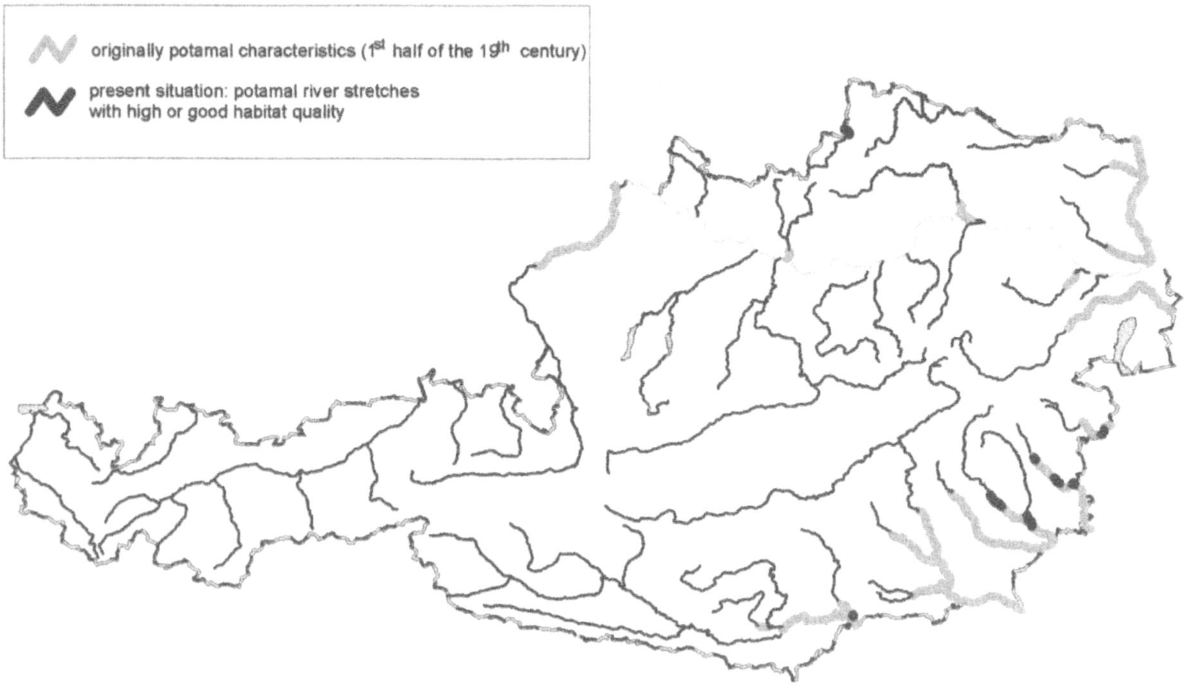

Figure 10b. (a) Loss of intact river stretches with rhithral characteristics. The thick gray lines show the spatial distribution of the original rhithral fish regions with intact physical environments (reference scale: 1st half of the 19th century); the thick black lines indicate all rhithral river stretches with high and good habitat quality (compare Figure 10b). (b) Loss of intact river stretches with potamal characteristics. The thick gray lines show the spatial distribution of the original potamal fish regions with intact physical environments (reference scale: 1st half of the 19th century); the thick black lines indicate all potamal river stretches with high and good habitat quality (compare Figure 10a).

conservation and environmental protection policy at the European level. As early as 1982, the Conference of Ministers of the European Council recommended that member governments preserve a European network of genetic reserves in the framework of the Bern Convention. This agreement legally binds its members to maintain and preserve wetlands and explicitly recognises the necessity for compiling species inventories and for evaluating representative areas. The International Convention on Biological Diversity, which was ratified by Austria's National Assembly in 1995, also legally anchored the 'establishment of a system of protected areas or areas that require special measures to preserve biodiversity'.

The EU's Water Framework Directive (Council Directive, EUROPEAN COMMISSION, 1999) determines the goals of an EC-wide water policy. One main objective is to improve the ecological quality of degraded running waters and to preserve high or good ecological quality. This Directive, formulated by representatives of all EU member states, defines both the qualitative and quantitative ecological status of running waters in terms of biological, chemical, hydromorphological and toxicological criteria on a 5-point scale ('high', 'good', 'fair', 'poor' and 'bad' water quality). The definitions applied in the present paper for river stretches with high and good habitat quality correspond to the respective categories of the Water Framework Directive. In addition to biological and chemical investigations, assessments, like the present study, that identify and document natural river stretches with original river typology are therefore a first step toward fulfilling the EU guidelines.

Acknowledgements

We would like to thank the many colleagues who participated in this study, in particular Martina Kainz und Michael Kaufmann. This paper draws on research supported by the Federal Ministry of Agriculture and Forestry and the Federal Ministry of Environment, Youth and Family Affairs. Michael Stachowitsch and Steven Weiss helped translate and improve the English.

References

Abe, J., W. Davis, T. Flanigan, A. Schwarz & M. McCarthy, 1992. Environmental indicators for surface water quality programs – Pilot Study. EPA-905-R-92-001. U.S. Environmental Protection Agency, Region 5, Chicago, Illinois and Office of Policy, Planning and Evaluation, Washington, D.C.

Allen, J. R. L., 1977. Changeable rivers: some aspects of their mechanics and sedimentation. In Gregory, K. J. (ed.), River Channel Changes. J. Wiley and Sons, Chichester: 15–45.

Bauer, H., 1992. Bewertungskriterien für Fließgewässer. Limnologie aktuell 3: 35–44.

Benke, A. C., 1990. A perspective on America's vanishing streams. J. n. am. Benthol. Soc. 9: 77–88.

Binder, W. & W. Kraier, 1999. Gewässerstrukturgütekartierung Bundesrepublik Deutschland – Stand und Ausblick. Wasserwirtschaft 89: 30–33.

Bohl, E., R. Lehmann & R. Obergruber, 1992. Entwicklung von Bewertungsverfahren für Bäche auf verschiedenen Planungsebenen. Limnologie aktuell 3: 417–444.

Boon, P. J., 2000. The development of integrated methods for assessing river conservation value. Hydrobiologia 422/423: 413–428.

Boon, P. J., N. T. Holmes, P. S. Maitland, T. A. Rowell & J. Davies, 1997. A System for evaluating rivers for conservation (SERCON): development, structure and function. In Boon, P. J. & D. J. Howell (eds), Freshwater Quality – Defining the Indefinable? The Stationery Office, Edinburgh, 299–326.

Böttger, K., 1986. Zur Bewertung der Fließgewässer aus der Sicht der Biologie und des Naturschutzes unter besonderer Berücksichtigung der Tieflandbäche. Landschaft und Stadt 18, 2: 77–82.

Bundi, U., A. Peter, A. Frutiger, M. Hütte, P. Liechti & U. Sieber, 2000. Scientific base and modular concept for comprehensive assessment of streams in Switzerland. Hydrobiologia 422/423: 477–487.

Dahl, T. E., 1990. Wetland losses in the U.S., 1780–1980s. Washington, D.C.: U.S. Department of the Interior. Fish and Wildlife Service, 124 pp.

Dynesius, M. & C. Nilsson, 1994. Fragmentation and Flow Regulation of the River Systems in the Northern Third of the World. Science 266: 753–762.

Ellenberg, H., 1996. Vegetation Mitteleuropas mit den Alpen. Eugen Ulmer Verlag, Stuttgart: 1095 pp.

EUROPEAN COMMISSION, 1999. Commission Proposal for a Council Directive Establishing a Framework für European Community Water Policy – 6404/99.

Galat, D. L. & R. Lipkin, 2000. Restoring ecological integrity of great rivers: historical hydrographs aid in defining reference conditions for the Missouri River. Hydrobiologia 422/423: 29–48.

Gepp, J., 1985. Die Auengewässer Österreichs, Bestandesanalyse einer minimierten Vielfalt. In Gepp. J. (ed.), Auengewässer als Ökozellen. Grüne Reihe des Bundesministeriums für Gesundheit und Umweltschutz (Wien) 4: 223–258.

Haidvogl, G. & H. Waidbacher, 1997. Ehemalige Fischfauna an ausgewählten österreichischen Fließgewässern. Abt. f. Hydrobiologie, Univ. f. Bodenkultur, Wien: 85 pp.

Hughes, R. M., S. G. Paulsen & J. Stoddard, 2000. EMAP-Surface Waters: a national, multiassemblage, probability survey of ecological integrity in the U.S.A. Hydrobiologia 422/423: 429–443.

Illies, J., 1961. Versuch einer allgemeinen biozönotischen Gliederung der Fließgewässer. Int. rev. Ges. Hydrobiol. 46, 2: 205–213.

Jungwirth, M., S. Muhar & S. Schmutz, 2000. Fundamentals of fish ecological integrity and their relation to the extended serial discontinuity concept. Hydrobiologia 422/423: 85–97.

Karr, J. R., P. R. Yant, K. D. Fausch & I. J. Schlosser, 1987. Spatial and Temporal Variability of the Index of Biotic Integrity in Three Midwestern Streams. Trans. am. Fish. Soc. 116 (1): 1–11.

Kaufmann, P. R. & E. G. Robinson, 1998. Physical habitat characterization. In Lazorchak, J. M., D. J. Klemm & D. V. Peck (eds), Field Operations and Methods for Measuring the Ecological Condition of Wadeable Streams. EPA/620/R-94/004F, U.S. Environmental Protection Agency, Corvallis, OR: 77–118.

Konold, W., 1992. Zur Bewertung von Fließgewässern auf ökologischer Grundlage. In Friedrich, G. & J. Lacombe (eds), Ökologische Bewertung von Fließgewässern. Limnologie aktuell. Gustav Fischer, Stuttgart 3: 19–33.

LAWA (Länderarbeitsgemeinschaft Wasser/Arbeitskreis 'Gewässerbewertung Fließgewässer'), 1993. Die Gewässerstrukturgütekarte der Bundesrepublik Deutschland. Verfahrensentwurf für kleine und mittelgroße Fließgewässer in der freien Landschaft.

Lazowski, W., 1997. Auen in Österreich. BMUJF Wien. Monographien 81: 240 pp.

LFU (Landesanstalt f. Umweltschutz Baden-Württemberg), 1994. Übersichtskartierung des morphologischen Zustands der Fließgewässer in Baden-Württemberg 1992/93. Handbuch Wasser 2: 43 pp.

LÖLF (Landesanstalt f. Ökologie, Landschaftsentwicklung und Forstplanung NW/Landesanstalt f. Wasser und Abfall NW (LWA)), 1985. Bewertung des ökologischen Zustandes von Fließgewässern. Düsseldorf. Teil I (26 pp.) und II (65 pp.).

Mader, H., T. Steidl & R. Wimmer, 1996. Abflußregime österreichischer Fließgewässer. Umweltbundesamt – Monographien: 82, Wien: 192 pp.

Mangelsdorf, J., K. Scheurmann & F. H. Weiß, 1990. River morphology. Springer Verlag, Berlin Heidelberg: 243 pp.

Martinet, F. & M. Dubost, 1992. Die letzten naturnahen Alpenflüsse. CIPRA, Kleine Schriften 11: 10–58.

Moog, O., 1993. Quantification of daily peak hydropower effects on aquatic fauna and management to minimize environmental impacts. Reg. Riv. 8: 5–14.

Moog, O., M. Jungwirth, S. Muhar & B. Schönbauer, 1993. Berücksichtigung ökologischer Gesichtspunkte bei der Wasserkraftnutzung durch Ausleitungskraftwerke. In Österr. Wasserwirtschaft 45, 7/8: 201–210.

Muhar, S. & M. Jungwirth, 1998. Habitat integrity of running waters – assessment criteria and their biological relevance. Hydrobiologia 386: 195–202.

Muhar, S., A. Muhar, S. Schmutz, R. Wimmer, H. Wiesbauer, B. Hozang, M. Jungwirth, G. Imhof & P. Tschemernig, 1993. Ausweisung naturnaher Fließgewässerabschnitte in Österreich. Blaue Reihe des Bundesministeriums f. Umwelt, Jugend und Familie, Wien, Bd 1: 175 pp.

Müller, N., 1991. Der Lech. Wandel einer Wildflußlandschaft. Augsburger Ökolog. Schriften 2: 174 pp.

Naiman, R. J., J. J. Magnuson, D. M. McKnight & J. A. Stanford, 1995. The Freshwater Imperative – A Research Agenda. Island Press, Washington, D.C.: 157 pp.

Naiman, R. J., T. J. Beechie, L. E. Benda, D. R. Berg, P. A. Bisson, L. H. Mac Donald, M. D. O'Connor, P. L. Olson & E. A. Steel, 1992. Fundamental Elements of Ecologically Healthy Watersheds in the Pacific Northwest Coastal Ecoregion. In Naiman, R. J. (ed.), Watershed Management: 127–188.

NRA, 1992. River Corridor Surveys. Methods and Procedures. Conservation Technical Handbook 1. Bristol.

Ohio EPA, 1988. Ohio water quality inventory, 1988 305 (b) report, Vol. 1, and Executive Summary. In Rankin, E. T., C. O. Yoder & D. A. Mishne (eds), Ohio EPA, Division of Water Quality Monitoring and Assessment. Columbus, OH.

Parde, M., 1947. Fleuves et Rivieres, 3rd edn. Paris.

Patzner, A., W. Herbst & E. Stüber, 1985. Methode einer ökologischen und landschaftlichen Bewertung von Fließgewässern. Natur und Landschaft 60, 11: 445–448.

Petts, G. E., 1989. Historical analysis of fluvial hydrosystems. In Petts, G. E., H. Möller & A. L. Roux (eds), Historical Change of Large Alluvial Rivers: Western Europe. J. Wiley and Sons, Chichester: 1–18.

Rankin, E. D., 1995. Habitat indices in water ressource quality assessments. In Davis, W. S. & Th. P. Simon (eds), Biological Assessment and Criteria. Tools for Water Resource Planning and Decision Making. Lewis Publishers, Boca Raton, London, Tokyo: 181–208.

Raven, P. J., N. T. H. Holmes, M. Naura & F. H. Dawson, 2000. Using river habitat survey for environmental assessment and catchment planning in the U.K. Hydrobiologia 422/423: 359–367.

Raven, P. J., P. Fox, M. Everard, N. T. H. Holmes & F. H. Dawson, 1997. River habitat survey: a new system for classifying rivers according to their habitat quality. In Boon, P. J. & D. J. Howell (eds), Freshwater Quality – Defining the Indefinable? The Stationery Office, Edinburgh: 215–234.

Rose, U., 1990. Beurteilung von Fließgewässerstrukturen aus ökologischer Sicht – Ergebnisse und Erfahrungen mit einer einfachen Methode. Wasserwirtschaft 80, 5: 236–242.

Schmutz, S., 1996. Anthropogene Eingriffe in Fließgewässer und deren Einfluß auf die Lebensgemeinschaften. Wasserwirtschaft 86, 5: 230–236.

Schmutz, S., M. Kaufmann, B. Vogel, M. Jungwirth & S. Muhar, 2000. A new multi-level concept for fish-based, river-type-specific assessment of ecological integrity. Hydrobiologia 422/423: 279–289.

Spiegler, A., G. Imhof, B. Pelikan & M. Katzmann, 1989. Strukturökologische Methode zur Bestandsaufnahme und Bewertung von Fließgewässern. BMLF, Wasserwirtschaftskataster, Wien 34 pp.

Stanford, J. A. & J. V. Ward, 1992. Management of aquatic resources in large catchments: recognizing Interactions Between Ecosystem Connectivity and Environmental Disturbance. In Naiman, R. J. (ed.), Watershed Management: 91–124.

Steidl, T., 1991. Typologie und Abflußverhalten österreichischer Fließgewässer. Hydrologisch-hydrographische Einteilung und regionale Gliederung. Master thesis, University of Agricultural Sciences, Vienna: 166 pp.

Stein, J. L., J. A. Stein & H. A. Nix, 1998. Identification of wild rivers – methodology and database development. Centre for Resource and Environmental Studies, Australien National University: 77 pp.

Strahler, A. N., 1957. Quantitative analysis of watershed geomorphology. Amer. Geophys. Union Trans. 38: 913–920.

Thienemann, A., 1925. Die Binnengewässer Mitteleuropas. Die Binnengewässer 1. Schweizerbartsche Verlagsbuchhandlung, Stuttgart.

Ward, J. V. & J. A. Stanford, 1995. The serial discontinuity concept. Extending the model to floodplain rivers. Reg. Riv.: Res. Manag. 10: 159–168.

Werth, W., 1987. Ökomorphologische Gewässerbewertung Österreichische Wasserwirtschaft 39: 122–128.

Wilhelmy, H., 1990. Geomorphologie in Stichworten – Exogene Morphodynamik. Verlag Ferdinand Hirt, Rendsburg: 192 pp.

Wimmer, R. & O. Moog, 1994. Flußordungszahlen österreichischer Fließgewässer. Umweltbundesamt – Monographien: 51 Wien: 581 pp.

Zauner, G., 1993. Fischökologische Studie untere Thaya. Studie im Auftrag der Wasserstraßendirektion: 182 pp.

Zauner, G. & S. Schmutz, 1994. Fischökologische Untersuchungen im Rahmen von Gewässerbetreuungskonzepten. In Stalzer, W. & M. Jungwirth (eds), Gewässerbetreuungskonzepte – Stand und Perspektiven. Wiener Mitteilungen 120: 323–359.

Hydrobiologia **422/423**: 359–367, 2000.
M. Jungwirth, S. Muhar & S. Schmutz (eds), Assessing the Ecological Integrity of Running Waters.

Using river habitat survey for environmental assessment and catchment planning in the U.K.

P. J. Raven[1], N. T. H. Holmes[2], M. Naura[3] & F. H. Dawson[4]
[1] *Environment Agency, Rio House, Waterside Drive, Aztec West, Almondsbury, Bristol BS32 4UD, U.K.*
[2]*Alconbury Environmental Consultants, The Almonds, 57 Ramsey Road, Warboys, Huntingdon, Cambridgeshire PE17 2RW, U.K.*
[3]*Environment Agency, North West Region, Richard Fairclough House, Knutsford Road, Warrington WA4 1HG, U.K.*
[4]*Natural Environment Research Council – Institute of Freshwater Ecology, River Laboratory, East Stoke, Wareham, Dorset BH20 6BB, U.K.*

Key words: habitats, rivers, catchment, environmental assessment

Abstract

This paper describes how the attributes of a method for determining the physical quality of rivers are well-placed for use in environmental assessment and catchment planning in the U.K. The system, known as River Habitat Survey (RHS), uses a standard field survey method with quality controls, a computer database for rapid analysis and includes outputs which quantify habitat quality and channel modification. Further refinement of the system is needed because there are limitations associated with the detail of data collected, with assumptions made about species' habitat requirements and the impact of channel modification. By taking full account of catchment characteristics, historical influences and geomorphological processes, the system can, together with other information, help in the conservation and rehabilitation of rivers.

Introduction

Maintaining or restoring river ecosystems requires an integrated approach to catchment management, addressing issues such as surface and groundwater pollution, the abstraction and transfer of water, the physical alteration of the river channels and the drainage of wetlands (Werritty, 1997). The forthcoming European Framework Directive on Water Policy is based on this approach and requires that physical, chemical and biological attributes of inland waters are measured to help determine whether 'high' or 'good' ecological status has been maintained or achieved (Commission for the European Communities, 1997). A system for measuring the physical structure and habitat quality of rivers is therefore needed to complement those which describe water quality based on chemical and biological data (Environment Agency, 1997). Using physical structure as a surrogate indicator of biodiversity can also help to assess the ecological integrity of rivers, leading to action to minimise or reverse habitat fragmentation which affects many streams and rivers in densely-populated countries such as the U.K. This paper considers how the attributes of such a system, developed in the U.K. and known as River Habitat Survey (RHS), can be used to protect and restore rivers.

Background

River habitat survey

The main driving force behind development of RHS was to provide an objective basis for determining the physical character of rivers which could then be used to derive a scheme for assessing habitat quality. RHS comprises several distinct components: a standard field survey method; a computer database for entering results and rapid comparison of information; various rule-based methods for assessing habitat quality; and a

Table 1. The main features recorded during an RHS survey

Valley form	Bank profile
Channel substrate	Bankside trees and associated features
Bank material	Channel habitat features
Flow type(s)	Artificial features
Channel and bank modifications	Features of special riparian interest
Bankface vegetation structure	Land use
Bankface vegetation types	

method to indicate the extent of artificial modification to the channel (Raven et al., 1997).

The method relies on a systematic collection of map and observational data based on a standard site comprising a 500 m length of river channel and its corridor extending 50 m outwards on either side. Observations of the presence, absence and, in some cases, extent of specific features in the channel and adjacent river corridor are entered onto a four-page field survey form. Predominant channel substrate, flow type, habitat features, aquatic vegetation types, bank vegetation structure and artificial modifications are recorded at 50 m intervals by way of *spot-check* transects; a checklist for the whole site is also completed, ensuring that features and modifications not occurring at the 10 spot-checks are also recorded (Table 1). Channel width, bank height and water depth are measured at a representative cross-section, and the number of riffles, pools and point bars also recorded. A full account of the development and testing of this method can be found in Fox et al. (1998).

Information such as geology, altitude and channel slope are obtained from maps and used with the field survey data to provide the necessary basis for describing the physical character of the river and its corridor, including the extent of artificial modification to the channel. However, the level of detail recorded reflects the need to obtain data cost-effectively, and provided there are no major access problems, a field survey should take no more than 1 h to complete. Surveyors are usually biologists but they have to be specially trained in the survey technique, be familiar with all the features recorded, and, for quality control, have to pass a written test. All the map-derived and field data are computerised using a double-entry verification system for input. The RHS database, which includes a photograph of each site, allows for storage and rapid analysis of the information.

To establish the full range of habitat features and channel modifications present in the U.K., a baseline survey comprising a geographically representative sample of rivers and streams was undertaken. The Ordnance Survey 10 × 10 km square grid was used for stratified random sampling of those rivers shown on 1:250 000 scale topographical maps and classified for water quality – approximately 85 000 km in all. During 1994–1996, each grid square was visited and more than 5600 sites surveyed, thereby establishing the necessary baseline dataset from which an objective system for determining quality could be established.

The occurrence of geomorphological features such as riffles, pools and point bars can be used as a basis for assessing habitat quality because, given suitable water quality, they help to determine the type, variety and abundance of plant and animal communities in rivers. In addition, riparian wetland features indicate intact hydrological links between the channel and river corridor. High habitat quality is determined by the presence of those features which occur in predominantly unmodified ecosystems. A classification can therefore be derived by comparing those features present at a site in the context of the range found on other rivers of similar type (Figure 1). Although the baseline survey provided the sample-based information necessary for this exercise, the quality of different river types needed to be calibrated by specially surveying a series of 'benchmark' sites of known high wildlife interest. Because most lowland rivers in the U.K. have been extensively modified, benchmark sites in these areas cannot be considered natural, but they nevertheless represent important calibration markers.

Habitat quality can be determined by the presence of one or more rare features at a site, or a rare combination of features which individually may not be scarce. In this case, the occurrence of features in the baseline survey determines rarity. By way of example, Table 2 shows those features occurring in 5% or less of upland and lowland sites. The RHS system also uses a scoring method for assessing habitat quality which is basically a broad measure of the diversity of features in the channel and river corridor. Habitat features are given a score and the higher the total at a site, the more diverse it is and the higher the habitat quality. Full details appear in Raven et al. (1998a). Developed primarily for use in England and Wales, further testing and refinement are needed to take account of geographical variations across the U.K. and beyond. Being a relative measure, it is important that habitat quality is assessed using rivers of similar char-

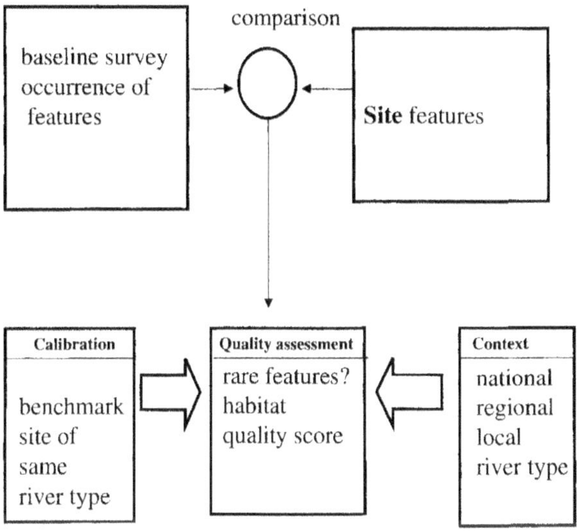

Figure 1. The process of assessing habitat quality.

Table 2. Habitat features with an occurrence of 5% or less in upland and lowland rivers in the U.K.

Feature	Upland	Lowland
Gorge	✓	✓
Braided/side channels		✓
Waterfalls more than 5 m high		✓
Extensive exposed bankside tree roots	✓	
Extensive underwater tree roots	✓	✓
Extensive coarse woody debris	✓	✓
Extensive fallen trees	✓	✓
Alder/willow carr	✓	✓
Extensive wetland on both banks	✓	✓

Extensive means at least one-third of channel length.

acter. This can be determined by defining river type on the basis of similar attributes such as altitude, geology, height of source, and size (Figure 1). Because of the variability of features along a river, broadly-defined types (e.g. upland, lowland) will mean lower levels of confidence when comparing sites regarding general characteristics or habitat quality. Determining river type is therefore crucial to the exercise. A recent and promising statistically-based approach, is to determine river type on a 'nearest neighbour' approach according to position on a principal components analysis plot (Jeffers, 1998).

Artificial modification to the river channel can also be expressed numerically using RHS data. Channel modification at a site is calculated from weighted scores which reflect different types of impact modification on habitat features. The system is also fully described in Raven et al. (1998a). Sites with a pristine unaltered channel score zero; the larger the score, the greater the extent of structural modification (Table 3).

In common with other survey methods which rely on field observations, the limitations of the data and the dangers of erroneously interpreting information beyond these limits need to be recognised. The consistency and quality of the data are dependent on the observational skills and training of individual surveyors. Intentionally, RHS only requires a basic understanding of river morphology, habitats and vegetation types – it does not require specialist geomorphological or botanical expertise. In the U.K., surveyors are trained to recognise features occurring in all river types and they must pass an RHS quality assurance test. Nevertheless, past channel management and historical land-use change may have significant impact on observational data. With the exception of experienced geomorphologists, most field observers will assume that a 'natural' looking bank and vegetation is unmodified. This may be wrong – historically re-aligned lowland rivers, may over time, develop a 'natural' form despite having not followed their original course for tens or even hundreds of years. Upland streams affected by historical mining works can exhibit similar 'recovery'. These subtleties will only be revealed by reference to historical records, photographs and maps. Interpretation of RHS data that link habitat features and quality to observed management must, therefore, be treated with appropriate caution, underlining the need for full use of all relevant sources of information when detailed, historically accurate assessment is needed.

Discussion

The baseline survey in context

Before the 1994–1996 baseline survey was carried out, the necessary data for objectively assessing the habitat quality of rivers on a U.K. basis were not available. The level and type of information gathered by RHS now allows the occurrence of specific individual features, or combinations of features on rivers of similar physical character to be compared. The extra physical dimension provided by RHS allows planners and river managers to address, in the proper context, environmental protection and improvement involving water pollution, water resources and habitat structure.

Table 3. The percentage of upland (>200 m) and lowland baseline survey sites in the U.K. classified according to channel modification (*Pristine is a subset of semi-natural*)

Category and habitat modification score	England & Wales		Scotland		Northern Ireland[a]
	Upland	Lowland	Upland	Lowland	Lowland
Pristine (0)	*38.9*	*13.6*	*47.3*	*28.0*	*10.1*
Semi-natural (0–2)	56.9	28.2	70.9	44.3	20.9
Predominantly unmodified (3–8)	26.0	20.9	16.5	23.8	20.9
Obviously modified (9–20)	12.1	21.3	10.4	15.1	20.5
Significantly modified (21–44)	4.0	25.7	1.6	14.9	32.6
Severely modified (45+)	1.0	3.7	0.0	1.7	5.0
Number of sites	404	4155	181	588	262

[a] Insufficient upland sample sites for meaningful comparison.

The 1994–1996 baseline survey has provided a representative picture of river habitat features and the extent of channel modification in the U.K. (Raven et al., 1998a). Results confirm that the impact of land and river management has been profound. For example, fallen trees and coarse woody debris which characterise river channels in unmanaged, wooded river catchments are rare (Table 2). Only 4.2% of sites less than 50 m in altitude have extensive floodplain wetlands along one or both banks, indicating how land drainage has allowed intensive agriculture, along with residential and industrial uses, to develop extensively in the lowlands. In lowland England and Wales, only 13.6% of survey sites have a pristine (untouched) channel, with 3.7% being severely modified (Table 3). Table 4 shows the percentage occurrence of road bridges, culverts and weirs, and extensively reinforced and reprofiled banks, all of which variously disrupt the natural character of rivers. Also indicated is the widespread occurrence of three invasive alien plant species, showing how widely established these now are throughout the U.K.

Using RHS for assessing quality and impacts

Since habitat features and channel modification differ geographically, RHS data can be used to provide an overview of the range and frequency of habitat quality and channel modification scores for different river types in the U.K. Figure 2 illustrates this using three examples. RHS data, together with other information, can also help to establish the relative importance of a site for use in environmental impact assessment and for deciding what management action (e.g. protection or enhancement) is required. The geographical representativeness of the 1994–1996 baseline survey across the whole of the U.K. allows the occurrence of individual features such as submerged tree roots, fallen trees and riparian wetland to be put into a national, regional or local context (Table 2). This information can be used in post-project appraisal to help predict and confirm the ecological consequences of river management, thus providing a consistent framework for determining, in broad terms, the likely environmental impact. An RHS survey will also establish whether more detailed information and analysis are required.

Structural modification of a river channel usually comprises one or more of the following: ***reinforcement*** (revetment in the form of concrete, steel piling, etc.); ***resectioning*** (mechanical reprofiling through dredging of the bed and banks); and ***regulation*** of flow by impoundment. These types of channel modification are widespread in the U.K., particularly in the lowlands (Table 4). Bearing in mind the abovementioned limitations to data interpretation, the RHS database allows the relationship between the habitat features and modifying factors at both individual spot-check transects and 500 m site levels to be analysed. This can be illustrated by way of example for specific features in riffle-dominated, small lowland rivers. This river type is defined by the following characteristics: altitude between 20 and 200 m; banktop width 2–15 m; height of source < 200 m; bedslope > 5°; no chalk geology upstream in the catchment. Extensive bank resectioning (i.e. affecting more than one-third of the site) adversely affects the occurrence of riffles, point bars and bankside trees when com-

Table 4. The percentage of upland (>200 m) and lowland baseline survey sites with specific modifying features

Modifying feature	England & Wales		Scotland		Northern Ireland[a]
	Upland	Lowland	Upland	Lowland	Lowland
Bridges	28.2	45.1	27.5	37.0	29.1
Culverts	5.2	10.5	0.0	4.8	3.1
Weirs	10.6	15.6	3.3	8.5	19.8
Extensive bank reinforcement	6.2	11.3	2.2	7.1	18.6
Extensive bank resectioning	3.7	33.2	7.7	19.4	36.8
Invasive bankside weeds					
Heracleum mantegazzianum	0.5	4.2	0.0	4.9	6.6
Impatiens glandulifera	0.2	15.8	0.5	4.4	14.0
Fallopia japonica	3.2	9.2	0.5	3.1	6.6
Number of sites	404	4155	181	588	262

Extensive means at least a third of bank length. [a] Insufficient upland sample sites for meaningful comparison.

pared to sites with unmodified channels (Figure 3). There is also a higher incidence of silt in the channel. Similar adverse effects are produced by extensive bank reinforcement and channel impoundment (Raven et al., 1998b). These types of analysis can be carried out for different features and river types.

Classifying rivers according to habitat quality and channel modification quality can help to determine the effectiveness and impact of proposed river management, although other information such as geomorphological influences, catchment history and species data must be used if a full environmental impact assessment is to be made. For example, RHS data is used as a one of many different sources of information to evaluate the overall conservation interest of a river using the System for Evaluating Rivers for Conservation (SERCON) (Boon et al., 1998).

Catchment plans

RHS is now being used in catchment management plans and provides a framework for describing the physical character of rivers and a basis for setting habitat targets. Although the national baseline survey provides a useful basis generally for describing river character, individual catchments will need extra survey sites depending on the detail required to provide a representative picture at regional or local level. A good example of this approach is the River Sankey, a small (179 km^2), heavily impacted catchment in north-west England. An inventory of habitat features has been produced by collating data from RHS and other surveys, maps, photographs and interviews with local interest groups, landowners and planning authorities. Additional data collected during the extra RHS surveys were used to determine stream power and determine geomorphological activity (Figure 4). As a result, a character profile of the river and its catchment has been produced, highlighting amongst other things landscape, water and habitat quality (Environment Agency, 1998). From this, recommendations for improvements to the water quality, habitat structure and general landscape character of the catchment – all of which will benefit local communities and fisheries interests – have been made. The emphasis on an integrated, catchment-wide approach has been fundamental, and the use of a Geographical Information System (GIS) to present the various sets of data and model the physical character of the catchment represents a major benefit (Webb et al., 1998). Illustrating graphically how various features found in a catchment can be put into a regional or national context has captured the imagination of local planning authorities, landowners and local communities. This support has helped river managers to identify imaginative long-term improvements in a heavily impacted river catchment.

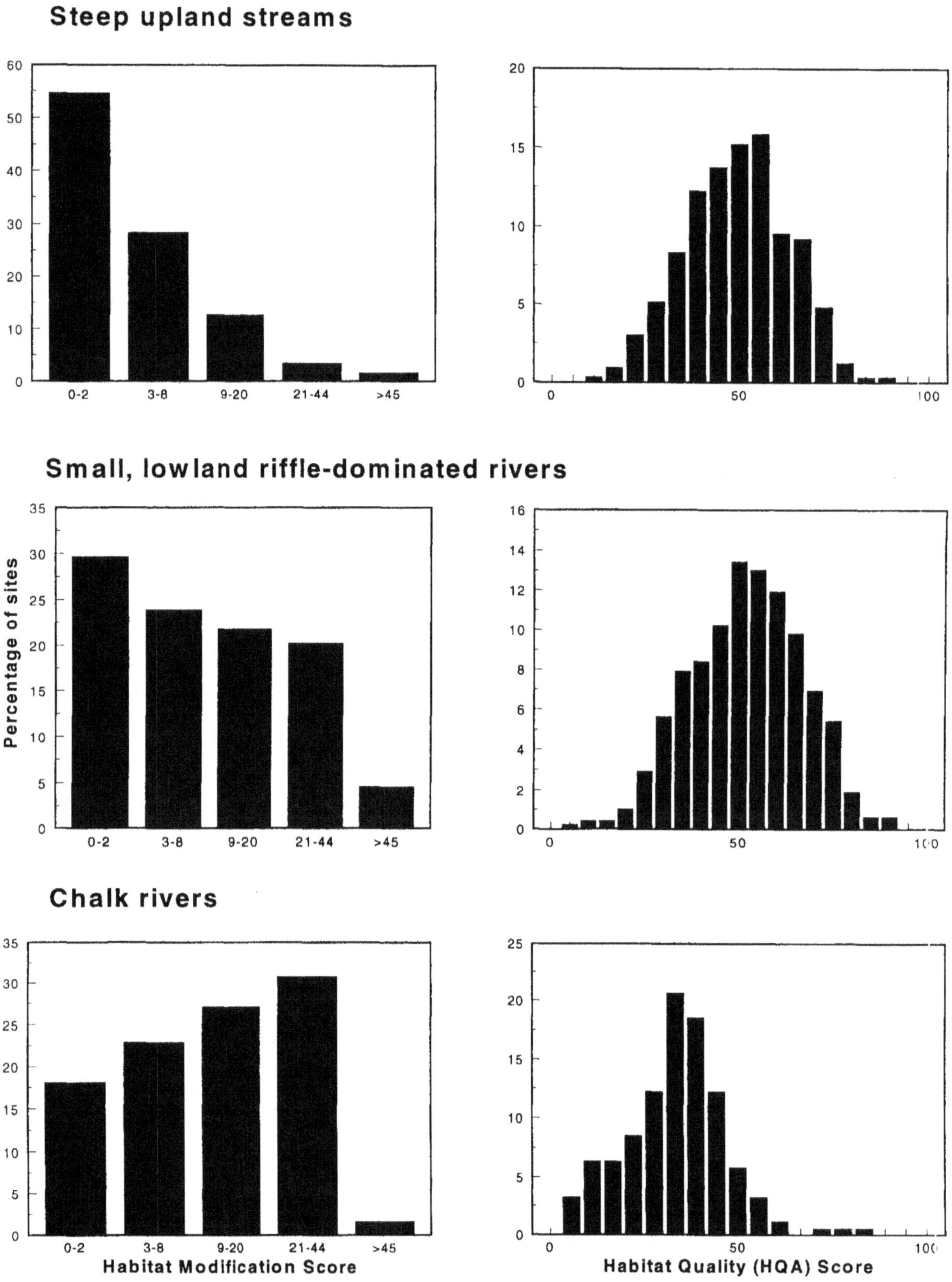

Figure 2. Distribution frequencies of habitat modification and habitat quality assessment scores for three different river types.

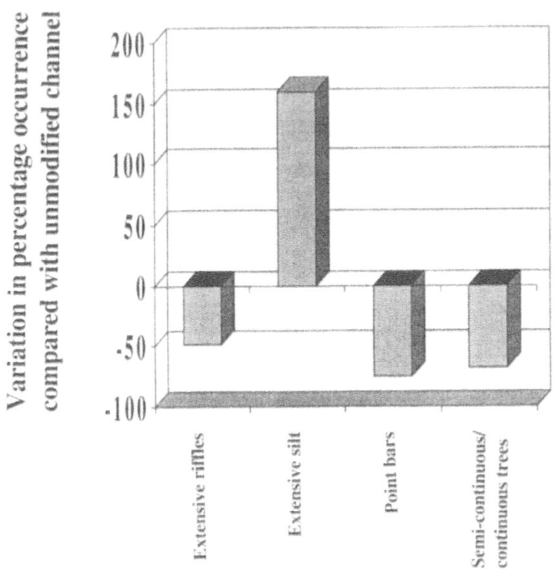

Figure 3. The impact of extensive bank resectioning on riffles, point bars, silt, substrate and bankside trees of riffle-dominated, small lowland rivers. Differences expressed as percentage variation in occurrences between extensively modified sites and those with unmodified channels. *Extensive means at least one-third of channel length.*

A wider context

Providing a national or regional context for individual sites or rivers is important because it can assist policy decisions on river management and land use, promote awareness of key issues and measure the performance of management actions. This is particularly relevant because assessing environmental quality is becoming increasingly important in benefit-cost and other economic assessments (Bowers, 1995). However, to be practicable, a system for describing rivers needs to be flexible enough to a variety of users. This can mean outputs ranging from instant snapshots for the general public and politicians, to sophisticated technical models for scientists involved in determining environmental impact assessment, diagnosis and prediction. Given its data attributes and options for generating outputs RHS can provide the basis for different levels of information (e.g. Figures 2 and 3). In the U.K. it is now possible, given appropriate data, to classify a river type on the basis of principal components analysis using map-derived variables (Jeffers, 1998) or aquatic plant community (Holmes et al., 1998a); its quality, based on habitat structure (Raven et al., 1998b) or aquatic invertebrate and fish communities (Mainstone et al., 1994; Wright et al., 1998); and its trophic status using macrophytes (Holmes et al., 1998b). The RHS database can also be used to help predict the occurrence of particular species, provided that water chemistry and other specific requirements are also fulfilled, although this approach is limited by the level of detail recorded by RHS and by assumptions made on species' habitat requirements (Buckton & Ormerod, 1997; Naura & Robinson, 1998).

Previous attempts worldwide to classify rivers with a view to conservation management have always encountered problems of scale, because different processes influence features from catchment to microhabitat level (e.g. Frissell et al., 1986; Rosgen, 1994). Classification schemes which impose arbitrary divisions in the continuum of a river system, therefore, have serious limitations (Kondolf, 1995). By necessity, analysis of the RHS database initially concentrated on site-based outputs, i.e. 500 m lengths of river. For a better understanding of the links between processes and habitat features, rivers need to be treated as a whole. The focus has now switched to a catchment basis, exploring further the geomorphological aspects of river channels (Newson et al., 1998a) and the links between seasonal flow types and habitats (Newson et al., 1998b). The catchment context is important because results from individual sampling sites cannot be considered entirely in isolation – they can be strongly influenced by channel and catchment processes upstream of the site, including factors such as land management.

Analysing sequences of RHS site data within an individual catchment will allow the relationship between habitat features, the primary 'driving' variables (e.g. gradient), and factors such as channel management and land use to be explored. A sound knowledge of how rivers behave is particularly important in the context of sustainable river rehabilitation schemes (Hansen et al., 1998). Using RHS and more detailed geomorphological information to establish channel character and predict behaviour should be able to reduce the failure risk of rehabilitation works. Bearing in mind the substantial costs involved, decisions on whether to restore features such as riffles, and if so where, can be better supported because the factors which maintain their occurrence in similar rivers elsewhere can be determined. Sustainable, cost-effective solutions such as in the Sankey catchment can be more confidently identified (Environment Agency, 1998).

The 1994–1996 baseline survey not only provides the basis for habitat quality assessment, but also a framework for monitoring changes in the features re-

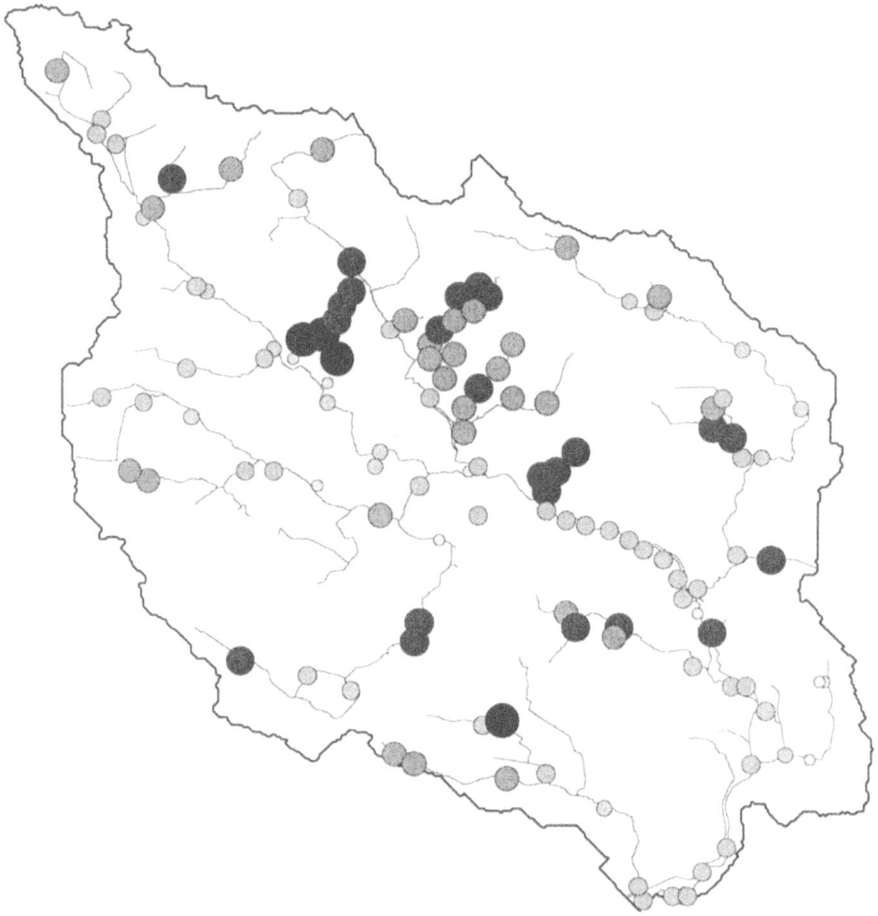

Figure 4. The River Sankey catchment, north-west England, showing relative stream power, as derived from RHS and additional geomorphological data. Stream power ranges from very low (lightest shading) to very high (darkest shaded circles).

corded. Future surveys can be used to reveal how changes in catchment and riparian land use affect river habitats and, in the context of climate change, how habitats might be altered as a result of reduced baseflows in groundwater-fed streams or increased flood peaks in spate rivers, caused respectively by the increased frequency of droughts and storms.

Although the RHS method has been developed specifically for the scale and management of rivers in the British Isles, it has also been tested successfully on rivers in Madeira, where altitude, channel slopes and geology differ quite significantly. The field method is not suitable for large (>100 m wide) or multi-thread rivers, but the underlying principles can be applied by adapting unit sample length, distance between spot-checks, vegetation and land-use categories to suit local conditions. On this basis, there is considerable potential for its application to environmental assessment and catchment management beyond the U.K. As the focus switches to a catchment context, there is significant scope for further developing the link RHS has already established between fluvial geomorphology and wildlife conservation, a valuable asset when the forthcoming European Framework Directive on Water Policy is implemented.

Acknowledgements

The authors would like to thank Prof. John Jeffers, Prof. Malcolm Newson, Prof. Mike Clark, Dr Phil Boon, Prof. Ron Edwards, Dr Mark Everard and Ian Fozzard for helping to develop the RHS system and its practical applications.

References

Boon, P. J., J. Wilkinson & J. Martin, 1998. The application of SERCON (System for Evaluating Rivers for Conservation) to a selection of rivers in the U.K. Aqua. Cons.: Mar. Freshwat. Eco. 8: 579–596.

Bowers, J., 1995. The interface between ecology and economics in catchment management. In Harper, D. M. & A. J. D. Ferguson (eds), The Ecological Basis for River Management. John Wiley, Chichester: 515–523.

Buckton S. T. & S. T. Ormerod, 1997. Use of a new standardised habitat survey for assessing the habitat preferences and distribution of upland river birds. Bird Study 44: 327–337.

Commission for the European Communities, 1997. Proposal for a Council Directive establishing a framework for Community action in the field of water policy. CEC COM(97)49. Official Journal, C184, 17 June 1997.

Environment Agency, 1997. The Quality of Rivers and Canals in England and Wales 1995. Bristol.

Environment Agency, 1998. A natural assets register for the Sankey catchment. Unpublished report, Environment Agency, North West Region, Warrington.

Fox, P. J. A, M. Naura & P. Scarlett, 1998. An account of the derivation and testing of a standard field method. River Habitat Survey. Aquat. Cons. Mar.: Freshwat. Eco. 8: 455–475.

Frissell, C. A., W. J. Liss, C. E. Warren & M. D. Hurley, 1986. A hierarchical approach to classifying stream habitat features: viewing streams in a watershed context. Envir. Mgmt. 10: 199–214.

Hansen, H. O., P. J. Boon, B. L. Madsen & T. M. Iverson (eds), 1998. River Restoration: The Physical Dimension. Aquat. Cons.: Mar. Freshwat. Eco. 8: 1–264.

Holmes, N. T. H., P. J. Boon & T. A. Rowell, 1998a. A revised classification system for British rivers based on their aquatic plant communities. Aquat. Cons.: Mar. Freshwat. Eco. 8: 555–578.

Holmes, N. T. H., J. R. Newman, F. H. Dawson, S. Chadd, K. Rouen & L. Sharp, 1998b. Mean trophic rank: a user's manual. Environment Agency R&D Technical Report E38, Bristol.

Jeffers, J. N. R., 1998. Characterisation of river habitats and prediction of habitat features using ordination techniques. Aquat. Cons.: Mar. Freshwat. Eco. 8: 529–540.

Kondolf, G. M., 1995. Geomorphological stream channel classification in aquatic habitat restoration: uses and limitations. Aquat. Cons.: Mar. Freshwat. Eco. 5: 127–141.

Mainstone, C. P., S. Barnard & R. J. Wyatt, 1994. The NRA National Fisheries Classification Scheme. A guide for users, R&D Note 206. National Rivers Authority, Bristol.

Naura, M. & M. Robinson, 1998. Principles of River Habitat Survey to predict the distribution of aquatic species: an example applied to the native, white-clawed crayfish *Austropotamobius pallipes*. Aquat. Cons.: Mar. Freshwat. Eco. 8: 515–528.

Newson, M. D., M. J. Clark, D. A. Sear & A. Brookes, 1998a. The geomorphological basis for classifying rivers. Aquat. Cons.: Mar. Freshwat. Eco. 8: 415–430.

Newson, M. D., D. M. Harper, C. L. Padmore, J. L. Kemp & B. Vogel, 1998b. A cost-effective approach for linking habitats, flow types and species requirements. Aquat. Cons.: Mar. Freshwat. Eco. 8: 431–446.

Raven, P. J., P. Fox, M. Everard, N. T. H. Holmes & F. H. Dawson, 1997. River Habitat Survey: a new system for classifying rivers according to their habitat quality. In Boon, P. J. & D. L. Howell (eds), Freshwater Quality: Defining the Indefinable? The Stationery Office, Edinburgh: 215–234.

Raven, P. J., N. T. H. Holmes, P. J. A. Fox, F. H. Dawson, M. Everard, I. Fozzard & K. Rouen, 1998a. River Habitat Quality: the physical character of rivers and streams in the U.K. the Isle of Man. Environment Agency, Bristol.

Raven, P. J., N. T. H. Holmes, F. H. Dawson & M. Everard, 1998b. Quality assessment using River Habitat Survey Data. Aqua. Cons.: Mar. Freshwat. Eco. 8: 477–499.

Raven, P. J., P. J. Boon, F. H. Dawson & A. J. D. Ferguson, 1998c. Towards an integrated approach to classifying and evaluating rivers in the U.K. Aquat. Cons.: Mar. Freshwat. Eco. 8: 383–393.

Rosgen, D. L., 1994. A classification of natural rivers. Catena 22: 169–199.

Webb, A. D., P. J. Bacon & M. Naura, 1998. Catchment stream surveys and the use of GIS for integrated management: DeeCAMP and the Deeside Rivers Survey. Aquat. Cons.: Mar. Freshwat. Eco. 8: 541–554.

Werritty, A., 1997. Conserving the quality of freshwater resources: the role of integrated catchment management. In Boon, P. J. & D. L. Howell (eds), Freshwater Quality: Defining the Indefinable? The Stationery Office, Edinburgh: 489–505.

Wright, J. F., M. T. Furse & D. Moss, 1998. River classification using invertebrates: RIVPACS applications. Aquat. Cons.: Mar. Freshwat. Eco. 8: 617–632.

Hydrobiologia **422/423**: 369–387, 2000.
M. Jungwirth, S. Muhar & S. Schmutz (eds), Assessing the Ecological Integrity of Running Waters.

Effects of catchment properties on hydrochemistry, habitat complexity and invertebrate community structure in a lowland river

Barbara Bis[1], Anna Zdanowicz[2] & Maciej Zalewski[1]
[1]*Department of Applied Ecology, Centre of Ecohydrological Studies, University of Łódź, 90-237 Łódź, Banacha 12/16, Poland*
[2]*Department of Water Resources, IMUZ, Falenty, 05-090 Raszyn, Poland*

Key words: GIS, ecohydrology, macroinvertebrates, biomonitoring, ecological integrity

Abstract

The main objective of this study was to quantify the relationship between taxonomic and functional parameters of macroinvertebrate assemblages and reach/landscape attributes of agricultural catchment. We aimed to analyse the relation between landscape patchiness and land-water ecotone structure, and their influence on benthic fauna composition, biomass and functional organization. A variety of catchment characteristics: land use patterns, surface geology (e.g. type of soils), elevation, hydrology database were incorporated into a geographical information system (GIS) to evaluate anthropogenic impact on landscape properties. The scenarios of potential transfer of nitrogen load along the gradient of the river valley were elaborated. The assessment of nutrient export focused on identifying the most effective nutrient barriers across spatial scales. The analysis of landscape and reach units demonstrated significant, negative correlations between the rate of nutrient transfer to surface/groundwaters and (i) the length of riparian ecotones, (ii) river width, and (iii) the areas of wetland and forested patches in the catchment. The trophic base (primary producers; CPOM; FPOM) in the stream channel was analysed as a factor modifying the structural and functional organization of macrobenthos. Canopy cover, and the consequent amount of solar input reaching the stream bed, significantly influenced food-related variables: leaf input and algal biomass. The factor that significantly affected food availability, retention and transport of benthic particulate organic matter (BPOM) and amount of chlorophyll *a* was stream discharge. Higher discharge contributed to low retention of BPOM. To obtain more information about environmental variability (56 variables) and macroinvertebrate functional organization, the entire data matrix was analysed using Principal Component Analysis. A significant correlation was shown in the case of shredder abundance (i) in the low-order station S1, (ii) high canopy cover complexity, and (iii) retention of benthic organic matter. Filtering collectors were significantly associated with the presence of suspended organic matter. Other faunal assemblages, highly related to stream hydraulics factors, were alternative gathering collectors/scrapers, much retained at fourth-order station S3. Stream discharge, depth and chlorophyll *a* significantly affected the presence of these organisms. This indicated the crucial role of habitat heterogeneity, which was attributable to local conditions (detrital inputs, hydraulics, morphological characteristics), and significantly influenced the trophic structure of biocommunities. In the regulated section of the river (S2) without overhanging vegetation, the abundance of scrapers was significantly correlated with open canopy. Domestic and agricultural impact, evidenced by high nutrient concentrations, significantly reduced species richness (stations S4/S4a). Macroinvertebrate communities were used to detect environmental stress in the river system. Assessment of water quality with biological indices directly revealed human-induced alterations (e.g. land use, lack of riparian vegetation, channelization).

Introduction

The biotic structure and water quality of streams and rivers reflect an integration of physical and biological processes occurring in a catchment. As a result, rivers are strongly influenced by the geology, valley contours and vegetation, and by human activities that alter land cover, hydrologic pathways and ultimately affect most of landscape-freshwater couplings (Allan & Johnson, 1997). Today, focusing on the ecological, geomorphologic and hydrological interactions between aquatic and terrestrial ecosystems has become standard procedure (e.g. Hynes, 1975; Vannote et al., 1980; Statzner & Higler, 1986; Frissell et al., 1986; Pringle et al., 1988; Naiman & Decamps, 1990; Boon et al., 1991; Hawkins et al., 1993). A refined understanding of the scale-dependent processes and hierarchy of linkages across the catchment and land-water ecotones is a crucial approach to ecologically sound water management (Statzner & Sperling, 1993; Carter et al., 1996; Allan et al., 1997; Allan & Johnson, 1997; Johnson & Gage, 1997). This was profoundly reinforced by development of the ecohydrology paradigm (Zalewski, 1999; Zalewski et al., 1997).

Many recent catchment studies, taking advantage of geographical information systems and/or multivariate statistics to quantify landscape structure, have revealed the longitudinal and lateral influences of the terrestrial ecosystems on water chemistry and stream biota (Verdonschot, 1990; Dolédec et al., 1996; Richards et al., 1996; Johnson et al., 1997; Johnson & Gage, 1997; Townsend et al., 1997). Moreover, the sensitivity of macroinvertebrate community structure as an indicator of anthropogenically induced changes in river ecosystems has been recently analysed in long-term, multiple-site based studies (Dolédec & Statzner, 1994; Statzner et al., 1994, 1997; Richards et al., 1997). These syntheses, using various metrics derived from taxonomic collections, provided further evidence that measurements of both catchment and local habitat have long been considered the 'templet' for local biotic diversity (Southwood, 1977); this has increasingly been developed by stream ecologists into habitat assessment protocols for environmental monitoring (e.g. Platts et al., 1983).

The assessment of ecological integrity in water resources systems (Karr et al., 1984; Karr, 1991), defined as the ability to maintain a balanced, integrated and adaptive community of autochthonous organisms, should encompass all factors affecting ecosystem stability (e.g. resistance and resilience). Refining our understanding of the catchment influences on richness and composition seems an important component in both invertebrate ecology and impact assessment.

The primary objective of this study was to quantify the relationship between a series of taxonomic and functional descriptors of macroinvertebrate assemblages and reach/landscape-scale attributes of agricultural catchment. Specifically, we aimed to analyse the relationships between landscape patchiness and land-water ecotone structure influence on benthic fauna composition, biomass and functional organization. This knowledge will be particularly relevant in developing macroinvertebrate indicators as monitoring and assessment tools for scale-dependent water resources management.

Materials

Study area

The study was conducted in the Grabia River catchment, encompassing 819.5 km^2 (Figure 1). This catchment was chosen because land use patterns within its component drainages ranges from heavily impacted agricultural to comparatively undisturbed areas. The river is 81 rectilinear km long. Its Quaternary geology is dominated by glacial remnants (sands, gravel, morainal clay and chalk rocks in the middle course). Annual average rainfall in this area is 549–600 mm, with precipitation being evenly distributed throughout the year. Annual average water temperature in July is 16–18 °C, and in January 2–3 °C. The catchment ranges in elevation from sea level to 270 m. A detailed description of the river basin was given by Maksymiuk (1970), Burchard & Maksymiuk (1985) and Zalewski et al. (1990).

Study design

Chemical measurements, analysis of in-stream habitat heterogeneity, temporal distribution of particulate organic matter (POM), as well as algae and benthic fauna collections were carried out from April 1995 to November 1996. Four stations were sampled (eight field campaigns). These stations were numbered by distance downstream from the stream source and represented common conditions along the river.

Catchment-scale measurements

A variety of catchment characteristics: land use patterns, type of soils, surface geology, elevation and hydrography database were incorporated into a geographical information system (GIS) to quantify anthropogenic effects on landscape properties. These factors were used as catchment-scale covariates to characterize the potential hydrological variation. All databases were transformed into a digital format and analysed in the ARC/INFO system.

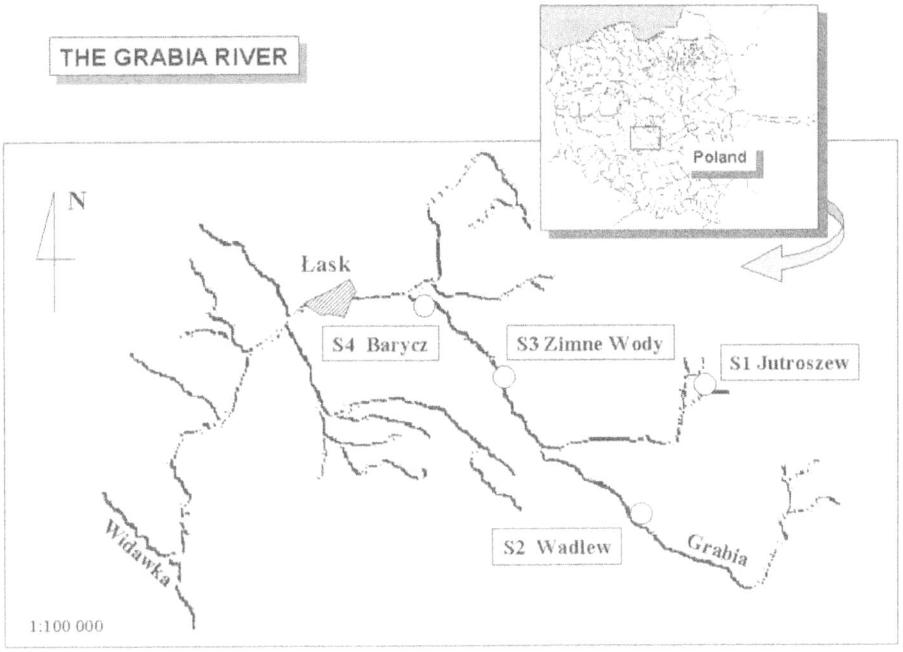

Figure 1. Location of sampling stations in the Grabia River (Central Poland).

Reach-(habitat-) scale measurements

Four stations were chosen to represent different geomorphologic and hydrological settings, as well as local conditions (e.g. altered land use, riparian zone management, channel characteristics).

The uppermost stations (S1, S2) were established near the headwaters (II order). At station 1 (S1 Jutroszew), with a natural, meandering bank line, the stream flows through wet meadows (*Molinietalia* Koch 1926), mixed forest (*Dicrano-Pinion* Libb. 1933) and field area (village, 200 m). It is a tributary of the Grabia River, heavily overshaded, with the dominant tree species *Alnus glutinosa* L., *Pinus silvestris* L. and *Betula pendula* Roth. Along the river course, peaty soil, clay and sand dominate. In the river channel, *Elodea canadensis* Rich. and free-floating *Lemna minor* L. appear. Allochthonous organic matter (detrital inputs, debris dams) composes an important part of the bottom cover.

Station 2 (S2 Wadlew) is the channelized section of the Grabia River, bordered by intensively farmed meadows and pastures. The riparian zone is well developed with communities of *Phalaridetum arundinaceae* (Koch 1926 n.n.) Libb. 1931 and *Filipendulo-Geranietum* Koch 1926. The sandy bottom lacks channel vegetation, but the bottom near the bank is fairly covered by *Elodea canadensis* Rich. and *Veronica beccabunga* L.

Stations S3 and S4 (IV order) are located in the middle course of the river. Station 3 (S3 Zimne Wody), with a meandering, natural stream profile (e.g. representative erosional and depositional habitats), has well-developed, large floodplain area (±160 m) with sand and organic soil (supporting infiltration processes). Near the village on the high plain (250 m from the river section), infiltration to the groundwater is highly limited by the substratum (marl debris). The riparian zone is formed by *Alnus glutinosa* L., *Padus avium* Mill., and communities of *Salici-Populetum* (R. Tx. 1931) Meijer Drees 1936, *Cuscuto-Convolvuletum sepium* R. Tx. 1937, *Phalaridetum arundinaceae* (Koch 1926 n.n.) Libb. 1931 and *Caricetum gracilis* (Graebn. & Hueck, 1931) R. Tx. 1937. The river bed does not have submerged macrophytes. The bottom near the bank is rarely covered by *Elodea canadensis* Rich., but bankside roots are exposed, forming optimal refugium for fish and benthic fauna.

Station 4 (S4/S4a Barycz) is located near a village (50 m) in a heavily impacted agricultural area. Because infiltration to the groundwater is minimal (clay-dominated soil type), the surface flow is relatively high. The predominant communities in the riparian zone are: *Arrhenatheretum medioeuropaeum*

(Br.-Bl. 1919) Oberd. 1952, *Salici-Populetum* (R. Tx. 1931) Meijer Drees 1936 and *Dicrano-Pinion* Libb. 1933. Water plants are scarce (*Elodea canadensis* Rich., *Potamogeton crispus* L.), but sandy bottom is covered regularly by fine particulate organic matter. In this river section, 2 stations were established: S4 – the main course of the river; S4a – the small side-arm of the studied section.

Physical characteristics

The physical characteristics of the stations over the study period are listed in Table 1. The following data were measured or classified in the field:

1. stream width – measured at 2–6 transects;
2. stream depth – measured across each cross-section;
3. stream velocity for discharge calculation;
4. substrate types;
5. incident light;
6. in-stream channel and riparian vegetation;
7. organic matter;
8. bank modifications (artificial features);
9. surrounding area (major impacts; land use within 100 m) (Table 1).

Sampling methods

Nutrients and other chemical properties related to water quality were assessed each month. The values shown in Table 2 represent ranges for the sampling stations. Water samples were collected in mid-channel, placed in pre-rinsed polyethylene bottles and kept in ice while being transported to the laboratory for analysis. The groundwater was sampled using a series of piezometers forming 2 transects at each station. Water levels were recorded and water samples were analysed (procedure as above) for the chemical nutrients presented in Table 2.

The bed sediment samples for qualitative and quantitative assessment of organic matter and substrate grain size were taken with a 100 mm diameter steel cylinder inserted into the bed to a depth of 30 cm. At each station, 3–5 samples were collected from different morphodynamic units (lotic, lenitic, depositional areas). Coarse particulate organic matter (CPOM) was sorted before drying in order to quantify the types of detritus (%). Distinctions were made between leaves, leaf fragments, woody debris and macrophytes because various types of detritus may support different macroinvertebrate functional feeding groups. Samples were oven-dried at 105 °C for 24 h to constant weight, before ashing at 550 °C for 5 h to determine ash-free dry weight (AFDW). Samples were mechanically shaken through a column of sieves (above 4.0; 2.0; 1.0; 0.5; 0.25; 0.125; 0.063 mm). The standard sieving time was 15 min., then the content of each sieve was weighed. The diameter of grains was calculated on semilogarithmic cumulative curves for 25, 50, 75% (Index Q_{25} Q_{50} Q_{75}) (Doeglas, 1968).

Total suspended matter (SS) was determinated by filtering water samples (1l) through preweighed Whatman GF/C filters. The filters were dried at 105 °C for 1 h and weighed to the nearest 0.01 mg. Chlorophyll *a* samples were filtered through Whatman GF/C filters and extracted in 90% acetone. The phytopigment concentration was assayed spectrophotometrically. Chlorophyll*a* represented the living biomass, and ash-free dry mass (AFDM) represented the total mass of organic matter including all dead material and allochthonous organic material.

Macroinvertebrates were quantified in 10–50 m study reaches by means of a micromacrofauna shovel (150 cm^2) (Tolkamp, 1980). Sites were mapped for a variety of physical variables, e.g. substrate types, aquatic and riparian vegetation, erosial or depositional areas, and the exact position of sampling microhabitats. At each reach, 8–15 samples were taken from random locations to a sediment depth of 3 cm every 10–12 weeks. Macroinvertebrates were counted and identified to the lowest possible taxonomic level. The collected individuals were preserved in ethanol (70–80%). For most collected species, the body length of all specimens was measured in 2 mm size classes to construct a life history pattern and for conversion to biomass (Smock, 1980; Meyer, 1989). From each sample, a group of representative individuals from each identified taxon was dried at 60 °C for 24 h and weighed. The animals were assigned to a particular functional feeding group according to the classification given by Cummins (1973) and Merritt & Cummins (1984).

All variables in all analyses were transformed when appropriate to approximate a normal distribution. Invertebrate densities and physicochemical parameters were typically $\log_{10}(x+1)$ transformed, while proportions and relative abundance were arcsin $\sqrt{x}$ transformed. Environmental variables were analyzed by principal component analysis (PCA) to investigate the relationships between variables. To establish species groups, the data were subjected to classification analysis by Ward's method. One-way ana-

Table 1. Main physical features of stations studied

	S1	S2	S3	S4	S4a
Stream order	II	II	IV	IV	IV
Distance from source (km)	3	15	40	50	50
Gradient (%)	3.3	1.7	0.8	0.8	0.8
Mean discharge (m^3 s^{-1})	0.12	0.35	0.46	0.47	0.25
Reach length (m)	50	10	20	30	6
Velocity (ms^{-1})	0.01–0.34	0.07–0.38	0–0.52	0–0.44	0–0.12
Depth (m)	0.01–0.55	0.25–0.50	0.38–0.57	0.15–0.59	0.07–0.42
Width (m)	0.75–2.5	1.6–2.3	9.3–14.5	7.8–13	2.3–5.7
Canopy cover (%)	30–75	0–50	10–70	0–30	30–60
Woody debris	present	none	present	none	none

Table 2. Chemical contents in the Grabia River

	Surface water					Groundwater			
	S1	S2	S3	S4	S4a	S1	S2	S3	S4
$P\text{–}PO_4^{3-}$ mg l^{-1}	0.002–0.474	0.017–2.228	0.003–0.656	0.014–0.759	0.023–2.546	0.022–1.677	0.010–2.380	0.010–0.439	0.001–1.676
$N\text{–}NO_3^-$ mg l^{-1}	0.001–1.210	0.009–4.190	0.007–1.150	0.002–1.250	0.006–1.540	0–5.600	0–4.003	0.011–5.480	0.001–6.060
$N\text{–}NH_4^+$ mg l^{-1}	0–0.359	0.016–1.309	0.017–0.288	0.010–0.400	0.033–0.374	0.020–1.649	0.010–1.340	0.018–1.630	0.015–3.040
Oxygen mg l^{-1}	3.00–8.40	6.40–8.40	4.10–8.50	5.30–9.10	3.70–7.80	1.30–8.60	1.30–5.70	1.20–9.10	1.20–9.30
Conductivity μS cm^{-1}	198–543	304–438	191–452	188–440	184–476	380–992	305–478	182–497	190–950
Acidity pH	7.10–8.40	6.90–7.50	7.10–8.20	7.04–7.76	7.15–8.00	5.70–8.47	6.90–7.90	6.14–8.60	5.20–9.30

lysis of variance (ANOVA) and Tukey's HSD test were used to determine the significance of differences between data (chemical parameters, food-related variables). All statistical analyses were performed using the STATISTICA package (version 5.1 PL; StatSoft. Inc., 1997).

Results

Water chemistry

The concentrations of most chemical variables, listed in Table 3, were related to land use. All nutrient concentrations ($P\text{–}PO_4^{3-}$, $N\text{–}NO_3^-$, $N\text{–}NH_4^+$) were higher at anthropogenically impacted stations (S2; S4). The concentration of $N\text{–}NH_4^+$ exhibited a significant seasonal increase at station 2 ($p = 0.023$, $df = 7$) (Table 2). Water quality trends were generally consistent with longer-term patterns along the gradient described by Zalewski et al. (1990).

Landscape properties

Land use categories were aggregated into four classes: 1. agricultural areas, including rowcrop and non-rowcrop agriculture, 2. urban areas, 3. forests, and 4. wetlands (Figure 2). These parameters reflected the extent of human impact in landscape changes.

Land use of the Grabia River catchment was dominated by agricultural areas, ranging from 49.7% (117 ha) to 86.3% (8784 ha). Forested land cover was less prevalent, ranging from 5.5% (2 ha) to 36.1% (2144 ha). Wetland habitats covered about 1.6% of total area. Surface geology was regionally patchy (loam, moor, sand, less loamy sand, loamy sand, sandy loam), but sandy substrate was the dominant soil type.

Landscape characteristics were elaborated and developed on the basis of data related to the structure and function of the river, including climate, catchment hydrology, topography, surface geology and hydrochemistry (surface/groundwaters). These input data, combined with selected variables such as travelling time of nitrogen (time required for nitrogen to reach groundwater table) or surplus of nitrogen (estimated from plant production, mineral fertilisers, cattle

Table 3. List of environmental variables used in PCA ordination

	Unit	Abbreviation	
Quantitative environmental variable			
Chemical variables	Water temperature	°C	TEMP
	Acidity		pH
	Electric conductivity	μS	Ec
	Oxygen concentration	mg l^{-1}	O2
	Ammonium	mg l^{-1}	NH4
	Nitrogen–NH4	mg l^{-1}	N–NH4
	Nitrate	mg l^{-1}	NO3
	Nitrogen–NO3	mg l^{-1}	N–NO3
	Nitrite	mg l^{-1}	NO2
	Nitrogen–NO2	mg l^{-1}	N–NO2
	Total phosphate	mg l^{-1}	T-P
	Ortho-phosphate	mg l^{-1}	P–PO4
Width of the river		m	WIDTH
Discharge		$m^3\ s^{-1}$	DISCH
Vegetation	algae	%	ALGAE
	floating	%	VEGEFLAT
	submerged	%	VEGESUBM
	emerged	%	VEGEEMER
Bottom substrate	silt	%	SILT
	sand	%	SAND
	gravel	%	GRAVEL
	stones	%	STONE
Habitat type	Riffle	%	RIF
	Run	%	RU
	Pool	%	PO
Total suspended solids	organic	ash-free	SOM
	inorganic	dry weight	SIM
Chlorophyll *a*		mg l^{-1}	CHL
Benthic organic matter	woody debris	%	WOOD
	leaves	%	LEAF
	coarse particulate organic matter	%	CPOM
	fine particulate organic matter	%	FPOM
Current velocity		m s^{-1}	VELOCITY
Depth		m	DEPTH
Nominal environmental variable			
Date	January	JAN	
Profile shape	line shaped regular	SHAPLSRE	
	line shaped irregular	SHAPLSIR	
	line shaped very irregular	SHAPLVIR	
Meandering	none	MEANDNT	
	moderate	MEANMOD	
	strong	MEAND	
Regulation	none	REGULANT	
	strong	REGULA	
Canopy cover	0%	SH 0	
	1–10%	SH 1–10	
	11–30%	SH 11–30	
	31–50%	SH 31–50	
	51–75%	SH 51–75	
	> 75%	SH >75	
Surrounding	urban	SURURB	
	forest	SURFOR	
	wooded bank	SURWOB	
	field	SURFIELD	
	meadow	SURMED	
Ripple marks		RIPPMARK	
Shifting sand		SHIFSAND	

selling), were used to quantify the potential nitrogen transport to groundwater in long-term scenarios (after 3 and 30 years). The model revealed notable differences in the potential nitrogen transfer to groundwater; thus, in the upper basin the nitrogen transfer was estimated at about 1–15 kg $ha^{-1}y^{-1}$, in the lower basin at about 20–35 kg $ha^{-1}y^{-1}$ (Figure 3).

Food sources

The variation in density and extent of the riparian canopy was significantly related to land use pattern, being greatest in the forested sections of the river (S1, S3). Canopy cover, and consequently the amount of solar input reaching the streambed, significantly influenced food-related variables ($p<0.05$): leaf input ($R = 0.58$, $F_{(1,14)} = 9.93$, $p = 0.01$) (Figure 4), and algal biomass ($R = 0.39$, $F_{(1,19)} = 4.36$, $p = 0.08$). Terrestrial litterfall (leaves) ranged from 10 gm^{-2} at station S4 (solar input = 870 $\mu E\ cm^{-2}\ s^{-1}$) to 615 gm^{-2} at station S1 (solar input = 65 $\mu E\ cm^{-2}\ s^{-1}$). Higher retention of plant litter at S1 was affected by the abundance of retention structures in the stream (debris dams).

The factor that significantly affected food availability, retention and transport of benthic particulate organic matter (BPOM), and amount of chlorophyll *a* was stream discharge. Higher discharge contributed to low retention of BPOM ($R = -0.58$, $F_{(1,18)} = 5.65$, $p = 0.02$) (Figure 5). Larger stream size and discharge were generally accompanied by significantly higher algal biomass ($R = 0.47$, $F_{(1,18)} = 5.28$, $p = 0.03$). The relationships between food-related variables (leaf inputs, algal biomass, BPOM) and discharge were estimated on the basis of data from the period 1990 to 1996.

Macroinvertebrate community

A hierarchical method (Ward's) of cluster analysis was performed on 85 benthic invertebrate taxa. The classification grouped similar items together, indicating relationships among groups. The results of the cluster analysis showed that seven species groups were present (Figure 6).

Group I was characterized by a number of taxa living on/in sandy substrate (e.g. *Armiger crista*) in shallow, moderate-flowing water with a rich riparian vegetation, abundant benthic particulate organic matter and/or woody debris dams (e.g. *Beraeodes minutus, Limnephilus flavicornis, Paraleptophlebia cincta*). Most of these taxa are common in unpolluted small streams, and their composition was attributed to S1. Group II was related to large water bodies, and had several taxa clearly bound to water flow (e.g. *Limnius volckmari, Goera pilosa*). Moderate canopy cover affected periphyton production and the presence of its consumers (e.g. *Ephemerella notata, Caenis horaria*). This species composition was representative for station S3. Group III was typified by few common coleopterans (e.g. *Platambus maculatus*) or Gastopoda, inhabiting slowly flowing small waters with sandy substrate and open canopy. These species were widespread in the stream with the following hydrological pattern: very low discharges in summer and high ones in winter (station S2). Group IV was represented by ubiquitous taxa commonly occurring in the first three groups (e.g. *Erpobdella monostriata, Gammarus fossarum*). Group V and group VI had several taxa which preferred slowly flowing/stagnant waters with silty substrate and dense vegetation (stations S4 and S4a). Many of them can be found in organically enriched conditions. Group VII contained common taxa resistant to organic pollution (*Chironomidae, Oligochaeta*) and occurring at all sites.

Macroinvertebrate distribution pattern

Average macroinvertebrate density and biomass (Figure 7A, B) were significantly lowest at station S2, located in the regulated section of the river. Clearing of riparian vegetation at this site was accompanied by a notable increase in the relative abundance of scrapers. Generally, macroinvertebrate densities gradually increased downstream. In contrast, average biomasses declined precipitously at station S2 and remained at comparatively low levels.

Macroinvertebrate trophic organization

The longitudinal distribution of macroinvertebrate functional groups in the Grabia River provided insight into the influence of downstream changes in food resources (quality and availability). The annual ranges and significant differences between density and biomass of functional feeding groups at each station are illustrated in Figure 7A, B. Maximum taxonomic richness, density, and biomass of shredders (e.g. *Gammarus fossarum, Limnephilus rhombicus, Potamophylax nigricornis, Nemoura* sp.) occurred at the low-order station S1. This was attributable to the high retention of particulate organic matter (leaf material) here. The total density of shredders at station S4a did not parallel the biomass, but shredders in the

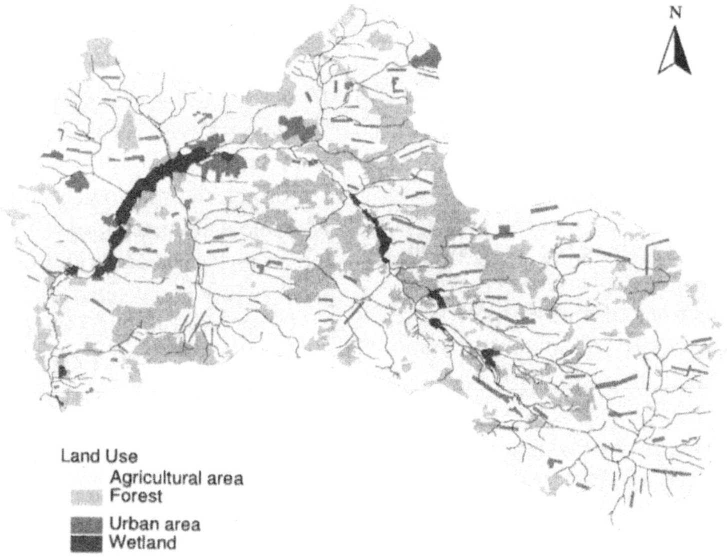

Figure 2. Map of the land use in the Grabia River catchment (scale 1: 250 000).

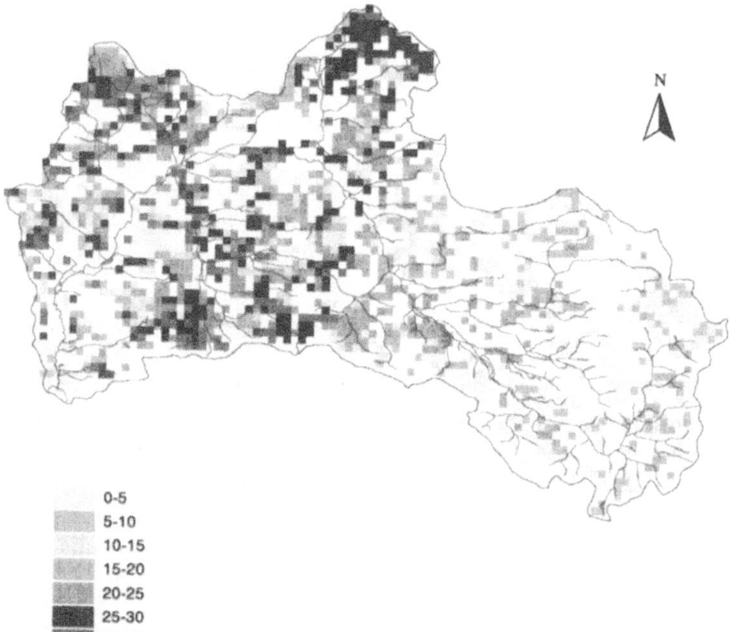

Figure 3. Map of the potential nitrogen transport to groundwater (kg ha^{-1} y^{-1}; after 3 years) in the Grabia River catchment (scale 1: 250 000). Note the diversity of the potential nitrogen transport in the upper and in the lower basin.

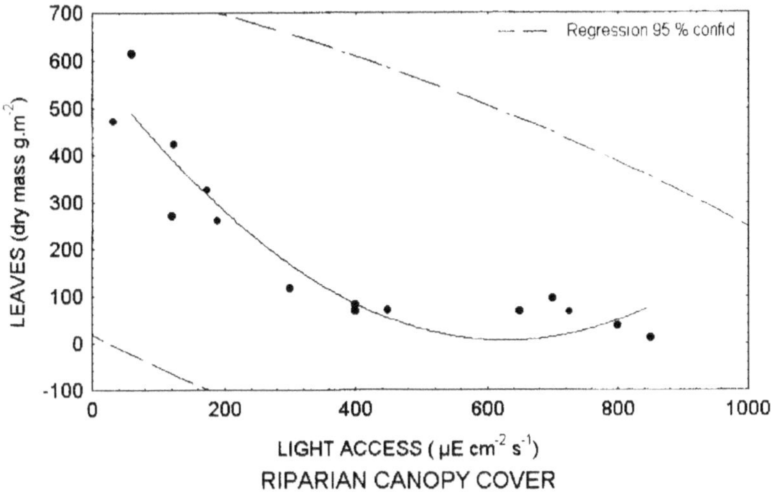

Figure 4. The relationship between canopy cover density (solar input) and leaf input–according to the negative exponentially-weighted smoothing procedure.

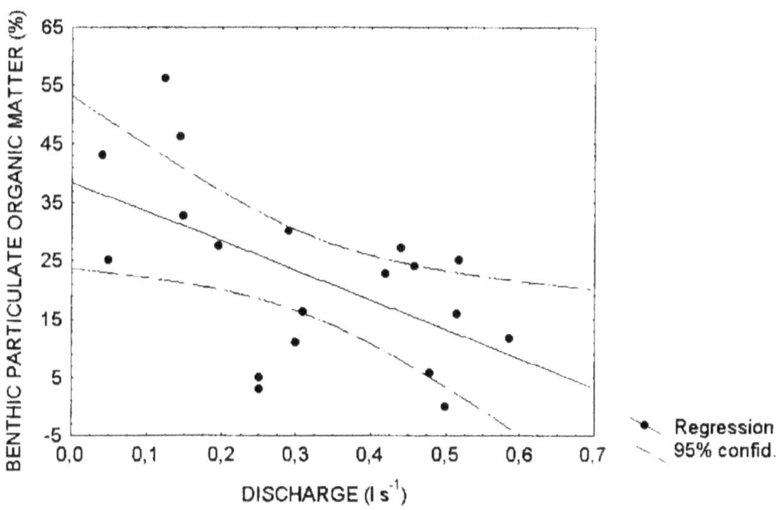

Figure 5. The relationship between stream discharge and BPOM retention.

side-arm were represented only by late instar larvae of *Anabolia nervosa*.

Gathering collectors were major components at all stations. These consumers were most abundant at stations S4 and S4a, with *Oligochaeta* and *Chironomidae* predominating, whereas gatherers at the low-order site S1 were dominated by the large-sized: *Ephemera vulgata* and *E. danica*, which significantly affected the high biomass values.

Filtering collectors were well represented at station 1, with widespread representatives of *Sphaeriidae* and *Hydropsychidae*. The species composition of these consumers changed along the river course, with increasing domination of *Simuliidae* downstream.

Scrapers (e.g. *Valvata naticina, Gyraulus albus, Anisus spirobis, Radix peregra, Limnius volckmari)* remained on comparable levels at each site, but significantly higher density and biomass were observed in logged, regulated section S2, situated in the channelized river valley.

Functional organization vs. environmental variability.

Faunal abundance and biomass, number of taxa, biotic indices and other similar faunal parameters are based on condensed data. Such data can give general community characteristics (e.g. annual fluctuations), but do not make maximum use of the information available on the faunal organization which might reveal the

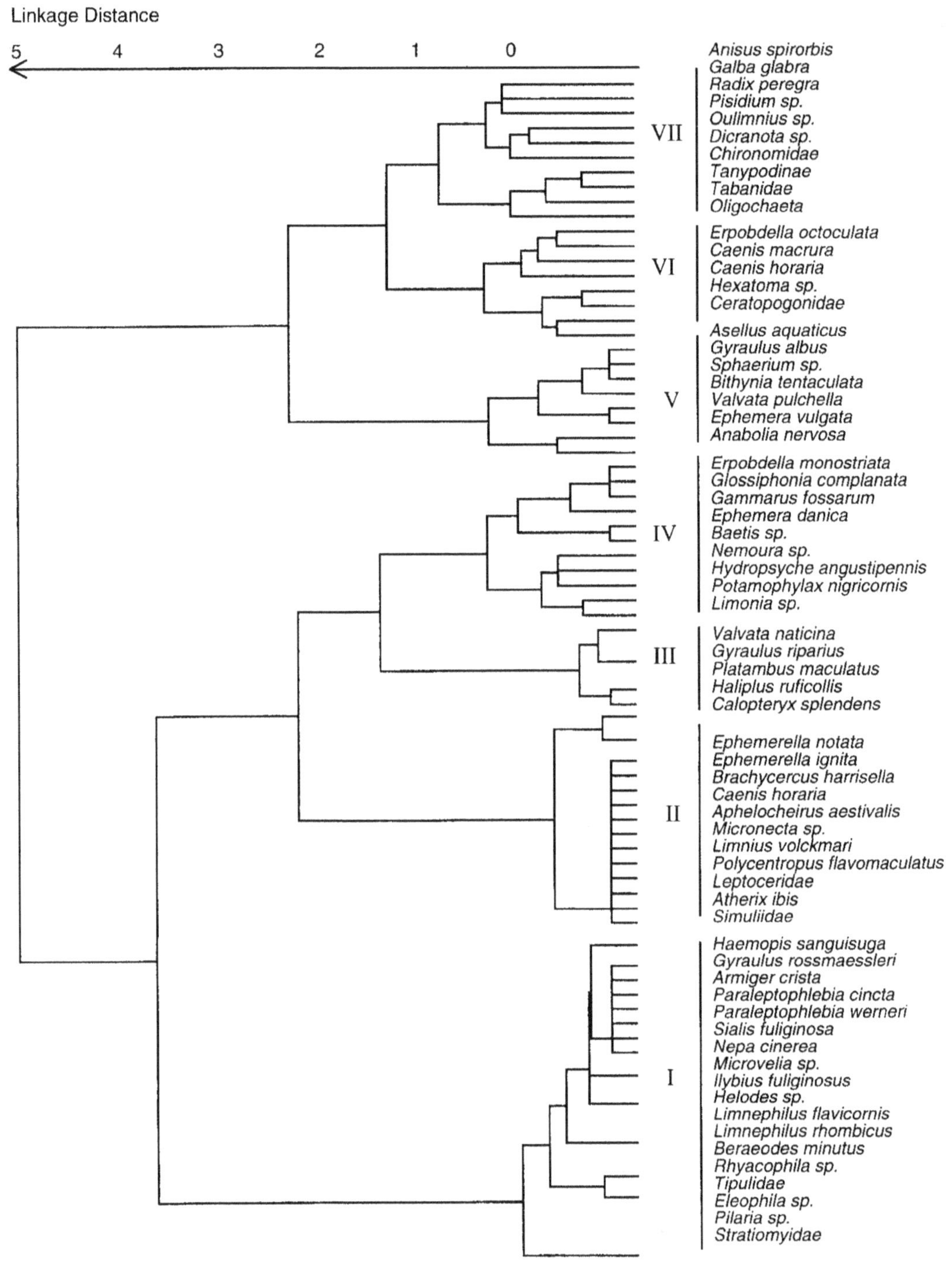

Figure 6. Classification of species assemblages. Ward's clustering methods based on disagreement percent.

more subtle effects of disturbance (Armitage & Gunn, 1996). To obtain more information about macroinvertebrate functional organization and environmental variability (56 variables: (physicochemistry, food-related variables), the entire data matrix was analysed using PCA (Principal Component Analysis) (Figure 8). The first seven PCA axes respectively account for 16.49%, 11.90%, 9.36%, 8.82%, 7.88%, 6,71% and 5. 86% of the variance (cumulative 67.10%). A significant correlation ($p<0.05$) was shown in the case of shredder abundance and 1. canopy cover complexity ($R=0.80$, $p=0.0004$) as well as 2. retention of benthic

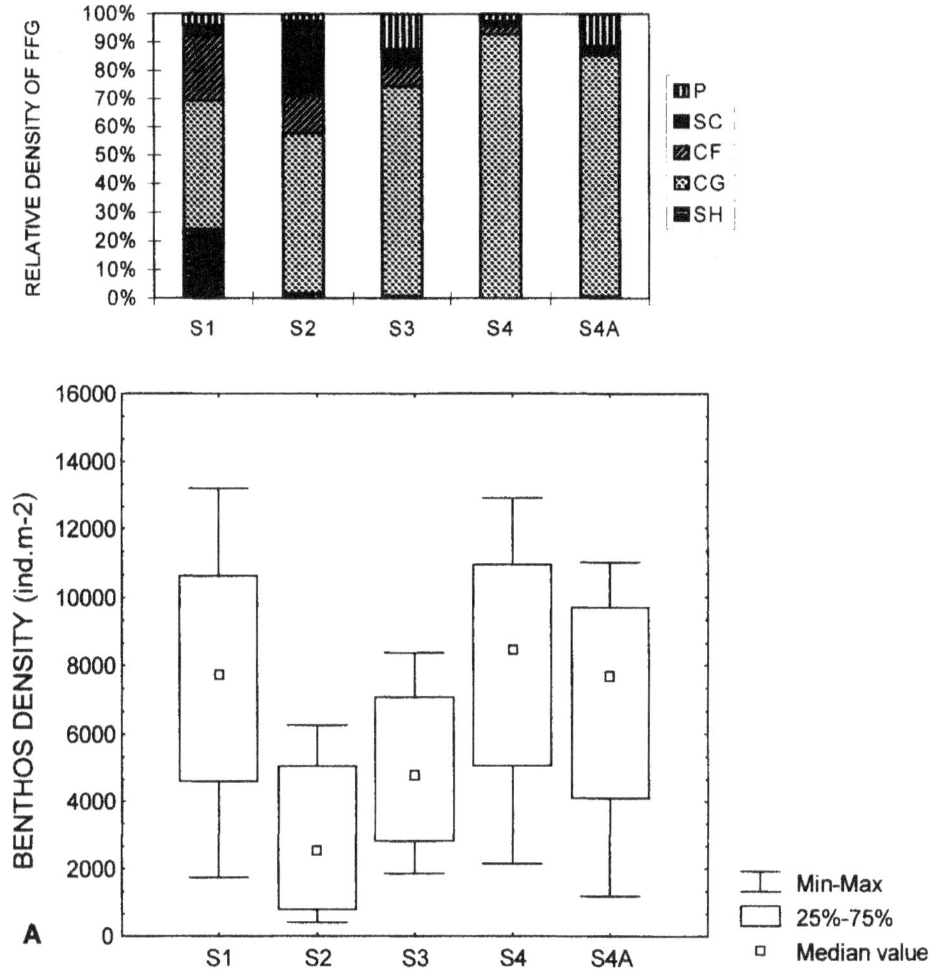

Figure 7a. Total density (A) and biomass (B) of the macroinvertebrate community in the Grabia River catchment at all stations. Upper schemes: relative contribution (by density and biomass) of macroinvertebrate functional feeding groups along the Grabia River longitudinal gradient: SH – shredders; CG – gathering collectors; CF – filtering collectors; SC – scrapers; P – predators.

organic matter (R= 0.49, p= 0.03). Filtering collectors were significantly associated with the presence of suspended organic matter (R= 0.48, p=0.01). Other faunal assemblages highly related to stream hydraulics factors were alternative gathering collectors/scrapers (e.g. *Ephemeroptera*), abundant at fourth-order station S3. Stream discharge (R = 0.51, p= 0.03), depth (R= 0.36, p=0.04), and chlorophyll *a* affected the presence of these organisms. It implied the crucial role of habitat heterogeneity, which was attributable to local conditions (detrital input, hydraulics, morphological characteristics), and significantly influenced the trophic structure of communities. In the regulated section of the river (S2) without overhanging vegetation, the abundance of scrapers was significantly correlated with open canopy (R = 0.63, p= 0.003). Domestic and agricultural impact, evidenced by high nutrient concentrations (e.g. P–PO_4^{3-} *v.* CG: R = 0.45, p= 0.04), significantly contributed to low species richness and dominance of gathering collectors (stations S4/S4a).

Biotic indices

The water quality assessment based on the analysis of macroinvertebrates reflected the natural and anthropogenic disturbances occurring in the river system. Information on the faunal parameters (number of taxa and biotic indices) is presented in Figure 9. The biotic indices showed significant differences between undisturbed (S1; S3) and anthropogenically impacted river reaches (S2; S4/S4a).

The wide range of BMWP scores (Biological Monitoring Working Party; Armitage et al., 1983) indicated strong seasonal variations and differences between sites. The BMWP recorded the best water

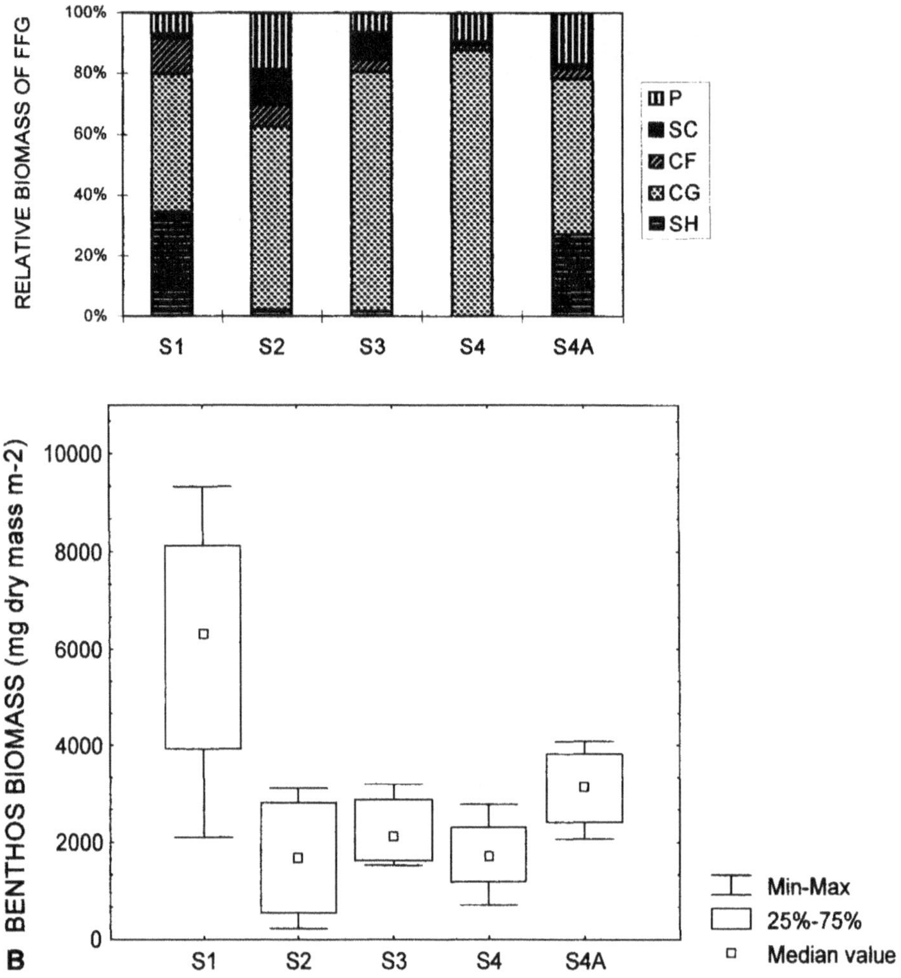

Figure 7b. Total density (A) and biomass (B) of the macroinvertebrate community in the Grabia River catchment at all stations. Upper schemes: relative contribution (by density and biomass) of macroinvertebrate functional feeding groups along the Grabia River longitudinal gradient: SH – shredders; CG – gathering collectors; CF – filtering collectors; SC – scrapers; P – predators.

quality at station 1: at this location the greatest total taxa number was observed. The ASPT scores (Average Score Per Taxon; Armitage et al., 1983), incorporating OQR values (Overall Quality Rating; Extence et al., 1987), indicated the highest biotic scores at station 3. In consequence, the rapid assessment techniques should be adapted to the regional faunal and environmental diversity (geographical, geological and catchment related).

Management implications

The relationships between landscape components and nutrient export to surface and groundwaters was examined using several metrics that described both landscape and reach properties in terms of the size, diversity, and position of patches within the catchment (Table 4). The classification of patch types, considered as effective nutrient barriers, was based on the Hillbricht-Ilkowska & Lawacz scheme (1984). Accordingly, three categories of landscape patches were distinguished: 1. 'strong' – wetlands; 2. 'moderate' – woodlots, forest fragments, natural and cultivated meadows, pastures; and 3. 'low' patches – ploughed arable land, urban sites.

Principal component analysis (Figure 10) showed that effective nutrient barrier capacities were a function of (a) river width ($N–NO_3^-$ $_{sur.}$ export: $R = -0.87$, $p = 0.24$), (b) 'strong' patch size ($P–PO_4^{3-}$ $_{sur.}$ export: $R = -0.90$, $p = 0.014$), (c) percentage of 'strong' cover in watershed ($N–NO_3^-$ $_{sur.}$ export: $R = -0.92$, $p = 0.008$), and (d) length of the riparian ecotone zone ($N–NO_3^-$ $_{sur.}$ export: $R = -0.91$, $p = 0.011$).

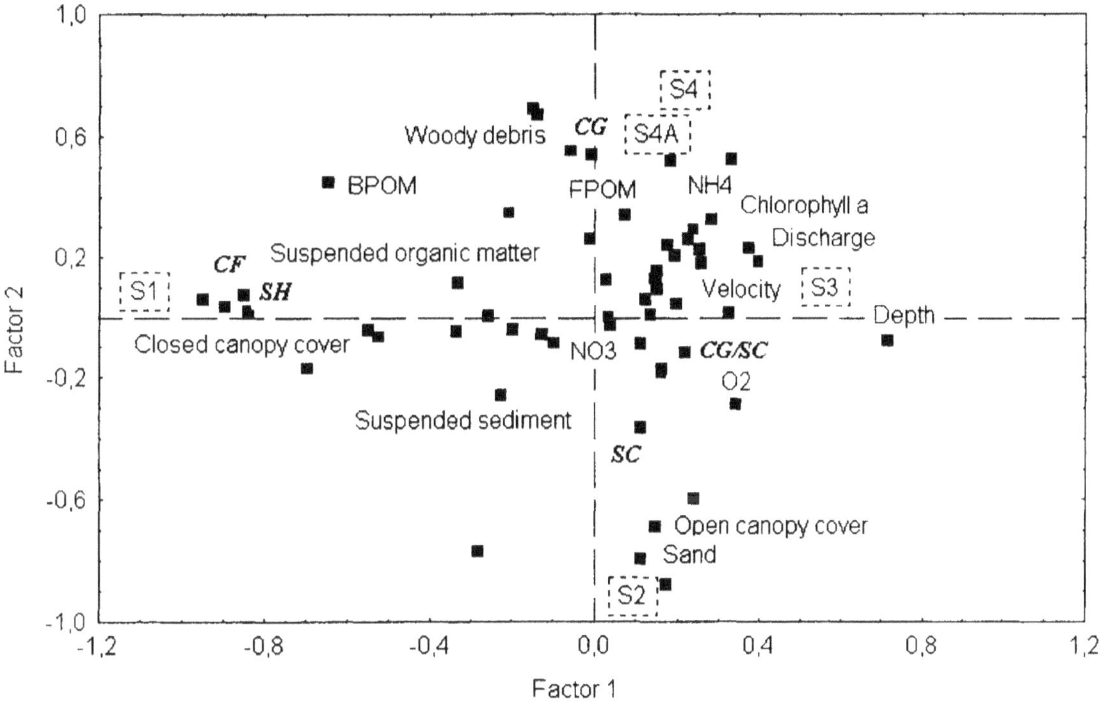

Figure 8. PCA plot (F1 × F2) of environmental variables and functional organization of macroinvertebrates (for abbreviations see Table 3 and Appendix 1).

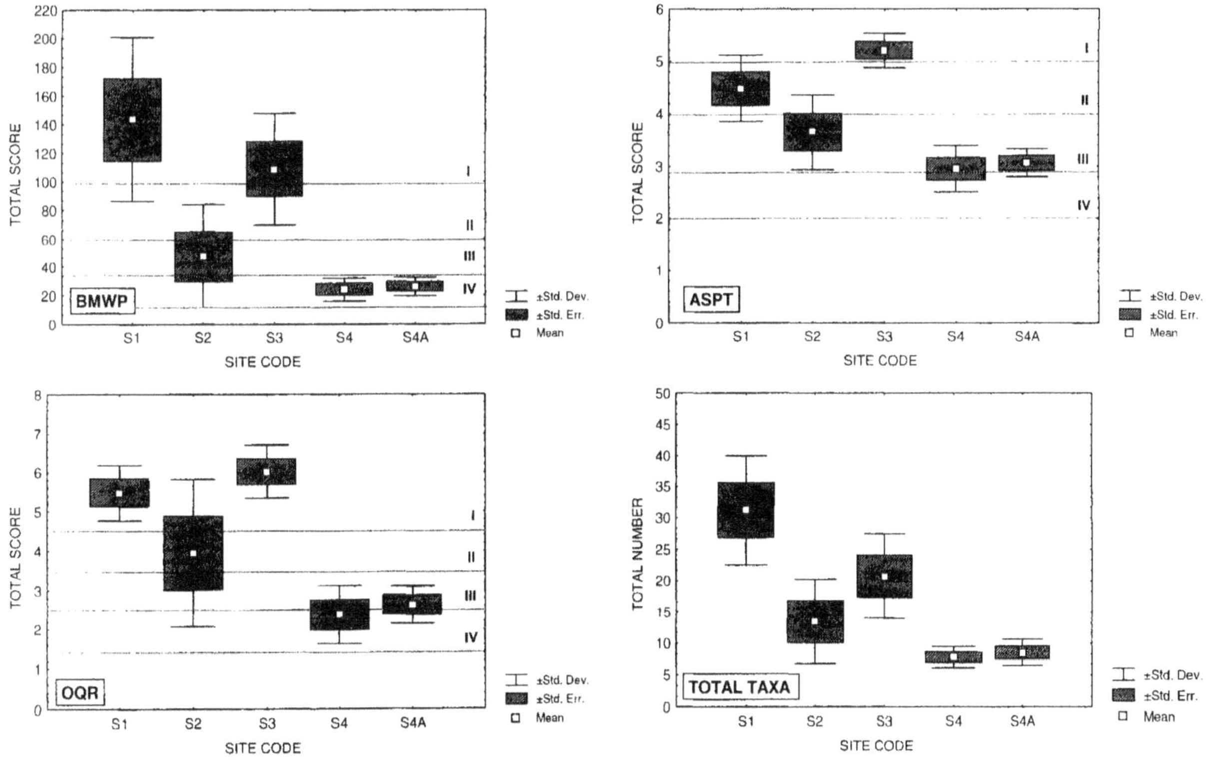

Figure 9. Assessment of water quality based on faunal parameters: number of taxa and biotic indices (BMWP, ASPT, OQR).

Table 4. Catchment properties in terms of size and dispersion of patches

Total area (ha)	AREA	*Average patch size (km^2)*	
Slope (%)	SLOPE	Size of wetlands area	PATCHSS
Density gradient ($km\ km^{-2}$)	STRNET	Size forested area	PATCHSM
Distance from source (km)	RIVL	Size arable/urban area	PATCHSL
River width (km)	RIVW	*% Cover in watershed*	
Length of ecotones (m)	ECOTL	% Cover of wetlands	%COVST
Width of ecotones (m)	ECOWI	% Cover of forested area	%COVM
Altitude (m a.s.l.)	ALT	% Cover of arable/urban area	%COVL
P–PO_4^{3-} export ($mg\ l^{-1}$)		*N–NO_3^- export ($mg\ l^{-1}$)*	
P–$PO^{3-}_{4\ surface.}$ export	P–PO4s	N–$NO^-_{3\ sur.}$ export	N–NO3s
P–$PO^{3-}_{4\ ground.}$ export	P–PO4g	N–$NO^-_{3\ sur.}$ export	N–NO3g

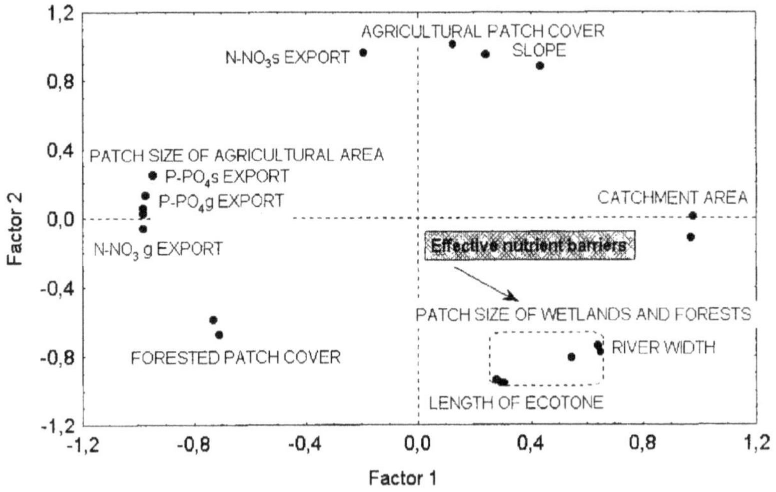

Figure 10. Ordination diagram (F1 × F2) of the catchment metrics in a two-dimensional space.

The first two eigenvalues of PCA analysis explained 95.79% (56.72; 39.06%) of the total variability within the landscape metrics. It confirmed that the ecological studies of freshwater ecosystems within a landscape framework could have important management implications.

Discussion

The hierarchical landscape analysis, driven by GIS techniques, provided useful descriptors of regional/local interactions between terrestrial and aquatic systems (Allan & Johnson, 1997; Johnson & Gage, 1997). These explanatory factors, developed as a number of metrics describing landscape properties in terms of the shape, diversity and position of patches, were very valuable for predicting ecological risk, ecosystem heterogeneity, biocommunities and potential responses to management practices (Baker & Cai, 1992; Hunsaker & Levine, 1995). The results of the present study, based on the application of GIS technologies to quantify landscape pattern and on multivariate statistical techniques to infer patterns of variation within data set (PCA), demonstrated that patch size (referred to land use categories), patch cover in watershed, the length of riparian ecotone and the river dimensions were the most important explanatory variables, significantly influencing stream conditions and nutrient dynamics in the catchment.

The land use pattern in the Grabia River basin was highly dominated by agriculture (70%). However, one of the major trends in land use changes was a gradual decline of forested and wetland areas. Therefore, in our study, GIS models also contributed to the analysis of long-term scenarios (Goodchild et al., 1996). As might be expected (Brunke & Gonser, 1997;

Richards et al., 1997), the potential nitrogen transfer to groundwaters was much higher in the lower part of the Grabia River basin, which is mostly covered by rowcrop agricultural areas. Additionally, the differences in nutrient delivery were significantly connected with elevation, soil infiltration coefficients, depth of groundwater table and management practices (Bis et al., unpublished).

Johnston et al.'s (1997) study of landscape influence on water chemistry indicated that catchment and reach-scale metrics were equally effective predictors only for certain water chemistry parameters (e.g. nitrate, orthophosphate, alkalinity). Other variables (total phosphorous, total suspended solids) were better explained by reach metrics, indicating the dominance of local controlling mechanisms; still others (ammonium, total dissolved solids) directly exhibited catchment-scale control. In our study, GIS data sets and multivariate statistics were used to identify the interactions between catchment/reach properties and nutrient export (nitrate, orthophosphate) to surface and groundwaters. The assessment focused on identifying the most effective nutrient barriers across spatial scales. The analysis of landscape and reach units (structure, configuration, spatial position within catchment) demonstrated significant, negative correlations between the rate of nutrient transfer to river/groundwaters and 1. the length of riparian ecotones, 2. river width, and 3. the size and relative cover of wetland and forested areas in the catchment. These results demonstrated that ecological studies of freshwater ecosystems within a landscape framework have important management implications. Statzner & Sperling (1993) emphasized that the riparian vegetation has an extremely high – in relation to its total area – influence on the effects of organic carbon and nutrient inputs into lowland streams, and can regulate many ecosystem processes (in-stream metabolism, self-purification potential).

Although the functions that riparian ecotones play in moderating water chemistry and in-stream biota composition are well described (e.g. Naiman & Decamps, 1990; Gregory et al., 1991; Zalewski et al., 1996, 1998), the relative influence of the riparian zone *versus* the catchment on stream conditions is poorly known. To date, only few studies (e.g. Carter et al., 1996; Richards et al., 1996, 1997) have used the large spatial scale approach to quantify the relationship between taxonomic and/or non-taxonomic descriptors of macroinvertebrate assemblages and landscape properties. The PCA analysis in the present study demonstrated that in the temperate, lowland river, 1. riparian canopy cover, 2. stream hydraulic processes, and 3. nutrient inflow were the most important variables affecting the structure and trophic organization of stream communities. As we expected, detrital inputs were a significant food source for shredding macroinvertebrates in the headwater stream with well-developed riparian cover (S1). The efficient use of energy at this site depended on retained organic matter particles (debris dams, meandering, instream/bank vegetation). In this undisturbed river section, the abundance of woody debris and its accumulation played an important role in providing channel roughness and improving retention processes. As a result, many heterogeneous (micro-) habitats formed excellent refugia for fish and macroinvertebrates. Additionally, we found a significant correlation not only between availability of benthic organic matter and shredders (*Gammaridae, Nemouridae, Limnephilidae*), but also between suspended organic matter and filter feeders (*Sphaeriidae, Hydropsychidae, Simuliidae*). This agreed with Roeding & Smock's (1989) study of low-gradient, sandy-bottomed streams with relatively poor organic matter retention capabilities. They confirmed that the abundance of debris dams significantly determined not only higher retention of the river, providing energetic stability in river dynamics, but also affected the temporal and spatial distribution of shredding macroinvertebrates. The fact that the filtering collectors *(Spheriidae, Hydropsychidae)* were abundant at site S1 was explained by high habitat complexity (enhanced OM retention) and an availability of suspended organic matter (discharge; adjacent ecotone zones). These results were consistent with Townsend et al.'s (1997) study, which described a high abundance of filter feeders in native forest streams. Additionally, the recent Johnson & Covich's (1997) study, undertaking a spatial analysis of the influence of riparian vegetation upon organic matter abundance, emphasized that control of coarse organic matter inputs needs to be placed in a hierarchical context, where distance downstream is the large-scale control, riparian forest cover the meso-scale control and retention the fine-scale mechanism regulating detritus distribution.

As expected (Armitage et al., 1987), deforestration and channelization had significant consequences, resulting in a totally different and less varied biocommunity (site S2). An increase of incident radiation enhanced algal productivity and significantly affected the occurrence of scrapers (*Gastropoda, Elmidae)* in the studied river section. Additionally, conversion of forested areas into agricultural land, and channel regulation, influenced stream water chemistry, characteristics of the channel and water temperature. Statzner & Sperling (1993) emphasized that sewer overflows into regulated streams have much greater synergetic effects on the biota (oxygen depletion, high suspended solids, increased NH_3, high flow and sewage) than in free-flowing lowland streams.

The analysis of environmental variables which control the structure and distribution of benthic fauna in a large-dimension river reach (site S3 – IV order) showed a significant role of stream hydraulic factors (discharge, depth). The annual hydrologic regime highly affected the presence of taxa clearly bound to water flow (Statzner & Higler, 1986) and also distinctly determined the biocommunity trophic structure. In our study, we conclude that the amounts of chlorophyll *a* and fine particulate organic matter were significantly correlated with alternative gathering collectors/scrapers (e.g. *Ephemerellidae, Caenidae, Baetidae*). Moreover, in this river reach, a relatively high species richness was recorded. According to Statzner & Higler (1986), the potentials of community stability and resilience in zones of major hydraulic and faunal change (transition zones; hydraulic inconstancy) are significantly different from zones upstream and downstream.

The fine particulate organic matter exported along the river gradient originates from in-stream microbial processing, physical abrasion and particularly from adjacent areas (Gregory et al., 1991). As a result, FPOM is an important source of energy and nutrients in the Grabia River and strongly influenced the functional organization of the macroinvertebrate community (Piechocki, 1969; Niesiolowski, 1980; Grzybkowska, 1993). A significantly higher density and biomass of gathering collectors (*Oligochaeta, Chironomidae*) was particularly observed in the river reach S4/S4a. Additionally, the effect of anthropogenic disturbances on benthic fauna composition – related to plant cover degradation, decrease of groundwater level and surface inflow – was significantly correlated with stream nutrient inputs. The detrimental effect of diffuse pollution was the most discriminating factor, drastically decreased the benthic fauna richness, biomass and functional diversity (collectors – 95%). These results show that (except landscape and reach metrics), protection of river channel environmental quality should focus primarily on the maintenance of habitat diversity. Armitage & Gunn (1996) have pointed out that increased knowledge of mesohabitat dynamics (spatial and temporal variation in occurrence, stability and distribution) in a range of river systems would greatly enhance both environmental assessment and provide data on the efficacy of habitat restorative measures (e.g. benthic fauna response to flow modification and other physical disturbances).

According to our results, biological assessment of water quality, based on the faunal parameters (number of taxa and biotic indices), did not reveal the multiple interactions with landscape and reach properties. In general, strict taxonomic-based community metrics do not provide insight into causal mechanisms of stream impairment (Resh & Jackson, 1992; Richards et al., 1997). Our results indicated that the large-scale properties (nutrient inputs, riparian cover, hydraulics) were strong predictors of macroinvertebrate functional organization. The biological and ecological species traits, based on the functional diversity of biocommunities, have been proposed by Statzner et al. (1994, 1997). This provided a new biomonitoring approach in defining mechanisms controlling the stream community assemblages. The predictive power of this proposal reflects the coupling of local environmental conditions and the set of adaptations among taxa. In consequence, this offers a unique dimension in assessing the natural and anthropogenic disturbances in freshwater ecosystems across large spatial scales.

Acknowledgements

This paper is a part of a research programme funded by the Polish National Research Committee (Grant PB 6PO4F 05312). GIS databases were developed thanks to the German Federal Environmental Agency (Umweltbundesamt) and the German Association for Water Resources and Land Improvement (DVWK). Special thanks to Prof. W. Mioduszewski (IMUZ) for his guidance in GIS data storage. We also thank Agnieszka Bednarek, Michal Niedzwiecki, Katarzyna Rubciak and Katarzyna Skretowska for crucial assistance in this project. The anonymous reviewers provided helpful comments and English corrections on an earlier version of this manuscript. All this help is gratefully acknowledged.

References

Allan, J. D. & L. B. Johnson, 1997. Catchment-scale analysis of aquatic ecosystems. Freshwat. Biol. 37, 1: 107–111.

Allan, J. D., D. L. Erickson & J. Fay, 1997. The influence of catchment use on stream integrity across multiple spatial scale. Freshwat. Biol. 37: 149–162.

Armitage, P. D. & R. J. M. Gunn, 1996. Differential response of benthos to natural and anthropogenic disturbances in 3 lowland streams. Int. Rev. ges. Hydrobiol. 81, 2: 161–181.

Armitage, P. D., D. Moss, J. F. Wright & M. T. Furse, 1983. The performance of a new biological water quality score system based on macroinvertebrate over a wide range of unpolluted running-water sites. Wat. Res. 17: 333–347.

Armitage, P. D., R. J. M. Gunn, M. T. Furse, J. F. Wright & D. Moss, 1987. The use of prediction to assess macroinvertebrate response to river regulation. Hydrobiologia 144: 25–32.

Baker, W. L. & Y. Cai, 1992. The r. le programmes for multiscale analysis of landscape structure using the GRASS geographical information system. Lands. Ecol. 7: 291–302.

Boon, P. J., P. Calow & G. E. Petts (eds), 1991. River Conservation and Management. John Wiley & Sons Ltd., Chichester, England: 470 pp.

Brunke, M. & T. Gonser, 1997. The ecological significance of exchange processes between rivers and groundwater. Freshwat. Biol. 37: 1–33.

Burchard, J. & Z. Maksymiuk, 1985. Przyrodnicze warunki kształtujace rezim hydrobiologiczny zlewni Grabi i Lubrzanki. In Zalewski, M. (ed.), Ochrona i Zagospodarowanie Malych Rzek. UL, PZW, Lódz-Warszawa: 9–35.

Carter, J. L., S. V. Fend & S. S. Kennelly, 1996. The relationships among three habitat scales and stream benthic invertebrate community structure. Freshwat. Biol. 35: 109–124.

Cummins, K. W., 1973. Trophic relations of aquatic insects. Ann. Rev. Ent.18: 183–206.

Doeglas, D. J., 1968. Grain-size indices, classification and environment. Sedimentology 10: 83–100.

Dolédec, S. & B. Statzner, 1994. Theoretical habitat templets, species traits and species richness: 548 plant and animal species in the Upper Rhône River and its floodplain. Freshwat. Biol. 31: 523–538.

Dolédec, S., D. Chessel, C. J. F. ter Braak & S. Champely, 1996. Matching species traits to environmental variables: a new three-table ordination method. Envir. Ecol. Statistics, 3: 143–166.

Extence, C. A., A. J. Bates, W. J. Forbes & P. J. Barham, 1987. Biologically based water quality management. Envir. Pollut. 45: 221–236.

Frissell, C. A., W. J. Liss, C. E. Warren & M. D. Hurley, 1986. A hierarchical framework for stream habitat classification: viewing streams in a watershed context. Envir. Manage. 10: 199–214.

Goodchild, M. F., L. T. Steyaert, B. O. Parks, C. Johnston, S. Maidment, M. Crane & S. Glendinning (eds), 1996. GIS and Environmental Modelling: Progress and Research Issues. GIS World, Inc., Fort Collins, CO: 486 pp.

Gregory, S. V., F. J. Swanson, W. A. McKee & K. W. Cummins, 1991. An ecosystem perspective of riparian zones. BioScience 41: 540–551.

Grzybkowska, M., 1993. *Chironomidae* w bentosie i dryfie odcinków rzek o róznej rzedowosci w Polsce Srodkowej. Acta Univ. Lodz: Lódz: 119 pp.

Hawkins, C. J., J. L. Kerschner, P. A. Bisson, M. D. Bryant, L. M. Decker, S. V. Gregory, D. A. McCullough, C. K. Overton, G. H. Reeves, R. J. Steedman & M. K. Young, 1993. A hierarchical approach to classifying stream habitat features. Fisheries 18: 3–11.

Hillbricht-Ilkowska, A. & W. Lawacz, 1984. Land impact – loading and dynamics of nutrients, biotic structure and processes in the lake system of the Jorka river watershed (Mazury, Lake Land, Poland). Ekol. Pol. 31: 539–585.

Hunsaker, C. T. & D. A. Levine, 1995. Hierarchical approaches to the study of water quality in rivers. BioScience 45: 193–203.

Hynes, H. B. N., 1975. The stream and its valley. Verh. int. Ver. Limnol. 19: 1–15.

Johnson, L. B. & S. H. Gage, 1997. Landscape approaches to the analyses of aquatic ecosystems. Freshwat. Biol. 37, 1: 113–132.

Johnson, L. B., C. Richards, G. H. Host & J. W. Arthur, 1997. Landscape influences on water chemistry in Midwestern stream ecosystems. Freshwat. Biol. 37: 193–208.

Johnson, S. L. & A. P. Covich, 1997. Scales of observation of riparian forests and distribution of suspended detritus in a prairie river. Freshwat. Biol. 37: 163–176.

Karr, J. R., 1991. Biological integrity: a long-neglected aspect of water resource management. Ecol. Appl. 1: 66–84.

Karr, J. R., P. R. Yant, K. D. Fausch & I. J. Schlosser, 1984. Evaluation of an index of biotic integrity: temporal variability and regional application in the Midwest. Environmental Research Brief, U.S. EPA. U.S. Goverment Printing Office: 7 pp.

Maksymiuk, Z., 1970. Hydrografia dorzecza Grabi. Acta Geogr. Lodz. 25: 10–33.

Merritt, R. W & K. W. Cummins, 1984. Aquatic Insects of North America. 2. Kendall, Hunt, Dubuque, IA: 441 pp.

Meyer, E., 1989. The relationship between body length parameters and dry mass in running water invertebrates. Arch. Hydrobiol. 117: 191–203.

Naiman, R. J. & H. Decamps (eds), 1990. The Ecology and Management of Aquatic-terrestrial Ecotones. Man and Biosphere Series, 4, UNESCO Paris and The Parthenon Publishing Group, 316 pp.

Niesiolowski, S., 1980. Black flies (*Simuliidae, Diptera*) of the Widawka and Grabia Rivers, Poland. Bull. Entomol. Pologne, 80: 413–462.

Piechocki, A., 1969. Mieczaki (*Mollusca*) rzeki Grabi i jej terenu zalewowego. Frag. Faun. 10: 111–197.

Platts, W. S., W. F. Megahan, & G. W. Minshall, 1983. Methods for evaluating stream, riparian and biotic conditions. U.S. Forest Service, General Technical Report No. INT-138. Ogden, UT: 70 pp.

Pringle, C. M., R. J. Naiman, G. Bretschko, J. R. Karr, M. W. Oswood, J. R. Webster, R. L. Welcome & M. J. Winterbourn, 1988. Patch dynamics in lotic ecosystems: the stream as a mosaics. J. n. am. Benthol. Soc. 7: 503–524.

Resh, V. H. & J. K. Jackson, 1992. Rapid assessment approaches to biomonitoring using benthic macroinvertebrates. In Rosenberg, D. M. & V. H. Resh (eds), Freshwater Biomonitoring and Benthic Macroinvertebrates. Chapman & Hall, New York and London: 195–233.

Richards, C., L. B. Johnston & G. E. Host, 1996. Assessing the influence of landscape-scale catchment features on physical habitat and stream biota. Can. J. Fish. aquat. Sci. 53: 295–311.

Richards, C., R. J. Haro, L. B. Johnston & G. E. Host, 1997. Catchment and reach-scale properties as indicators of macroinvertebrate species traits. Freshwat. Biol. 37: 219–230.

Roeding, C. E. & L. A. Smock, 1989. Ecology of macroinvertebrate shredders in a low-gradient sandy-bottomed stream. J. n. am. Benthol. Soc. 8, 2: 149–161.

Smock, L. A., 1980. Relationship between body size and biomass of aquatic insects. Freshwat. Biol. 10: 375–383.

Southwood, T. R. E., 1977. Habitat, the template for ecological strategies. J. anim. Ecol. 46: 337–365.

Statzner, B. & B. Higler, 1986. Stream hydraulics as a major determinant of benthic invertebrate zonation patterns. Freshwat. Biol. 16: 127–139.

Statzner, B. & F. Sperling, 1993. Potential contribution of system-specific knowledge (SSK) to stream-management decision: ecological and economic aspects. Freshwat. Biol. 29: 313–342.

Statzner B., V. H. Resh. & A. L. Roux, 1994. The synthesis of long-term ecological research in the context of concurrently developed ecological theory: design of a research strategy for the Upper Rhône River and its floodplain. Freshwat. Biol. 31: 253–263.

Statzner B., K. Hoppenhaus, M. F. Arens & P. Richoux, 1997. Reproductive traits, habitat use and templet theory: a synthesis of world-wide data on aquatic insects. Freshwat. Biol. 8: 109–135.

Tolkamp, H. H., 1980. Organism – substrate relationships in lowland streams. Ph.D. Thesis, Centre for Agricultural Publishing and Documentation, Wageningen: 211 pp.

Townsend, C. R., Ch. J. Arbuckle, T. A. Crowl & M. R. Scarsbrook, 1997. The relationship between land use and physicochemistry, food resources and macroinvertebrate communities in tributaries of the Taieri River, New Zealand: a hierarchically scaled approach. Freshwat. Biol. 37: 177–192.

Vannote, R. L., G. W. Minshall, K. W. Cummins, J. R. Sedell & C. W. Cushing, 1980. The river continuum concept. Can. J. Fish. aquat. Sci. 37: 130–137.

Verdonschot, P. F. M., 1990. Ecological characterization of surface waters in the province of Overijssel (The Netherlands). Ph.D. Thesis, Research Institute for Nature Management: 254 pp.

Zalewski, M., 1999. Minimising the risk and amplifying the opportunities for restoration of shallow reservoirs. Hydrobiologia 395/396: 107–114.

Zalewski, M., G. A. Janauer & G. Jolankai, 1997. Ecohydrology. A new paradigm for the sustainable use of aquatic resources. IHP - V Projects 2.3/2.4, UNESCO Paris: 60 pp.

Zalewski, M., B. Bis, M. Nowak, P. Frankiewicz & W. Puchalski, 1996. Land/water ecotones as a key component of the integrated strategy of conservation and management of freshwater ecosystem. In Smith, J. P. D., L. Holt & A. Wheeler (eds), Aspects of the Environment in Europe. Univ. Wolverhampton Publisher, U.K., Chapter 5: 66–87.

Zalewski, M., B. Bis, M. Lapinska, P. Frankiewicz & W. Puchalski, 1998. The importance of the riparian ecotone and river hydraulics for sustainable basin-scale restoration scenarios. Aquatic Conserv: Mar. Freshw. Ecosyst. 8: 287–307.

Zalewski, M., P. Frankiewicz, J. Przybylski, J. Banbura & M. Nowak, 1990. Structure and dynamics of fish communities in temperate rivers in relation to the abiotic-biotic regulatory continuum concept. Pol. Arch. Hydrobiol. 37: 151–168.

Appendix 1. List of aquatic macroinvertebrates considered in this analysis (functional feeding groups: SH – shredder, CG – gathering collector, CF – filtering collector, SC – scraper, P – predator)

TAXON	FFG	TAXON	FFG
NEMATODA		**Aphelocheiridae**	
Gordius aquaticus Dujardin	P	*Aphelocheirus aestivalis* (Fabricius)	P
OLIGOCHAETA	CG	**Veliidae**	
HIRUDINEA		*Microvelia* sp.	P
Glossiphonidae		*Micronecta* sp.	P
Glossiphonia complanata (L.)	P	**MEGALOPTERA**	
Glossiphonia sp.	P	**Sialidae**	
Erpobdellidae		*Sialis fuliginosa* Pictet	P
Erpobdella octoculata (L.)	P	*Sialis* sp.	P
Erpobdella monostriata (L.)	P	**COLEOPTERA**	
Hirudinidae		**Haliplidae**	
Haemopis sanguisuga (L.)	P	*Haliplus ruficollis* (De Geer)	P
MOLLUSCA		*Haliplus* sp.	P
GASTROPODA		**Dytiscidae**	
Lymnaeidae		*Ilybius fuliginosus* (Fabricius)	P
Lymnaea stagnalis (L.)	SC	*Platambus maculatus* (L.)	P
Galba glabra (Müller)	SC	*Agabus* sp.	P
Radix peregra (Müller)	SC	**Helodidae**	
Bithyniidae		*Helodes* sp.	SC
Bithynia leachi (Sheppard)	SC	**Elmidae**	
Bithynia tentaculata (L.)	SC	*Limnius volckmari* (Panzer)	SC
Valvatidae		*Oulimnius* sp.	SC
Valvata naticina Menke	SC	**TRICHOPTERA**	
Valvata pulchella Struder	SC	**Rhyacophilidae**	
Planorbidae		*Rhyacophila* sp.	P
Gyraulus albus (Müller)	SC	**Hydropsychidae**	
Gyraulus rossmaessleri (Auerswald)	SC	*Hydropsyche angustipennis* (Curtis)	CF
Gyraulus riparius (Westerlund)	SC	*Hydropsyche* sp.	CF

Appendix 1. contd.

Anisus spirorbis (L.)	SC	**Polycentropodidae**	
Armiger crista (L.)	SC	*Polycentropus flavomaculatus* (Curtis)	P
BIVALVIA		**Limnephilini**	
Sphaeriidae		**Limnephilidae**	
Pisidium spp.	CF	*Limnephilus flavicornis* (Fabricius)	SH
Sphaerium spp.	CF	*Limnephilus rhombicus* (L.)	SH
ISOPODA		*Limnephilus* sp.	SH
Asellidae		*Potamophylax nigricornis* (Pictet)	SH
Asellus aquaticus (L.)	SH	*Potamophylax* sp.	SH
AMPHIPODA		*Apatania* sp.	SC
Gammaridae		*Anabolia nervosa* (Curtis)	SH
Gammarus fossarum L.	SH	**Stenophylacini**	
EPHEMEROPTERA		*Halesus* sp.	SH
Baetidae		**Goeridae**	
Baetis spp.	CG/SC	*Goera pilosa* (Fabricius)	SC
Ephemerellidae		**Leptoceridae**	
Ephemerella ignita (Poda)	CG/SC	*Athripsodes* sp.	SC
Ephemerella notata Eaton	CG/SC	**Beraeidae**	
Caenidae		*Beraeodes minutus*	SH
Brachycercus harrisella Curtis	CG/SC	**DIPTERA**	
Caenis macrura Stephens	CG/SC	**Tipulidae**	SH
Caenis horaria (L.)	CG/SC	**Limoniidae**	
Caenis sp.	CG/SC	*Limonia* sp.	P
Leptophlebiidae		*Pilaria* sp.	CG
Leptophlebia vespertina (L.)	CG	*Dicranota* sp.	P
Leptophlebia sp.	CG	*Hexatoma* sp.	P
Paraleptophlebia cincta (Retzius)	CG	*Eleophila* sp.	P
Paraleptophlebia werneri Ulmer	CG	**Ptychopteridae**	CG
Ephemeridae		**Ceratopogonidae**	P
Ephemera danica Müller	CG	**Chironomidae**	CG
Ephemera vulgata L.	CG	*Tanypodinae*	P
Ephemera sp.	CG	**Tabanidae**	P
ODONATA		**Stratiomyidae**	CG
Calopterygidae		**Sciomyzidae**	P
Calopteryx splendens (Harris)	P	**Simuliidae**	CF
PLECOPTERA		**Athericidae**	
Nemouridae		*Atherix ibis* (Fabricius)	CG
Nemoura spp.	SH	**Empididae**	
HETEROPTERA		*Wiedemannia* sp.	P
Nepidae		**ARACHNOIDEA**	
Nepa cinerea L.	P	**Hydracarina**	P

Hydrobiologia **422/423:** 389–412, 2000.
M. Jungwirth, S. Muhar & S. Schmutz (eds), Assessing the Ecological Integrity of Running Waters.

Integrated ecological assessment methods as a basis for sustainable catchment management

Piet F.M. Verdonschot
Institute for Forestry and Nature Management, P.O. Box 23, 6700 AA Wageningen, The Netherlands

Abstract

During the 20th century, environmental problems have increased from sewage discharge in the first decennia towards climate change today. An increase in scale of threats implies an increase in scale of management and assessment. Successively, physico-chemical, biological and ecological assessment evolved but failed to stop deterioration. The development in techniques runs parallel to the evolution in assessment. Eight major groups of assessment techniques are distinguished: indices (saprobic, diversity, biotic), multimetrics and rapid techniques, physico-ecological, catchment scale, ecosystem components, assemblage/community, process, and non-taxonomic assessment. An increasing refinement is observed in objectives, measures and complexity, and a decreasing one in levels of scale, stream type and, more often, taxonomy. The applied algorithms were typically restricted to single summary measures. Multivariate analysis was introduced only recently. Assessment can do without a reference condition. Still, this condition is often used in assessment. Therefore, the role of the reference is discussed and it is concluded that, if used, it needs to be defined strictly. The role of nine concepts of lotic ecology in assessment is studied. Two major groups of concepts are distinguished. One is related to catchment-scale functioning of streams and the other to instream habitat-related processes and biodiversity. The 5-S-Model, a frame that divides the stream ecosystem into five major components: system conditions, stream hydrology, structures, substances and species (Verdonschot et al., 1998: The 5-S-Model, an integrated approach for stream rehabilitation. In H.O. Hansen & B.L. Madsen, River Restoration '96, Session lectures proceedings. National Environmental Research Institute, Denmark, International Conference arranged by the European Centre for River Restoration: 36–44), is a first attempt to comprise this knowledge in management. Finally, integrated ecological assessment is defined. It is based on three major approaches: an ecological typology approach, an ecological catchment approach and a societal approach. *Ecological typology* implies a nested multiple parameter approach based on regional ecological stream typology where types are scaled and different taxonomic groups are incorporated. Ecological typology sets the ecological demands necessary to make management sustainable. The *ecological catchment approach* implies a nested multiple scale approach which couples natural and anthropogenic features and dynamics of the catchment both in space and time. This approach sets the conditions within the catchment to facilitate sustainable catchment management. The *societal approach* adds human activities to the first two. It couples ecological demands and anthropogenic uses and supports those user choices that make management sustainable. The application of these three approaches is initiated to enable *sustainable catchment management*.

Introduction: what is the problem?

Hydrological changes, physical disturbance (habitat alteration, urban land use), and point and non-point sources of pollution (chemical contamination, surface runoff, agriculture) are examples of processes responsible for a broad-scale deterioration of lotic ecosystems (see Boon, 1992; Kristensen & Hansen, 1994; Petts, 1990). Deterioration is not only increasing at a local or regional scale but also at the global scale (Pringle, 1998). These examples indicate why control and assessment have become a priority for local, regional, national and international authorities, not least because of the economic losses (Fore et al., 1996; Statzner & Sperling, 1993). Furthermore, an increase in scale of threats implies an increase in scale of assessment.

For a long time, biological, biochemical and/or physico-chemical monitoring were the most appropriate means of detecting effects of human activities on aquatic communities. Traditionally this was only done when problems occurred (e.g., smell, visible pollution, fish kills; Ellis, 1937; Forbes & Richardson, 1913; Hynes, 1960; Kolkwitz & Marsson, 1908). Mostly, (bio-)chemical or uni-dimensional biological assessment of water quality was used (Cairns & Pratt, 1993; Hellawell, 1978; Matthews et al., 1982; Newmann, 1988; Sladecek, 1973). The methods were descriptive and all dealt with the component water and its dissolvents. Traditional physico-chemical, biochemical and biological assessment methods have increasingly become inadequate due to their restricted approach to one or a few aspects of the aquatic ecosystems (Knoben et al., 1995; Metcalfe, 1989; Miller et al., 1988). A number of human disturbances are not fundamentally chemical (Davis & Simon, 1995) or physical.

The traditional biological assessment methods appear to be low-sensitive, robust to natural variation, unpredictable, especially in the middle ranges of their measuring scale and confounded by the co-occurrence of other environmental disturbances. Furthermore, biotic indices (e.g., Armitage et al., 1983; Chandler, 1970; Woodiwiss, 1980) allow the condition of sites near the ends of the measuring scale to be simply judged, but are inadequate for moderately degraded sites (Tolkamp, 1985).

Therefore, assessment methods are becoming refined and broadened. They increasingly include other instream characters, the riparian zone or even the whole catchment. Sustainable stream management implies supporting and maintaining a balanced, integrated, adaptive community of organisms having a species composition, diversity and functional organisation comparable to that of the original or natural stream in the region (Cairns, 1975; Frey, 1975; Karr & Dudley, 1981). In a sustainable stream, the inherent potential is realised, its condition is stable, its capacity for self repair when perturbed is preserved, and minimal external support for management is needed (Karr et al., 1986). This sustainable management objective corresponds to the ecological integrity or health of streams and implies integrated ecological assessment (Jensen & Bourgeron, 1994; Jensen et al., 1996; Slocombe, 1993; Tockner & Schiemer, 1997).

In ecological assessment the overall environment is added to the biological aspect (Odum, 1971). The lotic system is then measured by using a variety of parameters which vary in sensitivity to different types of disturbance (e.g., habitat degradation, Nelson, 1990; landscape change, Roth et al., 1996; hydrological fluctuations, Richter et al., 1996). The goal of integrated ecological assessment is to assess the effect of human activities on ecological resources at integrated levels of scale in space and time; the results are then translated into management measures to be taken and evaluated. The primary objective should be to restore and maintain the ecological (hydrological, physical, chemical and biological) sustainability of streams. In this article I will evaluate certain key developments in and applications of theoretic and applied lotic ecology which enhance the use of integrated ecological assessment.

Lotic ecological assessment: what do we have?

Assessment methods

Indices assessment

The first and most traditional biological assessment system was the saprobic system, which focused on species presence in relation to organic pollution (Kolkwitz & Marsson, 1902, 1908/9; Liebmann, 1962). It was quantified by Pantle & Buck (1955) and Zelinka & Marvan (1961) and extended and reviewed by a number of European authors (see Knoben et al., 1995. Figure 2.4). Three techniques dominated in the first 85 years in Europe (Metcalfe, 1989; Newman, 1988; Sladecek, 1973), namely:

- Saprobic indices; use the difference in pollution tolerance of aquatic organisms. The tolerance is described in parameters of indicator values (1–5), weights (tolerance ranges) and species abundances (Sladecek, 1973).
- Diversity indices; use the decrease in species diversity under increasing disturbance/stress. Most widely used is the Shannon–Weaver formula (Shannon & Weaver, 1949), which is based on the number of species and their individual abundances. Amongst others, Washington (1984), Hellawell (1986) and Boyle et al. (1990) review and evaluate diversity indices.
- Biotic indices and scores; use both a saprobic index and a diversity measure and thus combine taxa richness and (mostly organic) pollution tolerance (BMWP, 1979; De Pauw & Vanhoren, 1983; Tuffery & Verneaux, 1968; Woodiwiss, 1964). They were considerably modified recently by Andersen et al. (1994). An overview is given by De Pauw et al. (1992) and Metcalfe (1989).

The relation of the indices with major five ecological catchment components as defined in the 5-S-Model (see paragraph 3.3; Verdonschot et al., 1998) is restricted to substances (organic components) and the intrinsic value of species, and all relate to instream features (Table 1).

Multimetrics and rapid assessment techniques

In recent years rapid assessment techniques and multimetrics have become popular in the US. The former emphasises a low-cost approach through reduced sampling and efficient data analysis. The multimetric approach is more advanced and complex; it uses a number of single metrics to assess environmental degradation (Karr et al., 1986). The Index of Biological Integrity (IBI) was restricted to fish (Karr, 1981). Later adaptations included the benthic macro-invertebrate assemblage (e.g., Invertebrate Community Index (ICI); Kerans & Karr, 1994; Ohio EPA, 1987; Plafkin et al., 1989), or the macrophytes (Nelson, 1990). Barbour et al. (1992, 1996) presented the conceptual basis for the multimetrics approach, in which community health is composed of community structure, community balance and functional feeding groups; in combination with habitat quality, an integrated assessment is obtained. Until now, the metrics remain based on ecological attributes of biological communities. Different groupings of metrics are possible. The following grouping, adapted after Resh & Jackson (1993) and Thorne & Williams (1997), only intends to indicate the metrics more and less often mentioned in multimetrics and rapid assessment techniques:

- Richness measures (e.g., number of taxa, number of EPT taxa, no. of Chironomidae taxa) and enumerations (e.g., number of individuals, percent of the total EPT taxa (sensitive) and chironomids (tolerant), percent dominant taxon, number of intolerant taxa, percent Oligochaeta, sediment tolerant taxa); in the past these metrics were often used to detect organic pollution, but often other sources of disturbances are measured too. Furthermore, these metrics often consider an increase in dominance of one or more taxa due to pollution or disturbance. Richness measures and enumerations, both the counting of something, do not differ.
- Tolerance/intolerance measures (presence of tolerants and intolerants)); these metrics rely on the assignment of (in-)tolerance values to taxa.
- Diversity measures (e.g., Shannon–Wiener Index, sequential comparison index); often these metrics are considered to decrease with increasing disturbance. Diversity measures are rarely used in multimetric indexes.
- Biotic indices (e.g., Hilsenhoff's family biotic index, BMWP score, ASPT score); these metrics use both the assignment of (in-)tolerance values to taxa and richness and/or diversity measures.
- Similarity/loss measures (e.g., number of taxa in common, community loss index, Bray–Curtis index); these metrics use comparisons between sites (reference versus disturbed sites) and are often mentioned and tested but rarely used in multimetric analyses.
- Functional measures (e.g., percent of functional feeding groups); these metrics use the alteration in feeding types under different types of disturbance. Their applicability is often doubted (e.g., Thorne & Williams, 1997).

Single metrics increase or decrease along an increase in disturbance. This behavior is tested and demonstrated for each finally selected and incorporated metric. Disparate quantitative scores of individual core metrics are converted to calculate a multimetric score on a quantitative scale (e.g., Barbour et al., 1996; Karr, 1981). The metrics sometimes lack sensitivity to contaminants, although these could provide additional information (Fore et al., 1995). Recently, examples of sensitivity to contaminats in terms of metrics which deal with deformities, lesions and tumors have become available.

Peeters et al. (1994) used indirect (thus biotic data) multivariate techniques to distinguish between six stream types in The Netherlands. Based upon all data they defined ten metrics related to current velocity, saproby, trophy, four substrate types and three functional feeding groups for each stream type. For each metric, indicator taxa (mainly family level) were given. For each stream type the metrics were divided into (five) different quality levels. The percentage of indicator taxa present in a new sample was scored at the metrics-scales of the relevant stream type. This application differs from the common use of a multimetric index in the sense that it combines stressor and reponse parameters in one approach. A key principle in the original multimetric approach is the use of only response parameters. Stressors themselves, however, can also indicate changes because their status changes when they become a stressor.

Table 1. The relation of assessment methods with the major components of the 5-S-Model (abbreviations see text)

		Indices	Multimetrics and rapid techniques	Physico-ecological	Catchment scale	Ecosystem components	Assemblage/community	Process and non-taxonomic
System conditions	*Land use*			RCS, RHS	CSA			
	Geology/ morphology			RHS, GSGK	CSA	SERCON	RIVPACS, AUSRIVAS, EKOO	
	Slope			QHEI	CSA			
Stream hydrology	*Hydrology*			HIS/HEP, IHA/RVA	CSA	AMOEBE		
	Flow			QHEI, RBP, RHS, GSGK, IFIM/PHABSIM, RCHARC		SERCON		
	Current		EBEOSWA				EKOO	
Structures	*Riparian zone*		Functional measures	HSI/HEP, RCS, QHEI, RBP, RHS, GSGK	CSA	SERCON		
	Channel			RCS, QHEI, RBP, RHS, GSGK	CSA	SERCON, AMOEBE	RIVPACS, AUSRIVAS, EKOO	
	Depth			HSI/HEP, IFIM/PHABSIM, RCHARC	CSA	AMOEBE		
	Substrates		EBEOSWA	HSI/HEP, QHEI, RBP, RHS, GSGK, IFIM/PHABSIM, RCHARC	CSA	SERCON, AMOEBE	EKOO	RIVPACS, AUSRIVAS,
	Cover			IFIM/PHABSIM, RCHARC	CSA			
Substances	*Pollutants*		Enumerations					
	Organic	Saprobic and biotic indices	(Richness measures), EBEOSWA	HSI/HEP				
	Nutrients		Functional measures, EBEOSWA					Process features
	Chemistry				CSA	SERCON, AMOEBE	RIVPACS, AUSRIVAS, EKOO	
Species	*Algae*	Biotic indices						
	Macrophytes			RCS, QHEI				
	Bank vegetation			QHEI, RBP, RHS, GSGK				
	Macrofauna	Biotic indices	Enumerations, richness, similarity and functional measures, EBEOSWA				RIVPACS, AUSRIVAS, EKOO	Functional feeding groups, species traits
	Fish		Enumerations, richness and similarity measures	HSI/HEP, IFIM/PHABSIM, RCHARC				
	All	Saprobic and diversity indices			CSA	SERCON, AMOEBE (target species)		Process features (functioning)
Spatial scale	*Instream*	Saprobic, biotic, diversity indices	enumerations, richness, similarity and functional measures, EBEOSWA	HSI/HEP, IFIM/PHABSIM, RCHARC, IHA/RVA			RIVPACS, AUSRIVAS, EKOO	Process features, functional feeding groups, species traits
	River corridor			RCS, QHEI, RBP, RHS, GSGK				
	Floodplain					AMOEBE		
	Catchment				CSA	SERCON		

The relation of major groups of metrics with major five ecological catchment components are restricted to substances (organic components and other pollutants) and the intrinsic value of the community (Table 1), and relate to instream features.

Physico-ecological assessment

The physical habitat of streams can be assessed with a number of methods, for example the 'River Corridor Survey' (RCS; NRA, 1992), the 'Habitat Evaluation Procedure' and 'Habitat Suitability Index' (HEP/HSI; US Fish and Wildlife Service, 1981), the 'Qualitative Habitat Evaluation Index' (QHEI; Ohio EPA, 1989), the Rapid Bioassessment Protocol (RBP; Plafkin et al., 1989), the River Habitat Survey (RHS; Raven et al., 1997) and the German stream structure assessment ('Gewässerstrukturgütekarte' (GSGK); Friedrich et al., 1993). They include metrics like substrate type, instream cover, channel morphology, riparian zone and bank erosion, pool/glide and riffle-run quality and gradient. In some approaches these categories are further separated into secondary categories. Habitat quality assessment evaluates physical stream characteristics which are controlled by fluvial processes and human interferences.

An important development in physical/hydrological assessment is the 'instream flow-habitat models'. Their primary application is the design of environmentally acceptable flow regimes. Most instream flow models oversimplify the complexity of stream ecosystems by describing flow regimes based on the average daily discharge or the mean annual flow (Gordon et al., 1992). More sophisticated are the 'Instream Flow Incremental Methodology' (IFIM; Bovee, 1982) and the 'Riverine Community Habitat Assessment & Restoration Concept' (RCHARC; Nestler et al., 1989). Herein, habitat preferences of fish are developed and used to predict habitat availability at different flow levels.

Richter et al. (1996, 1997) introduced the 'Range of Variability Approach' (RVA). RVA characterises ecologically relevant attributes of a flow regime and translates these into more simple, flow-based management targets.

The relation of physico-ecological assessment techniques with the five major ecological catchment components is mainly focused on stream hydrology, structures and bank vegetation (Table 1). Most techniques relate to morphological and physical within-channel conditions and features of the riparian zone.

Catchment-scale assessment

Catchment assessment mainly deals with the impact of land use on the stream system. Relations between agricultural versus forest land use were found by Jacobson et al. (1992), Hall et al. (1994), Martin (1996) and Johnson et al. (1997). Roth et al. (1996) reported that measures of land use were superior predictors of stream ecological integrity, in terms of IBI and HSI, than were most local measures.

Land use activities are often evaluated using geographic information systems and image processing (Geographic Information Systems (GIS); Angermeier & Bailey, 1992; IP; Allan & Johnson, 1997), whereby land use categories are scored (e.g., dominance > 50%, Hall et al., 1994). Hall et al. (1994) discussed chemical, biological, physical habitat and land use evaluations, but stopped short of integrating approaches over different scales. GIS and image processing technology allow quantitative assessment of lateral, longitudinal and vertical components of the landscapes that interact at a variety of spatial and temporal scales to influence streams (Johnson & Gage, 1997).

The use of GIS is meant to summarise landscape properties that contribute directly to the structure and functioning of aquatic systems, and to derive secondary data synthesising knowledge about the landscape over multiple scales. These hierarchically nested classification systems are useful for assessment. Catchment assessment uses extensive sampling designs, and data are analysed by using gradient analysis, regression and multivariate techniques that explain or predict species composition (or other response variables such as physical and chemical attributes) from gradients or patterns in environmental conditions.

The relation of catchment-scale assessment with the five major ecological catchment components is focused on all components (Table 1). Most techniques relate to features of the catchment although the scale is limited to the interaction between catchment and stream.

Ecosystem components assessment

The System for Evaluating Rivers for Conservation (SERCON) is a broad-based and consistent technique for stream evaluation and provides a simple way of communicating technical information to decision makers (Boon et al., 1997). SERCON uses attributes of physical descriptors, catchment characteristics, physical and chemical characteristics of channel and floodplain, biotic characteristics, and aquatic impacts to evaluate a catchment section. Hereby, six conservation criteria (physical diversity, naturalness, representativeness, rarity, species richness and special features) are used along with impacts. The attributes are classified and scores are assigned to each of these classes. The actual score is multiplied by a weight to obtain the criterion score. For Dutch water management, a general method of ecosystem description and assessment was developed (AMOEBE; Brink et al., 1991). The approach compares a reference condition (for the Rhine River defined as the historical condition of 1900–1930) with the actual state. Parameters include species of different organism groups (key species) and abiotic conditions representative of ecosystem compartments. The strengths are that it presents different ecosystem compartments and is a visually strong presentation tool.

The relation of ecosystem components assessment with major ecological catchment components is also focused on all components (Table 1), though mutual interactions are often less clear. Most techniques relate to different, isolated features of the floodplain or catchment.

Assemblage/community assessment

From the early 1980s, with the upcoming multivariate analysis techniques, ecologists started to explore relationships between taxa lists and accompanying environmental parameters.

Wright et al. (1984) used multivariate analysis techniques to classify unpolluted running water sites and to predict community types from environmental data. The results were used in the River Invertebrate Prediction and Classification System (RIVPACS). RIVPACS predicts the expected macro-invertebrate fauna at a given site from a small number of environmental parameters recorded. Comparing the observed fauna (at species or family level) with the expected or 'target' predicted fauna yields a measure of site quality (Wright et al., 1989). The key principle is comparable to the multimetrics although in RIVPACS species assemblages are used while in multimetrics individual species or taxa are used.

The Australian River Assessment Scheme (AUSRIVAS) is based on the RIVPACS model. The difference is that the major habitats are sampled and modelled separately. Furthermore, different models are used for different bio-regions in Australia.

Verdonschot (1990) conducted an extensive data collection and multivariate analysis (clustering and ordination) of macrofauna in undisturbed and disturbed surface waters in The Netherlands. He described macrofaunal site groups which are recognised on the basis of explanatory environmental variables and the abundances of organisms. Therefore no clear boundaries were described between the groups, only a recognisable centroid. Each group showed a limited internal variation. Each of these site groups (species assemblages) and their explanatory environments were defined as cenotypes. The cenotypes are mutually related in terms of key environmental factors which represent major ecological processes. The cenotypes and their mutual relationships form a web, in which the gradual transitions (e.g., quality or developmental gradients) between cenotypes are shown. The web offers an ecological basis for the daily practice of water and nature management (Verdonschot, 1991). The web allows the development of water quality objectives, provides a tool for monitoring and assessment, indicates targets, and guides the management and restoration of water bodies. The web is included in a software package named EKOO (Verdonschot & Nijboer, 1997). The main modules in this package are (1) the assignment of a newly sampled site to one of the cenotypes, (2) the characterization of a new sample in terms of diversity and biotic characteristics (comparable to metrics but not meant to value, merely to inform), and (3) the option to choose a target for a newly sampled site and to establish a set of measures to reach this target.

The relation of assemblage/community approaches with major ecological catchment components is focused on almost all components, and mutual interactions are clear (Table 1). All assemblage/community approaches relate to instream features.

Process and non-taxonomic assessment

The use of processes such as nutrient spiralling, P/R ratio (Odum, 1975), mineralisation/degradation rate (Aleksandrova et al., 1986), community respiration (Grimm & Fisher, 1984), primary production and others is very limited. Until now, they have not been used for water management purposes.

Non-taxonomic assessment is defined as assessment based on non-taxonomic characteristics. In fact the taxonomic entities are grouped into non-taxonomic categories. Two examples are the functional group and species trait assessment. Functional group assessment is based upon functional groups such as the macro-invertebrate functional feeding groups, but it can also be growth forms in macrophytes. Cummins & Wilzbach (1985) developed a key to the macro-invertebrate functional feeding groups and mutual score between pairs of groups. Groups can be scored for habitat-organic resource categories. Ratios can be calculated and related to general ranges in three groups of stream orders. Recently, the species traits approach was introduced by Southwood (1977, 1988) and applied in the Upper Rhone (Statzner et al., 1994). The value of functional feeding groups and related parameters is still quite controversial.

The relation of functional and non-taxonomic appr oaches with major ecological catchment components is focused on species (Table 1) and all approaches relate to instream features.

Assessment characteristics

General features

One of the most important features of all assessment methods is the objective for which they were developed. The different assessment methods reflect different objectives which became important over time (Table 2). They also reflect the deterioration issues that were important when these assessment methods were developed. Catchment scale, ecosystem components and assemblage/community assessment methods all show characteristics necessary for integrated ecological assessment.

Parallel to this, a development towards a smaller scale and specific stream types becomes apparent (Table 2). There is a general tendency to regionalise the approach. Ecosystem properties tend to vary across regions and thus influence the potential species assemblages. A second argument arrises from a major problem in using indicator organisms to determine reference communities. As a result of the biogeographical distribution of species and typological differences between streams, an optimal assessment can only be achieved through regional adaptations (Seager et al., 1992; Tolkamp, 1984; Verdonschot, 1990) for each stream type. Stream types can only be defined at a regional or smaller scale. For example, in using the saprobic index for different stream types in Germany, Braukmann (1997) showed the change in range of saprobic scores related to different stream types. This explains the large number of modifications in all assessment techniques. Ecoregions are the basic geographic unit for establishing reference conditions in North America. They comprise large, relatively homogeneous, climatological and geomorphological landscape ecological areas (Omernik, 1987). Knoben et al. (1995) note that these reflect political rather than ecological regions but also plea for the development of European ecoregions. From the above it can be argued that ecoregions are still too large units for integrated ecological assessment.

Most assessment methods use uni-dimensional measures except for the assemblage/community assessment. The latter is also the most complex method to develop for selected stream types in a certain region. Though almost all assessment techniques regard the biota as an indicator of environmental quality, they differ in the extent to which they respond to this environment (Table 3). Only the physico-ecological assessment does not use true biological indicators (except bank vegetation and instream macrophytes as a structure parameter). A number of assessment methods detect degradation, regardless of the factor(s) responsible for that degradation. In others the response ranges from substances (organic pollution) to the whole environment. The latter refers to those techniques listed in Table 1 which respond to all five major ecosystem components. Integrated ecological assessment asks for these broadly responding techniques as well as for the specific identification of each of the stressors/environmental parameters.

Use of algorithms

A number of statistical techniques that can be used for assessment are given by Hellawell (1986), Norris & Georgis (1993) and Johnson & Gage (1997). The applied techniques are mostly single summary measures like saprobic, biotic and diversity indices alongside gradient analysis, classification and clustering, path analysis, structural equation modelling, analysis of variance, statistics, and fuzzy logic. Norris & Georgis (1993) divide the analysis options into two groups: univariate and multivariate approaches. The insight that can be obtained from the univariate statistics is limited because different but compensating trends appear non-significant in composite variables, derived measures do not always change monotonically along gradients, and the natural variability of individual taxa is hard to quantify.

Multivariate approaches are able to detect subtle differences across taxa in space and time (Jongman et al., 1987). They allow patterns of variability in groups of taxa and/or environmental variables to be detected. It can be argued that these provide more information than does variation in single parameters (Norris & Georgis, 1993). The greatest shortcoming of such multivariate approaches is that the underlying ecological mechanisms (cause–effect) still remain unclear. One

Table 2. Some general differentiating features of groups of assessment techniques discussed

Assessment methods	Objective	Spatial scale/ Stream type	Taxonomic level	Scale of measurement	Complexity
Saprobic indices	Organic pollution	High/ None	Species	Uni-dimensional	Simple
Diversity indices	Water quality	High/ None	Species	Uni-dimensional	Simple
Biotic indices	(Organic) pollution	High/ None	Higher taxa	Uni-dimensional	Simple
Multimetrics	System quality	High/ Ecoregion	Higher taxa (species)	Uni-/multi-dimensional	Moderate
Physico-ecological assessment	Riparian quality	High/ None	–	Uni-dimensional	Moderate
Catchment scale assessment	Land use effect	Moderate/ None	Higher taxa	–	Moderate
Ecosystem components assessment	System quality	Moderate/ Main types	Species and higher taxa	Uni-/multi-dimensional	Moderate
Assemblage/community assessment	System state	Low/ Stream types	Species	Multi-dimensional	High

strength of univariate approaches over multivariates is that the univariates can give information to diagnose causes or relations often based on empirical patterns. Table 4 shows the major groups of statistics and their features.

Most assessment methods are limited to single summary measures or sometimes other univariate statistics (Table 5). Assemblage/community assessment uses all techniques within the group of multivariate statistics. This stresses the objectives of the assemblage/community approach: to detect subtle changes in taxa composition along complex gradients. Integrated ecological assessment needs this subtle approach.

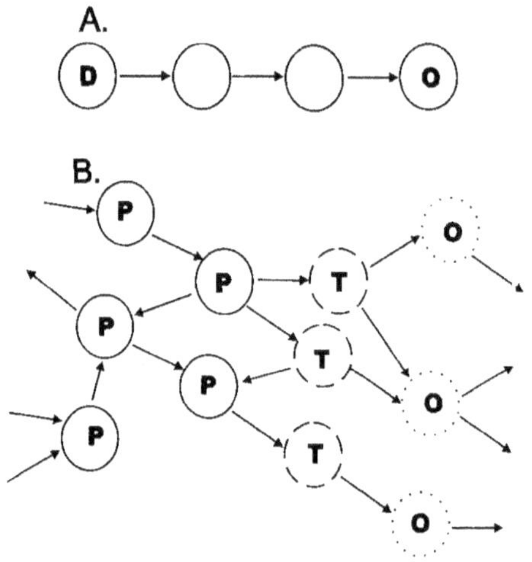

Figure 1. Reference conditions and scale of measurement approaches. (**A**). An assessment approach with fixed endpoints, O, for optimal or reference, and D for 'dead water'. It expresses a singular series of ecosystem developmental stages. (**B**). A reference framework with several developmental stages situated in different directions (a network). Stages are more or less defined as indicated by closed circles with a P (present), open ones with a T (target), and dotted ones with an O (optimal/reference).

Scale of measurement

Most traditional assessment systems – like metrics, saprobic and biotic indices – use a uni-dimensional series of conditions expressed in a singular measure (Figure 1A). The scale runs from the reference or optimum condition (O; Figure 1A) to the severely disturbed one (D; Figure 1A). Two major problems arise in using this uni-dimensional scale. Firstly, the reference condition is difficult to define, which does not mean that one should not use such a state; also, the state does not necessarily have to be absolutely pristine. Secondly, uni-dimensionality does not correspond to ecological reality and therefore does not address the multitude of interacting factors that management is confronted with in practice.

Therefore, it is necessary to describe different states in stream ecosystem development along gradients of decreasing disturbance. This refers not only all different stages between reference and severely dis-

Table 3. Response (*) of groups of assessment methods to the major components of the abiotic environment

Assessment method	System conditions	Stream hydrology	Structures	Substances
Saprobic indices				*
Diversity indices				
Biotic indices				*
Multimetric assessment			*	
Physico-ecological assessment		*	*	
Catchment scale assessment		*	*	*
Ecosystem components assessment		*	*	*
Assemblage/community assessment	*	*	*	*

turbed conditions along different gradients, but also different reference conditions, each with its own series of intermediate or less to more disturbed stages. The result is a web of developmental stages (Verdonschot, 1991), each described by its average community as well as by its ranges of abiotic variables. The list of abiotic parameters is even more important than a list of rare and/or characteristic species. These developmental stages are termed *cenotypes*, i.e., community types. The cenotypes are mutually related by key factors. Figure 1B illustrates a web of cenotypes (circles) and their interrelations (arrows),which can serve as a reference framework for management. The term reference is used here in the sense that one can refer to it, e.g., refer to the optimal stage or reference condition as well as to disturbed conditions.

Integrated ecological assessment needs such a web of cenotypes, which also implies a multi-dimensional scale of measurement.

What is a reference condition?

Under undisturbed conditions, a stream ecosystem is optimally developed and thus structures and functioning are as they would be under pristine circumstances (Verdonschot & Schot, 1987). Reference conditions are often defined in such terms. Note that this paragraph deals with the contents of the definition of reference and not with the goals society or management sets for future conditions.

Any anthropogenic disturbance will affect a stream and cause it to change, often set it back in development. It is important that management assess the magnitude and direction of these changes, not only between a previous and a present condition (assessment) but also between a present and a future one (prediction) (Armitage 1994; Bradshaw, 1988). Recovery is not simply the reverse disturbance; other developmental stages may also occur. Thus, reference in terms of a previous (historic) or future condition can differ.

Most methods of assessing stream condition compare a test stream with a (regional) reference stream (Hosper & Brink, 1989; Kerans & Karr, 1994; Wright et al., 1984). The reference stream is assumed to represent the best attainable condition of the test stream. Often, especially in Europe, natural or pristine reference streams are lacking. In such cases, water management uses a desired, pursued or almost 'natural' condition as a reference. Both examples above are subjective conceptions; none clearly defines reference. The result is confusion. Higler & Verdonschot (1992) discussed a number of definitions of a reference condition to show the different options of how reference was defined in the past:

(a) *Naturalness* and *succession* are conceptions that are often confused in defining reference. The 'natural' condition is often defined as self-maintaining, i.e., without human influences. But even then, certain streams would gradually become wetlands or floodplain swamps. Naturalness in the sense of the endpoint of a succession (a climax) cannot be the only criterion to define the

Table 4. Some features of assessment-related statistical analysis techniques (HG, hypothesis generating, HT, hypothesis testing)

	Group	Objective	Generation	Testing	Pattern analysis	Prediction	Major use
Univariate statistics		*Species <-> factor*					
	Single summary measures	'State'	HG	Yes		No	Quantify the state of an entity in a single measure
	Analysis of variance	Comparison	HT	Yes	Single	Yes	Test differences between entities mean and variance
	Multiple regression	Exploratory	HT	Yes	Single, Multiple	Yes	Quantifies explained variance and predict response of entities
Multivariate statistics		*Assemblage <-> complex*					
	Classification	Grouping	HG		Multiple	No	Groups similar entities
	Association	Comparison	HG(T)		Multiple	No	Compares entities
	Ordination	Exploratory	HG		Multiple	(Yes)	Groups similar entities and relates them to others

reference condition. A climax does not really exist for running waters.

(b) The stream current causes a dynamic or metastable balance in the sense of the *dynamic equilibrium* (Huston, 1979). On the other hand the current velocity can change quickly during periods of drought and spates. This disturbance can be more or less predictable but always sets the stream back in development (intermediate disturbance hypothesis; Poff & Ward, 1990; Ward & Stanford, 1983). Therefore, the use of a dynamic equilibrium as a reference is open to discussion.

(c) The reference condition in terms of the 'former' or 'original' condition can be known from *historic literature*. For example, for the Rhine River, the AMOEBE approach uses information available from the period 1900–1930 (Hosper & Brink, 1989). It can also be based on paleolimnological data. Although these historic conditions are often called natural, they were at that time also subject to human disturbances or different environmental circumstances (Dam, 1987).

(d) Each fully developed stream system has *unique* features which hamper an exact description of an overall reference condition. The reference condition should be described in dynamic terms (e.g., ranges of patterns and processes) and thus fit within a certain stream type. The question remains how wide a range is acceptable.

(e) The reference condition refers to the potential optimal condition, taking the present and *future conditions* into consideration. Altered human activities will lead to new environmental circumstances and thus induce new species combinations. Even a reduction in today's human disturbances or restoration of former conditions does not automatically imply the return of former communities. Irreversible processes are often involved. Thus it also makes sense to include such observations in defining a reference.

All arguments above show that there are no objective and unanimous criteria to define the reference condition. The reference is therefore better defined as the ecologically optimal developmentalcondition chosen. This directional process is described as ecosystem development (Verdonschot, 1991).

Taxonomic group and level

The use of certain taxonomic groups (Table 6) is related to the objective and scale of the study, the knowledge on and ecological response of the group, the presence in the respective water type, the expected information and the ease of handling (Table 7). Resh & Unzicker (1975) and Resh & McElravy (1993) argued the importance of using the species level in assessment. This increases the information content and the performance of analysis techniques. Thus, the more subtle the information that is required, the better it

Table 5. Statistical approaches used (*) in different assessment methods

Assessment methods	Univariate			Multivariate		
	Single	ANOVA	Regression	Classification	Association	Ordination
Saprobic indices	*					
Diversity indices	*	(*)				
Biotic indices	*					
Multimetric assessment	*	*			*	
Physico-ecological assessment	*					
Catchment scale assessment			*			
Ecosystem components assessment	*					
Assemblage/community assessment				*	*	*

Table 6. Number of times different organism groups are used in 44 saprobic, diversity and biotic indices (references from De Pauw et al., 1992)

	Macro-invertebrates	Diatoms and periphyton	Macrophytes	Fish	Macrofauna and fish
No. of times used	29	9*	1	2	2

*Plankton was also included once.

Table 7. Some major characteristics of biotic components in assessment

		Scale	Ease of handling	Present knowledge	Information about	
					Structure	Processes
Taxonomic group	it Algae	Low	No	Low	Low	Moderate
	Macrophytes	High	Yes	High	Good	Low/Moderate
	Macrofauna	Moderate	Yes/No	Moderate/High	High	Moderate/High
	Fishes	Moderate/High	Yes	High	Moderate	Low
Identification level	*High (species)*	Low	Yes/No	Moderate	High	High
	Low (family)	High	Yes	High	Low	Low/Moderate
Sample frequency	*Low*	Moderate	Yes	–	Moderate/Low	Moderate
	High	Low	No	–	High	High

is to use the species level. This level is the result of the speciation mechanisms that emphasised ecological specialisation in the past. On the other hand, the use of higher taxonomic levels is easier (keys are available, which is not always true for species), it reduces costs and handling time, and it is also effective when strong environmental differences or gradients occur. Many applications of indices and multimetrics use higher taxonomic levels (Tables 1 and 8). Though multivariate techniques often start at the species level, often information is lost when rare taxa are excluded from the analysis for mathematical reasons. The use of multiple taxonomic groups is restricted to the ecosystem components and assemblage/community assessment.

Combining the taxonomic identification level with the abiotic conditions (Table 3) used in the assessment also shows a gradient from biotic indices towards assemblage/community assessment (Table 9). The diagonal through table 9 runs parallel to the use of single summary measures (left upper corner) towards multivariate analysis techniques (right lower corner).

Lotic ecology: what do we know?

Scale and hierarchy

Large watersheds are comprised of tributaries and their catchments. Tributaries contain multiple stream reaches; each reach potentially includes riffles, pools and other habitat units, and these habitat units each contain multiple microhabitats (Frissell et al., 1986; Sedell et al., 1990). The multitude of processes that form stream systems exist within a hierarchical framework (Allan & Starr, 1982; Frissell et al., 1986). The hierarchy theory provides a framework for describing the components of an ecosystem and their scaled relations (O'Niell et al., 1986; Jensen et al., 1996):

(1) Every component of a stream is a whole and a part at the same time (the whole/part duality).
(2) Patterns, processes and their interactions can be defined at multiple spatial and temporal scales (Levin, 1992) and are interwoven (Hutchinson, 1953).
(3) The relationship between ecological processes (and the patterns they create) changes with spatial scale (Turner, 1990).
(4) Levels of ecosystem organisation at coarser scales bound the range of ecological properties that emerges at finer scales (Allen et al., 1984). At the same time, finer scales to a lesser extent affect coarser ones (e.g., 'top down' and 'bottom up' control of food webs; feedback loops, Costanza et al., 1993).
(5) No single scale of ecological organisation is correct for all purposes. This is important because more often only information at one or a limited number of scales is provided.
(6) The definition of an ecological hierarchy (component patterns and processes) is always dictated by the objectives of a study.

Implementing hierarchy theory in the description of stream ecosystems involves explicitly characterising the scaled relations between the patterns of interest, the ecological factors (processes) that determine these patterns, the spatial and temporal bounds of each, and the order in which they are nested.

The smallest resolvable area (grain), the area influenced by the phenomenon under study (extent), and the boundaries of the respective system influence the scale of observation in assessment. Most assessments focus on stream ecosystem patterns at one fixed spatial scale. Integrated ecological assessment should be conducted at multiple spatial and temporal scales (in order to include system dynamics), dependent on the type of (policy) questions, issues or objectives being addressed. Ecological data to perform assessment always should be collected at one scale lower than that at which the question is asked.

In conclusion, integrated ecological assessment at multiple scales facilitates catchment management decisions and places management strategies into a proper context. In 'top down' water management, regional assessment supports the development of management plans, while local assessment helps to evaluate the res-

Table 8. Number of groups of assessment techniques using different structure, process, taxonomic and non-taxonomic parameters

	Number of groups
Structure parameters	
Taxonomic	
Higher level	4
Species level	5
Assemblages/Community	1
Multiple groups	2
Non-taxonomic	
Functional feeding groups	1
Species traits	1
Process parameters	>4

Table 9. Relationships between taxonomic identification level and ecological level of groups of assessment methods

	ABIOTIC: →	None +	Factor +	Factor-complexes
Taxonomic				
↓		*Biological*	*Aut-ecological*	*Syn-ecological*
↓				
	Higher taxa	Biotic indices		
		Multimetrics		
			Catchment scale	
↓	*Species*	Saprobic indices		
		Diversity indices		
		Ecosystem components		
Syntaxo-	*Assemblages/*		Assemblage/	
nomic	*Communities*		Community	
	Ecosystems			Web approach
				Nested web approach

ults of these plans. 'Bottom up' water management implies the description of fine-scale stream typologies by local managers; this supports the daily practice of management. Aggregation of stream types sets the frame for regional and catchment management. Thus, integrated ecological assessment includes a 'bottom up' and a 'top down' approach.

Concepts

Hynes (1975) widened the view on streams by including the catchment as part of the system. Ward (1989) introduced the concept of the four-dimensional nature of stream ecosystems with a longitudinal, lateral, vertical and temporal component. In a catchment perspective these four dimensions need to be broadened and need to deal with ecological connectivity.

Most longitudinal interactions relate to upstream/downstream effects, either seen as a sequence of interlinked zones (Hawkes, 1975; Illies & Botosaneanu, 1963) or as a longitudinal continuum (Wallace et al., 1977; Vannote et al., 1980). Nonetheless, the effect of downstream activities on upstream stretches – such as the effect of downstream channelisation on discharge and erosion patterns up-stream (Schumm, 1977), or weir downstream on migration – should not be neglected.

The lateral component (Naiman & Décamps, 1990; Petersen et al., 1987) should be broadened and include the 'dry' area of the catchment. It should, for example, include the surface runoff and wind transport of silt and substances towards the stream.

The vertical component includes the groundwater flow (Brunke & Gonser, 1997) and the hyporheic community (Stanford & Ward, 1988) and can be broadened by including the exchange between stream and air, such as evaporation and deposition of substances (e.g., Kristensen & Hansen, 1994) and the terrestrial phase of insects.

The temporal component, such as the organism's life history, the processes of meandering (Boon, 1992), and patch dynamics (Townsend, 1989) should also include historic developments (Kondolf & Larson, 1995).

The concept of 'the four-dimensional nature of lotic systems' identifies in general terms all the interactions and the functioning of the stream as an integral part of the whole catchment; as such it can be used as a frame for integrated water management (Table 10). The next four concepts (Table 10) all deal with the effect of stream hydrology on the functioning of the stream and its direct surrounding, i.e., the longitudinal and lateral interactions as a whole. The Habitat Template and Patch Dynamics Concept deal with the species and their biotic and abiotic interactions at the habitat scale. Both deal with an important aspect of the species themselves (*diversity*) and support nature conservation. All concepts affect different scales or can be applied at different scales.

Table 10. Some characteristics of major ecological catchment and habitat concepts in lotic ecology

Ecological concept	No.	Key theme	Direction	Spatial scale	5-S-Model	Reference
Four-dimensional nature of lotic systems	1	Longitudinal, lateral, vertical and temporal exchange	Conceptual, four directions	Catchment	Scale, Hierarchy	Ward, 1989
River Continuum Concept	2	Longitudinal gradient	Longitudinal, lateral	Stream, valley	Stream hydrology	Vannote et al., 1980
Serial Discontinuity Concept	3	Discontinuity through human interference	Longitudinal	Stream	Stream hydrology	Ward & Stanford, 1983b
Nutrient Spiralling Concept	4	Longitudinal nutrient cycling	Longitudinal, lateral	Stream, valley	Substances	Wallace et al., 1977
Flood Pulse Concept	5	Lateral exchange of substances	Lateral, temporal	Lower reach, valley	Stream hydrology	Junk et al., 1989
Habitat Template Concept	6	r, K, A selection in space and time	Spotwise	Stream section	Species, Structures	Southwood, 1977
Patch Dynamics Concept	7	Competition versus disturbance	Spotwise	Stream section	Species, Structures	Townsend, 1989
Dynamic Equilibrium Model	8	Dynamic equilibrium in systems	Conceptual, temporal	Independent	Species	Huston, 1979
Intermediate Disturbance Hypotheses	9	Non equilibrium maximises diversity	Conceptual, temporal	Independent	Species	Ward & Stanford, 1983a

The concepts dealing with stream functioning as a whole imply a more or less gradual shift in species composition along the stream gradient. For water management, which cannot deal with a gradient in environmental circumstances, it is necessary to identify and arrange ecological types. Assessment, though based on sites, prefers to refer to larger, more or less homogeneous stream lengths or sections. Defining stream types yields entities that have a recognisable centroid but do not require clear boundaries, whereby each entity may show limited biotic and abiotic internal variation. The continuum is partitioned. If discontinuities occur, these mostly reflect 'natural' boundaries between types.

The concepts dealing with habitats and patches in streams can be applied in two ways. Firstly, there is a relationship between patch dynamics and diversity as well as between *K*-selective, *r*-selective and adverse taxa. Diversity depends on the relation between temporal and spatial variability. Secondly, these concepts most probably can be used at higher scales as well, e.g., landscapes. Surprisingly, this has not yet been tested.

In general, little has been done to implement knowledge on concepts of lotic ecology in ecological assessment. It seems as if theoretic and applied aquatic ecology have developed independently. Integrated ecological assessment could greatly profit by including concepts. This issue deserves much more attention.

5-S-Model

Making the proper choices in stream and catchment management requires understanding the functioning and interactions of the controlling factors. All the considerations on concepts, scale and hierarchy provide a conceptual basis for 'catchment ecology', which I define as stream ecology embedded into a landscape ecological frame. In order to implement 'catchment ecology' in water management the *5-S-Model* was formulated. This conceptual model provides guidelines

Table 11. Summary of results of a discussion by Dutch water and nature managers on ecological objectives and tools needed for the future

Framework

Differentiation = basis
- region
- stream/river type

Multiple scales = ecological catchment approach
- uniform system description

Multiple webs = ecological typology
- regional ecological typology
- target (standard) and reference conditions

Multiple activities = societal needs
- human activities in the catchment
- cause-effect relationships

Tools
- techniques to describe, develop, monitor, assess, evaluate and test
- knowledge on cause-effect relationships
- techniques to survey, prioritise, predict, aggregate and split
- techniques for trend-analysis
- knowledge- and expert-systems

Others
- coherency between tools
- support (pr. standardisation, presentation)

for assessment and management (Verdonschot et al., 1998). The five main components are:

(1) *System conditions* comprise the structures and processes related to climate (temperature, rainfall), geology and geomorphology (like slope, soil composition) and set the boundary conditions for stream ecosystem functioning at a high hierarchical level in space (the catchment) as well as in time (± 100 years). Generally, system conditions are not often changed by management.

(2) *Stream hydrology* comprises the hydrological processes of the catchment and the hydraulic processes of the stream and the habitat (Henry & Amoros, 1995). The two main directions of flow are one running from the boundary of the catchment towards the stream (lateral) and one running from source to mouth of the stream (longitudinal). Groundwater flow, precipitation and evaporation also play a role.

(3) *Structures* refer to the morphological features of the longitudinal and transversal stream bottom, banks and bed, as well as to the substrate patterns within. Cut off meanders, terrestrialization, sand deposits and other features of the stream valley are included here.

(4) *Substances* comprise the processes related to dissolved components like nutrients, organic matter, oxygen, major ions and contaminants. From catchment boundary towards the stream the amount of dissolved substances increases. This increase is also visible from source to mouth.

Stream hydrology, structures and substances together compose the group of controlling factors that directly determine how the stream community functions. They are set by the system conditions.

(5) *Species* are the response to the functioning of all above-mentioned groups of controlling factors. Species includes all taxonomic and non-taxonomic entities as well as biotic processes like production, respiration and so on. Species and their communities are the actual goal of ecological stream management and rehabilitation.

The five components mutually interact on different hierarchical levels and with different intensity. Structures, for example, can respond to the action of stream hydrology but can also reduce discharge fluctuations. Or species can be adapted to stream hydrology and at the same time (e.g., trees) can operate on stream hydrology and morphology. Despite a dom-

Table 12. Relation between framework and objectives in ecological water management in the Netherlands

OBJECTIVES	FRAMEWORK: needs *Ecological catchment approach*	*Ecological typology*	*Societal needs*
Monitor/describe		•	•
Assess/test/standardise		•	•
Develop/evaluate		•	•
Cause-analysis/prioritise	•		•
Survey/predict	(•)	•	•
Aggregate/split	•	•	
PR/integrate/present	•	•	•

inant hierarchical effect, a feedback is always present. Knowledge of the hierarchy in factors and processes acting in space and time in streams allows us also to infer the direction and magnitude of potential changes due to human activities (Naiman et al., 1992). Such changes can refer both to disturbance and restoration, or to the time involved/needed (Niemi et al., 1990). Human disturbances can be seen as a sixth 'S': the 'S' of *Steering*. Steering is not separately distinguished in the model but is part of all five other S's. The disturbance and restoration of streams is steered in a negative or positive direction. Integrated ecological assessment includes these aspects.

Towards integrated ecological assessment; what do we need?

A framework for integrated ecological assessment

Stream management has changed from responding to problems in the past to detecting them in an early stage and even to predicting and preventing them in the future. The increase in types and scale of threats led to an increase in management objectives (e.g., Raven et al., 1998) and thus in the need for more assessment techniques and ecological tools. Recently, the ecological objectives in Dutch water management were discussed. Tables 11 and 12 summarise these objectives, the need for information, and the fields for which ecological tools should be adapted or developed in The Netherlands. Three major approaches are deduced from these discussions (Figure 2). Firstly, integrated ecological assessment needs a frame in which scale, hierarchy and concepts of the catchment and diversity are integrated: an *ecological catchment approach*. Secondly, the abiota and biota should be made operational in an *ecological typology approach*; the entities must be applicable in the daily practice of water management. Thirdly, actual demands or consequences of society should be included: *a societal approach*. The overall implication is that assessment is merely one aspect of sustainable catchment management.

Ecological typology approach; a nested multiple parameter approach

Patterns of variability in groups of taxa related to groups of environmental variables provide subtle information on the ecosystem functioning. Ecological typology implies, for practical reasons, the recognition of such patterns in a continuum (Hawkes, 1975). In ecological typology, a type is described in terms of a general average state (comparable to a class) of a group of taxa with a certain range of biotic variation and under a certain range of (groups of) environmental factors (comparable to a zone), whereby types grade into one another (comparable to a continuum; Verdonschot, 1990). Ecological types can be put into a framework in which all developmental stages of stream ecosystems are represented and mutually related by the processes of change between them. This framework is based on the concept of Warren et al. (1979). Types and their interrelations in terms of variables compose a web. Different, more or less defined conditions and different directions of potential development from the actual condition should be available (Figure 1). The web serves as a reference framework describing different states of less and more optimal stream cenotypes: an ecological optimal or reference condition, intermediate or target conditions, and the present conditions. The target stage represents conditions in between the present and the reference condition. From a management point of view, these target stream cenotypes have an important practical use as a standard ('norm') that can be accomplished within a plan period of about 5–20 years (in contrast to the optimal type, which only can be reached in the long term).

Until now these webs were developed at one scale level and mostly for one taxonomic group (Verdonschot & Nijboer, 1997). Both should change in future applications:

(a) One goal is to aggregate types to higher scales or split them to lower ones. Therefore, it is necessary to start to define webs at the lowest scale,

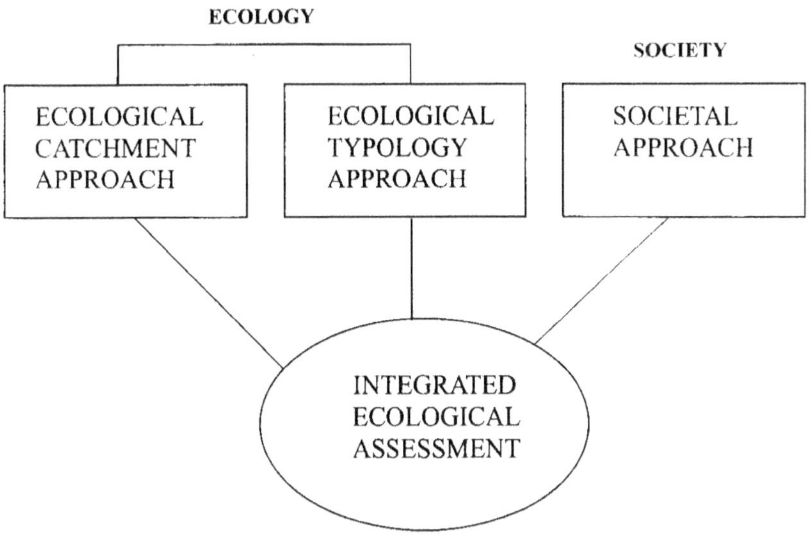

Figure 2. Nested ecological typology. Each higher scale type includes all lower ones.

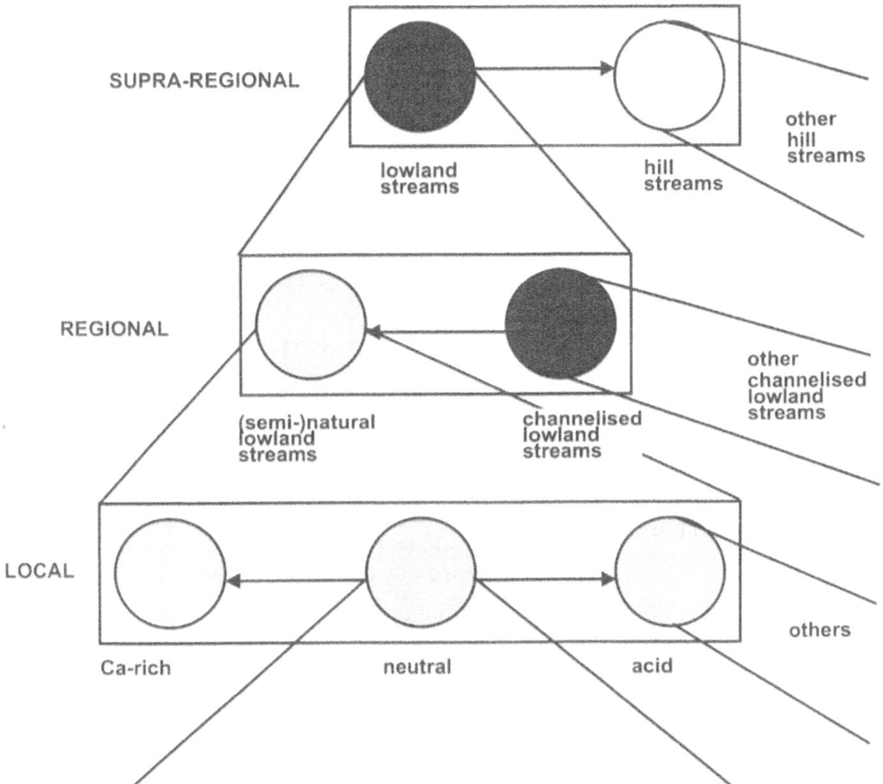

Figure 3. Tripod of tools in integrated ecological assessment.

i.e. the habitat level (Figure 3). These types can be aggregated at higher scales: a 'bottom up' approach. If types are described as assemblages on a habitat level, taking the above concepts (section "Concepts", p. 401) into consideration, tools can be developed for local management as well as, through aggregation, for regional and national policies. The higher scale aggregates the habitat types into stream-reach types, whole stream types and even catchment types. A type at a higher level can be viewed as a window through which the lower scale types are still visible. Types are mutually related as a hierarchically ordered, nested chain. The hierarchy in related types corresponds to syn-taxonomy in vegetational science, where syn-taxonomic units such as classes, orders and associations are nested and used (Braun-Blanquet, 1928). As a consequence dominant taxa differentiate between types at a higher scale, while rare taxa have this function at a lower scale. Finally, webs can only be used within the region for which they were defined.

(b) To date, the typology approach has mainly concerned macro-invertebrates. In the near future, more taxonomic groups should be included in an ecological typology. Types should be composed of at least macrofauna, diatoms, macrophytes, and fishes. This implies combining taxonomic groups acting at different scales into one type. The respective groups should therefore be incorporated in the types and applied as indicators at their relevant scale.

This so-called nested web approach (see bottom line in Table 9) meets the following ecological requirements: it includes (groups of) abiotic and biotic variables which are mutually related, it incorporates the variety of development stages and, if nested, considers a stream as part of its catchment.

The nested web approach is applicable in integrated ecological management because it is:

(1) Able to assess stream ecosystem quality and disturbance at different scales whereby interactions are taken into account

The web can be used as an assessment system and quality classes can be assigned to the cenotypes. The web is constructed out of the full range of disturbed and undisturbed conditions. Depending on the value a water manager attributes to a particular developmental condition, the quality for the type will be established. One can also assign or even refine a quality class by extracting and comparing several metrics from both the types and the new sample. To use types in assessment, both multivariate analysis techniques and multi-metrics are available and useful. This implies that at a high scale, single summary measures at higher taxonomic levels will suffice; at a regional scale both the taxonomic level and the algorithm should be more sophisticated. At the local level only refined and detailled approaches will be successful.

(2) Able to predict effects of management policies and applications at different scales

The web of types can be used to predict the effects of a particular management choice. In predict on, the direction and amount of change between a present and a future condition is established. Two approaches are available. Firstly, one decides to change an actual human activity which currently affects the stream ecosystem. The present condition as well as the predicted change in key factors is known. This change in key factors is projected on the web and the effects in terms of operational factors can be deduced. Secondly, one chooses a target type or target species. The operational key factors that must be changed to reach the target are extracted from the web.

(3) Able to evaluate stream ecosystem development at different scales

A decision-support system or expert system based on the knowledge gathered by using the 5-S-Model and the web can be used to set up a proper, integrative, habitat–stream–catchment management plan. For example, in the project EKOO (Verdonschot, 1990) such an expert system was developed at one scale for a region in The Netherlands. This supported choices of the key factors to be changed and the management options to be chosen. The expert system is computerised and consists of assignment techniques, metrics and a decision key. The decision key is used to obtain a proper set of measures to be taken to reach the chosen target.

(4) Able to support monitoring and evaluation of management undertaken at different scales

The web of types can be used as a monitoring system. Monitoring involves taking samples at different moments in time and comparing them with earlier samples. The difference provides information about the type, the direction and the nature and amount of change because individual environmental parameters are identified. Large changes become visible in the web as changes between types. Small changes are detected in terms of the direction and 'distance' of a sample to the average of the nearest type. Change in 'distance' indicates change at the respective sample within a type. Both the amount and direction of change can be compared to the chosen, desired change. In evaluation procedures, this step is important to determine whether management targets have been accomplished or not. It also informs about the necessity of additional measures or corrections in actual management procedures.

Ecological catchment approach; a nested multiple-scale approach

The joint management of land and water within entire catchments to ensure stream integrity is emerging as a response to the increasing degradation of stream systems (Allan et al., 1997). Much of the rationale for stream catchment management derives from the idea that a catchment is a topographically and hydrologically defined unit: the catchment approach (Hynes, 1975). Catchments are hierarchically organised (Frissell et al., 1986). However, there is limited understanding of the relative importance of local versus regional versus supra-regional factors. Certain processes are likely to be primarily under local control, while others depend on factors acting over larger areas. As an example, studies on buffer strips showed that riparian zones are important to the stream and that their width buffers the influence of human activities in the catchment (Osborne & Kovacic, 1993). Thus, land use becomes less important relative to riparian land use (Allan et al., 1997). On the other hand, Roth et al. (1996) and Allan et al. (1997) concluded that while local and riparian conditions are important, regional landscape conditions may be of even greater importance. The assessment they used related different metrics to a number of different scale parameters. Multiple regression showed the importance of regional land use. Local measurements added independent information.

Catchment approaches often focus on the spatial scale, which should be analysed and presented using Geographic Information Systems (GIS) and Image Processing (IP). A classification of geographical areas with comparable system conditions (ecoregions), supplemented by knowledge on all relevant key factors in the catchments, supports these techniques.

The temporal scale, especially system dynamics, is crucial. Here, statistical techniques support trend-analysis and deal with dynamic features like flow fluctuation, seasonality and erosion-sedimentation dynamics. One example which combines the ecological catchment approach with the ecological typology is presented for spring areas in the Netherlands by Verdonschot (1996).

Societal approach; a nested multiple functional approach

Boon (1992) added a fifth dimension to Ward's 'four-dimensional nature of lotic systems'. This fifth dimension is conceptual and addresses basic questions of philosophy, policy, and practice, i.e. questions on 'why', 'what' and 'how' from the societal point of view. Also, Naiman et al. (1992) argued for a new perspective on catchment management that recognises the need to find a balance between ecological, economic and social values within a long-term framework of sustainability and human use. This meets a number of problems such as co-operation between different governmental authorities and the co-ordination of spatially and temporal autonomic developments in human use. Previously, the tradition in stream assessment was to report on different societal aspects in isolation, such as on chemical water quality related to effluent discharges of sewage or industry, or bank erosion related to drainage intensity or urbanisation. A more multidisciplinary approach is needed based on catchment-level principles. Human uses, economic and societal values are part of integrated catchment management. This is the only manner in which catchments can be managed sustainably.

Integrated catchment management requires decision-support systems to weigh different interests of society including water quality, natural values and biodiversity. Through the use of GIS, in combination with knowledge on cause-effect relationships and weights added to human uses, the whole can be incorporated in integrated ecological management.

Application of integrated ecological assessment

The application of integrated ecological catchment management or the assessment of a new stream site involves the following steps (for further explanation see also Verdonschot & Nijboer, 1997):

(1) Ecological framework

(a) *Construct*: construct a nested web of stream types including target and/or reference conditions by the principles described in sections "What is a reference condition?", p. 397 and "A framework for integrated ecological assessment", p. 404.

(b) *Gather*: gather data on human activities in the catchment and weigh their impact on the stream system.
(c) *Map*: map spatial data on system conditions and human uses of the catchment at multiple scales.

(2) Application for a new site/sample

(a) *Establish*: establish the objectives and the relevant scale or multiple scales. The objectives determine part of the web of types (see 1.a) and the scales (see 1.b) to be used.
(b) *Measure*: measure by standardised methods or collect data on the actual abiotic and biotic key parameters at the relevant scale of the stream system, establish the relevant key factors in the catchment according the 5-S-Model, identify all human activities and quantify the most important ones in terms of operational parameters for the stream system.
(c) *Identify*: compare the actual condition with the web of types at the relevant scale(s), identify the actual type, the conditional and operational key factors, and actual factors (and causes) of disturbance.
(d) *Choose*: choose, at the scale of the objective, the target (and reference) type and thus the desired ecological development within the web of types, which will always be limited by the conditional key factors within the catchment.
(e) *Identify*: identify the operational factors that influence/disturb the site and identify the operational key factors to be managed under the given natural system conditions.
(f) *Plan*: plan the measures to be taken, based on the key factors to be changed and thus processes to be altered (steered), taking the system conditions and present and future human uses (see 1.c) of the catchment into account.
(g) *Execute*: execute the chosen measures. This calls for a detailed plan as well as careful guidance during execution.
(h) *Evaluate*: monitor, assess and evaluate the effects of the measures and the direction of development towards the target.

It must be clear that an exact instruction and detailed description of each of the steps mentioned above is necessary to make this frame applicable in the daily practice of water management. Methods need to be written down in protocols and techniques need to be tested prior to use.

Conclusions

Assessment methods which establish and report on the state of streams are required to meet a range of local, regional and (inter)national needs. An underlying consistency of approach – based on flexibility on different scales, hierarchy and subtle information on structure and functioning (processes) of ecological entities and their mutual interactions – meets monitoring requirements and can assimilate and communicate meaningful information to all interested parties. This is the prime concern of an integrated ecological assessment system (Raven et al., 1998). Integrated ecological assessment is a multivariate approach. In order to implement this scaled integrated ecological assessment, three major questions were dealt with:

(1) What do we have?

A total of eight major groups of assessment techniques were distinguished and analysed. They differ in objectives, operational scales and (group of) stream/river types assessed, algorithms and parameters (taxonomic groups and levels) used, and catchment components to which they respond.

(2) What do we know?

Ecological hypotheses showed that two major aspects must be included in assessment, namely hierarchy in components and multiple scales. Both are included in the 5-S-Model. Furthermore, the leading concepts in lotic ecology support water management (the patterns and processes and ecological connectivity in catchment concepts (River Continuum Concept, Nutrient Spiralling Concept, Flood Pulse Concept, Serial Discontinuity Concept)) and nature conservation in terms of biodiversity (species distribution/interactions and diversity (Dynamic Equilibrium Model, Intermediate Disturbance Hypothesis, Habitat Template Concept, Patch Dynamics Concept)).

(3) What do we need?

First, we need a *differentiation* into ecological *region* and stream and river *type* as an essential starting point. Second, stream ecosystems should be described in a catchment context, i.e., over multiple scales: an *ecological catchment approach*. These system descriptions should be uniform. Third, each stream and river type must be defined in ranges of abiotic and biotic terms at different scales: an *ecological typology*

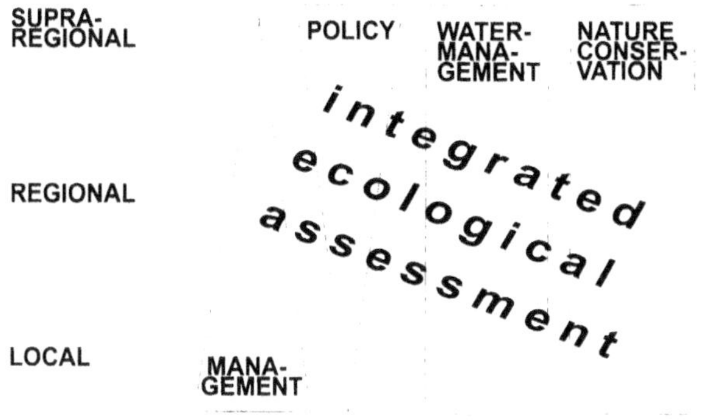

Figure 4. Integrated ecological assessment joins water management and nature conservation, and supports policies and practical management.

approach. Types are placed in a nested web with relations between types designated in terms of steering key factors (processes) and quantified. The web is extended with target (standard) and reference conditions. Fourth, all human activities in the catchment should be recognised and their impact on the stream/river type quantified: a *societal approach*. The latter requires knowledge on cause–effect relationships.

These four items bring water management and nature conservation together and offer a framework over multiple scales (Figure 4). The implementation of these needs necessitates techniques to describe, develop, monitor, assess, evaluate and test; knowledge on cause-effect relationships; techniques to survey, prioritise, predict, aggregate and split; techniques for trend-analysis and knowledge- and expert-systems. The coherency between these tools is vital: all should be supported by a standardisation of techniques and methods, a communicative presentation and a well thought-out public relations.

In final conclusion, integrated ecological assessment is not always everything everywhere. Although termed a catchment approach, it does not require knowing everything about the catchment, but rather knowing everything about the ecologically relevant interactions within the catchment in relation to the stream system functioning. Above all, integrated ecological assessment concerns interactions (ecological connectivity). It can better be defined as integrated ecological methods because it involves much more than how to produce a few colours on a map. It enables sustainable water management, which must not only assess but also monitor, evaluate, predict, restore, etc. The ecological assessment methods chosen are always objective-dependent, but should rely on multiple scales. The methods and techniques are constructed 'bottom up' and take advantage of existing tools. Finally, such a multiple scale and multidisciplinary approach must always be founded on co-operation.

References

Allan, D. J. & L. B. Johnson, 1997. Catchment-scale analysis of aquatic ecosystems. Freshwat. Biol. 37: 107–111.

Allan, T. F. H. & Starr, 1982. Hierarchy: Perspectives for Ecological Complexity. Univ. Chicago Press, Chicago.

Allan, D. J., D. L. Erickson & J. Fay, 1997. The influence of catchment land use on stream integrity across multiple spatial scales. Freshwat. Biol. 37: 149–161.

Allan, T. F. H., T. W. Hoekstra & R. V. O'Niell, 1984. Interlevel relations in ecological research and management: some working principles from hierarchy theory. Gen. Tech. Rep. RM-110, US Dept. of Agriculture, Forest Service, Rocky Mountain Research Station, Fort Collins, CO.

Aleksandrova, N. G., T. G. Moroz, V. S. Polishchuk & E. Y. Rossova, 1986. Combined evaluation of water quality of the lower Dnepr. Wat. Res. 4: 589–596.

Andersen, M. M., F. F. Riget & H. Sparholt, 1994. A modification of the Trent index for use in Denmark. Wat. Res. 18: 145–151.

Angermeier, P. L. & A. Bailey, 1992. Use of a geographic information system in the conservation of rivers in Virginia, USA. In: Boon, P. J., P. Calow & G. E. Petts (eds), River Conservation and Management. John Wiley & Sons, New York.

Armitage, P. D., 1994. Prediction of biological responses. In: Calow, P. & G.E. Petts (eds), The Rivers Handbook, Hydrological and Ecological Principles. Blackwell Sci. Publ. 2: 254–275.

Armitage, P. D., D. Moss, J. F. Wright & M. T. Furse, 1983. The performance of a new biological water quality score system based on macroinvertebrates over a wide range of unpolluted running water sites. Wat. Res. 17: 333–347.

Barbour, M. T., J. L. Plafkin, B. P. Bradley, C. G. Graves & R. W. Wisseman, 1992. Evaluation of EPA's rapid bioassessment benthic metrics: metric redundancy and variability among reference stream sites. Envir. Toxicol. Chem. 11: 437–449.

Barbour, M. T., J. Gerritsen, G. E. Griffith, R. Frydenborg, E. McCarron, J. S. White & M. L. Bastian, 1996. A framework for biological criteria for Florida streams using benthic macroinvertebrates. J. North am. Benthol. Soc. 15: 185–211.

BMWP, 1979. Biological monitoring working party. The 1978 national testing exercise. Technical Memorandum 19. Water Data Unit, Reading, UK.

Boon, P. J., 1992. Essential elements in the case for river conservation. In: Boon, P. J., P. Calow & G.E. Petts (eds), River Conservation and Management. Wiley & Sons, Chichester, UK: 11–34.

Boon, P. J., N. T. H. Holmes, P. S. Maitland, T. A. Rowell & J. Davies, 1997. A system for evaluating rivers for conservation (SERCON): development, structure and function. In: Boon, P. J. & D. L. Howell (eds), Freshwater Quality: Defining the Indefinable? The Stationary Office, Edinburgh, UK: 299–326.

Bovee, K. D., 1982. A guide to stream habitat analysis using the instream flow incremental methodologies. US Wildlife Service Instream Flow Information Paper no. 12. Report no. FWS/OBS-82/26. USFWS Instream Flow Group, Fort Collins, USA.

Boyle, T. P., G. M. Smillie, J. C. Anderson & D. R. Beeson, 1990. A sensitivity analysis of nine diversity and seven similarity indices. J. Wat. Poll. Contr. Fed. 62: 749–762.

Bradshaw, A. D., 1988. Alternative endpoints for reclamation. In: Cairns, J. J. (ed.), Rehabilitating Damaged Ecosystems. CRC Press, Boca Raton, FL. II: 69–85.

Braukmann, U., 1997. Zoocoenological and saprobiological contributions to a general regional typology of brooks. Arch. Hydrobiol. 26: 1–355.

Braun-Blanquet J., 1928. Pflanzensoziologie. Springer, Wien, 3e Aufl.

Brink, B. J. E. ten, S. H. Hosper & F. Colijn, 1991. A quantitative method for description and assessment of ecosystems: the AMOEBA approach. Mar. Poll. Bull. 23: 265–270.

Brunke, M. & T. Gonser, 1997. The ecological significance of exchange processes between rivers and groundwater. Freshwat. Biol. 37: 1–33.

Cairns, J. Jr., 1975. Quantification of biological integrity. In: Kusler, J. A., M. L. Quammen & G. Brooks (eds), Mitigation of Impacts and Losses. Proc. Nat. Wetland Symp. Berne: 276–282.

Cairns, J. Jr. & Pratt J. R., 1993. A history of biological monitoring using benthic macroinvertebrates. In: Rosenberg, D. M. & V. H. Resh (eds), Freshwater Biomonitoring and Benthic Macroinvertebrates. Chapman & Hall, New York: 10–27.

Chandler, R. J., 1970. A biological approach to water quality management. Wat. Poll. Contr. 69: 415–422.

Costanza, R., L. Wainger, C. Folke & K. Maler, 1993. Modeling complex ecological economic systems. BioScience 43: 545–555.

Cummins, K. W. & M. A. Wilzbach, 1985. Field procedures for analysis of functional feeding groups of stream macroinvertebrates. Appalachian Environmental Laboratory, Univ. Maryland, Frostburg, USA, 18 pp.

Dam, H. van, 1987. Acidification of moorland pools: a process in time. Thesis, Agricultural Univ. Wageningen, 175 pp.

Davis, W. S. & T. P. Simon, 1995. Biological Assessment and Criteria: Tools for Water Resource Planning and Decision Making. CRC Press, Boca Raton, FL.

De Pauw, N. & G. Vanhoren, 1983. Method for biological quality assessment of watercourses in Belgium. Hydrobiologia 100: 153–168.

De Pauw, N., P. F. Ghetti & D. P. Manzini, 1992. Biological assessment methods for running waters. In Newman et al. (eds), River Water Quality: Ecological Assessment and Control.

Ellis, M. M., 1937. Detection and measurement of stream pollution. Bull. US Bur. Fish. 48: 365–437.

Forbes, S. A. & R. E. Richardson, 1913. Studies on the biology of the upper Illinois river. Bull. Illinois State Lab. Nat. Hist. 9: 481–574.

Fore, L. S., Karr J. R. & R. W. Wisseman, 1996. Assessing invertebrate responses to human activities: Evaluating alternative approaches. J. N. am. Benthol. Soc. 15: 212–231.

Frey, D., 1975. Biological integrity of water: an historical perspective. In: Ballantine, R. K. & L.G. Guarraia (eds), The Integrity of Water. EPA, Washington, DC: 127–139.

Fore, S. A., S. I. Guttman, A. J. Bailer, D. J. Altfater & B.V. Counts, 1995. Exploratory analysis of population genetic assessment as water quality indicator. Ecotoxicol. Environ. Safe. 30: 36–46.

Friedrich, G., K. J. Hesse & J. Lacombe, 1993. Die ökologische Gewasserstrukturkarte. Wass. Abwass. 11, Kassel.

Frissell, C. A., W. J. Liss., C. E. Warren & M. D. Hurley, 1986. A hierarchical approach to classifying stream habitat features: viewing streams in a watershed context. Envir. Manage. 10: 199–214.

Gordon, N. D., T. A. McMahon & B. L. Finlayson, 1992. Stream Hydrology: An Introduction for Ecologists. John Wiley, Chichester.

Grimm, N. B. & S. G. Fisher, 1984. Exchange between interstitial and surface water: implication for stream metabolism and nutrient cycling. Hydrobiologia 111: 219–228.

Hall, L. W. Jr., S. A. Fisher, W. D. Jr. Killen, M. C. Scott, M. C. Ziegenfuss & R. D. Anderson, 1994. Status assessment in acid-sensitive and non-acid-sensitive Maryland coastal plain streams using an integrated biological, chemical, physical, and land use approach. J. aquat. Ecosyst. Health 3: 145–167.

Hawkes, H. A., 1975. River zonation and classification. In: Whitton, B. A. (ed.), River Ecology. Studies in Ecology. Univ. Calif. Press 2: 312–374.

Hellawell, J. M., 1978. Biological Surveillance of Rivers. A Biological Monitoring Handbook. NERC, Stevenage, 333 pp.

Hellawell, J. M., 1986. Biological Indicators of Freshwater Pollution and Environmental Management. Elsevier, London, 546 pp.

Henry, C. P. & C. Amoros, 1995. Restoration ecology of riverine wetlands: I. A scientific base. Envir. Manage. 19: 891–902.

Higler, L. W. G. & P. F. M. Verdonschot, 1992. Ökologische Bewertung von Fliessgewässern in den Niederlanden. Gustav Fischer Verlag, Stuttgart, Limnologie aktuell 3: 97–110.

Hosper, S. H. & B. J. E. ten Brink, 1989. Naar toetsbare ecologische doelstellingen voor het waterbeheer: de AMOEBE-benadering. H2O 22: 612–617.

Huston, M., 1979. A general hypothesis of species diversity. Am. Nat. 113: 81–101.

Hutchinson, G. E., 1953. The concept of pattern in ecology. Proc. Nat. Acad. Sci. USA 105: 1–12.

Hynes, H. B. N., 1960. Biology of Polluted Waters. Liverpool Univ Press, Liverpool, UK, 202 pp.

Hynes, H. B. N., 1975. The stream and its valley. Verh. int. Verein. Limnol. 19: 1–15.

Illies, J. & L. Botosaneanu, 1963. Problémes et méthodes de la classification et de la zonation écologique des eaux courantes. considereés surtout du point de vue faunistique. Mitt. int. Verein. Limnol. 12: 1–57.

Jacobson, R., P. Kazyak, D. Janicki, D. Wade, H. Wilson & R. P. Morgan, 1992. Feasability of using an index of biotic integrity (IBI) approach for synthesizing data from a Maryland biological stream survey. Rep. prep. by Versar Inc., Columbia, MD.

Jensen, M. E. & P. Bourgeron, 1994. Ecosystem management: principles and applications (Vol. II). Gen. Tech. Rep. PNW-GTR-318, US Department of Agriculture, Forest Service, Pacific Northwest Research Station, Portland.

Jensen, M. E., P. Bourgeron, R. Everett & I. Goodman, 1996. Ecosystem management: a landscape ecology perspective. Wat. Resources Bull. 32: 203–216.

Johnson, L. B. & S. H. Gage, 1997. Landscape approaches to the analysis of aquatic ecosystems. Freshwat. Biol. 37: 113–132.

Johnson, L. B., C. Richards, G. E. Host & J. W. Arthur, 1997. Landscape influences on water chemistry in midwestern stream ecosystems. Freshwat. Biol. 37: 193–208.

Jongman, R. H. G., C. J. F. ter Braak & O. F. R. van Tongeren, 1987. Data analysis in community and landscape ecology. Pudoc, Wageningen, 299 pp.

Junk, J. W., B. P. Bayley & E. R. Sparks, 1989. The flood pulse concept in river-floodplain systems. Canadian Special Publication of Fisheries and Aquatic Sciences.

Karr, J. R., 1981. Assessment of biotic integrity using fish communities. Fisheries 6: 21–27.

Karr, J. R. & D. R. Dudley, 1981. Ecological perspective on water quality goals. Envir. Manage. 5: 55–68.

Karr, J. R., K. D. Fausch, P. L. Angermeier, P. R. Yant & I. J. Schlosser, 1986. Assessing biological integrity in running waters: a method and its rationale. Special Publ. 5. Illinois Natural History survey, Urbana.

Kerans, B. L. & J. R. Karr, 1994. A benthic index of biotic integrity (B-IBI) for rivers of the Tennessee Valley. Ecol. Appl. 4 (4) 768–785.

Knoben, R. A. E., C. Roos & M. C. M. van Oirschot, 1995. Biological assessment methods for watercourses. UN/ECE task force on monitoring & assessment, Lelystad, Volume 3: 1–86.

Kolkwitz, R. & M. Marsson, 1902. Grundsätze für die biologische Beurteilung des Wassers nach seiner Flora und Fauna. Mitt. Aus d. Kgl. Prüfungsanstalt für Wasserversorgung u. Abwässerbeseitigung 1: 33–72.

Kolkwitz, R. & M. Marsson, 1908. Ökologie der planzlichen Saprobien. Ber. dtschen. bot. Ges. 26: 505–519.

Kolkwitz, R. & M. Marsson, 1909. Ökologie der tierischen Saprobien. Int. Rev. Hydrobiol. 2: 126–519.

Kondolf, G. M. & M. Larson, 1995. Historical channel analysis and its application to riparian and aquatic habitat restoration. Aquat. Conserv. Mar. Freshwat. Ecosyst. 5: 109–126.

Kristensen, P. & H. O. Hansen, 1994. European rivers and lakes. Assessment of their environmental state. European Environment Agency, EEA Environm. Monogr. 1, Copenhagen, 122 pp.

Levin, S. A., 1992. The problem of pattern and scale in ecology. Ecology 73: 1942–1968.

Liebmann, H., 1962. Handbuch der Frischwasser und Abwasserbiologie. Band I, R. Oldenburg, Munich, 588 pp.

Martin, D., 1996. Influence du mode d'utilisation du sol sur les caracteristiques chimiques de certaines sources aquifers. Faculte de Sciences Economiques de Potiers. Int. rapp. IBN, Leersum: 1–23.

Matthews, R. A., A. L. Buikema, J. Cairns & J. H. Rodgers, 1982. Biological monitoring: Part IIa: receiving system functional methods, relationships and indices. Wat. Res. 16: 129–139.

Metcalfe, J. L., 1989. Biological water quality assessment of running water based on macro-invertebrate communities: history and present status in Europe. Envir. Pollut. 60: 101–139.

Miller, K. L., P. M. Leonard, R. M. Hughes, J. R. Karr, P. B. Moyle, L. H. Schrader, B. A. Thompson, R. A. Daniels, K. D. Fausch, G. A. Fitzhugh, J. R. Gammon, D. B. Halliwell, P. L. Angermier & D. J. Orth, 1988. Regional applications of an index of biotic integrity for use in water resource management. Fisheries 13 (5): 12–20.

Naiman, R. J. & H. Décamps, 1990. The ecology and management of aquatic-terrestrial ecotones. MAB series, Volume 4, UNESCO, Paris. Parthenon Publishing Group, NJ.

Naiman, R. J., D. G. Lonzarich, T. J. Beechie & S. C. Ralph, 1992. General principles of classification and the assessment of conservation potential in rivers. In: Boon, P. J., P. Calow & G. E. Petts (eds), River Conservation and Management, Wiley & Sons, Chichester, UK: 93–124.

National Rivers Authority, 1992. River Corridor Surveys: Methods and Procedures. Bristol.

Nelson, W. G., 1990. Prospects for development of an index of biotic integrity for evaluating habitat degradation in coastal systems. Chem. Ecol. 4: 197–210.

Newman, P. J., 1988. Classification of surface water quality. Heinneman, Oxford.

Nestler, J. M., R. T. Milhous & J. B. Layzer, 1989. Instream habitat modelling techniques. In: Gore, J. A. & G. E. Petts, Alternatives in regulated river management. CRC Press, Boca Raton, FL: 295–315.

Niemi, G. J., P. DeVore, N. Detenbeck, D. Taylor, A. Lima, J. Pastor, J. D. Yount & R. J. Naiman, 1990. Overview of case studies on recovery of aquatic systems from disturbance. Envir. Manage. 14: 571–587.

Norris, R. H. & A. Georgis, 1993. Analysis and interpretation of benthic macroinvertebrate surveys. In: Rosenberg, D. M. & V. H. Resh (eds), Freshwater Biomonitoring and Benthic Macroinvertebrates. Chapman & Hall, New York: 234–286.

Odum, E. P., 1971. Fundamentals of Ecology. Saunders Company, Philadelphia, 574 pp.

Odum, E. P., 1975. Ecology, 2nd edn. Holt, Rinehart & Winston, London, 244 pp.

Ohio EPA, 1987/1989. Biological criteria for the protection of aquatic life. Vol. I, II, III. Ohio Environmental Protection Agency, Columbus, OH.

Omernik, J. M., 1987. Ecoregions of the conterminous United States. Ann. Assoc. Am. Geol. 77: 118–125.

O'Niell, R. V., D. L. DeAngelis, J. B. Waide & T. F. H. Allen, 1986. A hierarchical concept of the ecosystem. Princeton Univ. Press, Princeton, NJ.

Osborne, L. L. & D. A. Kovacic, 1993. Riparian vegetated buffer strips in water quality restoration and stream management. Freshwat. Biol. 29: 243–258.

Pantle, E. & H. Buck, 1955. Die biologische Überwachung der Gewässer und die Darstellung der Ergebnisse. Gas und Wasserfach, 96 (18), 604 pp.

Peeters, E. T. H. M., J. J. P. Gardeniers & H. T. Tolkamp, 1994. New methods to assess the ecological status of surface waters in the Netherlands. Part 1: Running waters. Verh. int. Verein Limnol.: 1914–1916.

Petersen, R. C., B. L. Madsen, M. A. Wilzbach, C. H. D. Magadza, A. Paarlberg, A. Kullberg & K. W. Cummins, 1987. Stream management: emerging global similarities. Ambio 16: 166–179.

Petts, G. E., 1990. Water, engineering and landscape: development, protection and restoration. In: Cosgrove, D. & G. E. Petts (eds), Water, Engineering and Landscape. Water Control and Landscape Transformation in the Modern Period. Belhaven Press, London: 188–208.

Plafkin, J. L., M. T. Barbour, K. D. Porter, S. K. Gros & R. M. Hughes, 1989. Rapid bioassessment protocols for use in streams and rivers: benthic macroinvertebrates and fish. EPA 444/4-89-001. US Environmental Protection Agency, Washington, DC.

Poff, N. L. & J. V. Ward, 1990. Physical habitat template of lotic ecosystems: recovery in the context of historical pattern of spatiotemporal heterogeneity. Envir. Manage. 14: 629–645.

Pringle, C. M., 1998. Managing riverine connectivity in complex landscapes to protect 'remnant natural areas'. Verh. int. Verein. Limnol. (in press).

Raven, P. J., N. T. H. Holmes, F. H. Dawson, P. J. A. Fox, M. Everard, I. R. Fozzard & K. J. Rouen, 1997. River Habitat Quality: the physical character of rivers and streams in the UK and Isle of Man. Environment Agency, Bristol.

Raven, P. J., P. J. Boon, F. H. Dawson & A. J. D. Ferguson, 1998. Towards an integrated approach to classifying and evaluating rivers in the UK. Aquat. Conserv. 8: 383–393.

Resh, V. H. & J. K. Jackson, 1993. Rapid assessment approaches to biomonitoring using benthic macroinvertebrates. In: Rosenberg, D. M. & V. H. Resh (eds), Freshwater Biomonitoring and Benthic Macroinvertebrates. Chapman and Hall, New York: 195–233.

Resh, V. H. & E. P. McElravy, 1993. Contemporary quantitative approaches to biomonitoring using benthic macroinvertebrates. In: Rosenberg, D. M. & V.H. Resh (eds), Freshwater Biomonitoring and Benthic Macroinvertebrates. Chapman & Hall, New York: 159–194.

Resh, V. H. & J. D. Unzicker, 1975. Water quality monitoring and aquatic organisms: the importance of species identification. J. Wat. Pollut. Contr. Fed. 47: 9–19.

Richter, B. D., J. V. Baumgartner, J. Powell & D. P. Braun, 1996. A method for assessing hydrologic alternation within ecosystems. Conserv. Biol. 10: 1163–1174.

Richter, B. D., J. V. Baumgartner, R. Wigington & Braun D. P., 1997. How much water does a river need? Freshwat. Biol. 37: 231–249.

Roth, N. E., J. D. Allan & D. L. Erickson, 1996. Landscape influences on stream biotic integrity assessed and multiple spatial scales. Landscape Ecol. 11: 141–156.

Schumm, S. A., 1977. The Fluvial System. John Wiley & Sons, New York.

Seager, J., I. Milne, G. Rutt & M. Crane, 1992. Integrated biological methods for river water quality. In: Newman et al. (eds), River Water Quality: Ecological Assessment and Control.

Sedell, J. R., G. H. Reeves, F. R. Hauer, J. A. Stanford & C. P. Hawkins, 1990. Role of refugia in recovery from disturbances: modern fragmented and disconnected river systems. Envir. Manage. 14: 711–724.

Shannon, C. E. & W. Weaver, 1949. The mathematical theory of communication. University of Illinois Press, Urbana.

Sládecek, V., 1973. System of water quality from the biological point of view. Ergebnisse der Limnologie 7: 1–128.

Slocombe, D. S., 1993. Implementing ecosystem-based management: development of theory, practice and research for planning and managing a region. BioScience 4: 612–622.

Southwood, T. R. E., 1977. Habitat, the templet for ecological strategies? J. anim. Ecol. 46: 337–365.

Southwood, T. R. E., 1988. Tactics, strategies and templets. Oikos 52: 3–18.

Stanford, J. A. & J. V. Ward, 1988. The hyporheic habitat of river ecosystems. Nature 335: 64–66.

Statzner, B. & F. Sperling, 1993. Potential contribution of system-specific knowledge (SSK) to stream-management decisions: ecological and economic aspects. Freshwat. Biol. 29: 313–342.

Statzner, B., V. H. Resh & A. L. Roux, 1994. The synthesis of long-term ecological research in the context of concurrently developed ecological theory: design of a research strategy for Upper Rhône River and its floodplain. Freshwat. Biol. 31: 253–263.

Thorne, R. S. T. J. & W. P. Williams, 1997. The response of benthic macroinvertebrates to pollution in developing countries: a multimetric system of bioassessment. Freshwat. Biol. 37: 671–686.

Tockner, K. & F. Schiemer, 1997. Ecological aspects of the restoration strategy for a riverfloodplain system of the Danube River in Austria. Global Ecol. Biogeography Lett. 6: 321–329.

Tolkamp, H. H., 1984. Biological assessment of water quality in running waters using macroinvertebrates: a case study for Limburg, The Netherlands. Wat. Sci. Techn. 17: 867–878.

Tolkamp, H. H., 1985. Using several indices for biological assessment of water quality in running water. Verh. int. Verein. Limnol. 22: 2281–2286.

Townsend, C. R., 1989. The patch dynamics concept of stream community ecology. J. N. am. Benthol. Soc. 8: 51–63.

Tuffery, G. & J. Verneaux, 1968. Methode de determination de la qualite biologique des eaux courantes. CERAFER, Paris, 21 pp.

Turner, M. G., 1990. Landscape changes in nine rural counties in Georgia. Photogramm. Eng. Remote Sensing 56: 379–386.

US Fish and Wildlife Service, 1981. Habitat Evaluation Procedures (HEP). ESM 102 USFWS, Washington, DC, 135 pp.

Vannote, R. L., G. W. Minshall, K. W. Cummins, J. R. Sedell & C. E. Cushing, 1980. The River Continuum Concept. Can. J. Fish. aquat. Sci. 37: 130–137.

Verdonschot, P. F. M., 1990. Ecological characterization of surface waters in the province of Overijssel (The Netherlands). Ph.D. Thesis, Wageningen, 1–255.

Verdonschot, P. F. M., 1991. The web-approach: a tool in water management. In: Ecological Water Management in Practice (Proceedings of the Technical Meeting held in Ede, The Netherlands, 3 October 1990). Proc. and Inf. CHO-TNO, 45: 59–76.

Verdonschot, P. F. M., 1996. Towards ecological spring management. Crunoecia 5: 183–194.

Verdonschot, P. F. M. & J. A. Schot, 1987. Macrofaunal community types in helocrene springs. Ann. Rep. Res. Inst. Nature Management, Leersum 1986: 85–103.

Verdonschot, P. F. M. & R. C. Nijboer, 1997. Typology of macrofaunal assemblages applied to water and nature management: a Dutch approach. The RIVPACS International Workshop, 16–18 September 1997, Oxford, UK.

Verdonschot P. F. M., J. M. C. Driessen, H. G. Mosterdijk & J. A. Schot, 1998. The 5-S-Model, an integrated approach for stream rehabilitation. In H.O. Hansen & B.L. Madsen, River Restoration '96, Session lectures proceedings. National Environmental Research Institute, Denmark, International Conference arranged by the European Centre for River Restoration: 36–44.

Vitousek, P. M. & W. A. Reiners, 1975. Ecosystem succession and nutrient retention: a hypothesis. Bioscience 25: 376–381.

Wallace, J. B., J. R. Webster & W. R. Woodall, 1977. The role of filter feeders in flowing waters. Arch. Hydrobiol. 79: 506–532.

Ward, J. V., 1989. The four dimensional nature of lotic ecosystems. J. North amer. Benth. Soc. 8 (1): 2–8.

Ward J. V. & J. A. Stanford, 1983a. The intermediate-disturbance hypothesis: an explanation for biotic diversity patterns in lotic ecosystems. In: Fontaine, T. D. & S. M. Bartell (eds), Dynamics of Lotic Ecosystems. Ann Arbor Science, Collingwood: 347–356.

Ward, J. V. & J. A. Stanford, 1983b. The serial discontinuity concept of lotic ecosystems. In: Fontaine, T. D. & S. M. Bartell (eds), Dynamics of Lotic Ecosystems. Ann Arbor Science, Collingwood: 29–42.

Warren, C. E., M. Allen & J. W. Haefner, 1979. Conceptual frameworks and the philosophical foundations of general living systems theory. Behav. Sci. 24: 296–310.

Washington, H. G., 1984. Diversity, biotic and similarity indices. A review with special relevance to aquatic ecosystems. Wat. Res. 18: 653–694.

Woodiwiss, F. S., 1964. The biological system of stream classification used by the Trent River Board. Chem. Indust. 11: 443–447.

Woodiwiss, F. S., 1980. Biological monitoring of surface water quality. Summary report, Commission of the European Communities. Severn Trent Water Authority, UK, 45 pp.

Wright, J. F., D. Moss, P. D. Armitage & M. T. Furse, 1984. A prelimnary classification of running-water sites in Great Britain based on macroinvertebrate species and the prediction of community type using environmental data. Freshwat. Biol. 14: 221–256.

Wright, J. F., P. D. Armitage & M. T. Furse, 1989. Prediction of invertebrate communities using stream measurements. Regul. Rivers Res. Manage. 4: 147–155.

Zelinka, M. & P. Marvan, 1961. Zur Präzisierung der biologischen Klassification der Reinheit fliessender Gewässer. Arch. Hydrobiol. 57: 389–407.

Hydrobiologia **422/423:** 413–428, 2000.
M. Jungwirth, S. Muhar & S. Schmutz (eds), Assessing the Ecological Integrity of Running Waters.

The development of integrated methods for assessing river conservation value

P. J. Boon
Scottish Natural Heritage, 2 Anderson Place, Edinburgh EH6 5NP, U.K.

Key words: rivers, conservation, integrated, evaluation, SERCON

Abstract

In recent years the conceptual barriers between river 'management' and river 'conservation' have begun to erode. As the environmental aspirations of water managers and conservationists broaden, there is a need for integrated methods of conservation assessment which can be of value to both groups. This paper considers the requirements inherent in such methods, and explores some of the problems in their design, using for illustration the development of SERCON (System for Evaluating Rivers for Conservation) in the UK. The design of integrated assessment methods, including the selection of attributes for evaluation, must be closely related to the objectives of any conservation activity. For example, selecting a representative series of rivers for statutory protection may require more focused assessment techniques than those needed for setting river conservation management objectives. In the European Union, the Habitats Directive provides a good example of the former approach, in which specific river types and species listed in the Directive must be evaluated for selecting Special Areas of Conservation. In contrast, the forthcoming Water Framework Directive will need to rely on a more comprehensive method of integrated assessment to ensure that 'good ecological quality' can be reliably defined, monitored and maintained. Attempts at developing integrated conservation assessment methods need to address six main issues: (i) differences in perception of conservation 'value'; (ii) defining conservation criteria (such as 'naturalness'), many of which are to an extent subjective; (iii) establishing weighting systems to differentiate important from less important conservation features; (iv) ensuring that methods are rigorous and consistent; (v) the need for extensive datasets to allow comprehensive assessment; and (vi) providing clear guidance on interpreting the outputs. This paper illustrates these problems using experience gained in the UK. It concludes that the inevitable search for pragmatism should not detract from the need for an inclusive approach to river conservation assessment.

The conservation ethic and its application to rivers

In a lecture presented at the Edinburgh International Science Festival in 1996, Martin Holdgate, President of the Zoological Society of London and former Director General of the World Conservation Union, stated that conservation needs a fundamentally new approach in which conserving biodiversity is seen as a mainstream social activity. This is a view that sees man not just as part of nature, but dependent on it (Holdgate, 1997).

In Britain, the conservation of nature has a long pedigree, moving from an essentially descriptive phase in the natural history studies of the 19th century, through to the establishment of a statutory body for nature conservation in 1949 (the Nature Conservancy, later the Nature Conservancy Council (NCC)), and culminating in the present conservation agencies for England, Wales and Scotland. Such bodies have undoubtedly achieved a great deal. On the other hand, it could be argued that making nature conservation a formal government responsibility (like education or the provision of health care) removes an individual sense of responsibility for conservation from the population at large. Nature conservation then becomes merely an activity undertaken by the minority, rather than an attitude influencing the majority (Boon, 1994).

In the 1990s, the British conservation agencies have worked hard to promote a conservation ethic that permeates all sectors of society. In part this has been achieved through broadening the role of the conservation bodies in Scotland and Wales to cover aspects of the countryside such as access and recreation as well as nature conservation. Yet even in England, where the body which replaced the NCC (English Nature) still has only a nature conservation remit, the emphasis has changed. The aim of the NCC was to ensure that "the national heritage of wild flora and fauna and geological and physiographic features remains as large and diverse as possible, so that society may use and appreciate its value to the fullest extent" (Nature Conservancy Council, 1984). English Nature now describes its role in the following way:

> Our philosophy of nature conservation sets improving wildlife and natural features in the context of natural, human-induced and global environmental change. We want soundly managed ecological and physical processes to operate.... We seek the sustainable management of special sites and high standards of environmental practice everywhere. We want to maintain healthy ecosystems, not botanical or zoological parks.... English Nature's leadership of the nature conservation agenda in England must be built on a full understanding of the roles and aspirations of others.... Together with our strategic allies, we aim to make big differences and jointly achieve lasting outcomes for wildlife and natural features (English Nature, 1997).

This represents a widening of the focus in which protected areas for conservation are seen very much in the context of the 'wider countryside' in which people are part and parcel of countryside management. It also signals a shift towards advocating 'sustainable use' of natural resources, and to 'partnership' approaches in seeking gains for nature.

Rivers are probably subjected to a greater diversity of human use and abuse than any other biotope on earth, providing a good illustration of the ways in which social, economic and ecological factors combine to shape the natural world. At the same time the relationship between rivers and people emphasizes the importance of an integrated approach to river conservation and management. In parts of the developing world, where environmental pressures are often acute, the link between socio-economics and conservation is especially important (Wishart et al., in press), yet is often missing. Dudgeon (1992), for example, catalogued a long list of problems on tropical Asian rivers and concluded that academic scientists must learn to communicate with developers and planners if river systems are to be conserved into the 21st century.

Ironically, it is sometimes in countries where river conservation has long been espoused that it has been hard to dislodge the separatist approach. For example, the structure of the Environment Agency in England and Wales is based on 'functions', one of which is 'conservation' which lies alongside other 'functions' such as navigation or flood defence. This position is changing, as part of a more fundamental shift which has taken place over the last 20 years in the water regulatory agencies in the UK. During the 1970s and 1980s, the central place of water chemistry in the assessment of water 'quality' began to be shared by biological assessment methods; with the development of techniques such as RIVPACS (Wright et al., 1989) they became standard tools for routine quality monitoring. In the 1990s, water chemistry and river biology have been further supplemented by assessments of physical habitat features (Raven et al., 1997, 1998) as the emphasis on pollution has expanded to take in other aspects of freshwater 'quality' (see Boon & Howell, 1997). All this reflects a gradual transition in which river conservation is now seen to be good not just for nature, but for economics as well. Thus, the progressive broadening of the role of the water regulators has begun to break down the conceptual barriers between what might be considered the province of river 'managers' and 'conservationists'.

Similar changes have been taking place in other countries. In the USA, the term 'biological integrity' is often used in routine monitoring of water quality. However, definitions of 'integrity', such as 'the capability of supporting and maintaining a balanced, integrated, adaptive community of organisms having a species composition, diversity, and functional organization comparable to that of the region' (see Angermeier & Karr, 1994) clearly incorporate many concepts (such as 'naturalness' and 'representativeness') traditionally considered the stock-in-trade of nature conservationists (e.g. Ratcliffe, 1977). Indeed, Angermeier & Karr (1994) consider that the primary goal of biological conservation is to protect biological integrity, whilst Cairns (1995) states that if integrity were seriously impaired then 'sustainable use' (another conservation concept) would be impossible.

As the perspectives of water managers and conservationists broaden, and their environmental aspirations move closer together, there is a need for integrated methods of conservation assessment which can be of value to both groups. It is important to note that the term 'integration' is used here to refer to a holistic approach in which nature conservation value is assessed against multiple criteria for habitats and biota, rather than relying on one or two traditional measures (such as the presence of rare species). It is not used to refer to the process in which human use of natural resources is set against inherent conservation value in determining policy on land use or development. Whilst this is an essential part of regional planning, there is merit in keeping inherent natural values and human use requirements separate up to the point of decision making.

This paper considers the need for integrated methods of assessing river conservation value and explores some of the problems in their design, using for illustration the development of SERCON (System for Evaluating Rivers for Conservation) (Boon et al., 1997). SERCON has been produced in the UK to help conservation agencies, river managers and others assess the conservation value of rivers in a systematic way. It does this by subjecting data on a wide range of riverine features to a weighted scoring system, composing a picture of river stretches (known as Evaluated Catchment Sections (ECSs)) in terms of generally accepted conservation criteria. Full details of the structure and function of SERCON are given in Boon et al. (1994, 1996, 1997, 1998). The system is under review at present, and it is anticipated that SERCON Version 2 will contain significant revisions.

River evaluation in relation to conservation objectives

Conservation assessment is not an end in itself, but is undertaken in response to a particular need. The objectives of specific river conservation programmes vary, so the methods of evaluation will need to vary to meet those objectives. Four rather different aims or scenarios are frequently encountered and are discussed below; each demands its own approach to conservation evaluation, which is not always the comprehensive and integrated method that forms the principal focus of this paper:

Aim 1: To conserve representative examples of all major river types within a defined geographical area (a country, state, county, province, etc.)

The principal conservation designation in Britain is the Site of Special Scientific Interest (SSSI). SSSIs are selected according to published guidelines (Nature Conservancy Council, 1989). The chapter on rivers describes the intention to select and 'notify' a national series of river SSSIs to represent the main river types throughout Britain (Boon, 1995). However, the reorganization of the NCC into separate bodies for the three countries in Britain has seen a divergence of policy on river SSSIs. Although an extensive programme of river notification has been completed by English Nature, far fewer rivers have been notified by the equivalent bodies in Wales and Scotland, due to a combination of resource constraints and some doubts (at least in Scotland) over the efficacy of the SSSI mechanism in bringing adequate protection to rivers.

The main evaluation requirement to meet this aim is an ability to assess the 'representativeness' of individual rivers against a classification of river types. The classification scheme used for conservation assessment in Britain is based on river plant communities (Holmes, 1983, 1989; Holmes et al., 1998) and divides rivers into 10 broad classes (River Community Types (RCTs)). By examining the occurrence of features such as the presence of dominant or commonly occurring species, it is possible not only to classify a stretch of river into one of the 10 RCTs but to see whether it is a good representative of its type. Rules for establishing the degree of representativeness in terms of macrophyte communities have now been incorporated within SERCON (Boon et al., 1996; see below).

Aim 2: To conserve rare or threatened riverine species, or examples of rare or threatened habitats

Although conservation often works best by focusing on habitats rather than species, conservation effort has frequently been directed at rare, threatened or declining species. In Britain, for example, the Wildlife and Countryside Act 1981 contains measures for protecting species listed in the Schedules of the Act. These Schedules, updated formally at 5-year intervals, list riverine species such as freshwater pearl mussel (*Margaritifera margaritifera*), Atlantic stream crayfish (*Austropotamobius pallipes*), allis shad (*Alosa alosa*), and otter (*Lutra lutra*). Recent introduction into national legislation of the measures in the EU Habitats Directive has added a new dimension with the requirement to designate Special Areas of Conservation (SACs) for species listed in Annex II of the Directive. Thus, in Britain, there are likely to be SACs for riverine species such as Atlantic salmon (*Salmo salar*) and sea lamprey (*Petromyzon marinus*).

This second scenario requires both autecological information on individual species considered import-

Table 1. Assessment criteria for selecting Special Areas of Conservation for species on Annex II of the EC Habitats Directive

• Size and density of the population of the species present on the site in relation to the populations present within national territory
• Degree of conservation of the features of the habitat which are important for the species concerned and restoration possibilities
• Degree of isolation of the population present on the site in relation to the natural range of the species
• Global assessment of the value of the site for conservation of the species concerned

ant for conservation as well as guidance on how rivers containing them are to be selected. For the Habitats Directive, Annex III provides some broad guidance on criteria for evaluating potential SACs (Table 1), but these need to be expanded further for selecting SACs for particular species. A good example is the freshwater pearl mussel, a species rapidly declining throughout its former holarctic range due to a combination of pollution, river engineering, and over-exploitation by pearl fishing. Scotland is now considered one of the few remaining strongholds for this species, although even here less than a third of the rivers where pearl mussels used to occur still harbour actively reproducing populations. Work on identifying possible SACs for this species has concentrated not only on population sizes, but also on the important criterion of age distribution, to ensure that all rivers selected contain actively reproducing pearl mussel populations.

Annex I of the Habitats Directive lists eight specific categories of European rivers for which SACs are to be designated (Table 2). Unfortunately, the classification in Annex I does not adequately cover the UK river resource, and only one of these categories – 'Floating vegetation of *Ranunculus* of plain, sub-mountainous rivers' – occurs in the UK. Selection of SACs for this feature was based on a combination of representativeness ('is it a good example of its type?'), the extent to which the river was still relatively free from human impact, and the coincidence of Annex II species within the same river.

Aim 3: To devise appropriate management strategies for specific rivers based on a broad assessment of their conservation value

The aim of this scenario is to provide planners and policy-makers with strategic guidance on appropriate management strategies for particular rivers. Depending on the perceived conservation value of each river, an appropriate strategy might be statutory protection, a careful limitation of proposed development in the catchment, or some form of rehabilitation work for degraded systems (Boon, 1992). This type of approach assesses conservation status together with the level of human impact, but leaves the assessment of priorities for future action firmly in the domain of the policy-maker. This can be especially useful where developers need to decide the location of new construction projects when faced with a choice of possible rivers. It was used, for instance, in South Africa to model the impact on conservation status of a range of dam construction and water allocation schemes using an 'expert system' known as the River Conservation System (O'Keeffe et al., 1987; O'Keeffe & Davies, 1991). The same general approach (but in this case with data only on riverine habitats) was applied to a series of proposed small-scale hydropower developments in North Wales (Litton, 1997), using a technique known as River Habitat Survey (Raven et al., 1997).

This third scenario aims for comprehensiveness. It takes as its starting point the philosophy that the assessment of conservation value should reflect the complexity of river systems, and thus treats comprehensiveness as a *sine qua non* in conservation evaluation. This reasoning has been applied to a range of river assessment systems (e.g. Angermeier & Karr, 1994; Blyth, 1983; Collier & McColl, 1992; Macmillan, 1983) and formed a foundation for the development of SERCON.

Table 2. Running water habitats listed in Annex I of the Habitats Directive ('Natural habitat types of community interest whose conservation requires the designation of Special Areas of Conservation')

- Alpine rivers and the herbaceous vegetation along their banks
- Alpine rivers and their ligneous vegetation with *Myricaria germanica*
- Alpine rivers and their ligneous vegetation with *Salix elaegnos*
- Constantly flowing Mediterranean rivers with *Glaucium flavum*
- Floating vegetation of *Ranunculus* of plain, sub-mountainous rivers
- *Chenopodietum rubri* of sub-mountainous rivers
- Constantly flowing Mediterranean rivers: *Paspalo-Agrostidion* and hanging curtains of *Salix* and *Populus alba*
- Intermittently flowing Mediterranean rivers

Aim 4: To ensure sustainability in the ecological structure and function of all rivers

This aim lies at the heart of the proposed EC Water Framework Directive (WFD). Although not a conservation Directive *per se*, and whilst it may not have the same focus as a 'traditional' nature conservation approach, the Directive provides a goal that 'conservation' and 'management' can both embrace. In essence, it brings with it the imperative to see rivers as whole systems rather than merely as containers and transporters of water. Instead of solely concentrating on trying to remove what is bad, or trying to maintain or enhance what is good, this approach aims for the sustainability of ecological processes and biological elements in all surface waters.

The riverine attributes for assessment under the WFD are summarized in Table 3. Within a given time-frame, Member States must aim to achieve 'good surface water status', of which part is described as 'good ecological status'. The latter is defined in Annex V of the Directive for each water type and for each quality element. The 'general definition' of 'good ecological status' for all waters is:

> 'The values of the biological quality elements for the surface water body type show low levels of distortion resulting from human activity, but deviate only slightly from those normally associated with the surface water body type under undisturbed conditions' (Council of the European Union, 1999).

The principle of comparing observed with predicted states forms a major plank of the assessment protocol. In other words, a river will be considered to have good ecological status if it contains habitats and species expected in a river of that 'type' in that locality, and if those features are relatively free from human disturbance. Thus, many of the variables listed in Annex V of the Directive could be considered to align with conservation criteria such as 'representativeness' and 'naturalness'.

Although the four scenarios described above are different from one another, each (to a greater or lesser extent) recognizes that rivers are complex systems, and any assessment of their 'integrity', their 'health', or their 'conservation value' must look beyond narrow indicators such as chemical water quality, the use of one group of organisms such as fish or invertebrates, or the success (or otherwise) of single, high-profile species such as salmon or otter.

Whilst approaches to sustainable river management and to river conservation occupy much common ground, there are still important differences between them. A nature conservation perspective highlights the special or extra-ordinary (the natural, the rare, the threatened, the diverse) yet still appreciates the ordinary. It lays stress on 'non-use' (e.g., aesthetic) values while realizing the importance of 'use' (e.g., economic) values. The river management perspective focuses on the ordinary but still recognizes the value of the uncommon. It emphasizes uses, but does not neglect non-use values. The WFD illustrates this differ-

Table 3. Quality elements for the classification of ecological status of rivers as described in the draft Water Framework Directive (Council of the European Union, 1999)

Biological elements
Composition and abundance of aquatic flora
Composition and abundance of benthic invertebrate fauna
Composition, abundance and age structure of fish fauna
Hydromorphological elements supporting the biological elements
Hydrological regime
quantity and dynamics of water flow
connection to groundwater bodies
River continuity
Morphological conditions
river depth and width variation
structure and substrate of the river bed
structure of the riparian zone
Chemical and physicochemical elements supporting the biological elements
General
Thermal conditions
Oxygenation conditions
Salinity
Acidification status
Nutrient conditions
Specific pollutants
Pollution by all priority substances identified as being discharged into the body of water
Pollution by other substances identified as being discharged in significant quantities into the body of water

ence well, having 'good ecological status' as the key target for rivers, but retaining a category of 'high ecological status' which represents 'type-specific', reference conditions. Conversely, bodies in Britain such as English Nature lay great stress on 'designated' rivers (Sites of Special Scientific Interest, Special Areas of Conservation) yet are at pains to emphasize the importance of sustainable river management in the 'wider countryside'.

Key issues in developing comprehensive methods for river conservation evaluation

During the 1980s and early 1990s, conservation evaluation of British rivers was primarily founded on an assessment of the aquatic plant communities. This had many advantages: the classification on which the assessment protocol was based was robust and well-tested (Holmes, 1983, 1989; Holmes et al., 1998), the aquatic plant flora of Britain is well described, and aquatic macrophyte surveys are relatively straightforward with a manageable number of species for identification. The principal disadvantage of evaluating rivers solely in terms of their aquatic plants is in failing to address the greater complexity of river structure and function. The development of SERCON in the mid-1990s (Boon et al., 1997) presented an opportunity to reconsider the methods for river conservation assessment, and specifically to address the question of complexity. Six key issues have emerged as central to this process:

What constitutes value?

There are some who are uneasy about the notion of assigning 'value' to natural systems. Part of the problem may be semantic, related to the rather confusing terminology which has grown up around the subject. In a discussion on freshwater nature conservation assessment in the Southern Hemisphere, O'Keeffe (1997) lists 10 terms that have been used by various authors to describe the environmental condition of rivers: conservation status, conservation importance, conservation values, conservation significance, conservation potential, natural value, ecological health, ecological integrity, biological integrity, and habitat integrity. Others could be added to this list; perhaps most notably in Britain the term 'conservation interest' or 'scientific interest' (as in 'Sites of Special Scientific Interest'). O'Keeffe argues that the proliferation of terms such as these does not help to persuade non-specialists (especially river managers and planners) to understand and embrace concepts of river conservation, and he recommends greater simplification. Although it is true that not all of these terms are taken to mean the same thing, it *is* the case that some such as 'conservation importance', 'conservation status', or 'conservation interest' are in essence synonymous.

Value judgements are an integral part of human society and sometimes arouse controversy. Some argue that there is no such thing as 'good' music or 'good' art, any more than there is 'good' food – all are simply a reflection of personal taste. Is there such a thing as a 'good' river, or is this, too, a matter or taste and perception? Of course, there will always be interest groups who will fight their corner – the engineer who sees perfection in a trapezoidal river channel; the angler for whom bankside trees are merely annoying obstacles that hinder his cast; the farmer who treats the river simply as a convenient source of water for irrigation; the salmocentric fisheries manager for whom a river is there for little else than catching salmon. This is nothing new. Rivers have always been used by people. Yet, from the early days of water quality classification, there has been a tacit and widespread acceptance that general terms such as 'bad', 'fair' and 'good' are appropriate for describing various aspects of river systems – a view emphasized in the Water Framework Directive's use of similar words to describe 'ecological status'.

If it is philosophically acceptable to ascribe high 'conservation value' to a river, the questions still remain – what constitutes value and who decides? In the development of SERCON, the Project Development Group (PDG) took the view that those with a scientific, technical or management role for rivers (e.g., in environment agencies, research institutes, conservation organizations, universities, etc.) were the most appropriate people to assist in determining the criteria for assessing value or quality. This approach mirrored that of O'Keeffe et al. (1987), who consulted an 'expert group' before designing the River Conservation System, and is line with the production of various assessment systems in North America (Angermeier et al., 1991; Karr, 1991).

Some authors have suggested that the general public should be encouraged to play a part in this type of work. Pollard and Huxham (in press) have recently argued that the public should be more involved in defining what constitutes 'good ecological status' under the WFD. Whilst it is unlikely that this suggestion will gain wide support, it is worth noting that studies on public perception of rivers have shown it to be more in tune with professional, ecological opinion than is often believed. In a study of two British rivers, Green & Tunstall (1992) concluded that the public desire and value what they regard as 'environmentally rich' river corridors, even if this is not necessarily what they personally find attractive. In summary, studies in Britain suggest that people value rivers that are unpolluted, with plentiful water and rich in wildlife (Tunstall et al., 1997). Similar work on public perception of coastal resources (Goodman et al., 1998) supports the conclusion that conservationists' views of what constitutes 'high conservation value' are often not dissimilar to those of the general public.

Perhaps this would not have been the case 20 years ago. Public perceptions change with time, and the impact of environmental education, especially through media such as television, cannot be underestimated. Angermeier et al. (1991) designed a 'multi-attribute judgement technique' to identify features of streams and rivers in Virginia that confer ecological value, and consulted fishery managers, aquatic biologists, and fish ecologists to draw up their protocol. However, they concluded that public perception was also important, and stated that management agencies should not merely reflect public attitudes but seek to be proactive by education.

When the views of a diverse set of studies are taken into account on what constitutes river or stream 'value', the picture that emerges indicates that not one but a suite of attributes is needed. Much of the early discussions by the SERCON PDG centred on which attributes should be selected for assessment. These remained remarkably constant throughout the development of the system. From an initial list of 57 features produced by the PDG, a structured list of 54 was sent for consultation to a panel comprising more than 160 individuals and organizations (Boon et al., 1997). Attributes were grouped in five categories – 'Physical descriptors' (10), 'Catchment characteristics' (5), 'Physical and chemical characteristics of

Table 4. Attributes assessed by SERCON (together with their standard abbreviations) grouped into six conservation criteria, and an impacts criterion

Physical Diversity (PDY)	**Species Richness (SR)**
PDY 1: Substrates	SR 1: Aquatic and Marginal Macrophytes
PDY 2: Fluvial Features	SR 2: Aquatic Invertebrates
PDY 3: Structure of Aquatic Vegetation	SR 3: Fish
	SR 4: Breeding Birds
Naturalness (NA) [1]	
NA 1: Channel Naturalness	**Special Features (SF)**
NA 2: Physical Features of the Bank	SF 1: Influence of Natural On-line Lakes
NA 3: Plant Assemblages on the Bank	SF 2: Extent and Character of Riparian Zone
NA 4: Riparian Zone	SF 3: Floodplain: Recreatable Water-dependent Habitats
NA 5: Aquatic and Marginal Macrophytes	SF 4: Floodplain: Unrecreatable Water-dependent Habitats
NA 6: Aquatic Invertebrates	SF 5: Invertebrates of River Margins and Banks
NA 7: Fish	SF 6: Amphibians
NA 8: Breeding Birds	SF 7: Wintering Birds on Floodplain of ECS
	SF 8: Mammals
Representativeness (RE)	
RE 1: Substrate Diversity	**Impacts (IM)**
RE 2: Fluvial Features	IM 1: Acidification
RE 3: Aquatic Macrophytes	IM 2: Toxic/Industrial/Agricultural Effluent
RE 4: Aquatic Invertebrates	IM 3: Sewage Effluent
RE 5: Fish	IM 4: Groundwater Abstraction
RE 6: Breeding Birds	IM 5: Surface Water Abstraction
Rarity (RA)	IM 6: Inter-river Transfers
RA 1: EC Habitats Directive/Bern Convention Species (and rare in UK)	IM 7: Channelization
RA 2: Scheduled Species	IM 8: Management for Flood Defence
RA 3: EC Habitats Directive Species (but not rare in UK)	IM 9: Man-made Structures
RA 4: Red Data Book/Nationally Scarce Macrophyte Species	IM 10: Recreational Pressures
RA 5: Red Data Book/Nationally Scarce Invertebrate Species	IM 11: Introduced Species
RA 6: Regionally Rare Macrophyte Species	

[1] Naturalness is sub-divided into NA (A) and NA (B), comprising NA 1-4 (physical attributes) and NA 5-8 (species attributes), respectively.

channel and floodplain (10), 'Biotic characteristics' (11), and 'Aquatic impacts' (18). Approval for including these attributes was generally high, ranging from 72 to 91% (mean, 87%). The final structure of SERCON comprises 62 attributes – 16 to provide the background data necessary to set an evaluation in context, 35 for assessing positive features important for conservation, and 11 for categorizing aquatic impacts (Table 4) (Boon et al., 1997).

A comparison of the river characteristics incorporated in other conservation assessment techniques shows many similarities with those in SERCON. In particular, there is a convergence of opinion that the evaluation of individual riverine *attributes* (e.g., intact riparian zone, unregulated flow) can also be used to give information on broader, generic, conservation *criteria*. For example, assessments of diversity, naturalness, and rarity feature in several techniques (e.g., Blyth, 1983; Harvey, 1981; Rabe & Savage, 1979). As the process of designing SERCON gathered momentum, it became clear that the many attributes required a structure. Thus, the 35 conservation attributes in SERCON were grouped under seven criteria: physical diversity, naturalness of physical features, species naturalness (presence of aliens), representativeness, rarity, species richness, and a category known as 'special features' (Boon et al., 1997). These were chosen to align with the framework set out in *A Nature Conservation Review* (Ratcliffe, 1977), which has formed the bedrock of conservation in Britain for the past two decades.

Are all values equal?

Agreeing that river conservation value can be described in terms of a suite of criteria is not the same as saying that all criteria are equally important, or that there is unanimity on which are more important than others. For example, the historical focus by conservationists on species considered rare, and the more recent prominence given to 'biodiversity' issues in the 1990s, may lead to an over-emphasis on criteria such as rarity and species richness. The study on Virginia streams by Angermeier et al. (1991) showed that both these criteria were considered to be major determinants of ecological value by the experts they consulted. The authors surmise that the homogeneity in responses probably reflected the exposure of those questioned to powerful arguments from scientists favouring species preservation, and to the pervasive influence that the maintenance of biological diversity has had in shaping natural resources policy in the USA (e.g., Endangered Species Act, 1983). The notion that high species richness always equals high conservation value was rejected in the development of SERCON. Guidance given in the user's manual (Boon et al., 1996) makes it clear, for example, that an undisturbed highland stream naturally low in nutrients and poor in species should be valued more highly than a disturbed system with lower naturalness but with higher species richness.

At the more detailed level, it is self-evident that not all *attributes* can be considered equal in terms of their value for (or their impact on) conservation. The presence in a river of one regionally scarce invertebrate species cannot, for example, be considered as important as the retention by a river of its natural planform, cross-section, unmodified banks, and riparian zone. Integrated assessment techniques therefore need a method of weighting to differentiate between attributes of differing importance. The difficulty is not so much in agreeing that weighting is necessary, but rather in determining how weights are to be derived and what those weights should be (Angermeier & Winston, 1997). The protocol for ranking the value of biotic resources in Virginia streams (Angermeier et al., 1993) assigned equal weight to their three components of value: 'rare biota, 'ecological value', and 'fishery value'. However, they recognized that weights might need to be adjusted later, and suggested that this might even be determined with reference to public rather than professional opinion.

Weights for SERCON attributes were set according to the combined views of the PDG, Steering Group, and the specialist panel. Clear instructions were issued to ensure that everybody followed the same procedure. For instance, the maximum weight allowed for any attribute was 5, with a minimum set at 1. The rationale for this was that a differential greater than 5 would suggest that some attributes were sufficiently unimportant that they did not even merit inclusion in SERCON. There was generally very good agreement on those attributes that should be weighted the highest (e.g., diversity of fluvial features, channel naturalness) and those the lowest (e.g., structural diversity of aquatic vegetation, naturalness of breeding bird communities), but in between there was a wide range of opinion. A series of 'ground rules' were adopted to allocate the extremes of the weighting scale to attributes in each criterion. A second set of rules then permitted the allocation of intermediate weights.

Setting weights entails value judgements, just as much as selecting conservation attributes. The SERCON PDG decided that in the interests of openness, the weights of each attribute would be clearly stated (even though the computer version uses the weights automatically to calculate conservation indices), and that for each evaluation unweighted indices would also be computed for comparison.

Problems of definition

The evaluation of fresh waters has often been hindered by confusion over definitions. The term 'quality' when applied to rivers and lakes was examined in depth by Boon & Howell (1997) in recognition that physical features, water chemistry, and biota all contribute to 'freshwater quality', and so, in a broader sense, do recreational, landscape, and economic attributes.

It is extremely important to arrive at clear definitions for conservation criteria. For example, the term 'biodiversity', which has now entered common parlance, is used in a variety of ways, even by biologists. Whilst a few equate biodiversity with species richness (e.g., Ward, 1998), most accept that biodiversity incorporates far more, but there is no universally agreed definition.

Most criteria in SERCON are readily defined. Physical diversity is assessed in terms of the richness and equitability of a standard list of substrate categories, river features such as gravel bars, eroding cliffs, and flow patterns, and the diversity of aquatic vegetation structure. With care, the definition of what is meant by 'species richness' or 'rarity' can also be made unequivocal. In SERCON, species richness is assessed with reference to standard UK lists of aquatic and marginal plants, invertebrates, fish, birds, and mammals, for comparison with the observed fauna and flora. Allowance is made for data on invertebrates where identifications are very often to family rather than species level. Rarity is defined in terms of a six-point hierarchy: (i) those species listed in the EC Habitats Directive or the Bern Convention, and which are rare in the UK; (ii) species protected under national legislation (the Wildlife and Countryside Act in Britain, and equivalent legislation in Northern Ireland); (iii) EC Habitats Directive species not rare in the UK; (iv) Red Data Book or nationally scarce macrophytes; (v) Red Data Book or nationally scarce invertebrates; (vi) regionally rare macrophytes.

Categories such as these are both easily defined and logically defensible. Defining what is meant by terms such as 'naturalness' or 'representativeness' is far more difficult – despite their widespread use, few people seem to have made the effort to tease apart the threads. Indeed, the copious minutes of the SERCON Project Development Group record the long, and at times arduous, debates within the Group to produce an agreed position on these concepts. It is easier to appreciate the usefulness of these discussions in retrospect; the same conclusion could undoubtedly be drawn from the development of SERCON in the UK as O'Keeffe et al. (1987) drew for the construction, evaluation, and testing of the RCS in South Africa. They described their work as "an extremely valuable exercise in its own right, forcing those who contributed to the system to examine their own intuitive judgements".

The task of defining these criteria is made still more complex owing to the relationship between criteria such as naturalness, representativeness, and species richness. The introductory sections of the SERCON manual (Boon et al., 1996) discussing naturalness offer the following guidance:

> The distinction between 'natural', 'semi-natural' and 'artificial' cannot be rigidly defined and arbitrary separations are required. Because of this, determining 'naturalness' requires considerable knowledge of habitats and their response to human intervention and to natural processes over time, and requires an element of professional judgement. Such judgements are specific to individual habitats and are intricately linked with the concepts of representativeness, diversity, etc. [covered by other parts of SERCON].

The present review of SERCON is examining the links between naturalness and representativeness, and some amalgamation of these criteria would seem to be a likely outcome.

The inter-relationship between criteria has been discussed in slightly different terms byAngermeier & Karr (1994), who state that many changes in species diversity can only be evaluated objectively on the basis of changes in integrity. For example, an increase in diversity may be brought about by artificial enrichment of a naturally oligotrophic system. Similarly, implicit in a definition of naturalness for rivers in Flanders (Schneiders et al., 1993) is the close link between naturalness and representativeness (i.e., between undisturbed habitats and communities showing characteristics expected for a particular river type and geographical location):

> 'Ecological naturalness of watercourses is reached when the biocoenosis occurring there by nature and belonging to the physical-geographical situation can survive permanently'.

The same concept is inherent in the Water Framework Directive, in which judgements are to be made about ecological features in the light of their naturalness (*'undisturbed conditions'*) and their representativeness (*'normally associated with the surface water body type'*).

Ensuring rigour and consistency

In the 1970s, the publication of *A Nature Conservation Review* (NCR) (Ratcliffe, 1977) provided a commentary on conservation criteria that would influence site evaluation procedures over the next 20 years. The second volume of the NCR comprises a series of descriptions of sites in Britain considered to be nationally important for nature conservation, framed in the light of the conservation criteria discussed in the first volume (size, diversity, naturalness, rarity, fragility, typicality, recorded history, position in an ecological/geographical unit, potential value, intrinsic appeal). Nevertheless, the site descriptions frequently contain value judgements with no clear substantiation. A lake is described as "the best example of an upland impermeable pool known in Britain" (Tarn Dub, Northern England), or a river as "a good example of the many eutrophic upland streams lying on the Old Red Sandstone of Caithness" (Burn of Latheronwheel, Northern Scotland). Yet there is no indication of why they were considered to be 'a good example' or 'the best example'.

Much of the impetus for designing SERCON came from a recognition that a more rigorous, repeatable, and transparent system of evaluation was needed both to set rivers in their conservation context and to help formulate management strategies. Three principal objectives were identified:

(1) the attributes and criteria for conservation assessment should be clearly described;
(2) a scoring system should be used to improve objectivity in assessment;
(3) the derivation and interpretation of scores should be readily understandable.

Scoring systems have been widely used in other freshwater conservation assessment protocols (e.g., Angermeier et al., 1993; Davies & Giesen, 1994; O'Keeffe et al., 1987; Rabe & Savage, 1979). Whilst scoring systems may be criticised for encouraging inappropriate, pseudo-mathematical treatment of data, their advantage in providing a framework for a rigorous and consistent comparison of one river with another outweighs the disadvantages. Scoring systems will fail, however, unless clear guidance is given on how scores are to be assigned. In the evaluation system for Virginia streams (Angermeier et al., 1993) the guidance for assigning scores (10, 20, 30 or 40) is in some cases extremely precise:

> 'Value of rare biota – score 20: stream contains historic habitat for Special Concern species, or supports anadromous populations of blueback herring (*Alosa aestivalis* or alewife (*Alosa pseudoharengus*)'.

However, the authors recognize that for some stream values, current knowledge is insufficient to provide any more than a subjective guide to scoring. For example:

> 'Habitat quality – score 20: stream exhibits fair habitat complexity and cover availability; channel or flow modifications are common, and some are severe.'

The immediate question is similar to the one discussed in relation to the NCR: 'what constitutes 'fair', 'common', and 'severe'?'

The development of SERCON raised similar problems. It is far easier, of course, to give clear guidance on scoring features that can be measured numerically (e.g., numbers of species, percent composition, percent cover) than those requiring a degree of judgement. There is also the danger that in an effort to be comprehensive, the process of assigning scores becomes too complex. For instance, in SERCON the presence of wintering birds on the floodplain (where one exists) is evaluated under the Special Features criterion. SERCON users refer to a table which lists 24 species in each of nine regions of the UK, and the percentage of 10 × 10-km squares in which each species has been recorded. The guidance for assigning a score (in this example, 2 on a 0–5 scale) states:

> 'All species found in >80% of the 10 × 10-km squares in the region present in the ECS, or at least 75% of those listed for >50% of 10 × 10-km squares present'.

It is possible that this will change as a result of the present review of SERCON, but it serves to illustrate the difficulties of using a scoring system when score allocation becomes over-complicated.

Complex systems require large datasets

There is a real dilemma in trying to construct a comprehensive evaluation system for rivers: on the one hand is the desire to ensure that it truly reflects the variety of features and attributes associated with rivers – on the other, the more complex the system the more information is needed, and time or money may not permit its collection.

Table 5 shows the percentage of attributes for which data are available for the 37 rivers (121 ECSs) in Britain so far evaluated using SERCON. The overall picture appears at first sight to be fairly encouraging, as 39 of the 46 attributes were scored in more than 80% of ECSs, and 26 of the 46 attributes were scored in 90–100% of ECSs. However, simply measuring the percentage of evaluations for which any given attribute is scored does not give the whole picture, as SERCON also requires a relative assessment to be made of *con-*

Table 5. The percentage of ECSs scored for each attribute for 37 rivers (121 ECSs) in Britain evaluated using SERCON (see Table 4 for an explanation of abbreviations)

Attribute	%	Attribute	%	Attribute	%	Attribute	%
PDY 1	93	RE 1	79	SR 1	95	IM 1	86
PDY 2	93	RE 2	79	SR 2	84	IM 2	99
PDY 3	93	RE 3	81	SR 3	81	IM 3	100
NA 1	99	RE 4	57	SR 4	98	IM 4	83
NA 2	93	RE 5	81	SF 1	100	IM 5	98
NA 3	93	RE 6	98	SF 2	86	IM 6	83
NA 4	93	RA 1	99	SF 3	7	IM 7	100
NA 5	95	RA 2	98	SF 4	4	IM 8	100
NA 6	79	RA 3	81	SF 5	96	IM 9	100
NA 7	81	RA 4	97	SF 6	88	IM 10	100
NA 8	98	RA 5	84	SF 7	94	IM 11	89
		RA 6	98	SF 8	55		

fidence in the data used for scoring (Boon et al., 1996). When the present review exercise is complete, it is likely to show that some attributes may be frequently scored, but with relatively low confidence. Moreover, there are a few attributes which do have much lower score rates than others. In particular, attributes SF 3 and SF 4, which relate to the extent of floodplain habitats, are hardly ever scored – not because of the demise of such habitats in the UK (a score of 0 is still assigned if there is no floodplain) but because detailed information on the extent of such habitats is difficult to obtain. Data for scoring some attributes in SERCON are taken from RHS surveys set up as a standard procedure prior to SERCON evaluation. It is not surprising, therefore, that most of these attributes (e.g., PDY 1–3, NA 1–4) can be scored almost every time. On the other hand, attributes requiring information on the status of particular species groups (e.g., mammals: SF 8) may be scored far less frequently, as for many rivers that information has simply never been systematically collected.

It is interesting to note that the problem of data availability was recognized at the very first Project Development Group meeting in May 1992, and it was agreed that the system should be capable of carrying out an evaluation at a fundamental level based on sparse data. By the sixth meeting in November 1992, the PDG considered that there might be merit in a system which allowed both a full evaluation when large datasets were available, as well as an evaluation based on a limited number of key attributes. Between then and the final (26th) PDG meeting in March 1995 the idea of a scaled-down version had been shelved, only to be revived recently in the review of SERCON!

Interpreting the results: the dangers of full integration

The process of 'integration' can be defined as 'to render whole or complete' and in that sense an attempt at producing a comprehensive conservation assessment system is an attempt at integration. However, the term is more commonly used to imply a combining, an interweaving, in which previously separate entities lose their identity in a composite whole. Superficially there may be much to be desired in a single conservation index for a river – one letter or number, much like a traditional water quality index. An end product such as this will be immediately understandable to a planner or to a river manager. One assessment technique that opted for this approach is the River Conservation System developed in South Africa by O'Keeffe et al. (1987). This consists of 41 attributes, divided between three categories – 'catchment', 'river', and 'biota' (Table 6). To derive an index of 'conservation status' for any or all of these categories requires the combination of a wide range of attributes. Moreover, some of the attributes are considered as 'negative' features and others as 'positive' and are differentially weighted either negatively or positively in calculating the indices. The output tends to mask valuable information on individual attributes; of more concern is that it combines 'good' with 'bad' features, so that a river with some elements of conservation importance, but

Table 6. The division of 'positive' and 'negative' attributes used in the River Conservation System (O'Keeffe et al., 1987) within three main categories, together with examples of each

Total no. of attributes		Positive attributes	Example	Negative attributes	Example
Catchment	12	3	% natural vegetation	9	% improved grazing
River	17	6	% unregulated	11	No. of mainstream dams
Biota	12	7	No. of endemic fish species	5	No. of introduced macrophyte species

also exposed to some negative impacts, is considered to be neither good nor bad, but a mediocre average.

The benefits of an integrated assessment method which still retains a clear distinction between individual criteria is illustrated by the following case study.

The benefits of integration with separation: case study of a SERCON evaluation of the River Douglas

The River Douglas is a small, second-order stream in north-west England. It rises at an altitude of 130 m, flows a distance of 36 km from Rivington Reservoir through the urban area of Wigan, and eventually enters the Ribble Estuary west of Preston. Although an urban landscape provides the setting for much of the upper river, further downstream it passes through agricultural land and improved grassland. Water quality is poor, although recent improvements to sewage works have brought some benefits.

The river was divided into three ECSs for the purpose of SERCON evaluation (Boon et al., 1996). A summary of the results is shown in Table 7. The physical diversity index for the upstream ECS 1 is high (in quality band A, on an A–E scale) owing to its diverse substrate with frequent cobbles, gravel and sand, and a range of fluvial features including unvegetated side bars, poached margins, runs and glides. Emergent broad and narrow-leaved herbs, especially reed canary-grass (*Phalaris arundinacea*), are very common. Although the planform of the channel is relatively natural, other elements of naturalness are moderately degraded. This places the ECS in quality band D both for physical attributes and for species attributes – the latter including the locally abundant alien macrophytes Indian balsam (*Impatiens glandulifera*) and Japanese knotweed (*Fallopia japonica*). Index values for representativeness, rarity, and species richness are also extremely low.

Table 7. River Douglas – SERCON criterion indices (0–100) and quality bands (A–E) (see Table 4 for an explanation of abbreviations)

ECS	1	2	3
PDY	82 (A)	44 (C)	56 (C)
NA (A)	37 (D)	26 (D)	28 (D)
NA (B)	36 (D)	57 (C)	57 (C)
RE	51 (D)	35 (D)	23 (E)
RA	14 (E)	0 (E)	0 (E)
SR	15 (E)	11 (E)	19 (E)
SF	32 (C)	42 (C)	25 (D)
IM	39 (C)	57 (D)	58 (D)

Note: For the purpose of this illustrative example, index values do not include suffixes indicating levels of data used in calculation (see Boon et al., 1997)

ECS 2 is characterized by extensive urban development and flows mainly through industrial and residential land. Physical diversity is considerably lower than in ECS 1 (band C), with few fluvial features. However, the structure of aquatic vegetation is relatively diverse, with abundant emergent herbs and filamentous algae, and with frequent floating, amphibious, and submerged broad-leaved herbs. This urban ECS has been extensively straightened and has bank reinforcements, resulting in a low index for physical naturalness (band D).

The ECS furthest downstream is characterized by a dramatic change in riparian land-use from urban to agricultural. The resectioned channel has few fluvial features apart from glides over a silt substrate. However, aquatic vegetation diversity scores highly, with abundant emergent herbs. Naturalness indices are very similar to ECS 2, with alien macrophytes Indian balsam and giant hogweed (*Heracleum mantegazzianum*) reducing the index for species naturalness. Indices for representativeness, rarity, species richness, and special features are all very low (bands E, E, E, D, respectively).

Table 8. Inappropriate integration within SERCON: an example of the River Douglas, north-west England (see Table 4 for an explanation of abbreviations)

Attribute	Index value	Weight	Weighted index value	Maximum possible index value	Maximum possible weighted index value
PDY	82	3	246	100	300
NA (A)	37	5	185	100	500
NA (B)	36	1.5	54	100	150
RE	51	3	153	100	300
RA	14	2	28	100	200
SR	15	2.5	37.5	100	250
SF	32	1	32	100	100
$\sum$			**735.5**		**1800**
Combined index value for ECS = 735.5 ÷ 1800 × 100 = **41**					

Although there are differences *between* ECSs in terms of conservation indices, one of the most notable contrasts is the relatively high value for physical diversity in ECS 1 compared with the levels of the remaining six indices within the same ECS (Table 7). Early in the development of SERCON there were plans to have a combined conservation index, in an attempt to provide a simpler indication of conservation value for the whole ECS. This idea was rejected. Even though individual criteria are not combined, weights for each criterion have been published (Boon et al., 1996, 1997) so that users can interpret the index values for any ECS in the knowledge that some are considered more important than others (Boon et al., 1998).

If, for the purpose of illustrating the problems with amalgamating information, the criterion indices are combined for ECS 1 of the River Douglas, this results in a rather mediocre index of 41 (on a 0–100 scale) (Table 8). The benefits to river management or rehabilitation of highlighting the value of the high structural diversity of this ECS, whilst at the same time contrasting this with low-scoring attributes such as species richness, are thus lost.

Conclusions

This paper has argued that river conservation assessment needs to be broadly based. It has also suggested that the most appropriate way of achieving this is to tease apart those elements that constitute conservation value. In a sense, this is a reductionist approach to rivers which stands the risk of being criticised for failing to see the wood for the trees. Whilst reductionism is helpful in highlighting the different characteristics that make a river important for conservation, it needs to be counter-balanced with holism.

The word 'holistic' is both over-used and mis-used – often to mean 'thorough', 'extensive', 'complete', or 'comprehensive'. Yet holism in its strict sense – 'the theory that a whole can be more than the sum of the parts' – is most certainly an appropriate concept to apply to rivers and their conservation. By reducing some rivers to their constituent values (conservation attributes), or by breaking them down into manageable lengths (e.g., SERCON ECSs), an overall impression of a river's uniqueness may be lost. For example, the River Hull in north-east England has long been designated as a Site of Special Scientific Interest. Yet an assessment of each section of the river in terms of criteria such as naturalness or species richness does not give the full picture, as the whole river is also considered to be important as the most northerly chalk stream in Britain (most are situated in southern and eastern England).

With this in mind, there are perhaps two principal challenges in the further development of river conservation assessment. One is to find ways of evaluation that combine the sense of holism with the clarity of reductionism. The other is to breathe some colour and life into the whole process of conservation assessment. The problem here is one of presentation: the need to demonstrate that what are essentially ecological concepts, such as 'naturalness', are not merely the domain of academic scientists – they are closely related to the unspoilt rivers rich in wildlife for which the public has obvious enthusiasm. Evaluation is not an end in itself; it is the first step in protecting and enhancing rivers and their catchments. Yet, unless the results of conservation evaluation can capture the enthusiasm and support of people at large, practical measures for conserving rivers are likely to be ineffective.

Acknowledgements

Data on SERCON attribute scoring were derived from the draft report on the SERCON review by Young Associates, Chester, who also undertook the evaluation of the River Douglas. The field survey of the Douglas was carried out by Michael Rose and Christopher Bates, under the direction of Jan Wilkinson (Young Associates). Other data for evaluating the river were provided by local offices of the Environment Agency (who also funded the project), Biological Records Centre, and the British Trust for Ornithology. I am also grateful to Nigel Holmes (Alconbury Environmental Consultants), Peter Maitland (Fish Conservation Centre), and two anonymous reviewers, for constructive comments on the manuscript.

References

Angermeier, P. L. & J. R. Karr, 1994. Biological integrity versus biological diversity as policy directives. BioScience 44: 690–697.

Angermeier, P. L. & M. R. Winston, 1997. Assessing conservation value of stream communities: a comparison of approaches based on centres of density and species richness. Freshwat. Biol. 37: 699–710.

Angermeier, P. L., R. J. Neves & L. A. Nielsen, 1991. Assessing stream values: perspectives of aquatic resource professionals. N. am. J. Fisheries Manage. 11: 1–10.

Angermeier, P. L., R. J. Neves & J. W. Kauffman, 1993. Protocol to rank value of biotic resources in Virginia streams. Rivers 4: 20–29.

Blyth, J. D., 1983. Rapid stream survey to assess conservation value and habitats available for invertebrates. In: Myers, K., C. R. Margules & I. Musto (eds), Survey Methods for Nature Conservation. Proceedings of a workshop held at Adelaide University, 31 August–2 September 1983. CSIRO Division of Water and Land Resources, Canberra, Australia: 1: 343–375.

Boon, P. J., 1992. Essential elements in the case for river conservation. In: Boon, P. J., P. Calow & G. E. Petts (eds), River Conservation and Management. John Wiley, Chichester, 11–33.

Boon, P. J., 1994. Nature conservation. In: Maitland, P. S., P. J. Boon & D. S. McLusky (eds), The Fresh Waters of Scotland: A National Resource of International Significance. John Wiley, Chichester: 555–576.

Boon, P. J., 1995. The relevance of ecology to the statutory protection of British rivers. In Harper, D. M. & A. J. D. Ferguson (eds), The Ecological Basis for River Management. John Wiley, Chichester, 239–250.

Boon, P. J. & D. L. Howell (eds), 1997. Freshwater Quality: Defining the Indefinable? The Stationery Office, Edinburgh, 552 pp.

Boon, P. J., N. T. H. Holmes, P. S. Maitland & T. A. Rowell, 1994. A system for evaluating rivers for conservation ('SERCON'): An outline of the underlying principles. Verh. int. Ver. Limnol. 25: 1510–1514.

Boon, P. J., N. T. H. Holmes, P. S. Maitland & T. A. Rowell, 1996. SERCON: System for Evaluating Rivers for Conservation. Version 1 Manual. Scottish Natural Heritage Research Survey and Monitoring Report No. 61. Scottish Natural Heritage, Edinburgh.

Boon, P. J., N. T. H. Holmes, P. S. Maitland, T. A. Rowell & J. Davies, 1997. A system for evaluating rivers for conservation (SERCON): development, structure and function. In: Boon, P. J. & D. L. Howell (eds), Freshwater Quality: Defining the Indefinable? The Stationery Office, Edinburgh: 299–326.

Boon, P. J., J. Wilkinson & J. Martin, 1998. The application of SERCON (System for Evaluating Rivers for Conservation) to a selection of rivers in Britain. In: Boon, P. J. & P. J. Raven (eds), The Application of Classification and Assessment Methods to River Management in the UK. Aquat. Conserv. Mar. Freshwat. Ecosyst. (special issue) 8: 597–616.

Cairns, J., 1995. Ecological integrity of aquatic systems. Regul. Rivers Res. Manage. 11: 313–323.

Collier, K. J. & R. H. S. McColl, 1992. Assessing the natural value of New Zealand rivers. In Boon, P. J., P. Calow & G. E. Petts (eds), River Conservation and Management. John Wiley, Chichester: 195–211.

Council of the European Union, 1999. Amended proposal for a Council Directive establishing a framework for Community action in the field of water policy, 19 February 1999, Brussels.

Davies, J. & W. Giesen, 1994. Towards a methodology for identifying tropical freshwater wetlands for protection. Mitt. int. Ver. Limnol. 24: 27–39.

Dudgeon, D. 1992. Endangered ecosystems: a review of the conservation status of tropical Asian rivers. Hydrobiologia 248: 167–191.

English Nature, 1997. Beyond 2000: English Nature's Strategy for Improving England's Wildlife and Natural Features, Peterborough.

Goodman, S. L., W. Seabrooke & S. A. Jaffry, 1998. Considering conservation value in economic appraisals of coastal resources. J. Envir. Plan. Manage. 41: 313–336.

Green, C. H. & S. M. Tunstall, 1992. The amenity and environmental value of river corridors in Britain. In: Boon, P. J., P. Calow & G. E. Petts (eds), River Conservation and Management, John Wiley, Chichester, 425–441.

Harvey, D. 1981. Conservation surveys – what do we need and how do we put it across? In: Harper, D. (ed.), Assessing the Conservation Value of Rivers. Department of Adult Education, University of Leicester: Leicester, 21–31.

Holdgate, M. 1997. What future for nature? Lecture at Edinburgh International Science Festival, 3 April 1996. Scottish Natural Heritage Occasional Paper No. 4, SNH, Edinburgh.

Holmes, N. T. H. 1983. Typing British rivers according to their flora. Focus on Nature Conservation 4. Nature Conservancy Council, Peterborough.

Holmes, N. T. H. 1989. British rivers – a working classification. Br. Wildlife 1: 20–36.

Holmes, N. T. H., P. J. Boon & T. A. Rowell, 1998. A revised classification system for British rivers based on their aquatic plant communities. In: Boon, P. J. & P. J. Raven (eds), The Application of Classification and Assessment Methods to River Management in the UK. Aquat. Conserv. Mar. Freshwat. Ecosyst. (special issue) 8: 555–578.

Karr, J. R. 1991. Biological Integrity: a long-neglected aspect of water resource management. Ecol. Appl. 1: 66–84.

Litton, G. 1997. Environmental impact of river and stream hydropower. Unpublished report to the Countryside Council for Wales, Bangor.

Macmillan, L. A. 1983. A method for identifying small streams of high conservation status. In Myers, K., C. R. Margules & I. Musto (eds), Survey Methods for Nature Conservation. Proceedings of a workshop held at Adelaide University, 31 August–2 September 1983, CSIRO Division of Water and Land Resources, Canberra, Australia: 1, 319–342.

Nature Conservancy Council, 1984. Nature Conservation in Great Britain. Peterborough.

Nature Conservancy Council, 1989. Guidelines for Selection of Biological SSSIs. Peterborough.

O'Keeffe, J. 1997. Methods of assessing conservation status for natural fresh waters in the Southern hemisphere. In: Boon, P. J. & D. L. Howell (eds), Freshwater Quality: Defining the Indefinable? The Stationery Office, Edinburgh: 369–386.

O'Keeffe, J. H. & B. R. Davies, 1991. Conservation and management of the rivers of the Kruger National Park: suggested methods for calculating instream flow needs. Aquat. Conserv. Mar. Freshwat. Ecosyst. 1: 55–71.

O'Keeffe, J. H., D. B. Danilewitz & J. A. Bradshaw, 1987. An 'expert system' approach to the assessment of the conservation status of rivers. Biol. Conserv. 40: 69–84.

Pollard, P. & M. Huxham, in press. The European Water Framework Directive: a new era in the management of aquatic ecosystem health? Aquat. Conserv. Mar. Freshwat. Ecosyst.

Rabe, F. W. & N. L. Savage, 1979. A methodology for the selection of aquatic natural areas. Biol. Conserv. 15: 291–300.

Ratcliffe, D. A. (ed.), 1977. A Nature Conservation Review, Cambridge University Press, Cambridge.

Raven, P. J., P. Fox, M. Everard, N. T. H. Holmes & F. H. Dawson, 1997. River Habitat Survey: a new system for classifying rivers according to their habitat quality. In: Boon, P. J. & D. L. Howell (eds), Freshwater Quality: Defining the Indefinable? The Stationery Office, Edinburgh: 215–234.

Raven, P. J., N. T. H. Holmes, F. H. Dawson & M. Everard, 1998. Quality assessment using River Habitat Survey data. In: Boon, P. J. & P. J. Raven (eds), The Application of Classification and Assessment Methods to River Management in the UK. Aquat. Conserv. Mar. Freshwat. Ecosyst. (special issue) 8: 477–499.

Schneiders, A., E. Verhaert, G. D. Blust, C. Wils, L. Bervoets, & R. F. Verheyen, 1993. Towards an ecological assessment of watercourses. J. Aquat. Ecosyst. Health 2: 29–38.

Tunstall, S., M. Fordham, C. Green, & M. House, 1997. Public perception of freshwater quality with particular reference to rivers in England and Wales. In: Boon, P. J. & D. L. Howell (eds), Freshwater Quality: Defining the Indefinable? The Stationery Office, Edinburgh: 39–58.

Ward, J. V. 1998. Riverine landscapes: biodiversity patterns, disturbance regimes, and aquatic conservation. Biol. Conserv. 83: 269–278.

Wishart, M. J., B. R. Davies, P. J. Boon & C. Pringle, in press. Global disparities in river conservation: 'first-world' values and 'third-world' realities. In: Boon, P. J., B. R. Davies & G. E Petts (eds), Global Perspectives on River Conservation: Science, Policy and Practice. John Wiley, Chichester.

Wright, J. F., P. D. Armitage, M. T. Furse & D. Moss, 1989. Prediction of invertebrate communities using stream measurements. Regul. Rivers Res. Manage. 4: 147–155.

Hydrobiologia **422/423:** 429–443, 2000.
M. Jungwirth, S. Muhar & S. Schmutz (eds), Assessing the Ecological Integrity of Running Waters.

EMAP-Surface Waters: a multiassemblage, probability survey of ecological integrity in the U.S.A.

R. M. Hughes[1], S .G. Paulsen[2] & J. L. Stoddard[2]
[1]*Dynamac, 200 SW 35th St., Corvallis, Oreg., U.S.A.*
E-mail: hughesb@mail.cor.epa.gov
[2]*USEPA, 200 SW 35th St., Corvallis, Oreg., U.S.A.*

Key words: sampling design, ecological indicators, reference condition

Abstract

Data analyzed at national, multistate or state scales often reveal more serious deterioration of various biological assemblages than that suspected from site specific studies simply because the impacts are observed regionally rather than locally. Unfortunately many regional assessments are based on data collected with differing sampling designs and methods, making their representativeness, accuracy and precision questionable. In 1989, the United States Environmental Protection Agency (USEPA) began EMAP-Surface Waters (EMAP-SW), a program for developing methods to monitor and assess status and trends in the nation's lakes and rivers in a statistically and ecologically rigorous manner. EMAP-SW has now conducted regional multistate pilots in streams and rivers throughout the U.S.A. in collaboration with State agencies, universities and EPA regional offices. We discuss EMAP's conceptual design, which focuses on biological integrity through use of multiple biological assemblages and includes physical and chemical habitat and landscape characteristics. When coupled with appropriate biological indicators, a probability-based design enabled us to provide accurate, precise and unbiased assessments of biological conditions, along with quantitative estimates of sampling uncertainty. Regional EMAP-SW surveys indicated the importance of assessing multiple biological assemblages because each assemblage was differentially sensitive to different stressors and at different spatial scales. Synthesizing multiple metrics from multiple assemblages allowed us to detect the effects of multiple anthropogenic disturbances. We also illustrate the value of using historical reconstruction and paleolimnological data for determining reference conditions where disturbance is extensive. We conclude that an EMAP approach to sampling design and indicators (recognizing natural ecoregional differences) has distinct advantages for monitoring and assessments that may be applicable to European Communities seeking to assess the ecological integrity of waters.

Introduction

Considerable deterioration of aquatic ecosystems at continental, national and regional scales has been reported worldwide (e.g., Petts et al., 1989; Williams et al., 1989; Hughes & Noss, 1992). Although there are multiple explanations for these reported declines (Miller et al., 1989; Petts et al., 1989; Allan & Flecker 1993), less attention has been focused on why detection of ecological changes across large spatial scales has been, until recently, limited.

There are at least five reasons why detection of these changes has been limited. (1) Often regional monitoring is not implemented until regional changes from natural conditions occur (Petts, 1989). Such changes are largely unavoidable because humans alter ecosystems markedly whenever they begin exploiting them (Marsh, 1885; Hughes, 1999). (2) Much monitoring of aquatic systems lacks a statistical sampling design, which precludes unbiased assessments of all of a continent's, nation's or state's waters (Paulsen et al., 1998). Focusing only on highly disturbed sites or pristine sites provides biased assessments of our waters, and subjective site selection hinders evaluation of characteristic conditions. (3) In addition, monitoring programs are fragmented among non-cooperating agencies or individuals with differing objectives, indicators, scales of interest and methods; therefore, data are difficult to combine in a quantitative manner (Schwaiger et al., 1994; Statzner et al., 1994; Yoder & Rankin, 1998). Although several excellent monitoring programs exist or have existed at state and basin scales, they are difficult to sustain because of annual fiscal and staffing uncertainties. (4) Long-term, intensive studies of particular sites on a network of ecological reserves are essential for understanding ecological processes. But the current design of these networks in the U.S.A. emphasizes selection of these sites for independent research questions and generally precludes aggregation or extrapolation to broader scales because of unknown representativeness (Hren et al., 1990).

The preceding four reasons are largely design-based, but there is also an indicator-based explanation for not detecting widespread ecological impairment. (5) Historically, the indicators have largely been chemical measures. This arose from an historical perspective that most of the stressors to aquatic systems were chemical in nature. Thus, while the objective may have been to assess the extent to which the systems supported aquatic life, chemical measures were used as surrogates. Frequently, this use of chemical surrogates for biota has been found wanting. For example, Karr et al. (1985) demonstrated that tertiary denitrification of sewage effluent alone had an insignificant effect on fish assemblages. Chovanec et al. (1995) showed that traditional assessment methods using measures of chemical water quality and saprobiological analyses grossly underestimated the degree of ecological alteration that had occurred in Austrian rivers when compared to assessments using a broader array of chemical, physical and biological measures. Yoder & Rankin (1998) found that use of chemical criteria failed to detect non-attainment of biological criteria for aquatic life in Ohio half the time. In addition, the proportion of Ohio stream miles in unacceptable condition rose from 9% to 44% when numerical biological criteria replaced qualitative biological indicators. They also reported enormous discrepancies among U.S.A. States in the percent of river miles assessed (5–100%), the percent fully attaining designated aquatic life uses (0–98%) and the number of river miles impaired by physical habitat modification (0–5500 miles). They concluded that these huge differences resulted largely from use of different indicators and reliance on chemical criteria and management action indicators (e.g., permits issued, management plans written, facilities constructed) rather than biological criteria. Although some interstate differences may represent environmental reality, without common indicators, sampling protocols, and sampling designs we cannot know this.

In response to shortcomings like these and recommendations from its Science Advisory Board (1988, 1990), the U.S. Environmental Protection Agency (USEPA) initiated EMAP (the Environmental Monitoring and Assessment Program) in 1990 (Messer et al., 1991). Initially, EMAP–Surface Waters (EMAP-SW) had three objectives (Paulsen & Linthurst, 1993): (1) Estimate, with known confidence, regional status and trends in condition of all lakes and rivers appearing on 1:100 000 scale maps. (2) Monitor status and trends and associations between indicators of condition and anthropogenic stressors. (3) Provide periodic assessments of ecological status and trends to managers and the public. The current EMAP-SW research strategy is focused on multi-state regional assessments; refinement of ecological indicators; improvement of nested, multi-tier monitoring designs; and development of systematic approaches for integrating ecological data to make regional assessments of ecological condition (EMAP, 1997).

EMAP differs from most other monitoring research programs in three ways: (1) It focuses on biological indicators to estimate condition of ecological resources and chemical, physical and biological indicators to evaluate stressors. It is not focused on human health or single anthropogenic stressors. (2) It selects sampling locations through use of a probability design to ensure that the set of sampling sites in a region is an unbiased representation of the entire population of waters of concern. (3) Its spatial focus is regional and national, while its temporal orientation ranges from years to decades.

Recently, Europe has also seen significant changes in monitoring programs. Concerns with the inadequacies of earlier monitoring designs in the United Kingdom, resulted in the development of the River Habitat Survey, which uses a stratified random sample of sites on rivers nationwide, but still focuses on physical habitat structure (Raven et al., 1998). Also, Sweden implemented a monitoring program employing both a probability sampling design and biological indicators to characterize lake condition (Wiederholm & Johnson, 1997).

The objective of this overview paper is to illustrate the advantages of a probability design in conjunction with a suite of biological, physical and chemical indicators for addressing a set of assessment questions across large areas by highlighting results of recent

EMAP pilot studies. As such, this is a broad summary rather than an incisive research paper; for detailed information the interested reader is encouraged to examine the cited literature.

Methods

A fundamental challenge in monitoring aquatic systems is that we cannot monitor everything, everywhere, all the time. Choices must be made on what to measure, how to measure it and where to measure it that are consistent with answering the assessment questions being posed. The choice of indicators and sampling design are critical to the success of any monitoring effort.

Indicators

The structures and functions of ecosystems are sensitive to an array of anthropogenic stresses. Different components of the system have different sensitivities to these stresses. Because of this, assessing their ecological condition requires a suite of indicators rather than a single, 'magic measurement'.

EMAP groups indicators into two broad classes, namely condition and stressor indicators (Figure 1; Paulsen & Linthurst, 1993). Condition indicators, often referred to as response indicators, are intended to be direct measures of important biological characteristics or important societal endpoints, such as biological integrity or biodiversity (important as evidenced by the Clean Water and Endangered Species Acts). These indicators should show the biological response to a host of environmental disturbances. They are most often measures of biological composition, structure, and processes from organism to ecosystem levels of organization (e.g., the size structure & species composition of a fish assemblage or the sediment metabolism rate of a stream site). The intent is that condition indicators be biological measures rather than chemical or physical surrogates. Sets of condition indicators should be responsive to a broad array of stressors and applicable across the U.S.A., but regionally calibrated. EMAP-Surface Waters examines multiple biotic assemblages (e.g., vertebrates, invertebrates, algae, microbes) to assess biological condition of rivers or lakes. Because of differences in responsiveness to different stressors, as well as differential sensitivity to the different spatial and temporal scales of stressors, no single assemblage is likely to adequately represent the condition of the others (O'Connor et al., 2000; Allen et al., 1999a).

Stressor indicators are essential for setting the biological expectations and for diagnosing probable causes of impaired biological conditions. Included are exposure, habitat and stressor indicators. They include measurements that are typically taken in the field, as well as measurements that are generated from satellite imagery, maps and other regionally available data sources. The term stressor indicator is perhaps a slight misnomer in that many of these are measures of characteristics that at a certain level are expected and necessary to support aquatic life. It is when they exist in extremely reduced amounts or levels beyond background that they serve as potential stressors to aquatic life. The types of measures that fall in this category include: temperature, fine sediments, hydrological and substrate conditions, riparian vegetation, large wood debris, nutrients, toxic chemicals, pH, climate, elevation, land use, human population density, chemical application and release rates, channel or flow modification and fisheries management. Certain of these indicators, such as catchment area, water body size or channel slope, are essential for calibrating candidate condition indicators.

Assemblage integration

Large, complex data sets like those generated by EMAP call for both ecological judgement and quantitative computer analyses. O'Connor et al. (2000) offered a four-phase process for integrating assessments of individual assemblages into a regional assessment of water bodies. (1) Control for natural covariates by examining residuals from maximum species richness or regression lines. For example, we know that river size affects species richness and composition of fish (e.g., Fausch et al., 1984), so fish data must be calibrated by such factors to detect effects of anthropogenic disturbances. (2) Rescale metrics by using z-scores, standardized ranks, or by dividing deviation from the metric median by the metric interquartile range (O'Connor et al., 2000). This gives all metrics a comparable range in values; upper and lower cutoffs can also be employed for extreme values. Re-scaling allows combining multiple metrics with great differences in ranges of values without having a wide-ranging metric distort the resulting index. (3) Standardize response signs where necessary so that all metrics decrease with increased environmental impact. This prevents negative and positive metrics from neut-

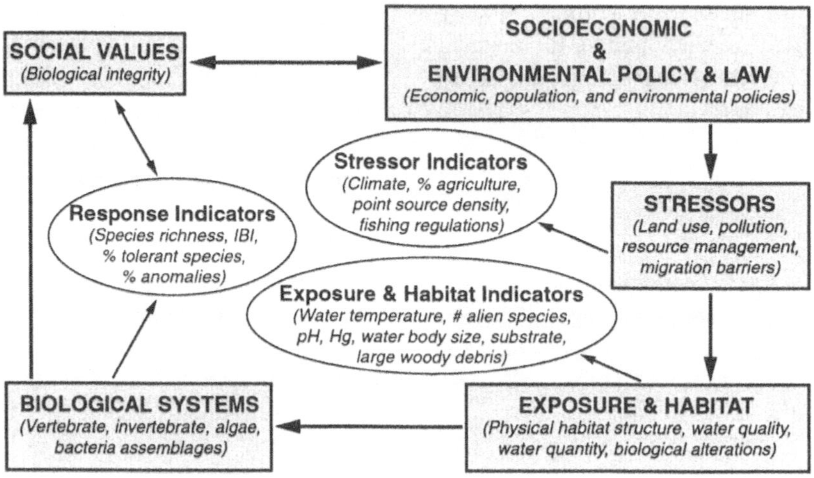

Figure 1. EMAP conceptual design linking surface water indicators, social values and resource policies (from Hughes, 1997).

ralizing each other. (4) Reduce the number of variables or metrics by calculating multivariate (Wright et al., 1984) and multimetric (Fausch et al., 1990) indices.

Reference condition

Perhaps the single biggest challenge in the development of indicators is addressing the concept of reference conditions. This reference condition is the yardstick, benchmark or standard against which to compare the waters being assessed. EMAP determines reference condition in three ways (regional reference sites, historical information and paleoecological data) as suggested by Hughes (1995). Regional reference sites are catchments or reaches that are representative of an ecoregion and minimally disturbed by humans; they were selected following thorough map and field analyses. Historical information included maps, text and data bases describing conditions preceding permanent European settlements. Paleoecological data were interpreted from lake sediment cores, mainly through the use of diatoms and pollen.

Field protocols

In monitoring, there is always a tradeoff between sampling a site in extensive detail and sampling a large number of sites. Our broad objective of assessing status and trends, gave relatively high priority to maximizing the number of sites visited and achieving trend detection by repeated estimates of the regional population of streams over time (Paulsen et al., 1991; Larsen et al., 1995; Urquhart et al., 1998). Thus, the indicators employed at a field site should be sampleable in less than a day. Crews ought to be small enough to fit in one vehicle, yet be large enough to provide a margin of field safety and sufficient manpower to haul gear (3–4 persons). We have found that 3–4 people can sample a large set of indicators in 8–10 h, allowing time to return to a motel or campsite before dark. It is useful to include indicators sensitive to immediate environmental conditions (e.g., algae), as well as those integrating larger scales and longer time periods (e.g., vertebrates). Paulsen et al. (1991) demonstrated that index variance (e.g., from annual, sampling period and measurement error) should be <50% of among-site variance if the concern in status estimates is with indicator scores lower than the 25th percentile. An indicator's ability to distinguish different sites or reaches (signal) should be at least three times as great as its combined sources of sampling error (noise) This allows explanation of a maximum of 50% of the variance (Allen et al., 1999a) and limits bias in the tenth percentile and the median to 40% and 5%, respectively (Paulsen et al., 1991).

Survey sampling design

The fundamental challenge in survey design, given that we cannot sample every lake and stream in the country, is to select a subset of sites that can be used to infer conditions for all aquatic systems, i.e., represent the entire population of lakes and streams/rivers. The approach with fewest assumptions is a probability-based sample survey design. Designs of this nature provide unbiased estimates of the parameters of in-

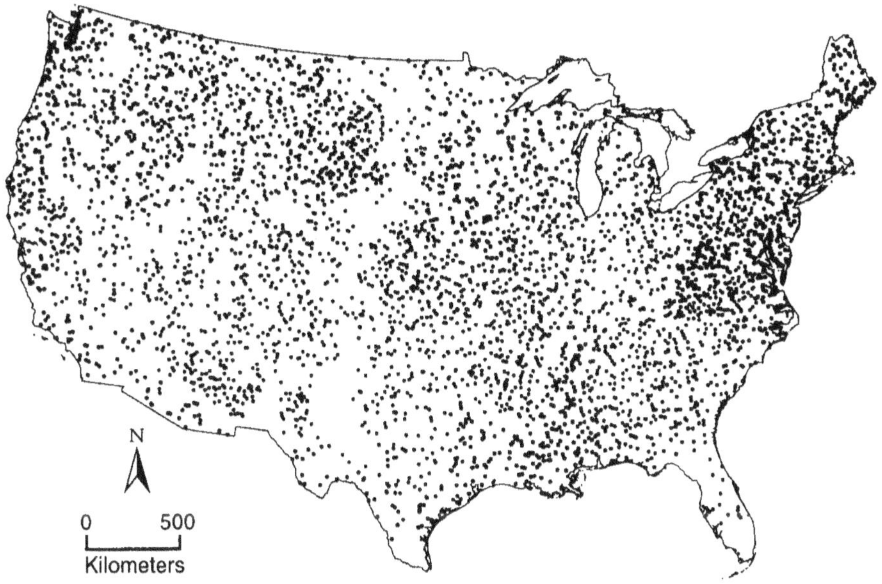

Figure 2. Distribution of sample points for a survey of the streams and rivers of the conterminous U.S.A. This site density would be sufficient for making national, regional and large state assessments. Note that the number of sites across the country can be easily increased (or decreased) simply by intensifying (or de-intensifying) the grid in any particular area (nation, region, state, basin). See Figure 3 for state boundaries.

terest for the population of lakes and streams/rivers with known confidence, i.e., an estimate of the uncertainty. The design should be flexible enough to allow focus on a specific problem in a small area or on particular types of streams or lakes. It should also permit statistical quantification of uncertainty and sources of variability for populations and sub-populations of waters. In the U.S.A., previous aquatic surveys using a probability design have focused on fish (Judy et al., 1984), lake acidification (Eilers et al., 1988; Landers et al., 1988; Baker et al., 1991) and stream acidification (Kaufmann et al., 1991; Baker et al., 1991). As explained by Overton et al. (1991), Urquhart et al. (1993) and Stevens (1997), the EMAP design facilitates regional ecological research and offers a clear alternative to regionally extrapolating from a small number of study sites with unknown representativeness. Although a probability design can apply to any spatial scale, EMAP is oriented toward national and regional issues (Figures 2 and 3).

Site selection by EMAP uses variable inclusion probabilities, a process somewhat akin to a stratified, random design (Stevens, 1994; Stevens & Olsen, 1999). EMAP minimizes pre-stratification of streams or rivers by type or region; however, it attempts to sample a sufficient number of sites to allow multiple post-stratifications by a wide range of classification schemes. Improper pre-stratification yields lower precision than does random sampling. Stevens (1994) found that pre-stratification confounds variance estimation, hinders detection of unanticipated problems and relationships, and produces unsatisfactory precision if $\geq$20% of sites among strata are misclassified. For example, it is unwise to pre-stratify on basin or region if subsequent sampling indicates that such classes account for small amounts of the variability present (Van Sickle & Hughes, 2000). Post-stratification was also employed in the national survey of 4000 Swedish lakes (Wiederholm & Johnson, 1997). To avoid swamping the sample with the much more numerous small water bodies (and failing to include enough large ones), EMAP selects small streams and ponds with a different frequency than would result from purely random selection. In addition, a grid network is employed in site selection to minimize clumping and to obtain broad spatial representation of all waters of interest (White et al., 1992).

Lakes and rivers are both selected in a similar manner. The electronic version of 1:100 000 topographic maps is used as a frame from which to select lakes, streams and rivers for sampling. A randomization process, utilizing a grid to force spatial coverage (Olsen et al., 1998; Sahr & White, 1998) provides a set of sampling locations that reflects the density of the resource in the region (Stevens & Olsen, 1999). Once the sampling sites are identified, the lakes and streams selected are checked on maps and in field visits to verify them and to gain site access.

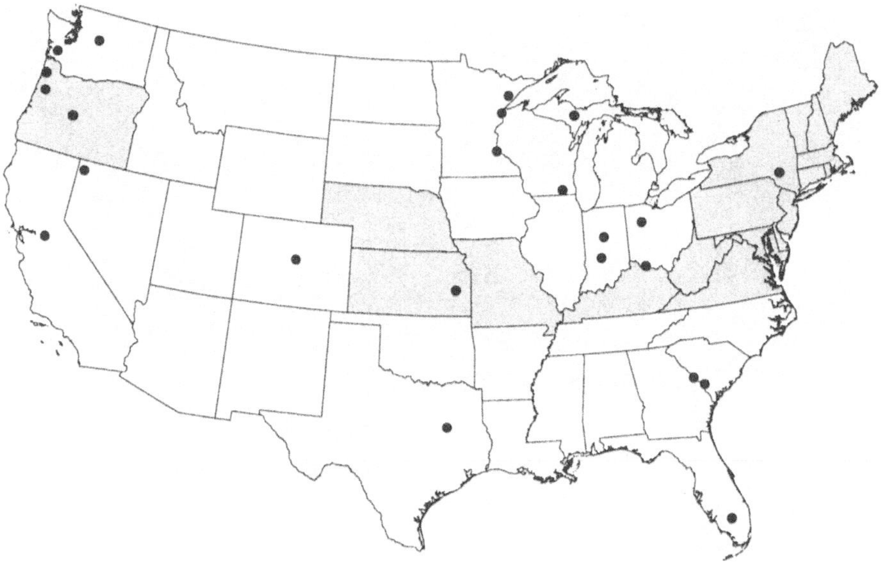

Figure 3. Locations of past and current EMAP-designed surveys in the conterminous U.S.A. Shaded areas represent state or multi-state surveys; dots represent basin or ecoregion

Since EMAP is concerned with both trend detection and status assessment, alternative temporal designs have also been evaluated (Urquhart et al., 1998). Sensitivity analysis demonstrated that several design alternatives have similar power for detecting trend; however, precise status estimates result from designs having the greatest number of sites. Thus, EMAP is using a design that selects four sets or panels of lakes/streams. Each set is visited in successive years with a revisit of the sets beginning in the fifth year. A subset of sites, generally 10–15%, is revisited every year.

Regional EMAP pilot studies

EMAP conducted a 1991–1996 pilot study on lakes in the northeastern U.S.A. and has also initiated multi-year regional pilots on streams in the Mid-Appalachians, eastern Corn Belt plains, central Great Plains, Rocky Mountains, California's Central Valley and the Pacific Northwest (Figure 3). More localized basin studies were begun by the states of Florida, Georgia, North Carolina, Virginia, West Virginia, Delaware, Maine, New York, New Hampshire, Vermont, Kentucky, Ohio, Indiana, Wisconsin, Minnesota, Kansas, Texas, Nevada, Oregon and Washington.

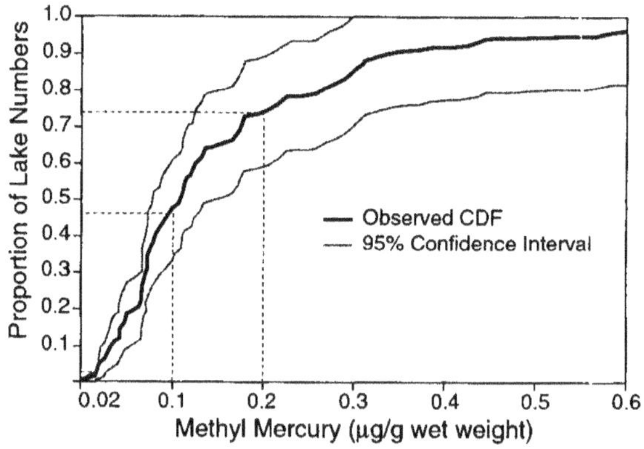

Figure 4. Piscivore risk as a function of methyl Hg concentrations in fish (from Yeardley et al., 1998). Methyl Hg concentrations of 0.02, 0.1 and 0.2 μg/g represent levels in fish that pose a risk to piscivorous birds, mustelids and humans, respectively.

Results

Making quantitative ecological assessments

Based on a survey of approximately 200 northeast U.S.A. lakes, EMAP made quantitative ecological assessments for the region's 9544 lakes. For example, Yeardley et al. (1998) reported that 26 ± 15% of the lakes in that region supported fish with methyl Hg levels exceeding a critical value of 0.2 μg/g for human consumers. For non-human consumers, 54 ± 13% and 98 ± 1% of the lakes had methyl Hg levels in

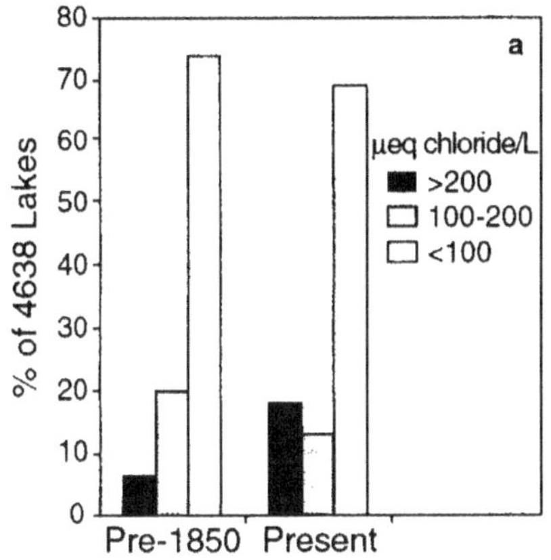

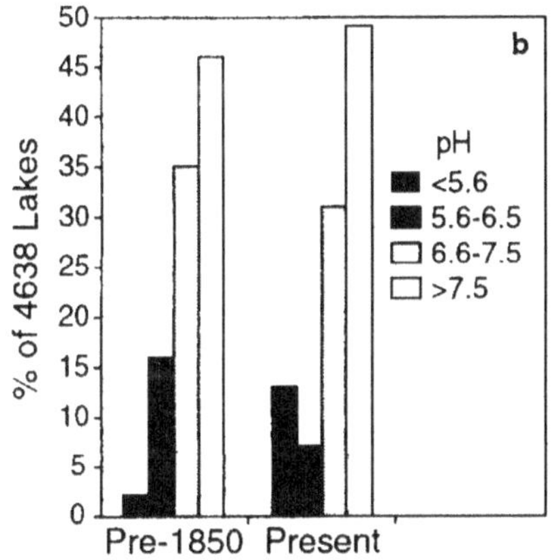

Figure 5. Changes in (a) lake chloride concentration and (b) lake pH as inferred from sediment diatoms taken from northeast U.S.A. lakes (from Dixit et al., 1999). Lake chloride concentrations > 200 μeq/L represent excessive landscape disturbance, while concentrations < 100 μeq/L represent background levels in this region. Both acidic (pH <5.6) and high pH (pH >7.5), another general disturbance indicator, lakes have increased since pre-industrial times.

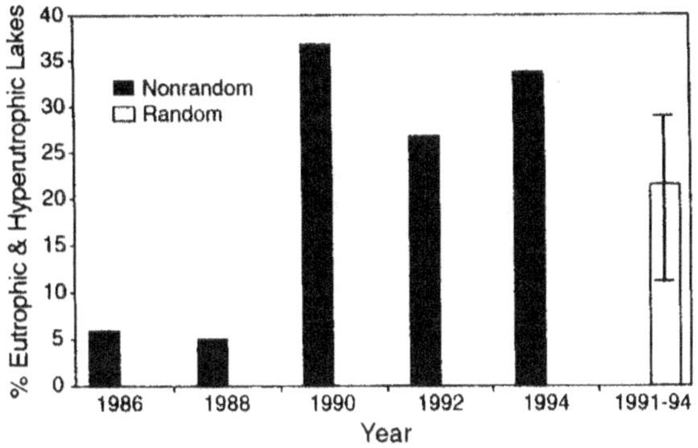

Figure 6. Percent of lakes estimated as eutrophic and hypereutrophic from nonrandom (305b) and random (EMAP) sampling designs (from Paulsen et al., 1998). Note that the random design also allows calculation of confidence bounds.

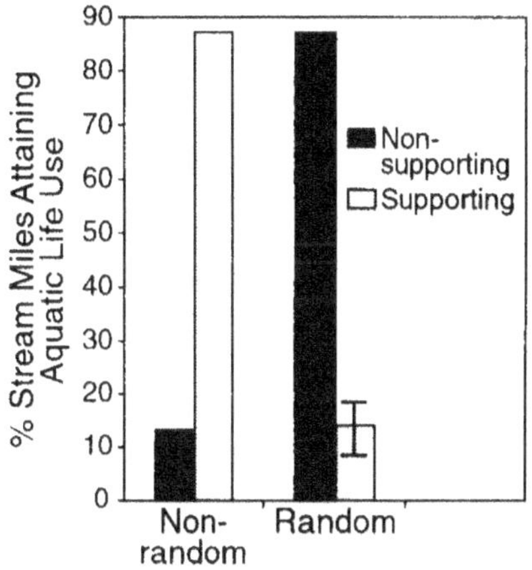

Figure 7. Percent of Delaware stream miles attaining aquatic life uses as predicted from random macrobenthos surveys versus nonrandom chemical samples (from Maxted, 1997; Paulsen et al., 1998).

fish tissue that placed piscivorous mammals and birds, respectively, at risk (Figure 4).

Since 1850, the percent of acidic lakes (pH < 5.6) in the northeast U.S.A. has increased by 11% and the percent of highly disturbed lakes (chloride concentration >200 μeq/L) has increased by 12% with a standard error of 6.8% (Figure 5; Dixit et al., 1999). The percent of acidic lakes is highest in the Adirondack (23%, 222 lakes) and Coastal Lowland & Plateau (18%, 200 lakes) regions.

Paulsen et al. (1998) found that the percent of eutrophic and hypereutrophic lakes in the northeast U.S.A. varied markedly among years when sites were subjectively selected, but patterns were much more consistent with a probability design (Figure 6). In only one year out of five was the nonrandom design comparable to the estimate and confidence bounds of the 4-year average obtained in a probability survey (23 ± 10% eutrophic and hypereutrophic lakes).

Probability designs have also proven useful at the state level for assessing river conditions. Maxted (1997) compared two assessments of aquatic life in Delaware streams. One that relied on a nonrandom survey and chemical indicators suggested that 87% of assessed Delaware streams fully supported aquatic life uses. In contrast, another employing a random survey and benthic macroinvertebrates indicated only 13 ± 5% fully supported aquatic life uses (Figure 7). This contradiction likely resulted from differences in indicators and sampling design.

The value of a random sampling design over a nonrandom approach was revealed in surveys of coastal coho salmon (*Oncorhynchus kisutch*) in Oregon, where the nonrandom sites led to a 3–5 fold overestimate (Jacobs & Cooney 1995; Figure 8). Soon

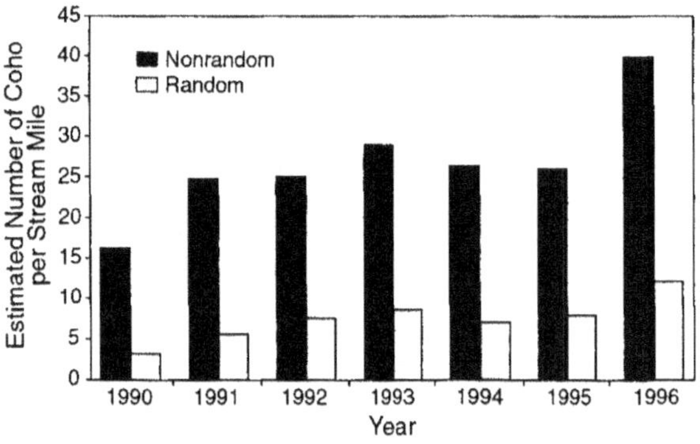

Figure 8. Coho salmon escapement in Oregon coastal streams as predicted from random and nonrandom sampling (from Jacobs & Cooney, 1995; Paulsen et al., 1998).

after coho population sizes were estimated from the probability sample, Oregon's coastal coho salmon were listed as threatened and coastal coho fisheries were closed. As indicated above, appropriate biological indicators – when coupled with a random sampling design – enable aquatic scientists to provide accurate, precise and unbiased assessments of biological conditions, along with quantitative estimates of sampling uncertainty.

Regional-scale biogeographic research

A probability design employing multi-assemblage indicators is especially powerful for conducting biogeographic research across large spatial scales. Although gradient-based designs are commonly used for such purposes, there are a large number of possible limiting factors to consider and the choice of factor gradients necessarily influences the effects observed. For example, a previously unsuspected scarcity of native minnow species was discovered in northeast U.S.A. lakes. The absence of minnows was associated with increased littoral piscivores, particularly bass (*Micropterus* spp.), and anthropogenic changes in catchments and nearshore areas (Figure 9; Whittier et al., 1997).

Allen et al. (1999b) concluded that anthropogenic factors were less important than natural factors for determining the richness of bird, fish, benthos, zooplankton and diatom assemblages (Table 1). As might be expected, but not commonly accepted, they observed only two, out of a possible 10, significant correlations in taxonomic richness for the five assemblages. These were between benthos and diatoms ($r = 0.37$) and zooplankton ($r = 0.31$). Although all but diatom richness were associated with lake area, the richness regression variables for each assemblage differed from the others

Table 1. Regression variables associated with species richness. Variance includes that for that variable plus those preceding it in the model (from Allen et al., 1999b)

Model	Variable	Variance
Benthos	Lake area	0.15
	Macrophytes	0.2
	Minimum temperature	0.25
	Minimum dissolved oxygen	0.34
Birds	Lake area	0.25
	Seasonal temperature range	0.29
Diatoms	pH	0.16
	pH^2	0.27
	Lake depth	0.37
	Shoreline human activity	0.39
Fish	Lake area	0.30
	Human population density	0.39
	Shoreline human activity^2	0.40
	Shoreline human activity	0.42
Zooplankton	Lake area	0.28
	pH^2	0.32
	Minimum dissolved oxygen	0.37
	Shoreline human activity	0.39
Richness PC1	Lake area	0.41
	Macrophytes	0.47
	pH^2	0.54
	Minimum temperature	0.58

(Table 1). That is, each assemblage responded to a different set of environmental variables. These regression results indicate the potential shortcomings of using species richness, as well as any single assemblage, when assessing the condition of an ecosystem. Others have also found fish species richness most highly correlated with natural variables when examined at large spatial scales (Oberdorff et al., 1995; Rathert et al., 1999). The fact that Allen et al. did not find consistent responses between human disturbances and species richness of various assemblages suggests that protecting diversity in one aquatic assemblage may not guarantee it will be protected in others. It is also possible that the indicators were too crude to detect some relationships.

Examining five biological assemblages of lakes through use of multiple canonical correspondence ana-

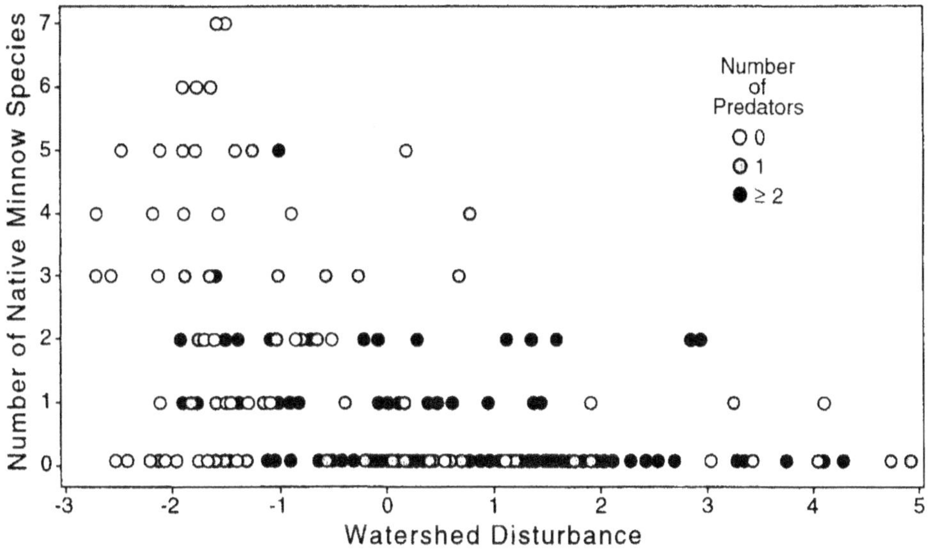

Figure 9. Minnow species richness as a function of human disturbance and littoral piscivorous fish (from Whittier et al., 1997).

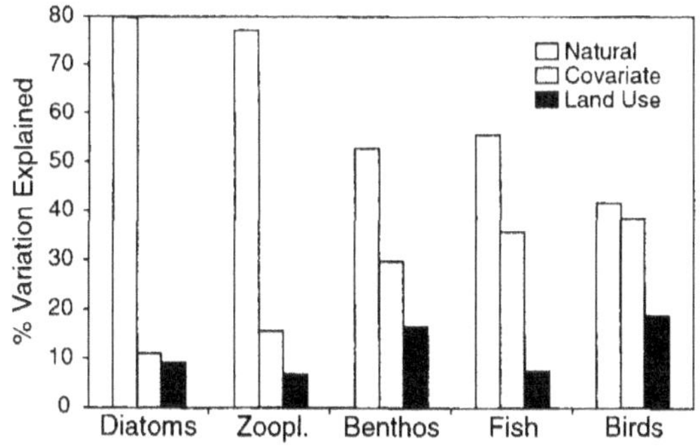

Figure 10. Variation explained by natural lake variables, covariate variables and land use variables for five biological assemblages of northeast U.S.A. lakes (from Allen et al., 1999a). Natural variables: climate, forest type, elevation, depth, area, minimum temperature, littoral substrate, aluminum, dissolved organic carbon and silica; covariate variables: calcium, chloride, chlorophyll-*a*, macrophyte cover, minimum dissolved oxygen, nitrogen, pH, phosphorus, sulphate, turbidity; land use variables: human density, % urban, % residential, % agriculture, road density, point source pollution density.

lyses, Allen et al. (1999a) detected two patterns possibly relevant to EC river assessments (Figure 10). First, variations in species composition corresponded most strongly with natural variables (lake area, lake depth, elevation, shoreline substrate,% wetland), *versus* land use variables (% agriculture,% urban, road density, human density, point source pollution) or natural variables co-varying with land use (climate, forest type, water quality, macrophyte density). Second, diatoms and zooplankton (small-bodied organisms) responded much more strongly than large bodied organisms to natural variables. This further supports the notion that large and small organisms use habitats differently and indicates the value of using both sizes when attempting rigorous community assessments. In addition, strong associations between biological assemblages and variables co-varying with land use may lead to overestimating the effects of either, and to erroneous conclusions about the ecological mechanisms involved. Separating the variation shared by land use and its natural co-variates requires long term monitoring, historical information, paleoecological research, or perhaps different analytical approaches and indicators.

Integrating multi-assemblage assessments

The papers discussed in the preceding two sections approached community analysis by comparing species richness and composition of individual assemblages separately. Another approach is to combine the community information into a single multivariate or multimetric index. The biological variables or metrics used are typically based on taxonomic, trophic, habitat and reproductive guilds, as well as abundance, evenness, life history and condition. For example, Moyle and Marchetti (1998) included both fish and amphibian metrics in their biointegrity index for California streams. O'Connor et al. (2000) applied principal components analysis (PCA) to reduce 29 biological metrics from five biological assemblages (birds, fish, benthos, zooplankton, diatoms) into a single community indicator (biological PCA-1). They also used

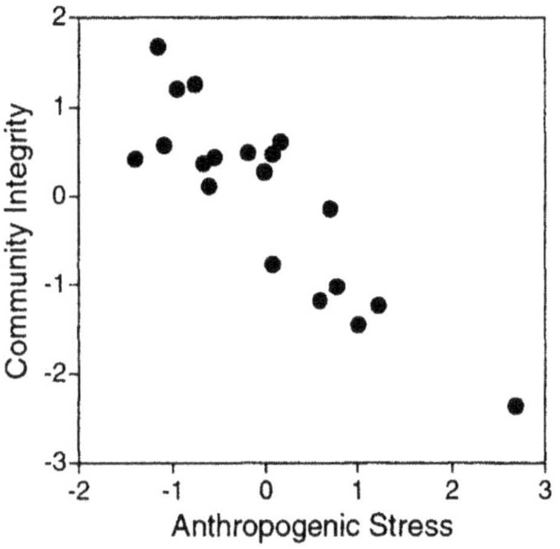

Figure 11. Community integrity versus anthropogenic stress (from O'Connor et al., 2000). Community integrity was indicated by PCA-1 scores of 29 biological metrics from five assemblages; anthropogenic stress was indexed by PCA-1 scores of 17 disturbance metrics. Spearman rank correlation is −0.78, $P<0.001$.

PCA to convert 17 land use and disturbance metrics into one anthropogenic stressor indicator (environmental PCA-1). Each assemblage but zooplankton had one or more metrics loading strongly on biological PCA-1, which had a strong negative correlation with environmental PCA-1 (Figure 11). The results from O'Connor et al. suggest that multivariate and multimetric indices integrating several biological variables, as opposed to species richness and composition, are more likely to reveal relationships between biological communities or assemblages and anthropogenic disturbances. This has also been an approach favored by Karr (1991) and his colleagues for a single assemblage.

Table 2. Key variables associated with diatom assemblage patterns at regional and ecoregional scales (from Pan et al., 1999). *r* is correlation with CCA axis 1

Region or Ecoregion	Variable	*r*
Mid-Atlantic Highlands Region	% forest	−0.68
	Incidence of riparian agriculture	0.41
	% stream bedrock	−0.33
	Chloride concentration	0.08
Valley Ecoregion	Channel slope	0.65
	% boulder	0.63
	Riparian vegetation cover	0.21
	Stream sinuosity	−0.02
Central Plateau Ecoregion	Turbidity	0.65
	Riparian disturbance	0.62
	Catchment area	−0.41
	Dissolved organic carbon	0.01
Western Plateau Ecoregion	% boulder	0.58
	Sulphate	−0.51
	Stream sinuosity	−0.35
	Stream slope	−0.06

Effect of spatial scale on indicators

A probability design coupled with multiple ecological indicators is particularly well suited for determining associations between various environmental stressors and various measures of biological response at different spatial scales. In a probability survey of 89 streams in the Mid-Atlantic Highlands, Pan et al. (1999; Table 2) reported that diatom assemblages were more strongly associated with land use indicators than with natural or natural/land use variables. However, when patterns were evaluated at the level of individual ecoregions, natural variables increased in importance.

Evaluating the same set of streams as Pan et al., Bryce et al. (1999) concluded that 60% of the total stream length was at high risk of disturbance (Figure 12). Within individual ecoregions, risks varied; 75% of the total stream length of the Western Appalachian Plateau was at high risk from mining, while 95% of the total stream length in the valleys was at high risk from agriculture. Higher risk index scores, calculated from landscape and riparian variables, were associated with increased nutrient richness, greater ionic strength and lowered relative abundances of mayflies, stoneflies and caddisflies.

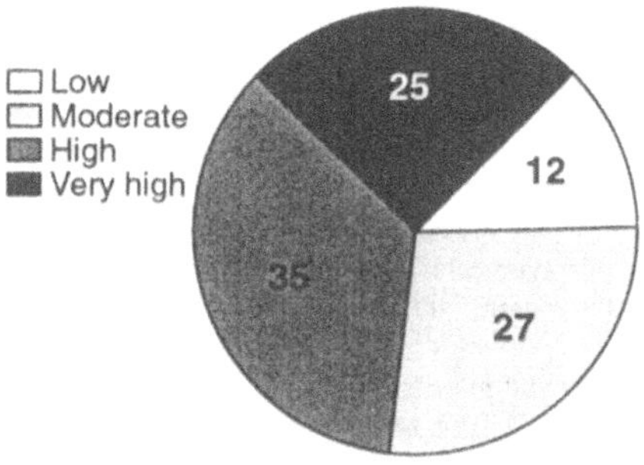

Figure 12. Percentages of Appalachian streams experiencing various levels of disturbance from human activities (from Bryce et al., 1999). Disturbances included mines, petroleum wells, roads, utility lines, residential development, logging, agriculture, livestock grazing, channelization, trash and point source pollution.

Understanding these associations between stressors and biological responses aids us in quantifying the relative importance of different stressors in different regions or at different geographic scales. This in turn should help managers and society to focus their restoration and mitigation efforts where they are likely to be most effective.

Reference condition

Bryce et al. (1999) found that only 14 of 102 randomly selected Appalachian streams qualified as minimally disturbed regional reference sites (Figure 12). Given that the sample included both large and small streams in three markedly different ecoregions, or a total of six classes (2 sizes × 3 ecoregions), this produces an average of two sites per class. This is too few to allow a reasonable assessment of their variances. A set of handpicked reference sites in that region was found to be more disturbed than those selected by Bryce et al. This further indicates the potential shortcomings of using regional reference sites in extensively disturbed regions.

In the intensively farmed Willamette Valley, Oregon, Hughes et al. (1998) also found few minimally disturbed catchments and therefore used historical land cover and fish distribution data to develop reference conditions for small streams. Using historical conditions as the reference, two of the 25 sites sampled were in comparable condition to the reference, whereas four could be considered comparable based on comparison with the least disturbed site in the sample. Where historical distribution information is fairly accurate, as with fish, it is becoming an increasingly powerful means to estimate potential species richness and status changes in individual species and to correlate those changes with environmental variables (Thurow et al., 1997; Rathert et al., 1999). Of course, this approach is limited to assemblage assessments for which historical distribution data exist.

EMAP has used paleolimnological data for determining reference conditions for lakes to assess changes in chemical conditions and diatom species richness (Dixit & Smol, 1994; Dixit et al., 1999; Figure 5). Its use in rivers appears largely restricted to oxbow lakes, sloughs or other habitat types where sediments may accumulate for hundreds of years. In lake rich regions, however, paleolimnological data are very useful for indicating the degree of regional chemical and physical change that has resulted from human land uses (Dixit et al., 1992; Smol, 1992).

Discussion

There have been numerous basin, state and regional EMAP pilot studies, and additional surveys are planned. For example, a national lake survey of fish tissue contaminants using the EMAP design and an EMAP stream and river survey of 12 conterminous western U.S.A. states began in 1999. At least 11 states (Alabama, Delaware, Indiana, Kentucky, Maryland, Minnesota, Nebraska, Oregon, South Carolina, Virginia, West Virginia) have incorporated probability surveys in their current stream monitoring programs; Florida, Kansas and Missouri are planning to do so. Other state and national agencies, however, have been slow to adopt long-term, statistically-designed and ecologically-rigorous monitoring of aquatic ecosystems – despite known shortcomings of current programs and existing data bases (Yoder & Rankin, 1998; Paulsen et al., 1998). There are many possible reasons why more agencies have not adopted such a monitoring program, but one key reason is an emphasis on specific water bodies. The State monitoring programs, like most programs, have limited resources and have used those resources to focus their efforts on water resource problems of which they are currently aware. Current lawsuits over lists of impaired waters (303d list) have resulted in an increased emphasis by States on specific water bodies, rather than on comprehensive assessment of all waters. Yet, PEER (1999) criticized USEPA and state monitoring programs for failing to accurately characterize the health of the nation's waters despite being required to do so since 1975 by the Clean Water Act.

In its assessments, EMAP detected fewer than expected fish species (Whittier et al., 1997; Figure 9), increases in highly disturbed lakes (Dixit et al., 1999; Figure 5) and mercury concentrations in fish tissue exceeding consumption criteria (Yeardley et al., 1998; Figure 4). These occurred across a multi-state region and had not been reported previously for the northeast U.S.A. In the case of fish species richness, the declines were observed only after natural variables were factored out, because at regional scales natural variation is often greater than, and co-varying with, anthropogenic stressors.

EMAP studies have also indicated the usefulness of monitoring multiple biological assemblages, as well as physical, chemical and landscape level indicators. As reported for lakes (Allen et al., 1999a; Figure 10), different assemblages respond differently to stressors. Croonquist & Brooks (1991) and Harding et al. (1998)

found similar differences between riparian birds and mammals and stream fish and invertebrates, respectively. This indicates that if assessment of ecological integrity is the concern, then multiple assemblages should be monitored.

Our work has demonstrated the value of a systematic random sample of sites (*versus* nonrandom site selection) for making estimates about entire populations of lakes and streams (Paulsen et al., 1998; Figures 6 and 8). Interestingly, the random designs have also revealed higher proportions of impaired waters than were suggested from non-randomly selected sites. Moreover, random designs can report information with known confidence bounds, an impossibility for non-random designs other than complete censuses.

Paulsen et al. (1998; Figure 7) indicated that biological indicators tend to reveal problems undetected or poorly characterized by physical and chemical indicators alone, although the indicator differences in this case were confounded by differences in sampling design. However, Chovanec et al. (1995), Karr (1991) and Yoder and Rankin (1998) have demonstrated the importance of using biological indicators for assessing biological integrity. Biological indicators are important largely because of the ability of resident biota to integrate temporal and spatial variations in many more abiotic stressors than we can monitor directly.

The usefulness of historical information for determining reference conditions was reported. Minimally disturbed reference sites were infrequently encountered in Appalachia and the Pacific Northwest – two mostly forested regions where the opposite might be expected (Hughes et al., 1998; Bryce et al., 1999; Figure 12). In areas long settled by humans or where agriculture and urbanization are extensive and intensive, this situation will be even more common. Historical, including paleoecological, information offers a proven and powerful complement to reference sites for estimating potential ecological integrity (Hughes et al., 1998; Dixit et al., 1999; Muhar et al., 2000), although such conditions are no longer likely to be attainable without major changes in human abundance and consumption. There is growing interest among states in the U.S.A. for using minimally-disturbed regional reference sites (even when these are quite disturbed), as opposed to historical conditions, to develop biological criteria and assess ecological condition. It would be very useful to compare the conclusions drawn from these two different approaches and to consider when and how to use one or the other or both.

To our knowledge, EMAP scientists have been the first to synthesize five biological assemblages (birds, fish, benthos, zooplankton and diatoms) into a single biological indicator of ecosystem integrity (O'Connor et al., 2000; Figure 11). Admittedly their index of community integrity was a crude product from a scientific perspective, but as Karr (1991) and Yoder and Rankin (1998) argued, it is useful to distill large amounts of complex biological data into a single index for the public and politicians. When reporting on ecological integrity, it seems wise to mimic economists and physicians, who integrate multiple indicators when they report on economic or individual health.

Conclusions

EMAP pilot surveys have demonstrated the value of probability designs, quantitative assessments of multiple biological assemblages, a suite of quantitative habitat indicators and the use of historical information. Field protocols have been developed and tested for collecting large amounts of pertinent physical, chemical and biological data by 3–4 persons in one day over a 3 month field season (Baker et al., 1997; Lazorchak et al., 1998; Averill & Peck 1999). Initially, data management and analysis and report writing required comparable amounts of labor–but this investment should decline once those processes are refined. In other words, monitoring and assessing regional ecological integrity are non-trivial issues, but they can be done and they do produce useful information for science and society.

Several scientific advisory groups have called for a national program to monitor surface waters in the U.S.A. (Paulsen et al., 1998). Such a program could cost \$5–10 million per year to assess all mapped U.S.A. streams and rivers by sampling 1000 reaches annually. This may sound like a considerable sum to biologists, but it equals what the U.S.A. government spent to subsidize U.S.A. popcorn and almond industries to advertise abroad in 1997. In addition, perhaps \$50–100 million are needed to bolster state aquatic monitoring programs (< \$0.5 per capita) for assessing local problems. An EMAP approach seems applicable for quantitatively assessing ecological status and trends at scales ranging from basins to continents. Based on the monitoring and assessment objectives outlined by the European Community (Commission of the European Communities, 1994), we conclude that an EMAP approach to sampling design and indicators (recognizing natural ecoregional differences) might be one method for meeting their needs.

Acknowledgements

The research results summarized herein were supported by the USEPA's Environmental Monitoring and Assessment Program and partially funded through contract 68-C6-0005 to Dynamac. An earlier draft received EPA's peer and administrative reviews and was cleared for publication. The draft benefited from critical reviews by J. P. Baker, R. T. Lackey, D. H. Landers, D. P. Larsen, G. E. Petts, C. O. Yoder and three anonymous reviewers. Suzanne Pierson ably prepared the final figures. We are especially grateful for the help and encouragement of Birgit Vogel during all phases of manuscript development.

References

Allan, J. D. & A. S. Flecker, 1993. Biodiversity conservation in running waters. BioScience 43: 32–43.

Allen, A. P., T. R. Whittier, P. R. Kaufmann, D. P. Larsen, R. J. O'Connor, R. M. Hughes, R. S. Stemberger, S. S. Dixit, R. O. Brinkhurst, A. T. Herlihy & S. G. Paulsen, 1999a. Concordance of taxonomic composition patterns across multiple lake assemblages: effects of scale, body size, and land use. Can. J. Fish. aquat. Sci. 56: 2029–2040.

Allen, A. P., T. R. Whittier, P. R. Kaufmann, D. P. Larsen, R. J. O'Connor, R. M. Hughes, R. S. Stemberger, S. S. Dixit, R. O. Brinkhurst, A. T. Herlihy & S. G. Paulsen, 1999b. Concordance of taxonomic richness patterns across multiple assemblages in lakes of the northeastern U.S.A. Can. J. Fish. aquat. Sci. 56: 739–747.

Averill, D. K. & D. V. Peck (eds), 1999. Field Operations and Methods for Measuring the Ecological Condition of Non-wadeable Rivers in Oregon. U. S. Environmental Protection Agency, Corvallis, OR: 219 pp.

Baker, J. R., D. V. Peck & D. W. Sutton, 1997. Field operations manual for lakes, EPA/620/R-97/001. U.S. Environmental Protection Agency, Corvallis, OR: 241 pp.

Baker, L. A., A. T. Herlihy, P. R. Kaufmann & J. M. Eilers, 1991. Acidic lakes and streams in the United States: the role of acidic deposition. Science 252: 1151–1154.

Bryce, S. A., D. P. Larsen, R. M. Hughes & P. R. Kaufmann, 1999. Assessing relative risks to aquatic ecosystems: a mid-Appalachian case study. J. Am. Wat. Res. Assoc. 35: 23–36.

Chovanec, A., V. Koller-Kreimel, O. Moog & S. Weiss, 1995. Assessment of the ecological integrity of running waters: the Austrian approach. Proceedings of the International Workshop on Assessment and Classification of Rivers, Düsseldorf, Germany.

Commission of the European Communities, 1994. Proposal for a council directive on the ecological quality of water. COM (93) 680 final, 15.06.1994, Office for Official Publications of the European Communities, Luxembourg.

Croonquist, M. J., & R. P. Brooks, 1991. Use of avian and mammalian guilds as indicators of cumulative impacts in riparian-wetland areas. Environ. Manage. 15: 701–714.

Dixit, S. S. & J. P. Smol, 1994. Diatoms as indicators in the Environmental Monitoring and Assessment Program–Surface Waters (EMAP-SW). Environ. Monitor. Assess. 31: 275–306.

Dixit, S. S., J. P. Smol, J. C. Kingston & D. F. Charles, 1992. Diatoms: powerful indicators of environmental change. Environ. Sci. Technol. 26: 22–33.

Dixit, S. S., J. P. Smol, D. F. Charles, R. M. Hughes, S. G. Paulsen & G. B. Collins, 1999. Assessing water quality changes in the lakes of the northeastern United States using sediment diatoms. Can. J. Fish. aquat. Sci. 56: 131–152.

Eilers, J. M., D. F. Brakke, D. H. Landers, and P. E. Kellar, 1988. Characteristics of lakes in mountainous areas of the western United States. Proceed. Internat. Assoc. Theo. Appl. Limnol. 23: 144–151.

EMAP (Environmental Monitoring and Assessment Program), 1997. Research Plan 1997. U.S. Environmental Protection Agency, Washington, DC: 138 pp.

Fausch, K. D., J. R. Karr & P. R. Yant, 1984. Regional application of an index of biotic integrity based on stream fish communities. Trans. Am. Fish. Soc. 113: 39–55.

Fausch, K. D., J. Lyons, J. R. Karr & P. L. Angermeier, 1990. Fish communities as indicators of environmental degradation. In Adams, S. M. (ed.), Biological Indicators of Stress in Fish. American Fisheries Society Symposium, Bethesda, MD 8: 123–144.

Harding, J. S., E. F. Benfield, P. V. Bolstad, G. S. Helfman & E. B. D. Jones III, 1998. Stream biodiversity: the ghost of land use past. Proc. natn. Acad. Sci. U.S.A. 95: 14843–14847.

Hren, J., C. J. Oblinger Childress, J. M. Norris, T. H. Chaney & D. N. Meyers, 1990. Regional water quality: evaluation of data for assessing conditions and trends. Envir. Sci. Technol. 24: 1122–1127.

Hughes, R. M., 1995. Defining acceptable biological status by comparing with reference conditions. In Davis, W. S. & T. P. Simon (eds), Biological Assessment and Criteria: Tools for Water Resource Planning and Decision Making. Lewis Publishers, Boca Raton, FL: 31–47.

Hughes, R. M., 1997. Do we need institutional change? In Stouder, D. J., P. A. Bisson & R. J. Naiman (eds), Pacific Salmon & Their Ecosystems: Status and Future Options. Chapman & Hall, New York: 559–568.

Hughes, R. M., 1999. Conservation of natural resources. In Alexander, D. E. & R. W. Fairbridge (eds), Encyclopedia of Environmental Science. Kluwer Academic Publishers, Dordrecht, NL: 90–95.

Hughes, R. M. & R. F. Noss, 1992. Biological diversity and biological integrity: current concerns for lakes and streams. Fisheries 17(3): 11–19.

Hughes, R. M., P. R. Kaufmann, A. T. Herlihy, T. M. Kincaid, L. Reynolds & D. P. Larsen, 1998. A process for developing and evaluating indices of fish assemblage integrity. Can. J. Fish. aquat. Sci. 55: 1618–1631.

Jacobs, S. E. & C. X. Cooney, 1995. Improvement of methods used to estimate the spawning escapement of Oregon coastal natural coho salmon. Oregon Department of Fish & Wildlife, Portland, OR: 24 pp.

Judy, R. D. Jr., P. N. Seeley, T. M. Murray, S. C. Svirsky, M. R. Whitworth & L. S. Ischinger, 1984. 1982 national fisheries survey. Volume I technical report: initial findings. U.S. Fish and Wildlife Service, Washington, DC: 140 pp.

Karr, J. R., 1991. Biological integrity: a long-neglected aspect of water resource management. Ecol. Appl. 1: 66–84.

Karr, J. R., R. C. Heidinger, and E. H. Helmer, 1985. Effects of chlorine and ammonia from wastewater treatment facilities on biotic integrity. J. Wat. Pollut. Cont. Fed. 57: 912–915.

Kaufmann, P. R., A. T. Herlihy, M. E. Mitch, J. J. Messer, and W. S. Overton, 1991. Stream chemistry in the eastern United States: synoptic survey design, acid-base status and regional patterns. Wat. Res. Resear. 27: 611–627.

Landers, D. H., W. S. Overton, R. A. Linthurst & D. F. Brakke, 1988. Eastern lake survey regional estimates of lake chemistry. Envir. Sci. Technol. 22: 128–135.

Larsen, D. P., N. S. Urquhart & D. L. Kugler. 1995. Regional scale trend monitoring of indicators of trophic condition of lakes. Wat Res. Bull. 31: 117–140.

Lazorchak, J. M., D. J. Klemm & D. V. Peck, 1998. Field operations and methods for measuring the ecological condition of wadeable streams, EPA/620/R-94/004. U.S. Environmental Protection Agency, Cincinnati, OH: 270 pp.

Marsh, G. P., 1885. The Earth as Modified by Human Action. Charles Scribner's Sons, New York: 629 pp.

Maxted, J. R., 1997. Biology, probability and the obvious. Human and Ecological Risk Assessment 3: 955–965.

Messer. J. J., R. A. Linthurst, & W. S. Overton, 1991. An EPA program for monitoring ecological status and trends. Envir. Monitor. Assess. 17: 67–78.

Miller, R. R., J. D. Williams & J. E. Williams, 1989. Extinctions of North American fishes during the past century. Fisheries 14(6): 22–38.

Moyle, P. B. & M. P. Marchetti, 1998. Applications of indices of biotic integrity to California streams and watersheds. In Simon, T. P. (ed.), Assessing the Sustainability and Biological Integrity of Water Resources Using Fish Communities. Lewis, Boca Raton, FL: 367–380.

Muhar, S., M. Schwarz, S. Schmutz & M. Jungwirth, 2000. Identification of rivers with high and good habitat quality: methodological approach and applications for Austria. Hydrobiologia 422/423: 343–358.

Oberdorff, T., J. Guegan & B. Hugueny, 1995. Global scale patterns of fish species richness in rivers. Ecography 18: 345–352.

O'Connor, R. J., T. E. Walls & R. M. Hughes, 2000. Using multiple taxonomic groups to index the ecological condition of lakes. Environ. Monitor. Assess. 61: 207–228.

Olsen, A. R., D. L. Stevens Jr. & D. White, 1998. Application of global grids in environmental sampling. In Weisberg, S. (ed.), Computing Science and Statistics, volume 30. Interface Foundation of North America, Fairfax Station, VA: 279–284.

Overton, W. S., D. White, and D. L. Stevens Jr., 1991. Design report for EMAP, the environmental monitoring and assessment program. EPA/600/3-91/053, U.S. Environmental Protection Agency, Washington, DC: 43 pp.

Pan, Y., R. J. Stevenson, B. H. Hill, P. R. Kaufmann & A. T. Herlihy, 1999. Spatial patterns and ecological determinants of benthic algal assemblages in mid-Atlantic Highlands streams. J. Phycol. 35: 460–468.

Paulsen, S. G. & R. A. Linthurst, 1993. Biological Monitoring in the Environmental Monitoring and Assessment Program. In Loeb, S. L. & A. Spacie (eds), Biological Monitoring of Aquatic Systems. Lewis, Boca Raton, FL: 297–322.

Paulsen, S. G., D. P. Larsen, P. R. Kaufmann, T. R. Whittier, J. R. Baker, D. V. Peck, J. McGue, R. M. Hughes, D. McMullen, D. Stevens, J. L. Stoddard, J. Lazorchak, W. Kinney, A. R. Selle & R. Hjort, 1991. EMAP-Surface Waters monitoring and research strategy: fiscal year 1991. EPA/600/3-91/022, U.S. Environmental Protection Agency, Corvallis, OR: 184 pp.

Paulsen, S. G., R. M. Hughes & D. P. Larsen, 1998. Critical elements in describing and understanding our nation's aquatic resources. J. Am. Wat. Res. Assoc. 34: 995–1005.

PEER, 1999. Murky waters. Public Employees for Environmental Responsibility, Washington, DC: 44 pp.

Petts, G. E., 1989. Historical analysis of fluvial hydrosystems. In Petts, G. E., H. Moller & A. L. Roux (eds), Historical Change of Large Alluvial Rivers: Western Europe. John Wiley & Sons, New York, NY: 1–17.

Petts, G. E., H. Moller & A. L. Roux (eds), 1989. Historical Change of Large Alluvial Rivers: Western Europe. John Wiley & Sons, New York, NY.

Rathert, D., D. White, J. C. Sifneos & R. M. Hughes, 1999. Environmental correlates of species richness for native freshwater fish in Oregon, U.S.A. J. Biogeogr. 26: 257–273.

Raven, P. J., N. T. H. Holmes, F. H. Dawson, P. J. A. Fox, M. Everard, I. R. Fozzard & K. L. Rouen. 1998. River habitat quality: the physical character of rivers and streams in the U.K. and Isle of Man. River Habitat Survey Report No. 2. Environment Agency, Bristol, England: 86 pp.

Sahr, K., & D. White, 1998. Discrete global grid systems. In Weisberg, S. (ed.), Computing Science and Statistics, volume 30. Interface Foundation of North America, Fairfax Station, VA: 269–278.

Schwaiger, K., J. Grath & A. Chovanec, 1994. Information needs for water quality: the Austrian context. In Advances in Water Quality Monitoring, Technical Reports in Hydrology & Water Resources No. 42. World Meteorological Organization, Geneva, Switzerland: 105–118.

Science Advisory Board, 1988. Future risk: research strategies of the 1990s. SAB-EC-88-040. U.S. Environmental Protection Agency, Washington, DC: 221 pp.

Science Advisory Board, 1990. Reducing risk: setting priorities and strategies for environmental protection. SAB-EC-90-021. U.S. Environmental Protection Agency, Washington, DC: 425 pp.

Smol, J.P., 1992. Paleolimnology: an important tool for effective ecosystem management. J. Aquat. Ecosyst. Health 1: 49–58.

Statzner, B., V. H. Resh & A. L. Roux, 1994. The synthesis of long-term ecological research in the context of concurrently developed ecological theory: design of a research strategy for the Upper Rhone River and its alluvial floodplain habitats. Freshwa. Biol. 31: 253–263.

Stevens, D. L. Jr., 1994. Implementation of a national monitoring program. J. environ. Manage. 42: 1–29.

Stevens, D. L. Jr., 1997. Variable density grid-based sampling designs for continuous spatial populations. Environmetrics 8: 167–195.

Stevens, D. L. Jr. & A. R. Olsen, 1999. Spatially restricted surveys over time for aquatic resources. J. Agri. Biol. Environ. Statistics 4: 415–428.

Thurow, R. F., D. C. Lee & B. E. Rieman, 1997. Distribution and status of seven native salmonids in the interior Columbia River basin and portions of the Klamath River and Great Basins. N. am. J. Fish. Mgmt. 17: 1094–1110.

Urquhart, N. S., W. S. Overton, and D. S. Birks, 1993. Comparing sampling designs for monitoring ecological status and trends: impact of temporal patterns. In Barnett, V. (ed.), Statistics in the Environment. Wiley & Sons, Sussex, England: 71–85.

Urquhart, N. S., S. G. Paulsen, and D. P. Larsen, 1998. Monitoring for policy-relevant regional trends over time. Ecol. Appl. 8: 246–257.

Van Sickle, J. & R. M. Hughes, 2000. Classification strengths of ecoregions, catchments and geographic clusters for aquatic vertebrates in Oregon. J. n. am. Benthol. Soc. 19.

White, D., A. J. Kimerling & W. S. Overton, 1992. Cartographic and geometric components of a global sampling design for environmental monitoring. Cart. Geogr. Info. Syst. 19: 5–22.

Whittier, T. R., D. B. Halliwell & S. G. Paulsen, 1997. Cyprinid distributions in northeast U.S.A. lakes: evidence of regional-scale minnow biodiversity losses. Can. J. Fish. aquat. Sci. 54: 1593–1607.

Wiederholm, T. & R. K. Johnson, 1997. Monitoring and assessment of lakes and watercourses in Sweden. In Ottens, J. J., F. A. M. Claessen, P. G. Stoks, J. G. Timmerman & R. C. Ward (eds), Monitoring Tailor Made II: Information Strategies in Water Management. Elsevier, New York, NY: 317–329.

Williams, J. E., J. E. Johnson, D. A. Hendrickson, S. Contreras-Balderas, J. D. Williams, M. Navarro-Mendoza, D. E. McAllister & J. E. Deacon, 1989. Fishes of North America endangered, threatened or of special concern: 1989. Fisheries 14(6): 2–20.

Wright, J. F., D. Moss, P. D. Armitage & M. T. Furse, 1984. A preliminary classification of running-water sites in Great Britain based on macroinvertebrate species and the prediction of community type using environmental data. Freshwat. Biol. 14: 221–256.

Yeardley, R. B. Jr., J. M. Lazorchak & S. G. Paulsen, 1998. Elemental fish tissue contamination in northeastern U.S. lakes: evaluation of an approach to regional assessment. Envir. Toxicol. Chem. 17: 1875–1884.

Yoder, C. O. & E. T. Rankin, 1998. The role of biological indicators in a state water quality management process. Envir. Monitor. Assess. 51: 61–88.

Hydrobiologia **422/423:** 445–452, 2000.
M. Jungwirth, S. Muhar & S. Schmutz (eds), Assessing the Ecological Integrity of Running Waters.

The Austrian way of assessing the ecological integrity of running waters: a contribution to the EU Water Framework Directive

A. Chovanec[1], P. Jäger[2], M. Jungwirth[3], V. Koller-Kreimel[4], O. Moog[3], S. Muhar[3] & St. Schmutz[3]
[1]*Federal Environment Agency, Spittelauer Laende 5, A-1090 Vienna, Austria*
[2]*Office of the Government of Salzburg, Ulrich Schreier-Strasse 18, PO Box 527, A-5010 Salzburg, Austria*
[3]*Department of Hydrobiology, Fisheries and Aquaculture, University of Agricultural Sciences, Max Emanuel-Straße 17, A-1180 Vienna, Austria*
[4]*Federal Ministry of Agriculture and Forestry, Marxergasse 2, A-1030 Vienna, Austria*

Key words: assessment, ecological integrity, concept, Water Framework Directive

Abstract

The term 'Oekologische Funktionsfaehigkeit', synonymously used as ecological integrity, has become a pivotal word within both Austrian water management and limnological research. Incorporated in the most important laws concerning water management issues, the term ecological integrity reflects the necessity of considering water bodies as ecological systems. Although ecological integrity of running waters is and must remain a holistic concept, the current Austrian methodology aims at assessing it by investigating a number of key components: hydro-morphological aspects (particularly habitat structures, flow regime, connectivity), physico-chemical parameters, macroinvertebrate and fish assemblages, biological water quality assessment and ecotoxicological evaluations. The classification is based on the assessment of these individual criteria by means of a comparison between a river-type-specific reference state and the current conditions. This approach, which has been laid down in the Austrian Standard M 6232 'Guidelines for the ecological study and assessment of rivers', also meets the general requirements of the classification of the ecological status of running waters discussed in the EU Water Framework Directive. Besides describing the Austrian way of assessing the ecological integrity of running waters, some nation-wide programmes concerning water quality, habitat assessment, and river typology are briefly presented.

Introduction

Rivers are complex systems characterised by intensive spatial and temporal dynamics of the biotic and abiotic conditions. Within the last decades, limnologists have aimed at describing the structure and function of streams; a short review of concepts in river ecology is given in Townsend (1996). As Karr et al. (1986) pointed out, the time is ripe for stream ecologists to use their science more effectively in protecting water resources. The demand was for a more integrated assessment of river systems that could evaluate the various and wide-reaching impacts of anthropogenic activities on the aquatic environment (e.g., Chovanec et al., 1994, 1997; Friedrich, 1998; Mauch, 1990; Moog & Chovanec, 1998; see also Cairns et al., 1993; Rapport, 1992; Schneider, 1992). These activities, including wastewater discharges, changes of habitat structures and connectivity aspects, as well as altered flow regimes, are often complex and difficult to describe directly in terms of their ecological repercussions. Therefore, it has been suggested that their combined effects should be assessed by focussing on the evaluation of the biotic community. This philosophy is also reflected by the requirements laid down in the EU Water Framework Directive (WFD; Council of the European Union, 1999). Analyses of physico-chemical and hydro-morphological parameters support the bioindication approach and are essential for an interpretation of disturbances.

The current Austrian methodology follows this logic and attempts to achieve a comprehensive assessment programme incorporating an array of both biotic and abiotic measurements. This approach of assessing the ecological integrity of running water strives to meet all the current socio-political demands of river management using, to the extent practical, current and standardised scientific methods. It is designed as an operational, multiparameter tool to assist the monitoring, maintenance, and restoration of the ecological integrity of rivers – a demand which is laid down in the Austrian Water Act. The close relationship between applied limnological research and the needs of river management is clear and, according to Grabherr (1995), this new approach for assessing ecological integrity is also a good example of conservation-oriented research.

In this paper the Austrian approach of assessing the ecological integrity of running waters and its legal basis are described; its function within Austrian water management and as a tool for the implementation of the WFD is discussed. Country-wide monitoring programmes, projects, and river typology inventories, which are an important basis for the implementation of the Directive, are presented.

Oekologische Funktionsfaehigkeit: the approach to assessing ecological integrity laid down in the Austrian Water Act

In 1985, the term 'Oekologische Funktionsfaehigkeit' – which can be synonymously used with ecological integrity – was incorporated into the Austrian Water Act [sec. 105 (1) lit.m]. Whereas this amendment was considered to be more symbolic than operational (Oberleitner, 1995), the broad use of ecological integrity in the Austrian Water Act of 1990 actually reflects the fact that the new approach is being taken into account in water resource management. All water bodies are now viewed as ecological systems and should be managed as such: sec. 30 (3) defines water protection as "the conservation of the naturalness of the waters and of the riparian areas necessary for the ecological integrity of the waters as well as the protection of groundwater resources". The term ecological integrity is also used in connection with regulations concerning the storage and use of hazardous substances (sec. 31) and the extent of water use (sec. 13).

The Austrian Standard (ÖNORM) M 6232 'Guidelines for the ecological study and assessment of rivers' (Oesterreichisches Normungsinstitut, 1997; first edition 1995) contains a methodological framework for the evaluation of running waters. It represents a general guideline for surveys which aim at assessing rivers according to ecological criteria in connection with river management plans, river water quality monitoring, river restoration, conservation measures, environmental impact assessment, etc.

The initial concept of evaluating the ecological integrity of running waters (Chovanec et al., 1994; Moog, 1994; Schmutz & Waidbacher, 1994) has been incorporated into this Austrian Standard. According to this guideline, ecological integrity can be defined as maintenance of all internal and external community processes and attributes, interacting with the environment in such a way that the biotic community corresponds to the natural state of the relevant aquatic habitat. This implies the preservation of regulation, resilience, and resistance (regulation: ability of an ecosystem to preserve its distinct characteristics; resilience: ability of an ecosystem to recover its distinct characteristics after short-term disturbances (ecological elasticity); resistence: ability of an ecosystem to withstand disturbances without incurring characteristic changes).

Thus, the ecological integrity of an aquatic system is fundamentally based on the ability of all naturally occurring aquatic species to maintain healthy and reproducing populations. The disturbance of this integrity is reflected in quantitative and qualitative changes in the assemblage of species which can eventually lead to either the disappearance of naturally occurring species or the appearance of species atypical of a given site. An acceptable and sustainable use of an aquatic environment would not overburden its capacity to maintain itself, and, hence, its ecological integrity would remain intact.

Working scheme

The assessment of ecological integrity in Austria represents a first step toward blending science and application; any classification of ecological integrity is a tool or a vehicle used to introduce the concept of thinking of a river at an ecosystem level to administrative authorities. Decision makers should start to regard rivers as functional entities and integrated systems.

Although ecological integrity is and must remain a holistic concept, a number of key components of the assessment procedure can be feasibly investigated and evaluated. The components used in this assessment system include hydrology, river morphology, physico-chemical parameters, ecotoxicological

Table 1. Criteria for assessing the ecological integrity of running waters (according to Chovanec et al., 1997, and Austrian Standard M 6232; modified)

Parameter/discipline	Examples of criteria, data requirements	Assessment scheme
Hydrology/hydraulics	Hydrological regime, velocity pattern, connectivity to groundwater, bed load regime	Verbal
River morphology	Morphological river type, habitat (instream and littoral) structures, substrate pattern longitudinal and lateral connectivity	Verbal
Physico-chemical parameters	Data on water quality	Threshold values; verbal
Ecotoxicology	Water quality data on hazardous substances; tests	Test results; NOEC; verbal
Biological water quality assessment	Saprobic system	Seven classes
Macroinvertebrate assemblages	Species inventory, dominance structure, abundance, functional feeding group classification, longitudinal zonation according the biocoenotic regions	Seven classes; verbal
Fish	River-type-specific species species with self-sustaining populations, fish region, number of guilds, guild composition population size, population age structure	Seven or five classes; verbal

Table 2. Seven-class evaluation of the ecological integrity of running waters

Class	Ecological integrity
1	Undisturbed
1–2	Slightly disturbed
2	Moderately disturbed
2–3	Significantly disturbed
3	Heavily disturbed
3–4	Very heavily disturbed
4	Completely disrupted

aspects, biological water quality assessment, as well as macroinvertebrate and fish communities (Table 1). The relevance of these components for an integrated river assessment is comprehensively discussed in Naiman et al. (1992), Moog & Chovanec (1998), and Muhar & Jungwirth (1998).

Based on an evaluation of these individual components (using either a classification process or verbal description and discussion) the disturbance of ecological integrity is assessed and classified within a seven-tiered system (Table 2). Component-specific ranks are not simply averaged to derive a final assessment. Those criteria indicating the highest anthropogenic disturbance have a strong influence on the final assessment; special interest is paid to the results obtained by the biological assessments. Focusing on the aquatic communities is crucial for determining ecosystem health. Cairns et al. (1993), Paulsen & Linthurst (1994), and Friedrich (1998) emphasise that biological criteria are essential components of a 'top down approach' of integrated river assessment.

The final classification presented in an expert's report is based on a comparison between a site-specific reference state and the current conditions (see also Hughes, 1995; Muhar et al., 2000; Schmutz et al., 2000). The reference state is defined as the state that has existed prior to human disturbances or at least without human influences that significantly alter the natural characteristics of the river. Friedrich (1992) and Mauch (1990) point out that an assessment of ecological integrity must relate to the degree of deviation from the natural state of the river. From the river management perspective, 'moderately disturbed' (which corresponds to the 'good status' of running waters in the evaluation scheme of the Water Framework Directive) is defined as the target for the ecological status of running waters. For all river stretches with worse evaluation results, mitigation measures are required by the Austrian Water Act as well as the Water Framework Directive.

Therefore, natural river stretches, which correspond to the original type of river, should be used whenever possible in defining reference conditions. When such stretches are not available they must be theoretically constructed using a combination of historical data (e.g., old maps or fish catch records) and models. In this connection, river classification plays a major role in defining reference states (Moog & Wimmer, 1990).

There were several reasons supporting the use of a seven-class system. First, the term 'significant disturbance' as laid down in the Austrian Water Act implies the necessity of a multi-tiered system. Second, seven-class systems of evaluation have a long tradition within river assessment, e.g., with the Saprobic System, and decision makers have learned to think along these categorical lines. And third, even minor positive or negative changes can be illustrated. A shift by one class can be a warning signal that current management practices are contributing to further degradation or can provide an incentive for remediation measures even in stretches where large-scale improvements are not possible. Assigning colours to each class will render consolidated data useful to administrative authorities. These coloured maps will allow a region-wide depiction of the status of ecological integrity across entire drainage basins. This can give water resource planning offices and policy makers a general idea of the current state of riverine conditions. The primary data and all detailed information on which the class assessment is based are available and help the assessment process to be as transparent and objective as possible.

Countrywide inventories, projects, and monitoring programmes

Austrian water management activities increasingly focus on countrywide river classification approaches and monitoring programmes. They can be recognised as the basis for a comprehensive and standardised river-type-specific assessment of the ecological integrity of Austrian running waters. These activities go back to the mid-1960s when the first maps on the biological water quality of the most important Austrian rivers were published. In the early 1990s a standardised water quality monitoring system was installed and various projects dealing with the evaluation of a countrywide river classification were initiated. The major target is creating a countrywide basis covering the disciplines listed in Table 1; this guarantees a standardised and comparable performance of river investigation projects on a high scientific level and helps implement the WFD.

River classification programmes

Stream orders of the Austrian running waters

The system of stream orders represents an ideal element within river classification, integrating a number of abiotic factors changing along a river's course (e.g., catchment area, water discharge, channel dimensions). All Austrian river segments were classified according to the system of Strahler (1957) by Wimmer & Moog (1994). This catalogue is the first countrywide evaluation of this key element of river classification and provides a sound basis for all future river assessment studies and typology programmes.

Classification of Austrian rivers according to their hydrological regime

The hydrological pattern of a river is an important element of its typological characteristics. It is determined by various components at the catchment level, in particular climate/precipitation, topography, geology, soils, vegetation, and land use. Aquatic communities are adapted to the hydrological properties of their habitat such as the ratio between low flow and high flow discharges, the time of low and high flows, their duration etc. An alteration of the flow regime can be caused, e.g., by water abstraction, hydroelectric power generation (reservoir, hydropeaking, water diversion), channelisation, changes in land use, or urbanisation. The hydrological condition in primarily responsible for all the depositional and erosional processes that form the diversity of habitable environments within a river's sediment and along its riparian and floodplain zones. A countrywide classification for a large number of Austrian rivers on the basis of their flow regime was developed by Mader et al. (1996).

Identification of Austrian rivers with high and good habitat quality The physical environment of rivers is fundamental for healthy riverine communities. In the framework of a national survey all larger Austrian rivers (except the Danube River) with a catchment area > 500 km^2 have been assessed in order to identify and document river stretches which still provide intact habitats of high or good integrity (see Muhar et al., this volume).

The assessment process focused on habitat features and processes like morphological–structural parameters, morphodynamics, hydrologic/hydraulic conditions, longitudinal and lateral connectivity, and anthropogenic changes related to associated wetlands.

The results of nearly 5000 investigated river kilometers provide baseline data for the characterisation of the large Austrian rivers and the definition and description of 'reference' conditions for different types of running waters; they also contribute to the identification of priority areas in the framework of habitat preservation programmes.

River typology of all Austrian rivers with a catchment size > 10 km²

A typological characterisation is been carried out according Annex II of the WFD. All Austrian rivers with a catchment size > 10 km^2 are characterised according the following typological features: stream orders, geology, catchment size, altitude, flow regime, and ecoregions (Wimmer & Chovanec, in press). This study represents an essential basis for the establishment of a surveillance monitoring according to the WFD, which has to cover all representative river types and also has to comprise type-specific reference sites.

The Austrian water quality monitoring system (AWQMS)

Recent legislation and administrative procedures regarding water pollution control in Austria, established by an amendment to the Austrian Water Act in 1990, have provided the basis for a new water quality monitoring system for running waters. Federal and provincial authorities as well as a large number of private laboratories are involved. The main goals of the monitoring system are to provide reliable and comparable water quality data for the most important rivers in order to check compliance with national and international threshold values and as information for decision makers, planning organs, and the general public. The standardisation efforts initiated in 1990 allow changes in water quality to be documented and the success of preventive or remediation measures to be assessed.

Water samples are collected monthly or bi-monthly at 245 sites. Sediments and biota (for the assessment of the biological water quality) are sampled once a year. At each sampling site and sampling date about 50 parameters are measured, including those needed for a general chemical characterisation of the water (e.g., nutrients, TOC, DOC), as well as organic micropollutants (e.g., pesticides; Chovanec, 1995) and heavy metals. The high number of parameters also enables the monitoring system to provide adequate information on the probability of ecotoxicological effects. Special attention is paid to countrywide standardisation (e.g., sampling methods, sampling cycles, analyses, data management, reporting) and analytical quality assurance (Chovanec & Winkler, 1994; Wasserwirtschaftskataster/Umweltbundesamt, 1997).

World Wide Web facilities are used within the Information System of the Federal Environment Agency for data administration, evaluation, and reporting. At the water-related homepage of the Federal Environment Agency, information and data of the monitoring system are available and comprise the annual report and the opportunity to query the data base (http://www.ubavie.gv.at).

The present monitoring network was designed in the late 1980s and early 1990s, focussing on surveying those running waters particularly stressed by waste water discharges. Because of Austria's obligations to meet the requirements arising from the WFD, the network has to be adapted: all representative types of running waters (independent of their pollution situation) have to be covered, and catchment area sizes have to be considered along with type-specific reference sites and sites of pressures (e.g., hydro-morphological impacts, pollution). This adaptation is currently carried out by elaborating the river typology mentioned in the previous chapter.

Biological assessment

Macroinvertebrate assemblages

A comprehensive classification catalogue of benthic macroinvertebrates in Austria regarding individual saprobic indices, their affiliation with functional feeding groups, and the distribution of longitudinal zonation within biocoenotic regions was evaluated by Moog (1995). All specifications occur at the species level since higher-level associations of autecological characteristics – with the exception of functional feeding groups – do not allow sufficiently meaningful conclusions. In addition to the ecological information, which is reflected in a 10-point scheme, the catalogue provides an overview of the indigenous limnofauna. Assessment procedures based on these criteria are an important element in evaluating the ecological integrity or the ecological status of running waters and can be supported by a computer programme ('Ecoprof').

Biological river water quality assessment

Biological river water quality assessment as a tool for water management in Austria dates back to the early 1960s and follows the principles of the traditional Saprobic System. It has recently been thoroughly revised in order to achieve countrywide harmonisation and comparability. An improved methodology published by the Bundesministerium für Land- und Forstwirtschaft (1998) and the Austrian Standard M 6232 provides guidelines on how to evaluate river water quality on four levels of accuracy varying according to sample effort, the number of indicator groups (macroinvertebrates, periphytic assemblages, protozoans) and the taxonomic precision. The basis of all levels is analysis of macroinvertebrates on the species level and a detailed assessment of the environmental factors of the sampling site ('site report') in the field. The final diagnosis is based on a stepwise discussion of the sub-disciplines (site-report, biological disciplines) and results in a verbal expression of the biological river water quality of a site as well as in a numerical value (see also Moog, 1995).

Establishment of type-specific reference communities
The target of a currently running project is to set up type-specific reference macroinvertebrate communities. This is recognised as an important element for the implementation of the WFD; the project will be completed in the year 2000. A comprehensive compilation and evaluation of faunistic data of relevant monitoring programmes (e.g., AWQMS), reports and scientific studies as well as the results of extensive expert consulting are used as a basis to define reference communities. An important part of the project is the additional investigation of pristine water sections. Until now those river sections were rarely subject to faunistic investigations. Each reference community will be described by a set of community-based parameters: e.g., longitudinal zonation, saprobiological reference condition, key species. A further target of the project is to develop a five-tiered classification scheme according to the WFD showing the deviation of macroinvertebrate assemblages from the type-specific reference condition. This system should be founded on the experience gained from the seven-tiered system.

Fish communities
For fish communities a multi-level concept is proposed (Schmutz et al., this volume). Out of five different biological organisation levels (fauna, community, guild, population, and individual), seven criteria are proposed: river-type-specific species, species with self-sustaining populations, fish region, number of guilds, guild composition, population size, and population age structure. Reference conditions have to be compiled for every distinct river type using historical fish and abiotic data, present river-type-specific reference sites and reference models. The final assessment procedure is done by comparing the assessment reach with the reference conditions using a five- or seven-tiered normative scheme and assigning that reach to the level of highest coincidence. The benefit of the multi-level concept is its potential for consistent sensitivity to low- and high-dose human alterations and, due to its general character, its adaptability to all river types.

Conclusion: the EU Water Framework Directive and its consequences for future scientific research in Austria

The new 'Council Directive establishing a Framework for Community Action in the Field of Water Policy' ('Water Framework Directive', WFD) will fundamentally change river management within the European Community. The term 'ecological status', which is laid down in the WFD and the way of assessing it by investigating aquatic communities (supported by the investigation of abiotic factors), corresponds to the general philosophy of an integrated approach of evaluating the ecological integrity of running waters in Austria (Koller-Kreimel & Chovanec, 1999).

From the WFD the following main obligations can be derived:

- to set up environmental objectives (good ecological and chemical status) in line with the type-specific characteristics of the river;
- to classify rivers according to typological features;
- to establish type-specific reference conditions;
- to classify the ecological and chemical status;
- to put emphasis on biological assessment methods supported by the evaluation of hydromorphological and chemical parameters;
- to implement a monitoring system;
- to classify the ecological potential of heavily modified and artificial water bodies;
- to define strategies against pollution (combined approach for point and diffuse sources);
- to establish river basin management plans.

With the country wide programmes listed above, essential baseline work for the implementation of the WFD has been done, particularly in the case of river typology, monitoring, and the development of biological assessment approaches.

As far as limnological research is concerned, future work has to be directed particularly towards:

- the definition and assessment of heavily modified and artificial water bodies as well as of the maximum ecological potential;
- the adaptation of different evaluation schemes to the five-class system of the WFD;
- limnological research, particularly concerning:

– methodological approaches to aggregate single biological assessment results, e.g., to integrate biological quality components such as phytoplankton, benthos, and fish;
– methodological approaches for the assessment of ecosystem processes, e.g., interactions river–floodplains;
– the investigation of sub-lethal toxicity as a consequence of chronic effects of pollutants;
– establishing biological databases at a European scale to harmonise assessment approaches.

Acknowledgements

The authors thank Michael Stachowitsch and Ulli Stärk for linguistic corrections.

References

Bundesministerium für Land- und Forstwirtschaft, 1998. Überarbeitete Richtlinie zur Bestimmung der saprobiologischen Gewässergüte von Fließgewässern (in press).

Cairns, J., P. V. McCormick & B. R. Niederlehner, 1993. A proposed framework for developing indicators of ecosystem health. Hydrobiologia 263: 1–44.

Chovanec, A., 1995. Pesticides in the aquatic environment – experiences from Austrian monitoring programmes. Toxicol. envir. Chem. 51: 205–220.

Chovanec, A. & G. Winkler, 1994. River monitoring and water quality assessment in Austria. Toxicol. envir. Chem. 44: 161–176.

Chovanec, A., H. Heger, V. Koller-Kreimel, O. Moog, T. Spindler & H. Waidbacher, 1994. Anforderungen an die Erhebung und Beurteilung der ökologischen Funktionsfähigkeit von Fließgewässern – eine Diskussionsgrundlage. Österr. Wasser- Abfallwirtschaft 46: 257–264.

Chovanec, A., O. Moog & V. Koller-Kreimel, 1997. Integrierte ökologische Bewertung – eine Vision? Stand der Diskussion in Österreich. In Lebensraum Gewässer – nachhaltiger Gewässerschutz im 21. Jahrhundert. Beiträge des Int. LAWA-Symposiums, Nov. 28–29, 1996, Heidelberg: 66–75.

Council of the European Union, 1999. Amended proposal for Council Directive establishing a framework for Community action in the field of water policy. ENV 68, PRO-CCOP 46, Interinstitutional File No 97/0067 (SYN), draft March 2, 1999.

Friedrich, G., 1992. Ökologische Bewertung von Fließgewässern – eine unlösbare Aufgabe? In G. Friedrich & J. Lacombe (eds), Ökologische Bewertung von Fließgewässern. Limnologie aktuell, Vol. 3. Gustav Fischer Verlag, Stuttgart, New York: 1–7.

Friedrich, G., 1998. Integrierte Bewertung der Fließgewässer – Möglichkeiten und Grenzen. In Bayerisches Landesamt für Wasserwirtschaft – Inst. f. Wasserforschung (ed.). Integrierte ökologische Gewässerbewertung – Inhalte und Möglichkeiten. Münchener Beiträge zur Abwasser-, Fischerei- und Flußbiologie, Band 51, München: 35–56.

Grabherr, G., 1995. Naturschutzforschung in Österreich. Natur Landschaft 70: 576–580.

Hughes, R. M., 1995. Defining acceptable biological status by comparing with reference conditions. In Davis, W. S. & T. P. Simon (eds). Biological Assessment and Criteria. Tools for Water Resource Planning and Decision Making. Lewis Publishers, Boca Raton, FL: 31–47.

Karr, J., K. D. Fausch, P. L. Angermeier, P. R. Yant & I. J. Schlosser, 1986. Assessing Biological Integrity in Running Waters, A Method and Its Rationale. Illinois Natural History Survey 5, 28 pp.

Koller-Kreimel, V. & A. Chovanec, 1999. Water Framework Directive – Ecological assessment of surface water status. International Conference on EU Water Management – Framework Directive and Danubian Countries, 21–23 June 1999, Bratislava, Proceedings: 93–110.

Mader, H., T. Steidl & R. Wimmer, 1996. Abflußregime österreichischer Fließgewässer. Monographien des Umweltbundesamtes, Wien. Band 82, 192 pp.

Mauch, E., 1990. Ein Verfahren zur gesamtökologischen Bewertung der Gewässer. Wasser Boden. 11: 763–767.

Moog, O., 1994. Ökologische Funktionsfähigkeit des aquatischen Lebensraumes. Wiener Mitteilungen, Band 120: 15–59.

Moog, O. (ed.), 1995. Fauna Aquatica Austriaca. Lieferung Mai/95. Wasserwirtschaftskataster, Bundesministerium für Land- und Forstwirtschaft, Wien.

Moog, O. & A. Chovanec, 1998. Die 'ökologische Funktionsfähigkeit' – ein Ansatz der integrierten Gewässerbewertung in Österreich. In Bayerisches Landesamt für Wasserwirtschaft – Inst. f. Wasserforschung (ed.). Integrierte Landesamt für Gewässerbewertung – Inhalte und Möglichkeiten. Münchener Beiträge zur Abwasser-, Fischerei- und Flußbiologie, Band 51, München: 57–118.

Moog, O. & R. Wimmer, 1990. Grundlagen zur typologischen Charakteristik österreichischer Fließgewässer. Wasser Abwasser 34: 55–211.

Muhar, S. & M. Jungwirth, 1998. Habitat integrity of running waters – assessment criteria and their biological relevance. Hydrobiologia 386: 195–202.

Muhar, S., M. Schwarz, S. Schmutz & M. Jungwirth, 2000. Identification of rivers with high and good habitat quality: methodological approach and applications in Austria. Hydrobiologia 422/423: 343–358.

Naiman, R. J., T. J. Beechie, L. E. Benda, D. R. Berg, P. A. Bisson, L. H. MacDonald, M. D. O'Connor, P. L. Olson & E. A. Steel, 1992. Fundamental elements of ecologically healthy watersheds in the pacific northwest coastal ecoregion. In Naiman, R. J. (ed.), Watershed Management: 127–188.

Oberleitner, F., 1995. Ökologische Anforderungen an Rechtsvorschriften – Das Prinzip der nachhaltigkeit. In Österreichische Gesellschaft für Natur- und Umweltschutz (ed.), Öko-Text 2: 43–52.

Österreichisches Normungsinstitut, 1997. Austrian Standard (ÖNORM) M 6232: Guidelines for the ecological study and assessment of rivers, Wien.

Paulsen, S. G. & R. A. Linthurst, 1994. Biological monitoring in the environmental monitoring and assessment programm. In Loeb, S. L. & A. Spacie: Biological Monitoring of Aquatic Systems. Lewis Publishers, Boca Raton, FL: 297–322.

Rapport, D. J., 1992. Evolution of indicators of ecosystem health. In McKenzie, D. H., D. E. Hyatt & V. J. McDonald (eds), Ecological Indicators, Vol. 1: 121–134.

Schmutz, S. & H. Waidbacher, 1994. Definition und Bewertung der fischökologischen Funktionsfähigkeit im Rahmen von Gewässerbetreuungskonzepten. Wiener Mitteilungen, Band 120: 61–88.

Schmutz, S., M. Kaufmann, B. Vogel, M. Jungwirth & S. Muhar, 2000. A new multi-level concept for fish-based, river-type-specific assessment of ecological integrity. Hydrobiologia 422/423: 279–289.

Schneider, E. D., 1992. Monitoring for ecological integrity: the state of the art. In McKenzie, D. H., D. E. Hyatt & V. J. McDonald (eds), Ecological Indicators, Vol. 2: 1403–1419.

Strahler, A. N., 1957. Quantitative analysis of watershed geomorphology. Am. Geophys. Union Trans. 38: 913–920.

Townsend, C., 1996. Concepts in river ecology: pattern and process in the catchment hierarchy. Arch. Hydrobiol. Suppl. 113, Large Rivers 10: 3–21.

Wasserwirtschaftskataster/Umweltbundesamt, 1997. Wassergüte in Österreich. Jahresbericht 1996. Bundesministerium für Land- und Forstwirtschaft, Wien, 231 pp.

Wimmer, R. & A. Chovanec, in press. Fließgewässertypen in Österreich als Grundlage eines Überwachungsnetzes im Sinne des Anhang II der EU Wasser-Rahmenrichtlinie. Bundesministerium für Land- und Forstwirtschaft, Wien.

Wimmer, R. & O. Moog, 1994. Flußordnungszahlen österreichischer Fließgewässer. Monographien des Umweltbundesamtes. Band 51, Wien, 631 pp.

Hydrobiologia **422/423:** 453–464, 2000.
M. Jungwirth, S. Muhar & S. Schmutz (eds), Assessing the Ecological Integrity of Running Waters.

Measuring the attainment of biological integrity in the USA: a critical element of ecological integrity

Michael T. Barbour[1], William F. Swietlik[2], Susan K. Jackson[2], David L. Courtemanch[3], Susan P. Davies[3] & Chris O. Yoder[4]
[1]*Tetra Tech, Inc., 10045 Red Run Boulevard, Suite 110, Owings Mills, MD 21117, U.S.A.*
[2]*US Environmental Protection Agency, Office of Science and Technology, MC# 4304, 401 M Street SW, Washington, DC 20460, U.S.A.*
[3]*Maine Department of Environmental Protection, State House #17, Augusta, ME 04333, U.S.A.*
[4]*Ohio Environmental Protection Agency, 1685 Westbelt Drive, Columbus, OH 43228, U.S.A.*

Key words: biological integrity, ecological integrity, bioassessment, biocriteria, water quality, biological indicators

Abstract

The concept of ecological integrity has become a worldwide phenomenon and is firmly entrenched into the regulatory structure of environmental law in the United States of America (USA). The attainment of ecological integrity requires the attainment of its three elements: physical, chemical, and biological integrity. In the USA, measures of chemical integrity were implemented first into monitoring programs and were effective in reducing pollutant loadings to the nation's surface waters. Because biological communities integrate the effects of different stressors such as reduced oxygen, excess nutrients, toxic chemicals, increased temperature, excessive sediment loading, and habitat degradation, the advent of bioassessment in regulatory programs has provided a more comprehensive and effective monitoring and assessment strategy. Measures of biological integrity clearly have become a priority in the USA. The development of biological criteria (biocriteria) within regulatory programs to serve as thresholds by which to judge the attainment of designated aquatic life conditions of surface waters is a major focus of states and Indian tribes within the USA. The derivation of reference conditions for the nation's surface waters (i.e., streams, rivers, lakes, wetlands, estuaries, and marine waters) across different physiographic regions is a critical element in the design of biocriteria and is currently a primary initiative in the USA. Nearly all state water resource agencies have developed bioassessment approaches for streams; 1600 to 75 000 km of streams require assessment in each state. Bioassessment development for other water body types is not as advanced to date. The US Environmental Protection Agency (USEPA) has produced technical guidance for developing effective bioassessment programs; they include crucial elements such as defining objectives, classifying water bodies according to expected biological attributes, deriving the reference condition of the site classes, developing standardized protocols for sampling and data analysis, and implementing a quality assurance plan. Approaches to bioassessment in the USA follow a basic design of incorporating various attributes of the elements and processes of the aquatic community, which is either an aggregation into a multimetric index or a series of multivariate analyses using the attributes as input variables. The Clean Water Act of 1972 and its subsequent amendments mandate maintaining, restoring, and protecting the ecological integrity of surface waters. Through use of robust bioassessments and other measures of ecological integrity, the USA has developed a strategic plan to establish priorities to meet this goal.

Introduction

The concept of ecological integrity is firmly entrenched in the regulatory structure of water law in the USA. The Clean Water Act (CWA) of 1972 identified as the fundamental, long-term goal of environmental protection for aquatic resources, the restoration and maintenance of ecological integrity, which is expressed in the law as physical, chemical, and biological integrity (Figure 1). The legislative record clearly documents the intent of the framers of the 1972 Act as well as the tensions between the ability to achieve

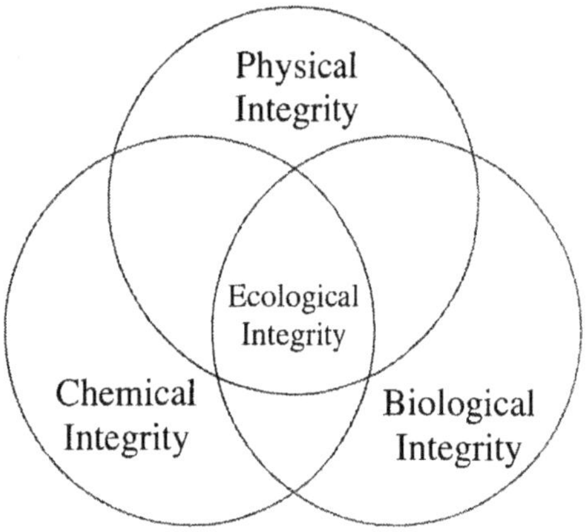

Figure 1. Ecological integrity and its three elements - physical, chemical, and biological integrity.

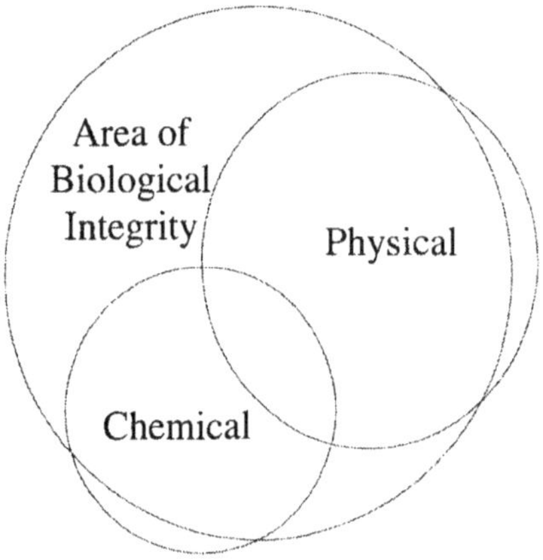

Figure 2. Alternative (dynamic) model of biological integrity (patterned after Yoder, 1995).

this visionary goal and the feasibility of implementation (USGPO, 1972a, b). This dichotomy has been at the heart of continuing debate and discourse as the USA has sought to implement the CWA (Jorling, 1977; Mackenthun, 1977). Karr (1991) argued that the model for ecological integrity (Figure 1) set forth by the USEPA was inadequate and not representative of the variable interaction of the three key components. Yoder (1995) proposed an alternative model (Figure 2) which shows that the overlapping influence of the three major components is both disproportionate and dynamic. The concept of biological integrity is all encompassing and, in essence, is the underpinnings of ecological integrity. This view is consistent with that of many European countries, e.g. Austria (Moog & Chovanec, 2000).

Several US federal agencies have different responsibilities associated with the mandates of the CWA. However, it is the US Environmental Protection Agency (USEPA) that, under the Act, is the primary arbiter of whether ecological integrity is being protected or restored in the nation's surface waters. The states and Indian tribes play an equally strong role to the USEPA through the implementation of their water quality standards, which are laws or regulations that the states and Indian tribes adopt to enhance and maintain water quality and to protect public health. These standards provide the foundation for accomplishing the goals and objectives of the CWA.

Although the broad goal of ecological integrity was set forth in federal law in 1972, implementation focused over the next two decades on engineering solutions to controlling toxic and municipal discharges (Adler, 1995). During the 1960s, public support called for more far-reaching, enforceable environmental legislation and regulation because of increasingly visible environmental catastrophes, e.g., widespread fish kills and occurrence of tumors and deformities, obnoxious odors and appearance of waters contaminated by raw sewage, and the loss of charismatic wildlife (such as birds) due to toxics and pesticides in streams and rivers. Chemical pollution, both organic and inorganic, was recognized as a major cause of impaired surface waters. The mobilization of public concern about the environment, culminating in the first Earth Day in 1970, directly led to the passage of the CWA. It was not surprising therefore that initial efforts to implement the CWA of 1972 focused on controlling industrial and municipal discharges into streams, rivers, and coastal waters.

Therefore, measures of chemical integrity were implemented first into monitoring programs. Measures of physical and biological integrity were slow to develop because of the perception that they yielded ambiguous results, were costly, and would add little to the existing monitoring techniques for direct measures of chemical pollution. Additionally, biological information was thought to be too variable in quality to be tolerated within the legalistic framework of public law (Courtemanch et al., 1989). Essentially, water pollution was regarded as an engineering problem that could be managed through engineered solutions to alter the water chemistry. Certain assumptions were made about what chemical criteria were appropriate to protect the aquatic ecosystem. They were based on response by representative test species that loosely acted as surrogates to resident aquatic biota (Cairns, 1980).

Chemical criteria, however, were effective in reducing pollutant loadings to the nation's surface waters. As a consequence, the concept of ecological integrity was narrowly interpreted to mean chemically clean water.

By the 1980s, engineering solutions had provided substantial progress in pollution abatement from point source discharges. However, the gains made in reducing the often catastrophic levels of chemical pollution in waters revealed more subtle and diffuse threats to the ecological integrity of aquatic resources, such as loss of habitat, reproductive defects from bioaccumulative chemicals, diversion of surface and ground waters, nutrient enrichment from stormwater runoff in both urban and agricultural environments, and colonization by nonindigenous species. Water resources, particularly their biological components, continued to be degraded at alarming rates (Karr, 1993). Water resource agencies began to shift their regulatory emphasis toward more complex problems, especially nonpoint sources (NPS). One of the main difficulties facing NPS pollution management over the years has been the lack of techniques to monitor and assess impacts (McCarron & Frydenborg, 1997). Traditional methods (i.e. water quality sampling of storm events) have remained largely inadequate due to the transient and unpredictable nature of NPS pollution (Barbour et al., 1996b). The need to accurately measure and diagnose the cumulative impacts of multiple stressors, not just individual pollutants, became a priority in water resource management.

Through trial and error, the USEPA began to move towards a more integrated, comprehensive approach to water resource protection with development of methods for measuring synergistic interactions of toxic chemicals, assessing biological and habitat indicators of ecosystem condition, and estimating ecological risk. In addition to strengthening the Agency's program to control toxic pollution, amendments to the Clean Water Act in 1977 and 1987 provided impetus for this shift by establishing geographically integrated programs, such as the Great Lakes Research, Chesapeake Bay, and National Estuary Programs, with an increasing emphasis on addressing nonpoint source pollution.

Direct measures and evaluation of the condition of the living system, or biota, are necessary to accurately assess the condition of the aquatic resource (Karr & Chu, 1999), particularly when impacted from multiple stressors. Biological communities integrate the effects of different stressors that water resource agencies are struggling to address, such as reduced oxygen, excess nutrients, toxic chemicals, increased temperature, excessive sediment loading and habitat degradation. This integration occurs through time, a dimension difficult to measure with chemical information. Thus, measures of biological condition provide a critically needed tool for water resource agencies to detect and evaluate the aggregate impact of the stressors.

During the 1980s and 1990s, federal and state resource management agencies benefitted from the development of integrative biological approaches with considerable potential to recognize and reverse the declining trends in aquatic resources (Davis & Simon, 1995). Biological monitoring programs were evolving and being expanded in regulatory agencies. This more ecologically realistic approach to water resource management has resulted in a renewed interest in assessment of the condition of aquatic resources and implementation of different monitoring tools, with a major shift in philosophy toward ambient monitoring of biological integrity to protect water resources (Karr, 1991; USEPA, 1988). Biological integrity has become an increasingly important focus within the USEPA for assessing the condition of the nation's surface waters and documenting the success of water resource restoration and protection (Jackson & Davis, 1994; USEPA, 1990). The assessment of the biological community using measures of species richness and composition, relative abundances of species or groups of species, and functional relationships among resident organisms (Karr, 1993) has become the standard in the USA for measuring progress towards achieving the CWA's goal of biological integrity.

Establishing biological criteria for judging attainment of biological integrity

To better protect aquatic ecological integrity, the long-range goal of the USEPA for water quality standards programs in all states and tribes over the next two decades is to:

> "...fully integrate biocriteria, nutrient criteria and microbial pathogen control with improved chemical-specific criteria, whole effluent toxicity methods and possible sedimentation, flow and wildlife criteria, into criteria and standards programs to better support watershed management for the protection of human health and the maintenance and improvement of the chemical, physical and biological integrity of the Nation's waters" (USEPA, 1998a).

The USEPA has developed a Water Quality Criteria and Standards Plan (currently an interim final document) that presents a 'vision' and strategy to better protect human health and enhance and maintain the quality of the nation's waterbodies. This interim final plan will guide USEPA and the states and tribes in their development and implementation of criteria and standards over the next decade and sets the goal that biological criteria and assessment methods are to be implemented. The interim final plan identifies the target date of 2005 for achieving this goal.

It is national policy that states and tribes designate aquatic life uses (i.e., environmental goals) for their waters that appropriately address biological integrity and adopt biological criteria necessary to protect those uses. To protect aquatic life, the biological component will dominate the other components because it is the product of the integrated interaction of the chemical and physical components (Yoder, 1995). Designated uses to support aquatic life can cover a broad range, or continuum, of biological conditions, with some waters being closer to an ideal of biological integrity than others. The attainable levels of biological integrity for any water is a state and/or tribal determination involving public participation.

Biocriteria are narrative descriptions or numerical values adopted into state or tribal water quality standards that can be used to factually and quantitatively describe a desired condition for the aquatic life in waters with a designated aquatic life use (USEPA, 1998a). The purpose of biocriteria is to establish enforceable standards based on biological characteristics related to biological integrity that can be used to direct water quality management. Biocriteria are developed by biologists and other natural resource scientists using accepted scientific principles to characterize the regional reference conditions for the different water bodies found within a state or tribal nation.

Defining the reference condition is a critical element in designing biocriteria. It expresses the objective and is the basis of all comparison and ultimately the basis on which attainment of goals is determined. Biocriteria determine the allowable range of condition for the aquatic community with respect to the reference condition. Ideally, reference conditions should represent pristine conditions. Because absolutely pristine sites do not exist (even the most remote lakes and streams are subject to atmospheric deposition), reference conditions typically describe the characteristics of waterbody segments that are minimally impacted by human activities. Acceptable reference conditions will differ among geographic regions because soil conditions, stream morphology, vegetation, and dominant land use differ. In heavily agricultural, industrial–commercial, or urbanized areas, reference conditions will need to be established based or historical records, simulation models, and expert judgement. Because of the extensive footprint of anthropogenic influence on the landscape, a common definition and nationally consistent approach has been elusive in the USA. However, the ideal reference condition must represent aspects of naturalness such as an aquatic ecosystem that is balanced, adaptive, and reflects natural evolutionary processes. Characteristics of such systems are those that support and maintain a community of organisms having a composition, diversity, and functional organization comparable to that of the natural habitat of a region (i.e., *regional reference conditions*).

The definition of reference needs to account for the conditions of large populations of general waterbody types (e.g., lakes, streams, wetlands), and needs to be acceptable and understandable both by persons most concerned with nature for its own sake and those who are concerned only with what it can provide humans (Hughes, 1995). The traditional upstream–downstream analysis has little value in the context of region-wide assessment programs and water quality standards that include biocriteria. Derivation of reference conditions may need to rely on many sources including empirical data from selected 'reference' sites, historical data, paleoecological data, laboratory studies, or quantitative models (Gibson et al., 1996; Hughes, 1995). Best professional judgement plays a vital role in establishing reference conditions. The initiation of any biological assessment program needs to focus on reference conditions applicable to the water quality program.

Elements of a bioassessment program

Any well-designed monitoring and assessment program is inherently anticipatory in that it will provide information for present needs and those not yet determined (Yoder & Rankin, 1995). Thus, implementation of a program must be proactive, because once the opportunity passes, there is no way to generate the needed data and information after the fact (Barbour & Yoder, in press). Measuring several attributes of the biological community ensures that future objectives will be addressed. Programs that are adaptable to immediate and future needs are also cost efficient.

The choice of a particular biological assessment design should be based on the monitoring objectives and ultimate decision process (Barbour et al., 1996a). The recent innovative rapid assessment techniques used in the USA (Barbour et al., 1999; Plafkin et al., 1989) offer cost efficiencies while maintaining data integrity, yet are not intended to replace more rigorous approaches. The constraint of resource investment and the requirement for quick 'turnaround' of biological data for decisions will dictate the nature of the study design. Regardless of approach, a certain amount of preparation is required. For instance, specific objectives must be defined, appropriate indicators selected, sites (or streams) classified into homogeneous classes, reference conditions established, protocols standardized and investigators trained, and a quality assurance plan developed. These elements are further explained, as follows:

Study objectives

To assess anthropogenic perturbations, objectives usually relate to determining the biological condition or status, demonstrating improvements following implementation of mitigation measures or restoration practices, or screening for problems or sensitive areas.

Selection of indicators

Environmental indicators must be relevant to the study objectives. Both technical and program considerations are important in identifying indicators. Keddy et al. (1993) provide guidance for selecting appropriate indicators that should be: (1) ecologically meaningful, i.e., closely related to maintenance of essential environmental processes and ecosystem functions (e.g., primary production); (2) macro-scale, i.e., indicate changes in entire communities rather than selected species; (3) general, i.e., can be measured meaningfully on different community types; (4) sensitive, i.e., quickly respond to stresses and perturbations; and (5) simple, i.e., are easy to measure. Four additional considerations, as articulated by McCarron & Frydenborg (1997), are necessary to incorporate bioassessment information in water management programs: (1) cost-effective, i.e., are within the financial and staffing resources of agencies; (2) scientifically and legally defensible, i.e., impairment caused by human activities must be distinguished from the natural variations in the biological condition; (3) integration into existing programs, i.e., better decision-making is accomplished by incorporating bioassessment information into regulatory and non-regulatory programs; and (4) effectively communicated, i.e., complicated scientific conclusions from monitoring efforts must be easy to understand for non-technical managers and the public; these scientific findings must tell understandable and compelling stories.

Site classification

A classification process,often based on regional attributes, may be useful to provide an ecological basis for identifying homogeneous areas from which reference conditions can be established. For example, the composition of the biota of coastal plain streams would be expected to be different from the streams of the Appalachian mountains. Site classes should not be constrained by political boundaries. Both the use of biological metrics and multivariate analyses of non-metric data have been shown to be effective classification tools.

Reference condition

Reference conditions are a critical element of assessing the quality or health of the aquatic system. The regional-based reference condition can be derived from an aggregate of reference sites or some empirical model of expectations that may include knowledge of historical condition or extrapolated from ecological principles. The establishment of reference condition is based on identification of minimally disturbed sites that represent the best physical, chemical, and biological condition attainable. The use of a regional-based reference is desired over individual site-specific references for broader application in state water resource programs.

Standardized protocols

Standardization of methods is crucial to establishing the reliability of biological data. Lack of standardization was a problem associated with early biological assessment studies, where diverse methods yielded conflicting information. It may be as important to standardize sampling bias as it is to minimize bias where comparison to a reference condition is required. Replication is necessary to establish the precision of the method. Thorough training of investigators enhances the ability to provide a consistent unit of effort, thus reducing the qualitative nature of the methods. Standardization is paramount when information will be used in later trend analysis.

Data analysis

For the bioassessment approach adopted by most state water resource agencies, metrics chosen for relevancy and sensitivity are calibrated for the site class and reference condition. This requires regional validation efforts. In aggregated indices, values of different scales are transformed to score categories; this is done from calibrated thresholds based on a selected percentile of the site class. Information from the individual metrics, as well as an aggregated index, allows judgement of the biological status of the assemblage.

Quality assurance (QA) plan

The development of an effective QA plan is paramount to the success of the program. It should assign responsibility, define personnel qualifications, establish protocols, define preventive and corrective action, provide information tracking, and ensure that study objectives are met. Quality assurance plans are especially important to regulate the performance and maintain data integrity.

Overview of state bioassessment programs in the USA

Much has been accomplished since the start of the biological assessment and criteria program in the USA in 1988. However, further progress is necessary to realize the full potential of bioassessments and biocriteria for managing water quality and protecting aquatic life.

The water resource programs in the USA have multiple objectives related to the sections of the Clean Water Act (Figure 3). The numerous programs of water resource agencies are best served by indicators that can be used for these multiple purposes. The ability to sustain a balanced biological community is one of the best indicators of the potential ecological integrity of a water resource (Karr, 1993). Bioassessment initially serves state and tribal monitoring needs through three primary functions or uses: (1) screening or initial assessment of conditions; (2) characterization of impairment and diagnosis; and (3) trend monitoring to evaluate improvements or further degradation (Barbour & Yoder, in press). Beyond monitoring and assessment functions, biological information can fill

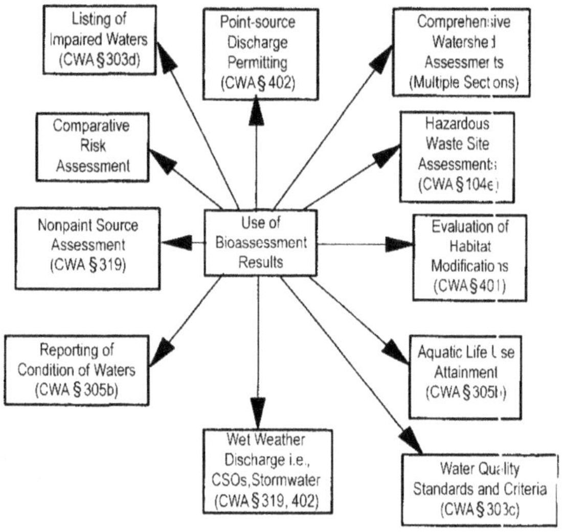

Figure 3. Applications of bioassessment data to water resource management programs in the USA (taken from Barbour & Yoder, in press).

an important gap in overall water resource management by indicating ecological condition (Courtemanch et al., 1989; Yoder, 1995). This critical transformation of utility for management occurs when states codify, into regulations, their expectations for acceptable biological conditions in a technically explicit way. These aquatic life expectations are referred to as biological criteria, and when codified in state law they are referred to as designated aquatic life uses. Biological assessment provides crucial water quality planning information for managing more complex water quality problems by providing a measure of aggregate impact (Courtemanch et al., 1989). This is valuable for state and tribal pollution control programs seeking to prioritize water quality problems, set restoration goals and new performance standards, and document results. Ohio and Maine, for example, have codified aquatic life uses in water quality standards. Here, such standards serve an 'action-forcing' function allowing water resource management decisions to be influenced by the extent to which actual ecological condition (as measured by the fish or benthic macroinvertebrate community) approaches aquatic life expectations.

The water resource agencies in the 50 states of the USA are in various stages of development and implementation of bioassessment. This is illustrated in maps updated by a report by the USEPA (Davis et al., 1996). With the publication of the USEPA Rapid Bioassessment Protocols (Plafkin et al., 1989) and the Program Guidance for Biocriteria (USEPA, 1990), the use of bioassessment by state water resource agencies in streams and wadeable rivers has advanced well beyond the other waterbody types (i.e., non-wadeable rivers, lakes and reservoirs, estuaries and coastal marine waters, and wetlands). Each state has from 1600 to 75 000 km of streams (exceptions are 423 km in Hawaii and 584 000 km in Alaska), whose ecosystem health needs to be assessed (Davis et al., 1996). Assessing the condition of so many stream kilometers is daunting and will require many years and substantial technical resources. Biological assessment approaches have therefore been designed to allow multiple-site investigations within a field season, be cost-effective, and yet be scientifically robust to judge biological condition (Barbour, 1997). All but three states (94%) have either developed, or are in the process of developing, bioassessment approaches for streams (Figure 4a). These three states, namely Nevada, South Dakota, and Utah, all have less than 20% of their stream kilometers characterized as perennial (Davis et al., 1996). Bioassessment approaches are lacking for intermittent streams, which are abundant in these typically arid states. State water resource agencies that have pioneered the development of bioassessment approaches in the USA include Ohio EPA (1987), Maine Department of Environmental (Courtemanch et al., 1989), and North Carolina Department of Environmental Management (Lenat, 1988). Common elements of these programs are the identification of reference conditions and the use of a suite of biological metrics to assess ecosystem health.

Bioassessment in most states is an effective tool to identify impairment from multiple sources of perturbation (i.e., chemical discharges and runoff, invasive land use to alter watershed characteristics and stream morphology, and air-borne deposition). However, the use of bioassessments for regulatory programs can apply to several mandates of the CWA (see Barbour & Yoder, in press, for a more detailed description of these applications). The primary application of bioassessment in state agencies is to evaluate the condition of the surface waters with respect to their designated aquatic life use. Approximately 60–70% of the states use bioassessment for this purpose (Figure 4b). They must report the condition of their surface waters to the US Congress under section 305b of the CWA.

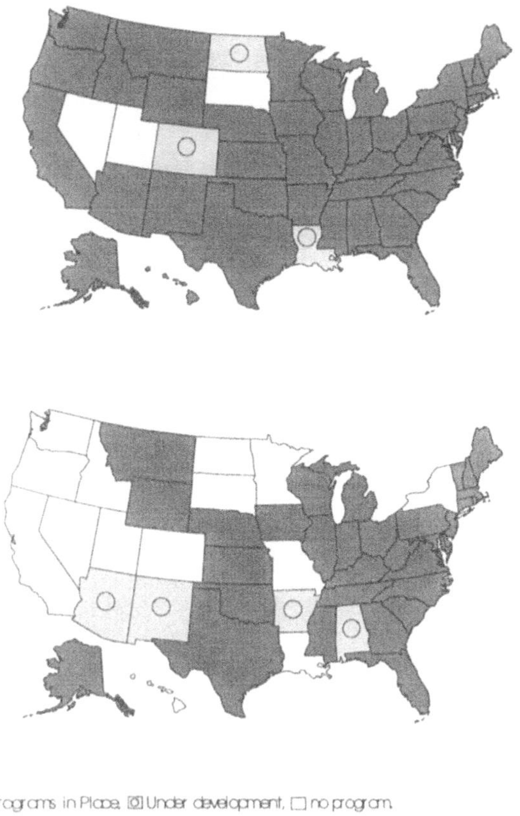

Figure 4. Map of the US illustrating use of stream bioassessment in state water resource programs. (a). Bioassessments used in water resource management (screening); (b). Bioassessments used to interpret aquatic life use attainment (reporting).

Selected bioassessment programs in the USA

Biological assessment programs have evolved to measure a range of attributes that represent various components of biological integrity. Biological integrity means more than achieving a high level of species diversity (Angermeier, 1997; Angermeier & Karr, 1994; Barbour & Yoder, in press). The bioassessment approach that relies on metrics derived from biological information is the technique most frequently used by the state water resource agencies. It provides a means of compiling and comparing ecological information relating to important elements and processes of naturally functioning aquatic communities and our current knowledge about those relationships (Barbour & Yoder, in press). The actual application of biological metrics in an assessment differs from state to state.

The most typical application of metrics is exemplified by Ohio, where the metrics are aggregated into

Table 1. Maine's water quality classification system for rivers, streams, and lakes, and associated biological standards (taken from Courtemanch, 1995)

Class	Management perspective	Biological standard
AA	High-quality water for recreation and ecological interests. No discharges or impoundments permitted	Habitat natural and free flowing. Aquatic life as naturally occurs
A	High-quality water with limited human interference. Discharges restricted to noncontact process water or highly treated wastewater equal to or better than the receiving water. Impoundments allowed	Habitat natural. Aquatic life as naturally occurs
B	Good-quality water. Discharge of well-treated effluent with ample dilution permitted	Habitat unimpaired. Ambient water quality sufficient to support life stages of all indigenous aquatic species. Only nondetrimental changes in community composition allowed
C	Lowest water quality. Maintains the interim goals of the Federal Water Quality Act (fishable/swimmable)	Ambient water quality sufficient to support life stages of all indigenous fish species. Change in community composition may occur but structure and function of the community must be maintained
GPA (lakes)	Protection of present high quality for recreation and ecological interests. No discharges allowed. Change of land use may not cause trophic state increase	Habitat natural. Aquatic life as naturally occurs. Maintain stable or decreasing trophic state

an index (Yoder & Rankin, 1995), patterned after the Index of Biotic Integrity (Karr, 1981; Karr et al., 1986). Ohio has developed an index for the benthic macroinvertebrate assemblage, the Invertebrate Community Index (ICI) (DeShon, 1995), and two for the fish assemblage, the Index of Biotic Integrity (IBI) and the Modified Index of Well-Being (MIwb) (Ohio EPA, 1987; Yoder & Rankin, 1995). The metrics represent various attributes of the assemblages, are calibrated to regional reference conditions, and transformed to unitless scores for aggregation into an index. Many papers describe the approach to aggregating metrics into an index (see for example, Barbour & Yoder, Barbour et al., 1995, Karr, 1991, 1993; Karr & Chu, 1999; Karr et al., 1986; Yoder & Rankin, 1995; in press). Some states, such as North Carolina, evaluate individual metrics and make decisions relevant to the response of those metrics (Lenat, 1988, 1993; Lenat & Barbour, 1994).

Maine has adopted an alternative approach using a probabilistic model to measure attainment of aquatic life goals. It follows the model developed by Reckhow. Reckhow (1994) developed a decision framework for complex environmental issues (e.g., development of biocriteria), recommending that management agencies should (1) clearly identify management objectives, (2) identify attributes to measure achievement of the objectives, and (3) identify feasible management options to achieve the objectives. A range of goals were identified for aquatic life condition from 'natural' to a maximum allowable change, described in four different water classes (see Table 1) and incorporated into state law. The law includes definitions of ecological terms that clarify management intent that can be translated into objective, measurable ecological attributes. Numerical biocriteria, using benthic macroinvertebrate assemblage metrics, were developed to quantify attainment of the attributes assigned to each classification. Maine's numeric criteria are based on a probability model for each of the four water quality classes for rivers and streams. A priori assemblage groups were identified by expert biologists' judgement of membership within the water quality classes. The models are linear discriminant functions that collectively use 23 variables, or metrics, relating to richness and abundance attributes of the benthic macroinvertebrate assemblage (Courtemanch, 1995). Constructing linear discriminant models yields advantages: (1) the analysis provides an objective means of identifying the most significant variables contributing to definition of the determined groups; (2) the weight of each variable used in classification is different and is determined by the analysis, contributing interpretive value; and (3) probabilities of fit can be calculated that provide information about the strength of a classification assignment when a new site is evaluated using the models (Courtemanch, 1995). The use of the decision analytic approach to establishing biological criteria offers many advantages to managers. Reckhow (1994) argues that careful decision-making about over-arching objectives for aquatic resources forces managers and scientists to ask key questions about (1) the characteristics of attributes that may be selected to measure achievement of those objectives (e.g., choice of biological metrics and indices), (2) the societal and ecological relevance of the selected measurable attributes, (3) the management options to more closely approach goals, and (4) the most technically appropriate assessment methods to quantify the selected attributes. Maine's use of probabilities offers certain advantages for implementation within the regulatory program for management of water quality. It is possible to use differing probabilities for regulatory and non-regulatory decisions. Assessment, which is used for status reporting, program management, and prioritization, can tolerate the use of weaker probabilities because these activities are administered continuously and interactively. Enforcement, on the other hand, is a discrete function, in which the action should not allow for much indecision in the model. Differing decision probabilities from the model can satisfy these management needs (Courtemanch, 1995). The New Brunswick, Canada Department of the Environment is proposing to adopt Maine's narrative aquatic life standards in their water classification regulation. Because Maine and the Province of New Brunswick share a largely overlapping physiographic region the Province is exploring the feasibility of also adopting Maine's numeric biocriteria model.

Each of the major approaches to implementing of bioassessment information has strengths and weaknesses relative to the four considerations listed by McCarron & Frydenborg (1997). Rigor of the methods varies with each approach and has to be weighed against the nature of information needed by program managers, the challenges it may evoke scientifically or legally, and the overall cost to the agency. The use of biological metrics is generally regarded as simple and easy to understand; it is a practical and very cost-effective screening tool to integrate water quality information at large numbers of sites. The decision analytic approach, utilizing multivariate techniques, while more complex in design may offer advantages for planning and adaptive management applications.

Case study of the value of bioassessment

A recent case demonstrates the potential for applying bioassessment information and associated biocriteria. Maine's Presumpscot River exhibits non-attainment of aquatic life standards in several locations from many potential sources of perturbation. In recent years, the USEPA has required that a Total Maximum Daily Load (TMDL) of pollutants be established for impaired river systems, such as the Presumpscot, where existing, required pollution controls are inadequate to attain applicable water quality standards. Recently, the USEPA approved Maine's TMDL assessment for the river. This approval was significant for several reasons. First, numeric biocriteria were used to evaluate the aquatic life standard and to determine that the biological condition was not attained. Second, bioassessment findings were used to identify the specific pollutant stressor. Of particular value, biological information served as the quantitative response variable or modeling endpoint used to develop a new pollutant discharge limit. This was accomplished by using data from other sites in the state where similar stressors were involved and where different levels of aquatic life attainment had been measured. The wastewater discharge license that resulted from this effort requires an initial 30% reduction in the discharge of total suspended solids and biochemical oxygen demand. Provisions are included in the license for phased reductions (up to 61%) if the benthic assemblage does not respond sufficiently to attain aquatic life standards. Maine DEP was able to apply this innovative approach to improving water quality and aquatic life conditions in the Presumpscot River because of the convergence of three factors: (1) the State had a sound legal basis for using biological monitoring findings to force action because clearly defined aquatic life standards exist in Maine's Water Quality Classification law, and scientifically defensible numeric criteria have been established by Maine DEP; (2) data, which were essential to model the recommended total suspended solids load reductions on the Presumpscot River, had been collected using standardized procedures that were consistent with studies performed in other parts of the State; and (3) teamwork and collaboration between Maine DEP water quality modelers and aquatic biologists resulted in an approach that integrated technical information and expertise from both disciplines; it enabled for Maine DEP to control the input of a stressor for which the State has no water quality standards.

This example demonstrates the use of biological information to directly manage multiple stressors affecting biological integrity in a complex river system. Biological response to these stressors was needed to identify and characterize impairment because performance standards for suspended solids (the primary stressor) did not exist. Interpreting the characteristics of the observed biological response and eliminating other plausible causes of impairment led investigators to focus on suspended solids from a specific industry. Extrapolation of biological assessment data from other similar sites in the State allowed managers to make predictions about allowable pollutant loads that could be incorporated into a new waste discharge permit. In Maine, biological assessment is directing ongoing adaptive management actions in that it serves as the measure of performance for the industry under its new license and as the overall barometer of the success of Maine's management of the resource.

USEPA's-long range goal of assessing and restoring the nation's surface waters

A number of key activities are necessary to meet USEPA's long-range goal of assessing and restoring surface waters (USEPA, 1998a). First, USEPA must complete the issuance of a series of biocriteria technical guidance manuals. These manuals provide states and tribes with scientific information and procedures used to conduct bioassessments and develop biocriteria. Bioassessment and biocriteria development manuals have been published for streams and small rivers (Barbour et al., 1999; Gibson et al., 1996; Plafkin et al., 1989) and lakes and reservoirs (USEPA, 1998b). A final draft manual is prepared for estuaries and near-coastal waters, and USEPA plans in the next 5 years to develop manuals for wetlands, large rivers and coral reefs.

Biocriteria are developed by the states and tribes by applying these methods to their respective ecological regions. Once the biocriteria technical guidance documents are complete, states and tribes will need additional resources and expertise to derive biocriteria and to incorporate biocriteria into their water quality standards. Policy, program, and legal issues often arise as the scientific and technical capabilities to implement biocriteria are developed. USEPA will assist the efforts of the states and tribes by providing technical assistance through funding, research and development, and policy and program guidance. Six long-term objectives of the USEPA for using bioassessments and biocriteria in the USA include: (1) all states use bioassessments to evaluate the health of aquatic life in all waterbodies; (2) bioassessment data are used to better define aquatic life uses of surface waters; (3) quantifiable biocriteria are in all state/tribal water quality standards to protect aquatic life uses; (4) biocriteria/bioassessments are used in ongoing regulatory programs; (5) biocriteria/ bioassessments are used to assess the effectiveness of water quality management efforts; and (6) bioassessment data and biocriteria are used to better communicate the health of the nation's waters.

References

Adler, R. W., 1995. Filling the gaps in water quality standards: Legal perspectives on biocriteria. In Davis, W. S. & T. P. Simon (eds), Biological Assessment and Criteria: Tools for Water Resource Planning and Decision Making. Lewis Publishers, Boca Raton, FL: 345–358.

Angermeier, P. L., 1997. Conceptual roles of biological integrity and diversity. In Williams, J. E., C. A. Wood & M. P Dombeck (eds). Watershed Restoration: Principles and Practices. American Fisheries Society: 49–65.

Angermeier, P. L. & J. R. Karr, 1994. Biological integrity versus biological diversity as policy directives: protecting biotic resources. BioScience 44: 690–697.

Barbour, M. T., 1997. The re-invention of biological assessment in the U.S. Hum. Ecol. Risk Assess. 3: 933–940.

Barbour, M. T. & C. O. Yoder, in press. The multimetric approach to bioassessment, as used in the United States. In Wright, J. F., D. W. Sutcliffe & M. T. Furse (eds). Assessing the Biological Quality of Freshwaters: RIVPACS and Similar Techniques. Freshwater Biological Association, London.

Barbour, M. T., J. B. Stribling & J. R. Karr, 1995. The multimetric approach for establishing biocriteria and measuring biological condition. In Davis, W. S. & T. P. Simon (eds), Biological Assessment and Criteria: Tools for Water Resource Planning and Decision making. CRC Press, Boca Raton, FL: 63–76.

Barbour, M. T., J. Diamond & C. Yoder, 1996a. Biological Assessment Strategies: applications and limitations. In Grothe, D. R., K. L. Dickson & D. K. Reed (eds). Whole-Effluent Toxicity Testing: an Evaluation of Methods and Predictability of Receiving System Responses. SETAC Publications, Pensacola, FL: 245–270.

Barbour, M. T., J. Gerritsen, G. E. Griffith, R. Frydenborg, E. McCarron, J. S. White & M. O. Bastian, 1996b. A framework for biological criteria for Florida streams using benthic macroinvertebrates. J. n. am. Benthol. Soc. 15: 185–211.

Barbour, M. T., J. Gerritsen, B. D. Snyder & J. B. Stribling, 1999. Rapid Bioassessment Protocols for Use in Streams and Wadeable Rivers: Periphyton, Benthic Macroinvertebrates and Fish. Second Edition. EPA/841-B-98-010. U.S. EPA, Office of Water, Washington, DC, 344 pp.

Cairns, J., Jr., 1980. Beyond single species toxicity testing. Mar. envir. Res. 3: 157–159.

Courtemanch, D. L., 1995. Merging the science of biological monitoring with water resource management policy: Criteria development. In Davis, W. S. & T. P. Simon (eds), Biological Assessment and Criteria: Tools for Water Resource Planning and Decision Making. Boca Raton, FL: 315–325.

Courtemanch, D. L., S. P. Davies & E. B. Laverty, 1989. Incorporation of biological information in water quality planning. Envir. Manage. 13(1): 35–41.

Davis, W. S. & T. P. Simon (eds), 1995. Biological Assessment and Criteria: Tools for Water Resource Planning and Decision Making. Lewis Publishers, Boca Raton, FL, 415 pp.

Davis, W. S., B. D. Snyder, J. B. Stribling & C. Stoughton, 1996. Summary of State biological assessment programs for streams and rivers. EPA 230-R-96-007. U.S. Environmental Protection Agency, Office of Planning, Policy, and Evaluation, Washington, DC, 159 pp.

DeShon, J. E., 1995. Development and application of the invertebrate community index (ICI). In Davis, W. S. & T. P. Simon (eds), Biological Assessment and Criteria: Tools for Water Resource Planning and Decision Making. Lewis Publishers, Boca Raton, FL: 217–244.

Hughes, R. M. 1995, Defining acceptable biological status by comparing with reference conditions. In Davis, W. S. & T. P. Simon (eds), Biological Assessment and Criteria: Tools for Water Resource Planning and Decision Making. Lewis Publishers, Boca Raton, FL: 31–47.

Gibson, G. R., M. T. Barbour, J. B. Stribling, J. Gerritsen & J. R. Karr, 1996. Biological criteria: Technical guidance for streams and small rivers (revised edition). EPA 822-B-96-001. U.S. Environmental Protection Agency, Office of Water, Washington, DC, 162 pp.

Jackson, S. & W. Davis, 1994. Meeting the goal of biological integrity in water-resource programs in the US Environmental Protection Agency. J. n. am. benthol. Soc. 13: 592–597.

Jorling, T., 1977. Incorporating ecological interpretation into basic statutes. In Ballentine, R. K. & L. J. Guarraia (eds), The Integrity of Water: A Symposium. U.S. Environmental Protection Agency, Washington, DC: 9–14.

Karr, J. R., 1981. Assessment of biotic integrity using fish communities. Fisheries 6(6): 21–27.

Karr, J. R., 1991. Biological integrity: a long-neglected aspect of water resource management. Ecol. Applic. 1: 66–84.

Karr, J. R., 1993. Defining and assessing ecological integrity beyond water quality. Envir. Tox. Chem. 12: 1521–1531.

Karr, J. R. & E. W. Chu. 1999. Restoring life in running waters: better biological monitoring. Island Press, Washington, DC, 206 pp.

Karr, J. R., K. D. Fausch, P. L. Angermeier, P. R. Yant & I. J. Schlosser, 1986. Assessing Biological Integrity in Running Waters: A Method and its Rationale. Special publication 5. Illinois Natural History Survey, Urbana, 28 pp.

Keddy, P. A., H. T. Lee & I. C. Wisheu, 1993. Choosing indicators of ecosystem integrity: Wetlands as a model system. In Woodley, S., J. Kay & G. Francis (eds), Ecological Integrity and the Management of Ecosystems. St. Lucie Press, Boca Raton, FL: 61–79.

Lenat, D. R., 1988. Water quality assessment of streams using a qualitative collection method for benthic macroinvertebrates. J. n. am. benthol. Soc. 7: 222–233.

Lenat, D. R., 1993. A biotic index for the southeastern United States derivation and list of tolerance values with criteria for assigning water quality rating. J. n. am. Benthol. Soc. 12: 279–290.

Lenat, D. R. & M. T. Barbour, 1994. Using benthic macroinvertebrate community structure for rapid, cost-effective, water quality monitoring: rapid bioassessment. In Loeb, S. L. & A. Spacie (eds), Biological Monitoring of Aquatic Systems. Lewis Publishers, Boca Raton, FL: 187–215.

Mackenthun, K. M., 1977. Legislative Requirements. In R. K. Ballentine & L. J. Guarraia (eds), The Integrity of Water: a Symposium. U.S. Environmental Protection Agency, Washington, DC: 5–7.

McCarron, E. & R. Frydenborg, 1997. The Florida bioassessment program: an agent for change. Hum. Ecol. Risk. Assess. 3: 967–977.

Moog, O. & A. Chovanec, 2000. Assessing the ecological integrity of rivers: walking the line among ecological, political and administrative interests. Hydrobiologia 422/423: 99–109.

Ohio Environmental Protection Agency, 1987. Biological Criteria for the Protection of Aquatic Life: Volumes I–III. Volume I: The role of biological data in water quality assessment. Volume II: Users manual for biological field assessment of Ohio surface waters. Volume III: Standardized biological field sampling and laboratory methods for assessing fish and macroinvertebrate communities. Ohio EPA, Division of Water Quality Monitoring and Assessment, Surface Water Section, Columbus (updated 1988, 1989), 287 pp.

Plafkin, J. L., M. T. Barbour, K. D. Porter, S. K. Gross & R. M. Hughes, 1989. Rapid bioassessment protocols for use in streams and rivers: benthic macroinvertebrates and fish. EPA 440-4-89-001. U.S. Environmental Protection Agency, Office of Water Regulations and Standards, Washington, DC, 177 pp.

Reckhow, K. H., 1994. A decision analytic framework for environmental analysis and simulation modeling. Envir. Toxicol. Chem. 17: 1901–1906.

USEPA, 1988. Proceedings of the first national workshop on biological criteria. Lincolnwood, Illinois, December 2–4, 1987. EPA 905-9-89-003. Region V, U.S. Environmental Protection Agency, Chicago, 47 pp.

USEPA, 1990. Biological Criteria: National Program Guidance for Surface Waters. EPA 440-5-90-004. Office of Water Regulations and Standards, U.S. Environmental Protection Agency, Washington, DC, 87 pp.

USEPA, 1998a. Water Quality Criteria and Standards Plan – Priorities for the Future. EPA 822-R-98-003. Office of Water, U.S. Environmental Protection Agency, Washington, DC, 54 pp.

USEPA, 1998b. Lake and Reservoir Bioassessment and Biocriteria: Technical Guidance Document. EPA 841-B-98-007. Office of Water, United States Environmental Protection Agency, 191 pp.

USGPO (U.S. Government Printing Office), 1972a. Report of the Committee on Public Works – United States House of Representatives with Additional and Supplemental Views on H.R. 11896 to Amend the Federal Water Pollution Control Act. House Report 92-911. 92nd Congress, 2nd session, March 11, 1972: P. 149.

USGPO (U.S. Government Printing Office), 1972b. Report of the Senate Public Works Committee – United States Senate. Federal Water Pollution Control Act Amendments of 1972. P.L. 92-500. Senate Report 92-414: P. 1468.

Yoder, C. O., 1995. Policy issues and management applications for biological criteria. In Davis, W. S. & T. P. Simon (eds), Biological Asessment and Criteria: Tools for Water Resource Planning and Decision Making. Lewis Publishers, Boca Raton, FL: 327–343.

Yoder, C. O. & E. T. Rankin, 1995. Biological criteria program development and implementation in Ohio. In Davis, W. S. & T. P. Simon (eds), Biological Assessment and Criteria: Tools for Water Resource Planning and Decision Making. Lewis Publishers, Boca Raton, FL: 109–144.

Hydrobiologia **422/423:** 465–475, 2000.
M. Jungwirth, S. Muhar & S. Schmutz (eds), Assessing the Ecological Integrity of Running Waters.

GIBSI – An integrated modelling system prototype for river basin management

A. N. Rousseau[1], A. Mailhot[1], R. Turcotte[1], M. Duchemin[1], C. Blanchette[1], M. Roux[1], N. Etong[1], J. Dupont[2] & J.-P. Villeneuve[1]
[1]*INRS-Eau, 2800 rue Einstein C.P. 7500, Sainte-Foy, Québec, Canada G1V 4C7*
E-mail: Alain_Rousseau@INRS-EAU.UQuebec.ca
[2]*Ministère de l'Environnement et de la Faune du Québec, 675, Boul. René-Lévesque Est, 7ième Étage, Québec, Canada G1R 5V7*

Key words: integrated modelling system, river basin management, hydrology, soil erosion, agricultural-chemical transport, water quality

Abstract

This paper introduces GIBSI, an integrated modelling system prototype designed to assist decision makers in their assessment of various river basin management scenarios in terms of standard water physical and chemical parameters and standards for various uses of the water. GIBSI runs on a personal computer and provides a user-friendly framework to examine the impacts of agricultural, industrial, and municipal management scenarios on water quality and yield. A database (including spatial and attribute data) and physically-based hydrological, soil erosion, agricultural-chemical transport and water quality models comprise the basic components of the system. A geographical information system and a relational database management system are also included for data management and system maintenance. This paper illustrates potential uses of GIBSI by presenting two sample applications applied to a 6680 km^2 complex river basin (63.2% forest, 17.2% agricultural land, 15.3% bush, 3.1% urban development and 1.2% surface water; population: 180 000) located in Québec, Canada: (i) a timber harvest scenario and (ii) a municipal clean water program scenario. Simulation results of the timber harvest scenario showed how clear-cut activities could lead to earlier and larger spring runoff than in the investigated reference state. Results of the municipal clean water scenario revealed that substantial reduction in coliform counts and total phosphorus could be made by constructing and operating wastewater treatment plants. These applications provide a framework of how GIBSI can be used as a decision support system and ultimately as a means of incorporating sound science into political decisions involving river basin management.

Introduction

In Québec, surface water accounts for more than 80% of the drinking water supply and, not surprisingly, plays an important role in the economy. Over the course of the last 40 years, the natural state of these waters, as well as those in the rest of Canada, has been affected by acid rain, surface runoff, and waste water discharges from urban, industrial, agricultural and rural sources. In certain regions of the country, these inputs have caused important changes to some aquatic ecosystems and, consequently, seriously impacted some traditional surface water uses such as sport fishing and swimming. On the other hand, during the same period, Canada experienced economic growth so that the population now enjoys one of the highest standards of living in the world. Unfortunately, as these facts indicate, industrial development and economic growth are almost always accompanied by a depreciation in the quality of natural water resources.

The above ecological-economic paradox is a motivation for river basin management. Because of this, project feasibility studies for industrial, forestry, urban, and agricultural developments need to balance economic growth with the preservation of ecological integrity, biological diversity, and social equity of all community members, as well as, future generations. Consequently, in addition to economic projections, such studies should include a complete investigation of the total impact of a project on the water quality and yield of a river basin.

The present state of the art of hydrological modelling and software engineering has reached a point where it is now possible to construct integrated modelling systems capable of simulating the impacts of various management practices (i.e. industrial, municipal and agricultural) on the water quality and yield of a river basin. These systems, which ought to be user-friendly for decision makers, must be based on sound hydrological modelling systems and efficient data management systems to better understand the processes causing changes and to predict the impacts of specific actions before problems occur. The integrated modelling system GIBSI (Gestion Intégrée des Bassins versants à l'aide d'un Système Informatisé) is a prototype designed to incorporate the above aspects (Villeneuve et al., 1998). For the decision maker, GIBSI offers the ability to predict the impacts of industrial, forestry, urban, and agricultural projects with respect to targeted water quality and yield standards for various uses of the water (i.e. agricultural, fish and wildlife, industrial, recreational). The scope of this paper is to present: (i) the basic features of GIBSI, and (ii) two sample applications on a complex Québec river basin: a timber harvest scenario and a municipal clean water program scenario.

The integrated modelling system

GIBSI proposes a scenario approach to evaluate the impacts of municipal, industrial, forestry, and agricultural projects on the water quality and yield of a river basin. Two types of users have been identified: the technical expert and the water resources manager. The former is responsible for setting up the system on a study river basin, while the latter elaborates scenarios and performs simulations.

Basic components of the system

GIBSI is comprised of a database, a Relational DataBase Management System (RDBMS), physically-based simulation models (hydrology, soil erosion, agricultural-chemical transport, and water quality), management models (point source, agricultural non-point source and reservoir management models) and a Geographic Information System (GIS). The GIS used is GRASSLAND (L.A.S., 1996), an open GIS built around GRASS (U.S. Army Corps of Engineers – CERL, 1988). MicrosoftAccessTM (Simpson, 1994) is the RDBMS of all relevant attribute data. The database includes spatial and attribute data (e.g. digital elevation model, meteorological data, soil database, gauge station locations, simulation results, quantitative and qualitative data associated with administrative units such as municipalities: livestock production, crop management, etc.) required for constructing a scenario and running a simulation. A complete description of the database is presented by Villeneuve et al. (1998). From a software engineering standpoint, GIBSI is modular; this structure allows, if necessary, for integration of new system components or replacement of existing ones (e.g. simulation models, RDBMS, GIS) with other commercial software. Interactions between users and system components are made through a single interface. Finally, GIBSI runs on a personal computer and does not require extensive computing equipment.

Simulation models, computational domains, and output variables

The hydrological model is HYDROTEL (Fortin et al., 1995), a distributed physically-based model compatible with remote sensing and GIS. Revised algorithms of the USLE (Wischmeier & Smith, 1978) and Yalin's (1963) sediment transport equation are used to model soil erosion and the inherent sediment transport. Modelling of agricultural-chemical transport and transformations are based on the nitrogen, phosphorus and pesticide transport algorithms of SWAT (Arnold & Williams, 1995) and EPIC (Williams, 1995). Finally, the water quality model is built around QUAL2E (Brown & Barnwell, 1987), a standard water quality model which simulates: (i) dispersion and advection of water contaminants; (ii) algal growth; (iii) nitrogen and phosphorus cycles; (iv) coliform decay; (v) re-aeration; and (vi) water temperature. Also, the sediment routing algorithm of SWAT (Arnold & Williams, 1995), which accounts for stream deposition and bed degradation, has been added to QUAL2E. A detailed description of the simulation models is presented by Villeneuve et al. (1998).

The hydrological simulation unit is an elementary drainage basin unit (i.e. area comprised of three soil layers of varying depths and drained by one reach). One-dimensional reaches of uniform hydraulics and biochemical characteristics are used for river flow routing and water quality modelling. The spatial simulation unit for modelling soil erosion and agricultural-chemical transport processes is the classified agricultural land use obtained from remote sensing information. The elementary basin units are traced using the basin's river network and the direction of flow from cell to cell obtained from a digital elevation model. The computational time step is the day. This time scale corresponds to the measurement frequency of meteorological data and is often associated with the application time scale of various agricultural management scenarios (e.g. fertilization dates). From a modelling point of view, the day represents the upper time scale limit for adequate forecasting of flooding events.

The list of output variables includes: river flows, sediment loads, coliform counts, water temperature, as well as loads and concentrations of dissolved oxygen, chlorophyll *a*, dissolved phosphorus and particulate organic phosphorus, particulate organic nitrogen, nitrate-nitrogen, nitrite-nitrogen, ammonia-nitrogen, BOD, and pesticides. Note that in GIBSI the absolute values of these variables may be interpreted in terms of various water quality and yield standards for various uses of the water (i.e. agricultural, fish and wildlife, industrial, recreational).

Scenarios, simulations, and interpretation of results

GIBSI offers three river basin management models: (i) a point source management model, (ii) an agricultural nonpoint source management model, (iii) and a reservoir management model. These management models, which are described in Villeneuve et al. (1998), allow users to implement various river basin management scenarios. Typical scenarios can be built by modifying:

(i) point sources, that is, waste water treatment plant discharge points (e.g., location, connected population, treatment type, load characteristics) and industrial discharge points (e.g. location, load characteristics);
(ii) agricultural production systems (e.g. number and type of livestock per administrative units, crop management practices and crop rotation systems, application dates and rates of chemical fertilizers, manure, and pesticides);
(iii) land use classes (e.g. global or discrete land use reclassification within a delineated area, that is, from one or several land use categories into another); and
(iv) hydraulics (e.g. addition or removal of reservoirs, hydraulic characteristics of dams).

Note that a scenario characterizes a specific state of the study river basin and that any states of the basin can be defined as a reference state. Once a management scenario is defined, the user must select a meteorological series before running a simulation (i.e. coupling of a scenario and meteorological data). Selection of meteorological data usually implies a simple statistical analysis of the selected data set so the simulation results can be linked to specific conditions (e.g. wet and dry periods, seasons or years). It is noteworthy that the simulation period corresponds to the time span of the selected meteorological data set. Simulation results are saved in the database along with a text file containing a complete description of the simulation context (e.g. comment lines describing the scenario and meteorological series). A particular value of GIBSI over other integrated modeling systems such as NELUP (O'Callaghan, 1995), WaterWare (Jamieson & Fedra, 1996), PÉGASE (Smitz et al., 1997), and BASINS (US EPA, 1997) is that it provides a user-friendly framework to examine the impacts of various agricultural, industrial, and municipal management scenarios on water quality and yield in terms of specific meteorological conditions, as simulation during wet versus dry periods can yield very different results.

Interpretation of results, using thematic maps and graphs, is based on comparing the reference state with results obtained either from combinations of different scenarios and one meteorological series or from combinations of one scenario and different meteorological series.

A prototype for connecting river water quality and flow regime to the ecological assessment of running waters

The simulated values of the physical and chemical parameters of water produced by GIBSI may be interpreted in terms of various water use standards. In this way, GIBSI provides a means to identify the river segments which may be suitable for specific water uses (e.g. source of drinking water, recreational boating or swimming, etc.). Additionally, if an *in situ* appraisal of the ecological integrity of several river segments is known, the simulated values of the physico-chemical parameters may be used to complement the ecological assessment of a watercourse. Following this line of reasoning, if there exists a reliable watershed database documenting the evolution of riparian vegetation, biotic community, land use, and hydrographic network (including identification of straightened out meanders and artificial river banks and floodplains), the prototype may be used to investigate the impact of anthropogenic disturbances on the natural state of relevant aquatic habitats. For example, for a given reference state and a given scenario characterizing two different time horizons of a watershed (baseline state and any other states), GIBSI could be used to estimate the river flow regime and the river water quality associated with the relevant aquatic ecosystems. In that sense GIBSI represents a valuable prototype for connecting water quality and flow regime of running waters to the *in situ* watershed assessments of the ecological integrity of running waters.

Sample applications

The Chaudière River basin was selected as the pilot basin for the development of GIBSI. It is located south of Québec City and covers an area of 6680 km^2. It has a population of around 180 000 and is characterized by various land use, agricultural, and industrial activities. This river basin has been designated by the Québec Government as the pilot river basin for a new administrative water management authority. The river basin is representative of most Québec basins, with around 63.2% forest, 17.2% agricultural land, 15.3% bush, 3.1% urban development and 1.2% surface water. Figure 1 presents the principal sub-basins of the Chaudière River basin. In order to illustrate some of the analytical functions of GIBSI, this section presents the results of two sample applications: (i) a timber harvest scenario and (ii) an implementation of a municipal clean water program scenario. In the developmental framework of GIBSI, these scenarios, although fictitious, were used to validate the hydrological and water quality models.

Model calibration

For this study, model calibration involved calibration of the hydrological and the water quality models. In the case of the latter model, only point sources of pollution were considered and several simplifications and assumptions were made since testing of the erosion and agricultural-chemical models is not completed at this point. Some such simplifications were that contaminant loads from combined sewer overflows, untreated industrial effluents, as well as leaking septic tanks and surface runoff in rural areas were neglected. In the case of the former model, default management parameters were assumed for reservoir operations. It is beyond the scope of this paper to present this calibration; a description of this exercise has been reported elsewhere (Turcotte et al., 1999).

Hydrological model

Because there is no gauge station at the Chaudière River basin outlet, calibration of the hydrological model was performed using two years of measured streamflows from the St. Lambert station, which covers 85% of the studied basin. Temporal validation was performed using data from two other years. Spatio-temporal validations were done as well on two different 700 km^2 sub-basins: a primarily forested basin, the Famine River basin, and a mostly agricultural basin, the Beaurivage River basin (see Figure 1). Nash-Sutcliffe coefficient values between 0.71 and 0.89 were obtained for all calibration and validation simulations (Turcotte et al., 1999). Moreover, simulated annual water yields were always within 15% of those observed, for both the calibration and validation periods.

Water quality model

Calibration of the water quality model was done using a year's worth of measured data obtained from 12 water quality monitoring stations located throughout the studied river basin for a total of 116 data points. The average absolute error between simulated and observed coliform counts (lumped in space) was centred around zero and close to 45% of the errors were less than 150 coliforms per 100 ml, which is quite acceptable considering that the standard for potability of untreated water is 1000 coliforms/100ml. For cases where simulated coliform counts were less than those measured, Turcotte et al. (1999) reported that the errors were most likely linked to the aforementioned assumptions on non-point and point sources of coliforms. For cases where simulated coliform counts were greater than those observed, larger than actual coliform loads and/or smaller than actual streamflows probably explain some of the observed errors (Turcotte et al., 1999).

Timber harvest scenario

This scenario examines the potential impact of a substantial timber harvest (covering nearly 200 km^2 of the

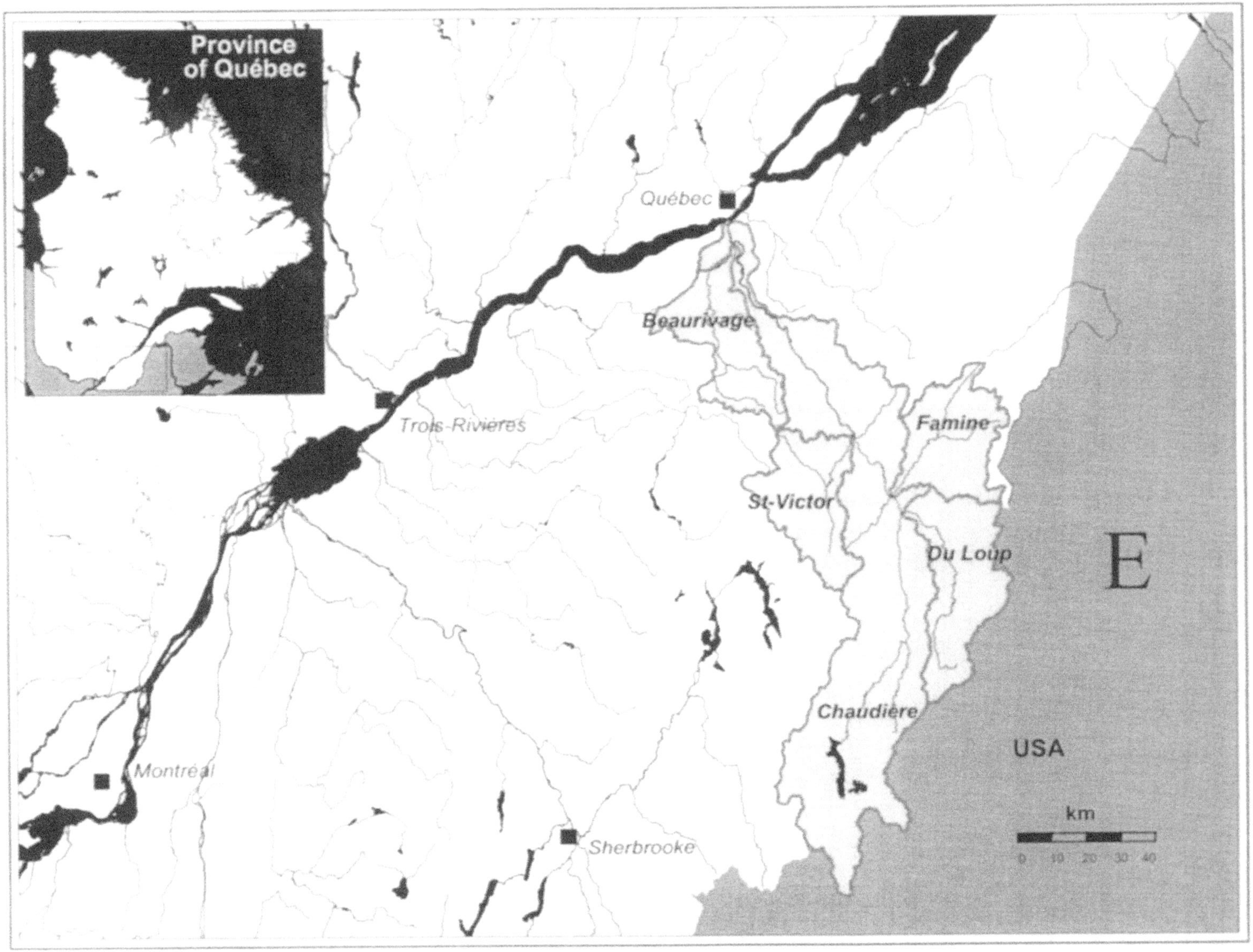

Figure 1. Principal sub-basins of the Chaudière River basin.

6680 km^2 Chaudière River basin) on the Famine River sub-basin (see Figure 2). For this scenario, the potential impacts were analysed in terms of two simulated responses at the sub-basin outlet: (i) the annual water yield and (ii) the spring runoff hydrograph.

As shown on Figure 2, coniferous forests, deciduous forests and bushes characterizing the reference state, and located in the northern portion of the Famine River sub-basin, were reclassified as clear-cut land (bare soil) – affecting 180 km^2 or 27% of the sub-basin area. Since GIBSI does not simulate forest growth, timber harvest was assumed to have taken place in the summer and the simulation period was that of the following year which extended from the current fall season to that of the following year (i.e. November of year 1 to November of year 2). To obtain a representative sample of potential impacts, eight meteorological series were used to generate eight sets of simulation results – where a set was made up of results from the current land use (the reference state) and those from the above clear-cut scenario (the timber harvest scenario). The one-year-long meteorological series were independent of one another and covered the period of November 1985 to November 1995. Figure 3 depicts a set of daily flows simulated with the 12-month meteorological series of November 1986 to November 1987. Only the simulated flows between March 15 and July 15, 1987, are considered for this sample application.

Table 1. Simulated annual water yields for the reference state and the timber harvest scenario

Simulation period (November to November)	Reference state [10^6 m^3]	Timber harvest [10^6 m^3]
1984–1985	248	315
1985–1986	370	440
1986–1987	226	298
1987–1988	233	307
1988–1989	218	293
1989–1990	483	559
1990–1991	319	392
1991–1992	249	321

Results and discussion

Annual water yield

Table 1 introduces the results of these simulations. On average (eight sets of simulation results), the timber harvest produced a 72 million m^3 increase in annual water yield when compared with the reference state (293 million m^3). Note that the annual increase in water yield varied between 66 and 76 million m^3 with a 3 million m^3 standard deviation. In terms of relative increase, the former corresponds to a 16% to 34% increase when compared with the reference state. This timber-harvest-related increase can be partly attributed to a drop in evapotranspiration, which increased the stored water volume and consequently storm runoff.

Spring runoff hydrograph

Figure 3 shows for the year following timber harvest an earlier spring runoff than that observed for the reference state. Moreover, not only was spring runoff early, but a faster time to peak, a larger spring runoff, and a shorter recession time were predicted. This behaviour characterized all sets of simulation results. Lack of tree shading partly explained the aforementioned changes in the spring runoff of the reference state. Consequently, clear-cut areas contributed to faster snowmelt and earlier spring runoff than those observed before timber harvesting.

Municipal clean water program scenario

Over the course of the last twenty years, the 'Programme d'assainissement des eaux du Québec' and the 'Programme d'assainissement des eaux municipales' have targeted treatment of municipal wastewater, treatment of effluents from pulp and paper production, and construction of manure storage (Gouvernement du Québec, 1997). With these programs, close to 98% of the Québec population connected to a sewer system will have their wastewater treated by year 2000 (compared with 2% in 1978 and 81% in 1995). One way of partially assessing the impact of these programs on surface water quality is to simulate the implementation of municipal wastewater treatment plants on the Chaudière River basin. This was accomplished by performing one set of simulations involving the following scenarios: (i) a first scenario depicting the 1997 conditions (the reference state) which is referred to as the 'post municipal clean water program (MCWP) scenario' and (ii) a second scenario illustrating the initial conditions – before implementation of municipal wastewater treatment plants – which is referred to as the 'pre-MCWP scenario'. These simulations did not consider contaminant loads from agricultural runoff, combined sewer overflows, or any industries not directly connected to a sewer system. Contaminant loads from septic systems in rural areas were neglected as well. The simulated river flows between March 15 and

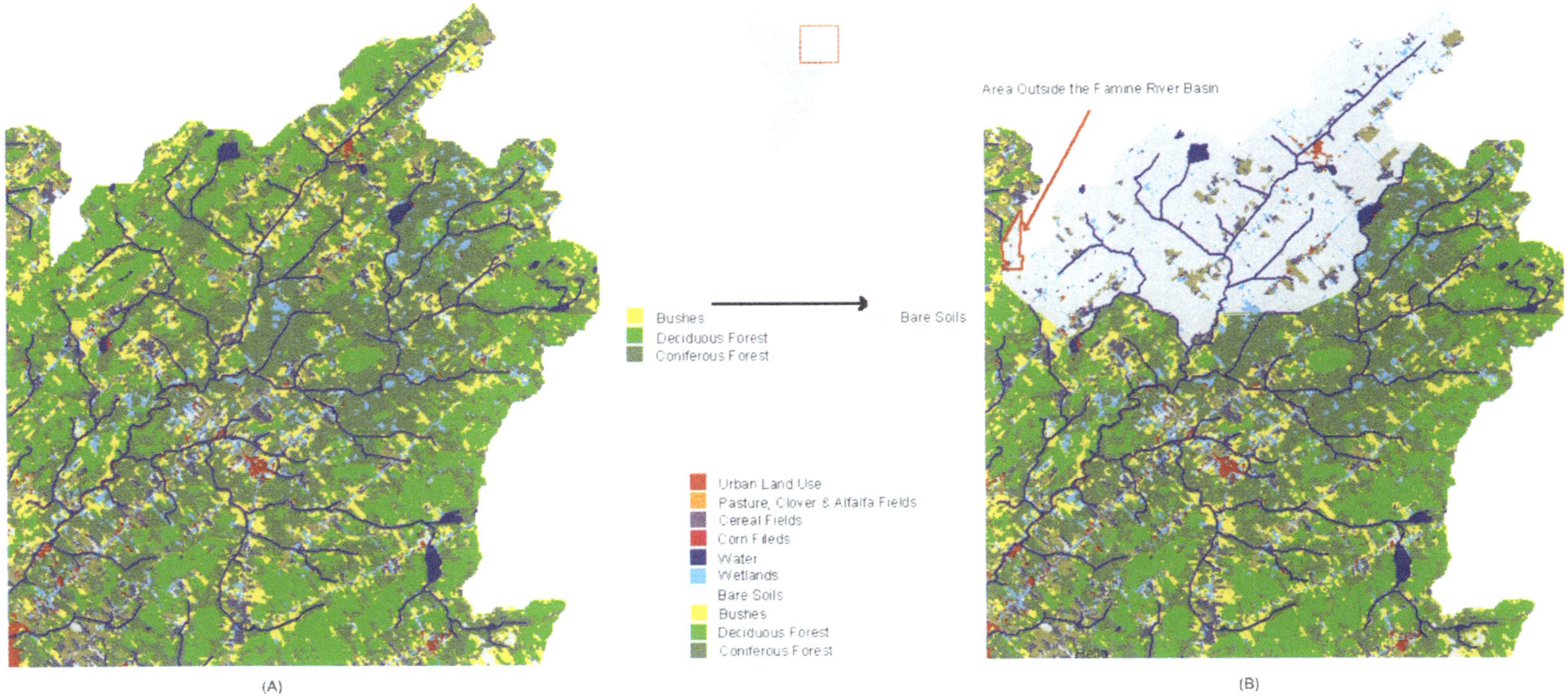

Figure 2. Distributed land use: (a) reference state and (b) timber harvest scenario.

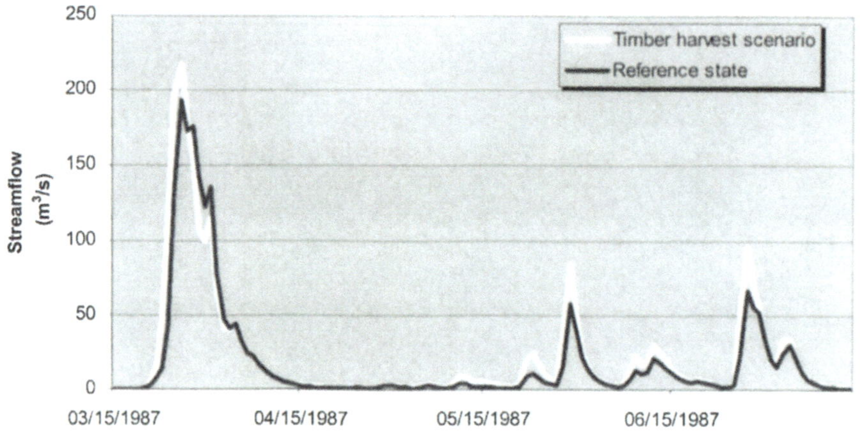

Figure 3. Daily simulated flows for the timber harvest scenario and the reference state at the outlet of the Famine River sub-basin for the 4-month period of March 15 to July 15, 1987.

July 15, 1987, are considered for this sample application. These flows correspond to those of the reference state defined in the above timber harvest application.

Figure 4 shows all surveyed municipal and industrial discharge points within the boundaries of the Chaudière River basin. In 1997, 66% of the Chaudière River basin's population was connected to a sewer system. Out of the 45 surveyed municipal discharge points, only seven did not have their effluent conveyed to a wastewater treatment plant. These points discharged the effluents of about 2% of the population connected to a sewer system. Table 2 presents the types of wastewater treatment plants in operation in the study river basin, their numbers, and the percentage of the population connected to these plants. All municipal loads (before and after plant operation) were estimated from statistical data compiled by the Québec Ministry of Environment and Wildlife (QMEW). The water quality variables considered for this application were concentrations of nitrogen, phosphorus, suspended solids and coliforms. However, for this presentation only total phosphorus and coliform concentrations are introduced and compared with their respective water quality standards set by the QMEW.

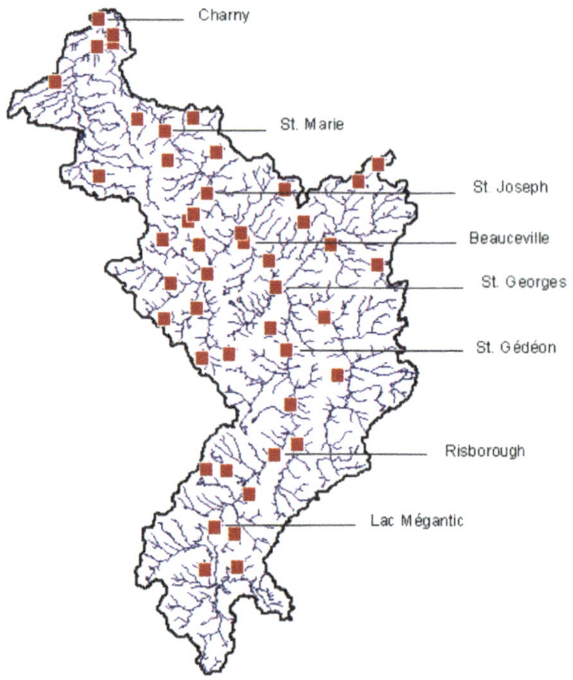

Figure 4. Surveyed municipal and industrial discharge points within the boundaries of the Chaudière River basin.

Results and discussion

Coliforms

Figure 5 presents simulated coliform concentrations along the Chaudière River between the outlet of Lake Mégantic and the river basin outlet. These concentrations were those for June 25, 1987 – the day associated with the lowest simulated flow at the outlet of the Chaudière River basin. This Figure clearly outlines all municipal discharge points located along this stretch of the Chaudière River. For the post-MCWP scenario, all concentrations are of the order of 10 coliform counts per 100 ml. This represents a major reduction in coliform concentrations compared with the pre-MCWP scenario. This Figure also shows that pre-MCWP concentrations rarely met the threshold concentration, less than 1000 coliform counts per 100 ml (Ministère de l'Environnement du Québec,

Table 2. Types of wastewater treatment plants in operation in the Chaudière River basin, their numbers, and the percentage of the population connected to these plants

Types of wastewater treatment plants	Number of treatment plants	Percentage of the population connected
Activated sludge	5	54.5
Aerated lagoon	26	43.5
Lagoon	4	1
Biodisk	1	0.5
Settling basin	2	0.5

1993), for using raw water as drinking water without requiring unconventional treatment. Finally, the post-MCWP coliform concentrations were in general lower than those measured and compiled by the QMEW (Simoneau et al., 1998). This is in line with the aforementioned assumptions. Moreover, it suggests that current water quality problems may be partially or totally attributed to agricultural runoff, combined sewer overflows, septic systems of rural housings, and/or inadequately treated discharges from industries not directly connected to a sewer system.

Total phosphorus

Figure 6 introduces the evolution of total phosphorus concentration at the river basin outlet. Similar to the above results for coliform counts, the post-MCWP concentrations for total phosphorus were below those for the pre-MCWP scenario. For the post-MCWP scenario, however, the total phosphorus concentration exceeded, for about 30 days of the 65-day simulation period, the threshold concentration for lake eutrophication, less than 0.02 and 0.03 mg/l for lakes and rivers, respectively (Ministère de l'Environnement du Québec, 1993). This finding indicates that there is still room for improvement and hence a need to reduce phosphorus load from the effluents of municipal wastewater treatment plants. Note that the post-MCWP total phosphorus concentrations were in general lower than those measured and compiled by the QMEW (Simoneau et al., 1998). Consequently, this does not exclude the probability that current water quality problems may be caused by agricultural runoff, combined sewer overflows, inefficient septic systems of rural housings, and/or any untreated effluents from industries not directly connected to a sewer system.

Conclusion

The sample applications presented in this paper have outlined the simulation and analytical capabilities of the integrated modelling system prototype GIBSI. In the timber harvest application, simulations showed how substantial clear-cut activities could lead to earlier and larger spring runoff than in the investigated reference state. Meanwhile, preliminary results of the municipal clean water scenario revealed that substantial gains in terms of reducing coliform counts and total phosphorus concentrations could be made by constructing and operating wastewater treatment plants. Furthermore, this application suggested that even if all other sources of pollution were controlled, total phosphorus loads from municipal plants could still represent a eutrophication threat to surface water during much of the summer.

Most importantly, this paper has demonstrated how an integrated modelling system prototype such as GIBSI can be used to incorporate sound science into the political decisions involving municipal clean water programs and timber harvest scenarios. Other scenarios, particularly those concerned with agricultural runoff, are already in place and being tested on the Chaudière River basin. Finally, future work will involve addition of an economic module for assessing the economic impact of different management scenarios. In a sustainable river basin management context, the use of GIBSI will thus provide decision makers with a complete tool for exploring a variety of integrated management programs. More importantly, the use of GIBSI may help answer issues such as: (i) How should all currently available regulatory and non-regulatory programs involving landowners and governing bodies (i.e. municipal, provincial and federal) be brought together to achieve effective river basin management? (ii) Who should pay for river basin management? and (iii) What should be done to meet clean water objectives?

Acknowledgements

The authors wish to thank J. Beauchemin, P. Bolduc, J.-F. Gagnon, D. Maheux, J.-F. Tremblay, and S. Trépanier, all of INRS-Eau, for their important contributions to the development of GIBSI. This work is largely funded by the Ministry of the Environment and Wildlife of Québec.

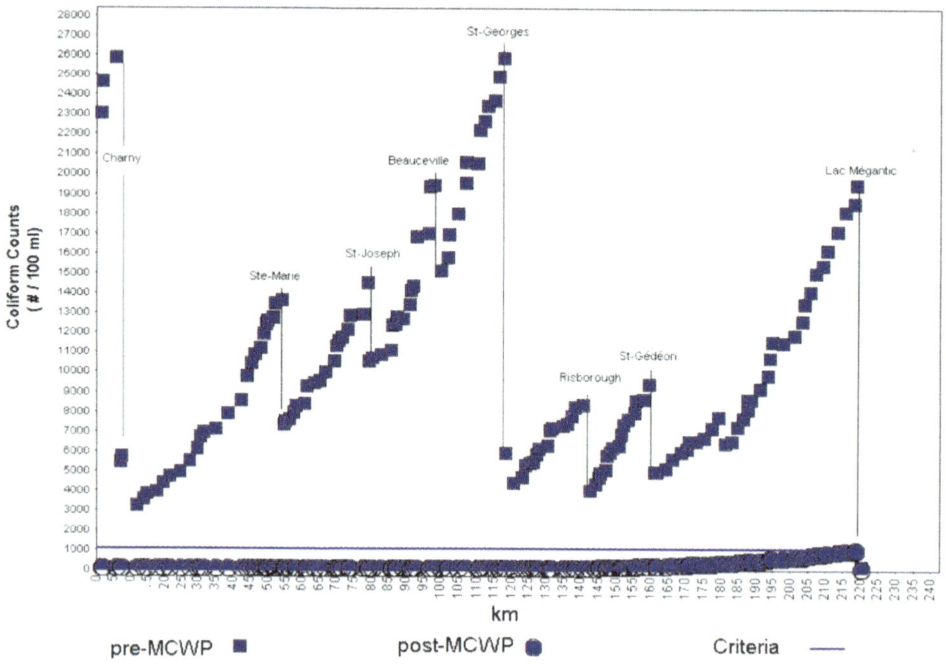

Figure 5. Simulated coliform concentrations along the Chaudière River between the outlet of Lake Mégantic and the river basin outlet for the post- and pre-MCWP scenarios and for June 25, 1987. N.B. Threshold concentration standard or criteria (less than 1000 coliform counts per 100 ml) for using raw water as drinking water without requiring unconventional treatments set by Québec Ministry of Environment and Wildlife (Ministère de l'Environnement du Québec, 1993).

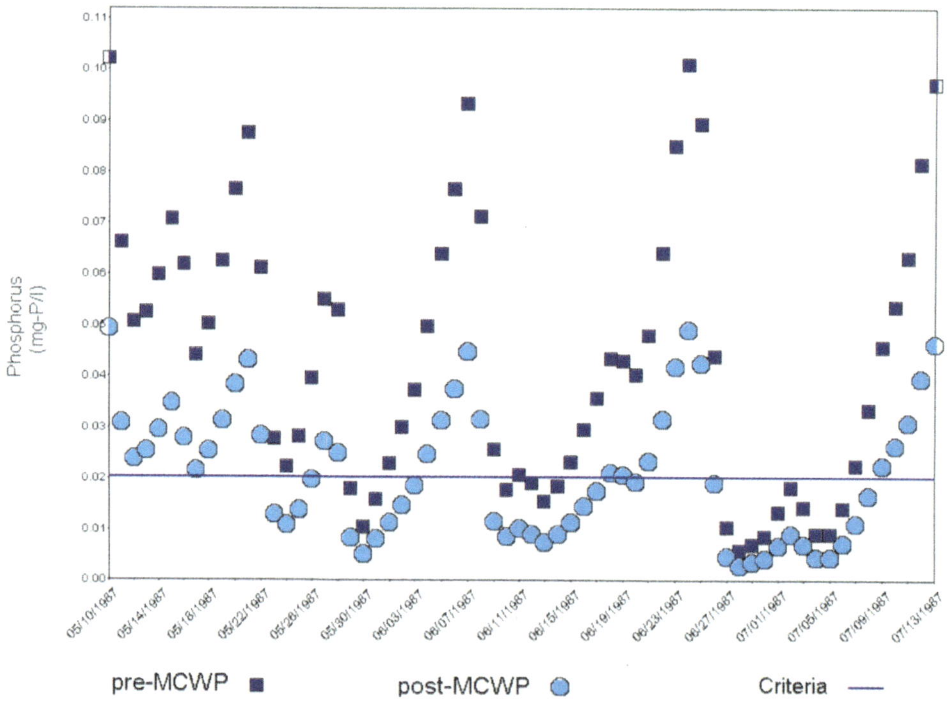

Figure 6. Evolution of simulated total phosphorus concentrations at the Chaudière River basin outlet for the post- and pre-MCWP scenarios and for the simulation period of March 15 to July 15, 1987. N.B. Threshold concentration standard or criteria (less than 0.02 and 0.03 mg/l for lakes and rivers, respectively) for preventing eutrophication set by Québec Ministry of Environment and Wildlife (Ministère de l'Environnement du Québec, 1993).

References

Arnold, J. G. & J. R. Williams, 1995. The EPIC model. In Singh, V. P. (ed.), Computer Models of Watershed Hydrology. Water Resources Publications, Highlands Ranch (CO): 847–908.

Brown, L. C. & T. O. Barnwell Jr., 1987. The Enhanced Stream Water Quality Models QUAL2 and QUAL2E-UNCAS: Documentation and User Manual. Athens, Georgia, U.S., Environmental Protection Agency, 189 pp.

Fortin, J.-P., R. Moussa, C. Bocquillon & J.-P. Villeneuve, 1995. HYDROTEL, un Modèle Hydrologique Distribué Pouvant Bénéficier des Données Fournies par la Télédétection et les Systèmes d'Information Géographique. Revue des Sciences de l'Eau, 8: 97–124.

Gouvernement du Québec, 1997. Symposium sur la Gestion de l'Eau au Québec, Document de Référence. Bibliothèque nationale du Québec, 59 pp.

Jamieson, D. G. & K. Fedra, 1996. The WaterWare decision-support system for river-basin planning. 1. Conceptual design. J. Hydrol., 177: 163–175.

L.A.S., 1996. Grassland User's guide for Windows95 and Windows NT, Version 1.0. Logiciels et Applications Scientifiques (L.A.S.) Inc., Montréal.

Ministère de l'Environnement du Québec, 1993. Critères de Qualité de l'Eau, 1990 (revised 1992). Service d'évaluation des rejets toxiques et Direction de la qualité des cours d'eau, ministère de l'Environnement du Québec, Québec, 425 pp.

O'Callaghan, J. R., 1995. NELUP: an introduction. J. Environ. Planning Manage. 38(1): 5–20.

Simoneau, M., L. Pelletier & N. Martel, 1998. Le bassin de la rivière Chaudière: profil géographique, sources de pollution et interventions d'assainisssement. In Ministère de l'Environnement et de la Faune (ed.), Le Bassin de la Rivière Chaudière: l'État de l'Écosystème Aquatique – 1996, Direction des Écosystèmes Aquatiques, Québec, envirodoq noEN980022.

Simpson, A, 1994. Microsoft Access 2. Sybex Inc., San Francisco.

Smitz, J., E. Everbecq, J. F. Deliège, J. P. Descy, R. Wollast & J. P. Vanderborght,, 1997. PÉGASE, une méthodologie et un outil de simulation prévisionnelle pour la gestion de la qualité des eaux de surface. Tribune de l'eau, No. 588 (4): 73.

Turcotte, R., A. N. Rousseau, A. Mailhot, J.-P. Villeneuve & J. Dupont (1999). GIBSI: an integrated modelling system for sustainable watershed management. In Proceedings of the Environmental Permitting Symposium, Air & Waste Management Association (AWMA) Symposium & Courses, Research Triangle Park, NC, February 17–19, 1999. AWMA, One Gateway Center, Third Floor, Pittsburgh, PA 15222.

U.S. Army Corps of Engineers – CERL, 1988. GRASS Reference Manual. Champaign, Ill.: U.S. Army Corps of Engineers – CERL.

US EPA, 1997. Basins Fact Sheet. United States Environmental Protection Agency, Office of Water (4505), Washington, DC 20460, EPA-823-F-97-0/0, October 1997.

Villeneuve, J.-P., C. Blanchette, M. Duchemin, J.-F. Gagnon, A. Mailhot, A. N. Rousseau, M. Roux, J.-F. Tremblay & R. Turcotte, 1998. Rapport Final du Projet GIBSI, Mars 1998. Rapport No. R-462. INRS-Eau, Sainte-Foy, Québec, 371 pp.

Williams, J. R., 1995. The EPIC model. In Singh, V. P. (ed.), Computer Models of Watershed Hydrology. Water Resources Publications, Highlands Ranch (CO): 909–1000.

Wischmeier, W. H. & D. D. Smith, 1978. Predicting Rainfall Erosion Losses – A Guide to Conservation Planning. Agricultural Handbook No. 537. U.S. Department of Agriculture, Washington, DC. 58 pp.

Yalin, Y. S., 1963. An Expression for Bed-Load Transportation. J. Hydraulic Division, ASCE, 89(HY3): 221–250.

Hydrobiologia **422/423:** 477–487, 2000.
M. Jungwirth, S. Muhar & S. Schmutz (eds), Assessing the Ecological Integrity of Running Waters.

Scientific base and modular concept for comprehensive assessment of streams in Switzerland

U. Bundi[1], A. Peter[2], A. Frutiger[1], M. Hütte[1], P. Liechti[3] & U. Sieber[3]
[1]*EAWAG, Swiss Federal Institute for Environmental Science and Technology, CH-8600 Dübendorf, Switzerland*
[2]*EAWAG, Swiss Federal Institute for Environmental Science and Technology, Limnological Research Center, CH-6047 Kastanienbaum, Switzerland*
[3]*BUWAL, Swiss Agency for The Environment, Forests and Landscape, CH-3003 Bern, Switzerland*

Key words: stream assessment, ecological integrity, restoration, rehabilitation

Abstract

There is a difficult polarity between reality and ecologically desirable goals of stream ecosystem management in Switzerland. Most streams are exposed to a variety of impacts. Due to a high intensity of anthropogenic activity there is a high pressure on land and water use. Essential water uses have to be guaranteed. On the other hand there is also a strong need to preserve or restore (bring back ecological integrity) or rehabilitate streams (bring back 'relative ecological integrity'). Stream assessment should, therefore, produce sound data suitable for characterizing the ecological condition of streams and for supporting their sustainable management. The methods should include a system approach as the basic unit and sound scientific principles of ecological integrity emphasizing habitat connectivities. The methods should allow: • the condition of streams to be rationally described and judged, • identification of different kind of impacts on a stream, • verification of the effects of water protection measures, • identification of suitable future actions in the context of a whole stream system. In order to cover the various requirements a modular concept for stream system-oriented analysis was developed. Altogether nine modules are elaborated. Each module deals with specific stream features (two hydrodynamic and ecomorphological modules, five biological modules, two chemical and ecotoxicological modules). Single modules, several modules, or all of them may be applied, depending on the purpose of the analysis. For each module a cost-effective survey procedure (rapid method) and a progressively more rigorous method providing more detailed information is suggested. Based on the result of the analysis, stream rehabilitation concepts can then be elaborated in cooperation with managers, river engineers and biologists.

Introduction

Situation of streams and rivers

Switzerland covers an area of 41 000 km^2. With an average rainfall of 1.5 m, the country is rich in water. The total length of the streams and rivers is 65 000 km. The population density (172 per km^2) shows large regional variation. Most of the population as well as the economic and agricultural activity is concentrated in the non-mountainous region of the Swiss Plateau, which covers about one third of the territory. 60% of the country's electric energy consumption is covered by hydropower. The pressure on land and water use has resulted in extensive physical degradation of rivers and streams. Figure 1 gives a synopsis of the development of river engineering works along major rivers and streams in the north-eastern part of Switzerland (Koblet, 1995).

Figure 2 shows the fractions of streams that have been lost in the Canton of Zurich, part of the Swiss Plateau, through putting them into culverts. This has been done for flood protection, to gain land for urban development and to improve farming conditions. About 35% of all streams in the Canton have been lost in this manner. First order streams are the most often involved, so that today much of the network of small streams is missing (Hütte et al., 1994). Of the

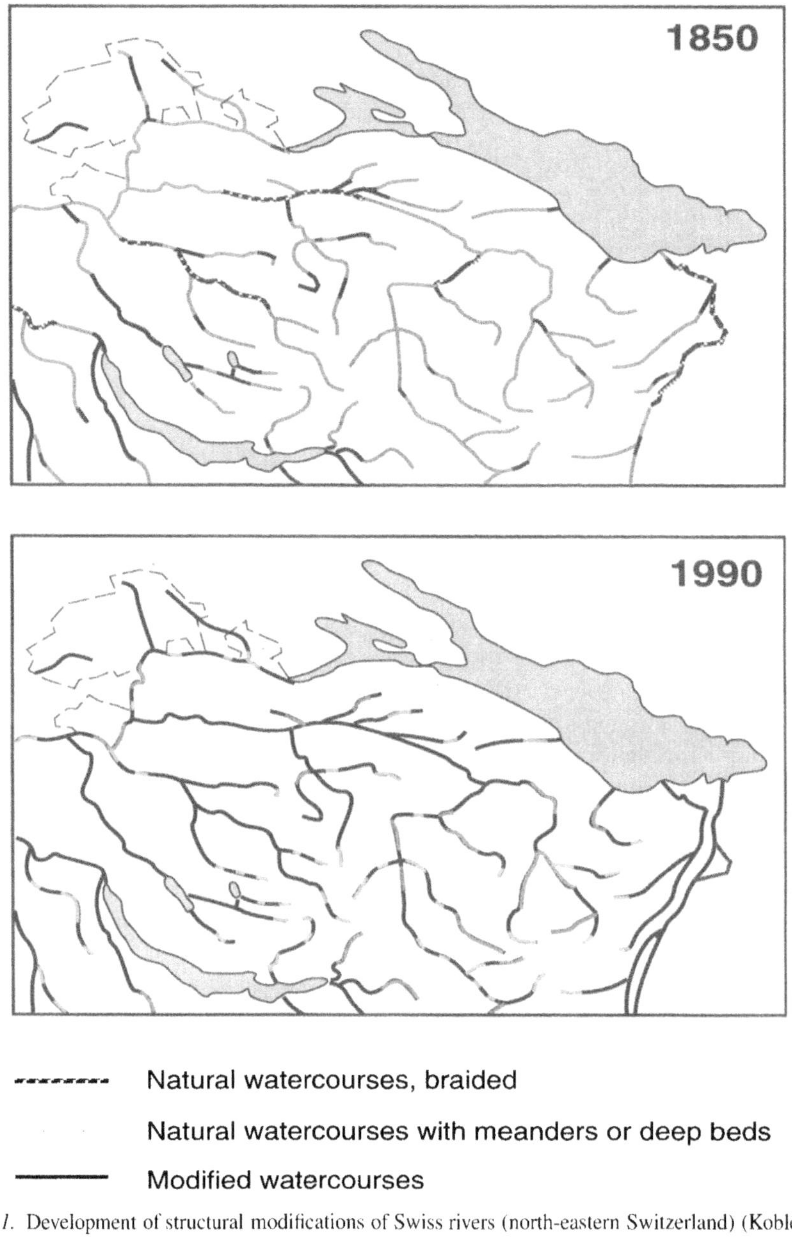

Figure 1. Development of structural modifications of Swiss rivers (north-eastern Switzerland) (Koblet, 1995).

remaining streams, a large majority has been physically altered in various ways. The situation in Canton Zurich reflects that in large parts of Switzerland. In the alpine regions there is less pressure from land use, but numerous streams suffer from the impact of hydropower generation (total or major withdrawal of water, temporary or seasonal changes of flow regime, thermal chaos, disruptions of longitudinal connectivity).

In all, only about 10% of the Swiss streams and rivers remain in a natural or near-natural state (BUWAL, 1997). In addition to the effects of physical changes, water quality must be considered. Although water quality – measured by conventional parameters like DOC, nitrogen compounds, and phosphorous – is usually good, ecotoxicological effects due to organic chemicals, metals and drugs still create a serious threat to aquatic organisms. Recently, widespread or-

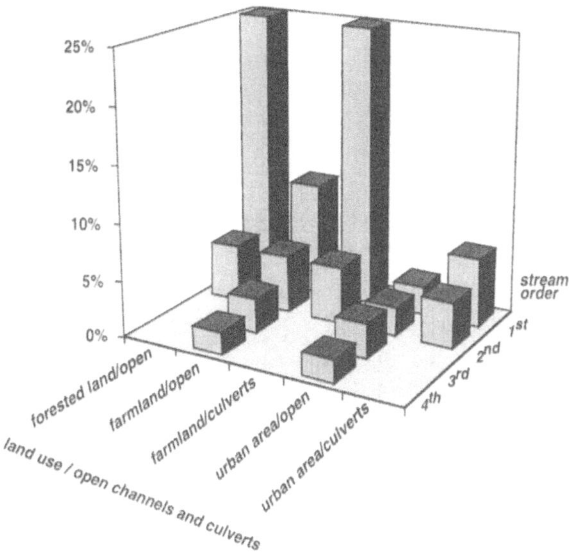

Figure 2. A broad classification of streams in Canton Zurich in terms of land-use and culverting (Hütte et al., 1994).

gan damage of fish and a decrease of fish catches was reported (Friedl & Staub, 1998). These effects are thought to have multifactorial causes, not only the presence of chemicals, but also changes in morphology, hydrology and temperature, along with fish management practices.

Political goals for stream management

Over the past 15 years some 50 km of streams have been ecologically improved every year (BUWAL, 1997), either by reopening of culverts or by rehabilitation of degraded streams. In 1992, new federal laws on Water Protection and on Water Correction Works, created the statutory basis for the protection and improvement of stream habitats. Additional regulations are found in the federal laws on the Protection of Nature and on Fisheries. The ongoing reform of the agricultural sector promotes ecologically sound farming and the creation of ecological compensation areas. The rehabilitation of removed buffer strips along streams will be possible in this context.

The Federal Water Protection Law aims at protecting water bodies against harmful effects and at the same time guaranteeing their different functions. In particular, the law serves:

- to ensure essential uses of water, like water supply and irrigation,
- to safeguard the health of humans, animals and plants,
- to preserve natural habitats for indigenous animals and plants and their communities,
- to preserve fish and fisheries,
- to preserve the water bodies as elements of the landscape,
- to maintain the natural functions of the water cycle.

The ecological goals for the surface waters are described as follows:

- The water bodies, and the surrounding area influenced by them, should maintain self-reproducing and self-regulating communities of animals, plants and microorganisms typical for a given location under near-natural conditions. The variety and frequency of species should be typical for water bodies of the same natural type which are exposed to little or no anthropogenic stress.
- The physical characteristics of water like morphology, sediment transport, flow regime and temperature regime shall have a near-natural variety and dynamics. In particular, unrestricted self-purification processes, a natural exchange of material between the water and the bottom, as well as unrestricted interactions between the water and its surroundings must be guaranteed.
- The chemical characteristics of the water must be such that
  - chemical compounds that may be harmful to aquatic life or to essential water uses do not accumulate to harmful concentrations, either in animals and plants or in suspended solids and sediments,
  - biological processes which regulate basic physiological processes of animals and plants like metabolism, reproduction and orientation by using the sense of smell, are not affected,
  - biological productivity does not exceed natural productivity.

The federal laws state that future engineering works may only be carried out if flood prevention cannot be achieved by either stream maintenance or by land use planning, or in cases where there is a major public economic interest. Unavoidable modifications should seek to retain near-natural conditions. Culverts have to be reopened when they need to be replaced. The Water Protection Law restricts water withdrawals, requiring that a certain residual flow be maintained, but this applies solely to new and renewed water use concessions.

What do the political goals mean?

Extensive action to improve the ecological quality of streams is necessary. Though some of the new legislation is progressive, restoration of most streams is impossible. By definition, restoration means bringing an ecosystem back to its original state. Under the given conditions in Switzerland, this can never be a realistic strategy. Human needs and exploitation of ecosystems most often come into conflict with maintaining natural or near-natural habitats and biodiversity. Striking a balance between conflicting interests is the only possibility.

What is the main goal for ecological improvement under these conditions, in terms of a scientifically sound concept? Obviously it cannot be ecological integrity, because this refers to original, undisturbed conditions and basically implies a catchment that is not – or only slightly – impaired. Preference should be given to the concept of sustainable development, which means economic development that satisfies both present and future human needs for resources and employment, while minimizing its impact on biological diversity (WRI/IUCN/UNEP, 1992). In this context, the integrity of ecosystems needs a new and broadened definition.

Scientific basis for stream assessment

The concepts of restoration and rehabilitation

'Preservation' is the maintenance of undisturbed ecosystems. It is crucial in habitat protection and conservation biology, but will not be discussed here. Stream 'restoration' aims at reestablishing ecological integrity, which is the sum of physical, chemical and biological integrity (Karr & Dudley, 1981). In its original definition 'biological integrity' refers to 'the capacity to support and maintain a balanced, integrated and adaptive biological system having the full range of elements and processes expected in a natural habitat of a region' (Frey, 1977; Angermeier & Karr, 1994).

For degraded streams with strongly modified catchments, the concept of restoration aiming at ecological integrity cannot be applied. A rational aim for ecological improvement of such streams could be to develop them 'in the right direction', i.e. towards a reference state of maximum naturalness attainable under the given conditions in a catchment. This process is 'rehabilitation' (improvement from a degraded state, Bradshaw, 1997). It implies focusing on attainable replacement of (artificial) elements and reestablishment of processes (Cooke & Jordan, 1995, cited in Angermeier, 1997). The ecosystem structure (species, habitat) is improved and ecosystem function will increase.

A rehabilitated ecosystem is different from a pristine ecosystem. It is, however, also an intact ecosystem and is characterized by complex habitat heterogeneity and the existence of longitudinal, lateral (riparian vegetation) and vertical connectivity. It possesses a certain potential for resistance (capacity to maintain natural structure and function after a natural disturbance) and for resilience (capacity to recover after a natural disturbance or following the cessation of an anthropogenic perturbation) (Kauffman et al., 1997).

Connectivity: longitudinal, lateral, vertical

Connectivity depends not only on environmental characteristics, but also on the behavior of organisms. Connectivity in a functional sense has to be distinguished from structural connection, which is the physical connection that permits exchange, including permanent or temporary connections between environmental units (Amoros et al., 1993). Connectivity is high if the physical connection is unimpaired and organisms can move in a longitudinal, lateral, or vertical direction. Fish depend on the physical connection of their habitat (spawning-feeding-wintering habitat) and fragmentation of habitat adversely affects the fish population. This is not only the case for potamodromous fish species but also for species that only carry out limited migrations (more or less resident fish species). Gravel spawning species like brown trout (*Salmo trutta fario* Linnaeus) and grayling (*Thymallus thymallus* Linnaeus) highly depend on the longitudinal, lateral and vertical dimensions of an ecosystem. They dig their redds in gravel (vertical dimension) and, after emergence, larvae depend on shelter and dead zones along the river margin (lateral dimension) (Sempeski & Gaudin, 1995). During their life cycle, individuals of both species need an intact habitat corridor and carry out migration (dispersal from the redd, feeding migration, remigration to the spawning grounds) in the longitudinal direction.

Ecologically important stream features

Hydrology/flow regime

Stream flow pattern is highly dynamic, varying not only between different sites of a stream but also over time. The catchment functions as a system which interacts with the surrounding environment. Hydrology is one of the major influences on the aquatic biocenosis.

Structure of stream bottom and instream structure

Most macroinvertebrates and many fish species live in close association with the stream bottom. Many macroinvertebrate taxa show some degree of substrate specialization and preference. Lithophilous taxa prefer stony substrate, xylophilous taxa live on woody debris and phytophilous taxa live in association with aquatic plants (Allan, 1995). In the same way, substrate preference plays an important role for spawning fish and for the time after hatching. Balon (1975) classified fish reproduction according to the site of spawning, the adaptations of the eggs and embryos and the degree of parental care.

Instream structure is mainly important for fish. It includes various kinds of overhead covers like woody debris or turbulence and submerged cover such as logs, boulders, rocks and vegetation. Instream structures transform streams into diverse, productive fish habitats.

Structure and vegetation of bank and surrounding area

Riparian zones (habitat along stream channels) are rich ecosystems in terms of biological structures and diversity. In addition they act as buffer strips, preventing fine sediment from entering the river channel. Moreover, bank vegetation provides shade, shelter and cover for organisms in the stream; it also regulates the water temperature. The transition zone between the stream and the terrestrial ecosystem consists of an ecotone which distinguishes itself by high species diversity. The well-being of continental water bodies depends largely on their terrestrial environment (Décamps & Naiman, 1990).

Disruptions of longitudinal connectivity

Longitudinal disruption may impede abiotic dynamic processes in the downstream direction (stream flow, bedload, suspended load) as well as downstream and upstream movements of organisms (Hütte et al., 1994), especially permanently aquatic organisms like fish, mussels and Gammaridae. Because aquatic insects emerge as winged animals they are less impeded by barriers. Barriers (weirs, culverts, etc.) block upstream migration and dams and turbines seriously interfere both with upstream and downstream migration. Longitudinal disruption disconnects stream reaches and splits them into small, isolated patches. This habitat fragmentation limits the dispersal and colonization for many species (e.g. *Cottus gobio* Linnaeus, Utzinger et al., 1998).

Temperature

Temperature sets limits on where a species can live. In contrast to the maximum summer temperature, low-temperature extremes are less critical. For example, brown trout do not occur in warm-water streams (maximum summer temperature above 25 °C).

Chemistry

Under pristine conditions the chemical composition of the water in streams is regulated by hydrogeochemical and hydrological processes and by natural input of organic matter. Human activity (waste water, agriculture, sealing of land surface, atmospheric pollution, etc.) leads to the addition of different kinds of chemical compounds and suspended solids. Many different physicochemical and biochemical processes are stimulated, leading to damage of organisms, loss of species and changes in the structure of communities. Important examples are:

- increased biological self-purification following the input of degradable organic compounds;
- nitrification and production of toxic nitrite;
- acute and chronic toxic effects of metals;
- toxic effects of numerous synthetic chemicals.

Nutrients (mainly P, N) may under certain conditions enhance biological productivity in streams.

Biological features and their temporal-spatial scales

The abiotic stream conditions play a pivotal role for the biological processes in streams. The different biological features represent different temporal-spatial scales (Table 1).

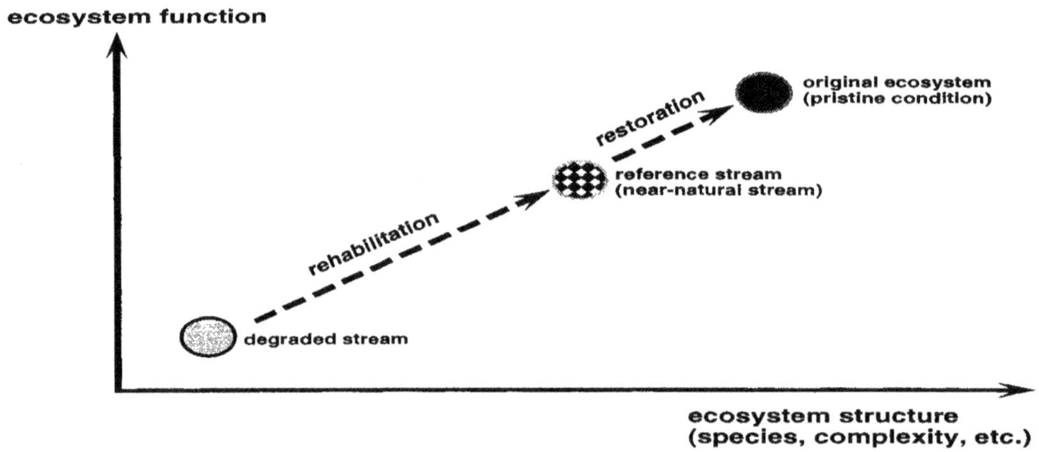

Figure 3. Ecosystem structure versus ecosystem function (modified from Bradshaw, 1997). Function and structure refer exclusively to biotic parameters.

Table 1. Temporal-spatial scales of different biological stream features

Biological feature	Temporal scale	Spatial scale (upper scale)
Algae	Days–weeks	m^2
Macroinvertebrates	Month–1 year	Several 100 m^2
Macrophytes	Years	Several 100 m^2
Fish	Years	km^2 (watershed)
Riparian vegetation	Decades	no scale (land–water interaction)

Approach to stream assessment in strongly modified catchments

Directing ecological improvement towards 'relative ecological integrity'

The governing concept for the ecological improvement of most Swiss streams must be rehabilitation. Rehabilitation measures should be directed towards a reference state (Figure 3). For anthropogenically modified catchments, we define the reference state in a pragmatic way as follows:

The reference state is the near-natural state of a stream in the given anthropogenically influenced environment. Certain marginal conditions of the catchment are accepted as being immutable, or subject to only limited change, e.g. draining of large areas of wetland. The reference state is characterized by:

- absence of engineering works, i.e. planform and cross-section are the result of the (anthropogenically modified) catchment morphology,
- no flow modification by hydropower or irrigation use,
- vegetation of banks and of adjacent areas typical for the given location,
- no or only very low anthropogenic load of chemical contaminants in the water, low alteration of water temperature.

A stream corresponding to the reference state is an ecologically intact system, even if it does not qualify for ecological integrity. An appropriate term to characterize the reference state might be 'relative ecological integrity'.

Aims of stream assessment

Methods for the assessment of streams should be in compliance with this concept. They should focus on the most important stream features and produce data suitable for characterizing the ecological condition of streams and for supporting their ecologically sound management. Since we are dealing with anthropogenically modified catchments, the methods used should always produce management-oriented data. They should include a system approach as the basic unit and be based on scientific principles of stream integrity emphasizing habitat connectivities. The methods should allow:

- the condition of a stream(system) to be rationally described and judged,

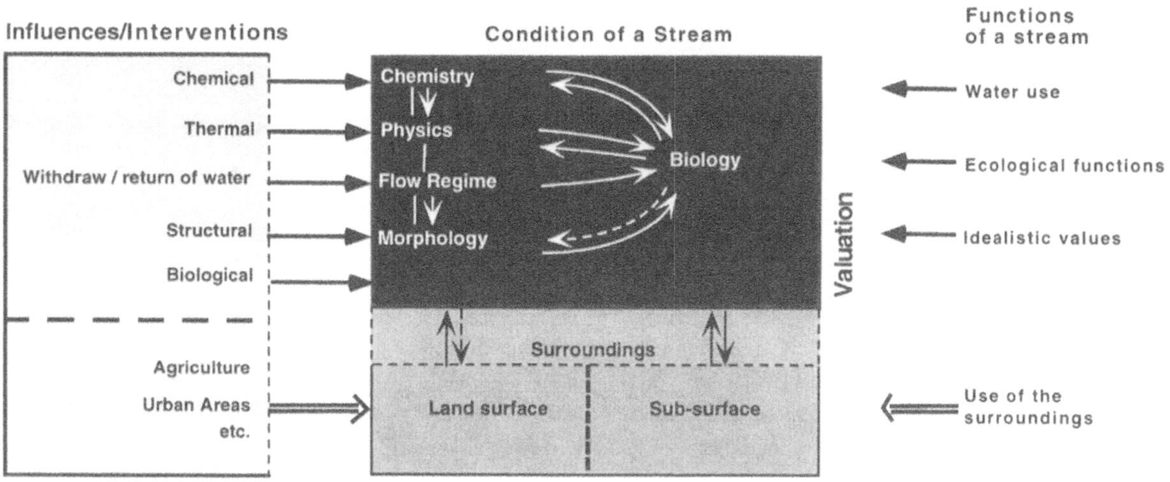

Figure 4. Factors determining the condition of a stream.

- identification at the different kinds of impact on a stream(system),
- verification of the effects of water protection measures,
- identification of suitable future actions in the context of a whole stream(system).

Parameters and processes to characterize stream condition

The condition of a stream is usually defined by quantitatively measurable values which attempt to describe the system or its constituents in detail (Wuhrmann, 1974). By considering the functions which a stream performs for people, an element of valuation or prioritization is introduced. Beside ecological functions, the functions of a stream may include different practical uses along with emotional and aesthetic values (Figure 4). Due to high intensity of human activity, streams usually have to fulfill several functions at the same time.

A comprehensive description of the condition of a stream taking into account all potentially important factors is not possible. For example, we are far from being able to understand the effects of thousands of interacting chemicals. Indeed, we are not even able to detect most of the chemicals and their metabolites, at least not with affordable expenditure. Thus, the description of stream condition must always be based on a limited set of parameters and processes.

Criteria for making a choice of parameters to measure include:

- relevance with respect to ecological pattern and functions, to the health of people and other organisms and to economic and recreational uses,
- availability of methods for measurement,
- cost of measurement.

The parameters and processes to be chosen should represent the ecologically most important stream features. The selected measures should allow a single feature to be described, e.g. the bank structure, or a set of combined features, e.g. morphology or oxygen balance, or a more comprehensive assessment to be made. Examples for such parameters and processes are:

- water flow: natural regime, direct modifications (withdrawals, returns) and indirect modifications (e.g. through land use),
- channel pattern: under pristine conditions, gradual changes through human influences,
- variability of bed width and water depth,
- natural and artificial falls,
- water temperature,
- oxygen content of water and sediment,
- concentrations and loads of oxygen-consuming compounds like DOC and ammonium,
- concentrations and loads of ecotoxicologically relevant chemicals like Cu, Pb, PCB, nonylphenol-compounds,
- occurrence of specific species, and diversity of fish, invertebrate and macrophyte communities,
- primary production of algae,
- ecotoxicological potential/effects of water samples.

Table 2. Definition of the survey levels with different degree of effort

	Level I Regional scale survey	II Catchment scale survey	III Extensive reach scale survey
Spatial scale	Region	Stream system/ catchment	Reach
Effort	Low	Medium	High
Aim	Overall view of stream conditions and ecological deficits, scoring of reaches	Detailed analysis of stream conditions, analysis of deficits, evaluation of measures	Specific analysis for specific questions

The parameters and processes represent different spatial scales (cm to km) and different temporal scales (days to thousands of years) of ecological processes (Ward, 1997). Different patterns of data collection (single time pairs, time series, etc.) and of statistical analysis, have to be applied for different parameters and processes, depending on their physical, chemical and biological dynamic and on the objective and the available resources for the assessment.

Other attempts at comprehensive evaluation systems and illustrations of the comprehensive approach to assessing river quality were published by Boon & Howell (1997). Boon et al. (1997) developed a system for evaluating rivers for conservation (SERCON) in the UK. SERCON is a broad-based technique for river evaluation, designed to be applied with greater consistency than other methods. Like our approach, SERCON has been developed principally for use at the scale of the catchment and subcatchment. There is a clear focus on the physical, chemical and biological features of river channels and banks, riparian zones and associated floodplains.

The Swiss modular concept for stream assessment

General principles

As an appropriate answer to the variety of the ecological processes to be assessed and the different requirements to be fulfilled by the assessment, a modular survey and evaluation concept for system-oriented analysis has been developed (Bundi et al., 1998). Each module deals with specific stream features. The different modules are designed to be complementary to each other. Single modules, several modules, or all of them may be applied, depending on the purpose of the analysis.

For each stream feature a cost-effective survey procedure (rapid method) and a progressively more rigorous method providing more detailed information is suggested. This step-by-step approach enables a better understanding of the complexity of a stream system and the human influences on it. Based on this, rehabilitation concepts can then be elaborated in cooperation with managers, river engineers and biologists.

The different levels

The methods of the first level (regional scale survey) are rapid, cost-effective, and provide information on the conditions of all streams within a certain area (e.g. river catchment area, Canton). The second level, the catchment scale survey, comprises modules which allow a more detailed analysis of whole stream systems including tributaries. Finally, the third level of effort, the extensive reach scale, includes an extensive, costly study of a short reach of a stream (0.1–1 km) in order to investigate specific questions or problems (minimum flow, survey after stream restoration, impact of sewage plants, etc.). Only general advice is given here for this level. Table 2 provides an overview of the three levels with different monitoring efforts.

Table 3. The modular concept with level I and II

	Hydrology and ecomorphology		Biology					Chemical and ecotoxicological effects	
Level	Hydrology	Ecomorphology	Riparian vegetation	Macrophytes	Algae	Macroinvertebrates	Fish	Water chemistry	Ecotoxicology
I	General characterization of flow discharge	main morphological impairment, longitudinal connectivity, assessment of naturalness		estimation of abundance of macrophytes	estimation of abundance, studies of diatoms	simple study on a taxonomic level that is suitable for field work	random sampling to detect species range	anthropogenetic contamination is identified by analysis of random samples	random sample, 2–3 simple test procedures, determination of NOEC levels
II	Systematic survey	deficit analysis on a catchment scale, evaluation of measures, setting of priorities		mapping of all species	detailed studies of diatoms	comparison of taxa present with reference sites	detailed studies of species present (diagnosis of population)	study of a larger number of contamination parameters	seasonal or frequent check of important point sources

Assessment of the survey result

Scoring in level I

For the different analyzed stream reaches a point scoring is carried out and the reaches are arranged according to their naturalness (natural stream properties) in four different classes (arbitrary scoring of each parameter with final calculation of the average).

Class 1: natural/near-natural, Class 2: little affected, Class 3: strongly affected, Class 4: artificial stream conditions. Class 1 combines the two classes 'high quality' and 'good quality' of the proposed classification by the European Community.

To present an overview, maps are drawn with different colors for the four classes. The maps are completed with additional information (e.g. natural and artificial barriers impeding upstream migration of fish, riparian conditions, etc.).

The classification of ecological status and presentation of the results are in accordance with the corresponding proposal of the European Community (Commission of the European Communities, 1998).

Deficit analysis in level II

At this level no point scoring is carried out. The actual condition of each stream reach is compared with a reference state as already defined. The reference state (ecomorphological reference) is derived from historical references/old maps, from reference sites, or from reconstruction based on 'best-available' information. The latter implies elements of arbitrary judgment, which is often unavoidable. The deficit analysis is followed by an evaluation of potential measures to improve the ecological conditions. The measures are prioritized and ordered according to an ecological benefit-cost ratio, whereas the benefits are measured in terms of the length of stream section made accessible to migrating organisms, or the area of adjacent terrestrial ecosystem networked with the stream.

The modules

Table 3 gives a synopsis on the whole modular concept for levels 1 and 2, each comprising 9 modules.

Level III contains detailed site-specific studies and no further directions are described.

Each module will be separately developed and can be used independently or combined with any other module.

Conclusion

The Swiss modular concept attempts to provide a rational, scientifically sound tool for stream assessment and development in anthropogenically intensively used catchments. In such catchments, restoring ecological integrity is not possible. Certain anthropogenic modifications of the environment have often to be accepted as immutable and dealing with conflicting interests is unavoidable. The will to achieve ecological improvement under these premises is the driving force for the Swiss modular concept. The direction of improvement (rehabilitation) is given by 'relative ecological integrity' defined as the near-natural state in the given anthropogenically influenced environment. Following this direction will lead to 'the best-possible-ecological solution' under the given circumstances.

Standardized methods will only be useful if they are accepted by the users, for example public authorities, private consultants and interest groups. Therefore, much emphasis was put on involving various users (and scientific disciplines) in the development of methods. This process included:

1. building the scientific basis of the modular concept with the support of multidisciplinary expert groups,
2. formulating the modular concept and developing the individual assessment methods in cooperation with representatives of public agencies and private consultants,
3. submitting drafts of methods to users for practical testing and comments,
4. holding hearings with major cantonal and federal agency players on the modular concepts and individual methods.

Involving users is very time consuming. The result in terms of support, however, is worth the effort.

Though rigorous application of the assessment concept will support major ecological improvements of the running waters, it has to be recognized that the complex problem of biological impairment resulting from the very complex mixture of industrial chemicals and drugs occurring in the environment cannot be solved on a catchment level. Nationally and internationally coordinated measures are needed. Awareness of ecotoxicological effects must be developed and scientific efforts must be made to gain a better understanding of these effects. This will support the political efforts needed to control chemical uses and inputs (including drugs).

Of the modules briefly described above, that on ecomorphology has been published (Hütte et al., 1998) and some are in progress. Most modules will be available within the next 2–3 years.

Acknowledgements

We thank U. Von Blücher, W. Meier, U. Kupper, P. Niederhauser, H. P. Willi and many colleagues for valuable comments.

References

Allan, J. D., 1995. Stream Ecology. Chapman & Hall, London: 388 pp.

Amoros, C., J. Gibert & M. Greenwood, 1993. Interactions entre unités de l'hydrosystème fluvial. In Amoros, C. & G. E. Petts (eds), Hydrosystèmes Fluviaux. Masson, Paris: 169–199

Angermeier, P. L., 1997. Conceptual roles of biological integrity and diversity. In Williams, J. E., C. A. Wood & M. P. Dombeck (eds), Watershed Restoration: Principles and Practices. American Fisheries Society, Bethesda, Maryland: 49–65.

Angermeier, P. L. & J. R. Karr, 1994. Biological integrity versus biological diversity as policy directives. BioScience 44: 690–697.

Balon, E. K., 1975. Reproductive guilds of fishes: a proposal and definition. J. Fish. Res. Bd Can. 32: 821–864.

Boon, P. J. & D. L. Howell, 1997. Defining the quality of fresh waters: theme and variations. In Boon, P. J. & D. L. Howell (eds), Freshwater Quality: Defining the Indefinable? The Stationery Office, Edinburgh.

Boon, P. J., N. T. H. Holmes, P. S. Maitland, T. A. Rowell & J. Davies, 1997. A system for evaluating rivers for conservation (SERCON): development, structure, function. In Boon, P. J. & D. L. Howell (eds), Freshwater Quality: Defining the Indefinable? The Stationery Office, Edinburgh: 299–326.

Bradshaw, A. D., 1997. What do we mean by restoration? In Urbanska, K. M., N. R. Webb & P. J. Edwards (eds), Restoration Ecology and Sustainable Development. Cambridge University Press, Cambridge: 8–14.

Bundi, U., A. Frutiger, C. Göldi, M. Hütte, U. Kupper, P. Liechti, W. Meier, P. Niederhauser, A. Peter, U. Sieber, U. Von Blücher & H.P. Willi, 1998. Methoden zur Untersuchung und Beurteilung der Fliessgewässer: Modul-Stufen-Konzept. Mitteilungen zum Gewässerschutz Nr. 26 (Vollzug Umwelt). Bundesamt für Umwelt, Wald und Landschaft (BUWAL): 43 pp.

BUWAL, Swiss Agency for the Environment, Forests and Landscape, 1997. Umwelt in der Schweiz, BUWAL, Bern: 372 pp.

Commission of the European Communities, 1998. Amended proposal for a Council Directive establishing a framework for Community action in the field of water policy. Brussels, 17.02.1998, COM (1998) 76 final.

Cooke, G. D. & W. R. Jordan, III., 1995. Ecosystem rehabilitation. In Using ecological restoration to meet Clean Water Act goals. U.S. Environmental Protection Agency, Chicago, Illinois: 1–4.

Décamps, H. & R. J. Naiman, 1990. Towards an ecotone perspective. In Naiman, R. J. & H. Décamps (eds), The Ecology and Management of Aquatic-terrestrial Ecotones. Man and the

Biosphere Series, Vol. 4. UNESCO Paris and the Parthenon Publishing Group: 1–5.

Frey, D. G., 1977. Biological integrity of water – a historical approach. In Ballentine, R. K. & L. J. Guarrida (eds), The Integrity of Water. U.S. Environmental Protection Agency, Washington, D.C.: 127–140.

Friedl, C. & E. Staub, 1998. Collapse of angling catches in running waters – a new phenomen in Switzerland and its possible causes. Poster presentation, International Conference on Assessing the Ecological Integrity of Running Waters. 9–11 November 1998, Vienna.

Hütte, M., U. Bundi & A. Peter, 1994. Konzept für die Bewertung und Entwicklung von Bächen und Bachsystemen im Kanton Zürich, EAWAG/Kanton Zürich: 133 pp.

Hütte, M., P. Niederhauser, U. Bundi, A. Frutiger, C. Göldi, U. Kupper, P. Liechti, W. Meier, P. Niederhauser A. Peter, U. Sieber, U. Von Blücher & H. P. Willi, 1998. Methoden zur Untersuchung und Beurteilung der Fliessgewässer in der Schweiz: Oekomorphologie – Stufe F. Mitteilungen zum Gewässerschutz Nr. 27 (Vollzug Umwelt). Bundesamt für Umwelt, Wald und Landschaft (BUWAL): 47 pp.

Karr, J. R. & D. R. Dudley, 1981. Ecological perspective on water quality goals. Envir. Manage. 5: 55–68.

Kauffman, J. B., R. L. Beschta, N. Otting & D. Lytjen, 1997. An ecological perspective of riparian and stream restoration in the Western United States. Fisheries 22, 5: 12–24.

Koblet, R., 1995. Entwicklung der Korrektionen an Fliessgewässern und Seen. Kartentafel 5.5 des Hydrologischen Atlas der Schweiz. Landeshydrologie und geologie, Bern.

Sempeski, P. & P. Gaudin, 1995. Size-related changes in diel distribution of young grayling (*Thymallus thymallus*). Can. J. Fish. and aquat. Sci. 52, 9: 1842–1848.

Utzinger, J., C. Roth & A. Peter, 1998. Effects of environmental parameters on the distribution of bullhead *Cottus gobio* with particular consideration of the effects of obstructions. J. appl. Ecol. 35: 882–892.

Ward, J. V., 1997. An Expansive Perspective of Riverine Landscapes: Pattern and Process Across Scales. GAIA 6: 52–60.

WRI/IUCN/UNEP, 1992. Global Biodiversity Strategy: Guidelines for actions to save, study, and use earth's biotic wealth sustainability and equitably. World Resources Institute, World Conservation Union, and the United Nations Environment Programme, Washington, D.C.

Wuhrmann, K. ,1974. Some problems and perspectives in applied limnology. Mitt. Int. Ver. Limnol. 20: 325–40.